Lecture Notes in Artificial Intelligence   12458

Subseries of Lecture Notes in Computer Science

More information about this subseries at http://www.springer.com/series/1244

Frank Hutter · Kristian Kersting ·
Jefrey Lijffijt · Isabel Valera (Eds.)

Machine Learning and Knowledge Discovery in Databases

European Conference, ECML PKDD 2020
Ghent, Belgium, September 14–18, 2020
Proceedings, Part II

 Springer

Editors
Frank Hutter 🆔
Albert-Ludwigs-Universität
Freiburg, Germany

Kristian Kersting 🆔
TU Darmstadt
Darmstadt, Germany

Jefrey Lijffijt 🆔
Ghent University
Ghent, Belgium

Isabel Valera 🆔
Saarland University
Saarbrücken, Germany

ISSN 0302-9743   ISSN 1611-3349 (electronic)
Lecture Notes in Artificial Intelligence
ISBN 978-3-030-67660-5   ISBN 978-3-030-67661-2   (eBook)
https://doi.org/10.1007/978-3-030-67661-2

LNCS Sublibrary: SL7 – Artificial Intelligence

This Springer imprint is published by the registered company Springer Nature Switzerland AG
The registered company address is: Gewerbestrasse 11, 6330 Cham, Switzerland

Preface

This edition of the European Conference on Machine Learning and Principles and Practice of Knowledge Discovery in Databases (ECML PKDD 2020) is one that we will not easily forget. Due to the emergence of a global pandemic, our lives changed, including many aspects of the conference. Because of this, we are perhaps more proud and happy than ever to present these proceedings to you.

ECML PKDD is an annual conference that provides an international forum for the latest research in all areas related to machine learning and knowledge discovery in databases, including innovative applications. It is the leading European machine learning and data mining conference and builds upon a very successful series of ECML PKDD conferences.

Scheduled to take place in Ghent, Belgium, due to the SARS-CoV-2 pandemic, ECML PKDD 2020 was the first edition to be held fully virtually, from the 14th to the 18th of September 2020. The conference attracted over 1000 participants from all over the world. New this year was a joint event with local industry on Thursday afternoon, the AI4Growth industry track. More generally, the conference received substantial attention from industry through sponsorship, participation, and the revived industry track at the conference.

The main conference programme consisted of presentations of 220 accepted papers and five keynote talks (in order of appearance): Max Welling (University of Amsterdam), Been Kim (Google Brain), Gemma Galdon-Clavell (Eticas Research & Consulting), Stephan Günnemann (Technical University of Munich), and Doina Precup (McGill University & DeepMind Montreal).

In addition, there were 23 workshops, nine tutorials, two combined workshop-tutorials, the PhD Forum, and a discovery challenge.

Papers presented during the three main conference days were organized in four different tracks:

- Research Track: research or methodology papers from all areas in machine learning, knowledge discovery, and data mining;
- Applied Data Science Track: papers on novel applications of machine learning, data mining, and knowledge discovery to solve real-world use cases, thereby bridging the gap between practice and current theory;
- Journal Track: papers that were published in special issues of the journals *Machine Learning* and *Data Mining and Knowledge Discovery*;
- Demo Track: short papers that introduce a new system that goes beyond the state of the art, accompanied with a video of the demo.

We received a record number of 687 and 235 submissions for the Research and Applied Data Science Tracks respectively. We accepted 130 (19%) and 65 (28%) of these. In addition, there were 25 papers from the Journal Track, and 10 demo papers (out of 25 submissions). All in all, the high-quality submissions allowed us to put together an exceptionally rich and exciting program.

The Awards Committee selected research papers that were considered to be of exceptional quality and worthy of special recognition:

- Data Mining best paper award: "Revisiting Wedge Sampling for Budgeted Maximum Inner Product Search", by Stephan S. Lorenzen and Ninh Pham.
- Data Mining best student paper award: "SpecGreedy: Unified Dense Subgraph Detection", by Wenjie Feng, Shenghua Liu, Danai Koutra, Huawei Shen, and Xueqi Cheng.
- Machine Learning best (student) paper award: "Robust Domain Adaptation: Representations, Weights and Inductive Bias", by Victor Bouvier, Philippe Very, Clément Chastagnol, Myriam Tami, and Céline Hudelot.
- Machine Learning best (student) paper runner-up award: "A Principle of Least Action for the Training of Neural Networks", by Skander Karkar, Ibrahim Ayed, Emmanuel de Bézenac, and Patrick Gallinari.
- Best Applied Data Science Track paper: "Learning to Simulate on Sparse Trajectory Data", by Hua Wei, Chacha Chen, Chang Liu, Guanjie Zheng, and Zhenhui Li.
- Best Applied Data Science Track paper runner-up: "Learning a Contextual and Topological Representation of Areas-of-Interest for On-Demand Delivery Application", by Mingxuan Yue, Tianshu Sun, Fan Wu, Lixia Wu, Yinghui Xu, and Cyrus Shahabi.
- Test of Time Award for highest-impact paper from ECML PKDD 2010: "Three Naive Bayes Approaches for Discrimination-Free Classification", by Toon Calders and Sicco Verwer.

We would like to wholeheartedly thank all participants, authors, PC members, area chairs, session chairs, volunteers, co-organizers, and organizers of workshops and tutorials for their contributions that helped make ECML PKDD 2020 a great success. Special thanks go to Vicky, Inge, and Eneko, and the volunteer and virtual conference platform chairs from the UGent AIDA group, who did an amazing job to make the online event feasible. We would also like to thank the ECML PKDD Steering Committee and all sponsors.

October 2020

Tijl De Bie
Craig Saunders
Dunja Mladenić
Yuxiao Dong
Frank Hutter
Isabel Valera
Jefrey Lijffijt
Kristian Kersting
Georgiana Ifrim
Sofie Van Hoecke

Organization

General Chair

Tijl De Bie — Ghent University, Belgium

Research Track Program Chairs

Frank Hutter — University of Freiburg & Bosch Center for AI, Germany
Isabel Valera — Max Planck Institute for Intelligent Systems, Germany
Jefrey Lijffijt — Ghent University, Belgium
Kristian Kersting — TU Darmstadt, Germany

Applied Data Science Track Program Chairs

Craig Saunders — Amazon Alexa Knowledge, UK
Dunja Mladenić — Jožef Stefan Institute, Slovenia
Yuxiao Dong — Microsoft Research, USA

Journal Track Chairs

Aristides Gionis — KTH, Sweden
Carlotta Domeniconi — George Mason University, USA
Eyke Hüllermeier — Paderborn University, Germany
Ira Assent — Aarhus University, Denmark

Discovery Challenge Chair

Andreas Hotho — University of Würzburg, Germany

Workshop and Tutorial Chairs

Myra Spiliopoulou — Otto von Guericke University Magdeburg, Germany
Willem Waegeman — Ghent University, Belgium

Demonstration Chairs

Georgiana Ifrim — University College Dublin, Ireland
Sofie Van Hoecke — Ghent University, Belgium

Nectar Track Chairs

Jie Tang	Tsinghua University, China
Siegfried Nijssen	Université catholique de Louvain, Belgium
Yizhou Sun	University of California, Los Angeles, USA

Industry Track Chairs

Alexander Ypma	ASML, the Netherlands
Arindam Mallik	imec, Belgium
Luis Moreira-Matias	Kreditech, Germany

PhD Forum Chairs

Marinka Zitnik	Harvard University, USA
Robert West	EPFL, Switzerland

Publicity and Public Relations Chairs

Albrecht Zimmermann	Université de Caen Normandie, France
Samantha Monty	Universität Würzburg, Germany

Awards Chairs

Danai Koutra	University of Michigan, USA
José Hernández-Orallo	Universitat Politècnica de València, Spain

Inclusion and Diversity Chairs

Peter Steinbach	Helmholtz-Zentrum Dresden-Rossendorf, Germany
Heidi Seibold	Ludwig-Maximilians-Universität München, Germany
Oliver Guhr	Hochschule für Technik und Wirtschaft Dresden, Germany
Michele Berlingerio	Novartis, Ireland

Local Chairs

Eneko Illarramendi Lerchundi	Ghent University, Belgium
Inge Lason	Ghent University, Belgium
Vicky Wandels	Ghent University, Belgium

Proceedings Chair

Wouter Duivesteijn	Technische Universiteit Eindhoven, the Netherlands

Sponsorship Chairs

Luis Moreira-Matias	Kreditech, Germany
Vicky Wandels	Ghent University, Belgium

Volunteering Chairs

Junning Deng	Ghent University, Belgium
Len Vande Veire	Ghent University, Belgium
Maarten Buyl	Ghent University, Belgium
Raphaël Romero	Ghent University, Belgium
Robin Vandaele	Ghent University, Belgium
Xi Chen	Ghent University, Belgium

Virtual Conference Platform Chairs

Ahmad Mel	Ghent University, Belgium
Alexandru Cristian Mara	Ghent University, Belgium
Bo Kang	Ghent University, Belgium
Dieter De Witte	Ghent University, Belgium
Yoosof Mashayekhi	Ghent University, Belgium

Web Chair

Bo Kang	Ghent University, Belgium

ECML PKDD Steering Committee

Andrea Passerini	University of Trento, Italy
Francesco Bonchi	ISI Foundation, Italy
Albert Bifet	Télécom Paris, France
Sašo Džeroski	Jožef Stefan Institute, Slovenia
Katharina Morik	TU Dortmund, Germany
Arno Siebes	Utrecht University, the Netherlands
Siegfried Nijssen	Université catholique de Louvain, Belgium
Michelangelo Ceci	University of Bari Aldo Moro, Italy
Myra Spiliopoulou	Otto von Guericke University Magdeburg, Germany
Jaakko Hollmen	Aalto University, Finland
Georgiana Ifrim	University College Dublin, Ireland
Thomas Gärtner	University of Nottinghem, UK
Neil Hurley	University College Dublin, Ireland
Michele Berlingerio	IBM Research, Ireland
Elisa Fromont	Université de Rennes 1, France
Arno Knobbe	Universiteit Leiden, the Netherlands
Ulf Brefeld	Leuphana Universität Luneburg, Germany
Andreas Hotho	Julius-Maximilians-Universität Würzburg, Germany

Program Committees

Guest Editorial Board, Journal Track

Michael Kamp	Monash University
Mehdi Kaytoue	Infologic
Marius Kloft	TU Kaiserslautern
Dragi Kocev	Jožef Stefan Institute
Peer Kröger	Ludwig-Maximilians-Universität Munich
Meelis Kull	University of Tartu
Ondrej Kuzelka	KU Leuven
Mark Last	Ben-Gurion University of the Negev
Matthijs van Leeuwen	Leiden University
Marco Lippi	University of Modena and Reggio Emilia
Claudio Lucchese	Ca' Foscari University of Venice
Brian Mac Namee	University College Dublin
Gjorgji Madjarov	Ss. Cyril and Methodius University of Skopje
Fabrizio Maria Maggi	Free University of Bozen-Bolzano
Giuseppe Manco	ICAR-CNR
Ernestina Menasalvas	Universidad Politécnica de Madrid
Aditya Menon	Google Research
Katharina Morik	TU Dortmund
Davide Mottin	Aarhus University
Animesh Mukherjee	Indian Institute of Technology Kharagpur
Amedeo Napoli	LORIA
Siegfried Nijssen	Université catholique de Louvain
Eirini Ntoutsi	Leibniz University Hannover
Bruno Ordozgoiti	Aalto University
Panče Panov	Jožef Stefan Institute
Panagiotis Papapetrou	Stockholm University
Srinivasan Parthasarathy	Ohio State University
Andrea Passerini	University of Trento
Mykola Pechenizkiy	Technische Universiteit Eindhoven
Charlotte Pelletier	Univ. Bretagne Sud/IRISA
Ruggero Pensa	University of Turin
Francois Petitjean	Monash University
Nico Piatkowski	TU Dortmund
Evaggelia Pitoura	Univ. of Ioannina
Marc Plantevit	Claude Bernard University Lyon 1
Kai Puolamäki	University of Helsinki
Chedy Raïssi	Inria
Matteo Riondato	Amherst College
Joerg Sander	University of Alberta
Pierre Schaus	UCLouvain
Lars Schmidt-Thieme	University of Hildesheim
Matthias Schubert	LMU Munich
Thomas Seidl	LMU Munich
Gerasimos Spanakis	Maastricht University
Myra Spiliopoulou	Otto von Guericke University Magdeburg
Jerzy Stefanowski	Poznań University of Technology

Giovanni Stilo	Università degli Studi dell'Aquila
Mahito Sugiyama	National Institute of Informatics
Andrea Tagarelli	University of Calabria
Chang Wei Tan	Monash University
Nikolaj Tatti	University of Helsinki
Alexandre Termier	Univ. Rennes 1
Marc Tommasi	University of Lille
Ivor Tsang	University of Technology Sydney
Panayiotis Tsaparas	University of Ioannina
Steffen Udluft	Siemens
Celine Vens	KU Leuven
Antonio Vergari	University of California, Los Angeles
Michalis Vlachos	University of Lausanne
Christel Vrain	LIFO, Université d'Orléans
Jilles Vreeken	Helmholtz Center for Information Security
Willem Waegeman	Ghent University
Marcel Wever	Paderborn University
Stefan Wrobel	Univ. Bonn and Fraunhofer IAIS
Yinchong Yang	Siemens AG
Guoxian Yu	Southwest University
Bianca Zadrozny	IBM
Ye Zhu	Monash University
Arthur Zimek	University of Southern Denmark
Albrecht Zimmermann	Université de Caen Normandie
Marinka Zitnik	Harvard University

Area Chairs, Research Track

Cuneyt Gurcan Akcora	The University of Texas at Dallas
Carlos M. Alaíz	Universidad Autónoma de Madrid
Fabrizio Angiulli	University of Calabria
Georgios Arvanitidis	Max Planck Institute for Intelligent Systems
Roberto Bayardo	Google
Michele Berlingerio	IBM
Michael Berthold	University of Konstanz
Albert Bifet	Télécom Paris
Hendrik Blockeel	Katholieke Universiteit Leuven
Mario Boley	MPI Informatics
Francesco Bonchi	Fondazione ISI
Ulf Brefeld	Leuphana Universität Lüneburg
Michelangelo Ceci	Università degli Studi di Bari Aldo Moro
Duen Horng Chau	Georgia Institute of Technology
Nicolas Courty	Université de Bretagne Sud/IRISA
Bruno Cremilleux	Université de Caen Normandie
Andre de Carvalho	University of São Paulo
Patrick De Causmaecker	Katholieke Universiteit Leuven

Nicola Di Mauro	Università degli Studi di Bari Aldo Moro
Tapio Elomaa	Tampere University
Amir-Massoud Farahmand	Vector Institute & University of Toronto
Ángela Fernández	Universidad Autónoma de Madrid
Germain Forestier	Université de Haute-Alsace
Elisa Fromont	Université de Rennes 1
Johannes Fürnkranz	Johannes Kepler University Linz
Patrick Gallinari	Sorbonne University
Joao Gama	University of Porto
Thomas Gärtner	TU Wien
Pierre Geurts	University of Liège
Manuel Gomez Rodriguez	MPI for Software Systems
Przemyslaw Grabowicz	University of Massachusetts Amherst
Stephan Günnemann	Technical University of Munich
Allan Hanbury	Vienna University of Technology
Daniel Hernández-Lobato	Universidad Autónoma de Madrid
Jose Hernandez-Orallo	Universitat Politècnica de València
Jaakko Hollmén	Aalto University
Andreas Hotho	University of Würzburg
Neil Hurley	University College Dublin
Georgiana Ifrim	University College Dublin
Alipio M. Jorge	University of Porto
Arno Knobbe	Universiteit Leiden
Dragi Kocev	Jožef Stefan Institute
Lars Kotthoff	University of Wyoming
Nick Koudas	University of Toronto
Stefan Kramer	Johannes Gutenberg University Mainz
Meelis Kull	University of Tartu
Niels Landwehr	University of Potsdam
Sébastien Lefèvre	Université de Bretagne Sud
Daniel Lemire	Université du Québec
Matthijs van Leeuwen	Leiden University
Marius Lindauer	Leibniz University Hannover
Jörg Lücke	University of Oldenburg
Donato Malerba	Università degli Studi di Bari "Aldo Moro"
Giuseppe Manco	ICAR-CNR
Pauli Miettinen	University of Eastern Finland
Anna Monreale	University of Pisa
Katharina Morik	TU Dortmund
Emmanuel Müller	University of Bonn
Sriraam Natarajan	Indiana University Bloomington
Alfredo Nazábal	The Alan Turing Institute
Siegfried Nijssen	Université catholique de Louvain
Barry O'Sullivan	University College Cork
Pablo Olmos	University Carlos III of Madrid
Panagiotis Papapetrou	Stockholm University

Andrea Passerini	University of Turin
Mykola Pechenizkiy	Technische Universiteit Eindhoven
Ruggero G. Pensa	University of Torino
Francois Petitjean	Monash University
Claudia Plant	University of Vienna
Marc Plantevit	Université Claude Bernard Lyon 1
Philippe Preux	Université de Lille
Rita Ribeiro	University of Porto
Celine Robardet	INSA Lyon
Elmar Rueckert	University of Lübeck
Marian Scuturici	LIRIS-INSA de Lyon
Michèle Sebag	Univ. Paris-Sud
Thomas Seidl	Ludwig-Maximilians-Universität Muenchen
Arno Siebes	Utrecht University
Alessandro Sperduti	University of Padua
Myra Spiliopoulou	Otto von Guericke University Magdeburg
Jerzy Stefanowski	Poznań University of Technology
Yizhou Sun	University of California, Los Angeles
Einoshin Suzuki	Kyushu University
Acar Tamersoy	Symantec Research Labs
Jie Tang	Tsinghua University
Grigorios Tsoumakas	Aristotle University of Thessaloniki
Celine Vens	KU Leuven
Antonio Vergari	University of California, Los Angeles
Herna Viktor	University of Ottawa
Christel Vrain	University of Orléans
Jilles Vreeken	Helmholtz Center for Information Security
Willem Waegeman	Ghent University
Wendy Hui Wang	Stevens Institute of Technology
Stefan Wrobel	Fraunhofer IAIS & Univ. of Bonn
Han-Jia Ye	Nanjing University
Guoxian Yu	Southwest University
Min-Ling Zhang	Southeast University
Albrecht Zimmermann	Université de Caen Normandie

Area Chairs, Applied Data Science Track

Michelangelo Ceci	Università degli Studi di Bari Aldo Moro
Tom Diethe	Amazon
Faisal Farooq	IBM
Johannes Fürnkranz	Johannes Kepler University Linz
Rayid Ghani	Carnegie Mellon University
Ahmed Hassan Awadallah	Microsoft
Xiangnan He	University of Science and Technology of China
Georgiana Ifrim	University College Dublin
Anne Kao	Boeing

Javier Latorre	Apple
Hao Ma	Facebook AI
Gabor Melli	Sony PlayStation
Luis Moreira-Matias	Kreditech
Alessandro Moschitti	Amazon
Kitsuchart Pasupa	King Mongkut's Institute of Technology Ladkrabang
Mykola Pechenizkiy	Technische Universiteit Eindhoven
Julien Perez	NAVER LABS Europe
Xing Xie	Microsoft
Chenyan Xiong	Microsoft Research
Yang Yang	Zhejiang University

Program Committee Members, Research Track

Moloud Abdar	Deakin University
Linara Adilova	Fraunhofer IAIS
Florian Adriaens	Ghent University
Zahra Ahmadi	Johannes Gutenberg University Mainz
M. Eren Akbiyik	IBM Germany Research and Development GmbH
Youhei Akimoto	University of Tsukuba
Ömer Deniz Akyildiz	University of Warwick and The Alan Turing Institute
Francesco Alesiani	NEC Laboratories Europe
Alexandre Alves	Universidade Federal de Uberlândia
Maryam Amir Haeri	Technische Universität Kaiserslautern
Alessandro Antonucci	IDSIA
Muhammad Umer Anwaar	Mercateo AG
Xiang Ao	Institute of Computing Technology, Chinese Academy of Sciences
Sunil Aryal	Deakin University
Thushari Atapattu	The University of Adelaide
Arthur Aubret	LIRIS
Julien Audiffren	Fribourg University
Murat Seckin Ayhan	Eberhard Karls Universität Tübingen
Dario Azzimonti	Istituto Dalle Molle di Studi sull'Intelligenza Artificiale
Behrouz Babaki	Polytechnique Montréal
Rohit Babbar	Aalto University
Housam Babiker	University of Alberta
Davide Bacciu	University of Pisa
Thomas Baeck	Leiden University
Abdelkader Baggag	Qatar Computing Research Institute
Zilong Bai	University of California, Davis
Jiyang Bai	Florida State University
Sambaran Bandyopadhyay	IBM
Mitra Baratchi	University of Twente
Christian Beecks	University of Münster
Anna Beer	Ludwig Maximilian University of Munich

Adnene Belfodil	Munic Car Data
Aimene Belfodil	INSA Lyon
Ines Ben Kraiem	UT2J-IRIT
Anes Bendimerad	LIRIS
Christoph Bergmeir	Monash University
Max Berrendorf	Ludwig Maximilian University of Munich
Louis Béthune	ENS de Lyon
Anton Björklund	University of Helsinki
Alexandre Blansché	Université de Lorraine
Laurens Bliek	Delft University of Technology
Isabelle Bloch	ENST - CNRS UMR 5141 LTCI
Gianluca Bontempi	Université Libre de Bruxelles
Felix Borutta	Ludwig-Maximilians-Universität München
Ahcène Boubekki	Leuphana Universität Lüneburg
Tanya Braun	University of Lübeck
Wieland Brendel	University of Tübingen
Klaus Brinker	Hamm-Lippstadt University of Applied Sciences
David Browne	Insight Centre for Data Analytics
Sebastian Bruckert	Otto Friedrich University Bamberg
Mirko Bunse	TU Dortmund University
Sophie Burkhardt	University of Mainz
Haipeng Cai	Washington State University
Lele Cao	Tsinghua University
Manliang Cao	Fudan University
Defu Cao	Peking University
Antonio Carta	University of Pisa
Remy Cazabet	Université Lyon 1
Abdulkadir Celikkanat	CentraleSupelec, Paris-Saclay University
Christophe Cerisara	LORIA
Carlos Cernuda	Mondragon University
Vitor Cerqueira	LIAAD-INESCTEC
Mattia Cerrato	Università di Torino
Ricardo Cerri	Federal University of São Carlos
Laetitia Chapel	IRISA
Vaggos Chatziafratis	Stanford University
El Vaigh Cheikh Brahim	Inria/IRISA Rennes
Yifei Chen	University of Groningen
Junyang Chen	University of Macau
Jiaoyan Chen	University of Oxford
Huiyuan Chen	Case Western Reserve University
Run-Qing Chen	Xiamen University
Tianyi Chen	Microsoft
Lingwei Chen	The Pennsylvania State University
Senpeng Chen	UESTC
Liheng Chen	Shanghai Jiao Tong University
Siming Chen	Frauenhofer IAIS

Liang Chen	Sun Yat-sen University
Dawei Cheng	Shanghai Jiao Tong University
Wei Cheng	NEC Labs America
Wen-Hao Chiang	Indiana University - Purdue University Indianapolis
Feng Chong	Beijing Institute of Technology
Pantelis Chronis	Athena Research Center
Victor W. Chu	The University of New South Wales
Xin Cong	Institute of Information Engineering, Chinese Academy of Sciences
Roberto Corizzo	UNIBA
Mustafa Coskun	Case Western Reserve University
Gustavo De Assis Costa	Instituto Federal de Educação, Ciência e Tecnologia de Goiás
Fabrizio Costa	University of Exeter
Miguel Couceiro	Inria
Shiyao Cui	Institute of Information Engineering, Chinese Academy of Sciences
Bertrand Cuissart	GREYC
Mohamad H. Danesh	Oregon State University
Thi-Bich-Hanh Dao	University of Orléans
Cedric De Boom	Ghent University
Marcos Luiz de Paula Bueno	Technische Universiteit Eindhoven
Matteo Dell'Amico	NortonLifeLock
Qi Deng	Shanghai University of Finance and Economics
Andreas Dengel	German Research Center for Artificial Intelligence
Sourya Dey	University of Southern California
Yao Di	Institute of Computing Technology, Chinese Academy of Sciences
Stefano Di Frischia	University of L'Aquila
Jilles Dibangoye	INSA Lyon
Felix Dietrich	Technical University of Munich
Jiahao Ding	University of Houston
Yao-Xiang Ding	Nanjing University
Tianyu Ding	Johns Hopkins University
Rui Ding	Microsoft
Thang Doan	McGill University
Carola Doerr	Sorbonne University, CNRS
Xiao Dong	The University of Queensland
Wei Du	University of Arkansas
Xin Du	Technische Universiteit Eindhoven
Yuntao Du	Nanjing University
Stefan Duffner	LIRIS
Sebastijan Dumancic	Katholieke Universiteit Leuven
Valentin Durand de Gevigney	IRISA

Saso Dzeroski	Jožef Stefan Institute
Mohamed Elati	Université d'Evry
Lukas Enderich	Robert Bosch GmbH
Dominik Endres	Philipps-Universität Marburg
Francisco Escolano	University of Alicante
Bjoern Eskofier	Friedrich-Alexander University Erlangen-Nürnberg
Roberto Esposito	Università di Torino
Georgios Exarchakis	Institut de la Vision
Melanie F. Pradier	Harvard University
Samuel G. Fadel	Universidade Estadual de Campinas
Evgeniy Faerman	Ludwig Maximilian University of Munich
Yujie Fan	Case Western Reserve University
Elaine Faria	Federal University of Uberlândia
Golnoosh Farnadi	Mila/University of Montreal
Fabio Fassetti	University of Calabria
Ad Feelders	Utrecht University
Yu Fei	Harbin Institute of Technology
Wenjie Feng	The Institute of Computing Technology, Chinese Academy of Sciences
Zunlei Feng	Zhejiang University
Cesar Ferri	Universitat Politècnica de València
Raul Fidalgo-Merino	European Commission Joint Research Centre
Murat Firat	Technische Universiteit Eindhoven
Francoise Fogelman-Soulié	Tianjin University
Vincent Fortuin	ETH Zurich
Iordanis Fostiropoulos	University of Southern California
Eibe Frank	University of Waikato
Benoît Frénay	Université de Namur
Nikolaos Freris	University of Science and Technology of China
Moshe Gabel	University of Toronto
Ricardo José Gabrielli Barreto Campello	University of Newcastle
Esther Galbrun	University of Eastern Finland
Claudio Gallicchio	University of Pisa
Yuanning Gao	Shanghai Jiao Tong University
Alberto Garcia-Duran	Ecole Polytechnique Fédérale de Lausanne
Eduardo Garrido	Universidad Autónoma de Madrid
Clément Gautrais	KU Leuven
Arne Gevaert	Ghent University
Giorgos Giannopoulos	IMSI, "Athena" Research Center
C. Lee Giles	The Pennsylvania State University
Ioana Giurgiu	IBM Research - Zurich
Thomas Goerttler	TU Berlin
Heitor Murilo Gomes	University of Waikato
Chen Gong	Shanghai Jiao Tong University
Zhiguo Gong	University of Macau

Hongyu Gong	University of Illinois at Urbana-Champaign
Pietro Gori	Télécom Paris
James Goulding	University of Nottingham
Kshitij Goyal	Katholieke Universiteit Leuven
Dmitry Grishchenko	Université Grenoble Alpes
Moritz Grosse-Wentrup	University of Vienna
Sebastian Gruber	Siemens AG
John Grundy	Monash University
Kang Gu	Dartmouth College
Jindong Gu	Siemens
Riccardo Guidotti	University of Pisa
Tias Guns	Vrije Universiteit Brussel
Ruocheng Guo	Arizona State University
Yiluan Guo	Singapore University of Technology and Design
Xiaobo Guo	University of Chinese Academy of Sciences
Thomas Guyet	IRISA
Jiawei Han	University of Illinois at Urbana-Champaign
Zhiwei Han	fortiss GmbH
Tom Hanika	University of Kassel
Shonosuke Harada	Kyoto University
Marwan Hassani	Technische Universiteit Eindhoven
Jianhao He	Sun Yat-sen University
Deniu He	Chongqing University of Posts and Telecommunications
Dongxiao He	Tianjin University
Stefan Heidekrueger	Technical University of Munich
Nandyala Hemachandra	Indian Institute of Technology Bombay
Till Hendrik Schulz	University of Bonn
Alexander Hepburn	University of Bristol
Sibylle Hess	Technische Universiteit Eindhoven
Javad Heydari	LG Electronics
Joyce Ho	Emory University
Shunsuke Horii	Waseda University
Tamas Horvath	University of Bonn and Fraunhofer IAIS
Mehran Hossein Zadeh Bazargani	University College Dublin
Robert Hu	University of Oxford
Weipeng Huang	Insight
Jun Huang	University of Tokyo
Haojie Huang	The University of New South Wales
Hong Huang	UGoe
Shenyang Huang	McGill University
Vân Anh Huynh-Thu	University of Liège
Dino Ienco	INRAE
Siohoi Ieng	Institut de la Vision
Angelo Impedovo	Università "Aldo Moro" degli studi di Bari

Muhammad Imran Razzak	Deakin University
Vasileios Iosifidis	Leibniz University Hannover
Joseph Isaac	Indian Institute of Technology Madras
Md Islam	Washington State University
Ziyu Jia	Beijing Jiaotong University
Lili Jiang	Umeå University
Yao Jiangchao	Alibaba
Tan Jianlong	Institute of Information Engineering, Chinese Academy of Sciences
Baihong Jin	University of California, Berkeley
Di Jin	Tianjin University
Wei Jing	Xi'an Jiaotong University
Jonathan Jouanne	ARIADNEXT
Ata Kaban	University of Birmingham
Tomasz Kajdanowicz	Wrocław University of Science and Technology
Sandesh Kamath	Chennai Mathematical Institute
Keegan Kang	Singapore University of Technology and Design
Bo Kang	Ghent University
Isak Karlsson	Stockholm University
Panagiotis Karras	Aarhus University
Nikos Katzouris	NCSR Demokritos
Uzay Kaymak	Technische Universiteit Eindhoven
Mehdi Kaytoue	Infologic
Pascal Kerschke	University of Münster
Jungtaek Kim	Pohang University of Science and Technology
Minyoung Kim	Samsung AI Center Cambridge
Masahiro Kimura	Ryukoku University
Uday Kiran	The University of Tokyo
Bogdan Kirillov	ITMO University
Péter Kiss	ELTE
Gerhard Klassen	Heinrich Heine University Düsseldorf
Dmitry Kobak	Eberhard Karls University of Tübingen
Masahiro Kohjima	NTT
Ziyi Kou	University of Rochester
Wouter Kouw	Technische Universiteit Eindhoven
Fumiya Kudo	Hitachi, Ltd.
Piotr Kulczycki	Systems Research Institute, Polish Academy of Sciences
Ilona Kulikovskikh	Samara State Aerospace University
Rajiv Kumar	IIT Bombay
Pawan Kumar	IIT Kanpur
Suhansanu Kumar	University of Illinois, Urbana-Champaign
Abhishek Kumar	University of Helsinki
Gautam Kunapuli	The University of Texas at Dallas
Takeshi Kurashima	NTT
Vladimir Kuzmanovski	Jožef Stefan Institute

Anisio Lacerda	Centro Federal de Educação Tecnologica de Minas Gerais
Patricia Ladret	GIPSA-lab
Fabrizio Lamberti	Politecnico di Torino
James Large	University of East Anglia
Duc-Trong Le	University of Engineering and Technology, VNU Hanoi
Trung Le	Monash University
Luce le Gorrec	University of Strathclyde
Antoine Ledent	TU Kaiserslautern
Kangwook Lee	University of Wisconsin-Madison
Felix Leibfried	PROWLER.io
Florian Lemmerich	RWTH Aachen University
Carson Leung	University of Manitoba
Edouard Leurent	Inria
Naiqi Li	Tsinghua-UC Berkeley Shenzhen Institute
Suyi Li	The Hong Kong University of Science and Technology
Jundong Li	University of Virginia
Yidong Li	Beijing Jiaotong University
Xiaoting Li	The Pennsylvania State University
Yaoman Li	CUHK
Rui Li	Inspur Group
Wenye Li	The Chinese University of Hong Kong (Shenzhen)
Mingming Li	Institute of Information Engineering, Chinese Academy of Sciences
Yexin Li	Hong Kong University of Science and Technology
Qinghua Li	Renmin University of China
Yaohang Li	Old Dominion University
Yuxuan Liang	National University of Singapore
Zhimin Liang	Institute of Computing Technology, Chinese Academy of Sciences
Hongwei Liang	Microsoft
Nengli Lim	Singapore University of Technology and Design
Suwen Lin	University of Notre Dame
Yangxin Lin	Peking University
Aldo Lipani	University College London
Marco Lippi	University of Modena and Reggio Emilia
Alexei Lisitsa	University of Liverpool
Lin Liu	Taiyuan University of Technology
Weiwen Liu	The Chinese University of Hong Kong
Yang Liu	JD
Huan Liu	Arizona State University
Tianbo Liu	Thomas Jefferson National Accelerator Facility
Tongliang Liu	The University of Sydney
Weidong Liu	Inner Mongolia University
Kai Liu	Colorado School of Mines

Shiwei Liu	Technische Universiteit Eindhoven
Shenghua Liu	Institute of Computing Technology, Chinese Academy of Sciences
Corrado Loglisci	University of Bari Aldo Moro
Andrey Lokhov	Los Alamos National Laboratory
Yijun Lu	Alibaba Cloud
Xuequan Lu	Deakin University
Szymon Lukasik	AGH University of Science and Technology
Phuc Luong	Deakin University
Jianming Lv	South China University of Technology
Gengyu Lyu	Beijing Jiaotong University
Vijaikumar M.	Indian Institute of Science
Jing Ma	Emory University
Nan Ma	Shanghai Jiao Tong University
Sebastian Mair	Leuphana University Lüneburg
Marjan Mansourvar	University of Southern Denmark
Vincent Margot	Advestis
Fernando Martínez-Plumed	Joint Research Centre - European Commission
Florent Masseglia	Inria
Romain Mathonat	Université de Lyon
Deepak Maurya	Indian Institute of Technology Madras
Christian Medeiros Adriano	Hasso-Plattner-Institut
Purvanshi Mehta	University of Rochester
Tobias Meisen	Bergische Universität Wuppertal
Luciano Melodia	Friedrich-Alexander Universität Erlangen-Nürnberg
Ernestina Menasalvas	Universidad Politécnica de Madrid
Vlado Menkovski	Technische Universiteit Eindhoven
Engelbert Mephu Nguifo	Université Clermont Auvergne
Alberto Maria Metelli	Politecnico di Milano
Donald Metzler	Google
Anke Meyer-Baese	Florida State University
Richard Meyes	University of Wuppertal
Haithem Mezni	University of Jendouba
Paolo Mignone	Università degli Studi di Bari Aldo Moro
Matej Mihelčić	University of Zagreb
Decebal Constantin Mocanu	University of Twente
Christoph Molnar	Ludwig Maximilian University of Munich
Lia Morra	Politecnico di Torino
Christopher Morris	TU Dortmund University
Tadeusz Morzy	Poznań University of Technology
Henry Moss	Lancaster University
Tetsuya Motokawa	University of Tsukuba
Mathilde Mougeot	Université Paris-Saclay
Tingting Mu	The University of Manchester
Andreas Mueller	NYU
Tanmoy Mukherjee	Queen Mary University of London

Ksenia Mukhina	ITMO University
Peter Müllner	Know-Center
Guido Muscioni	University of Illinois at Chicago
Waleed Mustafa	TU Kaiserslautern
Mohamed Nadif	University of Paris
Ankur Nahar	Indian Institute of Technology Jodhpur
Kei Nakagawa	Nomura Asset Management Co., Ltd.
Haïfa Nakouri	University of Tunis
Mirco Nanni	KDD-Lab ISTI-CNR Pisa
Nicolo' Navarin	University of Padova
Richi Nayak	Queensland University of Technology
Mojtaba Nayyeri	University of Bonn
Daniel Neider	MPI SWS
Nan Neng	Institute of Information Engineering, Chinese Academy of Sciences
Stefan Neumann	University of Vienna
Dang Nguyen	Deakin University
Kien Duy Nguyen	University of Southern California
Jingchao Ni	NEC Laboratories America
Vlad Niculae	Instituto de Telecomunicações
Sofia Maria Nikolakaki	Boston University
Kun Niu	Beijing University of Posts and Telecommunications
Ryo Nomura	Waseda University
Eirini Ntoutsi	Leibniz University Hannover
Andreas Nuernberger	Otto von Guericke University of Magdeburg
Tsuyoshi Okita	Kyushu Institute of Technology
Maria Oliver Parera	GIPSA-lab
Bruno Ordozgoiti	Aalto University
Sindhu Padakandla	Indian Institute of Science
Tapio Pahikkala	University of Turku
Joao Palotti	Qatar Computing Research Institute
Guansong Pang	The University of Adelaide
Pance Panov	Jožef Stefan Institute
Konstantinos Papangelou	The University of Manchester
Yulong Pei	Technische Universiteit Eindhoven
Nikos Pelekis	University of Piraeus
Thomas Pellegrini	Université Toulouse III - Paul Sabatier
Charlotte Pelletier	Univ. Bretagne Sud
Jaakko Peltonen	Aalto University and Tampere University
Shaowen Peng	Kyushu University
Siqi Peng	Kyoto University
Bo Peng	The Ohio State University
Lukas Pensel	Johannes Gutenberg University Mainz
Aritz Pérez Martínez	Basque Center for Applied Mathematics
Lorenzo Perini	KU Leuven
Matej Petković	Jožef Stefan Institute

Bernhard Pfahringer	University of Waikato
Weiguo Pian	Chongqing University
Francesco Piccialli	University of Naples Federico II
Sebastian Pineda Arango	University of Hildesheim
Gianvito Pio	University of Bari "Aldo Moro"
Giuseppe Pirrò	Sapienza University of Rome
Anastasia Podosinnikova	Massachusetts Institute of Technology
Sebastian Pölsterl	Ludwig Maximilian University of Munich
Vamsi Potluru	JP Morgan AI Research
Rafael Poyiadzi	University of Bristol
Surya Prakash	University of Canberra
Paul Prasse	University of Potsdam
Rameshwar Pratap	Indian Institute of Technology Mandi
Jonas Prellberg	University of Oldenburg
Hugo Proenca	Leiden Institute of Advanced Computer Science
Ricardo Prudencio	Federal University of Pernambuco
Petr Pulc	Institute of Computer Science of the Czech Academy of Sciences
Lei Qi	Iowa State University
Zhenyue Qin	The Australian National University
Rahul Ragesh	PES University
Tahrima Rahman	The University of Texas at Dallas
Zana Rashidi	York University
S. S. Ravi	University of Virginia and University at Albany – SUNY
Ambrish Rawat	IBM
Henry Reeve	University of Birmingham
Reza Refaei Afshar	Technische Universiteit Eindhoven
Navid Rekabsaz	Johannes Kepler University Linz
Yongjian Ren	Shandong University
Zhiyun Ren	The Ohio State University
Guohua Ren	LG Electronics
Yuxiang Ren	Florida State University
Xavier Renard	AXA
Martí Renedo Mirambell	Universitat Politècnica de Catalunya
Gavin Rens	Katholiek Universiteit Leuven
Matthias Renz	Christian-Albrechts-Universität zu Kiel
Guillaume Richard	EDF R&D
Matteo Riondato	Amherst College
Niklas Risse	Bielefeld University
Lars Rosenbaum	Robert Bosch GmbH
Celine Rouveirol	Université Sorbonne Paris Nord
Shoumik Roychoudhury	Temple University
Polina Rozenshtein	Aalto University
Peter Rubbens	Flanders Marine Institute (VLIZ)
David Ruegamer	LMU Munich

Matteo Ruffini	ToolsGroup
Ellen Rushe	Insight Centre for Data Analytics
Amal Saadallah	TU Dortmund
Yogish Sabharwal	IBM Research - India
Mandana Saebi	University of Notre Dame
Aadirupa Saha	IISc
Seyed Erfan Sajjadi	Brunel University
Durgesh Samariya	Federation University
Md Samiullah	Monash University
Mark Sandler	Google
Raul Santos-Rodriguez	University of Bristol
Yucel Saygin	Sabancı University
Pierre Schaus	UCLouvain
Fabian Scheipl	Ludwig Maximilian University of Munich
Katerina Schindlerova	University of Vienna
Ute Schmid	University of Bamberg
Daniel Schmidt	Monash University
Sebastian Schmoll	Ludwig Maximilian University of Munich
Johannes Schneider	University of Liechtenstein
Marc Schoenauer	Inria Saclay Île-de-France
Jonas Schouterden	Katholieke Universiteit Leuven
Leo Schwinn	Friedrich-Alexander-Universität Erlangen-Nürnberg
Florian Seiffarth	University of Bonn
Nan Serra	NEC Laboratories Europe GmbH
Rowland Seymour	Univeristy of Nottingham
Ammar Shaker	NEC Laboratories Europe
Ali Shakiba	Vali-e-Asr University of Rafsanjan
Junming Shao	University of Science and Technology of China
Zhou Shao	Tsinghua University
Manali Sharma	Samsung Semiconductor Inc.
Jiaming Shen	University of Illinois at Urbana-Champaign
Ying Shen	Sun Yat-sen University
Hao Shen	fortiss GmbH
Tao Shen	University of Technology Sydney
Ge Shi	Beijing Institute of Technology
Ziqiang Shi	Fujitsu Research & Development Center
Masumi Shirakawa	hapicom Inc./Osaka University
Kai Shu	Arizona State University
Amila Silva	The University of Melbourne
Edwin Simpson	University of Bristol
Dinesh Singh	RIKEN Center for Advanced Intelligence Project
Jaspreet Singh	L3S Research Centre
Spiros Skiadopoulos	University of the Peloponnese
Gavin Smith	University of Nottingham
Miguel A. Solinas	CEA
Dongjin Song	NEC Labs America

Arnaud Soulet	Université de Tours
Marvin Ssemambo	Makerere University
Michiel Stock	Ghent University
Filipo Studzinski Perotto	Institut de Recherche en Informatique de Toulouse
Adisak Sukul	Iowa State University
Lijuan Sun	Beijing Jiaotong University
Tao Sun	National University of Defense Technology
Ke Sun	Peking University
Yue Sun	Beijing Jiaotong University
Hari Sundaram	University of Illinois at Urbana-Champaign
Gero Szepannek	Stralsund University of Applied Sciences
Jacek Tabor	Jagiellonian University
Jianwei Tai	IIE, CAS
Naoya Takeishi	RIKEN Center for Advanced Intelligence Project
Chang Wei Tan	Monash University
Jinghua Tan	Southwestern University of Finance and Economics
Zeeshan Tariq	Ulster University
Bouadi Tassadit	IRISA-Université de Rennes 1
Maryam Tavakol	TU Dortmund
Romain Tavenard	Univ. Rennes 2/LETG-COSTEL/IRISA-OBELIX
Alexandre Termier	Université de Rennes 1
Janek Thomas	Fraunhofer Institute for Integrated Circuits IIS
Manoj Thulasidas	Singapore Management University
Hao Tian	Syracuse University
Hiroyuki Toda	NTT
Jussi Tohka	University of Eastern Finland
Ricardo Torres	Norwegian University of Science and Technology
Isaac Triguero Velázquez	University of Nottingham
Sandhya Tripathi	Indian Institute of Technology Bombay
Holger Trittenbach	Karlsruhe Institute of Technology
Peter van der Putten	Leiden University & Pegasystems
Elia Van Wolputte	KU Leuven
Fabio Vandin	University of Padova
Titouan Vayer	IRISA
Ashish Verma	IBM Research - US
Bouvier Victor	Sidetrade MICS
Julia Vogt	University of Basel
Tim Vor der Brück	Lucerne University of Applied Sciences and Arts
Yb W.	Chongqing University
Krishna Wadhwani	Indian Institute of Technology Bombay
Huaiyu Wan	Beijing Jiaotong University
Qunbo Wang	Beihang University
Beilun Wang	Southeast University
Yiwei Wang	National University of Singapore
Bin Wang	Xiaomi AI Lab

Jiong Wang	Institute of Information Engineering, Chinese Academy of Sciences
Xiaobao Wang	Tianjin University
Shuheng Wang	Nanjing University of Science and Technology
Jihu Wang	Shandong University
Haobo Wang	Zhejiang University
Xianzhi Wang	University of Technology Sydney
Chao Wang	Shanghai Jiao Tong University
Jun Wang	Southwest University
Jing Wang	Beijing Jiaotong University
Di Wang	Nanyang Technological University
Yashen Wang	China Academy of Electronics and Information Technology of CETC
Qinglong Wang	McGill University
Sen Wang	University of Queensland
Di Wang	State University of New York at Buffalo
Qing Wang	Information Science Research Centre
Guoyin Wang	Chongqing University of Posts and Telecommunications
Thomas Weber	Ludwig-Maximilians-Universität München
Lingwei Wei	University of Chinese Academy of Sciences; Institute of Information Engineering, CAS
Tong Wei	Nanjing University
Pascal Welke	University of Bonn
Yang Wen	University of Science and Technology of China
Yanlong Wen	Nankai University
Paul Weng	UM-SJTU Joint Institute
Matthias Werner	ETAS GmbH, Bosch Group
Joerg Wicker	The University of Auckland
Uffe Wiil	University of Southern Denmark
Paul Wimmer	University of Lübeck; Robert Bosch GmbH
Martin Wistuba	University of Hildesheim
Feijie Wu	The Hong Kong Polytechnic University
Xian Wu	University of Notre Dame
Hang Wu	Georgia Institute of Technology
Yubao Wu	Georgia State University
Yichao Wu	SenseTime Group Limited
Xi-Zhu Wu	Nanjing University
Jia Wu	Macquarie University
Yang Xiaofei	Harbin Institute of Technology, Shenzhen
Yuan Xin	University of Science and Technology of China
Liu Xinshun	VIVO
Taufik Xu	Tsinghua University
Jinhui Xu	State University of New York at Buffalo
Depeng Xu	University of Arkansas
Peipei Xu	University of Liverpool

Yichen Xu	Beijing University of Posts and Telecommunications
Bo Xu	Donghua University
Hansheng Xue	Harbin Institute of Technology, Shenzhen
Naganand Yadati	Indian Institute of Science
Akihiro Yamaguchi	Toshiba Corporation
Haitian Yang	Institute of Information Engineering, Chinese Academy of Sciences
Hongxia Yang	Alibaba Group
Longqi Yang	HPCL
Xiaochen Yang	University College London
Yuhan Yang	Shanghai Jiao Tong University
Ya Zhou Yang	National University of Defense Technology
Feidiao Yang	Institute of Computing Technology, Chinese Academy of Sciences
Liu Yang	Tianjin University
Chaoqi Yang	University of Illinois at Urbana-Champaign
Carl Yang	University of Illinois at Urbana-Champaign
Guanyu Yang	Xi'an Jiaotong - Liverpool University
Yang Yang	Nanjing University
Weicheng Ye	Carnegie Mellon University
Wei Ye	Peking University
Yanfang Ye	Case Western Reserve University
Kejiang Ye	SIAT, Chinese Academy of Sciences
Florian Yger	Université Paris-Dauphine
Yunfei Yin	Chongqing University
Lu Yin	Technische Universiteit Eindhoven
Wang Yingkui	Tianjin University
Kristina Yordanova	University of Rostock
Tao You	Northwestern Polytechnical University
Hong Qing Yu	University of Bedfordshire
Bowen Yu	Institute of Information Engineering, Chinese Academy of Sciences
Donghan Yu	Carnegie Mellon University
Yipeng Yu	Tencent
Shujian Yu	NEC Laboratories Europe
Jiadi Yu	Shanghai Jiao Tong University
Wenchao Yu	University of California, Los Angeles
Feng Yuan	The University of New South Wales
Chunyuan Yuan	Institute of Information Engineering, Chinese Academy of Sciences
Sha Yuan	Tsinghua University
Farzad Zafarani	Purdue University
Marco Zaffalon	IDSIA
Nayyar Zaidi	Monash University
Tianzi Zang	Shanghai Jiao Tong University
Gerson Zaverucha	Federal University of Rio de Janeiro

Javier Zazo	Harvard University
Albin Zehe	University of Würzburg
Yuri Zelenkov	National Research University Higher School of Economics
Amber Zelvelder	Umeå University
Mingyu Zhai	NARI Group Corporation
Donglin Zhan	Sichuan University
Yu Zhang	Southeast University
Wenbin Zhang	University of Maryland
Qiuchen Zhang	Emory University
Tong Zhang	PKU
Jianfei Zhang	Case Western Reserve University
Nailong Zhang	MassMutual
Yi Zhang	Nanjing University
Xiangliang Zhang	King Abdullah University of Science and Technology
Ya Zhang	Shanghai Jiao Tong University
Zongzhang Zhang	Nanjing University
Lei Zhang	Institute of Information Engineering, Chinese Academy of Sciences
Jing Zhang	Renmin University of China
Xianchao Zhang	Dalian University of Technology
Jiangwei Zhang	National University of Singapore
Fengpan Zhao	Georgia State University
Lin Zhao	Institute of Information Engineering, Chinese Academy of Sciences
Long Zheng	Huazhong University of Science and Technology
Zuowu Zheng	Shanghai Jiao Tong University
Tongya Zheng	Zhejiang University
Runkai Zheng	Jinan University
Cheng Zheng	University of California, Los Angeles
Wenbo Zheng	Xi'an Jiaotong University
Zhiqiang Zhong	University of Luxembourg
Caiming Zhong	Ningbo University
Ding Zhou	Columbia University
Yilun Zhou	MIT
Ming Zhou	Shanghai Jiao Tong University
Yanqiao Zhu	Institute of Automation, Chinese Academy of Sciences
Wenfei Zhu	King
Wanzheng Zhu	University of Illinois at Urbana-Champaign
Fuqing Zhu	Institute of Information Engineering, Chinese Academy of Sciences
Markus Zopf	TU Darmstadt
Weidong Zou	Beijing Institute of Technology
Jingwei Zuo	UVSQ

Program Committee Members, Applied Data Science Track

Deepak Ajwani	Nokia Bell Labs
Nawaf Alharbi	Kansas State University
Rares Ambrus	Toyota Research Institute
Maryam Amir Haeri	Technische Universität Kaiserslautern
Jean-Marc Andreoli	Naverlabs Europe
Cecilio Angulo	Universitat Politècnica de Catalunya
Stefanos Antaris	KTH Royal Institute of Technology
Nino Antulov-Fantulin	ETH Zurich
Francisco Antunes	University of Coimbra
Muhammad Umer Anwaar	Technical University of Munich
Cristian Axenie	Audi Konfuzius-Institut Ingolstadt/Technical University of Ingolstadt
Mehmet Cem Aytekin	Sabancı University
Anthony Bagnall	University of East Anglia
Marco Baldan	Leibniz University Hannover
Maria Bampa	Stockholm University
Karin Becker	UFRGS
Swarup Ranjan Behera	Indian Institute of Technology Guwahati
Michael Berthold	University of Konstanz
Antonio Bevilacqua	Insight Centre for Data Analytics
Ananth Reddy Bhimireddy	Indiana University Purdue University - Indianapolis
Haixia Bi	University of Bristol
Wu Bin	Zhengzhou University
Thibault Blanc Beyne	INP Toulouse
Andrzej Bobyk	Maria Curie-Skłodowska University
Antonio Bonafonte	Amazon
Ludovico Boratto	Eurecat
Massimiliano Botticelli	Robert Bosch GmbH
Maria Brbic	Stanford University
Sebastian Buschjäger	TU Dortmund
Rui Camacho	University of Porto
Doina Caragea	Kansas State University
Nicolas Carrara	University of Toronto
Michele Catasta	Stanford University
Oded Cats	Delft University of Technology
Tania Cerquitelli	Politecnico di Torino
Fabricio Ceschin	Federal University of Paraná
Jeremy Charlier	University of Luxembourg
Anveshi Charuvaka	GE Global Research
Liang Chen	Sun Yat-sen University
Zhiyong Cheng	Shandong Artificial Intelligence Institute

Silvia Chiusano	Politecnico di Torino
Cristian Consonni	Eurecat - Centre Tecnòlogic de Catalunya
Laure Crochepierre	RTE
Henggang Cui	Uber ATG
Tiago Cunha	University of Porto
Elena Daraio	Politecnico di Torino
Hugo De Oliveira	HEVA/Mines Saint-Étienne
Tom Decroos	Katholieke Universiteit Leuven
Himel Dev	University of Illinois at Urbana-Champaign
Eustache Diemert	Criteo AI Lab
Nat Dilokthanakul	Vidyasirimedhi Institute of Science and Technology
Daizong Ding	Fudan University
Kaize Ding	ASU
Ming Ding	Tsinghua University
Xiaowen Dong	University of Oxford
Sourav Dutta	Huawei Research
Madeleine Ellis	University of Nottingham
Benjamin Evans	Brunel University London
Francesco Fabbri	Universitat Pompeu Fabra
Benjamin Fauber	Dell Technologies
Fuli Feng	National University of Singapore
Oluwaseyi Feyisetan	Amazon
Ferdinando Fioretto	Georgia Institute of Technology
Caio Flexa	Federal University of Pará
Germain Forestier	Université de Haute-Alsace
Blaz Fortuna	Qlector
Enrique Frias-Martinez	Telefónica Research and Development
Zuohui Fu	Rutgers University
Takahiro Fukushige	Nissan Motor Co., Ltd.
Chen Gao	Tsinghua University
Johan Garcia	Karlstad University
Marco Gärtler	ABB Corporate Research Center
Kanishka Ghosh Dastidar	Universität Passau
Biraja Ghoshal	Brunel University London
Lovedeep Gondara	Simon Fraser University
Severin Gsponer	Science Foundation Ireland
Xinyu Guan	Xi'an Jiaotong University
Karthik Gurumoorthy	Amazon
Marina Haliem	Purdue University
Massinissa Hamidi	Laboratoire LIPN-UMR CNRS 7030, Sorbonne Paris Cité
Junheng Hao	University of California, Los Angeles
Khadidja Henni	Université TÉLUQ

Martin Holena	Institute of Computer Science Academy of Sciences of the Czech Republic
Ziniu Hu	University of California, Los Angeles
Weihua Hu	Stanford University
Chao Huang	University of Notre Dame
Hong Huang	UGoe
Inhwan Hwang	Seoul National University
Chidubem Iddianozie	University College Dublin
Omid Isfahani Alamdari	University of Pisa
Guillaume Jacquet	Joint Research Centre - European Commission
Nishtha Jain	ADAPT Centre
Samyak Jain	NIT Karnataka, Surathkal
Mohsan Jameel	University of Hildesheim
Di Jiang	WeBank
Song Jiang	University of California, Los Angeles
Khiary Jihed	Johannes Kepler Universität Linz
Md. Rezaul Karim	Fraunhofer FIT
Siddhant Katyan	IIIT Hyderabad
Jin Kyu Kim	Facebook
Sundong Kim	Institute for Basic Science
Tomas Kliegr	Prague University of Economics and Business
Yun Sing Koh	The University of Auckland
Aljaz Kosmerlj	Jožef Stefan Institute
Jitin Krishnan	George Mason University
Alejandro Kuratomi	Stockholm University
Charlotte Laclau	Laboratoire Hubert Curien
Filipe Lauar	Federal University of Minas Gerais
Thach Le Nguyen	The Insight Centre for Data Analytics
Wenqiang Lei	National University of Singapore
Camelia Lemnaru	Universitatea Tehnică din Cluj-Napoca
Carson Leung	University of Manitoba
Meng Li	Ant Financial Services Group
Zeyu Li	University of California, Los Angeles
Pieter Libin	Vrije Universiteit Brussel
Tomislav Lipic	Ruđer Bošković Institut
Bowen Liu	Stanford University
Yin Lou	Ant Financial
Martin Lukac	Nazarbayev University
Brian Mac Namee	University College Dublin
Fragkiskos Malliaros	Université Paris-Saclay
Mirko Marras	University of Cagliari
Smit Marvaniya	IBM Research - India
Kseniia Melnikova	Samsung R&D Institute Russia

João Mendes-Moreira	University of Porto
Ioannis Mitros	Insight Centre for Data Analytics
Elena Mocanu	University of Twente
Hebatallah Mohamed	Free University of Bozen-Bolzano
Roghayeh Mojarad	Université Paris-Est Créteil
Mirco Nanni	KDD-Lab ISTI-CNR Pisa
Juggapong Natwichai	Chiang Mai University
Sasho Nedelkoski	TU Berlin
Kei Nemoto	The Graduate Center, City University of New York
Ba-Hung Nguyen	Japan Advanced Institute of Science and Technology
Tobias Nickchen	Paderborn University
Aastha Nigam	LinkedIn Inc
Inna Novalija	Jožef Stefan Institute
Francisco Ocegueda-Hernandez	National Oilwell Varco
Tsuyoshi Okita	Kyushu Institute of Technology
Oghenejokpeme Orhobor	The University of Manchester
Aomar Osmani	Université Sorbonne Paris Nord
Latifa Oukhellou	IFSTTAR
Rodolfo Palma	Inria Chile
Pankaj Pandey	Indian Institute of Technology Gandhinagar
Luca Pappalardo	University of Pisa, ISTI-CNR
Paulo Paraíso	INESC TEC
Namyong Park	Carnegie Mellon University
Chanyoung Park	University of Illinois at Urbana-Champaign
Miquel Perelló-Nieto	University of Bristol
Nicola Pezzotti	Philips Research
Tiziano Piccardi	Ecole Polytechnique Fédérale de Lausanne
Thom Pijnenburg	Elsevier
Valentina Poggioni	Università degli Studi di Perugia
Chuan Qin	University of Science and Technology of China
Jiezhong Qiu	Tsinghua University
Maria Ramirez-Loaiza	Intel Corporation
Manjusha Ravindranath	ASU
Zhaochun Ren	Shandong University
Antoine Richard	Georgia Institute of Technology
Kit Rodolfa	Carnegie Mellon University
Mark Patrick Roeling	Technical University of Delft
Soumyadeep Roy	Indian Institute of Technology Kharagpur
Ellen Rushe	Insight Centre for Data Analytics
Amal Saadallah	TU Dortmund
Carlos Salort Sanchez	Huawei
Eduardo Hugo Sanchez	IRT Saint Exupéry
Markus Schmitz	University of Erlangen-Nuremberg/BMW Group
Ayan Sengupta	Optum Global Analytics (India) Pvt. Ltd.
Ammar Shaker	NEC Laboratories Europe

Manali Sharma	Samsung Semiconductor Inc.
Jiaming Shen	University of Illinois at Urbana-Champaign
Dash Shi	LinkedIn
Ashish Sinha	IIT Roorkee
Yorick Spenrath	Technische Universiteit Eindhoven
Simon Stieber	University of Augsburg
Hendra Suryanto	Rich Data Corporation
Raunak Swarnkar	IIT Gandhinagar
Imen Trabelsi	National Engineering School of Tunis
Alexander Treiss	Karlsruhe Institute of Technology
Rahul Tripathi	Amazon
Dries Van Daele	Katholieke Universiteit Leuven
Ranga Raju Vatsavai	North Carolina State University
Vishnu Venkataraman	Credit Karma
Sergio Viademonte	Vale Institute of Technology, Vale SA
Yue Wang	Microsoft Research
Changzhou Wang	The Boeing Company
Xiang Wang	National University of Singapore
Hongwei Wang	Shanghai Jiao Tong University
Wenjie Wang	Emory University
Zirui Wang	Carnegie Mellon University
Shen Wang	University of Illinois at Chicago
Dingxian Wang	East China Normal University
Yoshikazu Washizawa	The University of Electro-Communications
Chrys Watson Ross	University of New Mexico
Dilusha Weeraddana	CSIRO
Ying Wei	The Hong Kong University of Science and Technology
Laksri Wijerathna	Monash University
Le Wu	Hefei University of Technology
Yikun Xian	Rutgers University
Jian Xu	Citadel
Haiqin Yang	Ping An Life
Yang Yang	Northwestern University
Carl Yang	University of Illinois at Urbana-Champaign
Chin-Chia Michael Yeh	Visa Research
Shujian Yu	NEC Laboratories Europe
Chung-Hsien Yu	University of Massachusetts Boston
Jun Yuan	The Boeing Company
Stella Zevio	LIPN
Hanwen Zha	University of California, Santa Barbara
Chuxu Zhang	University of Notre Dame
Fanjin Zhang	Tsinghua University
Xiaohan Zhang	Sony Interactive Entertainment
Xinyang Zhang	University of Illinois at Urbana-Champaign
Mia Zhao	Airbnb
Qi Zhu	University of Illinois at Urbana-Champaign

Hengshu Zhu	Baidu Inc.
Tommaso Zoppi	University of Florence
Lan Zou	Carnegie Mellon University

Program Committee Members, Demo Track

Deepak Ajwani	Nokia Bell Labs
Rares Ambrus	Toyota Research Institute
Jean-Marc Andreoli	NAVER LABS Europe
Ludovico Boratto	Eurecat
Nicolas Carrara	University of Toronto
Michelangelo Ceci	Università degli Studi di Bari Aldo Moro
Tania Cerquitelli	Politecnico di Torino
Liang Chen	Sun Yat-sen University
Jiawei Chen	Zhejiang University
Zhiyong Cheng	Shandong Artificial Intelligence Institute
Silvia Chiusano	Politecnico di Torino
Henggang Cui	Uber ATG
Tiago Cunha	University of Porto
Chris Develder	Ghent University
Nat Dilokthanakul	Vidyasirimedhi Institute of Science and Technology
Daizong Ding	Fudan University
Kaize Ding	ASU
Xiaowen Dong	University of Oxford
Fuli Feng	National University of Singapore
Enrique Frias-Martinez	Telefónica Research and Development
Zuohui Fu	Rutgers University
Chen Gao	Tsinghua University
Thomas Gärtner	TU Wien
Derek Greene	University College Dublin
Severin Gsponer	University College Dublin
Xinyu Guan	Xi'an Jiaotong University
Junheng Hao	University of California, Los Angeles
Ziniu Hu	University of California, Los Angeles
Chao Huang	University of Notre Dame
Hong Huang	UGoe
Neil Hurley	University College Dublin
Guillaume Jacquet	Joint Research Centre - European Commission
Di Jiang	WeBank
Song Jiang	University of California, Los Angeles
Jihed Khiari	Johannes Kepler Universität Linz
Mark Last	Ben-Gurion University of the Negev
Thach Le Nguyen	The Insight Centre for Data Analytics
Vincent Lemaire	Orange Labs
Camelia Lemnaru	Universitatea Tehnică din Cluj-Napoca
Bowen Liu	Stanford University

Sponsors

Contents – Part II

Adversarial Learning

Federated Learning

Transfer and Multi-task Learning

Bayesian Optimization and Few-Shot Learning

Deep Learning Optimization and Theory

ADMMiRNN: Training RNN with Stable Convergence via an Efficient ADMM Approach

Yu Tang[1], Zhigang Kan[1], Dequan Sun[1], Linbo Qiao[1(✉)], Jingjing Xiao[2], Zhiquan Lai[1], and Dongsheng Li[1(✉)]

[1] National University of Defense Technology, Changsha, China
qiao.linbo@nudt.edu.cn, lds1201@163.com
[2] Army Medical University (Third Military Medical University), Chongqing, China

Abstract. It is hard to train Recurrent Neural Network (RNN) with stable convergence and avoid gradient *vanishing* and *exploding*, as the weights in the recurrent unit are repeated from iteration to iteration. Moreover, RNN is sensitive to the initialization of weights and bias, which brings difficulty in the training phase. With the gradient-free feature and immunity to unsatisfactory conditions, the Alternating Direction Method of Multipliers (ADMM) has become a promising algorithm to train neural networks beyond traditional stochastic gradient algorithms. However, ADMM could not be applied to train RNN directly since the state in the recurrent unit is repetitively updated over timesteps. Therefore, this work builds a new framework named ADMMiRNN upon the unfolded form of RNN to address the above challenges simultaneously and provides novel update rules and theoretical convergence analysis. We explicitly specify essential update rules in the iterations of ADMMiRNN with deliberately constructed approximation techniques and solutions to each subproblem instead of vanilla ADMM. Numerical experiments are conducted on MNIST and text classification tasks, where ADMMiRNN achieves convergent results and outperforms compared baselines. Furthermore, ADMMiRNN trains RNN more stably without gradient vanishing or exploding compared to the stochastic gradient algorithms. Source code has been available at https://github.com/TonyTangYu/ADMMiRNN.

1 Introduction

Recurrent Neural Network (RNN) [4] has made great progress in various fields, namely language modelling, text classification [13], event extraction [17], and various real-world applications [10]. Although RNN models have been widely used, it is still difficult to train RNN models because of the *vanishing gradients* and *exploding gradients* problems[1]. Moreover, RNN models are sensitive to the

---

[1] More information about *vanishing gradients* and *vanishing gradients* could be found in [1].

Electronic supplementary material The online version of this chapter (https://doi.org/10.1007/978-3-030-67661-2_1) contains supplementary material, which is available to authorized users.

© Springer Nature Switzerland AG 2021
F. Hutter et al. (Eds.): ECML PKDD 2020, LNAI 12458, pp. 3–18, 2021.
https://doi.org/10.1007/978-3-030-67661-2_1

weights and biases [23], which may not converge with poor initialization. These problems still need a method to be solved simultaneously.

Nowadays, gradient-based training algorithms are widely used in deep learning [14], such as Stochastic Gradient Descent (SGD) [20], Adam [12], RMSProp [25]. However, they still suffer from *vanishing* or *exploding gradients*. Compared to the traditional gradient-based optimization algorithms, the Alternating Direction Method of Multipliers (ADMM) is a much more robust method for training neural networks. It has been recognized as a promising policy to alleviate *vanishing gradients* and *exploding gradients* problems and exert a tremendous fascination on researchers. Besides, ADMM is also immune to poor conditioning with gradient-free technique [24].

In light of these properties of ADMM and to alleviate the problems mentioned above in RNN simultaneously, we are motivated to train RNN models with ADMM. However, it is not easy to apply ADMM to RNNs directly due to the recurrent state compared with MLP and CNN. The recurrent states are updated over timesteps instead of iterations, which is not compatible with ADMM. In this paper, to tackle this problem, we propose an ADMMiRNN method with the theoretical analysis. Experimental comparisons between ADMMiRNN and some typical stochastic gradient algorithms, such as SGD and Adam, illustrate that ADMMiRNN avoids the *vanishing gradients* and *exploding gradients* problems and surpasses traditional stochastic gradient algorithms in term of stability and efficiency. The main contributions of this work are four-fold:

- We propose a new framework named ADMMiRNN for training RNN models via ADMM. ADMMiRNN is built upon the unfolded RNN unit, which is a remarkable feature of RNN, and could settle the problems of *gradient vanishing* or *exploding* and sensitive parameter initialization in RNN at the same time. Instead of using vanilla ADMM, some practical skills in our solution also help converge.
- To the best of our knowledge, we are the first to handle RNN training problems using ADMM, which is a gradient-free approach and brings extremely significant advantages on stability beyond traditional stochastic gradient algorithms.
- The update rules of ADMMiRNN is presented, and also the theoretical analysis of convergence property is given. Our analysis ensures that ADMMiRNN achieves an efficient and stable result. Moreover, the framework proposed in this work could be applied to various RNN-based tasks.
- Based on our theoretical analysis, numerical experiments are conducted on several real-world datasets. The experiment results demonstrate the efficiency of the proposed ADMMiRNN beyond some other typical optimizers. Experimental results further verify the stability of our approach.

2   Related Work

The fundamental research of Recurrent Neural Networks was published in the 1980s. RNNs are powerful to model problems with a defined order but no clear

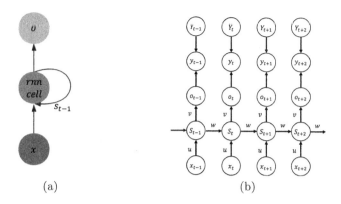

(a)   (b)

Fig. 1. Two different forms of RNN. **a**: The typical RNN cell. **b**: The unfolded form of (a), which is functionally identical to the original form [9].

concept of time, with a variant of Long short-term memory(LSTM) [11]. In [1], they argued that it is difficult to train RNN models due to the *vanishing gradients* and *exploding gradients*. Moreover, since RNN is sensitive to the initialization of weights and bias, those parameters should be initialized according to the input data [23]. In [18], they also state some difficulties in train RNNs. There is still a lack of a method to solve these above problems in RNN at the same time until now.

ADMM was first introduced in [6]. Its convergence was established in [5,7]. Since ADMM can decompose large problems with constraints into several small ones, it has been one of the most powerful optimization frameworks. It shows a multitude of well-performed properties in plenty of fields, such as machine learning [2], signal processing [22] and tensor decomposition [8].

In general, ADMM seeks to tackle the following problem:

$$\min_{x,y} f(x) + g(y), \quad \text{s.t.} \quad Ax + By = c. \tag{1}$$

Here, $f : \mathbb{R}^{n_1} \to \mathbb{R}$ and $g : \mathbb{R}^{n_2} \to \mathbb{R}$ are usually assumed to be convex functions. In Eq. (1), $A \in \mathbb{R}^{m \times n_1}$, $B \in \mathbb{R}^{m \times n_2}$, $c \in \mathbb{R}^m$, and $Ax + By = c$ is a linear constraint and n_1, n_2 are the dimensions of x, y respectively. It is solved by the Augmented Lagrangian Method which is formalized as:

$$\mathcal{L}_\beta(x, y, \lambda) = f(x) + g(y) + <\lambda, Ax + By - c> + \frac{\beta}{2}\|Ax + By - c\|^2, \tag{2}$$

where β is the penalty term, λ is the Lagrangian multiplier.

Since ADMM was first proposed, plenty of theoretical and practical works have been developed in recent years [16]. In 2016, [24] proposed a new method to train neural networks using ADMM. They abandoned traditional optimizers and adopted ADMM, which trains neural networks in a robust and parallel fashion. Furthermore, ADMM was applied to deep learning and obtained a remarkable result [26]. They provided a gradient-free method to train neural networks, which

Table 1. Important notations and corresponding descriptions.

Notations	Descriptions
t	The timestep
x_t	The input of RNN cells
o_t	The output of RNN cells
s_t	The state at timestep t
u	The weight corresponding to the input
w	The weight corresponding to the state
y_t	The prediction at timestep t
N	The cell numbers after unfolding
R	The loss function
$\Omega(w)$	The regularization term
θ	$\{u, w, b, a, s, v, c, o\}$
k	T he iteration count

gains convergent and great performance. Both of their works prove that ADMM is a powerful optimization method for neural networks because of its gradient-free property. However, RNNs are not as simple as a Multilayer Perceptron. The recurrent state brings many challenges to solving RNN with ADMM.

3   ADMM for RNNs

3.1   Notation

Before we dive into the ADMM methods for RNNs, we establish notations in this work. Considering a simple RNN cell as shown in Fig. 1(a), at timestep $t \geq 1$, x_t is the input of the RNN cell and o_t is the related output, RNNs could be expressed as:

$$
\begin{aligned}
&\Phi(x_1, x_2, \cdots, x_N, u, v, w, b, c) \\
&= vf(ux_N + w(vf(ux_{N-1} + w(\cdots vf(ux_0 + b) + c \cdots) + b) + c) + b) + c \quad (3) \\
&\quad + \Omega(W),
\end{aligned}
$$

where $f(\cdot)$ is an activation function and v, u, w, b and c are learnable parameters, $s_0 = 0$. These parameters are also unified. The recurrent state in RNNs varies over timesteps as well as iterations, which brings difficulties to applying ADMM into RNNs directly. We adopt an unfolding form of RNN unit shown in Fig. 1(b) and decouple these above parameters into three sub-problems. Normally, at timestep t, the updates are listed in the following:

$$
\begin{aligned}
a_t &= ux_t + ws_{t-1} + b, \\
s_t &= f(a_t), \quad (4) \\
o_t &= vs_t + c,
\end{aligned}
$$

whore $f(\cdot)$ io the activation function, ouch as RcLU or $tanh$, usually $tanh$ in RNNs. Important notations are summarized in Table 1. In this paper, we consider RNN in an unfolding form and present a theoretical analysis based on it.

For the sake of convenience, we define $\theta = \{u, w, b, a, s, v, c, o\}$ in the sequel. In term of applying ADMM into RNNs, assuming the RNN cell is unfolded into N continuous cells, we try to solve the mathematical problem as follows:

Problem 1.

$$\min_{\theta_t} \Phi(\theta_t) \equiv R(\theta_t) + \Omega(w),$$

$$\text{s.t.} \quad a_t = ux_t + ws_{t-1} + b, s_t = f(a_t), o_t = vs_t + c. \tag{5}$$

In Problem 1, $R(\theta_t)$ is the loss function which is convex and continuous, $\Omega(w)$ is the regularization term on the parameter w. It is also a convex and continuous function. Rather than solving Problem 1 directly, we can relax it by adding an l_2 penalty term and transform Eq. (5) into

Problem 2.

$$\min_{\theta_t} R(\theta_t) + \Omega(w) + \frac{\nu}{2} \sum_{t=1}^{N-1} (\|a_t - ux_t - ws_{t-1} - b\|^2$$

$$+ \|s_t - f(a_t)\|^2 + \|o_t - vs_t - c\|^2) \tag{6}$$

$$\text{s.t.} \quad a_N = ux_N + ws_{N-1} + b, s_N = f(a_N), o_N = vs_N + c,$$

where ν is a tuning parameter. Compared with Problem 1, Problem 2 is much easier to solve. According to [26], the solution of Problem 2 tends to be the solution of Problem 1 when $\nu \to \infty$. For simplicity and clarity, we often use $<\cdot, \cdot>$ to denote the inner product and $\tilde{k} = k + 1$. For a positive semidefinite matrix G, we define the $G-$norm of a vector as $\|x\|_G = \|G^{1/2}x\|_2 = \sqrt{x^T Gx}$.

3.2 ADMM Solver for RNN

As aforementioned in Sect. 2, we explain that ADMM utilizes the Augmented Lagrangian Method to solve problems like Eq. (2). Similarly, we adopt the same way and present the corresponding Lagrangian function of Eq. (6), namely Eq. (7):

$$\mathcal{L}_{\rho_1, \rho_2, \rho_3}(\theta) = R(o) + \Omega(w) + \phi(\theta_t), \tag{7}$$

where $\phi(\theta_t)$ is defined in Eq. (8).

$$\phi(\theta_t) = \frac{\nu}{2} \sum_{t=1}^{N-1} (\|a_t - ux_t - ws_{t-1} - b\|^2 + \|s_t - f(a_t)\|^2 + \|o_t - vs_t - c\|^2)$$

$$+ <\lambda_1, a_N - ux_N - ws_{N-1} - b> + <\lambda_2, s_N - f(a_N)> + <\lambda_3,$$

$$o_N - vs_N - c> + \frac{\rho_1}{2}\|a_N - ux_N - ws_{N-1} - b\|^2 + \frac{\rho_2}{2}\|s_N - f(a_N)\|^2$$

$$+ \frac{\rho_3}{2}\|o_N - vs_N - c\|^2. \tag{8}$$

Problem 2 is separated into eight subproblems and could be solved through the updates of these parameters in θ_t. Note that u, w, b, v, c in θ are not changed over timestep t. Consequently, these parameters are supposed to update over iterations. To make it clear, we only describe the typical update rules for u, a and s in the following subsections because there are some useful and typical skills in these subproblems while analysis of the other parameters detailed in Appendix A is similar.

Update u. We begin with the update of u in Eq. (7) at iteration k. In Eq. (8), u and x_t are coupled. As a result, we need to calculate the pseudo-inverse of the (rectangular) matrix x_t, making it harder for the training process. In order to solve this problem, we define $\mathbf{G} = rI_d - \rho_1 x_t^T x_t$ and replace it with Eq. (9).

$$
\begin{aligned}
u^{\tilde{k}} \leftarrow \arg\min \frac{\nu}{2} \sum_{t=1}^{N-1} \|a_t - ux_t - ws_{t-1} - b\|^2 + \frac{\rho_1}{2} \|a_N \\
- ux_N - ws_{N-1} - b - \lambda_1/\rho_1\|^2 + \frac{N}{2} \|u - u^k\|_{\mathbf{G}}^2.
\end{aligned}
\tag{9}
$$

It is equivalent to the linearized proximal point method inspired by [21]:

$$
\begin{aligned}
u^{\tilde{k}} \leftarrow \arg\min \frac{Nr}{2} \|u - u^k\|^2 + \nu(u - u^k)^T \sum_{t=1}^{N-1} [(x_t^k)^T \\
(a_t - u^k x_t^k - w^k s_{t-1}^k - b^k)] + \rho_1 (u - u^k)^T [(x_N^k)^T \\
(a_N^k - u^k x_N^k - w^k s_{N-1}^k - b^k - \lambda_1^k/\rho_1)].
\end{aligned}
\tag{10}
$$

In this way, the update of u is greatly sped up than the vanilla ADMM. It is worth noting that r needs to be set properly and r could also affect the performance of ADMMiRNN.

Update a. Adding a proximal term similar to that in Sect. 3.2, if $t < N$, this could be done by

$$
\begin{aligned}
a_t^{\tilde{k}} \leftarrow \arg\min \frac{r}{2} \|a_t - a_t^k\|^2 + \nu(a_t - a_t^k)^T (a_t^k - u^k x_t^k \\
- w^k s_{t-1}^k - b^k) + \frac{\nu}{2} \|s_t - f(a_t)\|^2.
\end{aligned}
\tag{11}
$$

When $t = N$,

$$
\begin{aligned}
a_N^{\tilde{k}} \leftarrow \arg\min \frac{r}{2} \|a_N - a_N^k\|^2 + \rho_1 (a_N - a_N^k)^T (a_N^k - u^k x_N^k - w^k s_{N-1}^k \\
- b^k - \lambda_1/\rho_1) + \frac{\rho_2}{2} \|s_N - f(a_N) - \lambda_2/\rho_2\|^2.
\end{aligned}
\tag{12}
$$

Here is a **trick**: If a_t is small enough, we have $f(a_t) = a_t$ as a result of the property of *tanh* function. In this way, we could simplify the calculation of a_t.

Update s. The parameter s represents the hidden state in the RNN cell shown in Fig. 1(a). With regard to the update of s, there are s_{t-1} and s_t in Eq. (8). However, we only consider s_t in the RNN model. It is because s_{t-1} has been updated in last unit and would cause calculation redundancy in the updating process. This is another **trick** in our solution. Besides, s_t also needs to be decoupled with w. If $t < N$, we could update s_t through

$$s_t^{\tilde{k}} \leftarrow \arg\min \frac{r}{2}\|s_t - s_t^k\|^2 + \nu(s_t - s_t^k)^T[(v^k)^T(o_t^k - v^k s_t^k - c^k)] + \frac{\nu}{2}\|s_t - f(a_t)\|^2.$$
(13)

And when $t = N$,

$$s_N^{\tilde{k}} \leftarrow \arg\min \frac{r}{2}\|s_N - s_N^k\|^2 + \rho_3(s_N - s_N^k)^T[(v^k)^T(o_N^k - v^k s_N^k - c^k - \lambda_3^k/\rho_3)].$$
(14)

Update Lagrangian Multipliers. Similar to the parameters update, λ_1, λ_2 and λ_3 are updated as follows respectively:

$$\lambda_1^{\tilde{k}} = \lambda_1^k + \rho_1(a_N - ux_N - ws_{N-1} - b),$$
(15a)

$$\lambda_2^{\tilde{k}} = \lambda_2^k + \rho_2(s_N - f(a_N)),$$
(15b)

$$\lambda_3^{\tilde{k}} = \lambda_3^k + \rho_3(o_N - vs_N - c).$$
(15c)

Algorithm. Generally, we update the above parameters in two steps. First, these parameters are update in a backward way, namely $o \rightarrow c \rightarrow v \rightarrow s \rightarrow a \rightarrow b \rightarrow w \rightarrow u$. Afterwards, ADMMiRNN reverses the update direction in $u \rightarrow w \rightarrow b \rightarrow a \rightarrow s \rightarrow v \rightarrow c \rightarrow o$. After all those variables in an RNN cell update, the Lagrangian multipliers then update. Proceeding with the above steps, we could arrive at the algorithms for ADMMiRNN, which is outlined in Algorithm 1.

3.3   Convergence Analysis

In this section, we present the convergent analysis of ADMMiRNN. For convenience, we define $\rho = \{\rho_1, \rho_2, \rho_3\}$. First, we give some mild assumptions as follows:

Assumption 1. *The gradient of R is H-Lipschitz continuous, i.e., $\|\nabla R(o_1) - \nabla R(o_2)\| \leq H\|o_1 - o_2\|$, $H \geq 0$ and is called the Lipschitz constant. This is equivalent to $R(o_1) \leq R(o_2) + \nabla R(o_2) \cdot (o_1 - o_2) + H/2\|o_1 - o_2\|^2$;*

Assumption 2. *The gradient of the objective function $\mathcal{L}_\rho$ is bounded, i.e., there exists a constant C such that $\nabla \mathcal{L}_\rho \leq C$;*

Algorithm 1. The training algorithm for ADMMiRNN.

Input: iteration K, input x, timestep N.
Parameter: u, w, b, v, c, s_0, λ_1, λ_2, and λ_3
Output: u, w, b, v, c,

1: Initialize $k = 0$, u, w, b, v, c, s_0, λ_1, λ_2, and λ_3.
2: **for** $k = 1, 2, \cdots, K$ **do**
3:   **for** $t = 1, 2, \cdots, N$ **do**
4:     **if** $t < N$ **then**
5:       Update $o_t^{\tilde{k}}$ in Eq. (20).
6:     **else if** $t = N$ **then**
7:       Update $o_N^{\tilde{k}}$ in Eq. (21).
8:     **end if**
9:     Update $c^{\tilde{k}}$ in Eq. (19).
10:    Update $v^{\tilde{k}}$ in Eq. (18).
11:    **if** $t < N$ **then**
12:      Update $s_t^{\tilde{k}}$ in Eq. (13).
13:      Update $a_t^{\tilde{k}}$ in Eq. (11).
14:    **else if** $t = N$ **then**
15:      Update $s_N^{\tilde{k}}$ in Eq. (14).
16:      Update $a_N^{\tilde{k}}$ in Eq. (12).
17:    **end if**
18:    Update $b^{\tilde{k}}$ in Eq. (17).
19:    Update $w^{\tilde{k}}$ in Eq. (16).
20:    Update $u^{\tilde{k}}$ in Eq. (10).
21:    Update $u^{\tilde{k}}$ in Eq. (10).
22:    Update $w^{\tilde{k}}$ in Eq. (16).
23:    Update $b^{\tilde{k}}$ in Eq. (17).
24:    **if** $t < N$ **then**
25:      Update $a_t^{\tilde{k}}$ in Eq. (11).
26:      Update $s_t^{\tilde{k}}$ in Eq. (13).
27:    **else if** $t = N$ **then**
28:      Update $a_N^{\tilde{k}}$ in Eq. (12).
29:      Update $s_N^{\tilde{k}}$ in Eq. (14).
30:    **end if**
31:    Update $v^{\tilde{k}}$ in Eq. (18).
32:    Update $c^{\tilde{k}}$ in Eq. (19).
33:    **if** $t < N$ **then**
34:      Update $o_t^{\tilde{k}}$ in Eq. (20).
35:    **else if** $t = N$ **then**
36:      Update $o_N^{\tilde{k}}$ in Eq. (21).
37:    **end if**
38: **end for**
39: Update $\lambda_1^{\tilde{k}}$ in Eq. (15a).
40: Update $\lambda_2^{\tilde{k}}$ in Eq. (15b).
41: Update $\lambda_3^{\tilde{k}}$ in Eq. (15c).
42: **end for**
43: **return** u, w, b, v, c,

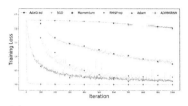

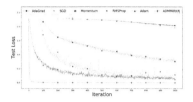

(a) training loss versus iterations.   (b) test loss versus iterations.

Fig. 2. Training loss and test loss versus iterations of RNN via ADMM, SGD, AdaGrad, Momentum, RMSprop, and Adam. ADMMiRNN achieves the best performance against other optimizers on MNIST.

Assumption 3. *The second-order moment of the gradient g_t is uniformly upper-bounded, that is to say $\mathbb{E}\|g_t\|^2 \leq C$.*

Such assumptions are typically used in [28]. Under these assumptions, we will have the properties [27] shown in the supplementary materials. Then we can prove that ADMMiRNN converges under the following theorems.

Theorem 1. *If $\rho_i > 2H$ $(i = 1, 2, 3)$ and **Assumption 1-3** hold, then **Property 1−3** in the supplementary materials hold.*

Theorem 2. *If $\rho_i > 2H$ $(i = 1, 2, 3)$, for the variables $(\theta, \lambda_1, \lambda_2, \lambda_3)$ in Problem 2, starting from any $(\theta^0, \lambda_1^0, \lambda_2^0, \lambda_3^0)$, it at least has a limit point $(\theta^*, \lambda_1^*, \lambda_2^*, \lambda_3^*)$ and any limit point $(\theta^*, \lambda_1^*, \lambda_2^*, \lambda_3^*)$ is a critical point of Problem 2. In other words, $0 \in \partial \mathcal{L}_{\rho_1, \rho_2, \rho_3}(\theta^*)$.*

Theorem 2 concludes that ADMMiRNN has a global convergence.

Theorem 3. *For a sequence θ generated by Algorithm 1, define $m_k = \min\limits_{0 \leq t \leq k} (\|\theta^{\tilde{k}} - \theta^k\|_2^2)$, the convergence rate of m_k is $O(1/k)$.*

Theorem 3 concludes that ADMMiRNN converges globally at a rate of $O(1/T)$. The convergence rate is consistent with the current work of ADMM [27]. Due to space limited, the proofs of the above theorems are also omitted in the supplementary materials.

4   Experiments

4.1   Setup

We train a RNN model shown in Fig. 1(a) on MNIST [15]. This is achieved by NumPy and those parameters are updated in a manner of Algorithm 1. The MNIST dataset has 55,000 training samples and 10,000 test samples and was first introduced in [15] to train handwritten-digit image recognition. All the experiments related to MNIST are conducted in 1000 iterations on a 64-bit Ubuntu 16.04 system.

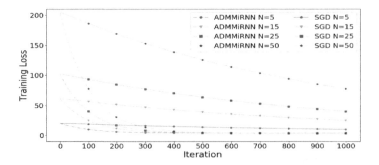

Fig. 3. The results of ADMMiRNN and SGD on different input sequence length. In this figure, N represents the length.

Furthermore, our experiments are also conducted on a text. The text could also be accessed from our open-source code repository. Training on a text is a typical RNN task. We achieved a typical RNN model and unfolded it to N cells with NumPy and N is also the length of the input sequence. In our experiments, we adopt a kind of smooth loss. These experiments are performed on a Macbook Pro with an Intel 3.1 GHz Core i5 Processor and 8 GB Memory.

In all of our experiments, we utilize a fixed value strategy for these hyperparameters, such as ρ_1, ρ_2 and ρ_3.

4.2   Convergence Results

Results on MNIST. We train the simple RNN model shown in Fig. 1(a) through different optimizers, including SGD, Adam, Momentum [19], RMSProp [25] and AdaGrad [3]. We compare our ADMMiRNN with these commonly-used optimizers in training loss and test loss and display our experimental results on MNIST in Fig. 2(a) and Fig. 2(b) respectively. Figure 2(a) and Fig. 2(b) indicate that ADMMiRNN converges faster than the other optimizers. ADMMiRNN gets a more smooth loss curve while the loss curves of other optimizers shake a lot. This means ADMMiRNN trains models in a relatively stable process. Besides, ADMMiRNN gets a much more promising training loss and test loss. These results not only prove that ADMMiRNN could converge in RNN tasks but also confirm that ADMMiRNN is a much powerful tool than traditional gradient-based optimizers in deep learning.

Results on Text Data. Besides experiments on MNIST, we also explore how ADMMiRNN performs in text classification tasks. One critical shortcoming of current RNN models is that they are sensitive to the length of the input sequence because the longer the input sequence is, the worse training results are. To investigate the sensitivity of ADMMiRNN to the input length, we measure the performance of ADMMiRNN and SGD on the text data with different input sequence length. The results are displayed in Fig. 3. Here, we adopt the average

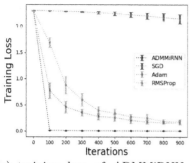 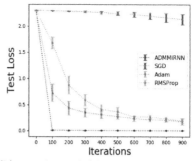

(a) training loss of ADMMiRNN and some typical optimizers.

(b) test loss of ADMMiRNN and some typical optimizers.

Fig. 4. The comparison of stability among ADMMiRNN, SGD, Adam and RMSProp. For each optimization method, we repeated experiments 10 times to obtain the mean and variance of the training loss and test loss against iterations on MNIST.

Table 2. Training loss and test loss under different hyperparameter settings. All of these values are obtained after 20 iterations.

ρ_1	ρ_2	ρ_3	r	Training loss	Test loss
1	1	1	1	5.045×10^{-2}	5.046×10^{-2}
0.1	1	1	1	5.339×10^{-2}	5.338×10^{-2}
1	0.1	1	1	5.338×10^{-2}	5.340×10^{-2}
1	1	0.1	1	3.776×10^{-4}	3.776×10^{-4}
1	1	10	1	0.9984	0.9985
1	1	1	10	5.339×10^{-2}	5.338×10^{-2}
1	1	10	10	0.9987	0.9986

loss of the input sequence as our target. From Fig. 3, we have evidence that ADMMiRNN always produces a remarkable result and is nearly immune to the length, which performs much more impressive than SGD regardless of the length of the input sequence.

4.3   Stability

As aforementioned, the initialization of weights and biases is critical in RNN models. In this section, we mainly compare ADMM with some different optimizers and explore its stability for RNN. In brief, we compare ADMMiRNN with SGD, Adam, and RMSProp and repeat each scheme ten times independently. The experimental results are displayed in Fig. 4. The blocks in Fig. 4(a) and Fig. 4(b) represent the standard deviation of the samples drawn from the training and testing process. The smaller the blocks are, the more stable the method is. From Fig. 4(a) and Fig. 4(b), we observe that at the beginning, SGD has a small fluctuation. But as the training progresses, the fluctuation gets more

Table 3. In this table, we mainly display the needed iteration counts in the training and test process when the accuracy reaches 100.0. This is a representation of the convergence speed under different hyperparameters.

ρ_1	ρ_2	ρ_3	r	Training iterations	Test iterations
1	1	1	1	2	2
0.1	1	1	1	2	2
1	0.1	1	1	2	2
1	1	0.1	1	2	2
1	1	10	1	3	3
1	1	10	10	11	11
1	1	1	10	2	2

and sharper, which means that SGD is tending to be unstable. As for Adam and RMSProp, their variance is getting smaller, but still large with regard to ADMMiRNN. According to different initialization of weights and biases, these optimizers may cause different results within a big gap between them. Specifically, ADMMiRNN has a relatively small variance from beginning to end compared with SGD, Adam and RMSProp, which is too small to show clearly in Fig. 4(a) and Fig. 4(b), which indicates that ADMMiRNN is immune to the initialization of weights and biases and settle the sensitivity of RNN models to initialization. No matter how the initialization changes, ADMMiRNN always gives a stable training process and promising results. The results demonstrate that ADMMiRNN is a more stable training algorithm for RNN models than stochastic gradient algorithms.

4.4   Different Hyperparameters

Varying ρs. In vanilla ADMM, the value of the penalty term is critical, and it may have negative effects on convergence. In this subsection, we mainly try different hyperparameters in ADMMiRNN and evaluate how they influence the training process of ADMMiRNN. These results are summarized in Table 2 and Table 3. Table 2 implies that ρ_3 determines the best result in ADMMiRNN. More precisely, larger ρ_3 delays the convergence speed in ADMMiRNN. However, if ρ_3 is too large, it may produce non-convergent results. In Table 3, we present the needed iteration count when the accuracy is 100.0 in the training and test process. When ρ_3 is 100, we need 11 iterations for the accuracy reaches 100.0 but we only need 2 iterations when $\rho_3 = 1$ or $\rho_3 = 10$. Furthermore, it turns out that ρ_1 and ρ_2 account less in ADMMiRNN while ρ_3 plays a much more crucial role with regard to the property of convergence and its convergence speed.

Varying ν. In this subsection, we investigate the influence of ν in Eq. 8. In our experiments on a text data, we fix all the hyperparameters other than ν and set it 10^{-2}, 10^{-3}, 10^{-4}, 10^{-6}, 10^{-8} respectively. We display the curves corresponding to different values of ν in Fig. 5. Figure 5 suggests that larger ν produces a relatively worse convergence result in ADMMiRNN. Small ν can not only lead to a small loss but is also able to push the training process to converge fast. However, when ν is small enough, the influence on the convergence rate and convergent result is not obvious.

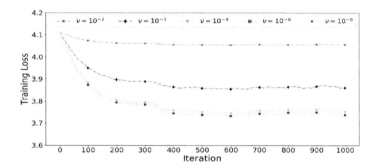

Fig. 5. The training loss V.S. iterations of ADMMiRNN on a text classification task with different ν.

4.5 Extension

In this subsection, we conduct another extensional experiment on MNIST to explore how the weight coefficient influences the performance and show it in Fig. 6. Figure 6(a) and Fig. 6(b) illustrate that if we adopt a large ν, the convergence would be delayed and also be more quivering than small ones. But we observe that the influence is not as apparent as training accuracy and validation accuracy with respect to the loss shown in Fig. 6(c).

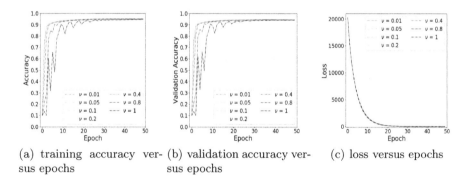

(a) training accuracy versus epochs   (b) validation accuracy versus epochs   (c) loss versus epochs

Fig. 6. Training RNN on MNIST via ADMMiRNN regarding training accuracy, validation accuracy and training loss V.S. epochs.

5   Conclusion

In this paper, we proposed a new framework to train RNN tasks, namely ADM-MiRNN. Since it is challenging to train RNNs with ADMM directly, we set up ADMMiRNN on the foundation of the expanded form of RNNs. The convergence analysis of ADMMiRNN is presented, and ADMMiRNN could achieve a convergence rate of $O(1/T)$. We further conduct several experiments on real-world datasets based on our theoretical analysis. Experimental results of comparisons regarding ADMMiRNN and several popular optimizers manifest that ADMMiRNN converges faster than these gradient-based optimizers. Besides, it presents a much more stable process than them. To the best of our knowledge, we are the first to apply ADMM into RNN tasks, and present theoretical analysis and ADMMiRNN is the first to alleviate the *vanishing* and *exploding* gradients problem and the sensitivity of RNN models to initializations at the same time. In conclusion, ADMMiRNN is a promising tool to train RNN models. In our future work, we will explore how to decide the best penalty parameter for ADMMiRNN in the training process rather than adopting a fixed value. Moreover, we will try to train ADMMiRNN in parallel as ADMM is an excellent parallel optimization method.

Acknowledgement. This work is partially sponsored by the National Natural Science Foundation of China under Grant No. 61806216, 61702533, 61932001, 61872376 and 61701506.

A   Appendix

A.1   Update w

Similar as the update of u in Sect. 3.2, we also define $\mathbf{G} = rI_d - \rho_1 s_{t-1}^T s_{t-1}$ and use with linearized proximal point method, then the update of w is transformed into

$$
\begin{aligned}
w^{\tilde{k}} \leftarrow \arg\min \Omega(w) \frac{Nr}{2} \|w - w^k\|^2 + \nu(w - w^k)^T \sum_{t=1}^{N-1} [(s_{t-1}^k)^T \\
(a_t^k - u^k x_t^k - w^k s_{t-1}^k - b^k)] + \rho_1 (w - w^k)^T [(s_{N-1}^k)^T \\
(a_N^k - u^k x_N^k - w^k s_{N-1}^k - b^k - \lambda_1^k/\rho_1)].
\end{aligned}
\tag{16}
$$

A.2   Update b

As far as b is concerned, it is updated by

$$
b^{\tilde{k}} \leftarrow \arg\min \phi(u^{\tilde{k}}, w^{\tilde{k}}, b, a^k, s^k, v^k, c^k, o^k).
\tag{17}
$$

A.3 Update v

Similar as aforementioned, the update rule for v is

$$
\begin{aligned}
v^{\tilde{k}} \leftarrow \arg\min \frac{Nr}{2}\|v - v^k\|^2 &+ \nu(v - v^k)^T \sum_{t=1}^{N-1} [(s_t^k)^T (o_t^k - v^k s_t^k - c^k)] \\
&+ \rho_3 (v - v^k)^T [(s_N^k)^T (o_N^k - v^k s_N^k - c^k - \lambda_1^k/\rho_3)].
\end{aligned}
\tag{18}
$$

A.4 Update c

The parameter c is quite simple, which is updated as follows:

$$
c^{\tilde{k}} \leftarrow \arg\min \phi(u^{\tilde{k}}, w^{\tilde{k}}, b^{\tilde{k}}, a^{\tilde{k}}, s^{\tilde{k}}, v^{\tilde{k}}, c, o^k).
\tag{19}
$$

A.5 Update o

Finally, we update o_t. It has to be noted that each o_t is also updated separably. If $t < N$,

$$
o_t^{\tilde{k}} \leftarrow \arg\min R(o) + \frac{\nu}{2}\|o_t^k - v^k s_t^k - c^k\|^2.
\tag{20}
$$

If $t = N$,

$$
o_N^{\tilde{k}} \leftarrow \arg\min R(o) + \frac{\rho_3}{2}\|o_N^k - v^k s_N^k - c^k - \lambda_3^k/\rho_3\|^2.
\tag{21}
$$

References

1. Bengio, Y., Simard, P., Frasconi, P., et al.: Learning long-term dependencies with gradient descent is difficult. IEEE Trans. Neural Netw. **5**(2), 157–166 (1994)
2. Boyd, S., Parikh, N., Chu, E., Peleato, B., Eckstein, J., et al.: Distributed optimization and statistical learning via the alternating direction method of multipliers. Found. Trends® Mach. Learn. **3**(1), 1–122 (2011)
3. Duchi, J., Hazan, E., Singer, Y.: Adaptive subgradient methods for online learning and stochastic optimization. J. Mach. Learn. Res. **12**(Jul), 2121–2159 (2011)
4. Elman, J.L.: Finding structure in time. Cogn. Sci. **14**(2), 179–211 (1990)
5. Gabay, D.: Augmented Lagrangian methods: applications to the solution of boundary-value problems, chapter applications of the method of multipliers to variational inequalities, vol. 3, p. 4. North-Holland, Amsterdam (1983)
6. Gabay, D., Mercier, B.: A dual algorithm for the solution of nonlinear variational problems via finite element approximation. Comput. Math. Appl. **2**(1), 17–40 (1976)
7. Glowinski, R., Le Tallec, P.: Augmented Lagrangian and operator-splitting methods in nonlinear mechanics, vol. 9. SIAM (1989)
8. Goldfarb, D., Qin, Z.: Robust low-rank tensor recovery: models and algorithms. SIAM J. Matrix Anal. Appl. **35**(1), 225–253 (2014)
9. Goodfellow, I., Bengio, Y., Courville, A.: Deep Learning. MIT Press (2016)

10. Graves, A., Fernández, S., Schmidhuber, J.: Multi-dimensional recurrent neural networks. In: de Sá, J.M., Alexandre, L.A., Duch, W., Mandic, D. (eds.) ICANN 2007. LNCS, vol. 4668, pp. 549–558. Springer, Heidelberg (2007). https://doi.org/10.1007/978-3-540-74690-4_56
11. Hochreiter, S., Schmidhuber, J.: Long short-term memory. Neural Comput. **9**(8), 1735–1780 (1997)
12. Kingma, D., Ba, J.: Adam: a method for stochastic optimization. Comput. Sci. (2014)
13. Lai, S., Xu, L., Liu, K., Zhao, J.: Recurrent convolutional neural networks for text classification. In: AAAI, vol. 333, pp. 2267–2273 (2015)
14. LeCun, Y., Bengio, Y., Hinton, G.: Deep learning. Nature **521**(7553), 436–444 (2015)
15. LeCun, Y., Bottou, L., Bengio, Y., Haffner, P., et al.: Gradient-based learning applied to document recognition. Proc. IEEE **86**(11), 2278–2324 (1998)
16. Monteiro, R.D., Svaiter, B.F.: Iteration-complexity of block-decomposition algorithms and the alternating minimization augmented Lagrangian method. Manuscript, School of Industrial and Systems Engineering, Georgia Institute of Technology, Atlanta, GA, pp. 30332–0205 (2010)
17. Nguyen, T.H., Cho, K., Grishman, R.: Joint event extraction via recurrent neural networks. In: Proceedings of the 2016 Conference of the North American Chapter of the Association for Computational Linguistics: Human Language Technologies, pp. 300–309 (2016)
18. Pascanu, R., Mikolov, T., Bengio, Y.: On the difficulty of training recurrent neural networks. In: International Conference on Machine Learning, pp. 1310–1318 (2013)
19. Qian, N.: On the momentum term in gradient descent learning algorithms. Neural Netw. **12**(1), 145–151 (1999)
20. Robbins, H., Monro, S.: A stochastic approximation method. Ann. Math. Stat. 400–407 (1951)
21. Rockafellar, R.T.: Monotone operators and the proximal point algorithm. SIAM J. Control Optim. **14**(5), 877–898 (1976)
22. Sun, T., Jiang, H., Cheng, L., Zhu, W.: Iteratively linearized reweighted alternating direction method of multipliers for a class of nonconvex problems. IEEE Trans. Signal Process. **66**(20), 5380–5391 (2018)
23. Sutskever, I., Martens, J., Dahl, G., Hinton, G.: On the importance of initialization and momentum in deep learning. In: International Conference Machine Learning, pp. 1139–1147 (2013)
24. Taylor, G., Burmeister, R., Xu, Z., Singh, B., Patel, A., Goldstein, T.: Training neural networks without gradients: a scalable ADMM approach. In: International Conference on Machine Learning, pp. 2722–2731 (2016)
25. Tieleman, T., Hinton, G.: Lecture 6.5-rmsprop, coursera: neural networks for machine learning. University of Toronto, Technical Report (2012)
26. Wang, J., Yu, F., Chen, X., Zhao, L.: ADMM for efficient deep learning with global convergence. In: Proceedings of the 25th ACM SIGKDD International Conference on Knowledge Discovery & Data Mining, pp. 111–119 (2019)
27. Wang, J., Zhao, L., Wu, L.: Multi-convex inequality-constrained alternating direction method of multipliers. arXiv preprint arXiv:1902.10882 (2019)
28. Zou, F., Shen, L., Jie, Z., Zhang, W., Liu, W.: A sufficient condition for convergences of ADAM and RMSPROP. In: Proceedings of the IEEE Conference on Computer Vision and Pattern Recognition, pp. 11127–11135 (2019)

Exponential Convergence of Gradient Methods in Concave Network Zero-Sum Games

Amit Kadan and Hu Fu$^{(\boxtimes)}$

University of British Columbia, Vancouver, BC V6T 1Z4, Canada
{amitkad,hufu}@cs.ubc.ca

Abstract. Motivated by Generative Adverserial Networks, we study the computation of Nash equilibrium in concave *network zero-sum games* (NZSGs), a multiplayer generalization of two-player zero-sum games first proposed with linear payoffs. Extending previous results, we show that various game theoretic properties of convex-concave two-player zero-sum games are preserved in this generalization. We then generalize last iterate convergence results obtained previously in two-player zero-sum games. We analyze convergence rates when players update their strategies using Gradient Ascent, and its variant, Optimistic Gradient Ascent, showing last iterate convergence in three settings—when the payoffs of players are linear, strongly concave and Lipschitz, and strongly concave and smooth. We provide experimental results that support these theoretical findings.

Keywords: Network zero-sum games · Last iterate convergence · Convergence of gradient ascent · Generative adversarial networks

1 Introduction

Connections between game theory and learning had long been known, before interest resurged recently in the machine learning community, largely due to the success of Generative Adversarial Networks (GANs), a novel framework for learning generative models [16]. A GAN is formulated as a two-player zero-sum game between two neural networks, a generator and a discriminator. The generator attempts to fool the discriminator by mapping random noise to images that look similar to samples from a target distribution, while the discriminator learns to distinguish the generator's output from real samples from the target distribution. Theoretically, at the equilibrium of this game, the generator outputs

The research was funded by an NSERC Discovery Grant, an NSERC Discovery Acceleration Grant, and Canadian Research Chair stipend.

Electronic supplementary material The online version of this chapter (https://doi.org/10.1007/978-3-030-67661-2_2) contains supplementary material, which is available to authorized users.

© Springer Nature Switzerland AG 2021
F. Hutter et al. (Eds.): ECML PKDD 2020, LNAI 12458, pp. 19–34, 2021.
https://doi.org/10.1007/978-3-030-67661-2_2

the target distribution. In practice, GANs produce promising results on a number of tasks including image generation, semantic segmentation, and text-to-image synthesis [15].

Among many theoretical questions opened up by GANs, that of *last iterate* convergence has attracted much attention and seen exciting progress. Classical results show that, when players use *no-regret online learning* algorithms to play a two-player zero-sum game, the *time average* of their strategies converge to a Nash equilibrium—a point where neither player can make gains by unilaterally deviating from their current strategy. In GANs, strategies correspond to parameters of neural networks; averaging strategies makes little sense. It is therefore desirable that the players' strategies, from iteration to iteration, should converge to an equilibrium. This is known as *last iterate* convergence, which is not implied by classical results. A number of simple algorithms have been shown to give rise to such convergence in two-player zero-sum games, with exponential convergence rates in various settings [14, 18, 21] (see Sect. 1.1 for more details).

On the other hand, many recently proposed extensions of GANs go beyond the two-player zero-sum framework, either to address challenges faced by the original GAN, or to make it more versatile. In particular, many models introduce more agents (neural networks) to the game. For example, Hoang et al. [17] proposed using an ensemble of generators to address *mode collapse*, a common problem of the classical GAN, where the generator captures only one or few modes of the data distribution. Other architectures incorporate a third classifying network, which is in direct competition with either the generator [22] or the discriminator [8]; such architectures are often built for semi-supervised learning. Lastly, some architectures incorporate an additional encoding network which, like the generator, competes with the discriminator, and allows for sampling from a latent distribution that encodes additional information about the data distribution [4, 7, 12]. Results on two-player zero-sum games do not apply to these architectures with more than two agents. It is also well known that two-player zero-sum games have many properties not extensible to games with more players or non zero-sum payoffs.

We observe that the extensions above all give rise to *network zero-sum games* (NZSGs), a class of games first proposed and studied by Cai et al. [5]. An NZSG is structured by a graph, where each node corresponds to a player, and along each edge a game is played between its two node players. A player chooses one strategy to be used in all the games in which she is engaged; the sum of all players' payoffs is always zero. Since players cannot choose different strategies for different games, an NZSG is not a simple parallelization of multiple two-player zero-sum games. However, Cai et al. [5] showed that NZSGs with linear payoffs preserve certain properties from two-player zero-sum games. In particular a Nash in an NZSG can be computed via a linear program.

We first generalize results of [5] on the tractability of equilibrium for NZSGs (Sect. 2); we show that in an NZSG with concave payoffs, a Nash can be computed via no-regret learning. Then, as our main result, we show last iterate convergence results for NZSGs with several classes of payoffs (Sect. 3), when players adopt simple learning rules used in practice, such as *Gradient Ascent* (GA) and *Optimistic Gradient Ascent* (OGA). GA is the most ubiquitous optimization algorithm. It may be seen as a smoothed best response, and so it may not be surprising that it produces dynamics that diverge from the equilibrium in two-player zero-sum games with linear payoffs [11,18]. We show that this phenomenon persists in NZSGs with linear payoffs. OGA, on the other hand, incorporates some minimal memory, and uses information from one step before. This small tweak has been shown to induce last iterate convergence in two-player zero-sum games with either linear payoffs or strongly concave payoffs that are smooth in various senses [9,18,21]. We extend these to NZSGs, showing comparable convergence performance. For two-player zero-sum games with strongly concave payoffs, GA is known to induce last iterate convergence; we generalize this as well.

We use two sets of tools. Our main tool for NZSGs with linear payoffs is dynamical systems. Strategies played in a repeated game give rise to a dynamical system; techniques for analyzing such systems naturally can be used to analyze various update algorithms [10,11,18]. Our results on both the divergence of GA and convergence of OGA dynamics are built on linear algebraic techniques used to analyze the corresponding dynamical systems. Crucial to the arguments is an algebraic property we show for NZSGs; namely, that a Hessian matrix associated with the payoff functions is antisymmetric everywhere.

We use Lyapunov-style convergence proofs to show results in NZSGs with strongly concave and smooth payoffs. Apart from existing arguments for two-player zero-sum games, our proof exploits a structural lemma (Lemma 1), which may be of independent interest.

In Sect. 4, we provide experiments that validate our theoretical findings.

1.1   Related Work

Cai et al. [5] introduced the class of network zero-sum games, and showed that a Nash equilibrium of an NZSG can be computed by a linear programming when each player's strategy is a distribution over a finite number of actions.

A few papers study convergence in n-player games. The most closely related work to ours is Azizian et al. [1]. They show that various gradient-based algorithms, including OGA, converge at an exponential rate to the Nash in a class of smooth and "monotone" n-player games. With slight modification explained in the technical sections, our results on the OGA dynamics in NZSGs with strongly concave and smooth payoffs or with linear payoffs can be obtained by showing these games to be smooth and monotone. Our proofs in these settings may be viewed as alternative approaches to showing these results. An advantage of our approach is that it is readily modified to apply for games with Lipschitz payoffs, as we demonstrate in Sect. 3.3.

Balduzzi et al. [2] study two classes of n-player games, Hamiltonian games and potential games, both of which are specific instances of NZSGs. They show that, when players use a continuous-time version of GA to update their strategies in a Hamiltonian game, the dynamics circle perpetually around the Nash of the game. They propose *Symplectic Gradient Adjustment* (SGA) and show it to converge in last iterate for both Hamiltonian and potential games. Balduzzi et al. [3] study another class of games called Smooth Market Games, which consist of payoffs that are pairwise zero-sum. They show that a continuous time version of GA converges in last iterate to the Nash of a game when payoffs are strictly concave in players' strategies.

A number of papers study last iterate convergence in concave two-player zero-sum games. Liang and Stokes [18] use tools from dynamical systems to show exponential convergence of the last iterate in bilinear games when players use OGA. They also show exponential convergence of the last iterate in games with smooth and strongly concave payoffs when players use GA. Mokhtari et al. [21] show exponential convergence of the last iterate in games with bilinear, or smooth and strongly concave payoffs when the players use OGA, by interpreting OGA as an approximation of the *Proximal Point Method*. Gidel et al. [14] use a variational inequality perspective to show exponential convergence of a variant of OGA in constrained two-player zero-sum games with smooth and strongly concave payoffs. Merkitopolous et al. [20] use similar tools to show last iterate, but not exponential convergence for Mirror Descent and Optimistic Mirror Descent when payoffs are strongly concave, and for Optimistic Mirror Descent when payoffs are linear.

1.2   Notations and Mathematical Conventions

Vectors in $\mathbb{R}^k$ are denoted by boldface, and scalars by lowercase. Time indices are denoted by superscripts, while players are identified by subscripts. For a square matrix A we denote the set of its eigenvalues by $\lambda(A)$. I_m denotes the $m \times m$ identity matrix.

Definition 1. *Given $U \subset \mathbb{R}^k$, and concave function $f : U \to \mathbb{R}$. $\mathbf{q} \in \mathbb{R}^k$ is a supergradient of f at $\mathbf{u}$ if $\forall \mathbf{u}' \in U$, $f(\mathbf{u}') \leq f(\mathbf{u}) + \langle \mathbf{q}, \mathbf{u}' - \mathbf{u} \rangle$. The set of supergradients of f at a point $\mathbf{u}$ is denoted by $\partial f(\mathbf{u})$.*

Definition 2. *For $\alpha > 0$, a function $f : U \to \mathbb{R}$ is α-strongly concave if $\forall \mathbf{u}, \mathbf{u}' \in U$ and $\mathbf{q} \in \partial f(\mathbf{u})$,*

$$f(\mathbf{u}') \leq f(\mathbf{u}) + \langle \mathbf{q}, \mathbf{u}' - \mathbf{u} \rangle - \frac{\alpha}{2} \| \mathbf{u} - \mathbf{u}' \|^2.$$

A function $g : U \times V \to \mathbb{R}$ is α-strongly concave in $\mathbf{u}$ if for any $\mathbf{v} \in V$, $h(\mathbf{u}) := g(\mathbf{u}, \mathbf{v})$ is α-strongly concave.

2   Network Zero-Sum Games Basics

In this section we extend network zero-sum games as defined by Cai et al. [5] to allow continuous action spaces. We then show that in games with concave payoff functions, an equilibrium can be efficiently computed with no-regret dynamics.

Definition 3. *A network game $\mathcal{G}$ consists of the following:*

- *a finite set $V = \{1, ..., n\}$ of players, and a set E of edges which are unordered pairs of players $[i, j], i \neq j$;*
- *for each player $i \in V$, a convex set $X_i \subseteq \mathbb{R}^{d_i}$, the strategy set for player i;*
- *for each edge $[i, j] \in E$, a two-person game (p_{ij}, p_{ji}), where $p_{ij} : X_i \times X_j \to \mathbb{R}$, and $p_{ji} : X_j \times X_i \to \mathbb{R}$.*

Given a strategy profile $\boldsymbol{x} = (\boldsymbol{x}_1, ..., \boldsymbol{x}_n) \in \prod_{j \in V} X_j$, player i's payoff is $p_i(\boldsymbol{x}) := \sum_{[i,j] \in E} p_{ij}(\boldsymbol{x}_i, \boldsymbol{x}_j)$.

A network game is a network zero-sum game *(NZSG) if for all strategy profiles $\boldsymbol{x} \in \prod_{j \in V} X_j$, $\sum_{i \in V} p_i(\boldsymbol{x}) = 0$.*

We let X denote $\prod_{i \in V} X_i$, and $d := \sum_{i \in V} d_i$. *Two-player zero-sum games* are special cases of NZSGs, where V has two nodes, connected by one edge.

In a *concave NZSG*, each $p_{ij}(\boldsymbol{x}_i, \boldsymbol{x}_j)$ is concave in $\boldsymbol{x}_i$. An NZSG is *linear* if each $p_{ij}(\boldsymbol{x}_i, \boldsymbol{x}_j)$ is linear in both $\boldsymbol{x}_i$ and $\boldsymbol{x}_j$.

Let $\boldsymbol{x}_{-i}$ denote the strategy profile without player i's strategy, i.e., $\boldsymbol{x}_{-i} = (\boldsymbol{x}_1, ..., \boldsymbol{x}_{i-1}, \boldsymbol{x}_{i+1}, ..., \boldsymbol{x}_n)$.

Definition 4. *A strategy profile $\boldsymbol{x}^*$ is a* Nash equillibrium *for an NZSG if for each player i, for any strategy $\boldsymbol{x}_i \in X_i$, $p_i(\boldsymbol{x}^*) \geq p_i(\boldsymbol{x}_i, \boldsymbol{x}^*_{-i})$.*

It can be shown via a fixed point argument that, in a concave NZSG where each player's strategy space X_i is convex and compact, a Nash equilibrium always exists [19]. Cai et al. [5] showed that for linear NZSGs where each player's strategy set is a simplex, a Nash can be computed efficiently by a linear program.

As a warm-up, we show that another classical technique for computing equilibrium in two-player zero-sum games, namely, no-regret learning algorithms, can be used to find an approximate Nash in general concave NZSGs.

Given an NZSG with compact strategy sets, consider the players playing it repeatedly. Let $\boldsymbol{x}_i^s$ and $\boldsymbol{x}_{-i}^s$ denote, respectively, player i's and the other players' strategies at time step s of the game. Each player should only respond to the past strategies of her opponents; i.e., $\boldsymbol{x}_i^s$ may depend only on $\boldsymbol{x}_{-i}^1, ..., \boldsymbol{x}_{-i}^{s-1}$.

Definition 5. *In a repeated game, a player's* regret *at time t, $r_i(t)$, is*

$$r_i(t) = \max_{\boldsymbol{x}_i \in X_i} \sum_{s=1}^{t} \left[p_i(\boldsymbol{x}_i, \boldsymbol{x}_{-i}^s) - p_i(\boldsymbol{x}_i^s, \boldsymbol{x}_{-i}^s) \right].$$

A player i's strategy $(\boldsymbol{x}_i^s)_s$ has no-regret *if for all t, $\frac{1}{t} r_i(t) \leq \epsilon(t)$ for some $\epsilon(t) \to 0$ as $t \to \infty$. An algorithm that produces no-regret strategies is a* no-regret algorithm.

It is well known that efficient no-regret algorithms exist [6], and that in a two-player zero-sum game, if players use no-regret dynamics, the *time average* of their strategies converges to a Nash equilibrium [6]. We show this phenomenon generalizes to NZSGs with concave payoffs.

Proposition 1. *In a concave NZSG with compact strategy sets, if each player uses strategies that have no-regret, then the strategy profile where each player plays her time-average strategy converges to a Nash equilibrium.*

A key step in the proof of Proposition 1 is the following property of NZSGs. We will make repeated use of this property later in the paper.

Lemma 1. *In an NZSG, for any two strategy profiles $\boldsymbol{x}$ and $\boldsymbol{x}^*$, we have*

$$\sum_i p_i(\boldsymbol{x}_i, \boldsymbol{x}^*_{-i}) = -\sum_i p_i(\boldsymbol{x}^*_i, \boldsymbol{x}_{-i}).$$

As we discussed in the Introduction, Proposition 1 is not adequate for applications where strategies are parameters of neural networks, since taking averages over strategies makes little sense in such settings. Following much recent literature, we shift the focus to last iterate convergence.

3   Last Iterate Convergence in NZSGs

In this section we present our main results on last-iterate convergence in NZSGs when players use gradient style updates. In this section we assume that the strategy spaces are unconstrained, i.e., $X_i = \mathbb{R}^{d_i}$ for each i.

We first formally define the two update rules we focus on. Recall that we use $\boldsymbol{x}_i^t$ to denote player i's strategy at time t. A player using *Gradient Ascent* (GA) modifies her strategy by

$$\boldsymbol{x}_i^{t+1} = \boldsymbol{x}_i^t + \eta \nabla_{\boldsymbol{x}_i} p_i(\boldsymbol{x}^t), \tag{GA}$$

where $\eta > 0$ is a fixed step size. A player using *Optimistic Gradient Ascent* (OGA) updates her strategy by

$$\boldsymbol{x}_i^{t+1} = \boldsymbol{x}_i^t + 2\eta \nabla_{\boldsymbol{x}_i} p_i(\boldsymbol{x}^t) - \eta \nabla_{\boldsymbol{x}_i} p_i(\boldsymbol{x}^{t-1}), \tag{OGA}$$

where $\eta > 0$ again is a fixed step size.

3.1   Linear NZSGs

Even in a two-player zero-sum bilinear game, i.e., $p_1(\boldsymbol{x}_1, \boldsymbol{x}_2) = -p_2(\boldsymbol{x}_2, \boldsymbol{x}_1) = p(\boldsymbol{x}_1, \boldsymbol{x}_2) = \boldsymbol{x}_1^\top C \boldsymbol{x}_2$, where C is a $d_1 \times d_2$ matrix, if each player uses GA, over time the players' strategies diverge from the set of Nash [18]. If, instead, players use OGA, their strategies converge to a Nash of the game [9,18,20]. We show that these phenomena continue to hold for linear NZSGs.

To state the rates of convergence and divergence, we need to introduce a matrix H for a linear NZSG, which we motivate later. Given an NZSG and a strategy profile $\boldsymbol{x}$, the *Hessian* $H(\boldsymbol{x})$ is a $d \times d$ block matrix with the $(i,j)^{th}$ block given by

$$H_{ij}(\boldsymbol{x}) = \nabla^2_{\boldsymbol{x}_j, \boldsymbol{x}_i} p_i(\boldsymbol{x}).$$

Denote the smallest nonzero modulus of an eigenvalue of H by $\omega(H)$, and denote the largest modulus of an eigenvalue of H by $\rho(H)$. Denote the distance to a set by $d(\mathbf{u}, S) := \min_{\mathbf{s} \in S} \|\mathbf{u} - \mathbf{s}\|$.

Theorem 1. *Consider an unconstrained, linear NZSG. Let X^* denote the set of Nash of the game. Assume each player uses GA to update her strategy at each time step. Assume $d(\boldsymbol{x}^0, X^*) \geq R$, for some $R > 0$. Then at each time step t,*

$$d(\boldsymbol{x}^t, X^*)^2 \geq (1 + \eta^2 \omega(H)^2)^t R^2.$$

Theorem 2. *Consider an unconstrained, linear NZSG. Assume that each player uses OGA as her update rule. Let X^* denote the set of Nash of the game. If $H(\boldsymbol{x})$ is diagonalizable for all $\boldsymbol{x}$, and if $d(\boldsymbol{x}^0, X^*) \leq r, d(\boldsymbol{x}^1, X^*) \leq r$ for some $r > 0$. Then setting $\eta = 1/2\rho(H)$, at each time step t,*

$$d(\boldsymbol{x}^{t+1}, X^*)^2 \leq \left(\frac{1}{2} + \frac{1}{2} \left(1 - \left(\frac{\omega(H)}{\rho(H)} \right)^2 \right)^{\frac{1}{2}} \right)^t r^2.$$

We sketch the proof ideas and relegate details to the supplementary file. We formulate the behavior of GA and OGA as trajectories of *dynamical systems*; this view has been taken in several previous works, which also analyze the behaviors of updating algorithms using tools from dynamical systems [10,11,18].

Definition 6. *A relation of the form $\boldsymbol{x}^{t+1} = \boldsymbol{g}(\boldsymbol{x}^t)$, also written as $\boldsymbol{x} \mapsto \boldsymbol{g}(\boldsymbol{x})$, is a discrete time dynamical system with update rule $\boldsymbol{g} : \mathbb{R}^k \to \mathbb{R}^k$. A point $\boldsymbol{z}$ is a* fixed point *of $\boldsymbol{g}$ if $\boldsymbol{g}(\boldsymbol{z}) = \boldsymbol{z}$.*

If players use GA, the strategies evolve according to the dynamical system $\boldsymbol{x}^{t+1} = (I_d + \eta H)\boldsymbol{x}^t$, where H is the Hessian matrix defined above. It is not hard to show that the set of Nash equilibria is precisely the set of fixed points of this dynamical system. Note that, when $\boldsymbol{g}$ is a linear function, as is the case for the GA dynamics, a point is its fixed point if and only if it is in the eigenspace of $\boldsymbol{g}$ for eigenvalue 1.

For a dynamical system with update rule $\boldsymbol{g} : \mathbb{R}^k \to \mathbb{R}^k$, the *Jacobian* is the matrix with its (i, j)-th entry $J_{ij} = \frac{\partial g_i}{\partial x_j}$. The eigenvalues of the Jacobian J at a fixed point $\boldsymbol{z}$ describe the behavior of the dynamics around $\boldsymbol{z}$. Roughly speaking, if all eigenvalues of J have modulus greater than 1, then in a neighborhood around $\boldsymbol{z}$, the dynamics diverges from $\boldsymbol{z}$; conversely, if all eigenvalues of J have modulus smaller than 1, in a neighborhood of $\boldsymbol{z}$ the dynamics converges to $\boldsymbol{z}$. When $\boldsymbol{g}$ is linear, this characterization of convergence/divergence extends to the entire space (beyond neighborhoods around $\boldsymbol{z}$), and allows some eigenvalues to be 1.

Proposition 2. *Let Z denote the set of fixed points of a dynamical system with linear update rule: $\boldsymbol{g}(\boldsymbol{z}) = J\boldsymbol{z}$, where J is diagonalizable. Let $d(\boldsymbol{x}, Z)$ denote $\min_{\boldsymbol{z} \in Z} \|\boldsymbol{x} - \boldsymbol{z}\|$.*

(a) If $\forall \lambda \in \lambda(J)$ either $|\lambda| < 1$ or $\lambda = 1$, then letting $\sigma_{\max_{<1}}(J)$ denote the largest modulus of any eigenvalue of J not equal to 1, $\forall x^0 \in X, d(x^{t+1}, Z) \leq (\sigma_{\max_{<1}}(J))^t d(x^0, Z)$.

(b) If $\forall \lambda \in \lambda(J)$ either $|\lambda| > 1$ or $\lambda = 1$, then letting $\sigma_{\min_{>1}}(J)$ denote the smallest modulus of any eigenvalue of J not equal to 1, $\forall x^0 \in X, d(x^{t+1}, Z) \geq (\sigma_{\min_{>1}}(J))^t d(x^0, Z)$.

To show Theorem 1, therefore, it suffices to analyze the eigenvalues of the matrix $J = I_d + \eta H$. The crucial observation is that, for NZSGs, the Hessian H is an antisymmetric matrix of the form

$$
H = \begin{bmatrix}
0 & C_{12} & \dots & C_{1n} \\
-C_{12}^\top & 0 & \dots & C_{2n} \\
\vdots & \vdots & \ddots & \vdots \\
-C_{1n}^\top & -C_{2n}^\top & \dots & 0
\end{bmatrix}.
$$

This is a consequence of the following lemma on NZSGs in general:

Lemma 2. *In an NZSG, if each p_i has continuous second partial derivatives, then*

$$
\nabla^2_{x_i, x_j} p_{ji}(x) = -(\nabla^2_{x_j, x_i} p_{ij}(x))^\top.
$$

As a result, for the GA dynamics in a linear NZSG, all eigenvalues of H are imaginary, and therefore all the eigenvalues of $I_d + \eta H$ are of the form $1 + i\eta\lambda$ for some $\lambda \in \mathbb{R}$. Part (b) of Proposition 2 indicates a diverging dynamics.

The antisymmetry of the Hessian H is also a crucial step in the proof of Theorem 2. We first need to augment the state space to allow the memory from a previous step to be passed as part of the state. Following Daskalakis and Panageas [11], we consider a dynamical system with the following update rule $g : \mathbb{R}^{2d} \to \mathbb{R}^{2d}$, defining $\hat{p}_i(x, x') := p_i(x)$:

$$
\begin{aligned}
g(x, x') &= (g^1(x, x'), g^2(x, x')), \\
g_i^1(x, x') &= x_i + 2\eta \nabla_{x_i} \hat{p}_i(x, x') - \eta \nabla_{x_i'} \hat{p}_i(x, x'), \\
g_i^2(x, x') &= x_i.
\end{aligned}
\tag{1}
$$

More explicitly, for the OGA update rule, we have the relation $(x^{t+1}, x^t) = g(x^t, x^{t-1})$. We make use of a connection established by [11] between the GA dynamics and the OGA dynamics (Proposition 3). Besides another application of the antisymmetry of H, we also use an expression for the determinant of a 2×2 block matrix (Lemma 3).

Proposition 3 ([11]). *Let z be a fixed point of the GA dynamics. Then, (z, z) is a fixed point of the OGA dynamics, and for each $\mu \in \lambda(J_{GA})$ we have two eigenvalues in $\lambda(J_{OGA})$ that are the roots of the quadratic equation*

$$
\lambda^2 - (2\mu - 1)\lambda + (\mu - 1) = 0.
$$

Lemma 3 ([13]). *Let A be a block matrix of the following form*

$$A = \begin{bmatrix} M_1 & M_2 \\ M_3 & M_4 \end{bmatrix},$$

where each M_i is a square matrix, and M_4 is invertible. Then the determinant of A is equal to the determinant of its Schur Complement:

$$\det(A) = \det(M_1 - M_2(M_4)^{-1}M_3)\det(M_4).$$

In order to be able to apply Proposition 2, we make an additional diagonalizability assumption on H. This is not a restrictive assumption; for any linear function, there is an arbitrarily small perturbation that makes its Hessian diagonalizable; in fact, the set of nondiagonalizable matrices over $\mathbb{C}$ has Lebesgue measure 0. In comparison, Azizian et al. [1] show exponential convergence of OGA in linear games with the assumption that the Hessian is invertible.

3.2 Smooth and Strongly Concave Payoffs

A payoff function p_{ij} is said to be β-smooth, for $\beta > 0$, if for all $\boldsymbol{x}_i, \boldsymbol{x}_i' \in X_i, \boldsymbol{x}_j, \boldsymbol{x}_j' \in X_j$,

$$\|\nabla_{\boldsymbol{x}_i} p_{ij}(\boldsymbol{x}_i, \boldsymbol{x}_j) - \nabla_{\boldsymbol{x}_i} p_i(\boldsymbol{x}_i', \boldsymbol{x}_j)\| \le \beta\|\boldsymbol{x}_i - \boldsymbol{x}_i'\|;$$
$$\|\nabla_{\boldsymbol{x}_i} p_{ij}(\boldsymbol{x}_i, \boldsymbol{x}_j) - \nabla_{\boldsymbol{x}_i} p_i(\boldsymbol{x}_i, \boldsymbol{x}_j')\| \le \beta\|\boldsymbol{x}_j - \boldsymbol{x}_j'\|. \tag{2}$$

An NZSG is said to have β-smooth payoffs if each payoff function p_{ij} is β-smooth for every $[i, j] \in E$. The game is said to be have α-strongly concave payoffs if each p_i is α-strongly concave in $\boldsymbol{x}_i$. In this section, we show that when players use GA and OGA to update their strategies in a game with payoffs that are α-strongly concave and β-smooth, their strategies converge to a Nash at an exponential rate. Throughout this section, we assume that for each player i, p_i is twice continuously differentiable. Since each p_i is differentiable, it has a unique supergradient, $\nabla_{\boldsymbol{x}_i} p_i(\boldsymbol{x})$ at a point $\boldsymbol{x}$.

Before stating our main results, we remark on the existence and uniqueness of Nash. Since we consider unconstrained NZSGs, Proposition 1 does not apply. Unlike linear NZSGs in Sect. 3.1, where $\boldsymbol{x} = \boldsymbol{0}$ is always a Nash, in general, Nash may not exist when the strategy spaces are not compact. With α-strong concavity, however, we do get uniqueness of Nash when one exists.

Lemma 4. *In an NZSG with α-strongly concave payoffs for $\alpha > 0$, if a Nash equilibrium exists, it is unique.*

For applications such as GANs, where strategies are parameters of neural networks, strategy spaces are practically compact, and a Nash equilibrium is guaranteed by Proposition 1 to exist.

We now state the main results of this section.

Theorem 3. *Consider an unconstrained NZSG with payoffs that are twice continuously differentiable, α-strongly concave and β-smooth for $\alpha, \beta > 0$. Assume the existence of a Nash, $\boldsymbol{x}^*$. Let $\boldsymbol{x}^0 \in X$ be such that $\forall i \in V, \|\boldsymbol{x}_i^0 - \boldsymbol{x}_i^*\| \leq r$ for $r > 0$. If each player uses GA, with $\eta = \frac{\alpha}{2n\beta^2}$, then at each time step t,*

$$\sum_i \|\boldsymbol{x}_i^t - \boldsymbol{x}_i^*\|^2 \leq \left(1 - \frac{\alpha^2}{4n\beta^2}\right)^t nr^2.$$

Theorem 4. *Consider an unconstrained NZSG with payoff functions that are twice continuously differentiable, α-strongly concave and β-smooth, for $\alpha, \beta > 0$. Assume the existence of a Nash, $\boldsymbol{x}^*$. Let $\boldsymbol{x}^0, \boldsymbol{x}^1 \in X$ be such that $\forall i \in V, \|\boldsymbol{x}_i^0 - \boldsymbol{x}_i^*\| \leq r, \|\boldsymbol{x}_i^1 - \boldsymbol{x}_i^*\| \leq r$ for $r > 0$. If each player uses OGA, with $\eta = \frac{1}{2n\beta}$, then at each time step t,*

$$\sum_i \|\boldsymbol{x}_i^{t+1} - \boldsymbol{x}_i^*\|^2 \leq \left(1 - \frac{\alpha}{4n\beta}\right)^t (n+1)2r^2.$$

In order show convergence for GA, we use a Lyapunov-style convergence argument. For two-player zero-sum games with strongly-concave and smooth payoffs, Liang and Stokes [18] show that, when players use GA to update their strategies, the strategies converge to the Nash of the game at an exponential rate. The key that allows us to extend the result to NZSGs is Lemma 1, which causes terms that are introduced by the strong concavity condition to vanish.

For the OGA update rule, we make use of writing OGA as a two step update, so that the second iterate results in a GA style update,

$$\boldsymbol{w}_i^t = \boldsymbol{w}_i^{t-1} + \eta \nabla_{\boldsymbol{x}_i} p_i(\boldsymbol{x}^t), \qquad \text{(OGA}')$$
$$\boldsymbol{x}_i^{t+1} = \boldsymbol{w}_i^t + \eta \nabla_{\boldsymbol{x}_i} p_i(\boldsymbol{x}^t).$$

Plugging in for $\boldsymbol{w}_i^t, \boldsymbol{w}_i^{t-1}$ in terms of $\boldsymbol{x}_i^t, \boldsymbol{x}_i^{t-1}$ gives us the original OGA update.

Mokhtari et al. [21] show that in a two-player zero-sum game with smooth and strongly concave payoffs, if each player uses the OGA update, the strategies converge to a Nash exponentially fast. Lemma 1 again plays a key role in our extension of the result to network zero-sum games.

Azizian et al. [1] show exponential convergence to a Nash when players use the OGA update strategy in a game with smooth payoffs and "strongly monotone" dynamics. We show in the supplementary file that NZSGs with strongly concave payoffs are in fact strongly monotone; this constitutes an alternative derivation of exponential convergence of the OGA dynamics.

3.3   Lipschitz and Strongly Concave Payoffs

In this section, we show that if players use GA or OGA to update their strategies in an NZSG where payoffs are α-strongly concave and L-Lipschitz, for $\alpha, L > 0$, then, given appropriate step sizes, their strategies converge to the unique Nash of the game. We assume that for each player i, p_i is continuously differentiable. If each p_i is L-Lipschitz, then for each player i,

$$\forall \boldsymbol{x} \in X, \|\nabla_{\boldsymbol{x}_i} p_i(\boldsymbol{x})\| \leq L. \tag{3}$$

Theorem 5. *Consider an unconstrained NZSG that is played for T rounds. Assume each p_i is α-strongly concave in X_i and L-Lipschitz for $\alpha, L > 0$. Assume the existence of a Nash, $\boldsymbol{x}^*$. Let $\boldsymbol{x}_i^0$ be such that, for each player i, $\|\boldsymbol{x}_i^0 - \boldsymbol{x}_i^*\| \leq r$ for $r > 0$. If each player uses GA with variable step size $\eta_s > 0$ at each time step s, then at each time step t,*

$$\sum_i \|\boldsymbol{x}_i^t - \boldsymbol{x}_i^*\|^2 \leq L^2 n \sum_{s=1}^t \eta_s^2 + nr^2 \prod_{s=1}^t (1 - \eta_s \alpha).$$

In particular, if $\eta_s = T^{-(0.5+\epsilon)}$ for $\epsilon \in (0, 0.5)$, then

$$\lim_{T \to \infty} \sum_i \|\boldsymbol{x}_i^T - \boldsymbol{x}_i^*\|^2 = 0.$$

Theorem 6. *Consider an unconstrained NZSG that is played for T rounds. Assume each p_i is α-strongly concave in X_i and L-Lipschitz for $\alpha, L > 0$. Assume the existence of a Nash, $\boldsymbol{x}^*$. Let $\boldsymbol{x}_i^0$ be such that for each player i, $\|\boldsymbol{x}_i^0 - \boldsymbol{x}_i^*\| \leq r$ for $r > 0$. Then if each player uses OGA with nonincreasing step size $\eta_s > 0$,*

$$\sum_i \|\boldsymbol{x}_i^t - \boldsymbol{x}_i^*\|^2 \leq 4nL^2 \sum_{s=1}^t \eta_s \eta_{s-1} + nr^2 \prod_{s=1}^t (1 - (\eta_s + \eta_{s-1})\alpha).$$

In particular, if $\eta_s = T^{-(0.5+\epsilon)}$ for $\epsilon \in (0, 0.5)$, then

$$\lim_{T \to \infty} \sum_i \|\boldsymbol{x}_i^T - \boldsymbol{x}_i^*\|^2 = 0.$$

Our proofs for these theorems resemble those from Sect. 3.2, with Lemma 1 facilitating the generalization to NZSGs. We note that the proof fails to achieve exponential convergence, due to the lack of smoothness in the game. Furthermore, the algorithm designer needs to know in advance the time horizon T, the number of time steps the game is to be played, in order to choose a learning schedule that allows for guaranteed last-iterate convergence.

4   Experiments

In this section, we provide examples validating our results. We first show convergence in the simplest setting—a game with three players where a zero-sum game is played between each pair. We provide experiments showing convergence in a game with linear payoffs, and a game with smooth and strongly concave payoffs. We then provide an experiment showing the effect that increasing the number of players has on convergence rate. For each experiment, we show the performance of both GA and OGA.

4.1   Three Player Game with Linear Payoffs

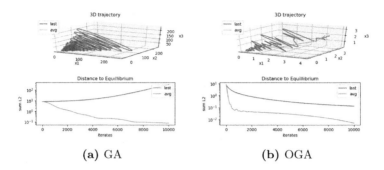

(a) GA   (b) OGA

Fig. 1. In the linear game, GA last iterate diverges, while GA average iterate converges. For OGA, both last and average iterate converge.

We provide experiments validating our theoretical results for a three player game with linear payoffs. The payoff of the players can be expressed as

$$
\begin{bmatrix} p_1(\boldsymbol{x}) \\ p_2(\boldsymbol{x}) \\ p_3(\boldsymbol{x}) \end{bmatrix} = \begin{bmatrix} \boldsymbol{x}_1^\top \ \boldsymbol{x}_2^\top \ \boldsymbol{x}_3^\top \end{bmatrix} \begin{bmatrix} 0 & C_{12} & C_{13} \\ -C_{12}^\top & 0 & C_{23} \\ -C_{13}^\top & -C_{23}^\top & 0 \end{bmatrix} \begin{bmatrix} \boldsymbol{x}_1 \\ \boldsymbol{x}_2 \\ \boldsymbol{x}_3 \end{bmatrix}.
$$

To track convergence, it is convenient if the game has a unique Nash. The linear game will have a unique Nash at $\mathbf{0}$ as long as the fixed point of dynamics is the singleton $\{\mathbf{0}\}$. This will occur if the Hessian of payoffs, H, has no eigenvalues equal to 0. Since H is antisymmetric, its eigenvalues come in complex pairs, and if d is even dimension, H will have an eigenvalue equal to 0 if its determinant is 0. If we sample entries of the C_i's i.i.d from the uniform distribution this will happen with probability 0.

We let $C_{ij} \in \mathbb{R}^{10 \times 10}$, and initialize the entries by sampling i.i.d. from the uniform distribution on $[0, 1]$. We initialize the coordinates of $\boldsymbol{x}_1^0, \boldsymbol{x}_2^0, \boldsymbol{x}_3^0$ i.i.d from the uniform distribution on $[-1, 1]$. For GA we set $\eta = 0.003$, to allow us to visualize the convergence of the average iterate and the divergence of the last iterate on the same plot. For OGA we let $\eta = 0.05$ for fastest convergence. We plot the trajectory of a single representative game simulation.

We demonstrate the performance of GA and OGA by plotting the trajectories of players' strategies; to show this in $\mathbb{R}^3$, we take the ℓ_2-norm of each player's strategies to form a three dimensional vector. We also plot the the sum of the squares of the ℓ_2 distance of player strategies from the origin on a log scale. This is shown in Fig. 1. From our results, it can be seen that GA diverges from the unique Nash of the game, while OGA converges to the unique Nash in last iterate. Notice that although the last iterate of OGA converges, it does so at a slower speed than the average iterate. The convergence in last iterate of OGA is not quite linear, but is upper bounded by a linear function, and hence does not contradict our theory.

4.2   Three Player Game with Smooth and Strongly Concave Payoffs

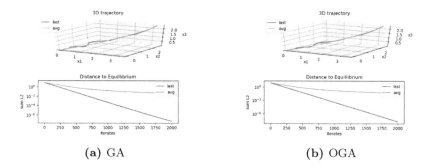

(a) GA   (b) OGA

Fig. 2. In the smooth and strongly concave game, last iterate and average iterate converges for both algorithms.

We next provide experiments showing convergence in a three player game with smooth and strongly concave payoffs. We set the payoffs for each player as follows:

$$p_{ij}(\boldsymbol{x}) = -\frac{1}{2}\|\boldsymbol{x}_i\|^2 + \boldsymbol{x}_i^\top C_{ij}\boldsymbol{x}_j + \frac{1}{2}\|\boldsymbol{x}_j\|^2$$

$$p_i(\boldsymbol{x}) = \sum_{j \in V\setminus\{i\}} p_{ij}(\boldsymbol{x}) \tag{4}$$

Like in the game with linear payoffs, this game has a unique Nash at $\mathbf{0}$ if and only if the determinant of H is nonzero, which we can guarantee by sampling entries of the C_i's uniformly at random.

As in the linear game, we initialize the entries of $C_{ij} \in \mathbb{R}^{10 \times 10}$ by sampling i.i.d. from the uniform distribution on $[0, 1]$, and the coordinates of $\boldsymbol{x}_1^0, \boldsymbol{x}_2^0, \boldsymbol{x}_3^0$ i.i.d from the uniform distribution on $[-1, 1]$. We set $\eta = 0.005$ for both GA and OGA. The results are shown in Fig. 2.

From our plots, it can be seen that both GA and OGA last iterates converge for the smooth and strongly concave game. Although OGA converges for both the game with linear payoffs and the game with smooth and strongly concave payoffs, the trajectory of GA and OGA take a more direct path to the Nash in the smooth and strongly concave game, as can be seen in the 3d trajectory. Furthermore, the last iterate of both GA and OGA follow a linear trend in the log scale, as predicted by our theory.

4.3   Effect of Number of Players on Convergence

In this section, we provide experimental results showing the effect of varying n in a NZSG of n players. For the smooth and strongly concave game, our theoretical upper bound has a linear dependence on the number of players in the game for both GA and OGA, and thus we test the dependence on players only in this setting.

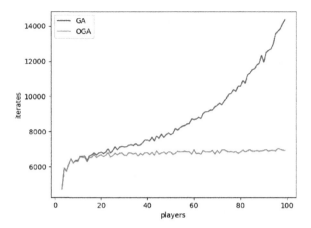

Fig. 3. In the smooth and strongly concave game, the performance of GA decays as players increase, while the performance of OGA plateaus.

We study a game with smooth and strongly concave payoffs, using the same payoffs as in Sect. 4.2 (see Eq. (4)). We perform the same initializations as in Sect. 4.2 - initializing the entries of $C_{ij} \in \mathbb{R}^{10 \times 10}$ by sampling i.i.d. from the uniform distribution on $[0, 1]$, and the coordinates of x_i^0 i.i.d from the uniform distribution on $[-1, 1]$. We set $\eta = 0.001$ for both GA and OGA. We let the number of players range from 3 to 100, plotting convergence for each setting of players. We track convergence by plotting the number of iterates it takes for the the sum of the squares of the ℓ_2 distance of player strategies from the origin to dip below 0.00001. For each fixed number of players, we run ten trials to convergence and plot the average. The results are shown in Fig. 3.

From this plot, we can see that the number of players affects the convergence rate of GA. However, for OGA, the effect of players on convergence disappears after enough players are introduced into the game. This suggests that the convergence rate for OGA in the smooth and strongly concave case may not be tight. This is an open question for future research.

5   Conclusion

In this paper, we studied the convergence of player strategies to equilibria in Network Zero-sum Games, a class of games that generalizes two-player zero-sum games and arises naturally in learning architectures that extend GANs. We show that many results in two-player zero-sum games on the convergence and divergence of these algorithms extend to NZSGs. We believe these results may guide practitioners working on extensions of GANs that involve more than two agents. Our results also shed some light on why existing extensions of GANs that employ more than two agents are successful in achieving convergent behaviour. Future research may search for models with more relaxed game theoretic assumptions where convergence can still be shown for reasonable algorithms. For example, the zero-sum assumption is absent from certain successful architectures, e.g. Wasserstein-GAN with Gradient Penalty [23].

References

1. Azizian, W., Scieur, D., Mitliagkas, I., Lacoste-Julien, S., Gidel, G.: A tight and unified analysis of gradient-based methods for a whole spectrum of differentiable games. In: The 23rd International Conference on Artificial Intelligence and Statistics (2020)
2. Balduzzi, D., Racanière, S., Martens, J., Foerster, J.N., Tuyls, K., Graepel, T.: The mechanics of n-player differentiable games. In: 35th International Conference on Machine Learning, pp. 363–372 (2018)
3. Balduzzi, D., et al.: Smooth markets: a basic mechanism for organizing gradient-based learners. In: International Conference on Learning Representations (2020)
4. Brock, A., Lim, T., Ritchie, J., Weston, N.: Neural photo editing with introspective adversarial networks. In: International Conference on Learning Representations, pp. 1–15 (2017)
5. Cai, Y., Candogan, O., Daskalakis, C., Papadimitriou, C.: Zero-sum polymatrix games: a generalization of minmax. Math. Oper. Res. **41**(2), 648–655 (2016)
6. Cesa-Bianchi, N., Lugosi, G.: Prediction, Learning, and Games. Cambridge University Press, Cambridge (2006)
7. Che, T., Li, Y., Jacob, A.P., Bengio, Y., Li, W.: Mode regularized generative adversarial networks. In: International Conference on Learning Representations (2017)
8. Chongxuan, L., Xu, T., Zhu, J., Zhang, B.: Triple generative adversarial nets. In: Advances in Neural Information Processing Systems, pp. 4088–4098 (2017)
9. Daskalakis, C., Ilyas, A., Syrgkanis, V., Zeng, H.: Training GANs with optimism. In: International Conference on Learning Representations (2018)
10. Daskalakis, C., Panageas, I.: Last-iterate convergence: zero-sum games and constrained min-max optimization. In: 10th Innovations in Theoretical Computer Science Conference (ITCS 2019) , vol. 124, pp. 27:1–27:18 (2018)
11. Daskalakis, C., Panageas, I.: The limit points of (optimistic) gradient descent in min-max optimization. In: Advances in Neural Information Processing Systems, pp. 9236–9246 (2018)
12. Donahue, J., Krähenbühl, P., Darrell, T.: Adversarial feature learning. In: International Conference on Learning Representations (2017)
13. Gallier, J.: The Schur complement and symmetric positive semidefinite (and definite) matrices. Penn Eng. 1–12 (2010)
14. Gidel, G., Berard, H., Vignoud, G., Vincent, P., Lacoste-Julien, S.: A variational inequality perspective on generative adversarial networks. In: International Conference on Learning Representations (2019)
15. Goodfellow, I.: NIPS 2016 tutorial: generative adversarial networks. arXiv preprint arXiv:1701.00160 (2016)
16. Goodfellow, I., et al.: Generative adversarial nets. In: Advances in Neural Information Processing Systems, pp. 2672–2680 (2014)
17. Hoang, Q., Nguyen, T.D., Le, T., Phung, D.Q.: MGAN: training generative adversarial nets with multiple generators. In: 6th International Conference on Learning Representations (2018)

18. Liang, T., Stokes, J.: Interaction matters: a note on non-asymptotic local convergence of generative adversarial networks. In: The 22nd International Conference on Artificial Intelligence and Statistics, pp. 907–915 (2019)
19. Menache, I., Ozdaglar, A.: Network games: theory, models, and dynamics. Synth. Lect. Commun. Netw. **4**(1), 1–159 (2011)
20. Mertikopoulos, P., Lecouat, B., Zenati, H., Foo, C.S., Chandrasekhar, V., Piliouras, G.: Optimistic mirror descent in saddle-point problems: going the extra(-gradient) mile. In: International Conference on Learning Representations (2019)
21. Mokhtari, A., Ozdaglar, A., Pattathil, S.: A unified analysis of extra-gradient and optimistic gradient methods for saddle point problems: proximal point approach. In: The 23rd International Conference on Artificial Intelligence and Statistics (2020)
22. Vandenhende, S., De Brabandere, B., Neven, D., Van Gool, L.: A three-player GAN: generating hard samples to improve classification networks. In: 16th International Conference on Machine Vision Applications (MVA), pp. 1–6 (2019)
23. Wei, X., Liu, Z., Wang, L., Gong, B.: Improving the improved training of Wasserstein GANs. In: International Conference on Learning Representations (2018)

Adaptive Momentum Coefficient
for Neural Network Optimization

Zana Rashidi[1]([✉]), Kasra Ahmadi K. A.[1], Aijun An[1], and Xiaogang Wang[2]

[1] Department of Electrical Engineering and Computer Science, York University,
Toronto, ON M3J 1P3, Canada
{zrashidi,kasraah,aan}@eecs.yorku.ca
[2] Department of Mathematics and Statistics, York University,
Toronto, ON M3J 1P3, Canada
stevenw@mathstat.yorku.ca

Abstract. We propose a novel and efficient momentum-based first-order algorithm for optimizing neural networks which uses an adaptive coefficient for the momentum term. Our algorithm, called *Adaptive Momentum Coefficient* (AMoC), utilizes the inner product of the gradient and the previous update to the parameters, to effectively control the amount of weight put on the momentum term based on the change of direction in the optimization path. The algorithm is easy to implement and its computational overhead over momentum methods is negligible. Extensive empirical results on both convex and neural network objectives show that AMoC performs well in practise and compares favourably with other first and second-order optimization algorithms. We also provide a convergence analysis and a convergence rate for AMoC, showing theoretical guarantees similar to those provided by other efficient first-order methods.

Keywords: Adaptive momentum · Neural networks · Optimization · Accelerated gradient descent · Convex optimization

1 Introduction

First order optimization methods such as Stochastic Gradient Descent (SGD) with Momentum [21] and their variants are the methods of choice for optimizing neural networks. While there has been extensive work on developing second-order methods such as Hessian-Free optimization [11] and Natural Gradients [1,12], such methods have not been successful in replacing the first-order ones due to their large per-iteration costs in time and memory.

Although Nesterov's accelerated gradient and its modifications have been very effective in deep neural network optimization [21], it has been shown that Nesterov's method might perform suboptimal for strongly convex functions [2] without looking at local geometry of the objective function. Further, in order to get the best of both worlds, search for optimization methods which combine the

© Springer Nature Switzerland AG 2021
F. Hutter et al. (Eds.): ECML PKDD 2020, LNAI 12458, pp. 35–51, 2021.
https://doi.org/10.1007/978-3-030-67661-2_3

efficiency of first-order methods and the effectiveness of second-order updates is still underway.

In this work, we introduce an adaptive coefficient for the momentum term in the Heavy Ball and Nesterov methods as an effort to combine first-order and second-order methods. We call our algorithm *Adaptive Momentum Coefficient* (AMoC). The adaptive coefficient effectively weights the momentum term based on the change in direction on the loss surface in the optimization process. The change in direction can contribute as implicit local curvature information without resorting to the expensive second-order information such as the Hessian or the Fisher Information Matrix.

Our experiments show the effectiveness of the adaptive coefficient on both strongly-convex functions with Lipschitz gradients and neural network objectives. AMoC can speed up the convergence process significantly in convex problems and performs well in the neural network experiments. AMoC has similar time-efficiency as first-order methods (e.g. Heavy Ball, Nesterov) while reaching lower errors.

The structure of the paper is as follows. In Sect. 2, we give a brief background on neural network optimization methods. In Sect. 3, we introduce our adaptive momentum coefficient algorithm and discuss its merits. In Sect. 4, we analyse the convergence of the algorithm and provide a convergence rate. In Sect. 4, we discuss some of the related work in the literature. Section 5 illustrates the algorithm's performance on convex and non-convex benchmarks. Proofs and details regarding the algorithm and the experiments can be found in the Appendix.

2   Background

We consider a neural network with a differentiable loss function $f : \mathbb{R}^D \to \mathbb{R}$ with the set of parameters θ. The objective is to minimize the loss function f with a set of iterative updates to the parameters θ. Gradient descent methods use the following update:

$$\theta_{t+1} = \theta_t - \epsilon \nabla f(\theta_t) \tag{1}$$

where ϵ is the learning rate. However, this update can be very slow and determining the learning rate ϵ can be hard. A large learning rate can cause oscillations and overshooting. A small learning rate can slow down the convergence drastically.

Heavy Ball Method. In order to speed up the convergence of gradient descent, one can add a momentum term [18].

$$d_{t+1} = \mu d_t - \epsilon \nabla f(\theta_t); \quad \theta_{t+1} = \theta_t + d_{t+1} \tag{2}$$

where d is the velocity and μ is the momentum parameter.

Algorithm 1: AMoC (ADAPTIVE MOMENTUM COEFFICIENT)

---

1 Initialize θ_1
2 Set $d_1 = \mathbf{0}$
3 **for** $t = 1$ **to** T **do**
4 $\quad\vert\quad$ Calculate the gradient $g_t = \nabla f(\theta_t)$
5 $\quad\vert\quad$ Calculate adaptive coefficient $\gamma_t = \mu\big(1 - \beta(\bar{g}_t \cdot \bar{d}_t)\big)$
6 $\quad\vert\quad$ Calculate update $d_{t+1} = \gamma_t d_t - \epsilon g_t$
7 $\quad\vert\quad$ Update parameters $\theta_{t+1} = \theta_t + d_{t+1}$

Nestrov's Method. Nesterov's accelerated gradient [16] can be rewritten as a momentum method [21]:

$$d_{t+1} = \mu d_t - \epsilon \nabla f(\theta_t + \mu d_t); \quad \theta_{t+1} = \theta_t + d_{t+1} \tag{3}$$

Nesterov's momentum is different from the heavy ball method only in where we take the gradient. Note that in both methods d_t is the previous update $\theta_t - \theta_{t-1}$.

3   Adaptive Momentum Coefficient

We propose Adaptive Momentum Coefficient (AMoC) as an alternative to the fixed momentum parameter in the Heavy Ball and Nesterov's algorithms. The adaptive coefficient utilizes the angle between the direction of the current gradient vector and the previous update to the parameters. This angle is characterized by the inner product between the normalized vectors of these two values, namely

$$\bar{g}_t \cdot \bar{d}_t = \cos(\pi - \phi_t) \quad \text{where} \quad \bar{g}_t = \frac{g_t}{\|g_t\|}; \quad \bar{d}_t = \frac{d_t}{\|d_t\|} \tag{4}$$

where ϕ_t is the angle between the negative gradient $-g_t$ and the previous update d_t. Our goal with the adaptive coefficient is to automatically reinforce the momentum term when these two directions align and gradually decrease this effect when they don't. Thus we propose the following coefficient for the momentum term:

$$\gamma_t = \mu\big(1 - \beta(\bar{g}_t \cdot \bar{d}_t)\big) \tag{5}$$

where $\mu \geq 0$ is the regular momentum parameter and $\beta \geq 0$ is a parameter controlling the amount of weight put on the inner product. With a large β, the algorithm will behave more aggressively, meaning, moving rapidly when directions align and bouncing back when they don't, which is suitable for convex functions. A small β however, leads to a more conservative behaviour, which is expected when optimizing non-convex objectives, e.g., neural networks (to avoid being trapped in local optima, for example). Since ϕ can range from 0 to π,

$$\mu(1 - \beta) \leq \gamma_t \leq \mu(1 + \beta) \tag{6}$$

Note that if we set β to zero, γ_t reduces to the regular momentum parameter for Heavy Ball, which can be seen as a special case of our algorithm. The adaptive coefficient embeds a notion of change of direction of the optimization path. This notion can be interpreted as implicit second-order information where it tells us how much the current gradient's direction is different from the previous gradients which is similar to what second-order information (e.g. the Hessian) provides intuitively:

$$H(\theta_t) = \nabla^2 f(\theta_t) \tag{7}$$

The Hessian provides the rate of change in the gradient of a function at a point while the gradient tells us the rate of change of the function itself. Since the previous update contains a running estimate of the past gradients, the dot product is basically comparing the current gradient against all aggregated past gradients.

The algorithm (AMoC) is shown in Algorithm 1. We also incorporate Nesterov's lookahead gradient into AMoC by simply taking the gradient at a further point, $g_t = \nabla f(\theta_t + \mu d_t)$, thus only changing line 4 of Algorithm 1. We call this version of the algorithm AMoC-N, with "N" standing for Nesterov.

4   Convergence Analysis

To analyse AMoC's convergence, consider a differentiable convex function f with Lipschitz gradients. On this class of functions, similar to the Heavy Ball algorithm, AMoC benefits from a convergence rate of order $1/T$ with a slightly different factor. To prove this proposition, we need to define the following sequence of coefficients:

Definition. *The sequence $\{\lambda_k\}$ is defined as:*

$$\lambda_{k+1} = \frac{\lambda_k}{\gamma_k} - 1, \qquad \lambda_1 = \frac{\mu(1-\beta)}{1 - \mu(1-\beta)} \tag{8}$$

Lemma. *For arbitrarily large integer T, there exists $\beta > 0$ such that the first $T + 1$ elements of the sequence $\{\lambda_k\}$ are positive.*

Proof. See Appendix A.

This lemma provides us with the proper β for the following theorem. This theorem guarantees the convergence of the algorithm, and provides an upper-bound for the convergence rate of a specific weighted average of all iterates.

Theorem. *For any differentiable convex function f with L-Lipschitz gradients, the sequence generated by AMoC with sufficiently small β, $\mu \in [0, \frac{1}{1+\beta})$, and $\epsilon \in (0, \frac{\mu(1+\beta)}{L\lambda_1})$ satisfies the following:*

$$f(\tilde{\theta}_T) - f(\theta^*) \leq \frac{\|\theta_1 - \theta^*\|^2}{2T(1 + \lambda_{T+1})} \left(\frac{1}{\epsilon} + \frac{\lambda_1^2 L}{\mu} \right) \tag{9}$$

where θ^* is the optimal point and $\tilde{\theta}_T = (\sum_{k=1}^{T} \frac{\lambda_k}{\gamma_k} \theta_k)/(\sum_{k=1}^{T} \frac{\lambda_k}{\gamma_k})$.

Proof. See Appendix A.

It can be easily verified that by setting β to 0, all the elements of the sequence $\{\lambda_k\}$ will be equal to $\frac{\mu}{1-\mu}$ and (9) becomes the same bound as the one for the Heavy Ball algorithm presented in [6].

5   Related Work

There has been extensive work on large-scale optimization techniques for neural networks in recent years. A good overview can be found in [3]. Here, we discuss some of the work more related to ours in three parts.

5.1   Gradient Descent Variants

Adagrad [5] is an optimization technique that extends gradient descent and adapts the learning rate according to the parameters. Adadelta [24] and RMSprop [22] improve upon Adagrad by reducing its aggressive deduction of the learning rate. Adam [9] improves upon the previous methods by keeping an additional average of the past gradients which is similar to what momentum does. Adaptive Restart [17] proposes to reset the momentum whenever rippling behaviour is observed in accelerated gradient schemes. AggMo [10] keeps several velocity vectors with distinct parameters in order to damp oscillations. AMS-Grad [19] on the other hand, keeps a longer memory of the past gradients to overcome the suboptimality of the Adam on simple convex problems.

5.2   Accelerated Methods

Several recent works have been focusing on acceleration for gradient descent methods. In [13], the authors propose an adaptive method to accelerate Nesterov's algorithm in order to close a small gap in its convergence rate for strongly convex functions with Lipschitz gradients adding a possibility of more than one gradient call per iteration. In [20], the authors propose a differential equation for modeling Nesterov inspired by the continuous version of gradient descent, a.k.a. gradient flow. The authors in [23] take this further and suggest that all accelerated methods have a continuous time equivalent defined by a Lagrangian functional, which they call the Bregman Lagrangian. Recently, in [4] the authors propose a differential geometric interpretation of Nesterov's method for strongly-convex functions with links to continuous time differential equations mentioned earlier and their Euler discretization.

5.3   Second-Order Methods

Second-order methods are desirable because of their fine convergence properties due to dealing with bad-conditioned curvature by using local second-order information. Hessian-Free optimization [11] is based on the truncated-Newton approach where the conjugate gradient algorithm is used to optimize the quadratic

Table 1. Number of iterations to convergence for the strongly-convex functions with Lipschitz gradients experiment. Note that in our experiments, Adam did not converge for the Smooth-BPDN and Ridge Regression problem.

	AMoC	AMoC-N	Heavy Ball	Nesterov	Adam
Anisotropic Bowl	78	**50**	1901	1749	888
Smooth-BPDN	**59**	77	110	115	–
Ridge Regression	**1160**	1817	>3000	>3000	–

approximation of the objective function. The natural gradient method [1] reformulates the gradient descent in the space of the prediction functions instead of the parameters. This space is then studied using concepts in differential geometry. K-FAC [12] approximates the Fisher information matrix which is based on the natural gradient method. Our method is different since we are not using explicit second-order information but rather implicitly deriving curvature information using the change in direction.

6 Experiments

We evaluated AMoC on strongly convex functions with Lipschitz gradients and neural network objectives including Autoencoders, Residual Networks and LSTMs. We compare our algorithm with Heavy-Ball, Nesterov, and Adam in addition with K-FAC in the Autoencoder experiment. Our neural network experiment setups closely follow [10], were tuned for best performance on the validation set and implemented in PyTorch (except K-FAC, where we used its official Tensorflow code[1] with the optimal parameters). See Appendix B for a detailed discussion of the inner product values during these experiments.

6.1 Strongly Convex and Lipschitz Functions

We borrow these three minimization problems from [13] where the authors try to accelerate Nesterov's method by using adaptive step sizes. The problems are Anisotropic Bowl, Ridge Regression and Smooth-BPDN. We fix the momentum parameter μ for all methods which is set to 0.99 for the Anisotropic Bowl and 0.9 for the other two problems. The learning rate ϵ for all methods is tuned for best performance. The parameter β in our algorithms is set to 1 for the Anisotropic Bowl for both AMoC and AMoC-N, 0.1 for Ridge Regression for both methods and 1 for AMoC in Smooth-BPDN and 0.1 for AMoC-N in the same problem. Results are shown in Fig. 1 and Table 1.

---

[1] https://github.com/tensorflow/kfac.

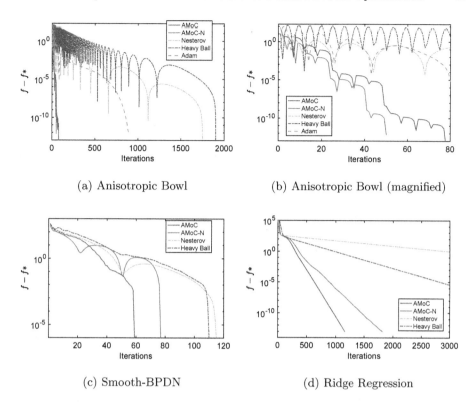

(a) Anisotropic Bowl   (b) Anisotropic Bowl (magnified)

(c) Smooth-BPDN   (d) Ridge Regression

Fig. 1. Results from experiments on strongly convex functions with Lipschitz gradients. All methods start from the same point. The vertical axes show the distance to the optimal value ($f - f^*$, where $f^* = f(\theta^*)$) and the horizontal axes show the number of iterations.

Anisotropic Bowl. The Anisotropic Bowl is a bowl-shaped function with a constraint to get Lipschitz continuous gradients:

$$f(\theta) = \sum_{i=1}^{n} i \cdot \theta_{(i)}^4 + \frac{1}{2}\|\theta\|_2^2, \qquad \text{subject to } \|\theta\|_2 \leq \tau \tag{10}$$

As in [13], we set $n = 500$, $\tau = 4$ and $\theta_0 = \frac{\tau}{\sqrt{n}}\mathbf{1}$. Figure 1a and 1b (magnified) show the convergence results for our algorithms and the baselines. The algorithms terminate when $f(\theta) - f(\theta^*) < 10^{-12}$. AMoC-N and AMoC take only 50 and 78 iterations to converge, while the closest result is that of Adam which takes 888 iteration to converge.

Ridge Regression. The Ridge Regression problem is a linear least squares function with Tikhonov regularization:

$$f(\theta) = \frac{1}{2}\|A\theta - b\|_2^2 + \frac{\lambda}{2}\|\theta\|_2^2 \tag{11}$$

where $A \in \mathbb{R}^{m \times n}$ is a measurement matrix, $b \in \mathbb{R}^m$ is the response vector and $\lambda > 0$ is the ridge parameter. The function $f(\theta)$ is a positive definite quadratic function with the unique solution of $\theta^* = (A^T A + \lambda I)^{-1} A^T b$.

Following [13], $m = 1200$, $n = 2000$ and $\lambda = 1$. A is generated from $U \Sigma V^T$ where $U \in \mathbb{R}^{m \times m}$ and $V \in \mathbb{R}^{n \times m}$ are random orthonormal matrices and $\Sigma \in \mathbb{R}^{m \times m}$ is diagonal with entries linearly distanced in $[100, 1]$ while $b = \text{randn}(m, 1)$ is drawn i.i.d. from the standard normal distribution. Figure 1d shows the results where AMoC and AMoC-N converge in 1160 and 1817 iterations respectively. The Nesterov version of the algorithm (AMoC-N) is not able to outperform the regular version as is the case with the original Heavy Ball and Nesterov algorithms. The tolerance is set to $f(\theta) - f(\theta^*) < 10^{-13}$. We have not included Adam since it was not able to perform well in this problem (also see [19]).

Smooth-BPDN. Smooth-BPDN is a smooth and strongly convex version of the BPDN (basis pursuit denoising) problem:

$$f(\theta) = \frac{1}{2} \|A\theta - b\|_2^2 + \lambda \|\theta\|_{\ell_1, \tau} + \frac{\rho}{2} \|\theta\|_2^2 \tag{12}$$

where:

$$\|\theta\|_{\ell_1, \tau} = \begin{cases} |\theta| - \frac{\tau}{2} & \text{if } |\theta| \geq \tau \\ \frac{1}{2\tau} \theta^2 & \text{if } |\theta| < \tau \end{cases}$$

and $\|\cdot\|_{\ell_1, \tau}$ is a smoothed version of the ℓ_1 norm also known as Huber penalty function with half-width of τ.

As in [13], we set $A = \frac{1}{\sqrt{n}} \cdot \text{randn}(m, 1)$ where $m = 800$ and $n = 2000$, $\lambda = 0.05$, $\tau = 0.0001$. The real signal is a random vector with 40 non-zero values and $b = A\theta^* + e$ where $e = 0.01 \frac{\|b\|_2}{\sqrt{m}} \cdot \text{randn}(m, 1)$ is Gaussian noise. Since we cannot find the solution analytically, Nesterov's method is used as an approximation to the solution $(f(\theta_N^*))$ and the tolerance is set to $f(\theta) - f(\theta_N^*) < 10^{-12}$. Figure 1c shows the results for the algorithms. AMoC-N and AMoC converge in 77 and 59 iterations respectively, outperforming all other methods. We observe the weakness of the lookahead gradient in this problem as well. Similarly, Adam was not able to perform well and hence not included it in the graph (also see [19]).

6.2   Deep Autoencoders

To evaluate the performance of AMoC, we apply it to the benchmark deep autoencoder problem first introduced in [8] on the MNIST dataset. We use the same network architectures as in [8] which is [1000 500 250 30 250 500 1000] except we use ReLU activation throughout the model. Our baselines are the Heavy Ball algorithm [18], SGD with Nesterov's Momentum [21] and K-FAC [12], a second-order method utilizing natural gradients using an approximation of the Fisher information matrix.

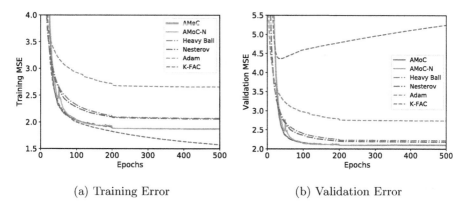

(a) Training Error (b) Validation Error

Fig. 2. Training and Validation Error during training for the MNIST Autoencoder experiment. The vertical axes show the Mean Squared Error (reconstruction error) and the horizontal axes show the number of epochs.

Table 2. Test errors from the MNIST Autoencoder experiment.

AMoC	AMoC-N	Heavy Ball	Nesterov	Adam	K-FAC
2.08	**2.08**	2.20	2.16	2.72	6.07

We use 90% of the training data for training and 10% for validation. All methods use the same parameter initialization scheme. All methods except K-FAC (which use a special learning rate and momentum schedule along with an increasing batch size schedule), use a fixed momentum parameter and we decay the learning rate by 0.1 at epochs 200 and 400. The parameter β is set to 0.1. For the momentum parameter μ, we did a search in $\{0.9, 0.99, 0.999\}$ and the learning rate in $\{0.1, 0.05, 0.01, 0.005, 0.001, 0.0005, 0.0001\}$. The minibatch size was set to 200. We set β_2 in Adam to 0.999 and searched over $\{0.9, 0.99, 0.999\}$ for β_1.

The training and validation results for 500 epochs of training are shown in Fig. 2 and the test set results are shown in Table 2. AMoC and AMoC-N outperform the baselines in terms of validation error and perform similarly to each other. K-FAC outperforms all methods in terms of training error but overfits to the dataset with the default parameters. Use of Nesterov's lookahead gradient does not affect the performance significantly in both Heavy Ball and AMoC. Adam performs poorly relative to other methods in this experiment. Further, AMoC and AMoC-N reach the lowest testing error among the algorithms followed by Nesterov and Heavy Ball.

6.3   Residual Networks

The classification experiments were done using 34-layer residual networks [7] on two datasets of CIFAR10 and CIFAR100. Results are shown in Fig. 3 and

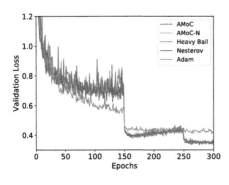
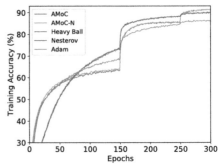

(a) Validation Loss for CIFAR10.   (b) Training Accuracy for CIFAR100.

Fig. 3. Results from the ResNet-34 experiments on CIFAR10 and CIFAR100. The vertical axis for the figure on the left shows the loss and of the figure on the right shows the accuracy. The horizontal axes show the number of epochs.

Table 3. Test accuracy (%) from the ResNet-34 experiments.

	AMoC	AMoC-N	Heavy Ball	Nesterov	Adam
CIFAR10	**89.37**	88.23	88.8	88.75	86.48
CIFAR100	68.05	**68.26**	64.04	64.04	63.58

Table 3. We trained each model for 300 epochs using 80% of the training data while holding out a random 20% for validation.

We used a batch size of 128 and searched for the learning rate in {0.1, 0.05, 0.01, 0.005, 0.001, 0.0005, 0.0001} while decaying by 0.1 at epochs 150 and 250. β_1, β_2 and μ were set similar to the Autoencoder experiment. β was set to 0.2 for CIFAR10 and 0.1 for CIFAR100. Batch normalization and a weight decay of 0.0005 were used in the models. Data augmentation was also limited to random resized cropping and horizontal flips.

In the CIFAR10 experiment, the algorithms AMoC, AMoC-N, Heavy Ball and Nesterov outperform Adam and perform similarly among themselves. However, AMoC achieves the highest accuracy on the test set. For CIFAR100, AMoC and AMoC-N outperform other methods by a noticeable margin, while the baselines perform similarly. Our algorithms also reach the highest accuracy on the test set with AMoC-N slightly higher.

6.4   LSTMs

We also experimented with LSTM word-level language models on the Penn Treebank dataset following the experimental setting of [14]. The LSTM model used has 3 layers each with 1150 hidden nodes with an embedding size of 400. Dropout is used with the probability of 0.1 on the embedding layer, 0.65 on the input embedding layer and 0.3 in the hidden layers. A weight decay of 1.2e-6 is used

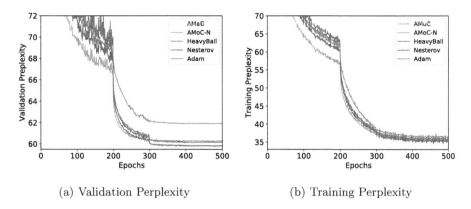

(a) Validation Perplexity   (b) Training Perplexity

Fig. 4. Training and Validation perplexity from the LSTM experiments on the Penn Treebank dataset. The vertical axes show the perplexity of the language model and the horizontal axes show the number of epochs.

Table 4. Test perplexity from the LSTM Penn Treebank experiment.

AMoC	AMoC-N	Heavy Ball	Nesterov	Adam
58.16	**57.61**	57.91	57.80	59.47

along with weight drop with a probability of 0.5 and temporal activation regularization with scaling of 1. L2 regularization is also used on the activations with a scale of 2.

We used a batch size of 80 and searched over {20, 15, 10, 5, 1, 0.5, 0.1, 0.05, 0.01, 0.005, 0.001} for the learning rate. We decay the learning rate by 0.1 at epochs 200, 300 and 400 while training for 500 epochs. β_1, β_2 and μ were set similar to the previous experiments. β in our algorithms was set to 0.2.

The results are shown in Fig. 4 and Table 4. All methods reach similar training perplexity with AMoC-N and Nesterov reaching lower validation perplexity followed by AMoC and Heavy Ball. AMoC-N achieves the lowest perplexity on the test set.

7   Conclusions

We proposed a novel and efficient momentum-based algorithm, AMoC, by introducing an adaptive coefficient for the momentum term. We compared AMoC against SGD with Nesterov's momentum, regular momentum (Heavy Ball), Adam and a recently proposed second-order method, K-FAC, on both convex functions and non-convex neural network objectives including Autoencoders, CNNs and RNNs. We saw that AMoC is able to perform well in various settings compared to popular first order and second-order methods. AMoC's implementation is straightforward and is computationally efficient. We also analyzed AMoC's convergence properties and proposed a convergence rate similar to that of the Heavy Ball algorithm. We believe that AMoC offers a new and promising direction in convex and non-convex optimization research and in particular, neural network optimization.

Acknowledgements. We would like to thank Ruth Urner for valuable discussions and insights. We would also like to thank the anonymous reviewers for their useful suggestions and constructive feedback.

A   Proofs

Lemma. *For arbitrarily large integer T, there exists $\beta > 0$ such that the first $T + 1$ elements of the sequence $\{\lambda_k\}$ are positive.*

Proof. Consider λ_1 to λ_k are positive and λ_{k+1} is negative. We have:

$$
\lambda_{j+1} - \frac{\mu(1+\beta)}{1 - \mu(1+\beta)} = \frac{\lambda_j}{\gamma_j} - \frac{1}{1 - \mu(1+\beta)}
$$

$$
\geq \frac{\lambda_j}{\mu(1+\beta)} - \frac{1}{1 - \mu(1+\beta)}
$$

$$
= \frac{\lambda_j - \frac{\mu(1+\beta)}{1-\mu(1+\beta)}}{\mu(1+\beta)}
$$

Combining all the inequalities for $j = 1, ..., k$ results:

$$
-\frac{\mu(1+\beta)}{1 - \mu(1+\beta)} \geq \lambda_{k+1} - \frac{\mu(1+\beta)}{1 - \mu(1+\beta)}
$$

$$
\geq \frac{\lambda_1 - \frac{\mu(1+\beta)}{1-\mu(1+\beta)}}{(\mu(1+\beta))^k}
$$

$$
= \frac{-2\mu\beta}{(1-\mu)^2 - \mu^2\beta^2} / (\mu(1+\beta))^k
$$

therefore:

$$
k \geq ln\left(\frac{\mu(1+\beta)(1-\mu(1-\beta))}{2\mu\beta}\right) / ln(\mu(1+\beta))
$$

The right-hand side of this inequality approaches $+\infty$ as β approaches 0. Thus, by choosing a small enough β, we achieve arbitrarily large number of positive terms.

Theorem. *For any differentiable convex function f with L-Lipschitz gradients, the sequence generated by AMoC with sufficiently small β, $\mu \in [0, \frac{1}{1+\beta})$, and $\epsilon \in (0, \frac{\mu(1+\beta)}{L\lambda_1})$ satisfies the following:*

$$
f(\tilde{\theta}_T) - f(\theta^*) \leq \frac{\|\theta_1 - \theta^*\|^2}{2T(1 + \lambda_{T+1})}\left(\frac{1}{\epsilon} + \frac{\lambda_1^2 L}{\mu}\right) \tag{13}
$$

where θ^* is the optimal point and $\tilde{\theta}_T = (\sum_{k=1}^{T} \frac{\lambda_k}{\gamma_k}\theta_k)/(\sum_{k=1}^{T}\frac{\lambda_k}{\gamma_k})$.

Proof. To prove this theorem, we follow a similar approach as [6]. By definition we have:

$$d_k = \theta_k - \theta_{k-1}$$
$$g_k = \nabla f(\theta_k)$$
$$\gamma_k = \mu \left(1 - \beta \bar{g}_k \cdot \bar{d}_k\right)$$
$$\theta_{k+1} = \theta_k - \epsilon g_k + \gamma_k d_k$$

therefore:

$$\theta_{k+1} + \lambda_{k+1} d_{k+1} = (\lambda_{k+1} + 1)\theta_{k+1} - \lambda_{k+1}\theta_k$$
$$= \theta_k - \frac{\epsilon \lambda_k}{\gamma_k} g_k + \lambda_k d_k$$

By subtracting θ^* and seting $\delta_k = \theta_k - \theta^*$, we get:

$$\|\delta_{k+1} + \lambda_{k+1} d_{k+1}\|^2 = \|\delta_k + \lambda_k d_k\|^2 + \left(\frac{\epsilon \lambda_k}{\gamma_k}\right)^2 \|g_k\|^2$$
$$- \frac{2\alpha \lambda_k}{\gamma_k} \delta_k \cdot g_k - \frac{2\epsilon \lambda_k^2}{\gamma_k} g_k \cdot d_k$$

(14)

According to [15, Theorem 2.1.5], $\delta_k \cdot g_k \geq f(\theta_k) - f(\theta^*) + \frac{1}{2L}\|g_k\|^2$ which combined with (14) results:

$$\|\delta_{k+1} + \lambda_{k+1} d_{k+1}\|^2 \leq \|\delta_k + \lambda_k d_k\|^2 + \left(\frac{\epsilon \lambda_k}{\gamma_k}\right)^2 \|g_k\|^2$$
$$- \frac{2\epsilon \lambda_k}{\gamma_k} \left(f(\theta_k) - f(\theta^*) + \frac{1}{2L}\|g_k\|^2\right)$$
$$- \frac{2\epsilon \lambda_k^2}{\gamma_k} g_k \cdot d_k$$
$$\leq \|\delta_k + \lambda_k d_k\|^2 - \frac{2\epsilon \lambda_k}{\gamma_k}(f(\theta_k) - f(\theta^*))$$
$$- \frac{2\epsilon \lambda_k^2}{\gamma_k} g_k \cdot d_k$$

Summing up all the inequalities for $k = 1, ..., T$ yields:

$$0 \leq \|\delta_{T+1} + \lambda_{T+1} d_{T+1}\|^2 \leq \|\delta_1\|^2 - 2\epsilon \sum_{k=1}^{T} \frac{\lambda_k}{\gamma_k}(f(\theta_k) - f(\theta^*))$$
$$- 2\epsilon \sum_{k=1}^{T} \frac{\lambda_k^2}{\gamma_k} g_k \cdot d_k$$
$$\leq \|\delta_1\|^2 - 2\epsilon \left(\sum_{k=1}^{T} \frac{\lambda_k}{\gamma_k}\right) \left(f(\tilde{\theta}_k) - f(\theta^*)\right)$$
$$- \frac{2\epsilon}{\mu} \sum_{k=1}^{T} \lambda_k^2 \|g_k\| \|d_k\| \frac{\bar{g}_k \cdot \bar{d}_k}{(1 - \beta \bar{g}_k \cdot \bar{d}_k)}$$

For $\beta < 1$, function $\frac{x}{1-\beta x}$ is convex for $x \in [-1, 1]$, thus:

$$0 \le \|\delta_1\|^2 - 2\epsilon(\sum_{k=1}^{T} \frac{\lambda_k}{\gamma_k}) \left(f(\tilde{\theta}_k) - f(\theta^*) \right)$$

$$-\frac{2\epsilon}{\mu} \frac{\sum_{k=1}^{T} \lambda_k^2 g_k \cdot d_k}{1 - \beta \frac{\sum_{k=1}^{T} \lambda_k^2 g_k \cdot d_k}{\sum_{k=1}^{T} \lambda_k^2 \|g_k\| \|d_k\|}}$$

Function $\frac{x}{1-\beta x}$ is also increasing, and $g_k \cdot d_k \ge f(\theta_k) - f(\theta_{k-1})$ [15, Theorem 2.1.5], therefore:

$$2\epsilon(\sum_{k=1}^{T} 1 + \lambda_{k+1}) \left(f(\tilde{\theta}_k) - f(\theta^*) \right) \le \|\delta_1\|^2 - \frac{2\epsilon}{\mu} \frac{\sum_{k=2}^{T} \lambda_k^2 \left(f(\theta_k) - f(\theta_{k-1}) \right)}{1 - \beta \frac{\sum_{k=2}^{T} \lambda_k^2 (f(\theta_k) - f(\theta_{k-1}))}{\sum_{k=1}^{T} \lambda_k^2 \|g_k\| \|d_k\|}}$$

Furthermore, easily one can show that sequence $\{\lambda_k\}$ is decreasing, and

$$\sum_{k=2}^{T} \lambda_k^2 \left(f(\theta_k) - f(\theta_{k-1}) \right) \ge -\lambda_1^2 \left(f(\theta_1) - f(\theta^*) \right)$$

Therefore:

$$2\epsilon T(1 + \lambda_{k+1}) \left(f(\tilde{\theta}_k) - f(\theta^*) \right) \le \|\delta_1\|^2 + \frac{2\epsilon}{\mu} \frac{\lambda_1^2 \left(f(\theta_1) - f(\theta^*) \right)}{1 + \beta \frac{\lambda_1^2 (f(\theta_1) - f(\theta^*))}{\sum_{k=1}^{T} \lambda_k^2 \|g_k\| \|d_k\|}}$$

$$\le \|\delta_1\|^2 + \frac{2\epsilon}{\mu} \lambda_1^2 \left(f(\theta_1) - f(\theta^*) \right)$$

$$\le \|\delta_1\|^2 \left(1 + \frac{\epsilon}{\mu} \lambda_1^2 L \right)$$

where the final inequality follows from $f(\theta_1) - f(\theta^*) \le \frac{L}{2} \|\delta_1\|^2$ [15, Theorem 2.1.5], and concludes the proof.

B   Inner Product Analysis

In this section we report the value of the inner product $(\bar{g}_t \cdot \bar{d}_t)$ for AMoC and AMoC-N per iteration/epoch for each of the experiments in Figs. 5 and 6. The results from the Anisotropic Bowl and the Smooth-BPDN experiments are particularly interesting. In the first case (Fig. 5a), with the inner product oscillating between positive and negative values, we can infer that the algorithm is crossing the optimum multiple times (without overshooting) but is able to bounce back and reach the optimum point eventually and in less iterations than the baselines. The second case (Fig. 5c) behaves in a similar way, except the algorithm seems to be moving close to the optimum but going up again and bouncing back several times (most likely in an oval-shaped trajectory) until it

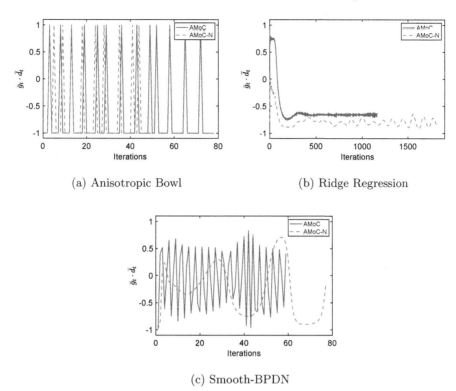

(a) Anisotropic Bowl   (b) Ridge Regression

(c) Smooth-BPDN

Fig. 5. Inner product, $\bar{g}_t \cdot \bar{d}_t = \cos(\pi - \phi_t)$, during optimization for both AMoC and AMoC-N for convex experiments in the paper.

gets close enough that it terminates. In the Ridge Regression problem (Fig. 5b), the inner product drops from values between 0.5 and 1 to values between -0.5 to -1, indicating that the algorithm is moving towards the optimum steadily, with the updates always keeping a low angle with the gradient. In the neural network experiments (Figs. 6a to 6d), the inner product gets closer to 0 by the end of training. We link this behaviour to the algorithm moving in a circular fashion around the optima with the direction of the negative gradient almost perpendicular ($\phi = \pi/2$) to the previous update.

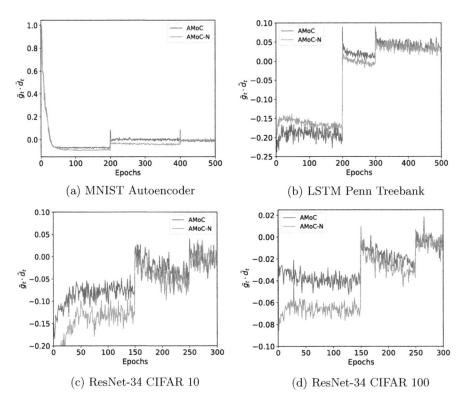

(a) MNIST Autoencoder

(b) LSTM Penn Treebank

(c) ResNet-34 CIFAR 10

(d) ResNet-34 CIFAR 100

Fig. 6. Inner product, $\bar{g}_t \cdot \bar{d}_t = \cos{(\pi - \phi_t)}$, during training for both AMoC and AMoC-N for NN experiments in the paper.

References

1. Amari, S.I.: Natural gradient works efficiently in learning. Neural Comput. **10**(2), 251–276 (1998)
2. Aujol, J.F., Rondepierre, A., Aujol, J., Dossal, C., et al.: Optimal convergence rates for Nesterov acceleration. arXiv preprint arXiv:1805.05719 (2018)
3. Bottou, L., Curtis, F.E., Nocedal, J.: Optimization methods for large-scale machine learning. SIAM Rev. **60**(2), 223–311 (2018)
4. Defazio, A.: On the curved geometry of accelerated optimization. arXiv preprint arXiv:1812.04634 (2018)
5. Duchi, J., Hazan, E., Singer, Y.: Adaptive subgradient methods for online learning and stochastic optimization. J. Mach. Learn. Res. **12**(Jul), 2121–2159 (2011)
6. Ghadimi, E., Feyzmahdavian, H.R., Johansson, M.: Global convergence of the heavy-ball method for convex optimization. In: 2015 European Control Conference (ECC), pp. 310–315. IEEE (2015)
7. He, K., Zhang, X., Ren, S., Sun, J.: Deep residual learning for image recognition. In: Proceedings of the IEEE Conference on Computer Vision and Pattern Recognition, pp. 770–778 (2016)
8. Hinton, G.E., Salakhutdinov, R.R.: Reducing the dimensionality of data with neural networks. Science **313**(5786), 504–507 (2006)

9. Kingma, D.P., Ba, J.: Adam: a method for stochastic optimization. arXiv preprint arXiv:1412.6980 (2014)
10. Lucas, J., Sun, S., Zemel, R., Grosse, R.: Aggregated momentum: stability through passive damping. arXiv preprint arXiv:1804.00325 (2018)
11. Martens, J.: Deep learning via hessian-free optimization. In: ICML, vol. 27, pp. 735–742 (2010)
12. Martens, J., Grosse, R.: Optimizing neural networks with Kronecker-Factored approximate curvature. In: International Conference on Machine Learning, pp. 2408–2417 (2015)
13. Meng, X., Chen, H.: Accelerating Nesterov's method for strongly convex functions with Lipschitz gradient. arXiv preprint arXiv:1109.6058 (2011)
14. Merity, S., Keskar, N.S., Socher, R.: Regularizing and optimizing LSTM language models. arXiv preprint arXiv:1708.02182 (2017)
15. Nesterov, Y.: Introductory Lectures on Convex Optimization: A Basic Course, vol. 87. Springer, Heidelberg (2013). https://doi.org/10.1007/978-1-4419-8853-9
16. Nesterov, Y.E.: A method for solving the convex programming problem with convergence rate o $(1/k^2)$. Dokl. akad. nauk Sssr. **269**, 543–547 (1983)
17. O'donoghue, B., Candes, E.: Adaptive restart for accelerated gradient schemes. Found. Comput. Math. **15**(3), 715–732 (2015)
18. Polyak, B.T.: Some methods of speeding up the convergence of iteration methods. USSR Comput. Math. Math. Phys. **4**(5), 1–17 (1964)
19. Reddi, S.J., Kale, S., Kumar, S.: On the convergence of ADAM and beyond (2018)
20. Su, W., Boyd, S., Candes, E.: A differential equation for modeling Nesterov's accelerated gradient method: theory and insights. In: Advances in Neural Information Processing Systems, pp. 2510–2518 (2014)
21. Sutskever, I., Martens, J., Dahl, G.E., Hinton, G.E.: On the importance of initialization and momentum in deep learning. In: ICML (3), vol. 28, pp. 1139–1147 (2013)
22. Tieleman, T., Hinton, G.: Lecture 6.5-rmsprop: divide the gradient by a running average of its recent magnitude. COURSERA: Neural Netw. Mach. Learn. **4**(2), 26–31 (2012)
23. Wibisono, A., Wilson, A.C., Jordan, M.I.: A variational perspective on accelerated methods in optimization. Proc. Natl. Acad. Sci. **113**(47), E7351–E7358 (2016)
24. Zeiler, M.D.: Adadelta: an adaptive learning rate method. arXiv preprint arXiv:1212.5701 (2012)

Squeezing Correlated Neurons for Resource-Efficient Deep Neural Networks

Elbruz Ozen$^{(\boxtimes)}$ (ID) and Alex Orailoglu

Department of Computer Science and Engineering,
University of California, San Diego, 9500 Gilman Dr, La Jolla, CA 92093, USA
elozen@eng.ucsd.edu, alex@cs.ucsd.edu

Abstract. DNNs are abundantly represented in real-life applications because of their accuracy in challenging problems, yet their demanding memory and computational costs challenge their applicability to resource-constrained environments. Taming computational costs has hitherto focused on first-order techniques, such as eliminating numerically insignificant neurons/filters through numerical contribution metric prioritizations, yielding passable improvements. Yet redundancy in DNNs extends well beyond the limits of numerical insignificance. Modern DNN layers exhibit a significant correlation among output activations; hence, the number of extracted orthogonal features at each layer rarely exceeds a small fraction of the layer size. The exploitation of this observation necessitates the quantification of information content at layer outputs. To this end, we employ practical data analysis techniques coupled with a novel feature elimination algorithm to identify a minimal set of computation units that capture the information content of the layer and squash the rest. Linear transformations on the subsequent layer ensure accuracy retention despite the removal of a significant portion of the computation units. The one-shot application of the outlined technique can shrink the VGG-16 model size 4.9× and speed up its execution by 3.4× with negligible accuracy loss while requiring **no additional fine-tuning**. The proposed approach, in addition to delivering results overwhelmingly superior to hitherto promulgated heuristics, furthermore promises to spearhead the design of more compact deep learning models through an improved understanding of DNN redundancy.

Keywords: Deep learning · Information redundancy · Pruning

1 Introduction

Deep learning is widely adopted in various fields, and its superb accuracy enables the design of applications for practical yet challenging tasks such as object detection, natural language processing, and even algorithms for autonomous vehicle control [4]. Deep learning enables designers to create complex systems by training the system solely with the labeled data. It consequently supersedes manual

© Springer Nature Switzerland AG 2021
F. Hutter et al. (Eds.): ECML PKDD 2020, LNAI 12458, pp. 52–68, 2021.
https://doi.org/10.1007/978-3-030-67661-2_4

feature extraction, a near-impossible task for such complex applications, while attaining a degree of precision rarely attainable even by humans. Although DNNs (deep neural networks) are the state-of-the-art machine learning algorithms in numerous practical applications, their memory and computational costs are excruciating since modern DNNs contain millions of parameters and require billions of operations for processing even a single quantum of data [32].

As numerous algorithmic and DNN architecture improvements (e.g., [15, 31]) have helped us to train more accurate DNN models efficiently, a significant portion of the improvements, particularly for challenging machine learning tasks, has been furnished by wider and deeper neural networks [10, 30]. However, the increasing computational complexity makes sustaining this trend a formidable task, particularly for systems deployed at the edge with limited resources.

It is widely recognized that the accurate performance of a DNN on a specific problem necessitates neither exact operations nor the majority of the network parameters embedded. The margin of inexact operations is commonly exploited by performing DNN inference in fixed-point format and quantizing fixed-point numbers into smaller bit-widths [21]. DNN pruning algorithms concentrate instead on the latter aspect of redundancy as they aim to deliver more compact DNN models with fewer parameters and operations. While pruning the parameters in an unstructured manner does lead to sparse models, further architectural support for sparsity (such as in [9]) is needed to savor the actual benefits delivered by the pruning algorithm. A *structured pruning algorithm* instead targets a group of parameters (i.e., all parameters of a neuron or convolution filter). The removal of an entire neuron or filter does not exacerbate irregular sparsity; therefore, the advantages can be reaped in any hardware platform with no demands on further memory compression or specialized hardware to skip computations.

It would be audacious to expect a pruning algorithm to deliver improvements not only in DNN compaction but in accuracy as well. As a matter of fact, the significant accuracy losses suffered by hitherto promulgated techniques have necessitated onerous fine-tuning (re-training) steps to retrieve baseline accuracy levels. The high computational cost of these techniques is further exacerbated by the need to implement timid pruning steps as even re-training fails to address the accuracy losses of bold pruning steps, otherwise. Iterative pruning algorithms attempt to minimize accuracy loss by pruning a small number of computation units followed up by fine-tuning at every iteration. The sub-optimal decisions of the pruning algorithm could be tolerated through frequent re-training, yet at significant costs in terms of time and computational resources. Data-center grade GPUs are necessary for training and comprehensively fine-tuning large DNN models; consequently, the associated costs and long computation times, frequently running into weeks, challenge the development process for practitioners with limited access to these resources. While a variety of pre-trained DNN models are readily available online [6], comprehensive fine-tuning necessitates access to a large amount of labeled data, furthermore. Data acquisition and management induce additional practical challenges. In contrast, precise redundancy removal techniques would minimize the need for comprehensive fine-tuning, thus facilitating the deployment of DNN models in resource-constrained applications.

Structured pruning algorithms target the computation units (i.e., a neuron or a filter) with the smallest contribution to network decisions, thus minimizing the impact on network accuracy. While the non-linear behavior of DNNs makes it challenging to identify these computation units, various heuristics [14,20,22, 24,26,34,35] deliver competitive results in neuron and filter ranking problems.

As these methods analyze the importance of the computation units within a reasonable precision, their analysis is restricted to an individual unit and fails to consider relationships among the units. A computation unit might seem essential when it is independently evaluated (i.e., due to its output contribution), yet its role might prove redundant when assessed within a group of computation units.

Let us clarify this argument with a somewhat simplistic yet motivational illustration. A botanical researcher gathers various types of measurements (temperature, humidity, altitude, and vegetation density) to determine an ideal location for a plant species. After observing that the altitude data exhibits small variation in the alternatives, the researcher discards the altitude data since it hardly delivers any useful information in the comparison. A careful investigation of the data furthermore reveals that temperature readings are strongly correlated with vegetation density, with the possible explanation of another non-measured variable, say the level of sunlight, affecting both. Although both variables are individually essential, the researcher ends up choosing only one of them since neither of the measurements provides significant additional information over the other. Analogously, if there are computation units with correlated outputs in a DNN layer, only a subset of the units should suffice to maintain the information content. The remaining units could be eliminated with minimal impact on accuracy if their magnitude contribution is properly expressed by the subset.

2   Related Work

Structured Pruning: Structured pruning algorithms aim to reduce the number of network parameters and operations without introducing additional sparsity in the DNN models. Previous techniques consider the properties of weight sets [11,12,20], or of layer outputs [14,24], or utilize Taylor approximation [22,26,34] or other back-propagated metrics [35] to estimate the neuron/filter significance in pruning. He *et al.* [13] utilize LASSO regression to sparsify and prune feature map channels while minimizing the construction error in the next layer. Zhuang *et al.* [37] consider construction error as well as discriminative power in channel pruning. Xiao *et al.* [33] enforce sparsity on auxiliary gate parameters to perform soft-pruning during the training. In attaining minimal accuracy loss, some of the listed work end up with fine-tuning stages as lengthy as the initial training phase in duration. Liu *et al.* [23] demonstrate that training the reduced architectures from scratch can outperform the pruned and fine-tuned models if an even larger number of training iterations can be afforded.

Weight/Activation Correlations: A limited subset of the extant literature utilizes pairwise weight [2] and activation [3] correlations to carry out computation unit elimination. Mariet *et al.* [25] pick a subset of diverse neurons through Determinantal Point Processes, and perform weight adjustments in the next layer to retain the magnitude contribution of the eliminated units. The analysis in [25], however, is limited to small models with fully connected layers only.

Tensor Decomposition: Tensor decomposition allows low-rank approximation of fully connected [7] and convolutional layers [16,36] with multiple smaller layers of the same type. Even though tensor decomposition provides significant reductions in the layer parameters and multiply-accumulate operations through a static investigation of the weight set properties, the expansion of the target layer into a sequence of layers might be problematic in terms of inference latency.

3   Information Redundancy in DNN Activations

While one could envision investigating the pairwise correlations among convolution filter outputs, as in Fig. 1, to this effect, relying solely on this approach fails to provide a complete picture of the redundancy since linear relationships can involve multiple computation units. We can characterize information redundancy more comprehensively by considering the computation unit output as a distinct vector for a set of predictions, and then identifying a minimal set of orthogonal base vectors that span the utilized layer output space.

Let us assume there are n_l computation units in layer l. We refer to the term *computation unit* to signify a neuron in the fully connected layers or a filter in the convolutional layers. We consider the outputs of a fully connected layer (after the non-linear activation function) as a two-dimensional matrix $A_{n_l \times m}$ where m is the number of examples that the inference is performed on. We restrict our analysis to fully connected layer outputs for simplicity; however, a similar analysis also applies to convolutional layers if each output feature map channel

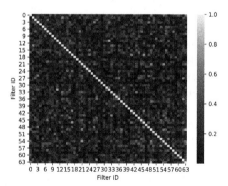

Fig. 1. Pairwise output correlation magnitudes in a convolutional layer (first layer of VGG-16 [30], model trained on CIFAR-10 [17])

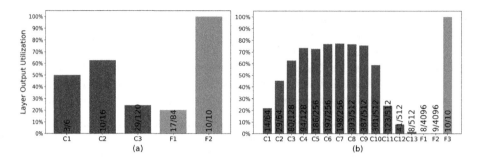

Fig. 2. Layer output utilization in (a) LeNet-5 [19] (on MNIST [19]) and (b) VGG-16 [30] (on CIFAR-10 [17]) architectures (C: convolutional layer, F: fully connected layer)

is flattened along the temporal dimension, forming a two-dimensional matrix where each row is allocated to the outputs of a single channel. If there are linear dependencies between the rows of $A_{n_l \times m}$, the activation matrix could be reduced into a smaller matrix with n_l' orthogonal rows through linear transformations.

While n_l' can be easily determined through the *Gram-Schmidt orthogonalization* [18] process, a rigid elimination approach will deliver only a minor reduction in the number of rows. The process will discard the row only if it exactly equals a linear combination of the other rows, yet such a condition is hard to satisfy for a matrix $A_{n_l \times m}$ when $m \gg n_l$ and in the presence of small numerical deviations.

A well-known data analysis tool, *PCA (Principal Component Analysis)* [29], seamlessly handles the effect of such inaccuracies. PCA expresses the input data with a smaller set of orthogonal variables that are known as principal components, which are ranked by the variance they explain in the data. As a preliminary analysis tool, we apply PCA to output activations and determine the minimum number of principal components that explain a certain amount of the output variance to assess the information redundancy in each layer.

Figure 2 presents the number of principal components that explain 95% of output variance in the layers of two DNNs: the LeNet-5 model trained on the MNIST dataset [19], and VGG-16 [30] trained on the CIFAR-10 dataset [17]. We observe that the rate of utilized output dimensions rarely exceeds 60% for LeNet-5 or 75% for VGG-16, excluding output layers. Moreover, certain convolutional and almost all fully connected layers are heavily under-utilized. For instance, two hidden fully-connected layers in VGG-16 consist of 4096 neurons each, yet 95% of their output variance could be approximated by 8 and 9 principal components. The utilization rates indicate an information redundancy in these models; thus, the proper layer transformations could drastically reduce the model size and the inference costs with minimal impact on accuracy.

4   Squeezing DNN Correlations with Feature Elimination

4.1   Method Description

Principal component analysis extracts a set of useful features from a set of observed variables. Nevertheless, the generated principal components are not directly useful for the computation unit elimination problem since the principal components can not be directly affiliated with a single computation unit. We design a novel algorithm to address this issue. We first construct an approximated orthogonal base for the layer output space by following a procedure closely related to Gram-Schmidt orthogonalization. Second, we utilize the orthogonal base to determine a minimal subset of computation units that can span the utilized output space. We conclude with the elimination of the remaining units and form a transformation matrix that will properly update the weights of the subsequent layer so that the magnitude contribution of the eliminated units can be retained.

We have already designated the activations produced by layer l as a matrix $A_{n_l \times m}$. We denote a particular row of A as $\boldsymbol{A}_i$ where $1 \leq i \leq n_l$. Let us define a few more quantities before proceeding into the details of the algorithm. The variance of each row could be calculated as follows:

$$\sigma^2{}_{(\boldsymbol{A}_i)} := \frac{1}{m} \sum_{j=1}^{m} (A_{i,j} - \mu_{(\boldsymbol{A}_i)})^2 \tag{1}$$

In (1), σ^2 denotes the variance and μ represents the mean value of the row. The average row variance of $A_{n_l \times m}$ can be calculated through (2):

$$\sigma^2{}_{avg} := \frac{1}{n_l} \sum_{i=1}^{n_l} \sigma^2{}_{(\boldsymbol{A}_i)} \tag{2}$$

We aim to find an approximated orthogonal base $U_{n_l' \times m} := [\boldsymbol{U}_1, \boldsymbol{U}_2, ..., \boldsymbol{U}_{n_l'}]$ for $A_{n_l \times m} := [\boldsymbol{A}_1, \boldsymbol{A}_2, ..., \boldsymbol{A}_{n_l}]$. A modified version of the Gram-Schmidt orthogonalization can be utilized for this purpose as detailed subsequently.

As a first step, we check to see if the first row of the original activation matrix contains sufficient variance through the condition in (3) where γ is an experimental tuning coefficient:

$$\gamma \cdot \sigma^2{}_{avg} \leq \sigma^2{}_{(\boldsymbol{A}_1)} \tag{3}$$

If (3) is satisfied, we set $\boldsymbol{A}_1$ as the first vector in the orthogonal base (4):

$$\boldsymbol{U}_1 := \boldsymbol{A}_1 \tag{4}$$

In the second step, we need to extract the sub-component of $\boldsymbol{A}_2$ that is orthogonal to $\boldsymbol{U}_1$. We first project $\boldsymbol{A}_2$ onto $\boldsymbol{U}_1$, then subtract the projection from $\boldsymbol{A}_2$ to obtain the orthogonal component $\boldsymbol{R}$:

$$R := A_2 - proj_{(U_1)}(A_2) = A_2 - \frac{(A_2 \cdot U_1)}{(U_1 \cdot U_1)}U_1 \tag{5}$$

If R contains sufficient variance (6), we include it in the orthogonal base (7):

$$\gamma \cdot \sigma^2_{avg} \leq \sigma^2_R \tag{6}$$

$$U_2 := R \tag{7}$$

Similar steps could be carried with the remaining rows of A by calculating the sub-component of the current row that is orthogonal to the base vectors obtained so far (8), and finally including the remaining vector R in the orthogonal base (9) if it still meets the constraints of sufficient variance expressed in Eq. (6):

$$R := A_i - \sum_{k=1}^{size(U)} proj_{(U_k)}(A_i) = A_i - \sum_{k=1}^{size(U)} \frac{(A_i \cdot U_k)}{(U_k \cdot U_k)}U_k \tag{8}$$

$$U_{size(U)+1} := R \tag{9}$$

Through the steps outlined, we construct an approximate orthogonal base $U_{n'_l \times m} := [U_1, U_2, ..., U_{n'_l}]$. As U is an approximate base for A, a linear transformation should construct the approximated version of A from U as in (10):

$$A_{n_l \times m} \approx C_{n_l \times n'_l} \times U_{n'_l \times m} \tag{10}$$

We refer to matrix C as the *composition matrix*. The composition matrix can be calculated by multiplying both sides of (10) with the right inverse of $U_{n'_l \times m}$:

$$A_{n_l \times m} \times (U_{n'_l \times m})^{-1} \approx C_{n_l \times n'_l} \times U_{n'_l \times m} \times (U_{n'_l \times m})^{-1} \tag{11}$$

$$C_{n_l \times n'_l} :\approx A_{n_l \times m} \times (U_{n'_l \times m})^{-1} \tag{12}$$

Since the rows of $U_{n'_l \times m}$ are orthogonal, $U_{n'_l \times m}$ is easily right-invertible by merely taking the transpose of $U_{n'_l \times m}$ and normalizing each column with the square of its magnitude. $U_{n'_l \times m}$ and $C_{n_l \times n'_l}$ share similarities with the Q and R matrices in *QR factorization* [18], yet the process differs since the described factorization is approximate, and the base vectors in $U_{n'_l \times m}$ are not normalized.

As a next step, we define another matrix $\tilde{A}_{n'_l \times m}$ that consists of the subset of the rows of the original activation matrix $A_{n_l \times m}$. We only pick the rows that deliver a distinct vector to $U_{n'_l \times m}$ after the projections (the rows in which (6) holds). It is straightforward to prove that the reduced form of (10) holds for $\tilde{A}_{n'_l \times m}$ if the corresponding rows of the selected features in the composition matrix are included in the reduced composition matrix $\tilde{C}_{n'_l \times n'_l}$:

$$\tilde{A}_{n'_l \times m} \approx \tilde{C}_{n'_l \times n'_l} \times U_{n'_l \times m} \tag{13}$$

We extract $U_{n'_l \times m}$ by left-multiplying both sides with $(\tilde{C}_{n'_l \times n'_l})^{-1}$:

$$(\tilde{C}_{n'_l \times n'_l})^{-1} \times \tilde{A}_{n'_l \times m} \approx (\tilde{C}_{n'_l \times n'_l})^{-1} \times \tilde{C}_{n'_l \times n'_l} \times U_{n'_l \times m} \tag{14}$$

$$U_{n'_l \times m} \approx (\tilde{C}_{n'_l \times n'_l})^{-1} \times \tilde{A}_{n'_l \times m} \tag{15}$$

Being a triangular matrix with no zero diagonal entries, $\tilde{C}_{n'_l \times n'_l}$ is always invertible. We embed $U_{n'_l \times m}$ in (10) to obtain:

$$A_{n_l \times m} \approx C_{n_l \times n'_l} \times (\tilde{C}_{n'_l \times n'_l})^{-1} \times \tilde{A}_{n'_l \times m} \tag{16}$$

We refer to $T_{n_l \times n'_l} := C_{n_l \times n'_l} \times (\tilde{C}_{n'_l \times n'_l})^{-1}$ as the *transformation matrix* which further reduces (16) into (17):

$$A_{n_l \times m} \approx T_{n_l \times n'_l} \times \tilde{A}_{n'_l \times m} \tag{17}$$

$T_{n_l \times n'_l}$ captures an information essential to our algorithm. It enables us to produce $\tilde{A}_{n'_l \times m}$ instead of $A_{n_l \times m}$ in the current layer by eliminating $n_l - n'_l$ computation units. We can still approximately construct $A_{n_l \times m}$ by multiplying with the transformation matrix $T_{n_l \times n'_l}$. Although it seems to create an additional computation stage at first glance, $T_{n_l \times n'_l}$ can be seamlessly merged with the subsequent layer thus further reducing the memory and computational costs of the subsequent layer since no non-linearity exists between these operations. In detail, the next layer has a weight matrix $W_{n_{l+1} \times n_l}$ and biases $B_{n_{l+1} \times 1}$. It accumulates pre-activations through (18) where the biases are broadcasted into rows:

$$Z_{n_{l+1} \times m} := (W_{n_{l+1} \times n_l} \times A_{n_l \times m}) + B_{n_{l+1} \times 1} \tag{18}$$

We can embed (17) into (18) to obtain:

$$Z_{n_{l+1} \times m} \approx (W_{n_{l+1} \times n_l} \times T_{n_l \times n'_l} \times \tilde{A}_{n'_l \times m}) + B_{n_{l+1} \times 1} \tag{19}$$

As a final step, we multiply $W_{n_{l+1} \times n_l}$ and $T_{n_l \times n'_l}$ to obtain a reduced weight matrix $\tilde{W}_{n_{l+1} \times n'_l}$ that operates on the remaining computation unit outputs (20):

$$Z_{n_{l+1} \times m} \approx \tilde{W}_{n_{l+1} \times n'_l} \times \tilde{A}_{n'_l \times m} + B_{n_{l+1} \times 1} \tag{20}$$

As a result of these design-time modifications, the subsequent layer accumulates the same partial sums even though the majority of the computation units have been eliminated in the current layer. It reduces the number of parameters and multiply-accumulate operations by a factor of n'_l / n_l in both layers and translates into remarkable memory footprint and performance improvements when this optimization is applied to all layers throughout the network.

The process could be naturally extended to convolutional layers as well. First, the spatial dimensions of the 4-dimensional output feature map ($A_{n_l \times s \times s \times m}$) need to be flattened along the temporal dimension to form a 2-dimensional input activation matrix $(A_{n_l \times s \cdot s \cdot m})$ where each row contains the outputs of a single channel, and s is the spatial dimension size of the feature map. The described steps could be carried out similarly to extract the transformation matrix (T). Finally, the modifications on the subsequent convolutional layer could be performed by first flattening the 4-dimensional weight tensor into two dimensions (e.g., from $W_{n_{l+1} \times r \times r \times n_l}$ to $W_{n_{l+1} \cdot r \cdot r \times n_l}$ where r is the spatial dimension size of the filter) so that all dimensions except the input channel dimension are merged, multiplying with the transformation matrix, and then un-flattening the resulting weight matrix rows back into the original dimensions to form the updated weight tensor $(W_{n_{l+1} \times r \times r \times n_{l'}})$.

4.2 Back-of-the-Envelope Analysis of the Algorithmic Complexity

The initial stage of our algorithm consists of the Gram-Schmidt orthogonalization process, which scales linearly with m and quadratically with the layer size (n_l), thus incurring total complexity of $O(m \times n_l{}^2)$. The derivation of the composition matrix $(C_{n_l \times n_l'})$ through the iterative matrix multiplication algorithm also requires $O(m \times n_l{}^2)$ complexity since n_l' is bounded by n_l in the worst case. The derivation of the inverse-reduced composition matrix $(\tilde{C}_{n_l' \times n_l'})^{-1}$ (with *Gauss-Jordan elimination* [18]) and the derivation of the transformation matrix (with iterative matrix multiplication) both incur $O(n_l{}^3)$ complexity. Finally, the update of the next layer's weight matrix through the iterative matrix multiplication requires $O(n_{l+1} \times n_l{}^2)$ complexity. In total, the run-time complexity scales in the order of $O((m \times n_l{}^2) + n_l{}^3 + (n_{l+1} \times n_l{}^2))$ in theory and frequently dominated by the first term in practice as m could be significantly large in certain scenarios. The number of columns in the profiled activation matrix (m) depends on both the number of predictions (linearly), as well as on the spatial dimensions of the output feature maps (quadratically). The feature map channels are flattened before the Gram-Schmidt process; therefore, an $s \times s$ feature map channel incurs s^2 columns for a single prediction in the profiled activation matrix $(A_{n_l \times m})$. The memory requirements scale in proportion with $O((m \times n_l) + n_l{}^2 + (n_{l+1} \times n_l))$ and are largely dominated by the first term due to analogous reasons.

4.3 Relationship with Low-Rank Tensor Decomposition

The fundamental operating principles of the algorithm share similarities with low-rank tensor decomposition techniques [7,16,36]. Let us explain this relationship on the fully connected layer whose pre-activations are accumulated through (21):

$$Z_{n_l \times m} := W_{n_l \times n_{l-1}} \times A_{n_{l-1} \times m} + B_{n_l \times 1} \tag{21}$$

Tensor decomposition expresses the layer weight matrix as a product of multiple lower-rank matrices. For instance, Girshick [7] decomposes the fully connected layer operation into two sequential matrix multiplications through *SVD (singular value decomposition)* [18] as follows:[1]

$$Z_{n_l \times m} = U_{n_l \times n_{l''}} \times (S_{n_{l''} \times n_{l''}} (V_{n_{l-1} \times n_{l''}})^T \times A_{n_{l-1} \times m}) + B_{n_l \times 1} \qquad (22)$$

If the number of utilized singular values ($n_{l''}$) is relatively small, tensor decomposition reduces the parameters and multiply-accumulate operations by a factor of $(n_{l''} \times (n_l + n_{l-1}))/(n_l \times n_{l-1})$ in the target layer. The reader will probably question when a small $n_{l''}$ allows a sufficient approximation of the original weight matrix. Neurons in a fully connected layer extract features from the input data, yet the derived reductions may overlap, resulting in feature correlation and consequent fair amount of redundant computations. Tensor decomposition allows a fully connected layer to extract only a small set of orthogonal features from the data (since SV^T is row-orthogonal). Then, each neuron forms its pre-activations by taking a linear mixture of the extracted features through the second matrix multiplication. Despite the benefits, the technique may introduce additional latency because single matrix multiplication is converted into two sequential ones. The reader will question the necessity of the second matrix multiplication since the orthogonal features are extracted in the first stage, and the second step is merely a static linear expansion of the data into larger output space.

Although $Z_{n_l \times m}$ is constructed through a pre-determined linear combination of $n_{l''}$ orthogonal features, such information can not be directly utilized for elimination of the linearly dependent rows because $A_{n_l \times m}$ is generated by processing $Z_{n_l \times m}$ with a non-linear function in the following step. Non-linear activation functions (i.e., tanh or ReLU) interfere with the dependencies among pre-activation rows, yet certain linear properties are still preserved due to the local (e.g., tanh) or piece-wise (e.g., ReLU) linear nature of these functions. First, we analyze the rows of $A_{n_l \times m}$ rather than the layer weights to account for the effect of the non-linearity. Second, we carry out a feature elimination procedure rather than feature extraction to be able to prune the computation units directly. As the magnitude contribution of the original layer still needs to be constructed, we avoid any depth expansion by embedding this operation into the next layer through weight updates. The procedure on the convolutional layers holds a similar relationship with the tensor decomposition method presented in [36].

5   Experimental Method

We utilize three different DNN models with three distinct datasets for the experiments: LeNet-5 on the MNIST dataset [19], VGG-16 [30] on the CIFAR-10 dataset [17], and ResNet-50 [10] on the ImageNet dataset [27]. Keras [6] with the

---

[1] S and V^T are statically multiplied to form weight matrix SV^T.

TensorFlow [1] backend is used to design the experiments. We train the LeNet-5 and VGG-16 base models from scratch and use the pre-trained weights for ResNet-50. LeNet-5 and VGG-16 training is performed with the SGD (Stochastic Gradient Descent) optimizer at a learning rate of 10^{-3} for 100 epochs. Model weights are initialized with the Glorot uniform initializer [8], and biases initialized to zeros. We utilize SCALE-Sim [28] to characterize the performance gains on a DNN accelerator model similar to Eyeriss [5] through cycle accurate simulations. We conduct our experiments on a moderate desktop system with Intel i5-8600K (6 core) CPU, 32 GB of memory, and NVIDIA GTX1060 (6GB) GPU.

The algorithm is tested with various coefficients (γ) to measure the accuracy drop after one-shot elimination. As γ is an algorithmic hyper-parameter, it does not directly reflect the pruned parameter percentage. Yet it allows us to match accuracy and footprint graphs in Sect. 6. The same pruning coefficient (γ) is used for all layers in LeNet-5 and VGG-16. ResNet-50 has four convolution layer groups where each layer contains the same number of filters in a group. Scaling the coefficients with a constant for each group (i.e., $k = [1, 1.5, 2, 2.5]$) helps to eliminate more filters from the layers with a large number of filters and maintains the accuracy better for less redundant networks. In addition, we omit the pruning of the ResNet-50 layers that merge with the residual connections to avoid any dimension mismatch. We note that the number of profiled examples (m) could be kept much smaller than the training set size with no adverse effect on precision. For instance, while the entire training set is profiled in the LeNet-5 experiments, we have profiled only 1000 training examples for VGG-16 and 500 training examples for ResNet-50 to overcome computational complexity and memory bottlenecks. We have also tried to down-sample the feature maps prior to flattening, yet forwent its adoption due to a noticeable accuracy loss.

We prune the same number of computation units with the previously suggested pruning heuristics in the literature by ranking the neurons and convolution filters with L1 and L2 norms (Li *et al.* [20]), output statistics such as APoZ[2] (Hu *et al.* [14]) and the standard deviation of the activations (described in Molchanov *et al.* [26]). We include the results of random pruning and the LeNet-5 and VGG-16 models trained from scratch (reduced network, initialized with [8]) for comparison. We have initially tried a gradient-based method similar to [26] as well. Although the local approximation obtained through the gradients is helpful in iterative pruning, it fails to produce satisfactory results when a significant portion of the units is pruned in one shot. We report the accuracy under various scenarios. First, we measure accuracy after one-shot elimination without any fine-tuning. We believe this is the most transparent metric to analyze the precision of an elimination algorithm since the insufficiency of the heuristic can be easily disguised through extensive fine-tuning.[3] Second, we fine-tune the network with minimal training for 5 epochs for LeNet-5 and VGG-16, and 0.1 epoch for ResNet-50, and then measure the accuracy. Third, we train the network more comprehensively until there is no improvement in validation accuracy for

---

[2] Evaluated only on VGG-16 and ResNet-50 with ReLU activation function.

[3] Unfortunately, this metric is rarely reported in prior literature.

LeNet-5 and VGG-16, and 1 epoch[4] for ResNet-50 before reporting the accuracy. We repeat experiments 10 times for LeNet-5, 5 times for VGG-16, and 1 time for ResNet-50 due to run-time constraints, and report average. We reduce the learning rate down to 10^{-6} during fine-tuning if necessary. Our method may require low learning rates in fine-tuning because of the effect of the weight transformations on the gradients. We report the number of network parameters, MAC (multiply-accumulate) operations to perform inference, and execution (clock) cycles spent on the accelerator.

6   Experimental Results and Discussion

The experimental results validate that our approach can dramatically shrink the DNNs with minimal impact on accuracy. Linear layer transformations allow accuracy retention even under extreme pruning scenarios, resulting in an extreme accuracy gap between our method and the other compared heuristics, particularly before any fine-tuning. Figure 3 and Fig. 5(a) demonstrate that we can eliminate 70.0% of the LeNet-5 parameters in one-shot, speed up the network by 2.2× on Eyeriss, and cause only a 3.9% accuracy loss without any fine-tuning. We further reduce the number of network parameters by 87.0% at around 9.7% accuracy loss with no fine-tuning while executing the network 3.7× faster. The best among the compared heuristics (*Activation Std.*) results in more than 42.0% accuracy loss at the same point. The highest accuracy difference between our method and the best of the heuristics even jumps to as large as 36.5% before any fine-tuning at certain pruning rates. The network pruned with our algorithm eradicates the accuracy loss after only a few epochs of fine-tuning (from 9.7% to 0.8%), and outperforms the competing heuristics in accuracy with only one fifth the amount of training when the networks are fine-tuned until accuracy improvement stops.

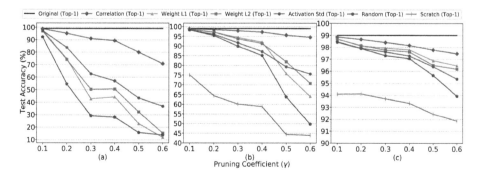

Fig. 3. LeNet-5 accuracy drop (a) after one-shot pruning (b) after pruning + fine-tuning (5 epochs) (c) after pruning + fine-tuning (until no improvement)

---

[4] It requires ~15 h of computation time.

Figure 4 and Fig. 5(b) demonstrate the performance of our method on VGG-16. We eliminate 79.7% of the parameters in one pruning step, delivering a 3.4× speedup with only a 1.8% accuracy loss and no fine-tuning. The comparison heuristics impair VGG-16 entirely even at lower pruning rates by causing the network to be stuck at a constant prediction while our accuracy loss is as small as 0.6% at the same pruning rate with no fine-tuning (79.4% accuracy gap). Fine-tuning delivers marginal benefits for all methods in VGG-16, and the networks pruned with some of the other techniques can not even be re-trained after they have been completely destroyed in the pruning stage. We observe a 55.3% accuracy gap between our method and the closest heuristic even after the comprehensive re-training stage in the most aggressive pruning case, as we eliminate 86.7% of the parameters and make the network 4.9× faster.

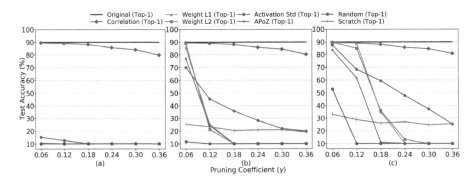

Fig. 4. VGG-16 accuracy drop (a) after one-shot pruning (b) after pruning + fine-tuning (5 epochs) (c) after pruning + fine-tuning (until no improvement)

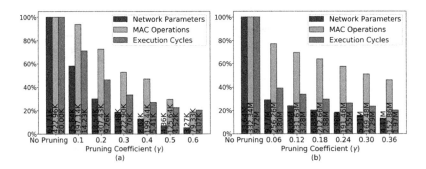

Fig. 5. LeNet-5 (a) and VGG16 (b) hardware footprint (remaining network parameters, required MAC (multiply-accumulate) operations to perform inference, and execution (clock) cycles spent on the accelerator) after elimination

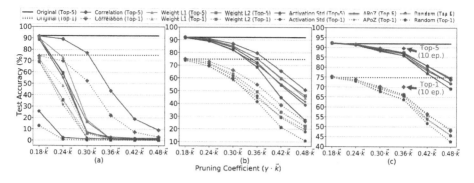

Fig. 6. ResNet-50 accuracy drop (a) after one-shot pruning (b) after pruning + fine-tuning (0.1 epoch) (c) after pruning + fine-tuning (1 epoch)

Our method further performs remarkably well on ResNet-50, as seen on Fig. 6 and Fig. 7 even though inherent redundancy is quite diminished relative to the previous examples. The relative absence of redundancy challenges the removal of computation units and accuracy retention with no fine-tuning. Yet we can still prune 21.3% of the parameters and reduce prediction times by 22.4% at the cost of only 2.9% top-5 accuracy loss with no fine-tuning. The closest method, with a top-5 accuracy comparable to the top-1 accuracy of the model pruned with our technique, diminishes top-5 accuracy by more than 19.2% at the same point. The top-5 accuracy difference between our technique and the closest heuristic surpasses 58.8% when no fine-tuning is applied, clearly demonstrating the superiority of our technique in accuracy retention. When complemented with fine-tuning our approach reduces the number of ResNet-50 parameters by 59.5% and multiply-accumulate operations by 52.7%; thus performance is boosted 2.14× with only 12.5%, 4.7%, and 2.3% top-5 accuracy drop after 0.1, 1, and 10 epoch(s) of fine-tuning, respectively. More fine-tuning epochs are expected to reduce the difference between the original and the pruned network further.

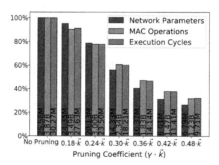

Fig. 7. ResNet-50 hardware footprint after elimination

While fine-tuning does not significantly improve VGG-16 accuracy, reduced ResNet-50 models notably benefit from fine-tuning. It is relatively challenging to fine-tune the plain networks (e.g., VGG-16) as pruning creates information bottlenecks in certain layers. Residual connections in ResNet-50 facilitate fine-tuning by providing an efficient flow for the gradients. As a result, despite their initial accuracy, the networks could reach similar accuracy values after sufficient fine-tuning. A good elimination heuristic still provides a head-start and allows the network to converge to final accuracy values with fewer iterations.

Finally, the execution time of the proposed algorithm fades in practice. It needs to be applied once, and takes $\sim$20 min for ResNet-50 (less than a minute per layer on average) in the most challenging case, which is a small fraction of the time required by a single epoch of fine-tuning (i.e., $\sim$15 h).

7   Conclusion

Memory and computational costs restrict the usage of DNNs in resource-limited applications. The number of extracted unique features rarely exceeds a fraction of the layer size in modern DNN models; therefore, the information content of the layer could be represented by employing only a small number of computation units. We design a novel procedure to carry-out feature elimination and perform successive modifications on the target model to compensate for the magnitude contribution of the eliminated units. We demonstrate the effectiveness of our approach as a powerful neuron/filter pruning technique that delivers results superior to the prior heuristics, yet requires only a minimal amount of fine-tuning. More importantly, we offer a principled way to understand and measure the redundancy in DNN computations, exposing an entirely new paradigm for the design of resource-efficient deep learning models for edge applications.

References

1. Abadi, M., et al.: TensorFlow: a system for large-scale machine learning. In: 12th USENIX Symposium on Operating Systems Design and Implementation (OSDI 2016), pp. 265–283 (2016)
2. Ayinde, B.O., Zurada, J.M.: Building efficient ConvNets using redundant feature pruning. arXiv:1802.07653 (2018)
3. Babaeizadeh, M., Smaragdis, P., Campbell, R.H.: NoiseOut: a simple way to prune neural networks. arXiv:1611.06211 (2016)
4. Chen, C., Seff, A., Kornhauser, A., Xiao, J.: DeepDriving: learning affordance for direct perception in autonomous driving. In: IEEE International Conference on Computer Vision, pp. 2722–2730 (2015)
5. Chen, Y.H., Emer, J., Sze, V.: Eyeriss: a spatial architecture for energy-efficient dataflow for convolutional neural networks. ACM SIGARCH Comput. Architect. News 44–3, 367–379 (2016)
6. Chollet, F., et al.: Keras (2015). https://keras.io
7. Girshick, R.: Fast R-CNN. In: Proceedings of the IEEE International Conference on Computer Vision, pp. 1440–1448 (2015)

8. Glorot, X., Bengio, Y.: Understanding the difficulty of training deep feedforward neural networks. In: Proceedings of the 13th International Conference on Artificial Intelligence and Statistics, pp. 249–256 (2010)
9. Han, S., et al.: EIE: efficient inference engine on compressed deep neural network. ACM SIGARCH Compu. Architect. News **44**(3), 243–254 (2016)
10. He, K., Zhang, X., Ren, S., Sun, J.: Deep residual learning for image recognition. In: IEEE Conference on Computer Vision and Pattern Recognition, pp. 770–778 (2016)
11. He, Y., Kang, G., Dong, X., Fu, Y., Yang, Y.: Soft filter pruning for accelerating deep convolutional neural networks. In: Proceedings of the 27th International Joint Conference on Artificial Intelligence, pp. 2234–2240 (2018)
12. He, Y., Liu, P., Wang, Z., Hu, Z., Yang, Y.: Filter pruning via geometric median for deep convolutional neural networks acceleration. In: Proceedings of the IEEE Conference on Computer Vision and Pattern Recognition, pp. 4340–4349 (2019)
13. He, Y., Zhang, X., Sun, J.: Channel pruning for accelerating very deep neural networks. In: Proceedings of the IEEE International Conference on Computer Vision, pp. 1389–1397 (2017)
14. Hu, H., Peng, R., Tai, Y.W., Tang, C.K.: Network trimming: a data-driven neuron pruning approach towards efficient deep architectures. arXiv:1607.03250 (2016)
15. Ioffe, S., Szegedy, C.: Batch normalization: accelerating deep network training by reducing internal covariate shift. In: International Conference on Machine Learning, pp. 448–456 (2015)
16. Kim, Y.D., et al.: Compression of deep convolutional neural networks for fast and low power mobile applications. In: International Conference on Learning Representations (2016)
17. Krizhevsky, A., Hinton, G.: Learning multiple layers of features from tiny images. Technical report. Citeseer (2009)
18. Lay, D.C., Lay, S.R., McDonald, J.J.: Linear Algebra and Its Applications, 5th edn. Pearson Publishing Co., USA (2016)
19. LeCun, Y., Bottou, L., Bengio, Y., Haffner, P.: Gradient-based learning applied to document recognition. Proc. IEEE **86**(11), 2278–2324 (1998)
20. Li, H., Kadav, A., Durdanovic, I., Samet, H., Graf, H.P.: Pruning filters for efficient ConvNets. In: International Conference on Learning Representations (2017)
21. Lin, D., Talathi, S., Annapureddy, S.: Fixed point quantization of deep convolutional networks. In: International Conference on Machine Learning, pp. 2849–2858 (2016)
22. Lin, S., et al.: Accelerating convolutional networks via global & dynamic filter pruning. In: Proceedings of the 27th International Joint Conference on Artificial Intelligence, pp. 2425–2432 (2018)
23. Liu, Z., Sun, M., Zhou, T., Huang, G., Darrell, T.: Rethinking the value of network pruning. In: International Conference on Learning Representations (2019)
24. Luo, J.H., Wu, J., Lin, W.: ThiNet: a filter level pruning method for deep neural network compression. In: Proceedings of the IEEE International Conference on Computer Vision, pp. 5058–5066 (2017)
25. Mariet, Z., Sra, S.: Diversity networks. In: International Conference on Learning Representations (2016)
26. Molchanov, P., Tyree, S., Karras, T., Aila, T., Kautz, J.: Pruning convolutional neural networks for resource efficient inference. In: International Conference on Learning Representations (2017)
27. Russakovsky, O., et al.: ImageNet large scale visual recognition challenge. Int. J. Comput. Vis. (IJCV) **115**(3), 211–252 (2015)

28. Samajdar, A., Zhu, Y., Whatmough, P., Mattina, M., Krishna, T.: SCALE-Sim: systolic CNN accelerator simulator. arXiv:1811.02883 (2018)
29. Shlens, J.: A tutorial on principal component analysis. arXiv preprint arXiv:1404.1100 (2014)
30. Simonyan, K., Zisserman, A.: Very deep convolutional networks for large-scale image recognition. In: International Conference on Learning Representations (2015)
31. Srivastava, N., Hinton, G., Krizhevsky, A., Sutskever, I., Salakhutdinov, R.: Dropout: a simple way to prevent neural networks from overfitting. J. Mach. Learn. Res. **15**(1), 1929–1958 (2014)
32. Sze, V., Chen, Y.H., Yang, T.J., Emer, J.S.: Efficient processing of deep neural networks: a tutorial and survey. Proc. IEEE **105**(12), 2295–2329 (2017)
33. Xiao, X., Wang, Z., Rajasekaran, S.: Autoprune: automatic network pruning by regularizing auxiliary parameters. In: Advances in Neural Information Processing Systems, pp. 13681–13691 (2019)
34. You, Z., Yan, K., Ye, J., Ma, M., Wang, P.: Gate decorator: global filter pruning method for accelerating deep convolutional neural networks. In: Advances in Neural Information Processing Systems, pp. 2130–2141 (2019)
35. Yu, R., et al.: NISP: pruning networks using neuron importance score propagation. In: Proceedings of the IEEE Conference on Computer Vision and Pattern Recognition, pp. 9194–9203 (2018)
36. Zhang, X., Zou, J., Ming, X., He, K., Sun, J.: Efficient and accurate approximations of nonlinear convolutional networks. In: Proceedings of the IEEE Conference on Computer Vision and Pattern Recognition, pp. 1984–1992 (2015)
37. Zhuang, Z., et al.: Discrimination-aware channel pruning for deep neural networks. In: Advances in Neural Information Processing Systems, pp. 875–886 (2018)

Activation Anomaly Analysis

Philip Sperl[1,2(✉)] [iD], Jan-Philipp Schulze[1,2(✉)] [iD], and Konstantin Böttinger[1] [iD]

[1] Fraunhofer Institute for Applied and Integrated Security, Garching, Germany
{philip.sperl,jan-philipp.schulze,
konstantin.boettinger}@aisec.fraunhofer.de
[2] Technical University of Munich, Munich, Germany

Abstract. Inspired by recent advances in coverage-guided analysis of neural networks, we propose a novel anomaly detection method. We show that the hidden activation values contain information useful to distinguish between normal and anomalous samples. Our approach combines three neural networks in a purely data-driven end-to-end model. Based on the activation values in the target network, the alarm network decides if the given sample is normal. Thanks to the anomaly network, our method even works in semi-supervised settings. Strong anomaly detection results are achieved on common data sets surpassing current baseline methods. Our semi-supervised anomaly detection method allows to inspect large amounts of data for anomalies across various applications.

Keywords: Anomaly detection · Deep learning · Intrusion detection · Semi-supervised learning · Coverage analysis · Data mining · IT security

1 Introduction

Anomaly detection is the task of identifying data points that differ in their behavior compared to the majority of samples. Reliable anomaly detection is of great interest in many real-life scenarios, especially in the context of security-sensitive systems. Here, anomalies can indicate attacks on the infrastructure, fraudulent behavior, or general points of interest. In recent years, the number of machine learning (ML) applications using deep learning (DL) concepts has steadily grown. DL methods allow to analyze highly complex data for patterns that are useful to minimize a certain loss function. Anomaly detection tasks are especially challenging for DL methods due to the inherent class imbalance. In research, anomaly detection is often only seen as an unsupervised task, thus ignoring the information gain when anomaly-related samples are available. In our work, we develop a new DL-based anomaly detection method showing superior results with only a handful of anomaly examples – and motivate that the very same method also works without any anomaly examples at all.

P. Sperl and J.-P. Schulze are co-first authors.

© Springer Nature Switzerland AG 2021
F. Hutter et al. (Eds.): ECML PKDD 2020, LNAI 12458, pp. 69–84, 2021.
https://doi.org/10.1007/978-3-030-67661-2_5

In DL-based anomaly detection, a popular idea is to use an autoencoder (AE) to preprocess or reconstruct the input. This type of neural network (NN) generates an output that is close to the given input under the constraint of small hidden dimensions. Intuitively, when trained on normal samples only, the AE will miss important features that distinguish anomalous samples, thus increasing the reconstruction error. Clearly, this method assumes that the overall error is large – however, anomalies may be too subtle to be detected based on the output only. In our paper, we consider the entire system context by analyzing more subtle patterns. We show that the hidden activations of AEs, but also other types of NNs, are useful to judge if the current input is normal or anomalous. By combining the information of three interrelated NNs, we achieve strong detection results even in semi-supervised settings.

During the conceptual phase, we were inspired by coverage-guided NN testing methods. In this promising new research direction, software testing concepts are transferred to DL models. The goal is to identify faulty regions in NNs responsible for unusual behavior, or errors during run-time. Pei et al. [24] first introduced the idea of neuron coverage to guide a testing process. Since then, further improvements and modifications have been proposed, e.g., by Ma et al. [16] and Sun et al. [39]. Recently, Sperl et al. [32] used this concept to detect adversarial examples fed to NNs. The authors analyze the activation values while processing benign and adversarial inputs. A second NN classifies if the recorded patterns resemble normal behavior or an attack. In this paper, we build upon this insight and further generalize the concept by adapting it to the constraints of anomaly detection. Whereas samples for benign as well as adversarial inputs are plentiful in adversarial ML, anomaly detection is a semi-supervised setting with only a few anomaly-related labels available. However, also here we assume that NNs behave in a special and distinguishable manner, when confronted with anomalous data. We show that this behavior is detectable by analyzing the activation values during run-time. When observing the neuron activations of NNs while processing normal inputs and synthetic anomalies, we train another NN to distinguish the nature of the analyzed data points. Our analysis shows that even artificial anomalies train an anomaly detection model, which performs well across multiple domains, and generalizes to yet unseen anomalies.

Applying our concept, we empirically show that anomalous samples cause different hidden activations compared to normal ones. We analyze the hidden layers of a so-called *target network* by an auxiliary network, called the *alarm network*. With our *anomaly network*, we automatically generate samples used during the training of the alarm network to distinguish between normal and anomalous samples. Our evaluation shows strong results on common data sets, and we report superior performance to common baseline methods. In summary, we make the following contributions:

- We propose a purely data-driven semi-supervised anomaly detection method based on the analysis of the hidden activations of NNs that we call A^3.
- Based on our thorough evaluation, we show that these patterns generalize to new anomaly types even with only a few anomaly examples available.

– We motivate that our method works in settings where no anomaly samples
  are available with a generative model as anomaly network.

2 Background and Related Work

Anomaly detection is a topic of active research with a wide range of use-cases
and methods. For instance, in network intrusion detection [18], power grids [31],
or industrial control systems [10], automated mechanisms improve the security
of the overall system. A good overview on anomaly detection in general and DL-
based systems in particular is given in the surveys [4] and [3], respectively. Note
that the term "semi-supervised" is often ambiguous in anomaly detection. We
follow the surveys' notation, thus calling any knowledge about the underlying
labels semi-supervised, e.g., also when the training data is assumed to be normal.
Famous unsupervised methods include OC-SVM [28], or Isolation Forest [15].

In DL-based anomaly detection systems, a popular choice are architectures
incorporating an AE, often used as feature extractor. AEs are combined with
classical ML classifiers like k-nearest neighbor [36], OC-SVMs for anomaly
detection [2], NN-based classifiers [26], or Gaussian Mixture Models [40]. Similarly, other feature extraction networks like recurrent NNs have been evaluated
[17,20,29]. To detect anomalies, AEs may also be used in their purest form: to
restore the input under the constraint of small hidden layers, similar to classical
dimensionality reduction methods like PCA. The reconstruction error of samples is used to discriminate between normal and anomalous data points [36,38].
Research has further analyzed how to improve the anomaly detection results,
e.g., by iteratively adding human feedback [9,35].

Over the past decades, computing power and data storage have steadily
risen. DL methods profit from the increased amount of training samples. Recent
research [21,22,27] has studied ways to incorporate known anomalies into DL-
based anomaly detection. They show that even a few anomaly labels, which
are usually available in practice, improve the overall detection performance. In
Sect. 4.2, we discuss state-of-the-art frameworks to which we compare A^3. With
our work, we show that the activations of NNs differ for normal and anomalous
samples. Using our method, we detect known and even new, yet unseen types
of anomalies in different scenarios with high confidence. With our method, we
significantly improve NN-based semi-supervised anomaly detection systems.

2.1 Nomenclature

NNs approximate an input-output mapping $f(\mathbf{x}; \boldsymbol{\theta}) = \hat{\mathbf{y}}$ based on the learned
parameters $\boldsymbol{\theta}$. The overall function $f = f_L \circ \ldots \circ f_1$ consists of multiple layers $f_i(\mathbf{x}; \boldsymbol{\theta}) = \mathbf{h}_i$. For easier readability, we summarize the input-output relation as $f : \mathbf{x} \mapsto \hat{\mathbf{y}}$. We denote the output, or *activation*, of the ith layer as
$\mathbf{h}_i = \sigma(\mathsf{W}_{i,i-1} \cdot \mathbf{h}_{i-1} + \mathbf{b}_i)$. Here $\sigma(\cdot)$ is a non-linear activation function, $\mathsf{W}_{i,j}$
and $\mathbf{b}_i$ the mapping parameters learned in layer i with respect to layer j. The

input corresponds to $\mathbf{x} = \mathbf{h}_0$. The weights $\boldsymbol{\theta}_i$ are determined by an optimization algorithm minimizing the expected loss between the desired and estimated output, $\mathcal{L}(\mathbf{y}, \hat{\mathbf{y}})$. Our data set $\mathcal{D} = \{(\mathbf{x}_k, \mathbf{y}_k)\}, \mathbf{x}_k \in \mathcal{X}, \mathbf{y}_k \in \mathcal{Y}$ is split into three parts: training, validation, and test. The network weights are adapted to the training set while evaluating the performance on the validation set. We consider categorical data, $\mathbf{y} \in \{1, 2, \dots |\mathcal{Y}|\}$, where $\mathcal{Y}$ denotes the set of available labels. Among other things, we evaluate the transferability of A^3, i.e., the performance of the model evaluated on more labels than it was trained on, $\mathcal{Y}_{\text{train}} \subset \mathcal{Y}_{\text{test}}$.

3   Activation Anomaly Analysis

We present the following hypothesis building the foundation of our paper:

> Evaluating the activations $\mathbf{h}_i$ of a neural network trained on the data set $\mathcal{D}_{\text{train}}$, we observe special patterns that allow to distinguish between classes the network has been trained on, and unknown classes $\mathbf{y}_i \notin \mathcal{Y}_{\text{train}}$.

This setting is analog to anomaly detection. An anomaly is defined as a sample different to normal data in some unspecified behavior. When $\mathcal{D}_{\text{train}}$ describes the normal data, then any point of a yet unknown class $\mathbf{y}_i \notin \mathcal{Y}_{\text{train}}$ defines an anomaly, i.e., a sample that does not belong to a normal class.

3.1   Architecture

Our new anomaly detection method A^3 comprises three parts. Whereas the target, and alarm network are closely related to the adversarial example detection method of Sperl et al. [32], we additionally add the anomaly network.

1. The *target network* performs a task unrelated to anomaly detection. In accordance to our assumption, the classes the target was trained on are considered normal. Several architectures of the target are possible – we evaluate fully-connected as well as convolutional autoencoders and classifiers.
2. The *anomaly network* generates counterexamples based on the inputs, used to train the alarm network. We show that even a random number generator as anomaly network gives state-of-the art results. Furthermore, we motivate that a generative model eliminates the need for anomaly-related labels.
3. The *alarm network* evaluates if the given sample is normal or anomalous by observing the hidden activations of the target network. While training, the activations caused by the inputs as well as synthetic anomalies by the anomaly network are considered.

All parts are combined to one connected architecture. In the scope of this paper, we fix the target network to its pretrained state. Our assumption is that the activations caused by the input show particular patterns for samples the target network was trained on (i.e., normal samples), and for other samples (i.e., anomalous samples). The alarm network analyzes the anomaly-related patterns in the target network's activations. A high level overview is given in Fig. 1.

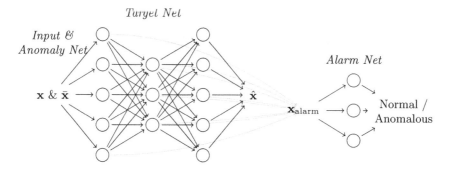

Fig. 1. A^3 consists of three connected parts: 1) a target network unrelated to anomaly detection (e.g., an autoencoder), 2) the anomaly network providing anomalous samples $\bar{\mathbf{x}}$, and 3) the alarm network judging if the input $\mathbf{x}$ is normal.

3.2 Training Phase

Target Network. The target network performs a task unrelated to anomaly detection on the input $\mathbf{x}$. We evaluate autoencoders, $f_{\text{target}} : \mathbf{x} \mapsto \hat{\mathbf{x}}$, and classifiers, $f_{\text{target}} : \mathbf{x} \mapsto \hat{y} \in \{1, \ldots, |\mathcal{Y}|\}$. According to our fundamental assumption, all samples used during training are considered normal.

Anomaly Network. The anomaly network transforms the input to a sample not resembling the normal data, $f_{\text{anomaly}} : \mathbf{x} \mapsto \bar{\mathbf{x}}$. In accordance with the semi-supervised setting, no information about the distribution of anomalous samples is necessary. The generated samples are fed in the target network during training. In our evaluation, we transform the input to a random realization of a normal distribution, i.e., $\bar{\mathbf{x}} \sim \mathcal{N}(\mu, \sigma^2)$. As further outlook, we use a variational autoencoder (VAE) [13], i.e., an autoencoder encoding the input to the inner states $\mathbf{h}_\mu$ and $\mathbf{h}_\sigma$ forming Gaussian posteriors. We multiply $\mathbf{h}_\sigma$ by a scaling factor drawn from $\mathcal{N}(0, 5)$ to sample from improbable regions of the learned distribution.

Alarm Network. The alarm network maps the input to an anomaly score. However, it does not operate on the input directly, but observes the target network's activations caused by the input. Hence, the alarm's output $\hat{y}$ is implicitly dependent on the input and the network weights of the target:

$$f_{\text{alarm}} : [\mathbf{h}_{\text{target},1}(\mathbf{x}; \boldsymbol{\theta}_{\text{target}}), \ldots, \mathbf{h}_{\text{target},L-1}(\mathbf{x}; \boldsymbol{\theta}_{\text{target}})] \mapsto \hat{y} \in [0, 1].$$

Optimization Objective. The alarm network is optimized on predicting the respective training labels, and classifying the anomaly network's output $\bar{\mathbf{x}} = f_{\text{anomaly}}(\mathbf{x})$ as anomalous. We fixed the weight parameter between these objectives to $\lambda = 1.0$. Note that only the alarm network's weights $\boldsymbol{\theta}_{\text{alarm}}$ are adapted while the target network's weights $\boldsymbol{\theta}^\star_{\text{target}}$ remain unchanged. Let $\mathcal{L}_x(y, \hat{y})$ denote the binary cross-entropy loss, the overall objective becomes:

$$\mathcal{L}(\mathbf{x}, y) = \mathcal{L}_x(y, \hat{y})|_{\hat{y}=f_{\text{alarm}}(\mathbf{x}; \boldsymbol{\theta}^\star_{\text{target}}, \boldsymbol{\theta}_{\text{alarm}})} + \lambda \cdot \mathcal{L}_x(1, \hat{y})|_{\hat{y}=f_{\text{alarm}}(\bar{\mathbf{x}}; \boldsymbol{\theta}^\star_{\text{target}}, \boldsymbol{\theta}_{\text{alarm}})}.$$

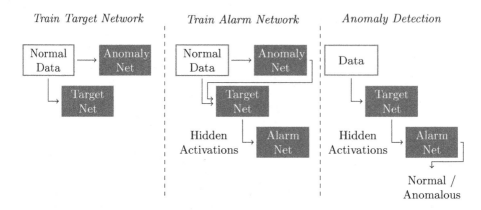

Fig. 2. We train the target network on the training data deemed as normal. The alarm network analyzes the target's hidden activations caused by the input as well as the anomaly network's output during training.

3.3   Prediction Phase

During the prediction phase, the target and the alarm network act as one combined system, mapping the input to an anomaly score: $f_{\text{detect}} : \mathbf{x} \mapsto \hat{y} \in [0, 1]$. The input is transformed by the very same pipeline: $\mathbf{x}$ gives rise to particular activations in the target, $\mathbf{h}_{\text{target},i}(\mathbf{x}; \boldsymbol{\theta}_{\text{target}})$. Based on the target's activations, the alarm network decides whether the input is more likely to be normal, or anomalous. Figure 2 gives an overview about the training and prediction phase.

4   Experiments

4.1   Data Sets

We evaluate the performance of A^3 on five publicly available and commonly used data sets. Our selection allows to compare the performance to related work and motivates that A^3 performs well even in complex scenarios.

1. *MNIST* [14]: common image data set for ML problems with 70 000 images showing ten handwritten digits.
2. *EMNIST* [8]: extension to MNIST with handwritten letters.
3. *NSL-KDD* [34]: common data set with around 150 000 samples. We use KDDTest$^+$ for testing containing new anomalies unseen in the training set.
4. *Credit Card Transactions* [25]: anonymized credit card data of around 285 000 transactions of which 492 being fraudulent.
5. *CSE-CIC-IDS2018* [30]: large network data set. We omit the DDoS data due to the high resource demands. Around five million samples remain.

We limited the preprocessing to minmax-scaling numerical and 1-Hot-encoding categorical data. Samples still containing non-numerical values were omitted. In the IDS data set, we discarded the IP addresses as well as the flow IDs. Generally, we took 80% of the data for training, 5% for validation, and 15% for testing. If a test set is given, we used it instead.

4.2   Baseline Methods

We compared the performance of A^3 to four common baseline methods. Note that we only considered baseline methods that scale to the large amount of data.

1. *Autoencoder Reconstruction Threshold:* When trained on normal data only, there is a measurable difference in the reconstruction compared to anomalous samples. We calculate the mean squared error, i.e., $\mathcal{L}(\mathbf{x}, \hat{\mathbf{x}}) = \|\hat{\mathbf{x}} - \mathbf{x}\|_2^2$, to quantify this difference. As some target networks use an AE architecture, we take the very same models for this baseline.
2. *Isolation Forest (IF):* IF is a common unsupervised anomaly detection method by Liu et al. [15]. Based on the given data, an ensemble of random trees is built. The average path length results in an anomaly score. We use the implementation provided by scikit-learn [23] along with the default parameters.
3. *Deep Autoencoding Gaussian Mixture Model (DAGMM):* DAGMM is a state-of-the-art unsupervised DL-based anomaly detection method by Zong et al. [40]. The authors combine information of an AE with a Gaussian mixture model. For all experiments, we use the implementation by Nakae [19], and the architecture recommended for the KDDCUP data set.
4. *Deviation Networks (DevNet):* DevNet is a state-of-the-art semi-supervised DL-based anomaly detection method by Pang et al. [21]. The anomaly detection is split between a feature learner and an anomaly score learner, both implemented as an NN. The anomaly scorer sets normal samples close to a Gaussian prior distribution, and enforces a minimum distance to anomalous samples. As recommended by the authors, we use their default architecture.

4.3   Anomaly Detection Constraints

Special constraints apply in the setting of anomaly detection. With our experiments, we show that A^3 performs well nonetheless.

1. *Scarcity of Anomaly Samples.* Generally, many samples are required for strong performance using DL frameworks. However, there is a natural imbalance in anomaly detection scenarios: most data samples are normal, only a few examples of anomalies were found manually. We show that A^3 performs well in this semi-supervised setting.
2. *Variable Extend of Abnormality.* Anomalous samples are not bound by a common behavior or magnitude of abnormality. By definition, the only difference is that anomalies do not resemble normal samples. We show the *transferability* of A^3, i.e., known anomalies during training also reveal yet unknown anomalies during testing.

Table 1. Experiments exploring the detection of known & unknown anomalies.

	Data	Normal	Train Anomaly	⊆ Test Anomaly
1a& 4a	MNIST	0, …, 5	6, 7	6, 7
1b& 4c	MNIST	4, …, 9	0, 1	0, 1
1c	NSL-KDD	Normal	DoS, Probe	DoS, Probe
1d	NSL-KDD	Normal	R2L, U2R	R2L, U2R
1e	IDS	Benign	BF, Web, DoS, Infil.	BF, Web, DoS, Infil.
1f	IDS	Benign	Bot, Infil., Web, DoS	Bot, Infil., Web, DoS
1g	CC	Normal	Fraudulent	Fraudulent
2a& 4b	MNIST	0, …, 5	6, 7	6, 7, 8, 9
2b& 4d	MNIST	4, …, 9	0, 1	0, 1, 2, 3
2c	NSL-KDD	Normal	DoS, Probe	DoS, Probe, R2L, U2R
2d	NSL-KDD	Normal	R2L, U2R	R2L, U2R, DoS, Probe
2e	IDS	Benign	BF, Web, DoS, Infil.	BF, Web, DoS, Infil., Bot
2f	IDS	Benign	Bot, Infil., Web, DoS	Bot, Infil., Web, DoS, BF
3a	(E-)MNIST	0, …, 9	A, B, C, D, E	A, B, C, D, E
3b	(E-)MNIST	0, …, 9	A, B, C, D, E	A, B, C, D, E, V, W, X, Y, Z
3c	(E-)MNIST	0, …, 9	V, W, X, Y, Z	V, W, X, Y, Z
3d	(E-)MNIST	0, …, 9	V, W, X, Y, Z	A, B, C, D, E, V, W, X, Y, Z

3. *Driven by Data, not Expert Knowledge.* A suitable anomaly detection algorithm should be applicable to multiple settings, even when no expert knowledge is available. Performance that is only achievable using domain knowledge may result in inferior results in other settings. We show that A^3 *generalizes* to other settings, i.e., uses the data itself to distinguish between normal and anomalous behavior.

4.4 Experimental Setup

We designed multiple experiments to show that A^3 works under all three constraints. An overview is given in Table 1.

1. *Experiment 1: Detection of Known Anomalies.* Considering constraint 1, we evaluated the fundamental assumption of our method, i.e.: the activation values of the target network contain information to distinguish between normal and anomalous samples. It is important to remember that only the alarm network, not the target, is trained on the anomaly detection task. We limited the anomaly samples to {5, 25, 50, 100} randomly selected instances during training in accordance with the semi-supervised setting. Note, this limitation may cause some classes not to be present during training.
2. *Experiment 2: Transferability to Unknown Anomalies.* Considering constraints 1 and 2, we evaluated the transferability of our fundamental assumption, i.e.: the activation values of the target network are inherently different for normal and anomalous samples. Similar to experiment 1, we evaluated the

Tablo 2. Dimensionality of the layers. All hidden layers are activated by ReLUs.

Data set	Target architecture	Alarm architecture
MNIST	According to [6]	1000, 500, 200, 75, 1
NSL-KDD	200, 100, 50, 25, 50, 100, 200	1000, 500, 200, 75, 1
IDS	150, 80, 40, 20, 40, 80, 150	1000, 500, 200, 75, 1
CreditCard	50, 25, 10, 5, 10, 25, 50	1000, 500, 200, 75, 1
MNIST & EMNIST	According to [5]	1000, 500, 200, 75, 1

detection performance bound by the scarcity of anomaly labels. Furthermore, the test data set contained more anomaly classes than the alarm network has been trained on. In other words, we tried to find anomalies that follow a different nature and data distribution than the one of the few known samples.

3. *Experiment 3: Generality of the Method.* Considering constraints 1, 2, and 3, we evaluated the generality of our fundamental assumption, i.e.: the activation values of any type of target network contain information to distinguish between normal and anomalous samples. We used a publicly available classifier as target, extracted the activation values, and tested whether these can be used to detect known as well as unknown anomalies. Hence, we motivate that our anomaly detection mechanism can be applied to already existing target networks and environments of any type.

4. *Experiment 4: Outlook to Unsupervised Anomaly Detection.* Considering the extreme case of constraint 1, we conducted first evaluations of the detection performance when no labeled anomalies are available during training. We used normal samples, as well as the output of a generative anomaly network.

4.5 Implementation Details

An overview about the architectures used for each experiment is given in Table 2. For the AE target models, we chose the first layer to be slightly larger than the dimension of the input vectors, whereas the hidden representation should be smaller. For the sake of simplicity, we used a common alarm model architecture throughout this paper. Note that for the MNIST-related experiments, we considered two publicly available architectures from Keras [7], i.e., a convolutional AE [6] extended by a dropout layer for experiment 1 and 4, as well as a CNN [5] for experiment 3. This underlines the generality of our method. All hidden layers are activated by ReLUs.

Parameter Choices. Based on a non-exhaustive parameter search on MNIST, we chose the following global optimizer settings: Adam [12] with a learning rate of 0.001 for the target network, and 0.00001 for the alarm network was used. The training was stopped after 30 and 60 epochs, respectively. No other regularizer than 10% dropout [33] before the last layer was used. We ran our experiments

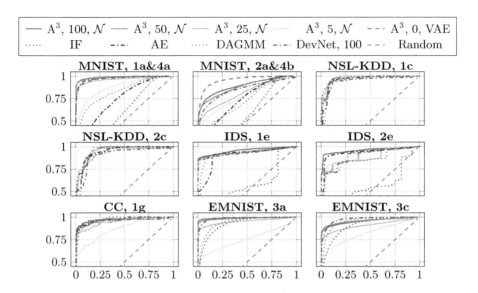

Fig. 3. ROC curves showing the *TPR vs. FPR* evaluated on the test set for several experiments. For A^3, we capped the amount of anomaly samples during training. We used a noise generator ($\mathcal{N}$) as anomaly network, and for experiment 4 a VAE.

on an Intel Xeon E5-2640 v4 server accelerated by an NVIDIA Titan X GPU. To support further research, we open-source our code[1].

Anomaly Network. For experiments 1, 2, & 3, we used a simple Gaussian noise generator as anomaly network. As all inputs are within $[0,1]$, we fixed the noise parameters to $\mathcal{N}(.5, 1)$. For experiment 4, we chose a VAE with the dense hidden layers $800, 400, 100, 25, 100, 400, 800$. During the training of the target network, it was adapted to reconstruct the normal samples.

5   Results and Evaluation

In the following, we present and evaluate the results measured during our experiments. To visualize our results, we show the receiver operating characteristic (ROC) curve in Fig. 3, i.e., the true positive rate (TPR) as a function of the false positive rate (FPR). As metrics, we chose the average precision (AP) and the area under the ROC curve (AUC) to be consistent with the related work [21]. Whereas the AP quantifies the trade-off between precision and recall, the AUC measures the trade-off between the TPR and FPR. Both metrics are independent of a detection threshold, and thus give a good overview about the general detection performance.

---

[1] Code available at: https://github.com/Fraunhofer-AISEC/A3.

Table 3. Test results given all normal, and 100 anomaly samples.

	A^3		AE		IF		DAGMM		DevNet	
	AUC	AP	AUC	AP	AUC	AP	AUC	AP	AUC	AP
1a	.99±.00	.99±.00	.70±.03	.44±.04	.57±.02	.28±.02	.85±.02	.70±.02	.98±.00	.95±.01
1b	1.0±.00	1.0±.00	.39±.04	.25±.03	.53±.01	.41±.02	.63±.04	.34±.03	.99±.00	.98±.00
1c	.97±.01	.97±.01	.96±.00	.96±.01	.97±.00	.98±.00	.94±.01	.91±.04	.95±.02	.95±.02
1d	.91±.03	.71±.08	.82±.03	.55±.06	.84±.00	.49±.02	.91±.01	.59±.03	.91±.03	.79±.02
1e	.95±.01	.92±.01	.88±.08	.75±.12	.47±.04	.17±.01	.90±.02	.68±.06	.93±.00	.90±.01
1f	.94±.01	.91±.01	.86±.07	.65±.13	.36±.03	.14±.01	.68±.12	.34±.18	.90±.01	.74±.04
1g	.97±.01	.77±.04	.97±.01	.54±.09	.97±.01	.14±.04	.96±.02	.35±.23	.97±.01	.77±.04
2a	.88±.01	.88±.01	.72±.02	.63±.02	.54±.01	.40±.01	.74±.02	.69±.02	.84±.04	.84±.03
2b	.92±.02	.91±.02	.64±.03	.62±.03	.64±.01	.60±.01	.71±.02	.61±.01	.90±.03	.90±.02
2c	.95±.01	.96±.01	.93±.01	.94±.01	.94±.00	.96±.00	.93±.01	.92±.03	.90±.02	.93±.01
2d	.94±.01	.94±.03	.94±.00	.94±.01	.94±.00	.96±.00	.93±.01	.92±.03	.88±.02	.89±.02
2e	.93±.02	.88±.03	.92±.02	.82±.07	.45±.03	.20±.01	.76±.08	.52±.10	.90±.00	.83±.01
2f	.95±.01	.93±.01	.89±.06	.79±.10	.45±.03	.20±.01	.80±.04	.56±.02	.92±.00	.81±.02
3a	.99±.00	.99±.00	-	-	.93±.00	.85±.01	.93±.01	.84±.03	.99±.00	.98±.00
3b	.96±.00	.97±.00	-	-	.91±.01	.89±.01	.95±.00	.93±.01	.97±.01	.97±.01
3c	.99±.01	.99±.01	-	-	.89±.01	.79±.02	.96±.00	.91±.01	.98±.01	.97±.01
3d	.97±.01	.97±.01	-	-	.91±.01	.89±.01	.95±.00	.93±.01	.96±.02	.96±.02

5.1   Test Results

In Fig. 3, we show the ROC curves of the first experiments. To simulate real-life conditions, where usually just a few anomaly samples are available, we restricted the known anomalies to {5, 25, 50, 100} for A^3. Note that these anomalies are sampled randomly from all available test anomaly samples – as a result, they might be restricted to a single class. We report that A^3 generally follows an intuitive behavior: the more known anomalies are used, the better the anomaly detection performance. Even for 5 samples an adequate performance is possible on most data sets. As little as 25 samples are needed to surpass the performance of the state-of-the-art unsupervised anomaly detection methods. Moreover, A^3 surpasses, or matches the performances of the state-of-the-art semi-supervised anomaly detection method. An important feature is the steep rise of the TPR. To choose a suitable detection threshold in practice, a very low FPR has to be tolerated using our method. This significantly reduces manual work, and builds trust in the detection results.

In Table 3, we summarize the results for A^3 and the baseline methods averaged over five passes using the test data sets. Note that not all baseline methods finished five runs on the IDS data set. To simulate real-life conditions, we limit the amount of available anomalies to 100 randomly chosen samples of the training set. We find that A^3 performs well even under this strict setting. It well surpasses the unsupervised baseline methods, and on most experiments also the semi-supervised baseline method DevNet. Especially for experiment 2, which poses the most competitive setting since new types of anomalies need

Table 4. Test result for exp. 4, where no anomaly samples were used to train A^3.

4a-AUC	4a-AP	4b-AUC	4b-AP	4c-AUC	4c-AP	4d-AUC	4d-AP
$.97 \pm .01$	$.94 \pm .03$	$.91 \pm .06$	$.90 \pm .06$	$.68 \pm .05$	$.49 \pm .03$	$.64 \pm .04$	$.61 \pm .05$

to be detected, we exceed state-of-the-art performance. Throughout all transfer experiments, A^3 achieves on average an improvement of 15% for the mean AUC and 36% for the mean AP compared to the unsupervised method DAGMM. For the semi-supervised method, DevNet, we report on average 4% higher AUC and 6% higher AP scores. Note that the scores are already close to 1.0, thus less improvement is expected. The performance remains strong across all data sets and experiments. With the results at hand, we can report that A^3 detects significantly more new types of anomalies than the chosen baseline methods. Finally, we conclude that our fundamental assumption, i.e., the hidden activations of NNs carry information useful to anomaly detection, is well supported.

We see strong evidence for the hypotheses made throughout the experiments. In experiment 1, we tried to find known anomalies. The test results show the highest results for this setting, thus our method very well identifies suitable patterns for this task. These patterns generalize well to yet unseen patterns as shown in experiment 2. Although only parts of the test anomalies are known during training, strong results are achieved. Whereas the aforementioned experiments used deep & convolution autoencoders as target network, we generalize the setting to deep classifiers in experiment 3. Also here, superior results are achieved. We conclude that A^3 is able to detect known and yet unknown anomalies with high confidence, and is flexible enough to adapt to a wide range of environments.

5.2 Outlook to Unsupervised Anomaly Detection

In experiment 4, we further motivate that A^3 performs well even when no anomaly samples are available, i.e., in a fully semi-supervised setting. For this, we chose a generative model as anomaly network, a VAE in our case. In this strict setting, A^3 does not require knowledge about anomalous behavior, and thus is closely related to unsupervised methods. Table 4 summarizes the results on the test set. For experiment 4d, the performance is comparable to the AE-based anomaly detection. For experiments 4a-c, we report very promising results outperforming the unsupervised baseline methods at the cost of a higher variance. Here, A^3 achieves an increase of 8% to 23% for the AUC and 19% to 45% for the AP score compared to the best unsupervised baseline method.

The ROC curves in Fig. 3 visualize the superior performance compared to the respective baseline methods. Sampling from the VAE's improbable distribution regions seems to be a suitable way to generate counterexamples useful to the alarm network. Future research may leverage these results to a more general, unsupervised setting. We are happy to report consistently strong results when using a simple noise generator as anomaly network along with a few labeled anomalies as shown in experiment 1, 2, & 3. With this setting, we provide a ready-to-use end-to-end anomaly detection framework achieving state-of-the-art performance even in strict semi-supervised environments for a variety of use-cases.

6   Discussion and Future Work

In this paper, we presented A^3, an anomaly detection method that analyzes the hidden activations of NNs. In A^3, the alarm network observes the context information of a target network during run-time whereas the anomaly network solves the natural class imbalance in anomaly detection tasks. Future research may leverage this concept to other use-cases, and further formalize the explored theory, which we motivated empirically. An additional in-depth analysis of the hidden activation values may reveal new aspects of NN-based anomaly detection. A^3 shows strong anomaly detection results for all five analyzed data sets across all experiments. We motivate that the activation analysis generalizes to yet unseen anomalies across different network architectures. Thanks to the modularity of our concept, various architectures may be used as target, alarm, or anomaly networks covering numerous types of data and use-cases. Future work may integrate other powerful architectures, e.g., generative adversarial networks [11] as already applied to other anomaly detection settings [1, 37]. We emphasize the real-life applicability by limiting the amount of anomaly samples during training. In practice, often a few known anomalies are available, e.g., by manual exploration or unsupervised methods. Good performance was already achieved with a handful of anomaly samples. During our research, we saw further improvements with a shifted output regularizer, e.g., $\mathcal{L}_{reg}(y) = \lambda \cdot |1 - y|$, favoring the detection of anomalies. With this set of regularizers, and the presented outlook to generative anomaly networks, we leave developers the ability to further tune the performance of A^3 in their implementation.

7   Conclusion

We introduce a novel approach for anomaly detection called A^3 based on the hidden activation patterns of NNs. Our architecture comprises three parts: a target network unrelated to the anomaly detection task, the anomaly network generating anomalous training samples, and the alarm network analyzing the resulting activations of the target. Our framework works under common assumptions and constraints typically found in anomaly detection tasks. We assume that anomalous training samples are scarce, and new types of anomalies exist during deployment. With our evaluation, we provide strong evidence that our method works on different target network architectures, and generalizes to yet unseen anomalous samples. Furthermore, we detect anomalies across different data types with just a few or even no labeled anomalies available during training. We present a valuable semi-supervised DL-based anomaly detection framework providing a purely data-driven solution for a variety of use-cases.

Acknowledgments. We would like to express our gratitude to all reviewers for their helpful remarks. This work was partly funded by the German Federal Ministry of Education and Research (BMBF) under the project *IUNO InSec* (16KIS0933K).

References

1. Akcay, S., Atapour-Abarghouei, A., Breckon, T.P.: GANomaly: semi-supervised anomaly detection via adversarial training. In: Jawahar, C.V., Li, H., Mori, G., Schindler, K. (eds.) ACCV 2018. LNCS, vol. 11363, pp. 622–637. Springer, Cham (2019). https://doi.org/10.1007/978-3-030-20893-6_39
2. Andrews, J.T.A., Morton, E.J., Griffin, L.D.: Detecting anomalous data using autoencoders. Int. J. Mach. Learn. Comput. **6**(1), 21–26 (2016). https://doi.org/10.18178/ijmlc.2016.6.1.565
3. Chalapathy, R., Chawla, S.: Deep Learning for Anomaly Detection: A Survey. arXiv:1901.03407 (2019)
4. Chandola, V., Banerjee, A., Kumar, V.: Anomaly detection: a survey. ACM Comput. Surv. **41**(3), 1–58 (2009). https://doi.org/10.1145/1541880.1541882
5. Chollet, F.: Keras Documentation (2015). https://keras.io/examples/mnist_cnn/
6. Chollet, F.: Building Autoencoders in Keras (2016). https://blog.keras.io/building-autoencoders-in-keras.html
7. Chollet, F., et al.: Keras (2015). https://keras.io
8. Cohen, G., Afshar, S., Tapson, J., Van Schaik, A.: EMNIST: extending MNIST to handwritten letters. In: 2017 International Joint Conference on Neural Networks (IJCNN), pp. 2921–2926 (2017). https://doi.org/10.1109/IJCNN.2017.7966217
9. Das, S., Wong, W.K., Dietterich, T., Fern, A., Emmott, A.: Incorporating expert feedback into active anomaly discovery. In: 2016 IEEE 16th International Conference on Data Mining (ICDM), pp. 853–858 (2017). https://doi.org/10.1109/icdm.2016.0102
10. Feng, C., Palleti, V.R., Mathur, A., Chana, D.: A systematic framework to generate invariants for anomaly detection in industrial control systems. In: Network and Distributed Systems Security (NDSS) Symposium 2019 (2019). https://doi.org/10.14722/ndss.2019.23265
11. Goodfellow, I.J., et al.: Generative adversarial nets. In: Advances in Neural Information Processing Systems 27, pp. 2672–2680. Curran Associates, Inc. (2014). http://papers.nips.cc/paper/5423-generative-adversarial-nets.pdf
12. Kingma, D.P., Ba, J.L.: Adam: A Method for Stochastic Optimization. arXiv:1412.6980 (2014)
13. Kingma, D.P., Welling, M.: Auto-Encoding Variational Bayes. arXiv:1312.6114 (2013)
14. Lecun, Y., Bottou, L., Bengio, Y., Haffner, P.: Gradient-based learning applied to document recognition. Proc. IEEE **86**(11), 2278–2324 (1998). https://doi.org/10.1109/5.726791
15. Liu, F.T., Ting, K.M., Zhou, Z.H.: Isolation forest. In: 2008 Eighth IEEE International Conference on Data Mining, pp. 413–422 (2008). https://doi.org/10.1109/ICDM.2008.17
16. Ma, L., et al.: DeepGauge: multi-granularity testing criteria for deep learning systems. In: Proceedings of the 33rd ACM/IEEE International Conference on Automated Software Engineering, pp. 120–131 (2018). https://doi.org/10.1145/3238147.3238202

17. Malhotra, P., Ramakrishnan, A., Anand, G., Vig, L., Agarwal, P., Shroff, G.: LSTM-based encoder-decoder for multi-sensor anomaly detection. In: ICML 2016 Anomaly Detection Workshop (2016). http://arxiv.org/abs/1607.00148

18. Mirsky, Y., Doitshman, T., Elovici, Y., Shabtai, A.: Kitsune: an ensemble of autoencoders for online network intrusion detection. In: Network and Distributed Systems Security (NDSS) Symposium 2018 (2018). https://doi.org/10.14722/ndss.2018.23204

19. Nakae, T.: DAGMM TF Implementation. https://github.com/tnakae/DAGMM

20. Nguyen, T.D., Marchal, S., Miettinen, M., Fereidooni, H., Asokan, N., Sadeghi, A.R.: DIoT: a federated self-learning anomaly detection system for IoT. In: Proceedings of the 39th IEEE International Conference on Distributed Computing Systems (ICDCS), Dallas, USA, July 2019. https://doi.org/10.1109/ICDCS.2019.00080

21. Pang, G., Shen, C., van den Hengel, A.: Deep anomaly detection with deviation networks. In: Proceedings of the ACM SIGKDD International Conference on Knowledge Discovery and Data Mining, pp. 353–362, November 2019. https://doi.org/10.1145/3292500.3330871

22. Pang, G., Shen, C., Jin, H., van den Hengel, A.: Deep Weakly-supervised Anomaly Detection. arXiv:1910.13601, October 2019

23. Pedregosa, F., et al.: Scikit-learn: machine learning in python. J. Mach. Learn. Res. **12**, 2825–2830 (2011)

24. Pei, K., Cao, Y., Yang, J., Jana, S.: Deepxplore: automated whitebox testing of deep learning systems. In: Proceedings of the 26th Symposium on Operating Systems Principles, pp. 1–18 (2017). https://doi.org/10.1145/3132747.3132785

25. Pozzolo, A.D., Caelen, O., Johnson, R.A., Bontempi, G.: Calibrating probability with undersampling for unbalanced classification. In: 2015 IEEE Symposium Series on Computational Intelligence, pp. 159–166. IEEE (2015). https://doi.org/10.1109/SSCI.2015.33

26. Qureshi, A.S., Khan, A., Shamim, N., Durad, M.H.: Intrusion detection using deep sparse auto-encoder and self-taught learning. Neural Comput. Appl. **32**(8), 3135–3147 (2019). https://doi.org/10.1007/s00521-019-04152-6

27. Ruff, L., et al.: Deep semi-supervised anomaly detection. In: International Conference on Learning Representations (ICLR) (2020). https://openreview.net/forum?id=HkgH0TEYwH

28. Schölkopf, B., Williamson, R., Smola, A., Shawe-Taylor, J., Piatt, J.: Support vector method for novelty detection. In: Advances in Neural Information Processing Systems, pp. 582–588 (2000). https://papers.nips.cc/paper/1723-support-vector-method-for-novelty-detection.pdf

29. Schulze, J.P., Mrowca, A., Ren, E., Loeliger, H.A., Böttinger, K.: Context by proxy: identifying contextual anomalies using an output proxy. In: Proceedings of the 25th ACM SIGKDD International Conference on Knowledge Discovery & Data Mining - KDD 2019, pp. 2059–2068 (2019). https://doi.org/10.1145/3292500.3330780

30. Sharafaldin, I., Lashkari, A.H., Ghorbani, A.A.: Toward generating a new intrusion detection dataset and intrusion traffic characterization. In: Proceedings of the 4th International Conference on Information Systems Security and Privacy, ICISSP 2018, pp. 108–116 (2018). https://doi.org/10.5220/0006639801080116

31. Shekari, T., Bayens, C., Cohen, M., Graber, L., Beyah, R.: RFDIDS: radio frequency-based distributed intrusion detection system for the power grid. In: Network and Distributed Systems Security (NDSS) Symposium 2019 (2019). https://doi.org/10.14722/ndss.2019.23462

32. Sperl, P., Kao, C.Y., Chen, P., Lei, X., Böttinger, K.: DLA: dense-layer-analysis for adversarial example detection. In: 5th IEEE European Symposium on Security and Privacy (2020). http://arxiv.org/abs/1911.01921

33. Srivastava, N., Hinton, G., Krizhevsky, A., Sutskever, I., Salakhutdinov, R.: Dropout: a simple way to prevent neural networks from overfitting. J. Mach. Learn. Res. **15**, 1929–1958 (2014)

34. Tavallaee, M., Bagheri, E., Lu, W., Ghorbani, A.A.: A detailed analysis of the KDD CUP 99 data set. In: 2009 IEEE Symposium on Computational Intelligence for Security and Defense Applications, pp. 1–6 (2009). https://doi.org/10.1109/CISDA.2009.5356528

35. Veeramachaneni, K., Arnaldo, I., Korrapati, V., Bassias, C., Li, K.: AI^2: training a big data machine to defend. In: 2016 IEEE 2nd International Conference on Big Data Security on Cloud (BigDataSecurity), IEEE International Conference on High Performance and Smart Computing (HPSC), and IEEE International Conference on Intelligent Data and Security (IDS), pp. 49–54 (2016). https://doi.org/10.1109/BigDataSecurity-HPSC-IDS.2016.79

36. Yousefi-Azar, M., Varadharajan, V., Hamey, L., Tupakula, U.: Autoencoder-based feature learning for cyber security applications. In: Proceedings of the International Joint Conference on Neural Networks, pp. 3854–3861 (2017). https://doi.org/10.1109/IJCNN.2017.7966342

37. Zenati, H., Romain, M., Foo, C.S., Lecouat, B., Chandrasekhar, V.R.: Adversarially learned anomaly detection. In: 2018 IEEE International Conference on Data Mining (ICDM), pp. 727–736 (2018). https://doi.org/10.1109/ICDM.2018.00088

38. Zhou, C., Paffenroth, R.C.: Anomaly detection with robust deep autoencoders. In: Proceedings of the 23rd ACM SIGKDD International Conference on Knowledge Discovery and Data Mining (2017). https://doi.org/10.1145/3097983.3098052

39. Zhou, Z., Zhou, W., Lv, X., Huang, X., Wang, X., Li, H.: Progressive Learning of Low-Precision Networks. arXiv:1905.11781 (2019)

40. Zong, B., et al.: Deep autoencoding Gaussian mixture model for unsupervised anomaly detection. In: International Conference on Learning Representations (2018). https://openreview.net/forum?id=BJJLHbb0-

Effective Version Space Reduction for Convolutional Neural Networks

Jiayu Liu[1(✉)], Ioannis Chiotellis[1], Rudolph Triebel[1,2], and Daniel Cremers[1]

[1] Department of Informatics, Technical University of Munich, Munich, Germany
{liuji,chiotell,triebel,cremers}@in.tum.de
[2] German Aerospace Center (DLR), Wessling, Germany

Abstract. In active learning, sampling bias could pose a serious inconsistency problem and hinder the algorithm from finding the optimal hypothesis. However, many methods for neural networks are hypothesis space agnostic and do not address this problem. We examine active learning with convolutional neural networks through the principled lens of version space reduction. We identify the connection between two approaches – prior mass reduction and diameter reduction – and propose a new diameter-based querying method – the minimum Gibbs-vote disagreement. By estimating version space diameter and bias, we illustrate how version space of neural networks evolves and examine the realizability assumption. With experiments on MNIST, Fashion-MNIST, SVHN and STL-10 datasets, we demonstrate that diameter reduction methods reduce the version space more effectively and perform better than prior mass reduction and other baselines, and that the Gibbs vote disagreement is on par with the best query method.

Keywords: Active learning · Deep learning · Version space · Diameter reduction

1 Introduction

Active learning is a supervised learning framework in which the learner is given access to a pool or stream of unlabeled samples and is allowed to selectively query labels from an oracle (e.g., a human annotator). In each query round, the learner queries the labels of some unlabeled samples and trains on the augmented labeled set to obtain new classifiers. The goal is to learn a good classifier or *hypothesis* using as few labels as possible. This setting is relevant in many real-world problems, where labeled data are scarce or expensive to obtain, but unlabeled data are cheap and abundant.

Many active learning methods for neural networks rely on measures of the "informativeness" of a query, in the form of classifier uncertainty, margin [10,17] or information gain [12,16,19]. Other methods capture the informativeness by representativeness of the query set using geometry-based [23] or discriminative [13] methods. However, most of these methods ignore the notion of the hypothesis

© Springer Nature Switzerland AG 2021
F. Hutter et al. (Eds.): ECML PKDD 2020, LNAI 12458, pp. 85–100, 2021.
https://doi.org/10.1007/978-3-030-67661-2_6

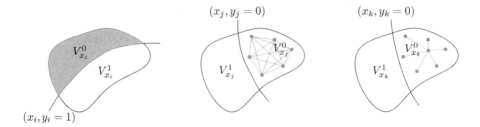

Fig. 1. Version space reduction for binary classification. Upon observing the label of x, the current version space V is split into subspaces V_x^0 and V_x^1, one of which will be removed and the other remains. **Left**: Prior mass reduction methods remove approximately half of the mass. **Middle**: Diameter reduction methods, like pairwise disagreement, query a sample that lead to sub-spaces of small diameter. **Right**: Proposed method, the *Gibbs-vote disagreement*, measures diameter by the expected distance between random hypotheses and their majority vote.

space and do not address the problem of sampling bias [8], which plague many active learning methods. Without carefully handling this problem, an active learning algorithm is not guaranteed to be *consistent*, i.e., capable of finding the optimal classifier in the hypothesis space.

We consider the hypothesis space of convolutional neural networks (ConvNets) and study version space reduction methods. Version space reduction works by removing hypotheses that are inconsistent with the observed labels from a predefined hypothesis space and maintaining the consistent sub-space, the *version space*. A key condition called the *realizability assumption* is that the hypothesis space contains the classifier that provides the ground truth – if not, there are no guarantees that the best hypothesis will not be removed, because a hypothesis might make mistakes on the queried samples but perform well on the data distribution.

For neural networks, the realizability assumption may not hold for all cases. For instance, no neural networks can achieve arbitrarily small test error on some classification datasets. A workaround is to consider the effective labelings on a set of i.i.d. pool samples. To avoid the problem of an unreasonably large effective hypothesis space, as implied by the result of [25], we only consider the labelings achievable by training on unaltered samples and correct labels. We examine experimentally whether the realizability holds with this restriction and analyze its implications on version space reduction methods.

Prior mass reduction [4,6,14] and diameter reduction [7,24] are two widely used version space reduction approaches. See Fig. 1 for illustration. However, prior mass reduction is not an appropriate objective for active learning [24] since any intermediate version spaces containing more than one hypothesis may still have a large diameter, i.e., large error rate in the worst-case scenario, despite having substantially reduced mass. We derive connections between prior mass and diameter reduction and introduce a new interpretation of diameter reduction as prior mass "reducibility reduction".

We propose a new diameter measure called the *Gibbs-vote disagreement*, which equals the expected distance between the random hypotheses and their majority vote classifier. We show its relation to a common diameter measure, the *pairwise disagreement*, and discuss under which situations the former may be advantageous. We show experimentally on four image classification datasets that diameter reduction methods perform better than all baselines and that prior mass reduction [4,6,14] and other baselines like [10,12,16,23] do not perform consistently better than random query and sometimes fail completely.

2 Related Work

A lot of research has been conducted to study the label complexity for active learning and optimality guarantees for greedy version space reduction. Hanneke [15] and Balcan et al. [1] prove upper-bounds on the label complexity in the realizable and non-realizable cases, using a parameter called the disagreement coefficient. Tosh and Dasgupta [24] propose a diameter-based active learning algorithm and characterize its sample complexity using a parameter called the splitting index. Dasgupta [6] shows that a greedy strategy maximizing the worst-case prior mass reduction is approximately as good as the optimal strategy. Golovin and Krause [14] show that the prior mass reduction utility function is adaptive submodular and a greedy algorithm is guaranteed to obtain near-optimal solutions in the average-case scenario. Cuong et al. [5] prove a worst-case optimality guarantee for pointwise submodular functions.

A variety of methods relying on the informativeness of a query have been proposed for neural networks. Gal et al. [12] use the Monte Carlo dropout to approximate the mutual information between predictions and model posterior [16] in a Bayesian setting. Kirsch et al. [19] extend [12,16] to a batch query method. Ducoffe and Precioso [10] use adversarial attacks to generate samples close to the decision boundaries. Sener and Savarese [23] adopt a core-set approach to select representative samples for query. Gissin and Shalev-Shwartz [13] use a discriminative method to select samples such that the labeled and the unlabeled set are indistinguishable. Pinsler et al. [22] formulate batch query as a sparse approximation to the expected complete data posterior of model parameters in a Bayesian setting. Beluch et al. [3] show that ensemble methods consistently outperform geometry-based [23] and the Monte Carlo dropout method [11,12].

3 Preliminaries

Let $\mathcal{X}$ be the input feature space and $\mathcal{Y}$ the label space. Let $\mathcal{H}$ be a hypothesis space of functions $h : \mathcal{X} \to \mathcal{Y}$ and assume a prior π over $\mathcal{H}$. A hypothesis randomly drawn from the prior is called a *Gibbs classifier*. Denote $S = \{(x_i, y_i)\}_{i=1}^n$

a pool of i.i.d. samples from the data distribution P_{XY} and $Q \subseteq S$ the set of queried labeled samples. Define the version space V corresponding to Q as

$$V := \{h \in \mathcal{H} : h(x) = y, \ \forall (x, y) \in Q\}. \tag{1}$$

Denote the subset of V that is consistent with x being labeled as y as

$$V_x^y := \{h \in \mathcal{H} : h(x) = y, \ h \in V\}, \tag{2}$$

and the pseudo-metric induced by the marginal distribution P_X as

$$d(h, h') := \text{Pr}_x(h(x) \neq h'(x)). \tag{3}$$

The disagreement and agreement region are defined as

$$\text{DIS}(V) := \{x \in \mathcal{X} : \exists h, h' \in V, h(x) \neq h'(x)\}, \tag{4}$$

$$\text{AGR}(V) := \mathcal{X} \setminus \text{DIS}(V). \tag{5}$$

4   Prior Mass Reduction

4.1   Gibbs Error

The Gibbs error [4] of an unlabeled sample x is the average-case relative prior mass reduction:

$$\text{GE}(x|V) := \mathbb{E}_y\left[1 - \text{Pr}_{h \sim \pi|_V}(h(x) = y)\right] = \mathbb{E}_y[1 - \pi|_V(V_x^y)], \tag{6}$$

where $\pi|_V(h) = \pi(h)/\pi(V)$ is the conditional distribution of $\mathcal{H}$ restricted to V. Gibbs error measures the proportion of inconsistent hypotheses taking expectation over all possible labelings of x, achievable by hypotheses in the version space. A greedy strategy that considers maximizing the average-case absolute prior mass reduction in each query can equivalently select the unlabeled sample that maximizes the Gibbs error

$$\arg\max_x \text{GE}(x|V). \tag{7}$$

Define the prior mass reduction utility function as

$$f(Q) := 1 - \text{Pr}\left(\{h \in \mathcal{H} : h(x) = y, \ \forall (x, y) \in Q\}\right) = 1 - \pi(V). \tag{8}$$

The optimization problem in (7) can be written, up to a scaling factor, as

$$\arg\max_x \pi(V)\text{GE}(x|V) = \arg\max_x \mathbb{E}_y[f(Q \cup \{(x, y)\}) - f(Q)] \tag{9}$$

$$= \arg\max_x \Delta_{\text{avg}}(x|Q), \tag{10}$$

where the notation $\Delta_{\text{avg}}(x|Q)$ denotes the expected marginal gain of x in terms of prior mass reduction given the labeled samples in Q.

A closely related objective for active learning is the label entropy given x. It can be shown that the Gibbs error lower bounds the entropy. However, a greedy strategy that maximizes the entropy is not guaranteed to be near-optimal in the adaptive case [5]. Furthermore, empirically this criterion performs similarly or worse than the maximum Gibbs error. For the sake of simplicity, we do not consider this method in this paper.

4.2   Variation Ratio

The variation ratio of an unlabeled sample x is the worst-case relative prior mass reduction upon the reveal of its label:

$$\mathrm{VR}(x|V) := \min_y \left[1 - \mathrm{Pr}_{h \sim \pi|_V}(h(x) = y)\right] = \min_y \left[1 - \pi|_V(V_x^y)\right]. \tag{11}$$

It measures the proportion of inconsistent hypotheses considering the worst-case labeling of x and is a lower bound on the Gibbs error. A greedy strategy that considers maximizing the worst-case absolute prior mass reduction in each query selects the unlabeled sample that maximizes the variation ratio

$$\arg\max_x \mathrm{VR}(x|V), \tag{12}$$

which can be expressed in terms of the prior mass reduction utility function, up to a scaling factor, as

$$\arg\max_x \pi(V)\mathrm{VR}(x|V) = \arg\max_x \min_y \left[f(Q \cup \{(x,y)\}) - f(Q)\right] \tag{13}$$

$$= \arg\max_x \Delta_{\mathrm{wc}}(x|Q), \tag{14}$$

where the notation $\Delta_{\mathrm{wc}}(x|Q)$ denotes the worst-case marginal gain of x in terms of prior mass reduction given the labeled samples in Q.

5   Diameter Reduction

5.1   Worst-Case Pairwise Disagreement

The size of the version space can be measured by the expected pairwise disagreement between hypotheses drawn from the conditional distribution:

$$\mathrm{PWD}(V) := \mathbb{E}_{h,h' \sim \pi|_V}\left[d(h, h')\right]. \tag{15}$$

It is the *average diameter* of the version space. A greedy strategy selects the unlabeled sample that minimizes the worst-case pairwise disagreement

$$\arg\min_x \max_y \mathrm{PWD}(V_x^y) = \arg\min_x \max_y \mathbb{E}_{h,h' \sim \pi|_{V_x^y}}\left[d(h, h')\right]. \tag{16}$$

Other measures of diameter based on the supremum distance [7,18] are not amenable to implementation because evaluation of such diameters involves optimization. The pairwise disagreement can be estimated from a finite set of sample hypotheses from the version space.

5.2   Worst-Case Gibbs-Vote Disagreement

We propose a new diameter measure called the *Gibbs-vote disagreement*. It is the expected disagreement between random hypotheses and their majority vote:

$$\text{GVD}(V) := \mathbb{E}_{h \sim \pi|_V}[d(h, h_{\text{vote}}|_V)], \tag{17}$$

where $h_{\text{vote}}|_V$ is the majority vote classifier of hypotheses from V. For each x, it induces a prediction

$$h_{\text{vote}}|_V(x) = \arg\max_y \mathbb{E}_{h \sim \pi|_V}[p(y|x; h)], \tag{18}$$

where $p(y|x; h)$ is the predicted probability of x belonging to class y given by a hypothesis h. The majority vote classifier is the deterministic classifier that has the smallest expected distance to the Gibbs classifier [9,18]:

$$\mathbb{E}_{h'}[d(h', h_{\text{vote}})] = \min_h \mathbb{E}_{h'}[d(h', h)]. \tag{19}$$

Hence the Gibbs-vote disagreement measures the size of the version space by the expected distance of the random hypotheses to their "center". Further, the following relation holds

$$\frac{1}{2}\text{PWD}(V) \leq \text{GVD}(V) \leq \text{PWD}(V) \tag{20}$$

We defer the proof to the appendix[1]. Essentially, Eq. (20) reveals that the Gibbs-vote disagreement is sandwiched between the average radius and diameter.

A greedy strategy selects the unlabeled sample that minimizes the worst-case Gibbs-vote disagreement

$$\arg\min_x \max_y \text{GVD}(V_x^y) = \arg\min_x \max_y \mathbb{E}_{h \sim \pi|_{V_x^y}}[d(h, h_{\text{vote}}|_{V_x^y})], \tag{21}$$

where $h_{\text{vote}}|_{V_x^y}$ is the majority vote of hypotheses from the subspace V_x^y.

5.3   Diameter Reduction as Reducibility Reduction

Pairwise disagreement shares a simple relation with Gibbs error – it is the expected Gibbs error:

$$\text{PWD}(V) = \mathbb{E}_{h,h' \sim \pi|_V}[\mathbb{E}_x[\mathbb{1}(h(x) \neq h'(x))]] \tag{22}$$

$$= \mathbb{E}_x[\mathbb{E}_{h \sim \pi|_V}[\text{Pr}_{h' \sim \pi|_V}(h(x) \neq h'(x))]] \tag{23}$$

$$= \mathbb{E}_x[\text{GE}(x|V)]. \tag{24}$$

A similar relation holds between Gibbs-vote disagreement and the variation ratio:

$$\text{GVD}(V) = \mathbb{E}_{h \sim \pi|_V}[\mathbb{E}_x[\mathbb{1}(h(x) \neq h_{\text{vote}}|_V(x))]] \tag{25}$$

$$= \mathbb{E}_x[\mathbb{E}_{h \sim \pi|_V}[\mathbb{1}(h(x) \neq h_{\text{vote}}|_V(x))]] \tag{26}$$

$$= \mathbb{E}_x[\text{VR}(x|V)], \tag{27}$$

---

[1] The paper with appendix is available at https://arxiv.org/abs/2006.12456.

where the last equality holds because the predictions of the majority vote classifier are always the worst-case labels for prior mass reduction. Diameter reduction selects samples such that, upon revealing their labels, the induced subspaces have minimum possibility to be further reduced by a potential random query. Thus, it can be thought of as reducing the expected prior mass "reducibility".

Prior mass reduction finds splits in directions that evenly partition the version space, but could result in version spaces that have irregular shapes, in the sense that the space can be whittled down finely in some directions while being under-split in others. The worst-case error rate of the resulted version space could still be large. Diameter reduction correctly resolve this issue. Figure 1 illustrates the differences between prior mass and diameter reduction.

5.4   Weighted Diameter Reduction

Tosh and Dasgupta [24] show that in general average diameter cannot be decreased at steady rate and propose to query the unlabeled samples that minimize the diameter weighted by the *squared* prior mass in the worst-case scenario

$$\arg \min_{x} \max_{y} \mathbb{E}_{h,h' \sim \pi}[\mathbb{1}(h, h' \in V_x^y) \, d(h, h')] \tag{28}$$

$$= \arg \min_{x} \max_{y} \pi(V_x^y)^2 \, \mathbb{E}_{h,h' \sim \pi|_{V_x^y}}[d(h, h')] . \tag{29}$$

The potential to be minimized is a surrogate for the "amount" of edges between hypotheses and is closely related to the splittablity of version space [7,24].

6   Realizability Assumption

Even though neural networks are capable of fitting an arbitrary pool set, we show experimentally that the version space obtained by training on a subset of the pool set with stochastic gradient descent – the "samplable" version space – is biased and not likely to contain the correct labeling of the pool set. Indeed, the distance from the *Bayes classifier*, which provides the ground truth labeling, to the "boundary" of the version space is non-negligible.

Let $h_\perp^*$ be the projection of the Bayes classifier h^* to the set of hypotheses $\tilde{V}$ that agree with V on $\mathrm{AGR}(V)$ (see the left plot of Fig. 2), i.e.,

$$h_\perp^* := \arg \min_{h \in \tilde{V}} d(h, h^*), \tag{30}$$

$$\tilde{V} := \{h : h(\mathrm{AGR(V)}) = h'(\mathrm{AGR(V)}), h' \in V\} . \tag{31}$$

It is easy to see that $h_\perp^*$ provides the ground truth on $\mathrm{DIS}(V)$ and predicts the same labels on $\mathrm{AGR}(V)$ as hypotheses in V do, hence

$$d(h_\perp^*, h^*) = d(h, h^*; \mathrm{AGR(V)}) = \mathbb{E}_x[\mathbb{1}(x \in \mathrm{AGV}(V))\mathbb{1}(h(x) \neq h^*(x))] , \, \forall h \in V. \tag{32}$$

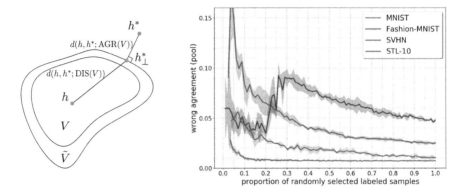

Fig. 2. Left: Projection of h^* to the samplable version space. **Right**: Wrong agreement of version spaces trained on random samples. Total numbers of samples are 400, 1000, 3000 and 2580 for MNIST, Fashion-MNIST, SVHN and STL-10 respectively.

where $d(h, h^*; \mathrm{AGR(V)})$ is the disagreement probability restricted to $\mathrm{AGR(V)}$, or equivalently the *wrong agreement* of hypotheses in V.

We show the evolution of wrong agreement in the right plot of Fig. 2. As more random samples are queried, the wrong agreement decreases for all datasets, but for some much slower than the others. In Fig. 3, we show for MNIST a 2-D embedding of version spaces using Multi-Dimensional Scaling (MDS) [20], which finds a low-dimensional representation of potentially high-dimensional data by preserving pairwise distances between the data points. The Bayes classifier is not contained in any of the samplable version spaces although the distances between them decrease steadily.

In general neural networks trained with a random subset do not automatically predict all labels in the pool set correctly, unless a relatively large proportion of samples are used for training. However, this fact does not render version space reduction inconsistent, because the samplable version space is not fixed, but it shifts towards the correct labeling and finally covers it when the whole pool set has been used.

We conjecture that the dynamics of active learning with neural networks have two major components: (1) shrinkage of the samplable version space, which is explicitly optimized by the learning algorithm and (2) reduction of bias, which is not directly controllable. Empirical evidence is provided in the next section.

7   Evaluation

Datasets and Architectures. We conduct active learning experiments[2] on four image classification datasets: MNIST, Fashion-MNIST, SVHN and STL-10. Neural network architectures are chosen to be competent for each dataset but as

---

[2] Source code is available at https://github.com/jiayu-liu/effective-version-space-reduction-for-convnets.

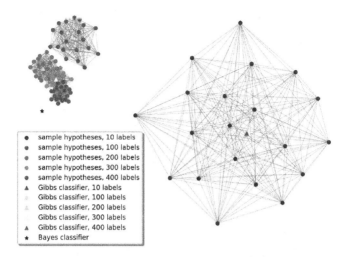

Fig. 3. Embedding of version spaces on MNIST using MDS. As more random samples are used for training, the samplable version spaces move closer to the Bayes classifier but hardly cover it.

simple as possible in the hope of controlling the model complexity and mitigating the effect of overfitting. See Table 1 for the complete experiment settings.

Active Learning Methods. We compare nine querying methods: Random, variation ratio (VR), Gibbs error (GE), Bayesian Active Learning by Disagreement with Monte Carlo dropout (BALD-MCD) [12,16], Core-Set [23], Deep-Fool Active Learning (DFAL) [10], pairwise disagreement (PWD), Gibbs-vote disagreement (GVD), and double-weighted pairwise disagreement (M^2-PWD) [24]. For each method on each dataset, at least three runs of active learning with different random balanced initial training set are performed.

Ensemble Size. We train networks multiple times from scratch to obtain sample hypotheses and use them for prior mass and diameter estimation. Since diameters are estimated by considering partitioned version spaces, the ensemble size should be at least in the order of number of classes. We set the size to 20. Larger ensemble improves estimation but at the cost of longer training time. In preliminary experiments, we tried larger ensembles (40) and did not observe

Table 1. Settings for each dataset used in the active learning experiments.

Dataset	Pool/Val/Test	Model	Ensemble size	Init/Query/Total	Runs
MNIST	45000/5000/10000	2-conv-layer ConvNet	20	10/5/400	4
Fashion-MNIST	55000/5000/10000	3-conv-layer ConvNet	20	10/10/1000	4
SVHN	40000/5000/15000	6-conv-layer ConvNet	20	100/20/3000	4
STL-10	4000/1000/8000	ResNet18	20	100/40/2580	3

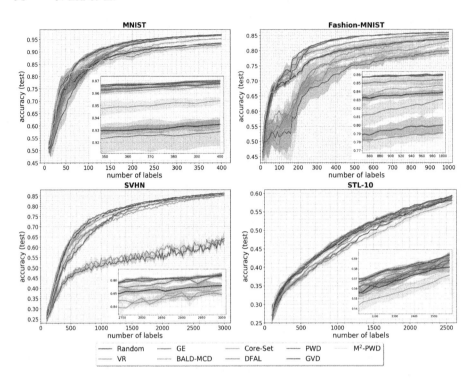

Fig. 4. Accuracy over number of queried labels on the test set. Direct diameter reduction methods PWD and GVD are consistently better than Random and are among the best methods. Weighted diameter reduction M²-PWD is on par with Random. Other baselines are effective on some datasets but inferior to Random on the others. Note that PWD, GVD and M²-PWD exhibit smaller variances than the others.

significant differences. Hence we do not include experiments on changing this hyper-parameter in the paper.

Query Size. We set a small query budget for each round to reduce the correlation between queries. Larger budget may alleviate the pressure of frequent retraining, but the effect of each query can not be estimated and examined reliably. We observed in preliminary experiments that using larger budget (one or two orders larger) hides the differences between methods.

7.1 Diameter Reduction Is More Effective Than Prior Mass Reduction

Figure 4 and Table 2 show that direct diameter reduction methods PWD and GVD are consistently better than Random and achieve higher accuracy than other baselines while weighted diameter reduction M2-PWD is on par with Random. Diameter reduction methods usually exhibit less variances because training on samples queried by PWD, GVD and M2-PWD yields version spaces with

Table 2. Accuracy on the test set in percentage.

#labels	MNIST	Fashion-MNIST	SVHN	STL-10
	400	1000	3000	2580
Random	93.47 ± 0.38	83.90 ± 0.38	85.60 ± 0.23	58.15 ± 0.54
VR	96.74 ± 0.15	83.05 ± 1.09	63.23 ± 1.99	59.13 ± 0.21
GE	96.79 ± 0.10	80.01 ± 0.94	64.08 ± 3.77	58.84 ± 0.34
BALD-MCD	96.51 ± 0.22	84.67 ± 0.41	85.26 ± 0.34	57.35 ± 0.64
Core-Set	95.38 ± 0.28	79.08 ± 0.82	84.91 ± 0.20	58.93 ± 0.33
DFAL	92.88 ± 1.19	85.38 ± 0.60	86.34 ± 0.33	58.81 ± 0.37
PWD	96.92 ± 0.12	85.92 ± 0.10	86.41 ± 0.12	**59.45 ± 0.11**
GVD	**97.02 ± 0.06**	**86.01 ± 0.15**	**86.44 ± 0.20**	59.33 ± 0.37
M²-PWD	93.24 ± 0.09	84.33 ± 0.03	85.42 ± 0.16	57.81 ± 0.20

Table 3. Diameter (pairwise disagreement) on the test set in percentage.

#labels	MNIST	Fashion-MNIST	SVHN	STL-10
	400	1000	3000	2580
Random	2.86 ± 0.18	7.55 ± 0.26	13.13 ± 0.29	32.88 ± 0.43
VR	2.27 ± 0.18	10.64 ± 0.72	46.88 ± 2.76	34.21 ± 0.08
GE	2.30 ± 0.04	11.38 ± 1.52	44.87 ± 4.25	34.25 ± 0.09
BALD-MCD	2.39 ± 0.15	8.11 ± 0.51	16.58 ± 0.42	33.55 ± 0.44
Core-Set	2.91 ± 0.18	10.79 ± 1.34	14.66 ± 0.47	33.13 ± 0.64
DFAL	3.79 ± 0.60	7.06 ± 0.60	13.98 ± 0.31	32.41 ± 0.27
PWD	**1.93 ± 0.04**	**6.91 ± 0.16**	**12.80 ± 0.08**	**32.25 ± 0.26**
GVD	1.98 ± 0.05	6.98 ± 0.26	12.88 ± 0.25	32.96 ± 0.48
M²-PWD	3.37 ± 0.13	7.22 ± 0.08	13.31 ± 0.13	33.23 ± 0.18

smaller diameters and less diverse sample hypotheses. Prior mass reduction is not always effective and even fails on SVHN. This failure is an example of prior mass reduction being incapable of reducing the diameter, and provides empirical evidence that it may not be an appropriate objective for active learning.

7.2 Comparison to Other Baselines

BALD-MCD, Core-Set and DFAL are not consistently better than Random although each of them achieves comparative test accuracy on certain dataset. Their inferiority to Random in terms of test accuracy usually correlates with higher diameter (See description in Fig. 5 and Table 3). BALD-MCD and DFAL are highly related to prior mass reduction methods in that BALD [16] seeks samples for which the model parameters under the posterior disagree the most about the prediction [16], and that DFAL, inspired by margin-based active learning [2], tries to locate the decision boundary with fewer labels which is essentially removing inconsistent hypotheses in the realizable case. However, none of them explicitly minimize the diameter, neither does Core-Set.

Note that for a fair comparison, we do not augment the training set by also adding the adversarial samples as the original DFAL paper [10] does. Samples with minimum adversarial perturbation are then verified reliably to be less effective than those lead to minimum diameter. The original Core-Set paper [23] uses a large query batch size (in the order of 1000). However, many baselines rely on greedy selection and do not perform any batch optimization. To reduce query correlation, we adopt as small batch size as possible. This allows reliable evaluation of the effectiveness of queried samples as in the online setting. We are therefore able to identify one major cause of inferiority to Random as failing to effectively reduce the version space diameter.

7.3   Evolution of Samplable Version Space and Its Implications

As shown in Fig. 5 and 3, the samplable version space shifts closer to the correct labeling while reducing its diameter as more labels are queried. These two processes together result in smaller test error.

No Direct Control Over Reduction of Version Space Bias. Interestingly, the Core-Set method, which queries representative samples from the pool set by solving a k-center problem in the feature space learned by neural networks, is incapable of achieving negligible wrong agreement on the learned version spaces. Indeed, it suffers larger version space bias than the direct diameter reduction methods. After all, random queries which are i.i.d. by assumption fail to achieve this goal as concluded in Sect. 6 and other attempts without augmenting the training data seem doomed.

Prior Mass Induced by Stochastic Gradient Descent May Not Be a Reliable Surrogate Measure. The continued decline in wrong agreement indicates that the distribution over labelings changes over time. This fact of shifting density over samplable labelings renders the notion of prior mass problematic, hence all notions relying on prior mass may not be well-defined. A direct consequence is that an estimate of the worst-case version space reduction would be more reliable than the average-case one. For example, VR provides a more reliable estimate of version space reduction than GE does.

Inferiority of Weighted Diameter Reduction Method. The estimation of weighted diameter involves estimating the prior mass. Hence, the inferiority of M^2-PWD to PWD and GVD can be attributed to the intrinsic difficulty of obtaining unbiased samplable version spaces and the resulted density shift. A supportive evidence can be seen by noting that on MNIST and Fashion-MNIST, where the wrong agreement is large (hence large density shift), the weighted variant performs worse, while on SVHN and STL-10, where the wrong agreement is small (hence small shift), the gap is less significant.

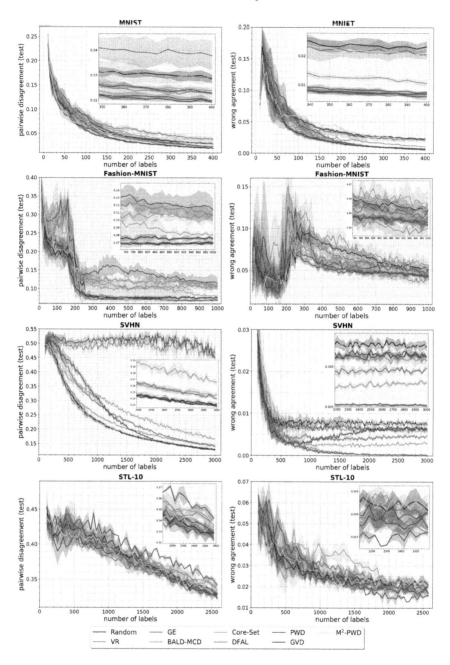

Fig. 5. Pairwise disagreement and wrong agreement over number of queried labels on the test set. Except direct diameter reduction methods PWD and GVD, other baselines are not consistently better than or on par with Random at reducing version space diameter. Performing worse than Random: GE, VR and BALD-MCD on datasets except MNIST, Core-Set on Fashion-MNIST and SVHN, and DFAL on MNIST and SVHN, and M^2-PWD on MNIST.

7.4   Gibbs-Vote Disagreement

The Gibbs-vote disagreement is among the best methods on all datasets, except for the early learning stage on SVHN. Its effectiveness can be ascribed to an interesting phenomenon – majority voting reduces mistakes. Although it need not necessarily be the case, this phenomenon occurs in many situations and the boost to accuracy depends on the variance of errors of Gibbs classifiers [21]. We show empirically that the majority vote classifier indeed has smaller error rate than random hypotheses in the version space in Fig. 6. Hence, optimiz-

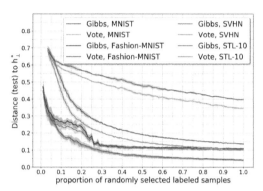

Fig. 6. Distance from the Gibbs and the majority vote classifier to the projection of h^*. On four datasets, the majority vote classifier has a smaller distance, hence smaller error rate. See description of Fig. 2 for total numbers of random samples.

ing the Gibbs-vote disagreement not only reduces the diameter but also implicitly moves the consistent hypotheses closer to the correct labeling, which is useful when the samplable version spaces are biased and do not contain the Bayes classifier.

8   Conclusion

In this work, we studied version space reduction for convolutional neural networks. We revealed the differences and connections between prior mass and diameter reduction methods and proposed the Gibbs-vote disagreement as a new effective diameter-reduction method. With experiments on four datasets, we shed light into how version space reduction works in the deep active learning setting and demonstrated the superiority of diameter reduction over prior mass reduction methods and other baselines.

Acknowledgments. This work was supported by the BMBF project MLWin and the Munich Center for Machine Learning (MCML).

References

1. Balcan, M.F., Beygelzimer, A., Langford, J.: Agnostic active learning. In: Proceedings of the 23rd International Conference on Machine Learning, pp. 65–72 (2006)
2. Balcan, M.-F., Broder, A., Zhang, T.: Margin based active learning. In: Bshouty, N.H., Gentile, C. (eds.) COLT 2007. LNCS (LNAI), vol. 4539, pp. 35–50. Springer, Heidelberg (2007). https://doi.org/10.1007/978-3-540-72927-3_5

3. Beluch, W.H., Genewein, T., Nürnberger, A., Köhler, J.M.: The power of ensembles for active learning in image classification. In: Proceedings of the IEEE Conference on Computer Vision and Pattern Recognition, pp. 9368–9377 (2018)
4. Cuong, N.V., Lee, W.S., Ye, N.: Near-optimal adaptive pool-based active learning with general loss. In: Proceedings of the 30th Conference on Uncertainty in Artificial Intelligence, pp. 122–131 (2014)
5. Cuong, N.V., Lee, W.S., Ye, N., Chai, K.M.A., Chieu, H.L.: Active learning for probabilistic hypotheses using the maximum Gibbs error criterion. In: Advances in Neural Information Processing Systems, pp. 1457–1465 (2013)
6. Dasgupta, S.: Analysis of a greedy active learning strategy. In: Advances in Neural Information Processing Systems, pp. 337–344 (2005)
7. Dasgupta, S.: Coarse sample complexity bounds for active learning. In: Advances in Neural Information Processing Systems, pp. 235–242 (2006)
8. Dasgupta, S.: Two faces of active learning. In: Proceedings of the 20th International Conference on Algorithmic Learning Theory, p. 1 (2009)
9. Devroye, L., Györfi, L., Lugosi, G.: A Probabilistic Theory of Pattern Recognition, vol. 31. Springer, Heidelberg (2013). https://doi.org/10.1007/978-1-4612-0711-5
10. Ducoffe, M., Precioso, F.: Adversarial active learning for deep networks: a margin based approach. arXiv preprint arXiv:1802.09841 (2018)
11. Gal, Y., Ghahramani, Z.: Dropout as a Bayesian approximation: representing model uncertainty in deep learning. In: Proceedings of the 33rd International Conference on Machine Learning, pp. 1050–1059 (2016)
12. Gal, Y., Islam, R., Ghahramani, Z.: Deep Bayesian active learning with image data. In: Proceedings of the 34th International Conference on Machine Learning, pp. 1183–1192 (2017)
13. Gissin, D., Shalev-Shwartz, S.: Discriminative active learning. arXiv preprint arXiv:1907.06347 (2019)
14. Golovin, D., Krause, A.: Adaptive submodularity: a new approach to active learning and stochastic optimization. J. Artif. Intell. Res. 427–486 (2011)
15. Hanneke, S.: A bound on the label complexity of agnostic active learning. In: Proceedings of the 24th International Conference on Machine Learning, pp. 353–360 (2007)
16. Houlsby, N., Huszár, F., Ghahramani, Z., Lengyel, M.: Bayesian active learning for classification and preference learning. Computing Research Repository abs/1112.5745 (2011)
17. Joshi, A.J., Porikli, F., Papanikolopoulos, N.: Multi-class active learning for image classification. In: Proceedings of the IEEE Conference on Computer Vision and Pattern Recognition, pp. 2372–2379 (2009)
18. Kääriäinen, M.: Generalization error bounds using unlabeled data. In: Auer, P., Meir, R. (eds.) COLT 2005. LNCS (LNAI), vol. 3559, pp. 127–142. Springer, Heidelberg (2005). https://doi.org/10.1007/11503415_9
19. Kirsch, A., van Amersfoort, J., Gal, Y.: BatchBALD: efficient and diverse batch acquisition for deep Bayesian active learning. In: Advances in Neural Information Processing Systems, pp. 7024–7035 (2019)
20. Kruskal, J.B.: Multidimensional Scaling, no. 11. Sage (1978)
21. Lacasse, A., Laviolette, F., Marchand, M., Germain, P., Usunier, N.: PAC-Bayes bounds for the risk of the majority vote and the variance of the Gibbs classifier. In: Advances in Neural Information Processing Systems, pp. 769–776 (2007)
22. Pinsler, R., Gordon, J., Nalisnick, E., Hernández-Lobato, J.M.: Bayesian batch active learning as sparse subset approximation. In: Advances in Neural Information Processing Systems, pp. 6356–6367 (2019)

23. Sener, O., Savarese, S.: Active learning for convolutional neural networks: a core-set approach. In: Proceedings of 6th the International Conference on Learning Representations (2018)
24. Tosh, C., Dasgupta, S.: Diameter-based active learning. In: Proceedings of the 34th International Conference on Machine Learning. pp. 3444–3452 (2017)
25. Zhang, C., Bengio, S., Hardt, M., Recht, B., Vinyals, O.: Understanding deep learning requires rethinking generalization. In: Proceedings of the 5th International Conference on Learning Representations (2017)

A Principle of Least Action
for the Training of Neural Networks

Skander Karkar[1]([⊠]), Ibrahim Ayed[2], Emmanuel de Bézenac[2],
and Patrick Gallinari[1,2]

[1] Criteo AI Lab, Criteo, Paris, France
as.karkar@criteo.com
[2] LIP6, Sorbonne Université, Paris, France
{ibrahim.ayed,emmanuel.de-bezenac,patrick.gallinari}@lip6.fr

Abstract. Neural networks have been achieving high generalization performance on many tasks despite being highly over-parameterized. Since classical statistical learning theory struggles to explain this behaviour, much effort has recently been focused on uncovering the mechanisms behind it, in the hope of developing a more adequate theoretical framework and having a better control over the trained models. In this work, we adopt an alternative perspective, viewing the neural network as a dynamical system displacing input particles over time. We conduct a series of experiments and, by analyzing the network's behaviour through its displacements, we show the presence of a low kinetic energy bias in the transport map of the network, and link this bias with generalization performance. From this observation, we reformulate the learning problem as follows: find neural networks that solve the task while transporting the data as efficiently as possible. This offers a novel formulation of the learning problem which allows us to provide regularity results for the solution network, based on Optimal Transport theory. From a practical viewpoint, this allows us to propose a new learning algorithm, which automatically adapts to the complexity of the task, and leads to networks with a high generalization ability even in low data regimes.

Keywords: Deep learning · Optimal Transport · Dynamical systems

1 Introduction

Deep neural networks (DNNs) have repeatedly shown their ability to solve a wide range of challenging tasks, while often having many more parameters than there are training samples. Such a performance of over-parametrized models is counter-intuitive. They seem to adapt their complexity to the given task, systematically achieving a low training error without suffering from over-fitting as

Electronic supplementary material The online version of this chapter (https://doi.org/10.1007/978-3-030-67661-2_7) contains supplementary material, which is available to authorized users.

© Springer Nature Switzerland AG 2021
F. Hutter et al. (Eds.): ECML PKDD 2020, LNAI 12458, pp. 101–117, 2021.
https://doi.org/10.1007/978-3-030-67661-2_7

could be expected [1,21,35]. This is in contradiction with the classical statistical practice of selecting a class of functions complex enough to represent the coherent patterns in the data, and simple enough to avoid spurious correlations [2,13]. Although this behavior has sparked much recent work towards explaining neural networks' success [8,17,22,24], it still remains poorly understood. Among the factors to consider are the implicit biases present in the choices made for the parametrization, the architecture, the parameter initialization and the optimization algorithm, and that contribute all to this success. Our aim in this work is to uncover some of these hidden biases and highlight their link with generalization performance through the lens of dynamical systems.

We will focus on residual networks (ResNets) [15,16], now ubiquitous in applications. This family of models has made it possible to learn very complex nonlinear functions by improving the trainability of very deep networks, and has thus improved generalization. Links have been derived between these networks and dynamical systems: a ResNet can be seen as a forward Euler scheme discretization of an associated ordinary differential equation (ODE) [30]:

$$x_{k+1} = x_k + v_k(x_k) \quad \longleftrightarrow \quad \partial_t x_t = v_t(x_t) \tag{1}$$

This link has yielded many exciting results, *e.g.* new architectures [20] and reversible networks [6]. Here, we make use of this analogy and analyze the behavior of residual networks by studying their associated differential flows. Adopting this dynamical point of view allows us to leverage the theories and mathematical tools developed to study, approximate and apply differential equations.

More specifically, we conduct experiments to observe how neural networks displace their inputs–seen as particles–through time. We measure a strong empirical correlation between good test performance and neural networks with low kinetic energy along their transport flow. From this, we reformulate the training problem as follows: retrieve the network which solves the task using the principle of least action, *i.e.* expending as little kinetic energy as possible. This problem, in its probabilistic formulation, is tightly linked with and inspired by the well-known problem of finding an optimal transportation map [27]. This yields new insights into neural networks' generalization capabilities, and provides a novel algorithm that automatically adapts to the complexity of the data and robustly improves the network's performance, including in low data regimes, without slowing down the training. To summarize, our contributions are the following:

- Through the dynamic viewpoint, we highlight the *low-energy bias* of ResNets.
- We formulate a Least Action Principle for the training of Neural Networks.
- We prove existence and regularity results for networks with minimal energy.
- We provide an algorithm for retrieving minimal energy networks compatible with different architectures, which leads to better generalization performance for different classification tasks, without complexifying the architecture.

We introduce in Sect. 2 some background on Optimal Transport (OT) and highlight the link between the dynamical formulation of OT and ResNets. We describe in Sect. 3 the general setting of our analysis. Section 4 provides empirical evidence illustrating our point. The formal framework of networks trained with minimized energy and a practical algorithm are described in Sect. 5. Experiments on standard classification tasks are provided in Sect. 6. The code is available online at http://github.com/skander-karkar/LAP.

2 Background

This section outlines the main elements of the formalism and reasoning of our work. Supplementary Material A gives more details about Optimal Transport.

2.1 Optimal Transport

The principle of least action is central to many fields in physics, mathematics and economics. It is found in classical and relativistic mechanics, thermodynamics, quantum mechanics [9–11], etc. It broadly states that the dynamical trajectory of a system between an initial and final configuration is one that makes a certain action associated with the system locally stationary [11]. One mathematical theory which can be associated with this general idea is the theory of Optimal Transport which was initially introduced as a way of finding a transportation map minimizing the cost of displacing mass from one configuration to another [27].

Formally, let α and β be absolutely continuous distributions compactly supported in $\mathbb{R}^d$, and $c : \mathbb{R}^d \times \mathbb{R}^d \to \mathbb{R}$ a cost function. Consider a transportation map $T : \mathbb{R}^d \to \mathbb{R}^d$ that satisfies $T_\sharp \alpha = \beta$, i.e. that pushes[1] α to β. The total cost of the transportation then depends on all the individual contributions of costs of transporting (infinitesimal) mass from each point x to $T(x)$, and finding the optimal transportation map amounts to solving:

$$\min_T \quad \mathcal{C}^{\text{stat}}(T) = \int_{\mathbb{R}^d} c(x, T(x)) \mathrm{d}\alpha(x)$$

$$\text{s.t.} \quad T_\sharp \alpha = \beta \tag{2}$$

A standard choice for c is the p-th power of a norm of $\mathbb{R}^d$, i.e. $c(x, y) = \|x - y\|^p$, but other costs can be used, defining different variants of the problem. This cost induces, through the p-th root of the minimal value of (2), a distance W_p between any two distributions α and β of finite $p-$th moment, called the p-Wasserstein distance [23].

In [3], the link between Optimal Transport and the principle of least action was made by showing that the static transportation can equivalently be viewed as a dynamical one that minimizes an action as it gradually displaces particles of mass in time. In other words, instead of directly pushing samples of α to β in $\mathbb{R}^d$ using T, we can displace mass from α according to a continuous flow with velocity $v_t : \mathbb{R}^d \to \mathbb{R}^d$. This implies that the density μ_t at time t satisfies the

---

[1] $T_\sharp \alpha$ is the *push-forward measure*: $T_\sharp \alpha(B) = \alpha(T^{-1}(B))$ for any measurable set B.

continuity equation $\partial_t \mu_t + \nabla \cdot (\mu_t v_t) = 0$, assuming that initial and final conditions are given respectively by $\mu_0 = \alpha$ and $\mu_1 = \beta$. In this case, the optimal displacement is the one that minimizes the action $\|v_t\|^p_{L^p(\mu_t)}$:

$$\min_v \quad \mathcal{C}^{\mathrm{dyn}}(v) = \int_0^1 \|v_t\|^p_{L^p(\mu_t)} \mathrm{d}t \tag{3}$$
$$\text{s.t.} \quad \partial_t \mu_t + \nabla \cdot (\mu_t v_t) = 0, \mu_0 = \alpha, \mu_1 = \beta$$

where $\|v_t\|^p_{L^p(\mu_t)} = \int_{\mathbb{R}^d} \|v_t(x)\|^p \mathrm{d}\mu_t(x)$ for costs $c(x, y) = \|x - y\|^p$ with $p > 1$. In this case, minimizers exist and the two transport costs are the same, *i.e.* $\mathcal{C}^{\mathrm{stat}}(T) = \mathcal{C}^{\mathrm{dyn}}(v)$ at the optimums. For $p = 2$ and the Euclidean norm, the dynamical cost $\mathcal{C}^{\mathrm{dyn}}(v)$ corresponds to the *kinetic energy*.

2.2   Link with Residual Networks

The dynamical formulation in (3) explicitly describes the evolution in time of the density μ_t, starting from an input distribution α. In this form, the link between deep residual networks and dynamical Optimal Transport is not clear. However, it is possible to adopt an alternate viewpoint which helps make it immediate. Instead of explicitly describing the density's evolution, we describe the paths $\phi^x : [0, 1] \to \mathbb{R}^d$, $t \mapsto \phi_t^x$ taken by particles from α at position x, when displaced along the flow v. The continuity equations can then equivalently be written as:

$$\partial_t \phi_t^x = v_t(\phi_t^x) \tag{4}$$

See Chaps. 4 and 5 of [27] for details. We can now note the resemblance between the residual network (1) and Eq. (4). Rewriting the conditions as necessary, the dynamical formulation (3) can equivalently be represented by:

$$\min_v \quad \mathcal{C}^{\mathrm{lag}}(v) = \int_0^1 \|v_t\|^p_{L^p((\phi_t)_\sharp \alpha)} \mathrm{d}t$$
$$\text{s.t.} \quad \partial_t \phi_t^x = v_t(\phi_t^x), \tag{5}$$
$$\phi_0 = \mathrm{id},$$
$$(\phi_1)_\sharp \alpha = \beta$$

where $\phi_t : x \in \mathbb{R}^d \mapsto \phi_t^x \in \mathbb{R}^d$ corresponds to the transport map induced by the flow, up until time t. As both formulations are equivalent, we have that for any flow v, $\mathcal{C}^{\mathrm{lag}}(v) = \mathcal{C}^{\mathrm{dyn}}(v)$. Moreover, optimal transportation plans in the static (2) and dynamical (5) cases coincide: if T and ϕ_t, are respectively solutions to (2) and (5), we have that $T = \phi_1$.

This link allows us to associate residual networks with a local action for each layer, which induces a global transportation cost $\mathcal{C}^{\mathrm{lag}}$, and taking $p = 2$ and the Euclidean norm allows us to refer to the network's kinetic energy.

3   General Setting

In order to better understand the inner workings of a DNN, it is essential to adopt a viewpoint in which the different driving mechanisms become apparent and are decoupled.

Decomposing a DNN. We consider the following model of a deep neural network f where computations are separated into the three steps, *i.e.* $f = F \circ T \circ \varphi$ (this is similar to [19] and corresponds to the general structure of recent deep models or to the structure of components of a deep model [16,31,34]):

1. **Dimensionality change:** Starting from an input distribution $\mathcal{D}$ in $\mathbb{R}^n$, a transformation φ is applied, transforming it into $\alpha = \varphi_\sharp \mathcal{D}$, a distribution in $\mathbb{R}^d$. This corresponds to the first few layers present in most recent architectures and represents a change of dimensionality. φ is known as the *encoder*.
2. **Data Transport:** Then α is transformed by a mapping $T : \mathbb{R}^d \to \mathbb{R}^d$, which we see as a transport map. Here, the dimensionality doesn't change and, if this part of the network is a sequence of residual blocks, T can be written as the discretized flow of an ODE.
3. **Task-specific final layers:** A final function $F : \mathbb{R}^d \to \mathcal{Y}$ is applied to $T_\sharp \alpha$ in order to compute the loss $\mathcal{L}$ associated with the task at hand, *e.g.* F could be a perceptron classifier. Like φ, F is typically made up of a few layers.

The focus of this work is on analyzing the second phase, Data Transport, and we assume that the encoder φ is pretrained and fixed (this will be relaxed in some experiments later). To solve a complex non-linear task for which a DNN is needed, the data has to be transformed in a non-trivial way, meaning that this is an essential phase, *e.g.* in the case of classification, $T_\sharp \alpha$ needs to be linearly separable if F is linear. This model is quite general, as many ResNet-based architectures [31,34] alternate modules that change the dimensionality (step 1) and transport modules that keep the dimensionality fixed (step 2) and according to [18], the transport modules have similar behaviour. The model can then be considered as a simplified ResNet, sometimes called a *single representation* ResNet. Note that [26] finds that networks that keep the same resolution remain competitive.

The Set of Admissible Targets. As recent neural architectures have systematically achieved near-zero training error [1,2,17,35], we place ourselves in this regime, which makes it possible to model this as a hard constraint. For some tasks, this constraint over T is obvious: in a generative setting for example, $T_\sharp \alpha$ must be equal to some prescribed distribution β which is the target of the generation process. But in general, T is less strictly constrained and the condition depends on F and $\mathcal{L}$. This leads us to define a *set of admissible targets* for the task:

$$S_{F,\mathcal{L}} = \{\beta \in \mathcal{P}(\mathbb{R}^d) \mid \mathcal{L}(F, \beta) = 0\} \tag{6}$$

with $\beta = T_\sharp \alpha$. In general, $\mathcal{L}$ is fixed while F is learned jointly with T. This set is supposed to be non-empty for some F and, in general, it will contain many distributions. The goal of the learning task can then be reformulated as:

$$\text{Find } (T, F) \text{ such that } T_\sharp \alpha \in S_{F, \mathcal{L}} \tag{7}$$

An important observation is that, even when $S_{F,\mathcal{L}}$ is reduced to a singleton, the problem is still strongly under-constrained and it is possible to obtain many such (T, F) that lead to poor generalization. One can then ask why this is not the case in practice, as good generalization performance is usually achieved.

The Case of Classification. Even though our framework is general, we focus our experiments on classification tasks, with $\mathcal{L}$ being the cross entropy loss. The task consists in separating N classes. Let us denote α_i the class distributions which are supposed to be distributions in $\mathbb{R}^d$ of mutually disjoint supports, meaning that there is no ambiguity in the class of data points, and such that $\alpha = \sum_i \alpha_i / N$. One wants to find a transformation T of these distributions such that all transported distributions can be correctly classified by a classifier F. When F is linear, $S_{F,\mathcal{L}}$ is the set of distributions which have N components that are linearly separated by F. Note that we place ourselves in a noiseless ideal setting where perfect classification is possible. The question we examine in this work is then twofold:

- What are the properties characterizing mappings reached by standard residual architectures with common hyper-parameters?
- Can we find a criteria to *automatically select* mappings with desirable properties in order to improve performance and robustness?

4   Empirical Analysis of Transport Dynamics in ResNets

Before introducing our framework, we conduct an exploratory analysis of the impact of the network's inner dynamics on generalization. We present below two experiments. The first one highlights how good generalization performance is closely related to low transport cost for classification tasks on MNIST and CIFAR10. This cost therefore appears as a natural characterisation of the complexity and disorder of a network. The second experiment, performed on a toy 2D dataset, visualizes the transport induced by the blocks of a ResNet.

We consider ResNets where, after encoding, a data point x_0 is transported by applying $x_{k+1} = x_k + v_k(x_k)$ for K residual blocks and then classified using F. We measure the disorder/complexity of a network by its transport cost which is the sum of the displacements induced by its residual blocks: $\mathcal{C}(v) = \sum_k \|v_k(x_k)\|_2^2$. This quantity corresponds to the kinetic energy of the total displacement.

Transportation Cost and Generalization on MNIST and CIFAR10. In order to study the correlation between the transport cost of a residual network and its generalization ability on image data, we train convolutional ResNets with 9 blocks and different initializations (orthogonal and normal with different gains), for 10-class classification tasks MNIST and CIFAR10. In Fig. 2, each point represents a trained network and gives the transport cost $\mathcal{C}$ as a function of the test accuracy of the network. This experiment clearly highlights the strong negative correlation between transport cost and good generalization. This illustrates the importance of the implicit initialization bias and motivates initialization schemes which favour a low kinetic energy. We believe a number of factors contribute to this low energy bias: small initialization gains tend to bias $\|v_k(x_k)\|_2^2$ towards small values, and training using gradient descent does not change this much.

Visualizing Network Dynamics on 2D Toy Data. This experiment provides a 2D visualization of the transport dynamics inside a network. The task is 2-class classification of a non-linearly separable dataset (two concentric circles, from `sklearn`) that contains 1000 points with a train-test split of 80%–20%, see Fig. 1 top left. The network is a ResNet with 9 residual blocks, followed by a fixed linear classifier. Each residual block contains two fully connected layers separated by a batch normalization and a ReLU activation.

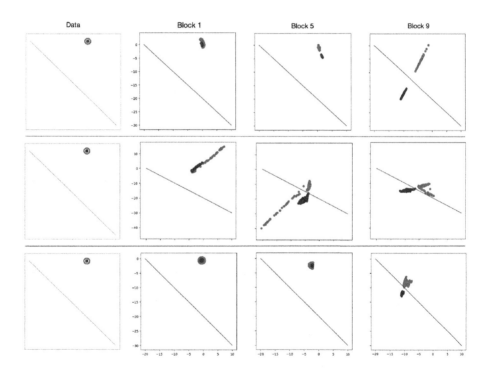

Fig. 1. Transformed circles testset by a ResNet9 after blocks 1, 5 and 9 after training; first row with good initialization; second row with a $\mathcal{N}(0,5)$ initialization; third row with a $\mathcal{N}(0,5)$ initialization and the transport cost added to the loss

With the cross-entropy loss alone, the behaviour of a well trained and carefully initialized network achieving 100% test accuracy is illustrated in the first row of Fig. 1. With a $\mathcal{N}(0,5)$ initialization, significantly bigger than an "optimal" initialization, the test accuracy drops to 98% (average of 100 runs) and the transport becomes chaotic (Fig. 1, second row). Adding the transport cost to the loss improves the test accuracy (99.7% on average) of this badly initialized network and the movement becomes more controlled (third row of Fig. 1). Thus, controlling transport improves the behavior and generalization ability of the network. This allows to explicitly control the network whereas implicit biases such as "good" initialization rely on heuristics. In Supplementary Material C.4, more experiments show that in other situations that deviate from the ideal setting where the task is perfectly solved, e.g. when using a network which is too large or too small, or a small training set, controlling the transport cost also improves generalization.

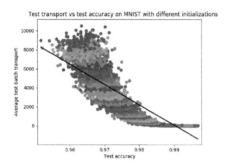 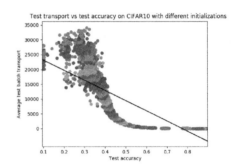

Fig. 2. Test transport against test accuracy of ResNet9 models on MNIST (left) and CIFAR10 (right) with fitted linear regressions, where each color indicates a different initialization (either orthogonal or normal with varying gains) (Color figure online)

5 Least Action Principle for Training Neural Networks

The previous section has shed some light on the low energy bias of networks as well as on its potential benefits on test accuracy. In this section, we take a step further and make this implicit bias explicit by considering a formulation for training that enforces minimal kinetic energy, closely related to the problem of Optimal Transport. This allows us to prove the existence of minimizers, and exhibit interesting regularity properties of the minimal energy neural networks which may explain good generalization performance.

5.1 Formulation

We consider costs $c(x, y) = \|x - y\|^p$ (where $\|.\|$ is a norm of $\mathbb{R}^d$), with $p > 1$, and suppose that $\alpha \in \mathcal{P}_p(\mathbb{R}^d)$ (the set of absolutely continuous measures on $\mathbb{R}^d$ with finite p-th moment). We assume that the space of classifiers is compact, that the loss $\mathcal{L}$ is continuous, that the set $\cup_{F \in \mathcal{F}} S_{F,\mathcal{L}}$ is at a finite p-Wasserstein distance W_p from α (in particular, it is non-empty) and that all its bounded subsets are totally bounded (*i.e.* can be covered by finitely many subsets of any fixed size). These properties depend on the choice of the loss $\mathcal{L}$ and of a class of functions $\mathcal{F}$ for the classifier F.

Returning to the transport problem as defined in Sect. 2.1, a natural way to select a robust model, given the empirical observations of Sect. 4, is to select, among the maps which transport α to $S_{F,\mathcal{L}}$ and thus solve the task, one with a minimal transport cost. This gives us the following optimization problem:

$$\inf_{F,T} \quad \mathcal{C}(T) = \int_{\mathbb{R}^d} c(x, T(x)) d\alpha(x) \tag{8}$$

$$\text{subject to} \quad T_\sharp \alpha \in S_{F,\mathcal{L}}$$

The equivalent dynamical version for $c(x, y) = \|x - y\|^p$ is, as per Sect. 2.2,

$$\inf_{F,v} \quad \int_0^1 \|v_t\|^p_{L^p((\phi_t)_\sharp \alpha)} \, dt$$

$$\text{subject to} \quad \partial_t \phi_t^x = v_t(\phi_t^x) \tag{9}$$

$$\phi_0^\cdot = \mathrm{id}$$

$$(\phi_1^\cdot)_\sharp \alpha \in S_{F,\mathcal{L}}$$

where $\|v_t\|^p_{L^p((\phi_t)_\sharp \alpha)} = \int_{\mathbb{R}^d} \|v_t\|^p d(\phi_t)_\sharp \alpha$. The result below shows that these two problems are equivalent and that the infima are realized as minima:

Theorem 1. *The infima of* (8) *and* (9) *are finite and are realized through a map T which is (or a velocity field v which induces) an optimal transportation map. When $c(x, y) = \|x - y\|^p$, then* (8) *and* (9) *are equivalent.*

Proof. From the hypothesis above, there exists $\beta \in S_{F,\mathcal{L}}$ at a finite distance from α. Taking any transport map between α and β, we see that the infima are finite.

Consider (8) and take a minimizing sequence $(F_i, T_i)_i$. Set $\beta_i = (T_i)_\sharp \alpha$. Then $(\mathcal{C}(T_i))_i$ converges to the infimum which is strictly bounded by $M > 0$. Then, by definition, for i large enough, $W_p^p(\alpha, \beta_i) \leq \mathcal{C}(T_i) \leq M$. So that $(\beta_i)_i$ is a bounded sequence in $\cup_F S_{F,\mathcal{L}}$. By the hypothesized total boundedness of bounded subsets and as $\mathcal{P}_p$ endowed with W_p is a complete metric space (see [4] for a proof), up to an extraction, $(\beta_i)_i$ converges to $\beta^\star$ in the closure of $\cup_F S_{F,\mathcal{L}}$. Moreover, up to an extraction, $(F_i)_i$ also converges to $F^\star$ by compactness of the class of classifiers. Taking $T^\star$ the OT map between α and $\beta^\star$ (see Supplementary Material A for existence results of OT maps), we then have, by continuity of $\mathcal{L}$,

$$T_\sharp^\star \alpha = \beta^\star \in S_{F^\star, \mathcal{L}}$$

and $\mathcal{C}(T^\star) \leq \lim \mathcal{C}(T_i)$ by optimality of $T^\star$, which means, since $(\mathcal{C}(T_i))_i$ is a minimizing sequence, that $\mathcal{C}(T^\star)$ minimizes (8). So $(F^\star, T^\star)$ is a minimizer and $T^\star$ is an OT map.

Finally, there exists, by dynamical OT theory (Supplementary Material A), a velocity field $v_t^\star$ inducing the OT map between α and $\beta^\star$ which then gives a minimizer $(F^\star, v^\star)$ for (9). By the same reasoning, taking a minimizing sequence $(F_i, v^{(i)})_i$ and the induced maps T_i shows that both problems are equivalent. $\square$

Note that uniqueness doesn't hold anymore, as the constraint $T_\sharp\alpha \in S_{F,\mathcal{L}}$ in (9) is looser than in standard OT. However, as we show in the following section, the fact that the optimization problems are solved by OT maps will give regularity properties for the models induced by these optimization problems.

5.2 Regularity

Intuitively, the fact that we minimize the energy of the transport map transforming the data is akin to the core idea of Occam's razor: among all the possible networks that correctly solve the task, the one transforming the data in the simplest way is selected. Moreover, it is possible to show that this optimal transformation is regular: our formulation provides an alternate view on generalization for modern deep learning architectures in the overparametrized regime.

Optimal maps can be as irregular as needed in order to fit the target distribution, however in much the same way as successfully trained DNNs, optimal maps are still surprisingly regular. In a way, they are as regular as possible given the constraints which is exactly the type of flexibility needed. However, the constraints in (8) and (9) are looser than in the standard definitions of Optimal Transport. Still, supposing that the input data distribution has a nicely behaved density, namely bounded and of compact support, with the same hypothesis as above, we have the following, which is mainly a corollary of Theorem 1:

Proposition 2. *Consider $T^\star$ the OT map induced by (8) (or (9)) given by Theorem 1. Take X, respectively Y, an open neighborhood of the support of α, respectively of $T_\sharp^\star\alpha$, then $T^\star$ is differentiable, except on a set of null α measure.*

Additionally, if $T^\star$ doesn't have singularities, there exists $\eta > 0$ and A, respectively B, relatively closed in X, respectively Y, such that $T^\star$ is η-Hölder continuous from $X \setminus A$ to $Y \setminus B$. Moreover, if the two densities are smooth, $T^\star$ is a diffeomorphism from $X \setminus A$ to $Y \setminus B$.

Proof. This is a consequence of Theorem 1, the hypothesis made in this section and the regularity theorems stated in Supplementary Material B.   □

There are two main results in Proposition 2: the first gives α-a.e. differentiability. This is already as strong as might be expected from a classifier: there are necessarily discontinuities at the frontiers between different classes. The second is even more interesting: it gives Hölder continuity over as large a domain as possible, and even a diffeomorphism if the data distribution is well-behaved enough. We recall that a function f is η-Hölder continuous for $\eta \in]0,1]$ if $\exists\, M > 0$ such that $\|f(x) - f(y)\| \leq M\|x-y\|^\eta$ for all x, y. η measures the smoothness of f, the higher its value, the better. In particular, in the case of classification, this means that the Hausdorff dimension along the frontiers between the different classes is scaled by less than a factor of $1/\eta$ in the transported domain. If the densities are smooth, the dimension even becomes provably smaller by this result.

Intuitively, this means that, in these models, the data is transported in a way that preserves and simplifies the patterns in the input distribution. In the following, we propose a practical algorithm implementing these models and use it for standard classification tasks, showing an improvement over standard models.

5.3 Practical Algorithm

We propose an algorithm for training ResNets using the least action principle by minimizing the kinetic energy. Starting from problem (9) with $p = 2$ and the Euclidean norm, we first discretize the differential equation via a forward Euler scheme, which yields $\phi_{k+1}^x = \phi_k^x + v_k(\phi_k^x)$. The discretized flow v_k is parameterized by a residual block, giving a standard residual architecture. The residual blocks, along with a classifier F, are parametrized by θ. Next, the constraint $(\phi_1)_\sharp\alpha \in S_{F,\mathcal{L}}$ is rewritten as $\mathcal{L}(F, (\phi_1)_\sharp\alpha) = 0$, denoted $\mathcal{L}(\theta) = 0$ below. Finally, as we only have access to a finite set $\mathcal{X}$ of samples x from α, we use a Monte-Carlo approximation of the integral $w.r.t$ the distributions $(\phi_t')_\sharp\alpha$, to obtain:

$$
\begin{aligned}
\min_\theta \quad & \mathcal{C}(\theta) = \sum_{x \in \mathcal{X}} \sum_{k=0}^{K-1} \|v_k(\phi_k^x)\|_2^2 \\
\text{s.t.} \quad & \phi_{k+1}^x = \phi_k^x + v_k(\phi_k^x), \\
& \phi_0^x = x, \ \forall \ x \in \mathcal{X}, \\
& \mathcal{L}(\theta) = 0
\end{aligned}
\tag{10}
$$

Is easy to see that the min-max problem $\min_\theta \max_{\lambda>0} \mathcal{C}(\theta) + \lambda\, \mathcal{L}(\theta)$ yields the same solution, as the first two constraints are satisfied trivially. If the constraint $\mathcal{L}(\theta) = 0$ corresponding to solving the task, which includes the classifier F, is not verified, this will cause the second term to grow unbounded, and the solution will thus be avoided by the minimization. This min-max problem can be solved using an iterative approach, starting from some initial λ_0 and θ_0:

$$
\begin{cases}
\theta_{i+1} = \arg\min_\theta \ \mathcal{C}(\theta) + \lambda_i\, \mathcal{L}(\theta) \\
\lambda_{i+1} = \lambda_i + \tau\, \mathcal{L}(\theta_{i+1})
\end{cases}
\tag{11}
$$

The minimization is done via SGD for a number of steps s, where a step means a batch, starting from the previous parameter value θ_i. This algorithm is similar to Uzawa's algorithm used in convex optimization [27]. In practice, it is more stable to divide the minimization objective in (11) by λ_i, yielding:

Algorithm: Training neural networks with Least Action Principle (LAP-Net)

Input: Training samples, step size τ, number of steps s, initial weight λ_0
Initialization: Initialize the parameters θ_0 and set $i = 0$
while *not converged* **do**
    1. Starting from θ_i, perform s steps of stochastic gradient descent:
       1.1. $\theta_{i+1}^0 = \theta_i$
       1.2. $\theta_{i+1}^l = \theta_{i+1}^{l-1} - \epsilon(\nabla\mathcal{C}(\theta_{i+1}^{l-1})/\lambda_i + \nabla\mathcal{L}(\theta_{i+1}^{l-1}))$ for l from 1 to s
       1.3. $\theta_{i+1} = \theta_{i+1}^s$
    2. Update the weight $\lambda_{i+1} = \lambda_i + \tau\, \mathcal{L}(\theta_{i+1})$ and increment $i \leftarrow i + 1$

Output: Learned parameters θ

While the high non-convexity makes it difficult to ensure exact optimality, we can still have some induced regularity when reaching a "good" local minimum:

Proposition 3. *Suppose* $(F^{\theta^*}, T^{\theta^*})$ *is reached by the optimization algorithm such that* T^{θ^*} *is an* $\epsilon - OT$ *map between* α *and its push-forward[2]. Then we have, with the same notations as in Proposition 2,*

$$\forall x, y \in X \setminus A, \|T^{\theta^*}(x) - T^{\theta^*}(y)\| \leq O(\epsilon + \|x - y\|^{\eta})$$

Proof. We simply write the decomposition:

$$T^{\theta^*}(x) - T^{\theta^*}(y) = T^{\theta^*}(x) - T^{\star}(x) + T^{\star}(x) - T^{\star}(y) + T^{\star}(y) - T^{\theta^*}(y)$$

and use the triangular inequality: the first and third terms are smaller than ϵ by hypothesis while Hölder continuity applies for the second by Proposition 2.   □

This shows that minimizing the transport cost still endows the model with some regularity, even in situations where the global minimum is not reached.

6   Experiments

MNIST Experiments. The base model is a ResNet with 9 residual blocks. Two convolutional layers first encode the image of shape $1 \times 28 \times 28$ into shape $32 \times 14 \times 14$. A residual block contains two convolutional layers, each preceded by a ReLU activation and batch normalization. The classifier is made up of two fully connected layers separated by batch normalization and a ReLU activation. We use an orthogonal initialization [28] with gain 0.01. This and all vanilla models and their training regimes are implemented by following closely the cited papers that first introduced them and our method is added over these training regimes. More implementation details are in Supplementary Material C.3.

When using the entire training set, the task is essentially solved (99.4% test accuracy). We penalize the transport cost as presented in Sect. 5.3, using $\lambda_0 = 5$, $\tau = 1$ and $s = 5$. The performance barely drops (99.3% test accuracy), and we can visualise the preservation of information from the point of view of a pre-trained autoencoder (see Supplementary Material C.1). From the experiments in two dimensions, we suspect that adding the transport cost helps when the training set is small. For performance comparisons, we average the highest test accuracy achieved over 30 training epochs (over random orthogonal weight initial-izations and random subsets of the complete training set). We find that adding the transport cost improves generalization when the training set is very small (Table 1). We see that the improvement becomes more important as the training set becomes smaller and reaches an increase of almost 14% points in the average test accuracy.

---

[2] By this, we mean that $\|T^{\theta^*} - T^{\star}\|_{\infty} \leq \epsilon$ where $T^{\star}$ is the OT map.

Table 1. Average highest test accuracy and 95% confidence interval of ResNet9 over 50 instances on MNIST with training sets of different sizes (in %)

Training set size	ResNet	LAP-ResNet (Ours)
500	90.8, [90.4, 91.2]	**90.9**, [90.7, 91.1]
400	88.4, [88.0, 88.8]	**88.4**, [88.0, 88.8]
300	83.5, [83.0, 84.1]	**86.2**, [85.8, 86.6]
200	74.9, [73.9, 75.9]	**82.0**, [81.5, 82.5]
100	56.4, [54.9, 58.0]	**70.0**, [69.0, 71.0]

CIFAR10 Experiments. We run the same experiments on CIFAR10. The architecture is exactly the same except that the encoder transforms the input which is of shape $3 \times 32 \times 32$ into shape $100 \times 16 \times 16$. For our method, we use $\lambda_0 = 0.1$, $\tau = 0.1$ and $s = 50$. We average the highest test accuracy achieved over 200 training epochs over random orthogonal weight initializations and random subsets of the complete trainset. Here, we find that adding the transport cost helps for all sizes of the trainset (which has 50 000 images in total). The increase in average precision becomes more important as the trainset becomes smaller (Table 2).

Table 2. Average highest test accuracy and 95% confidence interval of ResNet9 over 20 instances on CIFAR10 with training sets of different sizes (in %)

Training set size	ResNet	LAP-ResNet (Ours)
50 000	91.49, [91.40, 91.59]	**91.94**, [91.84, 92.04]
30 000	88.61, [88.47, 88.75]	**89.41**, [89.31, 89.50]
20 000	85.73, [85.59, 85.87]	**86.74**, [86.61, 86.87]
10 000	79.25, [79.00, 79.49]	**80.90**, [80.74, 81.06]
5 000	70.32, [70.00, 70.63]	**72.58**, [72.36, 72.79]
4 000	67.80, [67.55, 68.07]	**70.12**, [69.81, 70.42]

CIFAR100 Experiments. On CIFAR100, results using a ResNet are in Supplementary Material C.2. We also used the ResNeXt [31] architecture: the residual block of a ResNeXt applies $x + \sum_i w_i(x)$ with the functions w_i having the same architecture but independent weights, followed by a ReLU activation. We used the ResNeXt-50-32×4d architecture detailed in [31]. This is a much bigger and state-of-the-art network, as compared with the single representation ResNet used so far. It also extends the experimental results beyond the theoretical framework in three ways: the embedding dimension changes between the residual blocks, a block applies $x_{k+1} = \text{ReLU}(x_k + \sum_i w_{k,i}(x_k))$ and the encoder is no longer fixed. We found that penalizing $\sum_i w_{k,i}(x_k)$ or $x_{k+1} - x_k$ is essentially equivalent.

Table 3 shows consistent accuracy gains as our method (with $\lambda_0 = 1$, $\tau = 0.1$ and $s = 5$) corrects a slight overfitting of the bigger ResNeXt compared to ResNet.

Table 3. Average highest test accuracy and 95% confidence interval of ResNeXt50 over 10 instances on CIFAR100 with training sets of different sizes (in %)

Training set size	ResNeXt	LAP-ResNeXt (Ours)
50 000	72.97, [71.79, 74.14]	**76.11**, [75.32, 76.89]
25 000	62.55, [60.18, 64.92]	**64.11**, [62.25, 65.96]
12 500	45.90, [43.16, 48.67]	**48.23**, [46.39, 50.07]

An important observation is that adding the transport cost significantly reduces the variance in the results. This is expected as the model becomes more constrained and can be seen as an advantage, especially in cases where the results vary more with the initialization (*e.g.* transfer learning). This is illustrated by the width of the 95% confidence intervals in the tables above often becoming narrower when the transport cost is penalized. Finally, we could also have considered a relaxation of the optimization program by considering a fixed weight λ, which provides a simpler and quite competitive benchmark (see Supplementary Material C.2). The training's progress is shown there as well, and we see that the training is not slowed down by our method.

7   Related Work

That ResNets [15,16] are naturally biased towards minimally transforming their input, especially for later blocks and deeper networks, is already shown in [18], which found that earlier blocks learn new representations while later blocks only slowly refine those representations. [14] found that the deeper the network the more its blocks minimally move their input. Both were inspirations for this work.

The ODE point of view of ResNets has inspired new architectures [6,12,20, 25]. Others were inspired by numerical schemes to improve stability, e.g. [6] add a penalty term that encourages the weights to vary smoothly from layer to layer and [36] replicate an Euler scheme and study the effect of diminishing the discretization step-size. More recently, [32] accelerate the training of [7]'s model for generative tasks using the link with dynamical transport. But most often, regularization is achieved by penalization of the weights (e.g. spectral norm regularization [33], smoothly varying weights [6]).

OT theory was used in [29] to analyse deep gaussian denoising autoencoders (not necessarily implemented through residual networks) as transport systems. In the continuous limit, they are shown to transport the data distribution so as to decrease its entropy. Closer to this work, the dynamical formulation of OT is used in [5] for the problem of unsupervised domain translation.

8 Discussion and Conclusion

In this work, we have studied the behavior of ResNets by adopting a dynamical systems perspective. This viewpoint leverages the vast literature in this field.

More specifically, we have analyzed ResNets' complexity through the lens of the transport cost induced by the data displacement across the model's blocks. We find that due to a certain number of factors, this transport cost is biased towards small values. Moreover, this cost is negatively correlated to test accuracy, which has brought us to consider explicitly minimizing it. This leads us to present a novel generic formulation for training neural networks, based on the least action principle, closely related to the problem of Optimal Transport: amongst all the neural networks that correctly solve the task, select the one that transforms the data with the lowest cost. Note that even though we have only considered residual networks as they induce an ODE flow, this framework can be applied to any architecture by considering the static formulation (8) of the problem.

We have proven general results of existence and regularity for models trained within our framework, studied their behaviour in low-dimensional settings when compared to vanilla models and shown their efficiency on standard classification tasks. We also found that the training is stabilized in an adaptive fashion without being slowed down.

An important property of our method which is yet to be tested and is hinted at by the regularity results and by the lower variance in the performances is the robustness of the models, more specifically in adversarial contexts. This will be one important venue of future work. Another interesting avenue of research would be to experiment with alternative transportation costs.

References

1. Belkin, M., Hsu, D., Ma, S., Mandal, S.: Reconciling modern machine-learning practice and the classical bias–variance trade-off. PNAS **116**, 15849–15854 (2019)
2. Belkin, M., Ma, S., Mandal, S.: To understand deep learning we need to understand kernel learning. In: 35th International Conference on Machine Learning (2018)
3. Benamou, J., Brenier, Y.: A computational fluid mechanics solution to the Monge-Kantorovich mass transfer problem. Numerische Mathematik **84**, 375–393 (2000)
4. Bolley, F.: Separability and completeness for the Wasserstein distance. In: Donati-Martin, C., Émery, M., Rouault, A., Stricker, C. (eds.) Séminaire de Probabilités XLI. LNM, vol. 1934, pp. 371–377. Springer, Heidelberg (2008). https://doi.org/10.1007/978-3-540-77913-1_17
5. de Bézenac, E., Ayed, I., Gallinari, P.: Optimal unsupervised domain translation (2019)
6. Chang, B., et al.: Reversible architectures for arbitrarily deep residual neural networks. In: AAAI Conference on Artificial Intelligence (2018)
7. Chen, R., Rubanova, Y., Bettencourt, J., Duvenaud, D.K.: Neural ordinary differential equations. In: Advances in Neural Information Processing Systems (2018)
8. De Palma, G., Kiani, B., Lloyd, S.: Random deep neural networks are biased towards simple functions. In: Advances in Neural Information Processing Systems (2019)

9. Feynman, R.P.: The principle of least action in quantum mechanics. In: Feynman's Thesis - A New Approach to Quantum Theory. World Scientific Publishing (2005)
10. Garcia-Morales, V., Pellicer, J., Manzanares, J.: Thermodynamics based on the principle of least abbreviated action. Ann. Phys. **323**, 1844–1858 (2008)
11. Gray, C.G.: Principle of least action. Scholarpedia (2009)
12. Haber, E., Lensink, K., Treister, E., Ruthotto, L.: IMEXnet a forward stable deep neural network. In: 36th International Conference on Machine Learning (2019)
13. Hastie, T., Tibshirani, R., Friedman, J.: The Elements of Statistical Learning. SSS. Springer, New York (2009). https://doi.org/10.1007/978-0-387-84858-7
14. Hauser, M.: On residual networks learning a perturbation from identity (2019)
15. He, K., Zhang, X., Ren, S., Sun, J.: Identity mappings in deep residual networks. In: Leibe, B., Matas, J., Sebe, N., Welling, M. (eds.) ECCV 2016. LNCS, vol. 9908, pp. 630–645. Springer, Cham (2016). https://doi.org/10.1007/978-3-319-46493-0_38
16. He, K., Zhang, X., Ren, S., Sun, J.: Deep residual learning for image recognition. In: IEEE Conference on Computer Vision and Pattern Recognition (CVPR) (2016)
17. Jacot, A., Gabriel, F., Hongler, C.: Neural tangent kernel: convergence and generalization in neural networks. In: Advances in Neural Information Processing Systems (2018)
18. Jastrzebski, S., Arpit, D., Ballas, N., Verma, V., Che, T., Bengio, Y.: Residual connections encourage iterative inference. In: ICLR (2018)
19. Li, Q., Chen, L., Tai, C., Weinan, E.: Maximum principle based algorithms for deep learning. J. Mach. Learn. Res. **18**, 1–29 (2018)
20. Lu, Y., Zhong, A., Li, Q., Dong, B.: Beyond finite layer neural networks: bridging deep architectures and numerical differential equations. In: 35th International Conference on Machine Learning (2018)
21. Nakkiran, P., Kaplun, G., Bansal, Y., Yang, T., Barak, B., Sutskever, I.: Deep double descent: where bigger models and more data hurt. In: ICLR (2020)
22. Novak, R., Bahri, Y., Abolafia, D.A., Pennington, J., Sohl-Dickstein, J.: Sensitivity and generalization in neural networks: an empirical study. In: ICLR (2018)
23. Peyre, G., Cuturi, M.: Computational Optimal Transport. Now Publishers (2019)
24. Rahaman, N., et al.: On the spectral bias of neural networks. In: 36th International Conference on Machine Learning (2019)
25. Ruthotto, L., Haber, E.: Deep neural networks motivated by partial differential equations. J. Math. Imaging Vis. **62**(3), 352–364 (2019). https://doi.org/10.1007/s10851-019-00903-1
26. Sandler, M., Baccash, J., Zhmoginov, A., Howard, A.: Non-discriminative data or weak model? on the relative importance of data and model resolution. In: International Conference on Computer Vision Workshop (ICCVW) (2019)
27. Santambrogio, F.: Optimal transport for Applied Mathematicians. Birkhäuser (2015)
28. Saxe, A.M., Mcclelland, J.L., Ganguli, S.: Exact solutions to the nonlinear dynamics of learning in deep linear neural network. In: ICLR (2014)
29. Sonoda, S., Murata, N.: Transport analysis of infinitely deep neural network. J. Mach. Learn. Res. **20**, 31–81 (2019)
30. Weinan, E.: A proposal on machine learning via dynamical systems. Commun. Math. Stat. **5**(1), 1–11 (2017). https://doi.org/10.1007/s40304-017-0103-z
31. Xie, S., et al.: Aggregated residual transformations for deep neural networks. In: The IEEE Conference on Computer Vision and Pattern Recognition (CVPR) (2017)
32. Yan, H., Du, J., Tan, V., Feng, J.: On robustness of neural ordinary differential equations. In: ICLR (2020)

33. Yoshida, Y., Miyato, T.: Spectral norm regularization for improving the generalizability of deep learning (2017)
34. Zagoruyko, S., Komodakis, N.: Wide residual networks. In: Proceedings of the British Machine Vision Conference (BMVC). BMVA Press (2016)
35. Zhang, C., Bengio, S., Hardt, M., Recht, B., Vinyals, O.: Understanding deep learning requires rethinking generalization. In: ICLR (2017)
36. Zhang, J., et al.: Towards robust resnet: a small step but a giant leap. In: Twenty-Eighth International Joint Conference on Artificial Intelligence (IJCAI) (2019)

Active Learning

Tackling Noise in Active Semi-supervised Clustering

Jonas Soenen[1,2]($\boxtimes$)(iD), Sebastijan Dumančić[1,2](iD), Toon Van Craenendonck[3](iD), and Hendrik Blockeel[1,2](iD)

[1] Department of Computer Science, KU Leuven, Leuven, Belgium
{jonas.soenen,sebastijan.dumancic,hendrik.blockeel}@cs.kuleuven.be
[2] Leuven.AI, Leuven, Belgium
[3] VITO NV, Unit Health, Mol, Belgium

Abstract. Constraint-based clustering leverages user-provided constraints to produce a clustering that matches the user's expectation. In active constraint-based clustering, the algorithm selects the most informative constraints to query in order to produce good clusterings with as few constraints as possible. A major challenge in constraint-based clustering is handling noise: the majority of existing approaches assume that the provided constraints are correct, while that might not be the case. In this paper, we propose a method to identify and correct noisy constraints in active constraint-based clustering. Our approach reasons probabilistically about the correctness of the user's answers and asks additional constraints to corroborate or correct the suspicious answers. We demonstrate the method's effectiveness by incorporating it into COBRAS, a state-of-the-art method for active constraint-based clustering. Compared to COBRAS and other active-constraint-based clustering algorithms, the resulting system produces better clusterings in the presence of noise.

Keywords: Active learning · Clustering · Semi-supervised learning

1 Introduction

Despite being one of the fundamental data mining tasks, clustering is an inherently subjective problem [3,8]: different users often expect different clusterings of the same dataset. In contrast to a supervised learning setting, there is no way to select the clustering algorithm, its hyper-parameters and a similarity metric based on data only; the user has to try different settings until an informative clustering is found.

Semi-supervised clustering relies on a limited amount of supervision to guide the clustering process towards an informative clustering. Such supervision comes in the form of *pairwise constraints*: a *must-link constraint* between two instances indicates that the instances must be in the same cluster, while *a cannot-link constraint* indicates that the instances must be in different clusters (Fig. 1).

These pairwise constraints are often gathered in advance without knowing to which extent they are useful for the clustering process. Moreover, obtaining constraints usually requires human intervention and can be prohibitively expensive.

© Springer Nature Switzerland AG 2021
F. Hutter et al. (Eds.): ECML PKDD 2020, LNAI 12458, pp. 121–136, 2021.
https://doi.org/10.1007/978-3-030-67661-2_8

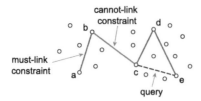

Fig. 1. In semi-supervised clustering, constraints between pairs of instances guide the clustering algorithm towards the desired solution. A must-link constraint indicates that two instances must be in the same cluster (e.g. a and b), a cannot-link constraint indicates that two instances must be in different clusters (e.g. b and c).

In this case, it is beneficial to let the algorithm actively query the user: the algorithm presents the user with a specific pair of instances (*a query*) and the user answers with a 'must-link' or 'cannot-link' *constraint*. This way, an algorithm can pose the most informative queries first and produce a high quality clustering with a substantially smaller number of constraints. This setting is known as *active semi-supervised clustering*.

The prevalent assumption among existing semi-supervised clustering approaches is that all user-provided constraints are correct. However, this assumption is often violated as the user might only have a vague idea of the clustering structure of the data or might simply make mistakes while answering queries. It is thus likely that a user provides inaccurate or even contradicting constraints. These noisy constraints can have a detrimental impact on the quality of the produced clustering. This is especially true for active methods which try to minimize the number of queries by asking only the most informative ones – wrongly answering such a query will have a significant impact on the final clustering.

In this work, we tackle the problem of handling noise in *active* semi-supervised clustering. The core idea behind our approach is to *intentionally introduce redundancy in the constraint set, by means of querying constraints that would form cycles in the available set.* Whether or not the resulting cycles are consistent will then be used to reason about noise in the constraints. For instance, answering a redundant query between instances c and e in Fig. 1 would make a cycle *cdec*. If the user answers this query with a cannot-link constraint, the new cycle is inconsistent: the newly provided constraint states that instances c and e should not be in the same cluster, while the available must-links ($c-b$ and $b-e$) state that they should. This is contradictory information implying the existence of a noisy constraint. On the other hand, if the user answers with a must-link constraint, the new constraint forms a consistent cycle with the must-link constraints (c, d) and (d, e) and, therefore, increases our confidence in the correctness of the constraints in this cycle. To reason about noise in a principled way, we use a probabilistic model that entails the reasoning illustrated above.

To show the effectiveness of our method, we integrate it into COBRAS [13], a state-of-the-art active semi-supervised clustering algorithm. COBRAS will select queries that are informative for clustering; our approach will complement these constraints by selecting queries that help to detect and correct the noisy constraints. We focus on COBRAS because it is one of the approaches most sensitive to noise. In our experiments, we show that with our approach COBRAS becomes significantly more robust to noise.

2 Related Work

While there is a substantial amount of research on constraint-based clustering, including active learning approaches, almost none of it explicitly deals with noisy constraints.

Most existing methods for semi-supervised clustering take pairwise constraints into account by optimising a loss function that includes a penalty for each violated constraint (e.g., PCK-means [1], MPCK-means [2], LCVQE [10] and COSC [11]). Thus, the constraints are interpreted as soft constraints. This makes these methods robust to noise in the constraints, but it does not really *counter* noise. The approach reflects the viewpoint that it may not be possible to satisfy all constraints, but without distinguishing two different reasons for this: because the constraint is simply wrong (noisy) or because a good solution that satisfies it could not be found. Ideally, when a constraint is deemed likely to be noisy, the penalty for violating it should be lower. Moreover, whether the constraint is likely noisy should not be assessed based on how well the constraint fits the inductive bias of the clustering system, as the purpose of these constraints is exactly to change that bias.

A system that does actively try to reduce the effect of noisy constraints is COP-RF [16]. Before the clustering process starts, COP-RF filters out 50% of the constraints that are the least similar to all other constraints of the same type. Besides this, the approach by Yang et al. [15] formulates a maximum entropy model that can learn from noisy pairwise constraints. This model can be used to learn a kernel matrix from the noisy pairwise constraints, which is then used to cluster the dataset using spectral clustering [12].

None of the above approaches are active learners: they consider the set of constraints as fixed, rather than actively looking for new constraints. In an active learning context, dealing with noise is even more important: to minimize the number of queries (instances pairs to be labeled by the user), active learners try to eliminate as much redundancy as possible in the constraints, when in fact such redundancy is crucial in a noisy context. In the context of active constraint-based clustering, the only noise-handling approach we are aware of is the work by Mazumbar and Saha [9]. They propose a method that theoretically guarantees that the correct ground truth is recovered with high probability. However, this method requires multiple constraints per instance in the dataset, so the number of queries becomes orders of magnitude larger than for typical active constraint-based clustering systems. This renders the method practically infeasible for large datasets.

What we propose in this paper, is a method that tries to identify noisy constraints in a way that is independent of the clustering system, and thus could in principle be combined with any of the existing constraint-based clustering methods (whether active or not, though non-active methods become active if this procedure is included). The method employs a small number of additional queries to that aim.

3   Reasoning About Noisy Constraints

In active semi-supervised clustering, the algorithm actively queries the user for the constraints between specific pairs of instances. The constraints obtained from the user are leveraged during clustering to produce high quality results. Usually, active semi-supervised clustering algorithms avoid asking redundant constraints. In a noiseless world, these redundant constraints do not give the algorithm any additional information. However, when there is noise, redundant constraints are crucial to reason about noisy constraints in a manner independent from clustering bias.

Our approach expects a set of constraints and will ask additional redundant queries to reason about the noise in these constraints. The end result is a corrected version of the given constraint set that is likely to be free of noisy constraints. This corrected set of constraints can then be used for clustering. Before we present the details of our approach, we first introduce the necessary terminology.

3.1   Terminology and Background

We will denote a must-link constraint between instances x and y as $ml(x, y)$ and a cannot-link constraint as $cl(x, y)$. Let S be the set of instance pairs for which the user has provided a constraint. Let $\mathcal{U}$ be the set of all constraints that are obtained from the user, such that $\mathcal{U}$ contains $ml(x, y)$ or $cl(x, y)$ for each $(x, y) \in S$. The set $\mathcal{G}$ stands for the set of ground truth constraints about the instance pairs in S. The constraints in $\mathcal{G}$ are unknown; our goal is to recover these constraints from $\mathcal{U}$. If a new query is posed, the query itself is added to S, the user's answer is added to $\mathcal{U}$ and $\mathcal{G}$ will contain the corresponding ground truth constraint. For notational convenience, $\mathcal{C}(x, y)$ will denote the constraint type (ml or cl) of the constraint in the constraint set $\mathcal{C}$ between the instances x and y. A **noisy constraint** is any user-provided constraint that differs from the corresponding ground truth constraint, i.e. $\mathcal{U}(x, y) \neq \mathcal{G}(x, y)$.

For example, suppose the user has provided the constraints in Fig. 1; then $\mathcal{U} = \{ml(a, b), cl(b, c), ml(c, d), ml(d, e)\}$. However, the constraint between d and e is actually a cannot-link constraint in the ground truth constraint set, thus $\mathcal{G} = \{ml(a, b), cl(b, c), ml(c, d), cl(d, e)\}$. In this case, the user-provided constraint $ml(d, e)$ is noisy (because $\mathcal{U}(d, e) = ml$ and $\mathcal{G}(d, e) = cl$).

Given a constraint set, we can deduce additional constraints that are implied by the constraints in the given set:

$$ml(x, y) \wedge ml(y, z) \Rightarrow ml(x, z) \qquad \text{(must-link transitivity)}$$
$$ml(x, y) \wedge cl(y, z) \Rightarrow cl(x, z) \qquad \text{(cannot-link entailment)}$$

We call a constraint set **consistent** if there is no instance pair (x, y) for which both a must-link and a cannot-link constraint are implied. In other words, a consistent constraint set does not contain any contradicting constraints.

If we represent a constraint set as a graph where nodes are instances and edges indicate the corresponding constraints (as in Fig. 1), then the following holds:

Proposition 1. *A constraint set is inconsistent if and only if its graph contains a cycle with exactly one cannot-link edge. We call such a cycle an **inconsistent cycle**.*

Proof. The *if* part in the above claim is trivial. To explain the *only if* part, note that the propagation rules can only derive both ml and cl for the same instance pair (x, y) if there are at least two paths between x and y, one of which is ml-only (hence the ml derivation) and the other contains exactly one cl (two cls would break the derivation chain). These two paths form a cycle with exactly one cannot-link edge.

If the user-provided constraint set $\mathcal{U}$ is inconsistent, there is no clustering where all constraints from $\mathcal{U}$ are satisfied and thus there must be at least one constraint in $\mathcal{U}$ that is noisy. Moreover, there is at least one noisy constraint in each inconsistent cycle in $\mathcal{U}$.

3.2   Overview

The goal of our work is to find a constraint set $\mathcal{C}$, derived from the user-provided constraints $\mathcal{U}$, such that we are reasonably confident that $\mathcal{C}$ corresponds to ground truth $\mathcal{G}$. To reason about the correctness of a constraint set $\mathcal{C}$, we define a probabilistic model that quantifies $P(\mathcal{G} = \mathcal{C} \mid \mathcal{U})$, i.e. the probability that the ground truth is equal to a set of constraints $\mathcal{C}$ given a set of user constraints $\mathcal{U}$. Our goal is to find a constraint set $\mathcal{C}$ such that $P(\mathcal{G} = \mathcal{C} \mid \mathcal{U}) \geq \alpha$ with α a predefined threshold. In the following, we call $P(\mathcal{G} = \mathcal{C} \mid \mathcal{U})$ the **confidence in** $\mathcal{C}$.

When the user-provided constraint set $\mathcal{U}$ is consistent, the most likely constraint set $\mathcal{C}$ is equal to $\mathcal{U}$. When $\mathcal{U}$ is inconsistent, there is noise in $\mathcal{U}$. In this case, the most likely constraint set $\mathcal{C}$ is the consistent constraint set that differs from $\mathcal{U}$ in as few constraints as possible.[1]

However, $\mathcal{C}$ does not automatically satisfy the confidence threshold α. There might not be enough redundancy in the constraint set to support $\mathcal{C}$. Therefore, we select additional redundant queries that complete cycles among the constraints in $\mathcal{U}$ until the confidence of the most likely constraint set $\mathcal{C}$ reaches the threshold. Adding the new constraint to $\mathcal{U}$ changes the most-likely constraint set $\mathcal{C}$. The confidence of the new $\mathcal{C}$ might be higher than that of the previous $\mathcal{C}$, e.g. the new constraint is consistent with the previous $\mathcal{C}$, or lower, e.g. a new inconsistent cycle is detected. By carefully selecting which queries we ask the user, we aim to

---

[1] there might be multiple most likely constraint sets with equal likelihood.

Algorithm 1: Verification procedure

Function verify($\mathcal{U}$):

$\quad \mathcal{C} \leftarrow most\_likely(\mathcal{U})$

$\quad$ **while** $confidence(\mathcal{C}, \mathcal{U}) < \alpha$ **do**

$\qquad new\_con \leftarrow$ ask_informative_redundant_query($\mathcal{U}$)

$\qquad \mathcal{U} \leftarrow \mathcal{U} \cup \{new\_con\}$

$\qquad \mathcal{C} \leftarrow most\_likely(\mathcal{U})$

$\quad$ **return** $\mathcal{C}$

---

reach the confidence threshold with as few queries as possible. Pseudocode for this procedure can be found in Algorithm 1.

In the remainder of this section, we present the probabilistic model used to calculate the confidence of a constraint set. However, because evaluating this model entails summing over all possible constraint sets, exact computation of the confidence is computationally infeasible. Therefore, we introduce a simple approximation of the actual confidence by focusing on a subset of all possible constraint sets and provide a procedure to efficiently enumerate this subset. Lastly, the way we select informative redundant queries is explained.

3.3 Defining the Confidence of a Constraint Set

We develop a probabilistic approach where the value of all user constraints $\mathcal{U}(x, y)$ and the corresponding ground truth constraints $\mathcal{G}(x, y)$ are considered random variables. We use a Bayesian approach that computes $P(\mathcal{G} = \mathcal{C} \mid \mathcal{U})$ using the Bayes rule:

$$P(\mathcal{G} = \mathcal{C} \mid \mathcal{U}) = \frac{P(\mathcal{U} \mid \mathcal{G} = \mathcal{C})P(\mathcal{G} = \mathcal{C})}{\sum_{\mathcal{C}' \in \mathbb{C}} P(\mathcal{U} \mid \mathcal{G} = \mathcal{C}')P(\mathcal{G} = \mathcal{C}')} \tag{1}$$

Where $\mathbb{C}$ is the set of all possible constraint sets about the pairs in S.

For the prior $P(\mathcal{G} = \mathcal{C})$, we assume a uniform distribution over all *consistent* constraint sets $\mathcal{C}$. This implies that all inconsistent constraint sets have a probability of 0, thus we can sum over all consistent constraint sets instead of all constraint sets. As $P(\mathcal{G} = \mathcal{C})$ is equal for every consistent constraint set, it can be ignored (assuming $\mathcal{C}$ is consistent, otherwise $P(\mathcal{G} = \mathcal{C} \mid \mathcal{U})$ is equal to 0).

We assume i.i.d. noise: there is a fixed probability ν that the user answers a query incorrectly. Consequently, for each pair (x, y) and label c, $P(\mathcal{U}(x, y) \neq c \mid \mathcal{C}(x, y) = c) = \nu$. With this assumption in place, the likelihood $P(\mathcal{U} \mid \mathcal{G} = \mathcal{C})$ can be written in function of n, the number of constraints in $\mathcal{U}$, and d, the number of constraints where $\mathcal{U}$ and $\mathcal{C}$ disagree.

$$P(\mathcal{U} \mid \mathcal{G} = \mathcal{C}) = \nu^d (1 - \nu)^{n-d} \tag{2}$$

Note that the assumptions behind the probabilistic model are likely to be violated in practice; for instance, some pairs may be easier to correctly label than others. Despite being violated in practical cases, the model allows us to set formal foundations for tackling noise.

3.4 Approximating the Confidence of a Constraint Set

To calculate the confidence of a certain constraint set $\mathcal{C}$, we have to sum over all consistent constraint sets $\mathcal{C}'$ (see denominator of Eq. 1). Naively enumerating all consistent constraint sets is not practically feasible as the number of consistent constraint sets is, in the worst case, exponential in the number of constraints. Therefore, we introduce an approximation that relies on a subset of the consistent constraint sets that contribute the most to the confidence and provide a procedure that finds these constraint sets efficiently.

High-Likelihood Constraint Sets. Given our noise model, the contribution of a consistent constraint set $\mathcal{C}'$ to the denominator of the confidence is equal to $\nu^d(1-\nu)^{n-d}$ with n the total number of constraints in $\mathcal{U}$ and d the number of pairs for which $\mathcal{U}$ and $\mathcal{C}$ disagree (Eq. 1, 2). For small ν, terms with high d contribute little to the sum. We therefore only sum over consistent constraint sets $\mathcal{C}'$ with $d \leq k$, for some parameter k called the **approximation order**. Higher values for k result in more accurate approximations, but increase the execution time of the algorithm considerably, as the size of the search space is exponential in k.

This approximation is accurate if it captures most of the probability mass, that is, when there is a high probability that the number of noisy constraint in $\mathcal{U}$ is smaller than the approximation order k. From our noise assumption follows that the number of noisy constraints in $\mathcal{U}$ follows a binomial distribution distributed with $|\mathcal{U}|$ trials and success probability ν.

$$P(\#\text{noisy constraint in } \mathcal{U} \leq k) = \sum_{d=0}^{k} \binom{|\mathcal{U}|}{d} \nu^d (1-\nu)^{|\mathcal{U}|-d} \tag{3}$$

If k is fixed, $P(\#\text{noisy constraints in } \mathcal{U} \leq k)$ increases when the size of $\mathcal{U}$ decreases or the noise probability ν decreases. Thus our approximation is most accurate for small constraint sets and low noise probabilities. If the approximation order k is too small relative to $|\mathcal{U}|$ and ν, the approximation overestimates the confidence and our approach will ask less redundant constraints than actually necessary to reach the desired confidence level.

To make our approximation more accurate for bigger constraint sets we employ two tricks: we use a dynamically growing *effective approximation order* and verify the user-provided constraints in *batches*.

When using a *static* approximation order k, the approximation becomes less accurate as the number of identified noisy constraints increases. After identifying k noisy constraints, all consistent constraint sets with more than k noisy constraints are not considered in the approximation of the confidence. This results in a highly overestimated confidence and causes the procedure to never identify more than k noisy constraints.

In order to resolve this, we use a *dynamically growing* **effective approximation order** k' instead of the static approximation order k. The effective approximation order k' is determined as the sum of the number of noisy constraints in the current most likely constraint set and the approximation order k. The effective approximation order thus grows each time a noisy constraint is detected in the user-provided constraints $\mathcal{U}$. This ensures that the approximation can always reason about k additional noisy constraints on top of the noisy constraints that have already been identified.

Additionally, the user-provided constraints can be verified in batches to enhance the accuracy of the approximation; instead of verifying all user-provided constraints $\mathcal{U}$ in one single go, $\mathcal{U}$ is divided in several smaller batches. One after the other, each of these batches is added to a constraint set $\mathcal{U}'$. After adding a batch to $\mathcal{U}'$, the full set is verified. Because the preexisting constraints in $\mathcal{U}'$ have been verified in the previous iteration, it is unlikely that these preexisting constraints contain undetected noisy constraints. The number of noisy constraints in the newly added batch $\mathcal{B}$ follows the same distribution as in Eq. 3 but proportional to the $|\mathcal{B}|$ instead of $|\mathcal{U}|$, with $|\mathcal{B}| < |\mathcal{U}|$. Therefore, the accuracy of the approximation will mostly depend on the size of the newly added batch.

For several active semi-supervised clustering approaches it is actually natural to verify the constraints in batches. NPU [14] and COBRAS [13] produce a new intermediate clustering every couple of queries. Therefore, it makes sense to verify the constraints right before each intermediate clustering is produced as to make sure that none of these intermediate clusterings is based on a noisy constraint.

Enumerating Consistent Constraint Sets. Algorithm 2 outlines a simple yet efficient recursive procedure to enumerate all consistent constraint sets up to the effective approximation order k'. These consistent constraints sets are used to approximate the confidence of a constraint set, determine the most likely constraint set and select redundant queries. In each recursive call, one constraint in a constraint set $\mathcal{A}$ is flipped until it reaches the desired effective approximation order. If the constraint set $\mathcal{A}$ is consistent, each of the constraints in $\mathcal{A}$ is considered as a candidate to be flipped next. If the constraint set $\mathcal{A}$ contains an inconsistent cycle, the only way this constraint set can become consistent is to flip one of the constraints involved in the inconsistent cycle. Therefore, only the constraints involved in the inconsistent cycle have to be considered as candidates to be flipped next. Every consistent constraint set that is encountered during this procedure is stored. To ensure that every consistent constraint set is only encountered once, a set of constraints *flipped* is maintained that keeps track of all the constraints that have already been flipped and should not be flipped again. Figure 2 shows part of the search tree explored by Algorithm 2 for an example scenario.

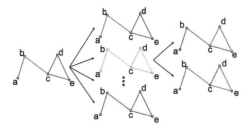

Fig. 2. An example of an incomplete search tree to find consistent constraint sets. The starting constraint set is consistent, hence every constraint is a candidate for flipping. The second constraint set at depth 1 is inconsistent, only the constraints between instance pairs (c, d) and (c, e) need to be considered as candidates $((d, e)$ was flipped in a previous recursive call).

Finding Inconsistent Cycles. To be able to use Algorithm 2, we need to detect when a constraint set is inconsistent and, if so, find an inconsistent cycle in the constraint set.

To check if a constraint set is inconsistent, we use a procedure similar to the feasibility test with must-link and cannot-link constraints of Davidson and Ravi [4]. The procedure maintains a set of must-link components, groups of instances that are interconnected by must-link constraints, and checks whether there are two instances within the same must-link component that are connected by a cannot-link constraint. If such a pair of instances is found, this cannot-link forms an inconsistent cycle with some must-links from this must-link component and thus the constraint set is inconsistent. To reconstruct the inconsistent cycle, we start from the two instances connected by the cannot-link and search for a path between them that only contains must-link constraints.

3.5 Selecting Redundant Queries

Now that we have all pieces to calculate the confidence of the most likely constraint set $\mathcal{C}$, we still have to select an informative query that will help us to reach the confidence threshold.

We distinguish two different scenarios in which a redundant query needs to be selected.

1. If there is only one most likely constraint set $\mathcal{C}$, we aim to increase the confidence of this constraint set. Therefore, we query the pair (x, y) for which the total likelihood of consistent constraint sets that imply a different constraint value for (x, y) than $\mathcal{C}$ is maximal. As before, we only use the consistent constraint sets up to the current effective approximation order k'.
2. If there are multiple most likely constraint sets, we query a pair that will reduce the amount of most likely constraint sets. As such, we aim to query a pair (x, y) for which the amount of most likely constraint sets that imply

---

Algorithm 2: Procedure to enumerate consistent assignments

Function solve($\mathcal{A}$, *flipped*):

    if *unsat_user_constraints($\mathcal{A}$)* $> k'$ **then**
        | **return** $\emptyset$

    solutions $\leftarrow \emptyset$
    if $\mathcal{A}$ *is consistent* **then**
        | *solutions* $\leftarrow$ *solutions* $\cup \{\mathcal{A}\}$
        | *candidates* $\leftarrow \mathcal{U} \setminus flipped$
    else
        | *cycle* $\leftarrow$ find_inconsistent_cycle($\mathcal{A}$)
        | *candidates* $\leftarrow$ *cycle* $\setminus flipped$
    for *con* **in** *candidates* **do**
        | *flipped* $\leftarrow$ *flipped* $\cup \{con\}$
        | $\mathcal{A}' \leftarrow \mathcal{A} \setminus con \cup \{con.flip()\}$
        | *solutions* $\leftarrow$ *solutions* $\cup$ solve($\mathcal{A}'$, *flipped*)
    return *solutions*

---

must-link is approximately equal to the amount of most likely constraint sets that imply cannot-link. The answer to this query will thus agree with approximately half of these constraint sets and disagree with the other half. Adding the answer of this query to $\mathcal{U}$ causes the confidence of the agreeing constraint sets to increase and the confidence of the disagreeing constraint sets to decrease. This gives rise to a binary-search-like procedure which is repeated until only one most likely constraint set remains.

It is possible that, at some point, no interesting queries are left (e.g. all possible instance pair within a single must-link component have been queried) and the confidence of the most likely constraint set is still lower than α. In this case, the most likely constraint set (or one of the most likely constraint sets if there are multiple) is assumed to be correct.

4   Integration in COBRAS

COBRAS [13] is a state-of-the-art active semi-supervised clustering algorithm. Its strengths are that it is time-efficient, query-efficient, and that it provides intermediate results. Its query-efficiency, however, naturally makes it more sensitive to noise. Moreover, COBRAS has no noise handling mechanism, as the algorithm aims to satisfy all of the user-provided constraints. Therefore, it is a good candidate for testing the effectiveness of our noise correcting approach.

COBRAS is an iterative algorithm where each iteration consists of a splitting phase and a merging phase. In its splitting phase, COBRAS identifies so-called

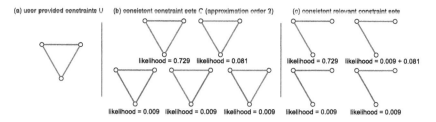

Fig. 3. Illustration of how the consistent constraints sets are marginalized over the irrelevant constraints (dotted lines) to only retain the relevant constraints (full lines). (a) shows the given user-provided constraints (b) shows all consistent constraint sets and their likelihood derived from the user-provided constraints with approximation order 2 (c) shows the marginalized constraint sets with their likelihood.

super-instances, which are groups of instances that are assumed to belong to the same cluster. During splitting, a small amount of queries are used to estimate the number of super-instances that are needed. The new super-instances (groups of instances) are then merged into clusters by querying constraints between different super-instances. Two super-instances are merged into the same cluster if the constraint between them is a must-link; two super-instances are not merged together if the constraint between them is a cannot-link. It is after this phase that our noise-correction approach is deployed.

After each merging phase, our approach tries to recover the correct constraints $\mathcal{C}$ from the user-provided constraints $\mathcal{U}$. If no noisy constraints are detected (i.e. $\mathcal{C} = \mathcal{U}$), COBRAS can safely continue with the next iteration. Otherwise, the current clustering is based on at least one noisy constraint and has to be corrected. Therefore, the merging phase of COBRAS is repeated with the corrected constraint set $\mathcal{C}$ instead of $\mathcal{U}$. We refer to the thus obtained noise-robust version of COBRAS as nCOBRAS.

Reasoning on a Subset of Relevant Constraints. In COBRAS, only a subset of all the constraints provided by the user is used to construct the next intermediate clustering (i.e. only the constraints between the current super-instances are used). Thus, it makes sense to only verify the constraints from this relevant subset and not waste queries on obtaining the correct values of the irrelevant constraints. However, the irrelevant constraints might still complete cycles with relevant constraints and thus might still provide useful information. Therefore, to calculate the confidence of a subset of relevant constraints $\mathcal{C}_{rel}$, we reason on the full set of constraints and marginalize over the irrelevant constraints $\mathcal{C}_{irrel}$.

$$P(\mathcal{G}_{rel} = \mathcal{C}_{rel} \mid \mathcal{U}) = \sum_{\mathcal{C}_{irrel} \in \mathbb{C}_{irrel}} P(\mathcal{G} = \mathcal{C}_{rel} \cup \mathcal{C}_{irrel} \mid \mathcal{U}) \qquad (4)$$

Where $\mathbb{C}_{irrel}$ is the set of all possible constraint sets involving the same pairs as the irrelevant constraints

To ensure that the redundant query selection procedure only selects queries that are informative to verify the relevant constraints, all consistent constraint sets enumerated by Algorithm 2 are marginalized over the irrelevant constraints. The active query selection procedure is then executed on the marginalized constraint sets that only contain the relevant constraints (Fig. 3).

5   Experiments

We compare the clustering performance of several active semi-supervised clustering approaches in the presence of varying amounts of noise. We are especially interested to investigate **(Q1)** whether nCOBRAS is indeed more noise robust than COBRAS and **(Q2)** how nCOBRAS performs compared to existing active constraint-based clustering approaches.

5.1   Algorithms

We compare nCOBRAS[2] to the following active semi-supervised clustering algorithms:

- COBRAS [13], the original COBRAS algorithm with no noise handling mechanism. We use the code provided by the authors.[3]
- NPU [14] is an active query selection scheme that can be used with any semi-supervised clustering algorithm, we use it with:
  - MPCK-means [2], a constrained modification of k-means that uses a modified objective function and allows constraint violation. We use the implementation in the wekaUT package[4].
  - COSC [11], a constrained extension of spectral clustering that optimizes a modified objective function. We use the fast implementation of the code provided by the authors[5],[6].

NPU with MPCK-means (NPU_MPCK) and NPU with COSC (NPU_Cosc) require the desired number of clusters K before clustering. In our experiments the true value of K (as indicated by the class labels) is given to these algorithms. This gives NPU-MPCK and NPU-Cosc an advantage over COBRAS and nCOBRAS, which do not require the number of clusters.

nCOBRAS needs three hyperparameters to be set: an estimated noise probability ν, the confidence threshold α and the approximation order k. In our experiments, we provide nCOBRAS with the true noise probability for ν, except when there is no noise: there, we set ν to 0.05 instead of 0, so that we can get an idea of the cost of noise-handling when it is in fact unnecessary. The value of the confidence threshold α is set to 0.95 and an approximation order of 3 is used.

---

[2] https://github.com/magicalJohn/noise_robust_cobras.

[3] https://dtai.cs.kuleuven.be/software/cobras/.

[4] https://www.cs.utexas.edu/users/ml/risc/code/.

[5] https://www.ml.uni-saarland.de/code/cosc/cosc.htm.

[6] In some cases the code provided by the authors fails to produce a clustering; if this happens, we use the clustering from the previous NPU iteration instead.

5.2 Experimental Methodology

We perform 3 times 10-fold cross-validation and report the average clustering quality. The algorithms cluster the full dataset, but only query constraints
between instances that are both in the training set. The clustering quality is
determined by calculating the adjusted rand index (ARI) [7] on the instances in
the test set. The ARI measures the similarity between the produced clustering
and the ground truth clustering. A random clustering has an expected ARI of
0, while a clustering identical to the ground truth clustering has an ARI of 1.

To guarantee that COBRAS and nCOBRAS do not query any test instances
during clustering, we ensure that all super-instance representatives are in the
training set. For NPU, the selection of the most informative instance x^* is modified to exclude points not in the training set.

We also report the average aligned rank (AAR) [5,6], which ranks the overall
performance of each of the algorithms. The lower the AAR the better the algorithm performs comparatively to the other algorithms. To calculate the average
aligned rank, for each dataset d and each algorithm a, compute the difference
between the average ARI of algorithm a on dataset d and the average ARI
achieved by all algorithms on dataset d. The resulting differences are ranked
from high to low. The AAR for an algorithm is the average of the positions of
its entries in the sorted list.

The noisy constraints are generated according to the noise assumption presented in Sect. 3: for an experiment with $x\%$ of noise, each of the queries has a
probability of $x\%$ that the query is answered incorrectly.

5.3 Datasets

For our experiments, we use 21 clustering tasks defined over 17 datasets as in Van
Craenendonck et al. [13]. These datasets include 15 UCI classification datasets:
iris, wine, dermatology, hepatitis, glass, ionosphere, optdigits389, ecoli, breast-
cancer-wisconsin, segmentation, column 2C, parkinsons, spambase, sonar and
yeast. For these datasets, the class labels indicate the target clustering. Additionally, we consider 4 clusterings of the CMU faces dataset and cluster 2 different
subsets of the 20 newsgroup text dataset. For details about the preprocessing
and the clustering tasks, we refer to Van Craenendonck et al. [13].

5.4 Results

We ran each of the algorithms for 200 queries with varying amounts of noise
(0%, 5% and 10%). For each noise value, we calculate the average ARI and the
average aligned rank of each algorithm over all clustering tasks. The results are
shown in Fig. 4 and Fig. 5.

(Q1) The results (Fig. 4) confirm that nCOBRAS is less sensitive to
noise than COBRAS. The average clustering quality of COBRAS significantly
decreases when the amount of noise increases: with 10% of noise, its ARI
decreases by half. In contrast, the average clustering quality of nCOBRAS is

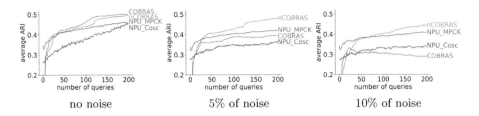

Fig. 4. Average ARI comparison of the different algorithms in the presence of varying amounts of noise (higher is better).

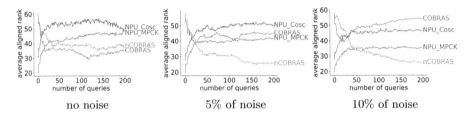

Fig. 5. Average aligned rank of the different algorithms in the presence of varying amounts of noise (lower is better).

relatively stable over all amounts of noise. Noise causes nCOBRAS' improvement to slightly slow down: with more noise, nCOBRAS has to ask more redundant queries to verify the constraints. Moreover, with no noise nCOBRAS' performance is very close to that of COBRAS which does not reason about the correctness of the constraints.

(**Q2**) The results also indicate that nCOBRAS outperforms both competitors, NPU-Cosc and NPU-MPCK, for all levels of noise. NPU-MPCK outperforms nCOBRAS for small amounts of constraints; this is due to the combined unsupervised and semi-supervised objectives of MPCK-means which allows it to get to a better initial clustering than nCOBRAS. Both NPU-MPCK and NPU-Cosc are less sensitive to noise than COBRAS.

5.5   Evaluation of Noise Detection

To show that our approach actually finds and corrects the majority of the noisy constraints, we calculate the average precision and average recall achieved by our noise detection algorithm when correcting a relevant constraint set with 5% of noise. For the majority of the datasets, the average recall is higher than 93% and the average precision is higher than 95%. This means our approach identifies almost all noisy constraints in the relevant constraints, and almost all constraints flagged by our approach are indeed noisy. For datasets where the average batch size is high, the precision and recall are typically lower; in this case our approximation fails to produce a reasonable estimate of the confidence. For details on the experimental set-up and full results, we refer to the appendix.

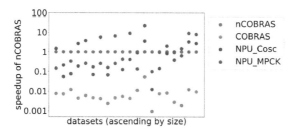

Fig. 6. Ratio of competitor to nCOBRAS average runtime for each of the 21 clustering tasks (sorted ascending by size) when there is 5% noise and the algorithms are run for 200 queries.

5.6 Runtime

Figure 6 shows the ratio of the average run time of each competitor to the average runtime of nCOBRAS for each of the 21 clustering tasks with 5% of noise for 200 queries. COBRAS is in both cases by far the fastest algorithm. nCOBRAS' noise identification procedure makes it substantially slower than COBRAS, but it is still about as fast as the other systems.

6 Conclusion

In this paper, we have presented a novel approach to handling noise in active semi-supervised clustering. Our method reasons probabilistically about the correctness of constraints and can correct noisy constraints with the help of additional redundant queries. We integrated our method into COBRAS and have shown that this makes COBRAS significantly more robust to noise. Under noisy conditions, the noise-robust variant of COBRAS produces better clusterings compared to two other active constraint-based clustering approaches.

Acknowledgments. This research received funding from the Flemish Government under the "Onderzoeksprogramma Artificiële Intelligentie (AI) Vlaanderen" programme. SD is supported by the Research Foundation-Flanders (FWO).

References

1. Basu, S., Banerjee, A., Mooney, R.J.: Active semi-supervision for pairwise constrained clustering. In: Proceedings of the 2004 SIAM International Conference on Data Mining, pp. 333–344 (2004)
2. Bilenko, M., Basu, S., Mooney, R.J.: Integrating constraints and metric learning in semi-supervised clustering. In: Twenty-First International Conference on Machine Learning, ICML 2004 (2004)
3. Caruana, R., Elhawary, M., Nguyen, N., Smith, C.: Meta clustering. In: Sixth International Conference on Data Mining (ICDM 2006), pp. 107–118 (2006)

4. Davidson, I., Ravi, S.S.: Agglomerative hierarchical clustering with constraints: theoretical and empirical results. In: Jorge, A.M., Torgo, L., Brazdil, P., Camacho, R., Gama, J. (eds.) PKDD 2005. LNCS (LNAI), vol. 3721, pp. 59–70. Springer, Heidelberg (2005). https://doi.org/10.1007/11564126_11

5. García, S., Fernández, A., Luengo, J., Herrera, F.: Advanced nonparametric tests for multiple comparisons in the design of experiments in computational intelligence and data mining: experimental analysis of power. Inf. Sci. **180**(10), 2044–2064 (2010)

6. Hodges, J., Lehmann, E.L., et al.: Rank methods for combination of independent experiments in analysis of variance. Ann. Math. Stat. **33**(2), 482–497 (1962)

7. Hubert, L., Arabie, P.: Comparing partitions. J. Classif. **2**, 193–218 (1985)

8. von Luxburg, U., Williamson, R.C., Guyon, I.: Clustering: science or art? In: Proceedings of ICML Workshop on Unsupervised and Transfer Learning. In: Proceedings of Machine Learning Research, vol. 27, pp. 65–79 (2012)

9. Mazumdar, A., Saha, B.: Clustering with noisy queries. In: Advances in Neural Information Processing Systems, pp. 5788–5799 (2017)

10. Pelleg, D., Baras, D.: K-means with large and noisy constraint sets. In: Kok, J.N., Koronacki, J., Mantaras, R.L., Matwin, S., Mladenič, D., Skowron, A. (eds.) ECML 2007. LNCS (LNAI), vol. 4701, pp. 674–682. Springer, Heidelberg (2007). https://doi.org/10.1007/978-3-540-74958-5_67

11. Rangapuram, S.S., Hein, M.: Constrained 1-spectral clustering. In: AISTATS, vol. 30, p. 90 (2012)

12. Shi, J., Malik, J.: Normalized cuts and image segmentation. Departmental Papers (CIS), p. 107 (2000)

13. Van Craenendonck, T., Dumančić, S., Van Wolputte, E., Blockeel, H.: COBRAS: interactive clustering with pairwise queries. In: Duivesteijn, W., Siebes, A., Ukkonen, A. (eds.) IDA 2018. LNCS, vol. 11191, pp. 353–366. Springer, Cham (2018). https://doi.org/10.1007/978-3-030-01768-2_29

14. Xiong, S., Azimi, J., Fern, X.Z.: Active learning of constraints for semi-supervised clustering. IEEE Trans. Knowl. Data Eng. **26**(1), 43–54 (2013)

15. Yang, T., Jin, R., Jain, A.K.: Learning from noisy side information by generalized maximum entropy model. In: Proceedings of the 27th International Conference on Machine Learning, ICML 2010, pp. 1199–1206 (2010)

16. Zhu, X., Loy, C.C., Gong, S.: Constrained clustering with imperfect Oracles. IEEE Trans. Neural Netw. Learn. Syst. **27**(6), 1345–1357 (2016)

A Taxonomy of Interactive Online Machine Learning Strategies

Agnes Tegen[✉], Paul Davidsson, and Jan A. Persson

Internet of Things and People Research Center,
Department of Computer Science and Media Technology, Malmö University,
Malmö, Sweden
`agnes.tegen@mau.se`

Abstract. In interactive machine learning, human users and learning algorithms work together in order to solve challenging learning problems, e.g. with limited or no annotated data or trust issues. As annotating data can be costly, it is important to minimize the amount of annotated data needed for training while still getting a high classification accuracy. This is done by attempting to select the most informative data instances for training, where the amount of instances is limited by a labelling budget. In an online learning setting, the decision of whether or not to select an instance for labelling has to be done on-the-fly, as the data arrives in a sequential order and is only valid for a limited time period. We present a taxonomy of interactive online machine learning strategies. An interactive learning strategy determines which instances to label in an unlabelled dataset. In the taxonomy we differentiate between interactive learning strategies when the computer controls the learning process (active learning) and those when human users control the learning process (machine teaching). We then make a distinction between what triggers the learning: active learning could be triggered by uncertainty, time, or randomly, whereas machine teaching could be triggered by errors, state changes, time, or factors related to the user. We also illustrate the taxonomy by implementing versions of the different strategies and performing experiments on a benchmark dataset as well as on a synthetically generated dataset. The results show that the choice of interactive learning strategy affects performance, especially in the beginning of the online learning process, when there is a limited amount of labelled data.

Keywords: Interactive machine learning · Online learning · Active learning

1 Introduction

The performance of a machine learning method is dependent on the labelled data it is trained on. While the amount of data in the world is increasing at an accelerating rate, the act of labelling that data is often costly. In active learning, a limited number of data instances are chosen for labelling, as compared to

F. Hutter et al. (Eds.): ECML PKDD 2020, LNAI 12458, pp. 137–153, 2021.
https://doi.org/10.1007/978-3-030-67661-2_9

all of them [14]. The aim is to achieve an equivalent performance to when all instances are labelled, by choosing the limited number of instances wisely. The active learning strategy selects which instances should be included and an oracle is then queried to provide correct labels for those instances. In machine teaching, the goal is also to achieve a high performance with a limited number of labelled samples, but with the teacher selecting the labelled instances to train the learner on [18,19]. Often the oracle, in the case of active learning, or the teacher, in the case of machine teaching, is a human interacting with the system and we will henceforth adopt the term user to describe both.

In settings where data arrive as streams, the selection has to be done differently compared to a pool-based setting. In an online learning scenario, the data arrives in a specific sequential order and the current data instance is the only one that can be labelled at a given point in time. Furthermore, if the machine learning method is supposed to produce estimations in real-time, the computations might have to be done at the edge where algorithms often have restrictions on computational complexity. Most work where interactive machine learning strategies are employed focus on active learning, especially strategies based on the uncertainty of the learner. Several works compare different strategies, but do often include only one type of interactive machine learning strategies. Particularly, machine teaching in a setting with streaming data is an area which needs further exploration.

In this work we present a taxonomy of interactive online machine learning strategies. We include active learning strategies, where the learner queries the user, and machine teaching strategies, where the user provides labels by their own initiative according to a specific strategy, as well as hybrid versions. Some of the strategy types included in the taxonomy are abstractions of previously proposed strategies, while others are novel, at least in the given problem setting. We implement versions of the presented strategies and compare them through experiments on a benchmark dataset and a synthetically created dataset.

2   Related Work

Fu et al. surveyed and organized existing work into two main categories based on whether the active learning strategy solely is based on the uncertainty of independent and identically distributed instances or if it also takes into account instance correlations [5]. Experiments comparing time complexity of the strategies discussed were carried out. While active learning for streaming data was mentioned as an emerging application and the related challenges discussed, the strategies are presented in a pool-based setting, where all unlabelled data is available at the same time. The strategies categorized as only being based on uncertainty can apply to an online learning setting, but the strategies presented where the instance correlations are included tend to be less suitable. First, they generally have a higher time complexity, which often is not suitable when estimations have to be produced in real-time. Second, they take into account a data instance's correlation to other instances, which means that a batch of data has to be collected before the strategy can be properly employed.

Active learning within a streaming data setting was explored by Miu et al. [13]. The authors present an online active learning framework, where user-provided annotations can be collected in real-time and used for Human Activity Recognition tasks. The framework is compared to more naïve annotation methods and evaluated on benchmark datasets. User studies were also carried out, where the framework was implemented in a mobile app through which the participants could provide labels. The user cannot provide labels by their own initiative however, as in machine teaching, but only when queried by the active learning strategy which is based on uncertainty of the learner.

Lughofer investigated active learning for data streams using evolving fuzzy classifiers [10]. Two different strategies of active learning are presented and tested, based on *conflict* and *ignorance*. *Conflict* means that the new data sample is on the border of previously defined classes, while *ignorance* means that the new data sample is far outside the previously defined borders. Both *conflict* and *ignorance* based learning can be considered strategies that are based on the uncertainty of the learning model. The paper does not explore a user providing labels by their own initiative.

A user with proactive capacities was studied by Chen et al., as they explored the role of adaptivity in algorithmic machine teaching [3]. In the work, the authors studied the teaching of version space learners in an interactive setting through experiments. An adaptive teacher, where the learner's hypothesis at every time step can be observed, was compared to a non-adaptive teacher, where only the initial hypothesis of the learner is known i.e. no feedback is received during teaching. In the first one, the teacher can proactively adapt the information provided to the learner on-the-fly, based on the performance. This can be seen as an interactive learning strategy where the user provides a label when the estimation of the learner is incorrect. The latter one can be seen as a pool-based setting where all examples are constructed beforehand. In the given setup, the learner did not have the possibility to query the teacher for labels however, and can only train on the instances provided by the teacher.

3 Interactive Online Machine Learning Strategies

The different categories of sampling strategies presented below are all interactive online learning strategies. Interactive learning includes both active learning, where the learner queries the user, who in turn responds with a label and machine teaching, where the user proactively provides labels based on a selection strategy. Online learning means that the data stream is received in a single-pass manner, i.e. the strategy is processing one data instance at a time. Special cases of sampling strategies that are meant to be combined with specific machine learning methods (e.g.. evolving models [11]) are not included. In Fig. 1, a visualization of the taxonomy can be seen.

In interactive online machine learning, several issues have to be taken into consideration that might not be as relevant for interactive machine learning in a pool-based setting. The most prominent issue is the one-pass manner in which the data instances arrive. Within an online setting a decision must be made for each new instance as it appears, whether or not to query for (in the case of active learning) or provide (in the case of machine teaching) the label. Streaming data also needs another approach for dealing with labelling expenses compared to a pool-based setting, since the incoming data instances in theory could be infinite. In the experiments the labelling expenses, unless otherwise stated, are calculated based on the the labelling status of recent instances. Other suggestions to handle the calculation over of budget with streaming data are presented for instance by Kottke et al. [8] and Žliobaitė et al. [20].

For each type of strategy, an algorithm illustrates the structure in pseudo code. It is assumed in the following descriptions that the user never is fallible or reluctant, i.e. that the user always will provide a correct label in accordance with the interactive machine learning strategy in question. Each algorithm has a data stream $X = \{x_0, x_1, ...\}$, a labelling budget B and classifier Ψ as input. In the case of additional input, it is further described for that particular strategy. The data stream contains a possibly infinite sequence of data instances, where x_0 is the first instance received by the learning algorithm. The labelling budget B determines the maximum ratio of incoming samples that can be labelled, with the restriction $0 < B \leq 1$, where $B = 1$ means that all incoming instances can be labelled. The classifier Ψ specifies which machine learning method is used along with any parameter settings. The classifier in its initial state can be pre-trained, if applicable data exists, or it might not have been previously trained, in the case of a cold-start scenario. The aim for the classifier is to accurately classify a state y. The classifier produces an estimate of the state $\hat{y}_i$ for each data instance x_i. The user provides a label y_i in accordance with the interactive learning strategy.

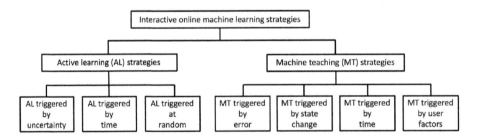

Fig. 1. Taxonomy of the interactive online machine learning strategies.

AL Triggered by Uncertainty. *AL triggered by uncertainty* is the most commonly used interactive learning strategy. The learner queries the user for labels for the data instances where it is least certain how to label [14]. Different implementations of the strategy exist and a lot of work in the area focus on comparing and developing strategies where the querying is triggered by uncertainty. As algorithm 1 shows, the decision whether to query or not is based on the output from an uncertainty measurement function $U(\hat{Y}_i, x_i)$, where $\hat{Y}_i$ is the possible classes and x_i is the current data sample, which is compared to a threshold value θ. Depending on which strategy is chosen for the function U and which classifier Ψ is used, $\hat{Y}_i$ might contain everything from only the most probable class $\hat{y}_i$ to all the possible classes ranked by probability. The implementation is especially straight-forward for probabilistic classifiers, but can be used with any classifier where an appropriate uncertainty measurement can be defined. Different versions of *AL triggered by uncertainty* include for example *least confident, margin sampling, entropy* and others where a region of uncertainty is explored [4]. In *least confident*, the learner queries when the probability of the prediction is low, *margin sampling* when the margin between the most probable and the second most probable class is low and in *entropy* when the combined probability of all possible classes is low.

Algorithm 1. AL triggered by uncertainty

Input: data stream $X = \{x_0, x_1, ...\}$, labelling budget B, classifier Ψ, threshold θ, step size s

$i \leftarrow 0$

repeat

    receive next data sample x_i

    $\hat{y}_i \leftarrow \Psi(x_i)$

    if $U(\hat{Y}_i, x_i) < \theta$ **and** $\hat{b} < B$ **then**

        ask user about label y_i for x_i

        $\Psi \leftarrow$ incremental learning procedure$((x_i, y_i))$

        $\theta \leftarrow \theta - s$

    else

        $\theta \leftarrow \theta + s$

    end if

    update labelling expenses $\hat{b}$

    $i \leftarrow i + 1$

until end of data stream

In the experiments described in Sect. 4 we include *least confident*, where the uncertainty function becomes $U(\hat{Y}_i, x_i) = 1 - P(\hat{y}_i \mid x_i)$. The inverted probability is compared to a threshold θ which determines whether the classifier is uncertain regarding its prediction. The value of the threshold is set initially, but can be time-variable [20], meaning that it can be increased or decreased with the step size s depending on whether or not a query was posed. Note that if $s = 0$, the threshold is constant.

A sliding window is used to calculate the current labelling expenses $\hat{b}$. The window contains the labelling status of the most recent data instances. Each element of the window is either a 0, if there was no label provided by the user, or 1, if the user did provide a label. The labelling expenses are calculated by computing the ratio of how many instances in the window has had an accompanying label provided by the user. This value is compared to the labelling budget B to decide whether the strategy can query the user. The labelling expenses are updated by shifting all elements of the window in accordance with a queue system, i.e. when a new element is entered the oldest one is discarded.

AL Triggered by Time. In *AL triggered by time*, the algorithm queries at a given time interval. The sampling rate is calculated based on the labelling budget B, as can be seen in Algorithm 2. A counter, c, is updated for every new data instance that arrives to keep track of when it is time to query. This strategy can be useful for instance as an option if another strategy is not using up the allowed budget or in a cold start scenario. In the latter case, the classifier has no access to any labelled data in the beginning, which means that *AL triggered by uncertainty* might not work, depending on how the uncertainty measurement is defined. Another scenario where this strategy would be useful is if concept drift is present [6]. Concept drift means that the underlying statistical properties of the data is changing over time. *AL triggered by uncertainty* might be inadequate in this case, since the measurement of uncertainty is based on older data.

Algorithm 2. AL triggered by time

Input: data stream $X = \{x_0, x_1, ...\}$, labelling budget B, classifier Ψ
$c \leftarrow 1$
$i \leftarrow 0$
repeat
   receive next data sample x_i
   $\hat{y}_i \leftarrow \Psi(x_i)$
   if $\frac{1}{B} \leq c$ **then**
      ask user about label y_i for x_i
      $\Psi \leftarrow$ incremental learning procedure$((x_i, y_i))$
      $c \leftarrow c - \frac{1}{B}$
   end if
   $c \leftarrow c + 1$
   $i \leftarrow i + 1$
until end of datastream

AL Triggered at Random. Similar to *AL triggered by time*, *AL triggered at random* does not take the data instance itself or the classifier into account when deciding whether or not to query. Unlike *AL triggered by time* however, the queries are not made with a set time interval, instead the time for querying is randomly chosen, as Algorithm 3 shows. First, a variable ζ is randomly generated from the uniform distribution $[0, 1]$. The value of ζ is then compared to the value of the labelling budget B, to decide whether to query or not. This type of strategy is often used as a baseline when evaluating performance of other interactive learning strategies [10, 13].

Algorithm 3. AL triggered at random

Input: data stream $X = \{x_0, x_1, ...\}$, labelling budget B, classifier Ψ
$i \leftarrow 0$
repeat
    receive next data sample x_i
    $\hat{y}_i \leftarrow \Psi(x_i)$
    generate a random variable $\zeta \in [0, 1]$
    if $\zeta \leq B$ **then**
        ask user about label y_i for x_i
        $\Psi \leftarrow$ incremental learning procedure$((x_i, y_i))$
    end if
    $i \leftarrow i + 1$
until end of datastream

MT Triggered by Error. In *MT triggered by error*, the decision of whether or not to provide a label is based on the estimation from the classifier. The user has access to the classifier's estimation of the current state. Whenever the estimation made by the classifier is incorrect, the user provides the correct label, given that the labelling expenses $\hat{b}$ does not exceed the labelling budget B. The labelling expenses are calculated and updated the same way as in *AL triggered by uncertainty*. Algorithm 4 showcases the process. There are previous strategies proposed where the user provides labels based on the output from the classifier [3], but they are typically not presented in the context of single-pass streaming data.

Algorithm 4. MT triggered by error

Input: data stream $X = \{x_0, x_1, ...\}$, labelling budget B, classifier Ψ
$\hat{b} \leftarrow 0$
$i \leftarrow 0$
repeat
    receive next data sample x_i
    $\hat{y}_i \leftarrow \Psi(x_i)$
    if $\hat{y}_i \neq y_i$ **and** $\hat{b} < B$ **then**
        label y_i for x_i is provided by user
        $\Psi \leftarrow$ incremental learning procedure$((x_i, y_i))$
    end if
    update labelling expenses $\hat{b}$
    $i \leftarrow i + 1$
until end of datastream

MT Triggered by State Change. In *MT triggered by state change*, the user will provide a label when the state y changes, given that the labelling expenses

$\hat{b}$ does not exceed the labelling budget B. The labelling expenses are calculated and updated the same way as in *AL triggered by uncertainty*. This strategy can be useful if there is a concern that all possible classes might not be properly represented in the labelled data used for training, e.g.. if the dataset is unbalanced. Unbalanced datasets can result in a high performance overall, even though the performance related to the less frequent classes is poor. For instance, if *AL triggered by time* is used, the learner will have many labelled instances from the frequently occurring classes, while much less from the rare classes. A user could counteract this by providing labels based on the changing status of the state. As illustrated in Algorithm 5, in *MT triggered by state change*, the user provides a label when the class of the current label changes. Depending on the scenario, the state might not change often, resulting in a low total number of labelled instances. However, as long as $\hat{b} < B$ holds, the user will provide a new label if the state changes. This means that while $\hat{b} < B$ and no new label is provided, the learner can assume the incoming instances have the same label as the last one provided by the user y_l and also add these to the collection of labelled data.

Algorithm 5. MT triggered by state change

Input: data stream $X = \{x_0, x_1, ...\}$, labelling budget B, classifier Ψ
$\hat{b} \leftarrow 0$
$i \leftarrow 0$
repeat
  receive next data sample x_i
  $\hat{y}_i \leftarrow \Psi(x_i)$
  if $\hat{b} < B$ **then**
    if $y_i \neq y_{i-1}$ **or** $i = 0$ **then**
      label y_i for x_i is provided by user
      $y_l \leftarrow y_i$
      $\Psi \leftarrow$ incremental learning procedure$((x_i, y_i))$
    else
      $\Psi \leftarrow$ incremental learning procedure$((x_i, y_l))$
    end if
  end if
  update labelling expenses $\hat{b}$
  $i \leftarrow i + 1$
until end of data stream

MT Triggered by Time. The user can by their own initiative also provide labels with a given frequency. The algorithm for this strategy will almost be identical to the one for *AL triggered by time*, presented in Algorithm 2. The only difference is that the user is not queried by an active learning strategy and instead keeps track of time themselves. Depending on the scenario, the user might be less exact compared to an active learning strategy that keeps track of time.

In our setting however we assume that the user does always provide a label in accordance with the strategy, in which case it does not affect the implementation or the performance of the strategy. Therefore, the results from *AL triggered by time* are representing both *AL triggered by time* and *MT triggered by time*.

MT Triggered by User Factors. The user can be triggered by factors related to themselves. These factors include for instance the user's internal state, e.g.. their current stress level, and characteristics, e.g. the knowledge level on what they are supposed to provide labels for. A user might provide fewer labels if they are stressed compared to when they are not. If a user only wants to provide labels when they are certain of the label, a knowledgeable user could provide labels for more data instances compared to a less knowledgeable user. The decision function of when a user will provide a label will vary depending on which factor is triggering the user. Because of the broad spectrum of possible strategies and the additional information needed, this strategy was not included in the experiments.

Hybrid Strategies. The strategies presented above do not necessarily have to be employed separately, but can be merged in different types of combinations to create hybrid strategies. There can be several reasons for creating a hybrid strategy. For instance, if an *MT triggered by error* does not use up it's budget, it can be combined with *AL triggered by time* which will guarantee an influx of new labelled instances. Another reason can be to benefit from the strength of two separate strategies. For example, Žliobaitė et al. present the Split strategy as a combination of *AL triggered by uncertainty* and *AL triggered at random* which can be useful especially if concept drift is occurring [20].

4   Experimental Setup

To illustrate how the different types of interactive machine learning strategies compare, we evaluated implementations of them in experiments on a benchmark dataset and a synthetically generated dataset with the user interaction simulated.[1] In all of the following experiments the learner started with no labelled data and had to incrementally learn as the labelled data was gradually accumulated over time. The evaluation was done in a test-then-train fashion. First, each data instance is used for testing, then the instance is added to the training data of the machine learning algorithm if a label is provided. The results show an average of 20 separate runs.

---

[1] See the following link for synthetic dataset and code: https://github.com/ategen..

4.1   Datasets

The mHealth Dataset. The mHealth dataset contains recordings of activities within a health application [1,2]. The total length of the recordings are 98304-161280 data instances, but the unlabelled data was excluded, resulting in data recordings of 32205-35532 instances. Wearable sensors, accelerometer, gyroscope, magnetometer and electrocardiogram sensor, were used for the recordings. Ten different subjects each perform a sequence of 12 different physical exercises in one recording instance. The exercises are performed one at a time, which means that when the interval of one type of exercise is over, it is not repeated in the same recording. In the experiments, the recordings were concatenated to create one longer sequence and the different physical exercises were repeated 10 times. The order of the recordings was randomly generated for each run.

The Synthetic Dataset. The synthetic dataset contains 5 classes where the mean values for each class were randomly generated in a 2D space. Ten thousand samples were then drawn for each class from a normal distribution with the given mean value and standard deviation. The total of 50000 generated samples constitutes the synthetic dataset. In all experiments, the values are the same but the order of the values in the data stream is shuffled for each run. How long the interval is for each class is also generated by sampling from a normal distribution. Figure 2 displays a visualisation of the distributions of the classes of the dataset.

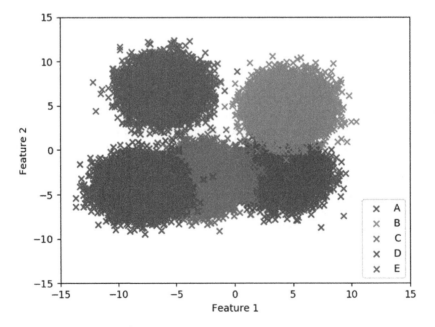

Fig. 2. A visualisation of the distributions of the classes of the synthetic dataset.

4.2   Machine Learning Algorithms

To compare the interactive learning strategies they were tested in combination with three different machine learning algorithms, Support Vector Machine with a polynomial kernel (SVM), k-Nearest Neighbor with $k = 3$ (k-NN) and Naïve Bayes classifier with assumption of Gaussian distributions of the features. The machine learning algorithms included were chosen because they are suitable for online learning, real-time classification and frequently employed in similar settings [7,9,12].

5   Results and Discussion

Figures 3a, 3c and 3e display the accumulated accuracy given the labelling budget for the mHealth dataset for the Naïve Bayes classifier, SVM and k-NN, respectively. In the Figs. 3b, 3d and 3f the accumulated accuracy over the amount of samples estimated, with a labelling budget of 3%. Figures 4a, 4c and 4e show the accumulated accuracy given the labelling budget for the synthetic dataset for the Naïve Bayes classifier, SVM and k-NN respectively. Similarly as for the mHealth dataset, Figs. 4b, 4d and 4f display the accumulated accuracy over the amount of estimated samples, with a labelling budget of 0.5%.

The left column of Fig. 3 (panels 3a, 3c and 3e) shows the final accumulated accuracy given the labelling budget. The figures show that when starting from a very low labelling budget, the performance substantially improves by only increasing the budget slightly. After a while however, there is only a small increase in performance, if any, even though the budget continues to increase. In the right column of the figure (panels 3b, 3d and 3f) the accumulated accuracy over number of samples can be found for a labelling budget of 3%. The best performing strategies, especially at the start, are *MT triggered by error* and *MT triggered by state change.*

In Fig. 4, the results from the experiments on the synthetic dataset are presented. The left column of Fig. 4 (panels 4a, 4c and 4e) displays the accumulated accuracy over labelling budget. Compared to Fig. 3, it does not display the final accumulated accuracy, but the accumulated accuracy after 5000 samples, or a tenth of the total amount of samples, has been processed. Since this dataset does not contain any concept drift, the performance of the different strategies will approach each other as the number of samples increases. After a while, the amount of labelled data is enough regardless of strategy to perform at a similar level. Consequently, it is interesting to analyze when the amount of samples is still relatively low. As the experiments where done using a cold start setup, the performance at the start of the data stream gives an indication of how the strategies perform with a limited amount of labelled data. Also here, the overall best performing strategies are *MT triggered by error* and *MT triggered by state change.*

It is the two machine teaching strategies *MT triggered by error* and *MT triggered by state change* that generally perform best in the experiments. The results are in line with our previous work [16,17] and give an indication that

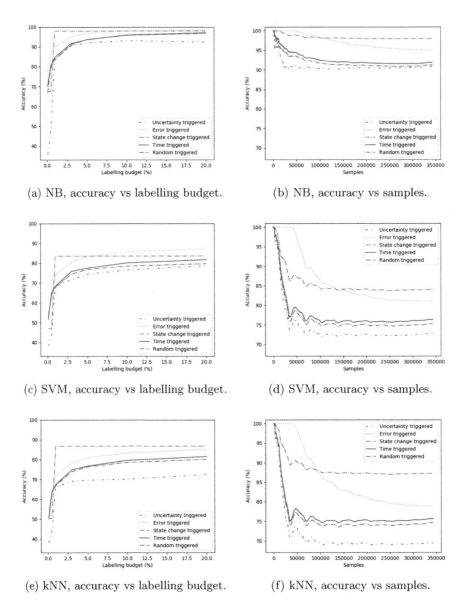

Fig. 3. The results from the experiments on the mHealth dataset. The left column (a, c and e) displays the final accumulated accuracy over labelling budget for Naïve Bayes (NB), Support Vector Machine (SVM) and k-Nearest Neighbor (kNN) respectively. The right column (b, d and f) displays accumulated accuracy over number of samples for a labelling budget of 3% for Naïve Bayes (NB), Support Vector Machine (SVM) and k-Nearest Neighbor (kNN) respectively.

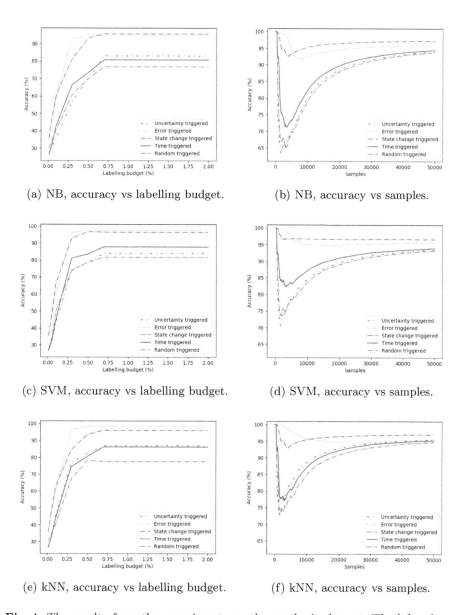

(a) NB, accuracy vs labelling budget.

(b) NB, accuracy vs samples.

(c) SVM, accuracy vs labelling budget.

(d) SVM, accuracy vs samples.

(e) kNN, accuracy vs labelling budget.

(f) kNN, accuracy vs samples.

Fig. 4. The results from the experiments on the synthetic dataset. The left column (a, c and e) displays accumulated accuracy after 5000 samples over labelling budget for Naïve Bayes (NB), Support Vector Machine (SVM) and k-Nearest Neighbor (kNN) respectively. The right column (b, d and f) displays accumulated accuracy over number of samples for a labelling budget of 0.5% for Naïve Bayes (NB), Support Vector Machine (SVM) and k-Nearest Neighbor (kNN) respectively.

letting the user be more proactive can be beneficial on performance. This is noteworthy since most interactive online machine learning strategies employed in the literature fall under the active learning subcategory. Online machine teaching is an area that currently deserves further exploration.

The reason for the high performance at the start in the right-hand side column of Figs. 3 and 4, contrasting the accuracy over samples, is the cold start scenario. The figures are displaying the accumulated accuracy for each new incoming sample. For the very first estimation, there will only be one labelled data instance collected. As the learner only has been introduced to one label at this point in time, it will continue to estimate this label until another option is presented. In many types of data streams it is probable that data instances with the same label follow consecutively for a period of time (e.g.. data streams collected in an activity recognition setting). At the very beginning, for a short period of time, this leads to an accumulated accuracy of 100%. When other labels are introduced to the learner, the performance goes down. As more labelled data instances are gathered for each class, the performance reaches a level that is more representative of the learner in the long run.

We chose to include experiments on both a dataset consisting of actual recordings, the mHealth dataset, and a synthetically generated dataset. The mHealth dataset gives an example of how the strategies perform in a real-life setting, where for instance data can be noisy and there might be concept drift. The synthetic dataset on the other hand, is not meant to represent a realistic scenario for data collection, but it provides a framework for comparing the different strategies. While we have included experiments on two different datasets, further experiments on a variety of datasets would be useful to get an exhaustive comparison of the strategies.

The high performance of *MT triggered by state change* is because it can collect a larger amount of labelled data to use as training data, while not having to involve the user all the time. The strategy expects the user to provide a label when the state is changing and if the labelling expenses still are below the allowed labelling budget, the learner assumes that the state is the same until the user provides a new label. The learner can thus continue to collect labelled data instances, without any effort from the user, as long as the labelling expenses are below the labelling budget. The current labelling expenses are calculated based on how many times the user has provided labelled data. The labelling expenses can thus be kept low while the learner continues to gather labelled data.

While the overall performance is high for *MT triggered by state change*, it relies on the assumptions that the user will always provide a correct label in accordance with the strategy as long as there is labelling budget. These assumptions are found in most work on interactive machine learning but whether they are realistic is rarely discussed. While there exists scenarios where these are reasonable assumptions (e.g.. a medical doctor labelling data within their expertise), in many scenarios the assumptions are simplifications of the actual setting. An interactive machine learning strategy that performs well given the assumptions on the user might not necessarily perform as well if the user did not always provide a label or sometimes provided an incorrect label. We plan to explore how relaxing these assumptions on the user can affect performance in future work. Even though we kept the assumptions in the experiments, the taxonomy also covers scenarios were they are relaxed.

We assume in the experiments that there is only one user, or possibly multiple users, but with the same behavior, but there exists several settings with multiple users that all have individual characteristics. For instance in crowdsourcing, there are multiple users with the possibility to provide labels, but they might not always be available or willing to provide a label [15]. When they do provide a label, it might be incorrect. We also assume that only one label can be provided for each instance. In many scenarios where the classes are disjoint this is a reasonable assumption (e.g.. is the room empty or not), but in other scenarios there might be a value in allowing multiple labels for one instance.

MT triggered by user factors was not included in the experiments because of the lack of necessary data. These types of factors are highly dependent on the user and strategies based upon them are difficult to model. The pattern of when the user is providing labels might not be known to an outside observer or even to the user themselves. Nevertheless, in many real-life scenarios they do have an impact on the result. For instance, a user might be more probable to provide a label when he or she is in a good mood and less likely to provide labels when in a bad mood or when they are stressed. Our aim is to explore these factors further in future work.

6   Conclusion and Future Work

In this work we have presented a taxonomy of interactive online machine learning strategies. We have also done implementations of the strategies presented in the taxonomy and performed experiments on one benchmark dataset and one synthetically generated dataset. The experiments show that the strategies where the user is triggered to provide labels when the state is changing or when an estimation is incorrect are overall better performing than when an active learning strategy queries the user. The difference in performance is especially noticeable when there is a lower amount of labelled data. The results gives an indication that giving the user a more proactive role in labelling, unlike typical active learning, can be beneficial on performance.

In future work we plan to further validate our conclusions and further explore the taxonomy through experiments. Our aim is to test the taxonomy on a variety of datasets and other machine learning algorithms. We also aim to explore machine teaching that is triggered by user related factors further. We plan to test the robustness of interactive online machine learning strategies if the user does not always respond with a correct label.

References

1. Banos, O., et al.: mHealthDroid: a novel framework for agile development of mobile health applications. In: Pecchia, L., Chen, L.L., Nugent, C., Bravo, J. (eds.) IWAAL 2014. LNCS, vol. 8868, pp. 91–98. Springer, Cham (2014). https://doi.org/10.1007/978-3-319-13105-4_14

2. Banos, O., et al.: Design, implementation and validation of a novel open framework for agile development of mobile health applications. Biomed. Eng. Online **14**(2), S6 (2015)

3. Chen, Y., Singla, A., Mac Aodha, O., Perona, P., Yue, Y.: Understanding the role of adaptivity in machine teaching: the case of version space learners. In: Advances in Neural Information Processing Systems, pp. 1476–1486 (2018)

4. Cohn, D., Atlas, L., Ladner, R.: Improving generalization with active learning. Mach. Learn. **15**(2), 201–221 (1994)

5. Fu, Y., Zhu, X., Li, B.: A survey on instance selection for active learning. Knowl. Inf. Syst. **35**(2), 249–283 (2013)

6. Gama, J., Žliobaitė, I., Bifet, A., Pechenizkiy, M., Bouchachia, A.: A survey on concept drift adaptation. ACM Comput. Surv. (CSUR) **46**(4), 1–37 (2014)

7. Khan, Z.A., Samad, A.: A study of machine learning in wireless sensor network. Int. J. Comput. Netw. Appl. **4**, 105–112 (2017)

8. Kottke, D., Krempl, G., Spiliopoulou, M.: Probabilistic active learning in datastreams. In: Fromont, E., De Bie, T., van Leeuwen, M. (eds.) IDA 2015. LNCS, vol. 9385, pp. 145–157. Springer, Cham (2015). https://doi.org/10.1007/978-3-319-24465-5_13

9. Krawczyk, B.: Active and adaptive ensemble learning for online activity recognition from data streams. Knowl.-Based Syst. **138**, 69–78 (2017)

10. Lughofer, E.: Single-pass active learning with conflict and ignorance. Evol. Syst. **3**(4), 251–271 (2012)

11. Lughofer, E.: On-line active learning: a new paradigm to improve practical useability of data stream modeling methods. Inf. Sci. **415**, 356–376 (2017)

12. Mahdavinejad, M.S., Rezvan, M., Barekatain, M., Adibi, P., Barnaghi, P., Sheth, A.P.: Machine learning for internet of things data analysis: a survey. Digit. Commun. Netw. **4**(3), 161–175 (2018)

13. Miu, T., Missier, P., Plötz, T.: Bootstrapping personalised human activity recognition models using online active learning. In: 2015 IEEE International Conference on Computer and Information Technology; Ubiquitous Computing and Communications; Dependable, Autonomic and Secure Computing; Pervasive Intelligence and Computing, pp. 1138–1147. IEEE (2015)

14. Settles, B.: Active learning literature survey. Technical report, University of Wisconsin-Madison Department of Computer Sciences (2009)

15. Shickel, B., Rashidi, P.: ART: an availability-aware active learning framework for data streams. In: The Twenty-Ninth International Flairs Conference (2016)

16. Tegen, A., Davidsson, P., Mihailescu, R.C., Persson, J.A.: Collaborative sensing with interactive learning using dynamic intelligent virtual sensors. Sensors **19**(3), 477 (2019)

17. Tegen, A., Davidsson, P., Persson, J.A.: Activity recognition through interactive machine learning in a dynamic sensor setting. Pers. Ubiquit. Comput. 1–14 (2020). https://doi.org/10.1007/s00779-020-01414-2

18. Zhu, X.: Machine teaching: an inverse problem to machine learning and an approach toward optimal education. In: Twenty-Ninth AAAI Conference on Artificial Intelligence (2015)

19. Zhu, X., Singla, A., Zilles, S., Rafferty, A.N.: An overview of machine teaching. arXiv preprint arXiv:1801.05927 (2018)
20. Žliobaitė, I., Bifet, A., Pfahringer, B., Holmes, G.: Active learning with drifting streaming data. IEEE Trans. Neural Netw. Learn. Syst. **25**(1), 27–39 (2013)

Knowledge Elicitation Using Deep Metric Learning and Psychometric Testing

Lu Yin[(⊠)], Vlado Menkovski, and Mykola Pechenizkiy

Eindhoven University of Technology, Eindhoven 5600, MB, Netherlands
{l.yin,V.Menkovski,m.pechenizkiy}@tue.nl

Abstract. Knowledge present in a domain is well expressed as relationships between corresponding concepts. For example, in zoology, animal species form complex hierarchies; in genomics, the different (parts of) molecules are organized in groups and subgroups based on their functions; plants, molecules, and astronomical objects all form complex taxonomies. Nevertheless, when applying supervised machine learning (ML) in such domains, we commonly reduce the complex and rich knowledge to a fixed set of labels, and induce a model shows good generalization performance with respect to these labels. The main reason for such a reductionist approach is the difficulty in eliciting the domain knowledge from the experts. Developing a label structure with sufficient fidelity and providing comprehensive multi-label annotation can be exceedingly labor-intensive in many real-world applications. In this paper, we provide a method for efficient hierarchical knowledge elicitation (HKE) from experts working with high-dimensional data such as images or videos. Our method is based on psychometric testing and active deep metric learning. The developed models embed the high-dimensional data in a metric space where distances are semantically meaningful, and the data can be organized in a hierarchical structure. We provide empirical evidence with a series of experiments on a synthetically generated dataset of simple shapes, and Cifar 10 and Fashion-MNIST benchmarks that our method is indeed successful in uncovering hierarchical structures.

Keywords: Hierarchical knowledge elicitation · Psychometric testing · Deep metric learning · Active learning

1 Introduction

Supervised learning models specified as a map from the data-space to a fixed set of labels is the panacea of machine learning (ML) applications. However, our goal is often to 'solve' a problem rather than to predict the labels. For example, consider a collection of texts or images – we may ask people to tag them, but our goal is not necessarily to predict tags, rather to understand the 'latent' taxonomy behind. Or one can imagine a scenario of a medical diagnosis involving images (e.g.. diabetic retinopathy) and a set of diagnostic images and a set of labels denoting the severity of the disease (e.g.. 'no disease', 'severity 1', …

© Springer Nature Switzerland AG 2021
F. Hutter et al. (Eds.): ECML PKDD 2020, LNAI 12458, pp. 154–169, 2021.
https://doi.org/10.1007/978-3-030-67661-2_10

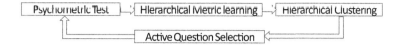

Fig. 1. Proposed hierarchy knowledge elicitation framework.

'severity 4'). From the very beginning, the domain expert is forced to project her comprehensive knowledge on the five given discrete values. This limits our model both from its performance as it cannot learn from the full depth of knowledge as well from its interpretability as projecting to a fixed set of points further contributes to the 'black-box-ness' of our model. In contrast, adding a more rich hierarchical structure to the data adds a significantly larger understanding of the finer subtleties of the data stemming from the relationships between the concepts in the domain, which provide us a taxonomy of different concepts relevant to decision making.

The reason why so many solutions are formulated as mapping of datapoints (commonly high dimensional) to a fixed set of labels is that eliciting the knowledge from the domain and forming a training dataset is a difficult task. A data point can have a rich set of properties associated with it and eliciting an expert to provide an exhaustive annotation for each datapoint is prohibitively unscalable.

In this paper, we propose an approach that addresses this challenge by first developing a hierarchical representation of the data that facilitates more efficient annotation of data. Our method for knowledge elicitation consists of psychometric testing that captures the perceived relative distance of the datapoints for a given query (context) and a deep metric learning embedding algorithm. The embedded data are then structured in a hierarchical fashion that allows assigning properties and annotations to a large number of datapoints in a scalable way.

We evaluate the method empirically by showing we can capture the hierarchical knowledge structure in virtual responders for a given latent structure. Furthermore, using our method, we confirm the shape bias in the responses of a human participant.

2   The Proposed Approach

We present an approach that consists of four components: psychometric testing, hierarchical metric learning, active question selection, and hierarchical clustering. The overall framework is shown in Fig. 1.

First, in order to capture the user's latent representation, we adopt a psychometric testing procedure [4] that relies on discriminative testing. The test collects information about the perceived relative differences to presented stimuli such as images or videos [2,17]. Measuring the perceived difference with a psychometric test is significantly more accurate [4] then directly quantifying a perceived value. On the other hand, we also do not need to present the expert with target labels at this stage, so they can express their knowledge without mapping it to a predefined set of concepts.

The captured responses from the psychometric testing are used to develop a distance metric using a deep metric learning method. The goal of the metric learning component is to develop an embedding of the data in a space where distances reflect the representation of the expert. In other words, images that lie closer together are perceived as more similar than images that are further apart in the embedding. To achieve this, we extend existing metric learning techniques and introduce dual-triplet loss with an adaptive margin. Unlike triplet loss in [15], in dual-triplet loss, we do not distinguish from anchor image and positive image and apply a symmetrical loss structure to align with psychometric testing and fully take advantage of every sample in a loss function. We explained it in detail in the proposed approach section.

Psychometric testing offers many advantages. However, the number of all possible discriminative tests is k combination of n, where k is 3, and n is the number of datapoints in the dataset. Asking all possible questions is typically not feasible. Nevertheless, to achieve a good embedding, we need only a small fraction of all possible questions. However, the quality of the embedding depends significantly on the selected questions as not all questions as equally informative. To address this, we develop a question selection approach using a Bayesian-based active learning method that selects questions with high uncertainty and high utility.

The embedding space is useful for many downstream tasks such as search and retrieval, but it also allows for efficient annotations. As now we have a metric to measure similarity, we can apply a label not only to a single datapoint but also to a region on the space covering multiple datapoints. Furthermore, if we can do this in a hierarchical fashion where labels at a different level of the hierarchy can be applied and then propagated to all lower levels. To achieve this, we combine the active question selection with a hierarchical clustering algorithm, such that we iteratively clusters and sub-clusters of the data to form a hierarchy and focus the question selection on sub-regions of the space. In this way, the approach starts first by forming the global distances and in turn organization of the data and then focuses on finer and finer differentiation as we go lower and lower in the hierarchy.

2.1 Psychometric Test

Psychometric testing is typically used to measure the subjectively perceived quantity of stimuli [4]. Different psychometric testing procedures are available, our method is a discriminative testing procedure, and a variation of the two-alternative forced-choice approaches. These methods typically present two alternatives to the participant that they are forced to choose from. One example of such set up would be to present two audio stimuli to a participant and ask which one is perceived as louder. To scale the perception of loudness, the experiment would consist of two pairs of stimuli, and the participant would need to answer which pair has a larger difference in loudness. Such experiments were also developed for estimating the quality of multimedia content [11]. Another option is to present three stimuli and ask to discriminate between the relative differcncc between the two pairs formed by the three examples. However, many of these questions can be ambiguous, as the images may seem to have similar distances, or they may present different aspects that are not directly comparable in a pair-wise fashion [8].

To deal with this, we adapt the psychometric testing such that the participants are presented three objects a_1, a_2, a_3, and are forced to choose the most dissimilar one among them (Fig. 2). By carrying out this simple test, we can extract hierarchical knowledge from the data by simple ternary decision. The differences between these three objects are expressed as distances in our neural network model. The choice is based on the annotator's personal perception. Therefore, different annotators may have different choices when facing the same question, and different hierarchical trees are created. We examine in detail how the method deals with these conditions in the experiment section.

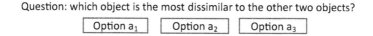

Fig. 2. Three-alternative-forced choice interface.

2.2   Hierarchical Metric Learning

To successfully project the data in an embedded space that captures the latent representation of the expert, we train a model using a variation of the triplet loss function. The triplet loss given in Eq. 1 and its training process was introduced in [15], consists of training using triplet of datapoints (x_a^i, x_n^i, x_p^i),

$$L = \sum_{i=1}^{N} \left[d(x_a^i, x_p^i) - d(x_a^i, x_n^i), +m \right]_+ \qquad (1)$$

where N is number of the possible triplets, $d(x_a^i, x_p^i) = ||(f(x_a^i) - f(x_p^i)||_2^2$ and $d(x_a^i, x_n^i) = ||(f(x_a^i) - f(x_n^i)||_2^2$, $f(z)$ is the representation of the data point z in embedding space and f is our model or more precisely f_θ, where θ represent all the model parameters. hinge function $[\cdot]_+$ indicate $max[0, \cdot]$ When the data is annotation with a fixed set, metric learning is typically implemented such that the triplets take the following roles: x_a^i is the anchor image, which has same label with positive image x_p^i and different label with negative image x_n^i. Therefore, the triplet loss produces a loss value when two images with the same label are further apart than two images with different labels.

Note that the form of triplet loss in Eq. 1 naturally fits with how we defined the 3AFC psychometric test. Based on the participant's response, we can select the most different image as the negative sample, and the other two images as positive and anchor images.

However, the anchor and the positive image are not interchangeable in the triplet loss term as the distance of the anchor to the positive and negative image are compared. One can imagine a scenario where selecting one of the images as an anchor results in a gradient for the model and selecting the other one does not, or results limited gradient, as depicted in Fig. 3.

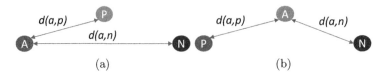

(a)   (b)

Fig. 3. (a) and (b) are different cases of choosing the anchor image. A represent the anchor image, N denotes negative and P for positive. Negative is chosen in 3AFC test and fixed, and we need to select the anchor image from the rest two. If $d(a, n) - d(a, p)$ is too large there will be limited gradients, and if it is larger than a margin value there will be no gradient. Since in case (a) the anchor is farther away from the negative image than in case (b), it is more likely that case(a) results in limited or zero gradients while case (b) results in more gradients.

To deal with this, we propose a dual-triplet loss function in which there is no specific differentiation between the anchor and the positive image but rather define two positive and one negative image. The participant gives an answer about which image is the furthest from the other two, and set it to be negative. The dual-triplet loss is defined as,

$$L = \sum_{i=1}^{N} \left[d(x_{p1}^i, x_{p2}^i) - d(x_n^i, x_{p1}^i), +m_a^i \right]_+ + \left[d(x_{p1}^i, x_{p2}^i) - d(x_n^i, x_{p2}^i), +m_a^i \right]_+$$

(2)

where, x_{p1}^i and x_{p1}^i are two positive images, and x_n^i are the negative image chosen by annotator during 3AFC test, N is the number of sampled triplets, m_a^i is an adaptive margin which will be explained later. Now negative image is compared with each of the rest two images, and both situations in Fig. 3 are considered. Every sampled is used twice in one triplet function to produce more gradients and to prevent zero gradients situation happen.

The triple loss margin has a significant impact on the performance of the metric learning model. Different methods have been developed to determine the value of the margin. Some of these approaches [3] propose using a margin value that adaptively changes during the training process. Since we collect the responses in a hierarchical manner, the margin needs to adapt accordingly, from large to small, as the questions become more focused. For example, if the responses are collected over the whole dataset, the margin needs to be appropriately large, and if we descent and become more focused in the testing, the margin needs to adapt and be smaller, as shown in Fig. 4. Specifically we compute the margin m_a value based on the diversity between the three samples in a triplet i and is defined as,

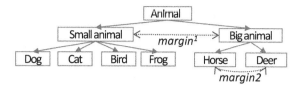

Fig. 4. Questions about small animal and big animal are more general and have a larger diversity than questions about horse and deer. So $margin^1$ should be larger than $margin^2$.

$$m_a^i = m_h^i + \gamma d_H^i \tag{3}$$

where m_h^i is a constant value that guarantees m_a^i is not zero and γd_H^i is the adaptive part. γ is a hyper-parameter that we set during training. d_H^i is a diversity factor of the triplet i. We calculate the diversity factor at a node i, by computing the average distance of the centroids of all sub-clusters of the cluster associated with the node i, as given in Eq. 4,

$$d_H^i = \frac{1}{n^{i^2} - n^i} \sum_{c_p, c_k \in L^i} ||c_p - c_k||_2^2 \tag{4}$$

where L^i is a node where question i is sampled from in the hierarchy. n_i is the number of child nodes in L^i, c_p is the center of cluster p, and c_k is the center of cluster k.

2.3   Active Questions Selection

Given a data set containing B images, there are $\binom{B}{3}$ potential questions for the annotators to answer. It is impractical to answer all of them, and randomly selecting the questions is suboptimal with respect to the efficiency of the training process. To address this, we develop an active question selection scheme.

We start by randomly selecting m questions for the annotators to answer. With the answers, we train the model M, create a hierarchical tree H, and construct a set of knowledge pool containing the answered questions D. The method iteratively repeats the process of question selection and update of the model M as well as the hierarchical representation of the data H. To select the questions for the following iteration, our method takes two steps. In the first step, a set of questions is proposed uniformly sampled from each node in the hierarchy H, and in the second step, we reject some of the proposed questions for which we do not expect high utility.

By sampling from each level of the hierarchy in the first step, we maximize the probability that the model will receive information both about the global distribution of the data as well as the specific differences at a finer level of detail. In contrast, randomly selecting questions is sufficiently less efficient in achieving the same goal.

Then, we use a pool-based active-learning question rejection scheme. We consider not only the uncertainty of the questions but also its utility, that is, select questions with high uncertainty and high variance. High uncertainty means annotators are not so sure which option to choose, thus the question could be informative, and its answer has a higher probability of introducing information to our model. High variance means similar questions have been answered many times, and this is to avoid ambiguous questions (i.e., similar questions have occurred too many times, but annotators are still not sure how to answer).

In order to evaluate the uncertainty and utility, we need to calculate the distribution of answering possibility by Bayes' theorem. For a given question q_i consists of 3 samples, $\theta = (\theta_1, \theta_2, \theta_3)$ denote the possibility of choosing a_1, a_2, a_3. We assume it follow dirichlet distribution $Dir(\alpha_1, \alpha_2, \alpha_3)$, and it's density can be written as,

$$p(\theta|\alpha) = \frac{\Gamma(\alpha_1 + \alpha_2 + \alpha_3)}{\Gamma(\alpha_1)\Gamma(\alpha_2)\Gamma(\alpha_3)} \prod_{i=1}^{3} \theta_i^{\alpha_i - 1} (\theta_i \geq 0; \sum_{i=1}^{3} \theta_i = 1) \tag{5}$$

where Γ denotes the gamma function, $\alpha = (\alpha_1, \alpha_2, \alpha_3)$. $\alpha_1 = \alpha_2 = \alpha_3 = 1$ at beginning, means the possibility of choosing a_1, a_2 or a_3 is equal. The distribution is updated based on the previous answered questions as,

$$p(\theta|\alpha, D) = p(\theta|\alpha + m) \tag{6}$$

where D is the set of answered questions. We define the questions sampled from neighbour questions of a_1, a_2, a_3 as q_i's similar questions, and count the times of choosing a_1, a_2, a_3 when facing similar questions of q_i in D. The counting number are defined as $m = (m_1, m_2, m_3)$.

Now we evaluate the uncertainty by the maximum of the expected possibility of choosing any of a_1, a_2, a_3. It does not matter which one we choose, because if any of them has a high possibility to be chosen, we can conclude that the model is confident about this choice based on the previous answers, and this question has less uncertainty. So if the maximum expected possibility $e(q_i)$ of choosing a_1, a_2 or a_3 is higher than s_e we will reject this question. Similar, we can evaluate the utility by the sum of possibility variance $var(q_i)$. If it is too high, we can interpret it as after facing similar questions many times, the annotator still has an ambiguous answer, so we also reject it.

2.4 Hierarchical Clustering

After answering enough actively selected questions, we embedded the data using the deep metric learning method such that its semantic hierarchical information represented by the Euclidean distance in that space. The hierarchical structure is extracted by a hierarchical clustering algorithm. Generally, there are two categories of methods for this task, a top-down approach called "Agglomerative" and a top-down approach called "Divisive" [14]. Since we have to decide the threshold very carefully if applying the "Agglomerative" approach, in our experiment, we choose the "Divisive" approach K-means to perform the divisive cluster, and Silhouette Coefficient is applied to help us choose K.

3   Related Work

Our work is related to three lines of research (1) Building hierarchical image structure (2) Deep metric learning (3) Informative samples mining.

Building Hierarchical Image Structure. There are two branches explored in building a hierarchical image structure. One is with the help of language information. Such as WordNet is used to build a semantic hierarchic classifier [10]. However, the language-based similarity is not always what we want. When comparing whitefish, goldfish, and cat, the former two are closer in word semantic because they are all fish, but this could not suit our needs if we want to build a hierarchical pet data set based on their cuteness. On the other hand, hierarchies could also be formed based on image features that are extracted by SIFT, HOG [18], or deep convolutional network [22]. While able to group images based on their visual similarities, the hierarchies are not build based on the user's perception or knowledge, so it is also difficult to cater to a special user's preference (e.g.. expert domain).

Deep Metric Learning. Deep metric learning is a data-driven approach to learn the measurement of similarity. It learns a nonlinear mapping for the original data to an embedding. Contrastive loss [5] is proposed to encourage the distance of positive paired samples be closer than negative samples of a margin of m. It is extended to triplet loss [15], quadruplet loss [1], N-paired loss [16] where three, four, N samples are used in loss function, and lifted structure loss [12] which take advantage all the positive and negative pairs in a mini-batch. Although yield promising results in distances based computer vision applications such as face recognition [15], person re-identification [1], and few-shot learning [13], their studies focus on the flat distance which has a weaker hierarchical distances representation ability.

Informative Samples Mining. For pair-based deep metric learning tasks, there are $O(N^n)$ potential tuples for training, N is the sample number in training set, and n is sample number in a tuple. It is impractical to train with all of them due to the GPU memory or train time concern. Besides, a lot of them are redundant or less informative and contribute little gradients during training. A semi-hard mining scheme is used in FaceNet [15] to select informative tuples online inside a mini-batch. Harwood *et al.* considers the global structure in their smart mining [6] for producing effective samples with a low computational cost. A distance weighted sampling strategy [19] is proposed to select samplings uniformly based on their relative distance. While these strategies boost the model performance by learning a more discriminative flat representation of each category object, our work aims to map data to an embedding space where a hierarchical structure is also considered. Therefore, a hierarchical sampling method with Bayesian-based active learning is proposed in our paper to better cope with our needs.

4   Experiments

As a first in our empirical evaluation, we aim to confirm that our approach can elicitate the latent hierarchical structure of a single participant. To remove any other sources of variation, we design an experiment with a virtual participant that always responds according to its given latent hierarchical model. In this manner, we can objectively evaluate the precision of the elicitated model against the one given to the virtual participant.

Next, we expand on this using multiple virtual participants that have partial agreements of their latent models. In large scale human experiments, we do not expect that the internal hierarchical representations of people will fully agree with each other. Therefore, in this analysis we aim to assess whether our approach can extract the common hierarchical model of a group of participants that have models with partial agreements. Towards this we simulate multiple individual participants and test whether the model can elicit common hierarchy accordingly.

Then, we evaluate with a human participant. The goal of this experiment is to evaluate how well our method can deal with the additional variability involved in using responses from a person. In this setting, the additional challenge is that we cannot be precisely sure of the human's latent hierarchical model of the data. To deal with this, we specifically design an experiment that would confirm a well understood psychological bias that humans have, specifically the shape bias [9]. The shape bias states that shapes are more important than other properties of an object, such as colors when we aim to discriminate between them. We develop an image dataset that allows us to test is we can confirm this bias in hierarchical models that we elicit from the responses.

Finally, we develop an experiment on a natural image dataset, specifically the Fashion-Mnist [20]. Our aim here is to test in natural settings. In this case, our evaluation is limited to the flat set of labels that the dataset comes with.

In all experiments, two evaluation metrics are considered: (1) *The accuracy of each cluster*. Every cluster in the hierarchy should share a similar concept. We tag a cluster by the majority shared concept in it, and calculate the accuracy of that node by counting the number of correctly assigned images and dividing by overall image number in that node. (2) *The dendrogram purity between the extracted structure and the ground-truth*. Since we want the hierarchy able to cater for special users, we can only claim a hierarchy is good when structure formed according to the ground-truth, which is the user's latent perception. That can be evaluated by dendrogram purity [7]. Rather than the flat cluster purity, it is a holistic way to evaluate the whole hierarchical structure tree. The value of it ranges from 0 to 1, 1 means every sample fits perfectly for the ground truth and on the contrary for 0. Note that dendrogram purity is only applicable when the ground-truth hierarchy is given in advance, so we only calculate it in simulation experiments, and as we are applying hierarchy clustering in this paper, every node of the hierarchy is a cluster.

4.1 Single Virtual Participant on Cifar 10

With our proposed hierarchical knowledge elicitation method, we experiment on a snipped tiny Cifar 10. The data set containing 1000 images sampled randomly from the original Cifar 10 with 100 images in each category.

Since we might not be able to know a user's latent perception before the experiment, in order to evaluate how well our extracted structure matches the user's latent perception, we pre-define a specific hierarchy (see Fig. 5) and run the experiment by simulation. A virtual participant gives perfect responses based on the given hierarchical structure. 1000 questions are simulated for model training with a fixed margin of 0.4 at the beginning. 600 more questions are sampled per iteration and used for training with adaptive margin in the following 5 iterations.

After collecting 3400 automatically respond triplets, we show our results in Fig. 5. We can see that our model can elicitate a clear hierarchical structure, which matches the given one well. Note that the small and big animals do not split down to individual animal nodes. That may because animal images are similar to each other, and it is not easy for the model to get a clear cluster with limited animal training triplets. We calculate the accuracy of the extracted nodes which are all above 90%.

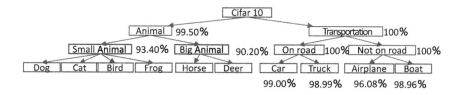

Fig. 5. Latent given hierarchy of a single virtual participant and the extracted result.

4.2 Multiple Virtual Participants on Cifar 10

We further evaluate our method in multiply-users situations when they have partial disagreements. Two more participants are simulated with different pre-designed hierarchies, as shown in Fig. 6. We follow the same experiment setup with single-user situation in Sect. 4.1.

We illustrate our results in Fig. 6, where we can see our model is able to extract varying hierarchical structures according to different given latent hierarchies. Individual animal nodes still can not be clustered well due to their higher similarity and limited training samples.

We also mixed the collected triplets from all three users, and extract a general hierarchy from all the triplets (See Fig. 6). Due to enough training triplets, individual animal nodes got extracted, and other nodes get better accuracies compared to single participant situations. We can also notice that both shared perception nodes (10 original Cifar 10 classes, animal and transportation) and

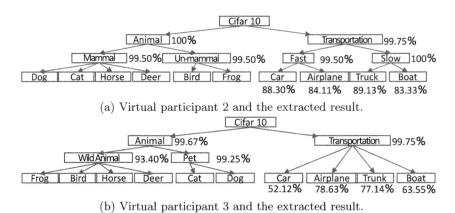

(a) Virtual participant 2 and the extracted result.

(b) Virtual participant 3 and the extracted result.

Fig. 6. Latent given hierarchy of multiple virtual participants and the extracted result.

individual perception nodes (big animal from participant 1, un-mammal from participant 2, pet from participant 3) can be seen in the structure. Therefore, our model can elicit a common structure that represents both shared and individual knowledge of all participants (Fig. 7).

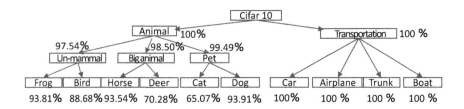

Fig. 7. Extracted common structure.

4.3   Real Participant on Synthetic Geometric Shape Data Set

The previous experiment has shown that our method is able to extract hierarchical structure by simulation. It is also crucial to test with real human responses. One challenge of real human involved experiments is how to evaluate our method since we don't know the latent hierarchical knowledge in advance. To deal with that, we test whether our method can uncover the well-studied shape bias in humans [9] on a synthetic geometric data set. The data set contains triangles, circles, rectangles 3 different shapes. Each shape contains 3 deformations, 5 colors, and 3 thicknesses (See the first layer in Fig. 8). Initially, we have the participant to answer 300 randomly chosen questions and train the model with a fixed margin of 0.2. During each following iteration, 300 actively sampled questions arc answered. After 5 iterations, 1500 more triplets are collected to train the model with an adaptive margin.

The extracted hierarchy are shown in Fig. 8. Because most of the nodes have 100% accuracy, we only mark the nodes which do not have perfect accuracies. We also notice there are few images located in the wrong clusters. Remember, all the questions are answered by real humans, so mistaken answers are inevitable, and that makes the dendrogram not perfect. The overall structure confirmed the human tend to believe shape is more important feature than other properties when they are asked to make choices.

4.4 Real Participant on Fashion-Mnist

Another real participant involved experiment is conducted on real-life image data set Fashion-Mnist [20]. Accuracy can be calculated when the node of the hierarchy containing the original flat labels. The ground truth latent perception is unknown before experiments and is discovered by our proposed method. We sample questions on a tiny Fashion-Mnist containing 900 images with 90 images for each class. During the first iteration, we randomly choose 600 questions, and train the model with a fixed margin of 0.4. In the following 2 iterations, we actively select 600 questions every iteration and train the model with an adaptive margin.

Figure 9 illustrates the extracted hierarchy. Note that Due to page size limitation, it is just a diagram, not the real hierarchical images. The first splits is based on cloth(left) and accessories(right) which further split down into tops (T-shirt, pullover, dress, coat, shirt)/trouser and bags/shoes(sneaker, boot, sandal). Tops continue down into long sleeves/short sleeves, which are organized by the length of their sleeves regardless of the original cloth category. Further splits of shoes are wrap shoes(sneakers, boots) and sandals. For the nodes composed of original Fashion-Mnist labels, we can calculate the accuracy of each node.

Further splits of sandals and bags can be seen in Fig. 10, which presents us with a finer granularity than the original labels.

4.5 Comparison and Ablation Study

We compare our method with three baselines and perform an ablation study to evaluate our proposed components on tiny Cifar 10. After training with 4000 triplets, dendrogram purities are calculated in different experiment settings and reported in Table 1. Trained curves are shown in Fig. 11.

Three benchmarks are chosen for comparison. First, pixel-level points are hierarchically clustered directly. Then, SIFT features are extracted and represented as BoVW [21] for hierarchical clustering. Besides, we train a CNN model with labels and hierarchically cluster on the last activation layer. For a fair comparison, we use the same CNN model as our HKE framework. From Table 1 and Fig. 11, we can see the dendrogram purities of pixel-level and SIFT BoVW are lower than 50%. For baseline CNN, the dendrogram purity is higher than 70%.

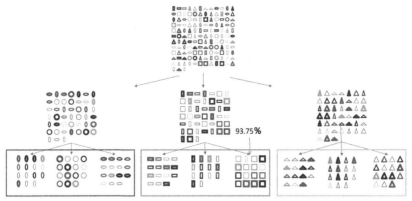

(a) The first three layers of the extracted hierarchy from the synthetic geometric shape data-set.

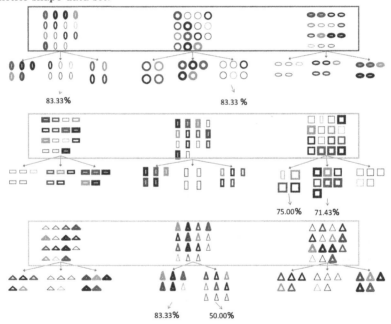

(b) The last layer of the extracted hierarchy from the synthetic geometric shape data-set.

Fig. 8. A four-layer hierarchy is extracted from the synthetic geometric shape data-set. The First 3 layers are shown in (a), and the last layer of the hierarchy is shown in (b) due to page size limitation. Clusters inside square-bound with the same color between (a) and (b) are the same. The first layer split into rectangles, circles, triangles three clusters, which is because the annotator thinks shape is the most discriminative property. In the second layer, shapes are again the discriminator; vertical stretc.hed, horizontally stretc.hed, and un-stretc.hed images are cluster together. When the hierarchical tree goes down, we can notice that images begin to be organized based on different thicknesses, which are also shapes. Note only not perfectly pure nodes are marked with accuracy.

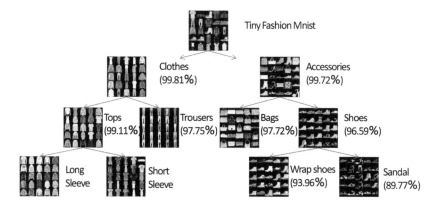

Fig. 9. Extracted hierarchy of tiny Fashion-Mnist.

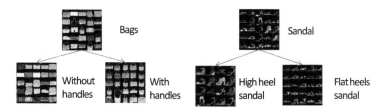

Fig. 10. Finer granularity of bags and sandals.

We also perform an ablation study to evaluate our proposed components. Experiment configuration are the same as in Sect. 4.2. In all the ablation experiments, dual-triplet is used to align with the 3AFC test. First, we sampling questions randomly and using a fixed margin of 0.4, the performance improves slightly in the first three iterations and keeps stuck even with more training data. Then, active question sampling (AQS) is applied to select questions with high-uncertainly and high utility, and performance gets a noticeable improvement. Finally, we integrate the adaptive margin (AM) and get the best performance.

When compare with HKE framework with baselines, we can notice that even with minimum components (Just 3AFC test, without AQS and AM), our proposed framework has better performance than all the baselines.

Table 1. Baselines and ablation study on Cifar 10 with 3 simulated participants.

Participant	Base line pixel level	Base line SIFT BoVW	Base line CNN	Random choosing + fixed margin	AQS + fixed margin	AQS+ AM
1	47.03	43.72	75.5	77.84	89.06	90.93
2	46.99	43.57	71.3	77.84	86.25	94.44
3	49.13	45.70	72.4	80.45	86.86	90.73

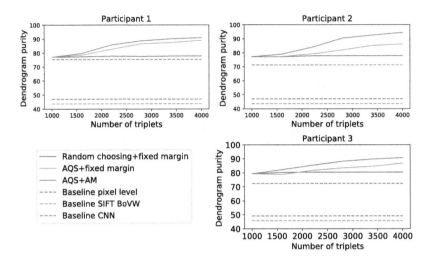

Fig. 11. Dendrogram purity training curves of different virtual responders. Our baselines are not trained with triplets, so they keep flat. Even with minimum components, our proposed HKE has better performance than baselines.

5   Conclusion

Developing a machine learning formulation for a given task usually involves compromises that are related to how much and what kind of data is available, but also very much around the quantity and kind of expert knowledge that can be collected. The later tends to lead to oversimplifications, such as forcing a fixed set of labels and disregarding any more precise descriptions of the data. In this work, we present a method for knowledge elicitation from experts that can be applied to a broad set of machine learning formulations of problems. The results show that the proposed 3AFC testing with dual metric learning can extract latent hierarchical representations of data in a scalable way. This opens possibilities for applications in many domains where expert knowledge has a rich and complex underlying structure as in medical domain, biology as well as many industrial applications.

References

1. Chen, W., Chen, X., Zhang, J., Huang, K.: Beyond triplet loss: a deep quadruplet network for person re-identification. In: Proceedings of the IEEE Conference on Computer Vision and Pattern Recognition, pp. 403–412 (2017)
2. Feng, H.C., Marcellin, M.W., Bilgin, A.: A methodology for visually lossless JPEG2000 compression of monochrome stereo images. IEEE Trans. Image Process. **24**(2), 560–572 (2014)

3. Ge, W., Huang, W., Dong, D., Scott, M.R.: Deep metric learning with hierarchical triplet loss. In: Ferrari, V., Hebert, M., Sminchisescu, C., Weiss, Y. (eds.) ECCV 2018. LNCS, vol. 11210, pp. 272–288. Springer, Cham (2018). https://doi.org/10.1007/978-3-030-01231-1_17
4. Gescheider, G.A.: Psychophysics: The Fundamentals. Psychology Press (2013)
5. Hadsell, R., Chopra, S., LeCun, Y.: Dimensionality reduction by learning an invariant mapping. In: 2006 IEEE Computer Society Conference on Computer Vision and Pattern Recognition (CVPR 2006), vol. 2, pp. 1735–1742. IEEE (2006)
6. Harwood, B., Kumar, B., Carneiro, G., Reid, I., Drummond, T., et al.: Smart mining for deep metric learning. In: Proceedings of the IEEE International Conference on Computer Vision, pp. 2821–2829 (2017)
7. Heller, K.A., Ghahramani, Z.: Bayesian hierarchical clustering. In: Proceedings of the 22nd International Conference on Machine Learning, pp. 297–304. ACM (2005)
8. Hellinga, N., Menkovski, V.: Hierarchical annotation of images with two-alternative-forced-choice metric learning. arXiv preprint arXiv:1905.09523 (2019)
9. Landau, B., Smith, L.B., Jones, S.S.: The importance of shape in early lexical learning. Cogn. Dev. **3**(3), 299–321 (1988)
10. Marszalek, M., Schmid, C.: Semantic hierarchies for visual object recognition. In: 2007 IEEE Conference on Computer Vision and Pattern Recognition, pp. 1–7. IEEE (2007)
11. Menkovski, V., Liotta, A.: Adaptive psychometric scaling for video quality assessment. Sig. Process. Image Commun. **27**(8), 788–799 (2012)
12. Oh Song, H., Xiang, Y., Jegelka, S., Savarese, S.: Deep metric learning via lifted structured feature embedding. In: Proceedings of the IEEE Conference on Computer Vision and Pattern Recognition, pp. 4004–4012 (2016)
13. Oreshkin, B., López, P.R., Lacoste, A.: TADAM: task dependent adaptive metric for improved few-shot learning. In: Advances in Neural Information Processing Systems, pp. 721–731 (2018)
14. Rokach, L., Maimon, O.: Clustering methods. In: Maimon, O., Rokach, L. (eds.) Data Mining and Knowledge Discovery Handbook, pp. 321–352. Springer, Heidelberg (2005). https://doi.org/10.1007/0-387-25465-X_15
15. Schroff, F., Kalenichenko, D., Philbin, J.: FaceNet: a unified embedding for face recognition and clustering. In: Proceedings of the IEEE Conference on Computer Vision and Pattern Recognition, pp. 815–823 (2015)
16. Sohn, K.: Improved deep metric learning with multi-class N-pair loss objective. In: Advances in Neural Information Processing Systems, pp. 1857–1865 (2016)
17. Son, I., Winslow, M., Yazici, B., Xu, X.: X-ray imaging optimization using virtual phantoms and computerized observer modelling. Phys. Med. Biol. **51**(17), 4289 (2006)
18. Wigness, M., Draper, B.A., Beveridge, J.R.: Efficient label collection for image datasets via hierarchical clustering. Int. J. Comput. Vis. **126**(1), 59–85 (2018)
19. Wu, C.Y., Manmatha, R., Smola, A.J., Krahenbuhl, P.: Sampling matters in deep embedding learning. In: Proceedings of the IEEE International Conference on Computer Vision, pp. 2840–2848 (2017)
20. Xiao, H., Rasul, K., Vollgraf, R.: Fashion-MNIST: a novel image dataset for benchmarking machine learning algorithms. arXiv preprint arXiv:1708.07747 (2017)
21. Yang, J., Jiang, Y.G., Hauptmann, A.G., Ngo, C.W.: Evaluating bag-of-visual-words representations in scene classification. In: Proceedings of the International Workshop on Multimedia Information Retrieval, pp. 197–206 (2007)
22. Zheng, Y., Fan, J., Zhang, J., Gao, X.: Hierarchical learning of multi-task sparse metrics for large-scale image classification. Pattern Recogn. **67**, 97–109 (2017)

Adversarial Learning

Adversarial Learned Molecular Graph Inference and Generation

Sebastian Pölsterl$^{(\boxtimes)}$ and Christian Wachinger

Artificial Intelligence in Medical Imaging (AI-Med),
Department of Child and Adolescent Psychiatry, Ludwig-Maximilians-Universität,
Munich, Germany
sebastian.poelsterl@med.uni-muenchen.de

Abstract. Recent methods for generating novel molecules use graph representations of molecules and employ various forms of graph convolutional neural networks for inference. However, training requires solving an expensive graph isomorphism problem, which previous approaches do not address or solve only approximately. In this work, we propose ALMGIG, a likelihood-free adversarial learning framework for inference and *de novo* molecule generation that avoids explicitly computing a reconstruction loss. Our approach extends generative adversarial networks by including an adversarial cycle-consistency loss to implicitly enforce the reconstruction property. To capture properties unique to molecules, such as valence, we extend the Graph Isomorphism Network to multi-graphs. To quantify the performance of models, we propose to compute the distance between distributions of physicochemical properties with the 1-Wasserstein distance. We demonstrate that ALMGIG more accurately learns the distribution over the space of molecules than all baselines. Moreover, it can be utilized for drug discovery by efficiently searching the space of molecules using molecules' continuous latent representation. Our code is available at https://github.com/ai-med/almgig.

1 Introduction

Deep generative models have been proven successful in generating high-quality samples in the domain of images, audio, and text, but it was only recently when models have been developed for *de novo* chemical design [8,16]. The goal of *de novo* chemical design is to map desirable properties of molecules, such as a drug being active against a certain biological target, to the space of molecules. This process – called inverse Quantitative Structure-Activity Relationship (QSAR) – is extremely challenging due to the vast size of the chemical space, which is estimated to contain in the order of 10^{33} drug-like molecules [27]. Searching this space efficiently is often hindered by the discrete nature of molecules, which prevents the use of gradient-based optimization. Thus, obtaining a continuous and differentiable representation of molecules is a desirable goal that could ease drug discovery. For *de novo* generation of molecules, it is important to produce chemically valid molecules that comply with the valence of atoms, i.e., how many

F. Hutter et al. (Eds.): ECML PKDD 2020, LNAI 12458, pp. 173–189, 2021.
https://doi.org/10.1007/978-3-030-67661-2_11

electron pairs an atom of a particular type can share. For instance, carbon has a valence of four and can form at most four single bonds. Therefore, any mapping from the continuous latent space of a model to the space of molecules should result in a chemically valid molecule.

The current state-of-the-art deep learning models are adversarial or variational autoencoders (AAE, VAE) that represent molecules as graphs and rely on graph convolutional neural networks (GCNs) [6,14,18,20,22,23,34,36,39]. The main obstacle is in defining a suitable reconstruction loss, which is challenging when inputs and outputs are graphs. Because there is no canonical form of a graph's adjacency matrix, two graphs can be identical despite having different adjacency matrices. Before the reconstruction loss can be computed, correspondences between nodes of the target and reconstructed graph need to be established, which requires solving a computationally expensive graph isomorphism problem. Existing graph-based VAEs have addressed this problem by either traversing nodes in a fixed order [14,22,34] or employing graph matching algorithms [36] to approximate the reconstruction loss.

We propose ALMGIG, a likelihood-free Generative Adversarial Network for inference and generation of molecular graphs (see Fig. 1). This is the first time that an inference and generative model of molecular graphs can be trained without computing a reconstruction loss. Our model consists of an encoder (inference model) and a decoder (generator) that are trained by implicitly imposing the reconstructing property via cycle-consistency, thus, avoiding the need to solve a computationally prohibitive graph isomorphism problem. To learn from graph-structered data, we base our encoder on the recently proposed Graph Isomorphism Network [37], which we extend to multi-graphs, and employ the Gumbel-softmax trick [13,24] to generate discrete molecular graphs. Finally, we explicitly incorporate domain knowledge such that generated graphs represent valid chemical structures. We will show that this enables us to perform efficient nearest neighbor search in the space of molecules.

In addition, we performed an extensive suite of benchmarks to accurately determine the strengths and weaknesses of models. We argue that summary statistics such as the percentage of valid, unique, and novel molecules used in previous studies, are poor proxies to determine whether generated molecules are chemically meaningful. We instead compare the distributions of 10 chemical properties and demonstrate that our proposed method is able to more accurately learn a distribution over the space of molecules than previous approaches.

2   Related Work

Graphs, where nodes represent atoms, and edges chemical bonds, are a natural representation of molecules, which has been explored in [6,14,18,20,22,23, 34,36,39]. Most methods rely on graph convolutional neural networks (GCNs) for inference, which can efficiently learn from the non-Euclidean structure of graphs [3,6,18,22,34,36,39]. Molecular graphs can be generated sequentially, adding single atoms or small fragments using an RNN-based architecture

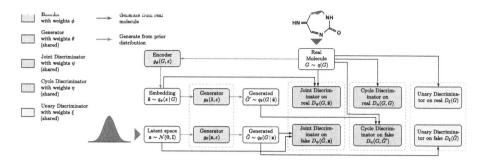

Fig. 1. Overview of the proposed ALMGIG model. Boxes with identical background color represent neural networks that share their weights. The joint discriminator plays a similar role as the discriminator in standard GANs. The cycle discriminator enforces the reconstruction property without explicitly computing a reconstruction loss. Molecules can be generated by transforming a sample from a simple prior distribution (blue path), or by embedding a real molecule into the latent space and reconstructing it (red path). (Color figure online)

$[3, 14, 18, 20, 22, 26, 34, 39]$, or in a single step $[6, 23, 36]$. Sequential generation has the advantage that partially generated graphs can be checked, e.g.., for valence violations $[22, 34]$. Molecules can also be represented as strings using SMILES encoding, for which previous work relied on recurrent neural networks for inference and generation $[2, 5, 8, 10, 16, 21, 25, 26, 29, 30, 35]$. However, producing valid SMILES strings is challenging, because models need to learn the underlying grammar of SMILES. Therefore, a considerable portion of generated SMILES tend to be invalid (15–80%) $[2, 10, 21, 30, 35]$ – unless constraints are built into the model $[5, 16, 18, 29]$. The biggest downside of the SMILES representation is that it does not capture molecular similarity: substituting a single character can alter the underlying molecule structure significantly or invalidate it. Therefore, partially generated SMILES strings cannot be validated and transitions in the latent space of such models may lead to abrupt changes in molecule structure $[14]$.

With respect to the generative model, most previous work use either VAEs or adversarial learning. VAEs and AAEs take a molecule representation as input and project it via an inference model (encoder) to a latent space, which is subsequently transformed by a decoder to produce a molecule representation $[2, 5, 8, 14-16, 21, 22, 34, 36]$. New molecules can be generated by drawing points from a simple prior distribution over the latent space (usually Gaussian) and feeding it to the decoder. AAEs $[2, 15]$ perform variational inference by adversarial learning, using a discriminator as in GANs, which allows using a more complex prior distribution over the latent space. Both VAEs and AAEs are trained to minimize a reconstruction loss, which is expensive to compute. Standard GANs for molecule generation lack an encoder and a reconstruction loss, and are trained via a two-player game between a generator and discriminator $[6, 10, 30]$. The generator transforms a simple input distribution into a distribution in the space of molecules, such that the discriminator is unable to distinguish between real molecules in the training data and generated molecules. Without an inference model, GANs cannot be used for neighborhood search in the latent space.

3   Methods

We represent molecules as graphs and propose our Adversarial Learned Molecular Graph Inference and Generation (ALMGIG) framework to learn a distribution in the space of molecules, generate novel molecules, and efficiently search the neighborhood of existing molecules. ALMGIG is a bidirectional GAN that has a GCN-based encoder that projects a graph into the latent space and a decoder that outputs a one-hot representation of atoms and an adjacency matrix defining atomic bonds. Hence, graph generation is performed in a single step, which allows considering global properties of a molecule. An undirected multi-graph $G = (\mathcal{V}, \mathcal{R}, \mathcal{E})$ is defined by its vertices $\mathcal{V} = \{v_1, \ldots, v_n\}$, relation types $\mathcal{R} = \{r_1, \ldots, r_m\}$, and typed edges (relations) $\mathcal{E} = \{(v_i, r_k, v_j) \,|\, v_i, v_j \in \mathcal{V}, r_k \in \mathcal{R}\}$. Here, we only allow vertices to be connected by at most one type of edge. We represent a multi-graph G by its adjacency tensor $\mathbf{A} \in \{0, 1\}^{n \times n \times m}$, where A_{ijk} is one if $(v_i, r_k, v_j) \in \mathcal{E}$ and zero otherwise, and the node feature matrix $\mathbf{X} = (\mathbf{x}_{v_1}, \ldots, \mathbf{x}_{v_n})^\top \in \mathbb{R}^{n \times d}$, where d is the number of features describing each node. Here, vertices are atoms, $\mathcal{R}$ is the set of bonds considered (single, double, triple), and node feature vectors $\mathbf{x}_{v_i}$ are one-hot encoded atom types (carbon, oxygen, ...), with d representing the total number of atom types. We do not model hydrogen atoms, but implicitly add hydrogens to match an atom's valence.

3.1   Adversarially Learned Inference

We first describe Adversarially Learned Inference with Conditional Entropy (ALICE) [17], which is at the core of ALMGIG and has not been used for *de novo* chemical design before. It allows training our encoder-decoder model fully adversarially and implicitly enforces the reconstruction property without the need to compute a reconstruction loss (see overview in Fig. 1). It has two generative models: (i) the *encoder* maps graphs into the latent space, and (ii) the *decoder* inverts this transformation by mapping latent codes to graphs. Training is performed by matching joint distributions over graphs $G \in \mathcal{G}$ and latent variables $\mathbf{z} \in \mathcal{Z}$. Let $q(G)$ be the marginal data distribution, from which we have samples, $\tilde{\mathbf{z}}$ a latent representation produced by the encoder, and $\tilde{G}$ a generated graph. The encoder generative model over the latent variables is $q_\phi(z \,|\, G)$ with parameters ϕ, and the decoder generative model over graphs is $q_\theta(G \,|\, z)$, parametrized by θ. Putting everything together, we obtain the *encoder joint distribution* $q_\phi(G, z) = q(G)q_\phi(z \,|\, G)$, and the *decoder joint distribution* $p_\theta(G, z) = p_z(z)q_\theta(G \,|\, z)$. The objective of Adversarially Learned Inference is to match the two joint distributions [7]. A discriminator network D_ψ

with parameters ψ is trained to distinguish samples $(G, \tilde{z}) \sim q_\phi(G, z)$ from $(\tilde{G}, \mathbf{z}) \sim p_\theta(G, z)$. Drawing samples $\tilde{z}$ and $\tilde{G}$ is made possible by specifying the encoder q_ϕ and decoder q_θ as neural networks using the change of variable technique: $\tilde{z} = g_\phi(G, \varepsilon)$, $\tilde{G} = g_\theta(\mathbf{z}, \varepsilon)$, where ε is some random source of noise. The objective then becomes the following min-max game:

$$
\min_{\theta, \phi} \max_{\psi} \quad \mathbb{E}_{G \sim q(G), \; \tilde{z} \sim q_\phi(z \,|\, G)}[\log \sigma(D_\psi(G, \tilde{z}))]
$$
$$
+ \mathbb{E}_{\tilde{G} \sim q_\theta(G \,|\, z), \; \mathbf{z} \sim p(z)}[\log(1 - \sigma(D_\psi(\tilde{G}, \mathbf{z})))]. \tag{1}
$$

where $\sigma(\cdot)$ denotes the sigmoid function. At the optimum of (1), the marginal distributions $q_\theta(G \,|\, z)$ and $q_\phi(z \,|\, G)$ will match, but their relationship can still be undesirable. For instance, the decoder $g_\theta(\mathbf{z}, \varepsilon)$ could map a given z to G_1 half of the time and to a distinct G_2 the other half of the time – the same applies to the encoder [17]. ALICE [17] solves this issue by including a cycle-consistency constraint via an additional adversarial loss. This encourages encoder and decoder to mimic the reconstruction property without explicitly solving a computationally demanding graph isomorphism problem. To this end, a second discriminator with parameters η is trained to distinguish a real from a reconstructed graph:

$$
\min_{\theta, \phi} \max_{\eta} \quad \mathbb{E}_{G \sim q(G)}[\log \sigma(D_\eta(G, G))]
$$
$$
+ \mathbb{E}_{\tilde{G}' \sim q_\theta(G \,|\, \tilde{z}), \; \tilde{z} \sim q_\phi(z \,|\, G)}[\log(1 - \sigma(D_\eta(G, \tilde{G}')))]. \tag{2}
$$

While encoder and decoder in ALICE have all the desired properties at the optimum, we rarely find the global optimum, because both are deep neural networks. Therefore, we extend ALICE by including a unary discriminator D_ξ only on graphs, as in standard GANs. This additional objective facilitates generator training when the joint distribution is difficult to learn:

$$
\min_{\theta, \phi} \max_{\xi} \quad \mathbb{E}_{G \sim q(G)}[\log \sigma(D_\xi(G))]
$$
$$
+ \mathbb{E}_{\tilde{G} \sim q_\theta(G \,|\, z), \; \mathbf{z} \sim p(z)}[\log(1 - \sigma(D_\xi(\tilde{G})))]. \tag{3}
$$

We will demonstrate in our ablation study that this is indeed essential.

During training, we concurrently optimize (1), (2), and (3) by first updating θ and ϕ, while keeping ψ, η and ξ fixed, and then the other way around. We use a higher learning rate when updating encoder and decoder weights θ, ϕ as proposed in [12]. Finally, we employ the 1-Lipschitz constraint in [11] such that discriminators D_ψ, D_η, and D_ξ are approximately 1-Lipschtiz continuous. Next, we will describe the architecture of the encoder, decoder, and discriminators.

3.2   Generator

Adjacency matrix (bonds) and node types (atoms) of a molecular graph are discrete structures. Therefore, the generator must define an implicit discrete distribution over edges and node types, which differs from traditional GANs that can only model continuous distributions. We overcome this issue by using the Gumbel-softmax trick similar to [6]. The generator network $g_\theta(\mathbf{z}, \varepsilon)$ takes a point $\mathbf{z}$ from latent space and noise ε, and outputs a discrete-valued and symmetric graph adjacency tensor $\mathbf{A}$ and a discrete-valued node feature matrix $\mathbf{X}$. We use an MLP with three hidden layers with 128, 256, 512 units and tanh activation, respectively. To facilitate better gradient flow, we employ skip-connections from $\mathbf{z}$ to layers of the MLP. First, the latent vector $\mathbf{z}$ is split into three equally-sized parts $\mathbf{z}_1, \mathbf{z}_2, \mathbf{z}_3$. The input to the first hidden layer is the concatenation $[\mathbf{z}_1, \varepsilon]$, its output is concatenated with $\mathbf{z}_2$ and fed to the second layer, and similarly for the third layer with $\mathbf{z}_3$.

We extend $\mathbf{A}$ and $\mathbf{X}$ to explicitly model the absence of edges and nodes by introducing a separate ghost-edge type and ghost-node type. This will enable us to encourage the generator to produce chemically valid molecular graphs as described below. Thus, we define $\tilde{\mathbf{A}} \in \{0,1\}^{n \times n \times (m+1)}$ and $\tilde{\mathbf{X}} \in \{0,1\}^{n \times (d+1)}$. Each vector $\tilde{\mathbf{A}}_{ij\bullet}$, representing generated edges between nodes i and j, needs to be a member of the simplex $\Delta^m = \{(y_0, y_1, \ldots, y_m) \mid y_k \in \{0,1\}, \sum_{k=0}^{m} y_k = 1\}$, because only none or a single edge between i and j is allowed. Here, we use the zero element to represent the absence of an edge. Similarly, each generated node feature vector $\tilde{\mathbf{x}}_{v_i}$ needs to be a member of the simplex Δ^d, where the zero element represents ghost nodes.

Gumbel-Softmax Trick. The generator is a neural network with two outputs, $\mathrm{MLP}_A(\mathbf{z}) \in \mathbb{R}^{n \times n \times (m+1)}$ and $\mathrm{MLP}_X(\mathbf{z}) \in \mathbb{R}^{n \times (d+1)}$, which are created by linearly projecting hidden units into a $n^2(m+1)$ and $n(d+1)$ dimensional space, respectively. Next, continuous outputs need to be transformed into discrete values according to the rules above to obtain tensors $\tilde{\mathbf{A}}$ and $\tilde{\mathbf{X}}$ representing a generated graph. Since the argmax operation is non-differentiable, we employ the Gumbel-softmax trick [13,24], which uses reparameterization to obtain a continuous relaxation of discrete states. Thus, we obtain an approximately discrete adjacency tensor $\tilde{\mathbf{A}}$ from $\mathrm{MLP}_A(\mathbf{z})$, and feature matrix $\tilde{\mathbf{X}}$ from $\mathrm{MLP}_X(\mathbf{z})$.

Node Connectivity and Valence Constraints. While this allows us to generate graphs with varying number of nodes, the generator could in principle generate graphs consisting of two or more separate connected components. In addition, generating molecules where atoms have the correct number of shared electron pairs (valence) is an important aspect the generator needs to consider, otherwise generated graphs would represent invalid molecules. Finally, we want to prohibit edges between any pair of ghost nodes. All of these issues can be addressed by incorporating regularization terms proposed in [23]. Multiple connected components can be avoided by generating graphs that have a path between every pair of non-ghost nodes. Using the generated tensor $\tilde{\mathbf{A}}$, which explicitly accounts for ghost edges, the number of paths between nodes i and j is given by

$$\tilde{\mathbf{B}}_{ij} = I(i = j) + \sum_{k=1}^{m} \sum_{p=1}^{n-1} \left(\tilde{\mathbf{A}}^p \right)_{ijk}. \tag{4}$$

The regularizer comprises two terms, the first term encourages non-ghost nodes i and j to be connected by a path, and the second term that a ghost node and non-ghost node remain disconnected:

$$\frac{\mu}{n^2} \sum_{i,j} \left[1 - (\tilde{\mathbf{x}}_{v_i})_0\right] \left[1 - (\tilde{\mathbf{x}}_{v_j})_0\right] \left[1 - \tilde{\mathbf{B}}_{ij}\right] + \frac{\mu}{n^2} (\tilde{\mathbf{x}}_{v_i})_0 (\tilde{\mathbf{x}}_{v_j})_0 \tilde{\mathbf{B}}_{ij}, \qquad (5)$$

where μ is a hyper-parameter, and $(\tilde{\mathbf{x}}_{v_i})_0 > 0$ if the i-th node is a ghost node.

To ensure atoms have valid valence, we enforce an upper bound – hydrogen atoms are modeled implicitly – on the number of edges of a node, depending on its type (e.g. four for carbon). Let $\mathbf{u} = (u_0, u_1, \ldots, u_d)^\top$ be a vector indicating the maximum capacity (number of bonding electron pairs) a node of a given type can have, where $u_0 = 0$ denotes the capacity of ghost nodes. The vector $\mathbf{b} = (b_0, b_{r_1}, \ldots, b_{r_m})$ denotes the capacity for each edge type with $b_0 = 0$ representing ghost edges. The actual capacity of a node v_i can be computed by $c_{v_i} = \sum_{j \neq i} \mathbf{b}^\top \tilde{\mathbf{A}}_{ij\bullet}$. If c_{v_i} exceeds the value in $\mathbf{u}$ corresponding to the node type of v_i, the generator incurs a penalty. The valence penalty with hyper-parameter $\nu > 0$ is defined as

$$\frac{\nu}{n} \sum_{i=1}^{n} \max(0, c_{v_i} - \mathbf{u}^\top \tilde{\mathbf{x}}_{v_i}). \qquad (6)$$

3.3   Encoder and Discriminators

The architecture of the encoder $g_\phi(G, \varepsilon)$, and the three discriminators $D_\psi(G, \mathbf{z})$, $D_\eta(G_1, G_2)$, and $D_\xi(G)$ are closely related, because they all take graphs as input. First, we extract node-level descriptors by stacking several GCN layers. Next, node descriptors are aggregated to obtain a graph-level descriptor, which forms the input to a MLP. Here, inputs are multi-graphs with m edge types, which we model by extending the Graph Isomorphism Network (GIN) architecture [37] to multi-graphs. Let $\mathbf{h}_{v_i}^{(l+1)}$ denote the descriptor of node v_i after the l-th GIN layer, with $\mathbf{h}_{v_i}^{(0)} = \mathbf{x}_{v_i}$, then node descriptors get updated as follows:

$$\mathbf{h}_{v_i}^{(l+1)} = \tanh\left[\sum_{k=1}^{m} \mathrm{MLP}_{r_k}^{(l)}\left((1 + \epsilon^{(l)})\mathbf{h}_{v_i}^{(l)} + \sum_{u \in \mathcal{N}_{r_k}(v_i)} \mathbf{h}_u^{(l)}\right)\right], \qquad (7)$$

where $\epsilon^{(l)} \in \mathbb{R}$ is a learnable weight, and $\mathcal{N}_{r_k}(v_i) = \{u \,|\, (u, r_k, v_i) \in \mathcal{E}\}$. Next, graph-level node aggregation is performed. We use skip connections [38] to aggregate node-level descriptors from all L GIN layers and soft attention [19] to allow the network to learn which node descriptors to use. The graph-level descriptor $\mathbf{h}_G$ is defined as

$$\mathbf{h}_{v_i}^c = \mathrm{CONCAT}(\mathbf{x}_{v_i}, \mathbf{h}_{v_i}^{(1)} \ldots, \mathbf{h}_{v_i}^{(L)}), \qquad (8)$$

$$\mathbf{h}_{v_i}^{c'} = \tanh(\mathbf{W}_1 \mathbf{h}_{v_i}^c + \mathbf{b}_1), \qquad \mathbf{h}_G = \sum_{v \in \mathcal{V}} \sigma(\mathbf{W}_2 \mathbf{h}_v^{c'} + \mathbf{b}_2) \odot \mathbf{h}_v^{c'}, \quad (9)$$

where $\mathbf{W}$ and $\mathbf{b}$ are parameters to be learned. Graph-level descriptors can be abstracted further by adding an additional MLP on top, yielding $\mathbf{h}_G'$. The discriminator $D_\xi(G)$ contains a single GIN module, $D_\eta(G_1, G_2)$ contains two GIN modules to extract descriptors $\mathbf{h}_{G_1}'$ and $\mathbf{h}_{G_2}'$, which are combined by component-wise multiplication and fed to a 2-layer MLP. $D_\psi(G, \mathbf{z})$ has a noise vector as second input, which is the input to an MLP whose output is concatenated with $\mathbf{h}_G'$

and linearly projected to form $\log[\sigma(D_\psi(G, \mathbf{z}))]$. The encoder $g_\phi(G, \varepsilon)$ is identical to $D_\psi(G, \mathbf{z})$, except that its output matches the dimensionality of $\mathbf{z}$.

4   Evaluation Metrics

To evaluate generated molecules, we adapted the metrics proposed in [4]: the proportion of valid, unique, and novel molecules, and a comparison of 10 physicochemical descriptors to assess whether new molecules have similar chemical properties as the reference data. *Validity* is the percentage of molecular graphs with a single connected component and correct valence for all its nodes. *Uniqueness* is the percentage of unique molecules within a set of N randomly sampled valid molecules. *Novelty* is the percentage of molecules not in the training data within a set of N unique randomly sampled molecules. Validity and novelty are percentages with respect to the set of valid and unique molecules, which we obtain by repeatedly sampling N molecules (up to 10 times). Models that do not generate N valid/unique molecules will be penalized. To measure to which extent a model is able to estimate the distribution of molecules from the training data, we first compute the distribution of 10 chemical descriptors d_k used in [4]. *Internal similarity* is a measure of diversity that is defined as the maximum Tanimoto similarity with respect to all other molecules in the dataset – using the binary Extended Connectivity Molecular Fingerprints with diameter 4 (ECFP4) [32]. The remaining descriptors measure physicochemical properties of molecules (see Fig. 4 for a full list).

For each descriptor, we compute the difference between the distribution with respect to generated molecules (Q) and the reference data (P) using the 1-Wasserstein distance. This is in contrast to [4], who use the Kullback-Leibler (KL) divergence $D_{\mathrm{KL}}(P \parallel Q)$. Using the KL divergence has several drawbacks: (i) it is undefined if the support of P and Q do not overlap, and (ii) it is not symmetric. Due to the lack of symmetry, it makes a big difference whether Q has a bigger support than P or the other way around. Most importantly, an element outside of the support of P is not penalized, therefore $D_{\mathrm{KL}}(P \parallel Q)$ will remain unchanged whether samples of Q are just outside of the support of P or far away.

The 1-Wasserstein distance, also called Earth Mover's Distance (EMD), is a valid metric and does not have these undesirable properties. It also offers more flexibility in how out-of-distribution samples are penalized by choosing an appropriate ground distance (e.g.. Euclidean distance for quadratic penalty, and Manhattan distance for linear penalty). We approximate the distribution on the reference data and the generated data using histograms $\mathbf{h}_k^{\mathrm{ref}}$ and $\mathbf{h}_k^{\mathrm{gen}}$, and define the ground distance $\mathbf{C}_{ij}(k)$ between the edges of the i-th and j-th bin as the Euclidean distance, normalized by the minimum of the standard deviation of descriptor d_k on the reference and the generated data. As overall measure of how well properties of generated molecules match those of molecules in the reference data, we compute the mEMD, defined as

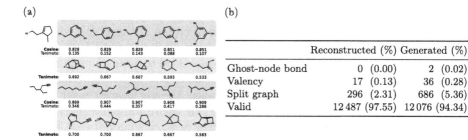

(a)

(b)

	Reconstructed (%)	Generated (%)
Ghost-node bond	0 (0.00)	2 (0.02)
Valency	17 (0.13)	36 (0.28)
Split graph	296 (2.31)	686 (5.36)
Valid	12 487 (97.55)	12 076 (94.34)

Fig. 2. (a) Nearest neighbors with respect to the molecule in the first column in embedding space (green rows) and according to Tanimoto similarity (gray rows). (b) Frequency of errors of ALMGIG on test set. (Color figure online)

$$\text{mEMD}(\mathbf{h}_k^{\text{ref}}, \mathbf{h}_k^{\text{gen}}) = \tfrac{1}{10} \sum_{k=1}^{10} \exp[-\text{EMD}(\mathbf{h}_k^{\text{ref}}, \mathbf{h}_k^{\text{gen}})], \qquad (10)$$

$$\text{EMD}(\mathbf{h}_k^{\text{ref}}, \mathbf{h}_k^{\text{gen}}) = \min_{\mathbf{P} \in \text{U}(\mathbf{h}_k^{\text{ref}}, \mathbf{h}_k^{\text{gen}})} \langle \mathbf{C}(k), \mathbf{P} \rangle, \qquad (11)$$

where $\mathbf{U}(\mathbf{a}, \mathbf{b})$ is the set of coupling matrices with $\mathbf{P}\mathbb{1} = \mathbf{a}$ and $\mathbf{P}^\top \mathbb{1} = \mathbf{b}$ [33]. A perfect model with EMD $= 0$ everywhere would obtain a mEMD score of 1.

5   Experiments

In our experiments, we use molecules from the QM9 dataset [31] with at most 9 heavy atoms. We consider $d = 4$ node types (atoms C, N, O, F), and $m = 3$ edge types (single, double, and triple bonds). After removal of molecules with non-zero formal charge, we retained 131 941 molecules, which we split into 80% for training, 10% for validation, and 10% for testing.

We extensively compare ALMGIG against three state-of-the-art VAEs: NeVAE [34] and CGVAE [22] are graph-based VAEs with validity constraints, while GrammarVAE [16] uses the SMILES representation. We also compare against MolGAN [6], which is a Wasserstein GAN without inference network, reconstruction loss, node connectivity, or valence constraints. Finally, we include a *random graph generation model*, which only enforces valence constraints during generation, similar to CGVAE and NeVAE, but selects node types ($\mathbf{X}$) and edges ($\mathbf{A}$) randomly. Note that generated graphs can have multiple connected components if valence constraints cannot be satisfied otherwise; we consider these to be invalid. For NeVAE, we use our own implementation, for the remaining methods, we use the authors' publicly available code. Further details are described in [28, sec. B].

5.1   Latent Space

In the first experiment, we investigate properties of the encoder and the associated latent space. We project molecules of the test set into the latent space, and perform a k nearest neighbor search to find the closest latent representation of a molecule in the training set in terms of cosine distance. We compare results against the nearest neighbors by Tanimoto similarity of ECFP4 fingerprints [32]. Figure 2a shows that the two approaches lead to quite different sets of nearest neighbors. Nearest neighbors based on molecules' latent representation usually differ by small substructures. For instance, the five nearest neighbor in the first row of Fig. 2a differ by the location of side chains around a shared ring structure. On the other hand, the topology of nearest neighbors by Tanimoto similarity (second row) differs considerably from the query. Moreover, all but one nearest neighbor contain nitrogen atoms, which are absent from the query. In the second example (last two rows), the nearest neighbors in latent space are all linear structures with one triple bond and nitrogen, whereas all nearest neighbors by Tanimoto similarity contain ring structures and only one contains a triple bond. Additional experiments with respect to interpolation in the latent space are in [28, sect. A.1].

5.2   Molecule Generation

Next, we evaluate the quality of generated molecules. We generate molecules from $N = 10000$ latent vectors, sampled from a unit sphere Gaussian and employ the metrics described in Sect. 4. Note that previous work defined uniqueness and novelty as the percentage with respect to all valid molecules rather than N, which is hard to interpret, because models with low validity would have high novelty. Hence, percentages reported here are considerably lower.

Figure 3a shows that ALMGIG generates molecules with high validity (94.9%) and is only outperformed by CGVAE, which by design is constrained to only generate valid molecules, but has ten times more parameters than ALMGIG (1.1M vs. 13M). Moreover, ALMGIG ranks second in novelty and third in uniqueness (excluding random); we will investigate the reason for this difference in detail in the next section on distribution learning. Graph-based NeVAE always generates molecules with correct valence, but often (88.2%) generates graphs with multiple connected components, which we regard as invalid. Its set of valid generated molecules has the highest uniqueness. We can also observe that graph-based models outperform the SMILES-based GrammarVAE, which is prone to generate invalid SMILES representation of molecules, which is a known problem [2,10,21,30,35]. MolGAN is a regular GAN without inference network. It is struggling to generate molecules with valid valence and only learned one particular mode of the distribution, which we will discuss in more detail in the next section. It is inferior to ALMGIG in all categories, in particular with respect to novelty and uniqueness of generated molecules.

Next, we inspect the reason for generated molecules being invalid to assess whether imposed node connectivity and valence constraints are effective. Molecules can be generated by (a) reconstructing the latent representation of a real molecule in the test set, or (b) by decoding a random latent representation draw from an isotropic Gaussian (see Fig. 1). Figure 2b reveals that most errors are due to multiple disconnected graphs being generated (2–5%). While

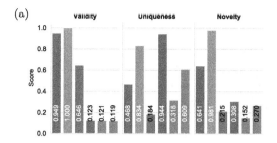
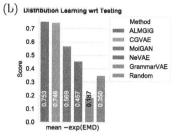

Fig. 3. (a) Overview of simple molecule generation statistics. (b) Comparison with respect to the proposed mEMD evaluation metric for distribution learning. See Fig. 4 and [28, Fig. A.2] for differences between individual distributions.

individual components do represent valid molecules, we treat them as erroneous molecular graphs. In these instances, 98–99% of graphs have 2 connected components and the remainder has 3. Less than 0.3% of molecules have atoms with improper valence and less than 0.1% of graphs have atomic bonds between ghost nodes. Therefore, we conclude that the constraints are highly effective.

Finally, we turn to the random graph generation model. It achieves a relatively high uniqueness of 60.9%, which ranks third, and has a higher novelty than MolGAN and GrammarVAE. Many generated graphs have multiple connected components, which yields a low validity. The fact that the random model cannot be clearly distinguished from trained models, indicates that validity, uniqueness, and novelty do not accurately capture what we are really interested in: *Can we generate chemically meaningful molecules with similar properties as in the training data?* We will investigate this question next.

5.3 Distribution Learning

While the simple overall statistics in the previous section can be useful rough indicators, they ignore the physicochemical properties of generated molecules and do not capture to which extent a model is able to estimate the distribution of molecules from the training data. Therefore, we compare the distribution of 10 chemical descriptors in terms of EMD (11). Distributions of individual descriptors are depicted in Fig. 4 and [28, Fig. A.2].

First of all, we want to highlight that using the proposed mEMD score, we can easily identify the random model (see Fig. 3b), which is not obvious from the simple summary statistics in Fig. 3a. In particular, from Fig. 3a we could have concluded that NeVAE is only marginally better than the random model. The proposed scheme clearly demonstrates that NeVAE is superior to the random model (overall score 0.457 vs 0.350). Considering differences between individual descriptors reveals that randomly generated molecules are not meaningful due to higher number of hydrogen acceptors, molecular weight, molecular complexity, and polar surface area (see [28, fig. A.2d]).

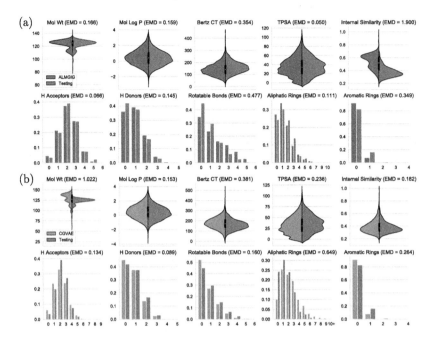

Fig. 4. Distribution of chemical properties of the test data and of 10,000 unique molecules generated by (a) ALMGIG, and (b) CGVAE [22]. Mol Wt: molecular weight. Mol Log P: water-octanol partition coefficient. Bertz CT: molecular complexity index. TPSA: molecular polar surface area. H Acceptors (Donors): Number of hydrogen acceptors (donors).

ALMGIG achieves the best overall mEMD score, with molecular weight and polar surface area best matching that of the test data by a large margin (EMD = 0.17 and 0.05, see Fig. 4a). The highest difference is due to the internal similarity of generated molecules (EMD = 1.90). Among all VAEs, CGVAE is performing best (see Fig. 4b). However, it notably produces big molecules with larger weight (EMD = 1.02) and lower polar surface area (EMD = 0.238). The former explains its high novelty value in Fig. 3a: by creating bigger molecules than in the data, a high percentage of generated molecules is novel and unique. Molecules generated by NeVAE (see [28, Fig. A.2a]) have issues similar to CGVAE: they have larger weight (EMD = 1.72) and higher polar surface area (EMD = 0.839), which benefits uniqueness. In contrast to ALMGIG and CGVAE, it also struggles to generate molecules with aromatic rings (EMD = 0.849). GrammarVAE in [28, Fig. A.2b] is far from capturing the data distribution, because it tends to generate long SMILES strings corresponding to heavy molecules (EMD = 5.66 for molecular weight). Molecules produced by MolGAN are characterized by a striking increase in internal similarity without any overlap with the reference distribution (EMD = 6.41; [28, Fig. A.2c]). This result is an example where comparison by KL divergence would be undefined. An interesting detail can be derived from the distribution of molecular weight: MolGAN has problems generating molecules with intermediate to low molecular weight (<115 g/mol), which highlights a common problem with GANs, where only one mode of the distribution is learned by the model (mode collapse). In contrast, ALMGIG can capture this mode, but still misses the smaller mode with molecular weight below 100 g/mol (see Fig. 4a). This mode contains only 2344 molecules (2.2%), which makes it challenging – for any model – to capture. This demonstrates that our proposed Wasserstein-based evaluation can reveal valuable insights that would have been missed when solely relying on validity, novelty, and uniqueness.

Table 1. Configurations evaluated in our ablation study. SC: Skip-connections.

	Penalties		Discriminators			Architecture		
	Conn.	Valence	Unary	Joint	Cycle	GIN SC	Gen. SC	Attention
ALMGIG	✓	✓	✓	✓	✓	✓	✓	✓
No connectivity	✗	✓	✓	✓	✓	✓	✓	✓
No valence	✓	✗	✓	✓	✓	✓	✓	✓
No Conn+Valence	✗	✗	✓	✓	✓	✓	✓	✓
No GIN SC	✓	✓	✓	✓	✓	✗	✓	✓
No generator SC	✓	✓	✓	✓	✓	✓	✗	✓
No attention	✓	✓	✓	✓	✓	✓	✓	✗
ALICE	✓	✓	✗	✓	✓	✓	✓	✓
ALI	✓	✓	✗	✓	✗	✓	✓	✓
(W)GAN	✓	✓	✓	✗	✗	✓	✓	✓

5.4   Ablation Study

Next, we evaluate a number of modeling choices by conducting an extensive ablation study (see Table 1). We first evaluate the impact of the connectivity (5) and valence penalty (6) during training. Next, we evaluate architectural choices, namely (i) skip connections in the GIN of encoder and discriminators, (ii) skip connections in the generator, and (iii) soft attention in the graph pooling layer (8). Finally, we use the proposed architecture and compared different adversarial learning schemes: ALI [7], ALICE [17], and a traditional WGAN without an encoder network $g_\phi(G, \varepsilon)$ [1,9]. Note that WGAN is an extension of MolGAN [6] with connectivity and valence penalties, and GIN-based architecture.

Our results are summarized in Fig. 5 (details in [28, Fig. A.2]). As expected, removing valence and connectivity penalties lowers the validity, whereas the drop when only removing the valence penalty is surprisingly small, but does lower the uniqueness considerably. Regarding our architectural choices, results demonstrate that only the proposed architecture of ALMGIG is able to generate a sufficient number of molecules with aromatic rings (EMD = 0.35), when removing the attention mechanism (EMD = 0.96), skip connections from the

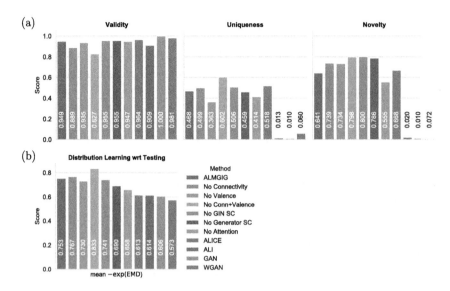

Fig. 5. Results of ablation study. (a) Simple molecule generation statistics. (b) Comparison with respect to the proposed mEMD evaluation metric for distribution learning.

GIN (EMD = 0.74), or from the generator (EMD = 1.26) the EMD increases at least two-fold. When comparing alternative adversarial learning schemes, we observe that several configurations resulted in mode collapse. It is most obvious for ALI, GAN, and WGAN, which can only generate a few molecules and thus have very low uniqueness and novelty. In particular, ALI and WGAN are unable to generate molecules with aromatic rings. ALICE is able to generate more diverse molecules, but is capturing only a single mode of the distribution: molecules with low to medium weight are absent. This demonstrates that including a unary discriminator, as in ALMGIG, is vital to capture the full distribution over chemical structures. Finally, it is noteworthy that WGAN, using our proposed combination of GIN architecture and penalties, outperforms MolGAN [6]. This demonstrates that our proposed architecture can already improve existing methods for molecule generation considerably. When also extending the adversarial learning framework, the results demonstrate that ALMGIG can capture the underlying distribution of molecules more accurately than any competing method.

6   Conclusion

We formulated generation and inference of molecular graphs as a likelihood-free adversarial learning task. Compared to previous work, it allows training without explicitly computing a reconstructing loss, which would require solving an expensive graph isomorphism problem. Moreover, we argued that the common validation metrics validity, novelty, and uniqueness are insufficient to properly assess the performance of algorithms for molecule generation, because they ignore physicochemical properties of generated molecules. Instead, we proposed to compute the 1-Wasserstein distance between distributions of physicochemical properties of molecules. We showed that the proposed adversarial learning framework for molecular graph inference and generation, ALMGIG, allows efficiently exploring the space of molecules via molecules' continuous latent representation, and that it more accurately represents the distribution over the space of molecules than previous methods.

Acknowledgements. This research was partially supported by the Bavarian State Ministry of Education, Science and the Arts in the framework of the Centre Digitisation.Bavaria (ZD.B), and the Federal Ministry of Education and Research in the call for Computational Life Sciences (DeepMentia, 031L0200A).

References

1. Arjovsky, M., Chintala, S., Bottou, L.: Wasserstein generative adversarial networks. In: 34th International Conference on Machine Learning, vol. 70, pp. 214–223 (2017)
2. Blaschke, T., Olivecrona, M., Engkvist, O., Bajorath, J., Chen, H.: Application of generative autoencoder in de Novo molecular design. Mol. Inf. **37**(1–2), 1700123 (2018)
3. Bradshaw, J., Paige, B., Kusner, M.J., Segler, M., Hernández-Lobato, J.M.: A model to search for synthesizable molecules. In: Advances in Neural Information Processing Systems, vol. 32, pp. 7937–7949 (2019)
4. Brown, N., Fiscato, M., Segler, M.H., Vaucher, A.C.: GuacaMol: benchmarking models for de Novo molecular design. J. Chem. Inf. Model. **59**(3), 1096–1108 (2019)
5. Dai, H., Tian, Y., Dai, B., Skiena, S., Song, L.: Syntax-directed variational autoencoder for structured data. In: 6th International Conference on Learning Representations (2018)
6. De Cao, N., Kipf, T.: MolGAN: an implicit generative model for small molecular graphs (2018). https://arxiv.org/abs/1805.11973
7. Dumoulin, V., et al.: Adversarially learned inference. In: 5th International Conference on Learning Representations (2017)
8. Gómez-Bombarelli, R., Wei, J.N., Duvenaud, D., Hernández-Lobato, J.M., Sánchez-Lengeling, B., et al.: Automatic chemical design using a data-driven continuous representation of molecules. ACS Cent. Sci. **4**(2), 268–276 (2018)
9. Goodfellow, I., et al.: Generative adversarial nets. In: Advances in Neural Information Processing Systems, vol. 27, pp. 2672–2680 (2014)
10. Guimaraes, G.L., Sanchez-Lengeling, B., Outeiral, C., Farias, P.L.C., Aspuru-Guzik, A.: Objective-reinforced generative adversarial networks (ORGAN) for sequence generation models (2017). https://arxiv.org/abs/1705.10843
11. Gulrajani, I., Ahmed, F., Arjovsky, M., Dumoulin, V., Courville, A.: Improved training of Wasserstein GANs. In: Advances in Neural Information Processing Systems, vol. 30, pp. 5767–5777 (2017)
12. Heusel, M., Ramsauer, H., Unterthiner, T., Nessler, B., et al.: GANs trained by a two time-scale update rule converge to a local nash equilibrium. In: Advances in Neural Information Processing Systems, vol. 30, pp. 6626–6637 (2017)
13. Jang, E., Gu, S., Poole, B.: Categorical reparameterization with Gumbel-softmax. In: 5th International Conference on Learning Representations (2017)

14. Jin, W., Barzilay, R., Jaakkola, T.: Junction tree variational autoencoder for molecular graph generation. In: 35th International Conference on Machine Learning, pp. 2323–2332 (2018)
15. Kadurin, A., Nikolenko, S., Khrabrov, K., Aliper, A., Zhavoronkov, A.: druGAN: an advanced generative adversarial autoencoder model for de novo generation of new molecules with desired molecular properties in silico. Mol. Pharm. **14**(9), 3098–3104 (2017)
16. Kusner, M.J., Paige, B., Hernández-Lobato, J.M.: Grammar variational autoencoder. In: 34th International Conference on Machine Learning, pp. 1945–1954 (2017)
17. Li, C., et al.: ALICE: towards understanding adversarial learning for joint distribution matching. In: Advances in Neural Information Processing Systems, vol. 30, pp. 5495–5503 (2017)
18. Li, Y., Zhang, L., Liu, Z.: Multi-objective de novo drug design with conditional graph generative model. J. Cheminform. **10**, 33 (2018)
19. Li, Y., Tarlow, D., Brockschmidt, M., Zemel, R.: Gated graph sequence neural networks. In: 4th International Conference on Learning Representations (2016)
20. Li, Y., Vinyals, O., Dyer, C., Pascanu, R., Battaglia, P.: Learning deep generative models of graphs (2018). https://arxiv.org/abs/1803.03324
21. Lim, J., Ryu, S., Kim, J.W., Kim, W.Y.: Molecular generative model based on conditional variational autoencoder for de novo molecular design. J. Cheminform. **10**, 31 (2018)
22. Liu, Q., Allamanis, M., Brockschmidt, M., Gaunt, A.: Constrained graph variational autoencoders for molecule design. In: Advances in Neural Information Processing Systems, vol. 31, pp. 7806–7815 (2018)
23. Ma, T., Chen, J., Xiao, C.: Constrained generation of semantically valid graphs via regularizing variational autoencoders. In: Advances in Neural Information Processing Systems, vol. 31, pp. 7113–7124 (2018)
24. Maddison, C.J., Mnih, A., Teh, Y.W.: The concrete distribution: a continuous relaxation of discrete random variables. In: 5th International Conference on Learning Representations (2017)
25. Olivecrona, M., Blaschke, T., Engkvist, O., Chen, H.: Molecular de-novo design through deep reinforcement learning. J. Cheminform. **9**, 48 (2017)
26. Podda, M., Bacciu, D., Micheli, A.: A deep generative model for fragment-based molecule generation. In: Proceedings of AISTATS (2020)
27. Polishchuk, P.G., Madzhidov, T.I., Varnek, A.: Estimation of the size of drug-like chemical space based on GDB-17 data. J. Comput. Aided Mol. Des. **27**(8), 675–679 (2013)
28. Pölsterl, S., Wachinger, C.: Adversarial learned molecular graph inference and generation (2020). https://arxiv.org/abs/1905.10310
29. Popova, M., Isayev, O., Tropsha, A.: Deep reinforcement learning for de novo drug design. Sci. Adv. **4**(7), eaap7885 (2018)
30. Putin, E., et al.: Adversarial threshold neural computer for molecular de novo design. Mol. Pharm. **15**(10), 4386–4397 (2018)
31. Ramakrishnan, R., Dral, P.O., Rupp, M., von Lilienfeld, O.A.: Quantum chemistry structures and properties of 134 kilo molecules. Sci. Data **1**, 1–7 (2014)
32. Rogers, D., Hahn, M.: Extended-connectivity fingerprints. J. Chem. Inf. Model. **50**(5), 742–754 (2010)
33. Rubner, Y., Tomasi, C., Guibas, L.J.: The earth mover's distance as a metric for image retrieval. Int. J. Comput. Vis. **40**(2), 99–121 (2000)

34. Samanta, B., De, A., Jana, G., Ganguly, N., Gomez-Rodriguez, M.: NeVAE: a deep generative model for molecular graphs. In: 33rd AAAI Conference on Artificial Intelligence, pp. 1110–1117 (2019)
35. Segler, M.H.S., Kogej, T., Tyrchan, C., Waller, M.P.: Generating focused molecule libraries for drug discovery with recurrent neural networks. ACS Cent. Sci. **4**(1), 120–131 (2018)
36. Simonovsky, M., Komodakis, N.: GraphVAE: towards generation of small graphs using variational autoencoders. In: Kůrková, V., Manolopoulos, Y., Hammer, B., Iliadis, L., Maglogiannis, I. (eds.) ICANN 2018. LNCS, vol. 11139, pp. 412–422. Springer, Cham (2018). https://doi.org/10.1007/978-3-030-01418-6_41
37. Xu, K., Hu, W., Leskovec, J., Jegelka, S.: How powerful are graph neural networks? In: 7th International Conference on Learning Representations (2019)
38. Xu, K., Li, C., Tian, Y., Sonobe, T., Kawarabayashi, K.I., Jegelka, S.: Representation Learning on Graphs with Jumping Knowledge Networks. In: 35th International Conference on Machine Learning, pp. 5453–5462 (2018)
39. You, J., Liu, B., Ying, R., Pande, V., Leskovec, J.: Graph convolutional policy network for goal-directed molecular graph generation. In: Advances in Neural Information Processing Systems, vol. 31, pp. 6412–6422 (2018)

A Generic and Model-Agnostic Exemplar Synthetization Framework for Explainable AI

Antonio Barbalau[1,3]($\boxtimes$), Adrian Cosma[2,3], Radu Tudor Ionescu[1], and Marius Popescu[1,3]

[1] University of Bucharest, Bucharest, Romania
`abarbalau@fmi.unibuc.ro`
[2] University Politehnica of Bucharest, Bucharest, Romania
[3] Sparktech Software, Bucharest, Romania

Abstract. With the growing complexity of deep learning methods adopted in practical applications, there is an increasing and stringent need to explain and interpret the decisions of such methods. In this work, we focus on explainable AI and propose a novel generic and model-agnostic framework for synthesizing input exemplars that maximize a desired response from a machine learning model. To this end, we use a generative model, which acts as a prior for generating data, and traverse its latent space using a novel evolutionary strategy with momentum updates. Our framework is generic because (*i*) it can employ any underlying generator, e.g.. Variational Auto-Encoders (VAEs) or Generative Adversarial Networks (GANs), and (*ii*) it can be applied to any input data, e.g. images, text samples or tabular data. Since we use a zero-order optimization method, our framework is model-agnostic, in the sense that the machine learning model that we aim to explain is a black-box. We stress out that our novel framework *does not* require access or knowledge of the internal structure or the training data of the black-box model. We conduct experiments with two generative models, VAEs and GANs, and synthesize exemplars for various data formats, image, text and tabular, demonstrating that our framework is generic. We also employ our prototype synthetization framework on various black-box models, for which we only know the input and the output formats, showing that it is model-agnostic. Moreover, we compare our framework (available at https://github.com/antoniobarbalau/exemplar) with a model-dependent approach based on gradient descent, proving that our framework obtains equally-good exemplars in a shorter time.

Keywords: Explainable AI · Black-box · Generative modelling · Evolutionary algorithm · Prototype synthetization · Exemplar generation

---

Electronic supplementary material The online version of this chapter (https://doi.org/10.1007/978-3-030-67661-2_12) contains supplementary material, which is available to authorized users.

© Springer Nature Switzerland AG 2021
F. Hutter et al. (Eds.): ECML PKDD 2020, LNAI 12458, pp. 190–205, 2021.
https://doi.org/10.1007/978-3-030-67661-2_12

1   Introduction

Due to the rise of deep learning [27] in recent years, scientists and engineers have developed solutions based on deep learning to solve almost every machine learning task in production-ready systems. While deep learning models obtain impressive accuracy levels [10,20,26,40], surpassing even human-level performance for many tasks [9,13], their inherent complexity transforms them into opaque decision systems. Critical processes that deal with potentially sensitive information in areas such as finance, medicine, security and justice have become essentially black boxes, with the underlying logic being too complex even for data scientists and inaccessible to the end-users. Deep learning models require training data usually generated and annotated by humans, thus containing various biases, including discriminatory views on race and gender. Hence, models trained on biased data will inherit these biases and, in turn, will make decisions that are unfair or socially unacceptable [5]. Another potential problem of such highly complex decision systems is the chance of inadvertently making correct decisions, but for the wrong reasons. A popular example here is an image of a wolf being classified correctly, but only because of the snowy background [41]. While this is a harmless example, the same kind of decisions resulting from spurious correlations from large amounts of data could potentially have a great negative impact on human lives.

In this context, explainable AI, a field that studies how artificial intelligence (AI) methods and techniques can be understood by human experts, gained a lot of attention recently. While there are many types of explanations that an explanatory method could provide [1,18], including rule extraction and outcome prediction, we chose to focus on explanation by exemplar generation. Prototypical examples (exemplars), which can describe a complex underlying data distribution, can offer meaningful insights about the behavior of a model, when a simple explanation is hard to extract. Prototype selection methods [2,7,19,32,48] return examples that are representative for a set of similar instances. An exemplar can be one of the instances observed in the training data set [2,19,48], or it can be an artificially-generated example in the data space [7,32].

Current explainable AI approaches mainly consider the glass-box scenario of highly-complex deep learning models, with no restriction towards accessing the models' weights. For example, Nguyen et al. [36] applied gradient descent to back-propagate through the model in order to synthesize the preferred inputs for neurons, which would not be possible without knowing the weights. We consider the more realistic case in which we have no information about the weights or other internal components of the model. Our framework is only allowed to inspect the input format and the output predictions. This strict definition of black-box models enables us to explain just about any machine learning model, not only deep learning models trained with gradient descent. In other words, our framework is *model-agnostic*.

While related methods obtain exemplars for image classification models [17, 36], we propose a more generic framework that it is not tied to a particular data modality – be it image, text, structured or tabular. Our framework is also agnostic with respect to the underlying generative model used to synthesize exemplars – be

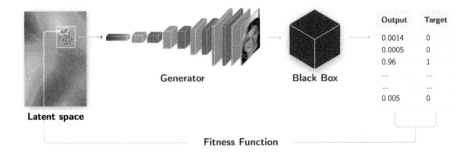

Fig. 1. General diagram of our exemplar generation method. Given a black-box classifier trained on some data set, we first train a generative model (VAE / GAN) on a disjoint data set, not necessarily containing the same class distribution. We then traverse the latent space of the generator, employing an evolutionary strategy, such that the black-box model provides a certain prediction (chosen beforehand) when the generated exemplar is given as input to the black-box model. Best viewed in color. Patent Pending No. 63053589.

it a Variational Auto-Encoders (VAE) [24] or a Generative Adversarial Network (GAN) [14]. To this end, we consider our framework as *generic*.

To our best knowledge, we are the first to propose a generic and model-agnostic explainable AI framework to synthesize exemplars that exhibit a high response with respect to the output of a black-box model. We exploit the structured latent space of the underlying generative model to progressively search for latent codes that can accurately explain a particular class or combination of classes, as learned by the model. We employ a novel evolutionary strategy with momentum updates as our search policy, as it was proven that evolutionary algorithms [42] represent an efficient way for black-box optimization, relieving us from the need to propagate gradients. Our framework is illustrated in Fig. 1.

We conduct experiments with two generative models, VAEs and GANs, to synthesize exemplars for various data formats, namely image, text and tabular, demonstrating that our framework is generic. We also employ our prototype synthetization framework on various black-box models, e.g.. Random Forest or neural networks, for which we only know the input and the output formats, showing that it is model-agnostic. We present experiments showing that our framework can also generalize to classes unseen by the generator. Moreover, we compare our framework with a model-dependent approach [36] based on gradient descent optimization, demonstrating that our framework converges to equally-good exemplars in a shorter amount of time.

2   Related Work

Explainable AI methods [1,18] for black-box models have gained significant attention in recent years, as bias in data and model training [3,5] have resulted in regulatory actions from the European Union [16], restricting the use of decision-making machine learning models without an explanatory component. Therefore, explainable AI is an area of research of utter importance, with open issues ranging from thorough testing and regulatory compliance [34] to finding what kind of explanations are best suited to answer questions of fairness.

Explainable AI methods can be classified into different taxonomies based on various criteria, e.g. as global or local methods, as model-specific or model-agnostic methods and so on [1,18]. We hereby focus on methods that are closer to our own, i.e. on model-agnostic or exemplar-based methods.

We note that deep neural networks are commonly regarded as complex models, their decisions being hard to understand due to the hierarchical non-linear structure. Local explainable AI methods [22,28,30,35,41,43,46,49] deal with explaining a particular decision, i.e. the decision provided for a certain input example. Some of these methods [30,41] employ a directly-interpretable surrogate model that is trained on the vicinity of an input sample, by modifying or occluding input features. Access to the network's weights enables the back-propagation of gradients, leading to saliency-based methods such as CAM [49], Grad-CAM [43] and Grad-CAM++ [6]. Unlike these local explainable AI methods, we propose a framework that does not look inside the models, i.e. the models are complete black boxes. Furthermore, our approach provides generic (non-local) explanations by synthesizing exemplars not tied to a certain data sample.

Exemplar-based methods [2,7,19,32,48], which offer a convenient way to communicate meaningful insights about the behavior of a model in situations where a direct explanation is hard to extract, are more closely related to our framework. Some exemplar-based methods select prototypical examples from the training data set [2,19,48]. For example, Gurumoorthy et al. [19] proposed a method aimed at describing the data distribution through case-based reasoning, while Chen et al. [7] presented an approach that selects relevant samples from the data set that have contributed to a decision of the model. Different from methods selecting exemplars from the data set, we introduce a framework that generates realistic examples without having access to the data set used to train the black box.

Synthesizing artificial (unrealistic) examples that provide maximal responses for a particular network component can shed light on the preferences and biases of a trained model. Indeed, methods [33,44] of visualizing a convolutional neural network represent a popular way of understanding its behavior. However, such methods are based on different variations of gradient ascent, requiring access to the internal weights of the model. Unlike such methods [33,44], we can produce realistic examples while treating the model as a black box.

We identified two works [17,36] that are very closely related to our approach. Our method is similar to that of Nguyen et al. [36] because our method, as much as theirs, requires a deep generative model to synthesize realistic images. Without a deep generative model to act as a realistic image prior, there is a high chance that preferred inputs could end up being unrealistic. In [37], the authors already proved that neural networks output high responses to texture-like images that were generated using genetic algorithms, with little to no resemblance to natural images. Nguyen et al. [36] applied gradient descent to back-propagate through the model in order to synthesize preferred inputs for neurons. Being base on gradient descent, their method requires access to the model's weights.

Different from Nguyen et al. [36], we consider the more realistic case in which we have no information about the weights or other internal components of the model. Our framework is only allowed to inspect the input format and the output predictions. This strict definition of black-box models enables us to explain just about any machine learning model, not only deep learning models trained with gradient descent. Without access to the gradients, our method generates exemplars through a novel evolutionary strategy with momentum updates.

Focusing on image classification, Guidotti et al. [17] presented an approach to explain the decisions of black-box models for a given input sample. Different from Guidotti et al. [17], we show that our method is applicable to different data types, namely to images, text samples and tabular data. We also show that our method works with various generators, namely VAEs and GANs. While Guidotti et al. [17] focus on explaining single instances, we focus on explaining output class probabilities, i.e. our exemplars are not tied to input data samples. All in all, we consider that there are significant differences between our method and that of Guidotti et al. [17].

3   Method

Given a black-box classification model $C : D \rightarrow \mathbb{R}^n$ and a generative model $G : \mathbb{R}^l \rightarrow D$ able to sample from a data set included in D, we aim to traverse the latent space of G using a gradient-free optimization method, namely an evolutionary strategy with momentum updates, in order to synthesize exemplars for which C provides a certain desired output $y \in \mathbb{R}^n$. Here, n is the number of classes, l is the size of the embedding space of the generator, and D is the data space, which depends on the input data type, e.g. for images $D = [0, 255]^{h \times w}$, where h and w are the height and the width of an input image, respectively. In our framework, we impose no restrictions upon the prediction model C. We require no access or knowledge of the internal structure of C, i.e. C is a black box. As generator, we can use any model that takes as input a noise vector z and outputs a corresponding data sample, including Variational Auto-Encoders, Generative Adversarial Networks and Auto-Regressive models. Given the target prediction $y \in \mathbb{R}^n$, we optimize an encoding $z \in \mathbb{R}^l$ such that $\hat{y} = C(G(z))$ is optimally close to y, i.e. $\hat{y} \approx y$. The objective for our optimization problem can be formally expressed as follows:

$$V(y, C, G) = \min_z \sum_{i=1}^{n} (C(G(z)) - y_i)^2 = \min_z \sum_{i=1}^{n} (\hat{y}_i - y_i)^2. \tag{1}$$

Nguyen et al. [36] employed gradient descent to optimize the objective defined in Eq. (1), by back-propagating though the classification model C. However, we assume that access to the internal structure or the weights of the model is not granted, i.e. the model C is a black box. Furthermore, we do not impose any architectural restrictions over the model, i.e. C needs not be a neural network. Even if access to the analytical gradients is not provided due to the black-box nature of the model C, one can still compute the numerical gradients and search for a z that minimizes the objective defined in Eq. (1), using gradient descent. However, since z belongs to an l-dimensional space, computing the numerical gradients for each component in z requires l forward passes through the model C, which is inefficient in comparison to our evolutionary approach. We show in our experiments that we can synthesize exemplars with a confidence greater than 95% with less than l model calls (forward passes) on average. Additionally, we show that our evolutionary approach provides better exemplars and converges faster than gradient descent, even when analytical gradients are available for the classification model, i.e. C becomes a glass-box model as in [36].

We hereby propose a novel evolutionary strategy based on optimization with momentum updates. We note that momentum is incorporated into a standard evolutionary strategy, i.e. the novelty consists in adding momentum updates. Our strategy is formally described in Algorithm 1. In steps 17–22, our algorithm starts by sampling t initial exemplars $z_i \in \mathbb{R}^l$ to form the initial population Z, such that $|Z| = t$. In step 20, each component of an exemplar z_i is sampled from an uniform distribution U over the interval $[-u, u]$. In the same time, we generate the set M of momentum vectors μ_i associated to exemplars z_i. The initial momentum vectors are zero vectors of l components, thus having the same size as the exemplars in Z.

After initializing the population, we perform a selection in step 23 by keeping the top k (elite) exemplars (and associated momentum vectors) that minimize our fitness function. The selection is performed inside the *select* function defined in steps 7–15. The fitness of each exemplar in the current population Z is computed in steps 8–11. Our fitness function is the sum squared error between the target output y and the predicted output $\hat{y}$ for an exemplar $z_i \in Z$:

$$\mathcal{L}(y, \hat{y}) = \sum_{j=1}^{n} (\hat{y}_j - y_j)^2, \qquad (2)$$

where $\hat{y} = C(G(z_i))$.

Until the fitness score f of our least fit exemplar in Z becomes smaller than f_{min}, we repeat steps 25–30. Inside the loop, each exemplar from the current population is duplicated and mutated m times. The mutation, performed inside the *mutate* function defined in steps 1–6, consists in adding a zero-centered Gaussian distributed velocity vector v to the exemplar z. The mutation applied to the exemplar z in step 5 can shift the new exemplar z' in the direction of the gradient. We note that, during exemplar selection, we will choose exemplars that minimize our fitness function defined in Eq. (2). Since the kept exemplars were likely shifted in the right direction during the previous mutation, we added a momentum component to the mutation operation, which leads to faster convergence. The momentum vector μ is added to the velocity v in a weighted sum computed in step 4, inside the *mutate* function. The momentum μ is the previous perturbation (velocity) applied on the exemplar z. The magnitude of the momentum μ with respect to the generated velocity vector v is controlled through the momentum rate α.

The current population together with the mutated duplicates form a new population that passes through the selection process in step 30 of Algorithm 1.

Algorithm 1: Evolutionary Optimization Algorithm with Momentum

---

Input   : A classifier C, a generator G, the target class probabilities y, a
               convergence threshold f_{min}, the initial population size t, a latent
               space boundary u, the number of elite exemplars k to keep during
               selection, the number of mutation operations m, a standard
               deviation s for velocity sampling, a momentum rate α.

Notations: $z \in Z$ – an exemplar z from a population Z; $\mu \in M$ – a momentum
               vector μ from a set M; l – the latent space size;
               v – an l-dimensional velocity vector (perturbation applied to an
               exemplar during mutation).

Output   : Synthesized exemplar z^*.

---

1 **Function** mutate(z,μ):
2 $\quad$ $v \sim N(0, s)$
3 $\quad$ **if** $\mu \neq \mathbf{0}_{1,l}$ **then**
4 $\quad\quad$ $v \leftarrow \alpha \cdot \mu + (1 - \alpha) \cdot v$
5 $\quad$ $z' \leftarrow z + v$
6 $\quad$ **return** (z', v)

7 **Function** select(Z, M):
8 $\quad$ $F \leftarrow \emptyset$
9 $\quad$ **for** $z_i \in Z$ **do**
10 $\quad\quad$ $f_i \leftarrow \mathcal{L}(y, C(G(z_i)))$
11 $\quad\quad$ $F \leftarrow F \cup \{f_i\}$
12 $\quad$ $I \leftarrow argsort(F)$
13 $\quad$ $Z \leftarrow \{z_{i_1}, z_{i_2}, ..., z_{i_k}\}$, where $i_j \in I$
14 $\quad$ $M \leftarrow \{\mu_{i_1}, \mu_{i_2}, ..., \mu_{i_k}\}$, where $i_j \in I$
15 $\quad$ **return** $(Z, M, \max_{f_i \in F}\{f_i\})$

16 **Algorithm** main()
17 $\quad$ $Z \leftarrow \emptyset$
18 $\quad$ $M \leftarrow \emptyset$
19 $\quad$ **for** $i \in \{1, 2, ..., t\}$ **do**
20 $\quad\quad$ $z_i \sim U(-u, u)$
21 $\quad\quad$ $Z \leftarrow Z \cup \{z_i\}$
22 $\quad\quad$ $M \leftarrow M \cup \{\mathbf{0}_{1,l}\}$
23 $\quad$ $Z, M, f \leftarrow select(Z, M)$
24 $\quad$ **while** $f > f_{min}$ **do**
25 $\quad\quad$ **for** $i \in \{1, 2, ..., k\}$ **do**
26 $\quad\quad\quad$ **for** $j \in \{1, 2, ..., m\}$ **do**
27 $\quad\quad\quad\quad$ $z', \mu' \leftarrow mutate(z_i, \mu_i)$
28 $\quad\quad\quad\quad$ $Z \leftarrow Z \cup \{z'\}$
29 $\quad\quad\quad\quad$ $M \leftarrow M \cup \{\mu'\}$
30 $\quad\quad$ $Z, M, f \leftarrow select(Z, M)$
31 $\quad$ $z^* \leftarrow z_1$
32 $\quad$ **return** z^*

By selecting only the top exemplars from every generation, we ensure that only the mutations that brought improvements are kept. Hence, only relevant perturbations are accumulated inside the momentum associated to an exemplar. As we will show in the experiments, the momentum component brings an increase of 19% in convergence speed compared to the plain version (that does not use momentum), subject to using the same hyperparameters. When the fitness of the least fit exemplar in Z goes under the threshold f_{min}, we store the best exemplar z_1 in z^* and return it as the output of our evolutionary algorithm.

4   Experiments

4.1   Data Sets

Adult Data Set. For tabular data, we present experiments on the *Adult Data Set* [25]. This is a binary classification data set for predicting the income of adults based on census information such as race, gender, marital status and level of education. It is composed of 48,842 samples with 14 features.

FER 2013. For image synthetization, we used the *Facial Expression Recognition* (FER) 2013 [15] data set that is comprised of grayscale images of faces representing 7 different classes of emotion. FER 2013 contains 28,709 training images, 3,589 validation images and 3,589 test images. The samples have a wide range of attributes, as they vary in illumination conditions, pose, gender, race and age.

Large Movie Review Dataset. We performed text synthetization experiments on the *Large Movie Review Dataset* [31]. The training set contains 25,000 movie reviews for binary sentiment classification: positive or negative. The test set is similar in size. Each review is highly polarised, neutral or close-to-neutral samples being absent.

4.2   Experimental Setup

For Algorithm 1, we used the same hyperparameters in all experiments: the initial population size is $t = 50$, the number of selected exemplars is $k = 10$, the latent space boundary is $u = 5$, the number of mutations per exemplar is $m = 2$, the standard deviation used for mutations is $s = 0.5$ and the momentum rate is $\alpha = 0.3$. We present results with other hyperparameters in the supplementary.

Setup for Tabular Data. To prove that our framework is truly model-agnostic, we employed a Random Forest (RF) classifier as the black box for the Adult Data Set, which attains an accuracy of 85.1% on the test set, while being trained on half of the training set. In the pre-processing step, 4 of the 6 numerical features (except for capital-gain and capital-loss) were normalized and each of the 8 categorical features was passed through a different embedding layer, generating a vector of two components for each categorical feature. The concatenation of all these features $(6 + 8 \cdot 2 = 22)$ gives us the final representation for the data samples. For data generation, we trained a VAE [11] on the other half of the training set (not used to train the RF classifier). We have synthesized prototype examples with regards to the output class probabilities of the RF classifier. The architecture of the VAE starts with two fully-connected layers having 64 and 128 neurons with batch normalization and Rectified Linear Unit (ReLU) activations, respectively. Finally, an 8-neuron dense layer determines the means and the standard deviations for a 4-dimensional encoding. During the reconstruction phase, embeddings are passed through two dense layers with 128 and 64 neurons, respectively. For the final output, there is an additional layer for numerical columns and an individual softmax layer for each categorical column. The loss is comprised of an L_2-distance component for numerical features and a categorical cross-entropy component for each categorical feature.

Setup for Image Data. For facial expression recognition, we used the VGG-16 [45] architecture, which yields an accuracy of 67.4% on the FER 2013 test set. As generators, we employed a Progressively Growing GAN [23] and a VAE trained with cyclical annealing [12]. The architectures and specifications for these networks are the ones specified in [23] and [12], respectively. We performed several experiments on this data modality. Firstly, we provide a quantitative comparison between exemplars synthesized using our framework and exemplars generated in the glass-box scenario, i.e. when access to the classifier structure and weights is available, as in [36]. Secondly, we provide a comparison of the convergence times of the two approaches, i.e. our evolutionary algorithm with momentum versus gradient descent (based on analytical gradients). In a set of preliminary trials, we noticed that the value of the gradient rapidly decreases within less than 5 iterations from values in the range of 10^{-1} to values in the range of 10^{-8}, which impedes the gradient descent optimization process. Therefore, we experimented using gradient descent with momentum. We measure converge times from two perspectives: the number of model calls until the generated sample is classified with a confidence greater than 95% and the duration of the optimization process in seconds. Thirdly, we show that our method is able to generate exemplars when the generator G and the model C are trained on the same data samples and on different data samples. Additionally, we prove that our method has the ability to generalize to previously unseen classes. To this end, we train the generator on all classes except one, e.g.. *surprised*, and successfully generate *surprised* exemplars even though the generator has never seen a surprised face before.

Setup for Text Data. Sentence generation from latent embeddings has been proposed in the past, both through VAEs [4,8] and through GANs [39]. In our approach, we used an LSTM VAE [4], with a latent dimension of 128 neurons and hidden size of 512 neurons for both encoder and decoder networks. We used GloVe embeddings [38] for the tokens processed by the encoder. The generator was trained for 120 epochs, with Kullback–Leibler annealing to avoid posterior collapse. The black-box classifier is a simple bidirectional LSTM, with hidden size of 256 neurons, with word embeddings trained alongside the final layer. The model achieves 85% accuracy on the test set. The generator and the classifier are trained on disjoint training sets. Our method for manipulating text resembles those of [21,47], since our classifier model acts as a sentiment discriminator.

Table 1. Exemplars synthesized by our evolutionary strategy with momentum for the Random Forest classifier trained on the Adult Data Set. There are two exemplars with *low income* and two exemplars with *high income*. Important features are highlighted in pale yellow.

Features	Low Income	Low Income	High Income	High Income
Age	28	19	48	38
Work Class	Private	Private	Private	Private
Final weight	315124	393950	105785	45519
Education	5th-6th	Some-college	Doctorate	Prof-School
Educational-num	0	10	19	15
Marital Status	Never-married	Never-married	Married	Married
Occupation	Other-service	Other-service	Prof-speciality	Prof-speciality
Relationship	Other-relative	Own-child	Husband	Husband
Race	White	White	White	White
Gender	Male	Female	Male	Male
Capital Gain	0	0	0	0
Capital Loss	0	0	0	0
Hours per Week	19	24	64	84
Native Country	Mexico	United-States	United-States	United-States

However, the classifier and generator networks are independent. Once they are established, we manipulate the generated text only by traversing the latent space.

4.3   Results on Tabular Data

Synthesized tabular exemplars are not only meaningful by themselves, but they additionally provide a clear picture of the model's decision process. Even though the RF model is treated as a complete black-box, with no knowledge of its type or internal structure, we are able to deduce the model's reasoning by observing the synthesized prototypes in Table 1. On the Adult Data Set, the features that influence the decision of the RF classifier seem to be the age, the level of education and the number of working hours per week. All *high-income* exemplars are older people, with significant academic achievements (typically PhD) and more than 40 working hours per week. On the other end, *low-income* exemplars have poor education, a young age and typically work part-time. Another type of *low-income* exemplars (not included in Table 1) features very old, retired and widowed people with poor education. Additionally, our analysis can reveal data set insights and model biases. In this scenario, the entirety of *high-income* exemplars are males born in the United States, while *low-income* exemplars are people born in Mexico. Hence, it seems that there is a bias towards classifying mexicans in the *low income* category, which can raise ethical concerns towards racism.

4.4   Results on Image Data

In the image synthetization scenario, we present the differences and strengths of our framework when compared to a gradient descent approach that works in

Table 2. Convergence comparison between gradient descent with momentum, a standard evolutionary strategy and our evolutionary strategy with momentum on the FER 2013 data set. Reported values represent the number of runs in which the algorithms converged, the average number of model calls (forward passes) and the average time required to produce an exemplar. The times are measured in seconds on an NVidia GeForce RTX 2080 GPU with 8 GB of RAM. Results are reported for 1000 runs.

Method	Converged (count)	Calls (average)	Time (seconds)
Gradient descent with momentum [36]	955	378	3.77
Evolutionary strategy (ours)	1000	323	0.14
Evolutionary strategy with momentum (ours)	1000	263	0.12

the glass-box scenario, while keeping the black-box scenario for our framework. For the comparison, we run both exemplar synthetization algorithms for 1000 times, generating 1000 exemplars in total.

Quantitative Analysis. Considering the convergence results presented in Table 2, we notice that both methods are able to synthesize exemplars that are classified with almost 100% confidence. However, there is a significant difference between the two methods in terms of convergence. While our evolutionary strategy is able to converge each and every time, the gradient descent approach is highly dependant on its starting point. We observe that, for both GAN and VAE generators, the gradient descent optimization does not always converge. We found that, out of 1000 runs, in 45 of them the gradient descent with momentum method failed to synthesize an exemplar with over 95% confidence, i.e. the algorithm got stuck in a non-optimal solution. This statement holds true when using both GANs and VAEs as generators.

Running Time. We measured the convergence times of the two exemplar generation approaches on an NVidia GeForce RTX 2080 GPU with 8GB of RAM. In Table 2, we present the number of model calls required to generate samples classified with more than 95% confidence by the classifier. We also present the amount of physical time required for convergence. We observe that our evolutionary strategy with momentum requires fewer model calls (forward passes) than the gradient descent with momentum. In terms of physical time, our evolutionary strategy is about 31× faster than gradient descent. We note that the gradient descent considered here is based on analytical gradients, which is faster than using numerical gradients. The remarkable difference in favor of our method can be explained by the following two factors: (i) our evolutionary strategy is able to make model calls in batches and (ii) it does not need to back-propagate gradients through the classifier or the generator. The experiments presented in Table 2 also show the benefit of introducing momentum in the evolutionary strategy. In terms of model calls, the speed up brought by momentum is 19%.

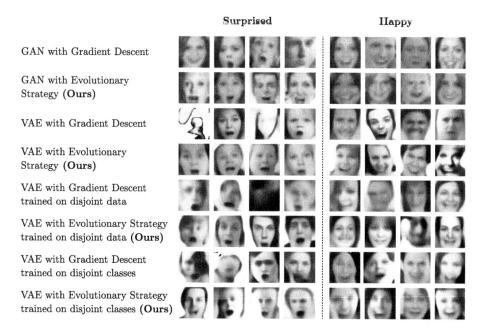

Fig. 2. Exemplars synthesized for the VGG-16 neural network trained on the FER 2013 data set. Exemplars are generated in various scenarios (different generators, same training data, disjoint training data, disjoint classes) using our evolutionary strategy with momentum or gradient descent with momentum [36]. In each scenario, we illustrate four exemplars for the *surprised* class and another four exemplars for the *happy* class.

Same Training Data. Since the GAN does not seem to produce realistic examples when its training data is not the same as that of the classifier, we present results in the context of using the same training data for both the generator and the classifier. We added this scenario to show that our method can produce exemplars with both GANs and VAEs. In Fig. 2, we present a subset of representative exemplars for two classes: *surprised* and *happy*. When using a GAN as generator (first two rows in Fig. 2), the exemplars present high quality visual features, irrespective of the synthetization algorithm. While the GAN exemplars are realistic, gradient descent does not always converge to a representative exemplar (the third *happy* exemplar on first row in Fig. 2 seems *neutral*). Moreover, gradient descent does not seem to always produce realistic exemplars for the VAE (see first and third exemplars for the *surprised* class on third row in Fig. 2).

Disjoint Training Data. We conducted experiments showing that our exemplar generation framework works well when the training data used for the generator is different from the training data used for the black-box classifier. The exemplars generated by our evolutionary strategy (sixth row in Fig. 2) are still realistic and representative for the *surprised* and *happy* classes. The exemplars

Table 3. Text exemplars synthesized by our evolutionary strategy with momentum for an LSTM classifier trained on the Large Movie Review Dataset. All samples are classified with high confidence (over 95%). Exemplars are provided for both polarity classes: *positive* (left hand-side) and *negative* (right hand-side).

Positive exemplars	Negative exemplars
"this is a great film and i recommend it to anyone"	*"one could have a cheap soap opera instead"*
"this is a great movie to watch and you will be a great time to watch it"	*"the film is not too long for the film to be a complete waste of time"*
"there is a lot of fun in this film and it is very well paced"	*"the acting is not much to save the entire movie"*
"i am a fan of the genre but this is one of the best films of all time"	*"the final scene in the movie is the worst of the year"*
"this is a very good movie for everyone but it is not perfect"	*"the film is not a complete waste of time"*
"he a terrific actor and he is great as the lead and the performances are absolutely perfect"	*"it is not a terrible movie but it is not a bad film"*
"it is one of the best movies i have seen in a long time"	*"the film is not that bad"*
"the film is well paced and it is not that good"	*"i mean it was not that bad"*
"it a good movie but it not worth a watch"	*"it was really worthless just below par"*

generated by gradient descent are not always realistic, and hard to interpret by humans (see third exemplar for the *surprised* class on fifth row in Fig. 2).

Disjoint Classes. We also conducted experiments to show that our exemplar synthetization framework generalizes to previously unseen classes. The images presented on the eighth row in Fig. 2 are generated with a VAE which was not trained on the respective classes, *surprised* and *happy*. Still, the generated images seem realistic and representative for these two classes. Some exemplars produced by the gradient descent (seventh row in Fig. 2) are less realistic.

Summary. Considering the overall results, we notice that our evolutionary strategy does not get stuck in non-optimal solutions, while converging faster than gradient descent. Non-optimal solutions are avoided because the evolutionary strategy employs multiple starting points and the velocity values (used instead of gradients) always stay within a reasonable range. Since our method relies on making small jumps in the latent space, while ignoring the gradients, it can easily escape saddle points. The benefits of our evolutionary framework are empirically demonstrated by the results on FER 2013. In summary, we conclude that our method is more robust than gradient descent, while treating the classifier as a black-box. Indeed, we showed that access to the gradients or the training data distribution of the classifier is not required.

4.5   Results on Text Data

In Table 3, we provide some selected exemplars generated for a simple LSTM text classifier, revealing the preferred inputs of the model for the *positive* and *negative* classes. We note that some generated reviews are realistic and representative for their class. Other reviews, especially the *negative* ones, indicate that the classification model outputs wrong class probabilities with high confidence when it encounters some specific words. For example, sentences containing words such as "good" or "great" are classified as positive reviews, even though they appear in negated form, e.g. "not that good". The classifier does not seem to understand contrasting transitions when evaluating the sentiment of reviews. These results are consistent with the problem of sentiment polarity classification observed by Li et al. [29]. Hence, even though the classifier has a relatively high test accuracy (85%), our method reveals that a naive training regime leads to sub-optimal results in real-world scenarios.

5   Conclusion

In this paper, we proposed a novel evolutionary strategy that incorporates momentum for generating exemplars for black-box models. Our framework requires an underlying generator, but it does not back-propagate gradients through the black-box model or the generator. We conducted experiments, showing that our approach can produce exemplars for three data types: image, text and tabular. Furthermore, our experiments indicate that our idea of incorporating momentum into a standard evolutionary strategy is useful, reducing the number of model calls by 19%. The empirical results demonstrate that our optimization algorithm converges faster than gradient descent with momentum, while providing similar or even more realistic exemplars. Given that our method does not require access to the weights or the training data of the black-box model, we believe it has a boarder applicability than gradient descent methods such as [36].

References

1. Adadi, A., Berrada, M.: Peeking inside the black-box: a survey on explainable artificial intelligence (XAI). IEEE Access **6**, 52138–52160 (2018)
2. Bien, J., Tibshirani, R.: Prototype selection for interpretable classification. Ann. Appl. Stat. 2403–2424 (2011)
3. Bolukbasi, T., Chang, K., Zou, J.Y., Saligrama, V., Kalai, A.: Man is to computer programmer as woman is to homemaker? Debiasing word embeddings. In: Proceedings of NIPS, pp. 4349–4357 (2016)
4. Bowman, S.R., Vilnis, L., Vinyals, O., Dai, A.M., Józefowicz, R., Bengio, S.: Generating sentences from a continuous space. In: Proceedings of CoNLL, pp. 10–21 (2016)
5. Caliskan, A., Bryson, J.J., Narayanan, A.: Semantics derived automatically from language corpora contain human-like biases. Science **356**(6334), 183–186 (2017)
6. Chattopadhay, A., Sarkar, A., Howlader, P., Balasubramanian, V.N.: Grad-CAM++: generalized gradient-based visual explanations for deep convolutional networks. In: Proceedings of WACV, pp. 839–847 (2018)
7. Chen, C., Li, O., Tao, D., Barnett, A., Rudin, C., Su, J.K.: This looks like that: deep learning for interpretable image recognition. In: Proceedings of NeurIPS, pp. 8928–8939 (2019)

8. Chung, J., Kastner, K., Dinh, L., Goel, K., Courville, A.C., Bengio, Y.: A recurrent latent variable model for sequential data. In: Proceedings of NIPS, pp. 2980–2988 (2015)

9. Cozma, M., Butnaru, A., Ionescu, R.T.: Automated essay scoring with string kernels and word embeddings. In: Proceedings of ACL (2018)

10. Devlin, J., Chang, M.W., Lee, K., Toutanova, K.: BERT: pre-training of deep bidirectional transformers for language understanding. In: Proceedings of NAACL, pp. 4171–4186 (2019)

11. Eduardo, S., Nazábal, A., Williams, C.K.I., Sutton, C.: Robust variational autoencoders for outlier detection and repair of mixed-type data. In: Proceedings of AISTATS (2020)

12. Fu, H., Li, C., Liu, X., Gao, J., Celikyilmaz, A., Carin, L.: Cyclical annealing schedule: a simple approach to mitigating KL vanishing. In: Proceedings of NAACL, pp. 240–250 (2019)

13. Georgescu, M.I., Ionescu, R.T., Popescu, M.: Local learning with deep and handcrafted features for facial expression recognition. IEEE Access **7**, 64827–64836 (2019)

14. Goodfellow, I., et al.: Generative adversarial nets. In: Proceedings of NIPS, pp. 2672–2680 (2014)

15. Goodfellow, I.J., et al.: Challenges in representation learning: a report on three machine learning contests. In: Lee, M., Hirose, A., Hou, Z.-G., Kil, R.M. (eds.) ICONIP 2013. LNCS, vol. 8228, pp. 117–124. Springer, Heidelberg (2013). https://doi.org/10.1007/978-3-642-42051-1_16

16. Goodman, B., Flaxman, S.: European union regulations on algorithmic decision making and a "right to explanation". AI Mag. **38**(3), 50–57 (2017)

17. Guidotti, R., Monreale, A., Matwin, S., Pedreschi, D.: Black box explanation by learning image exemplars in the latent feature space. In: Brefeld, U., Fromont, E., Hotho, A., Knobbe, A., Maathuis, M., Robardet, C. (eds.) ECML PKDD 2019. LNCS (LNAI), vol. 11906, pp. 189–205. Springer, Cham (2020). https://doi.org/10.1007/978-3-030-46150-8_12

18. Guidotti, R., Monreale, A., Ruggieri, S., Turini, F., Giannotti, F., Pedreschi, D.: A survey of methods for explaining black box models. ACM Comput. Surv. **51**(5), 1–42 (2018)

19. Gurumoorthy, K.S., Dhurandhar, A., Cecchi, G.A., Aggarwal, C.C.: In: Proceedings of ICMD, pp. 260–269 (2019)

20. He, K., Zhang, X., Ren, S., Sun, J.: Deep residual learning for image recognition. In: Proceedings of CVPR, pp. 770–778 (2016)

21. Hu, Z., Yang, Z., Liang, X., Salakhutdinov, R., Xing, E.P.: Toward controlled generation of text. In: Proceedings of ICML, pp. 1587–1596 (2017)

22. Karpathy, A., Johnson, J., Li, F.: Visualizing and understanding recurrent networks. In: Proceedings of ICLR (Workshop Track) (2016)

23. Karras, T., Aila, T., Laine, S., Lehtinen, J.: Progressive growing of GANs for improved quality, stability, and variation. In: Proceedings of ICLR (2018)

24. Kingma, D.P., Welling, M.: Auto-encoding variational bayes. In: Proceedings of ICLR (2014)

25. Kohavi, R.: Scaling up the accuracy of naive-bayes classifiers: a decision-tree hybrid. In: Proceedings of KDD, pp. 202–207 (1996)

26. Krizhevsky, A., Sutskever, I., Hinton, G.E.: ImageNet classification with deep convolutional neural networks. In: Proceedings of NIPS, pp. 1097–1105 (2012)

27. LeCun, Y., Bengio, Y., Hinton, G.: Deep learning. Nature **521**(7553), 436–444 (2015)

28. Li, J., Chen, X., Hovy, E., Jurafsky, D.: Visualizing and understanding neural models in NLP. In: Proceedings of NAACL, pp. 681–691 (2016)
29. Li, S., Huang, C.R.: Sentiment classification considering negation and contrast transition. In: Proceedings of PACLIC, pp. 307–316 (2009)
30. Lundberg, S.M., Lee, S.I.: A unified approach to interpreting model predictions. In: Proceedings of NIPS, pp. 4765–4774 (2017)
31. Maas, A.L., Daly, R.E., Pham, P.T., Huang, D., Ng, A.Y., Potts, C.: Learning word vectors for sentiment analysis. In: Proceedings of ACL, pp. 142–150 (2011)
32. Mahendran, A., Vedaldi, A.: Understanding deep image representations by inverting them. In: Proceedings of CVPR, pp. 5188–5196 (2015)
33. Mahendran, A., Vedaldi, A.: Visualizing deep convolutional neural networks using natural pre-images. Int. J. Comput. Vis. **120**(3), 233–255 (2016)
34. Mittelstadt, B., Russell, C., Wachter, S.: Explaining explanations in AI. In: Proceedings of FAccT, pp. 279–288. ACM (2019)
35. Mullenbach, J., Wiegreffe, S., Duke, J., Sun, J., Eisenstein, J.: Explainable prediction of medical codes from clinical text. In: Proceedings of NAACL, pp. 1101–1111 (2018)
36. Nguyen, A., Dosovitskiy, A., Yosinski, J., Brox, T., Clune, J.: Synthesizing the preferred inputs for neurons in neural networks via deep generator networks. In: Proceedings of NIPS, pp. 3387–3395 (2016)
37. Nguyen, A.M., Yosinski, J., Clune, J.: Deep neural networks are easily fooled: high confidence predictions for unrecognizable images, pp. 427–436 (2015)
38. Pennington, J., Socher, R., Manning, C.D.: GloVe: global vectors for word representation. In: Proceedings of EMNLP, pp. 1532–1543 (2014)
39. Rajeswar, S., Subramanian, S., Dutil, F., Pal, C.J., Courville, A.C.: Adversarial generation of natural language. In: Proceedings of RepL4NLP, pp. 241–251 (2017)
40. Ren, S., He, K., Girshick, R., Sun, J.: Faster R-CNN: towards real-time object detection with region proposal networks. In: Proceedings of NIPS, pp. 91–99 (2015)
41. Ribeiro, M.T., Singh, S., Guestrin, C.: "Why should i trust you?": explaining the predictions of any classifier. In: Proceedings of KDD, pp. 1135–1144 (2016)
42. Salimans, T., Ho, J., Chen, X., Sutskever, I.: Evolution strategies as a scalable alternative to reinforcement learning. ArXiv abs/1703.03864 (2017)
43. Selvaraju, R.R., Cogswell, M., Das, A., Vedantam, R., Parikh, D., Batra, D.: Grad-CAM: visual explanations from deep networks via gradient-based localization. In: Proceedings of ICCV, pp. 618–626 (2017)
44. Simonyan, K., Vedaldi, A., Zisserman, A.: Deep inside convolutional networks: visualising image classification models and saliency maps. In: Proceedings of ICLR (Workshop Track) (2014)
45. Simonyan, K., Zisserman, A.: Very deep convolutional networks for large-scale image recognition. In: Proceedings of ICLR (2014)
46. Wiegreffe, S., Pinter, Y.: Attention is not not explanation. In: Proceedings of EMNLP, pp. 11–20 (2019)
47. Yang, Z., Hu, Z., Dyer, C., Xing, E.P., Berg-Kirkpatrick, T.: Unsupervised text style transfer using language models as discriminators. In: Proceedings of NIPS, pp. 7287–7208 (2018)
48. Yeh, C.K., Kim, J., Yen, I.E.H., Ravikumar, P.K.: Representer point selection for explaining deep neural networks. In: Proceedings of NIPS, pp. 9291–9301 (2018)
49. Zhou, B., Khosla, A., Lapedriza, A., Oliva, A., Torralba, A.: Learning deep features for discriminative localization. In: Proceedings of CVPR, pp. 2921–2929 (2016)

Quality Guarantees for Autoencoders via Unsupervised Adversarial Attacks

Benedikt Böing[1]($\boxtimes$), Rajarshi Roy[2], Emmanuel Müller[1], and Daniel Neider[2]

[1] University of Bonn, Bonn, Germany
{boeing,mueller}@bit.uni-bonn.de
[2] Max Planck Institute for Software Systems, Saarbrücken, Germany
{rajarshi,neider}@mpi-sws.org

Abstract. Autoencoders are an essential concept in unsupervised learning. Currently, the quality of autoencoders is assessed either internally (e.g.. based on mean square error) or externally (e.g.. by classification performance). Yet, there is no possibility to prove that autoencoders generalize beyond the finite training data, and hence, they are not reliable for safety-critical applications that require formal guarantees also for unseen data.

To address this issue, we propose the first framework to bound the worst-case error of an autoencoder within a safety-critical region of an infinite value domain, as well as the definition of unsupervised adversarial examples that cause such worst-case errors. Technically, our framework reduces the infinite search space for a uniform error bound to checking satisfiability of logical formulas in Linear Real Arithmetic. This allows us to leverage highly-optimized SMT solvers, a strategy that is very successful in the context of deductive software verification. We demonstrate our ability to find unsupervised adversarial examples as well as formal quality guarantees both on synthetic and real-world data.

1   Introduction

Autoencoders are widely used for many unsupervised learning tasks such as cluster analysis [4], compression [14], anomaly detection [18], as well as a variety of pre-processing steps [10,14,17] in other machine learning pipelines. The general assumption is that data can be compressed into a lower dimensional latent space by an encoder function extracting the most relevant features of the data distribution. From this latent representation the decoder tries to reconstruct the original input. As the latent representation is an information bottleneck the autoencoders input deviates from its output. Typically the autoencoder reconstructs better in dense regions (i.e. regions with many training examples) than in regions with few training examples [18] giving rise to its application in anomaly detection. Moreover even the small errors in dense regions are a desirable property as they allow it to be used e.g. for denoising. At the same time it is necessary to control the error for all points in dense regions because otherwise the result - whether it is the latent representation or the reconstruction - is less useful. To this end

© Springer Nature Switzerland AG 2021
F. Hutter et al. (Eds.): ECML PKDD 2020, LNAI 12458, pp. 206–222, 2021.
https://doi.org/10.1007/978-3-030-67661-2_13

current approaches to assess autoencoders either measure internally the mean square error (MSE) on the unsupervised training data or external performance on some supervised application such as classification performance.

However, a major shortcoming of these approaches is that they cannot provide a formal guarantee in terms of the maximum deviation between input and output of the autoencoder as it is evaluated on training data only (i.e. with a finite number of inputs). We are not aware of any existing scheme to calculate the largest error of an autoencoder in an infinite input space. This lack of formal quality guarantees for autoencoders leads to a very limited applicability of such unsupervised learning schemes for safety-critical applications. For instance, it is particularly important to consider the maximum deviation when working with data containing clusters. In such situations the autoencoder should not mix up the clusters because otherwise the autoencoders results are meaningless. If the maximum deviation for the respective clusters are small enough though, the autoencoder is guaranteed to keep the clusters separated.

To address this and other shortcomings of unsupervised learning with autoencoders, we provide the first methodology to bound an autoencoder's worst-case error in a safety-critical region. As a first step towards this goal we define the notion of *unsupervised adversarial examples* which are inputs (not necessarily contained in the training data) on which the autoencoder's error exceeds a user-defined threshold. Then we define the worst-case error of an autoencoder as the largest error that can possibly manifest. Since we cannot expect to find a global maximum of the error (as there is no reason for the error itself to be bounded), we restrict our search to user-defined regions with an infinite value domain of the input space. We leave this region as a parameter to be provided by the user as it clearly depends on knowledge about the use case at hand, characteristics of the training data, or other domain-specific information.

Following a popular approach in the area of software verification, we reduce the problem of finding an unsupervised adversarial example to a satisfiability check of a formula in Real Arithmetic. This allows us to apply highly-optimized, off-the-shelf satisfiability modulo theory (SMT) solvers which can effectively reason about the infinite domains and, hence, can prove the existence or non-existence of unsupervised adversarial examples. Once we have found an unsupervised adversarial example, it serves as a lower bound for the worst-case error. Moreover, a simple binary search allows us to approximate worst-case error arbitrarily well. Note that naive approaches, such as sampling, cannot provide an upper bound on the worst-case error of an autoencoder as an exhaustive search of the input space is intractable. Moreover, our experimental evaluation shows that sampling often underestimates that worst-case error.

We demonstrate the effectiveness of our QUGA (QUality Guarantees for Autoencoders) approach and evaluate our quality guarantees for unsupervised learning on a synthetically created dataset as well as on a real dataset. In both cases we can find unsupervised adversarial examples as well as formal quality guarantees by lower and upper error bounds in safety-critical regions.

2   Related Work

Adversarials in Supervised vs. Unsupervised Learning. In the area of supervised learning, adversarial attacks have been widely studied [7,11,20]. While common definitions of adversarial attacks rely on the robust separation of class labels, we aim at unsupervised learning without given labels. Therefore supervised definitions do not cover the unsupervised learning case. Similarly existing approaches of adversarial attacks in unsupervised learning focus on a particular task such as clustering [6] or image retrieval [22] assuming that there is a notion of a wrong output. In contrast to these approaches we define adversarial attacks directly in terms of the intrinsic learning objective of autoencoders which is - as reflected by the loss function - approximating the identity function.

Empirical Quality Assessment vs. Formal Guarantees. Common evaluation schemes for autoencoders do an empirical quality assessment based on a given set of training data. The variety of quality measures ranges from simple average MSE to stability and robustness measures [13,15,21]. All of these measures have in common that they rely on the given training data. In contrast to such empirical evaluation, many safety-critical applications require formal guarantees explicitly also on unseen data. We propose such formal guarantees for trained autoencoders. Given a safety-critical data region, our method is able to either find an adversarial example or prove that such an example does not exist.

External vs. Internal Evaluation. Common external evaluation uses, for example, the classification quality of a down-stream step after the autoencoder as indirect measure of quality for the autoencoder. As such evaluation of multiple tasks is prone to the mix-up of fluctuating quality of individual tasks and dependency effects between these tasks. We belief that the modular evaluation of individual tasks is an additional requirement for safety-critical systems. Such a design-by-contract has been successfully established in modular software verification [3]. Similarly, we propose the first formal guarantee of an autoencoder (i.e., an upper bound on the maximal error on the entire data domain).

Verification of Neural Networks. Our work is related to formal methods and verification of neural networks in general (e.g., see [2,12]). However, most of the research in this area focuses on the problem of finding adversarial examples in supervised learning tasks and lacks formal insights for unsupervised learning. In contrast, our algorithm searches for unsupervised adversarial examples. It does so by reducing the problem to a series of satisfiability checks in a Real Arithmetic and applies a highly-optimized Satisfiability Module Theories (SMT) solver as computational back-end to perform these checks. We have implemented a prototype of our algorithm on top of the Z3 SMT solver [16] which provides a convenient API and is one of the most popular tool in the domain of software verification. For extremely large, real-world scenarios, however, one would clearly use a solver that is optimized for constraints arising from feed-forward networks, such as ReluPlex [12] or Planet [8].

Apart from constraint solving, other techniques from the area of deductive software verification have been used for finding adversarial examples in supervised learning and proving robustness properties of feed-forward neural nets. The perhaps most popular approach is abstract interpretation [9, 19]. However, abstract interpretation inherently overapproximates the behavior of the neural network and, hence, can only be used to prove safety properties. However, neither our unsupervised adversarial examples nor our worst-case error of an autoencoder can be achieved by their safety properties.

3 $QUGA$: Problem Statement

In general, an autoencoder tries to reproduce its input while propagating it through a latent space which typically has less dimensions than the input/ output space. This latent space serves as an information bottleneck and, hence, introduces errors to the identity function the autoencoder is supposed to learn. However, most applications of autoencoders rely on a good approximation of the identity function, and we are naturally interested in quantifying its error. More precisely, our goal is to give formal guarantees in terms of the maximum deviation from the identity function.

As a first step towards this goal, we define the notion of *adversarial examples of autoencoders*. Intuitively, such adversarial examples are inputs on which the "distance" between the input and the output of the autoencoder is larger than a (user-defined) threshold $\varepsilon > 0$. Given the lack of definitions for adversarials in unsupervised learning (and in particular for autoencoders), we define adversarial example based on an abstract distance function *dist* which maps two data points to a non-negative real number. However, we stress that the exact distance function is not important for our definition (e.g.., any L_p-norm could be used) because all autoencoders share the goal of reconstructing the input.

Definition 1 (ε-adversarial examples). *Let $f \colon \mathbb{R}^n \to \mathbb{R}^n$ be an autoencoder, $dist \colon \mathbb{R}^n \times \mathbb{R}^n \to \mathbb{R}_+$ a distance function, and $\varepsilon > 0$. An ε-adversarial example is a point $x \in \mathbb{R}^n$ such that*

$$dist(x, f(x)) > \varepsilon$$

(i.e., a point on which the input and output of f deviate more than ε).

Note that our definition of ε-adversarial examples is not restricted to inputs in the training or test sets but allows any input $x \in \mathbb{R}^n$. This property makes finding ε-adversarial examples a very challenging task, and in contrast to traditional internal evaluation (e.g.., mean square error) on training data, searching adversarial examples is a computationally hard problem.

In the context of safety-critical systems, however, it is not enough to identify individual ε-adversarial examples, but it is necessary to know the worst-case (i.e., maximum) error an autoencoder produces. Of course, we cannot expect to find a global maximum of the error as there is no reason for the error itself to be bounded. Therefore, we restrict the region for which we want to find a bound on the error. This region depends on knowledge about the use case at hand, characteristics of the training data, or other domain-specific information. Thus, we leave it as a parameter to be provided by the user.

Definition 2 (Worst-case error of autoencoders). *Let* $f\colon \mathbb{R}^n \to \mathbb{R}^n$ *be an autoencoder, dist* $\colon \mathbb{R}^n \times \mathbb{R}^n \to \mathbb{R}_+$ *a distance function, and* $A \subseteq \mathbb{R}^n$ *an (infinite) safety-critical region of inputs. Then, the* worst-case error *of* f *in* A *is defined as*

$$wce(f, A) = \sup \left\{ dist\big(x, f(x)\big) \in \mathbb{R}_+ \mid x \in A \right\}$$

(i.e., the largest deviation of an input in the region A *from the output).*

Definition 2 serves as our novel *quality criterion* for autoencoders that reflects how good the identity function is learned in the specific region of interest. Our *wce*-definition is inspired by many areas of reliable system design, including soft- and hardware verification, as $wce(f, A)$ guarantees that a system f employed in a safety-critical region A stays within its design parameters. Furthermore, our notion of *wce* overcomes limitations of classical quality metrics that are defined on finite training data only. We actively design *wce* for typically infinite data domains of safety-critical regions. In total, this leads us to the main problem statement, which we call *QUGA: QUality Guarantees for Autoencoders*.

Problem 1 (QUGA: Quality Guarantees for Autoencoders). Given an autoencoder $f\colon \mathbb{R}^n \to \mathbb{R}^n$, a distance function $dist\colon \mathbb{R}^n \times \mathbb{R}^n \to \mathbb{R}_+$, and a region $A \subseteq \mathbb{R}^n$, compute $wce(f, A)$.

In general, computing the worst-case error is a very challenging problem as it involves reasoning about an infinite number of inputs (not just training data) and does not make any assumption on the autoencoder, the distance function or the region. In the following section, we consider a restricted version of Problem 1 and show how a reduction to a series of constraint solving can be used to answer this restriction.

4   Solution Framework

In this section, we provide a framework for computing ε-adversarial examples and the worst-case error of autoencoders. To make these problems computationally tractable, we consider a restricted version of Problem 1. The following restrictions are designed in such a way that the solution framework remains applicable to a wide range of autoencoders used in practice:

1. We assume the the neurons of the autoencoder have linear or ReLU (Rectified Linear Units) activation functions.
2. We assume the distance function to be the L_1 or L_∞-norm.
3. We assume the safety-critical region A to be a finite union of convex compact polytopes (i.e. each polytope is an intersection of half-spaces of the $\mathbb{R}^n$).

4. We approximate the worst-case error up to a user-defined accuracy because our framework can find ε-adversarial examples for fixed ε only.

In the remainder of this section, we formally introduce autoencoders (Sect. 4.1) and the Satisfiability Modulo Theories (SMT) framework (Sect. 4.2). In Sect. 4.2, we then show how the existence of an ε-adversarial example can be phrased as a satisfiability problem in Linear Real Arithmetic, one of the theories supported by the SMT framework. This allows us to use highly-optimized SMT solvers to do a symbolic search on the (potentially infinite) input space. In Sect. 4.4, we finally provide an effective method to approximate the worst-case error of an autoencoder by repeatedly solving the easier problem of determining the existence of ε-adversarial attacks for different values of ε.

4.1   Autoencoders

Intuitively, an *autoencoder*—like most feed-forward networks—is a collection of neurons (or nodes) arranged sequentially in layers. Each neuron (except input neurons) is connected to neurons of the previous layer by edges carrying weights (e.g.., see Fig. 1 on Page 9). Functionally, an autoencoder evaluates a function $f : \mathbb{R}^n \to \mathbb{R}^n$, where internally it uses the neurons to propagate information through its layers.

For an autoencoder f, we use N to denote the number of layers and $l_k \in \mathbb{N} \setminus \{0\}$ with $k \in \{1, \ldots, N\}$ to represent the number of neurons in Layer k. Layer 1 is called *input layer*, Layer N is called *output layer*, and the remaining layers are called *hidden layers*. The *topology* of an autoencoder is a tuple $(l_1, l_2, \ldots, l_N)$ denoting the number $l_k \in \mathbb{N} \setminus \{0\}$ of neurons in each layer $k \in \{1, \ldots, N\}$. In contrast to general feed-forward networks, autoencoders have an equal number $n \in \mathbb{N} \setminus \{0\}$ of neurons in the input and output layers (i.e., $l_1 = l_N = n$). Furthermore there is a special layer that separates the autoencoder into the encoder and the decoder. The neurons in this layer span the latent space which typically has less dimensions than the input space.

For each layer $k \in \{2, \ldots, N\}$, the autoencoder has a weight matrix W^k of dimension $l_{k-1} \times l_k$, containing the weights of the connections between layer $k - 1$ and k. Moreover, each layer has a so-called *bias vector* $b_k \in \mathbb{R}^{l_k}$ associated with it that contains the bias for each neuron in Layer k.

The output of each neuron is calculated by taking a linear combination of the output of the previous layer and then applying an *activation function* (typically a non-linear function) on the result. We consider the linear (trivial) and the ReLU activation function, giving rise to linear and ReLU neurons. The output of Neuron j in Layer k is then $x_{k,j} = \sum_{i=1}^{l_{k-1}} x_{k-1,i} W_{i,j}^k + b_j^k$ if it is a linear neuron and $x_{k,j} = \max \{0, \sum_{i=1}^{l_{k-1}} x_{k-1,i} W_{i,j}^k + b_j^k\}$ if it is a ReLU neuron.

4.2   Satisfiability Modulo Theories (SMT)

Various problems in computer science, especially in the area of formal verification, can be solved by reducing them to constraint satisfaction problems in a suitable logic. Although propositional logic is a popular choice for many such problems, some of them require a more expressive logic: first-order logic. A formula in first-order logic is formed using constants, variables, function and predicate symbols, logical connectives, and quantifiers. In this paper, however, we require only a specific first-order logic, namely the the *quantifier-free fragment of linear real arithmetic (LRA)*, which we introduce next.

First, let $\mathcal{X} = \{x_0, x_1, \ldots\}$ be a set of *variables* which range over values in $\mathbb{R}$. Then, we define *terms* as follows: a term is either a constant $c \in \mathbb{R}$, a variable $x \in \mathcal{X}$, or a function application $t_1 \circ t_2$, where $\circ \in \{+, \cdot\}$ and t_1, t_2 are two terms. For instance, 5, x, and $3 \cdot x + 2 \cdot y$ are terms. To reflect the usual notation, we often drop the multiplication sign.

An *atomic formula* is a predicate symbol applied to terms. In LRA, we allow the usual binary predicates $<, \leq, =, \geq$, and $>$. For example, $3x + 2y > 5$ is an atomic formula. Moreover, a *formula* is inductively defined as follows: a formula is either an atomic formula, the negation $\neg\varphi$ of a formula φ, or the disjunction $\varphi_1 \vee \varphi_2$ of two formulas φ_1, φ_2. We also add syntactic sugar and allow the formulas $\varphi_1 \wedge \varphi_2$, $\varphi_1 \rightarrow \varphi_2$, and $\varphi_1 \leftrightarrow \varphi_2$, which are defined as usual.

To assign meaning to formulas, we introduce the concept of *interpretations*. An interpretation is a mapping $\mathcal{I}\colon \mathcal{X} \rightarrow \mathbb{R}$ which assigns to each variable a real value. Interpretations can easily be lifted to terms in the usual way, and we write $\mathcal{I}(t)$ for the interpretation (i.e., the value) of the term t under $\mathcal{I}$. Finally, we can define when an interpretation $\mathcal{I}$ *satisfies* a formula φ which we denote by $\mathcal{I} \models \varphi$: we have $\mathcal{I} \models t_1 \diamond t_2$ for $\diamond \in \{<, \leq, =, \geq, >\}$ if and only if $\mathcal{I}(t_1) \diamond \mathcal{I}(t_2)$ is true, $\mathcal{I} \models \neg\varphi$ if $\mathcal{I} \not\models \varphi$, and $\mathcal{I} \models \varphi_1 \vee \varphi_2$ if and only if $\mathcal{I} \models \varphi_1$ or $\mathcal{I} \models \varphi_2$. We say that a formula φ is *satisfiable* if an interpretation $\mathcal{I}$ with $\mathcal{I} \models \varphi$ exists.

Highly-optimized procedures for deciding satisfiability of formulas in LRA have been implemented in a framework called *Satisfiability Modulo Theories (SMT)* [1], which not only allows to check the satisfiability of formulas in LRA but also in many other (usually quantifier-free) fragments of first-order logic, called theories. Moreover, SMT solvers typically return an interpretation if the given formula is satisfiable. In the following, we exploit this property for our search for ε-adversarial attacks and to approximate the worst-case error of an autoencoder.

4.3   Identifying ε-Adversarial Examples

Let us now describe how to translate the problem of finding an ε-adversarial example of an autoencoder f into LRA. At its core is a formula φ^f that encodes the function computed by f in LRA. Moreover we add further constraints in the form of formulas φ^A and $\varphi_\varepsilon^{dist}$ which encode the input region A and the distance function (including the existence of an ε-adversarial example), respectively. The resulting encoding is then the conjunction $\varphi_\varepsilon^{ae} := \varphi^f \wedge \varphi^A \wedge \varphi_\varepsilon^{dist}$ which is satisfiable if and only if an ε-adversarial example exists. Moreover, a satisfying interpretation of φ_ε^{ae} carries sufficient information to extract such an ε-adversarial example. Let us now describe these formulas in detail.

Encoding the Autoencoder: To encode the function computed by an autoencoder f in LRA, we introduce variables $x_{k,j}$ for each layer $k \in \{1, \ldots, N\}$ and each neuron $j \in \{1, \ldots, l_k\}$ in Layer k. Intuitively, each such variable captures the output of a neuron and is used as the input for other neurons. Correspondingly, variables $x_{1,1}, \ldots, x_{1,l_1}$ represent the input to the autoencoder, while variables $x_{N,1}, \ldots, x_{N,l_N}$ represents the output of the autoencoder. To ensure that the variables $x_{k,j}$ actually have the desired meaning, we introduce constraints that describe the computation of each neuron. For a linear neuron we construct

$$\psi_{k,j} := \Big[x_{k,j} = [\sum_{i=1}^{l_{k-1}} W_{i,j}^k x_{k-1,i}] + b_j^k \Big].$$

On the other hand, for a ReLU neuron, we construct the constraint

$$\psi_{k,j} := \Big[[S_{k,j} = \sum_{i=1}^{l_{k-1}} W_{i,j}^k x_{k-1,i} + b_i^k] \wedge [x_{k,j} = \mathsf{ite}(S_{k,j} < 0, 0, S_{k,j})] \Big],$$

where ite (short for "if-then-else") is syntactic sugar for a conditional evaluation of terms, which is supported by virtually all SMT solvers.

Finally, we define

$$\varphi^f := \bigwedge_{2 \leq k \leq N} \bigwedge_{1 \leq j \leq l_k} \psi_{k,j},$$

which collects the constraints for all individual neurons. By construction φ^f completely encodes the autoencoder f in the sense that $f(\mathcal{I}(x_{1,1}), \ldots, \mathcal{I}(x_{1,l_1})) = (\mathcal{I}(x_{N,1}), \ldots, \mathcal{I}(x_{N,l_N}))$ holds for all satisfying interpretations $\mathcal{I} \models \varphi^f$.

Encoding the Region: Recall that we assume that the safety-critical region A in which to search for ε-adversarial examples is provided as a finite union of compact convex polytopes. Formally, a convex polytope $\mathcal{P}$ is the finite intersection of half-spaces $\mathcal{H}_i$ of the form $\sum_{j=1}^{l_1} a_{i,j} x_j \leq c_i$ for $a_{i,j}, c_i \in \mathbb{R}$, and we write $A = \{\mathcal{P}_1, \ldots, \mathcal{P}_\ell\}$ for the sake of brevity. Thus, restricting the search space for ε-adversarial examples to a convex polytope $\mathcal{P}$ consisting of m half-spaces can simply be achieved by the formula

$$\psi_{\mathcal{P}} := \bigwedge_{1 \leq i \leq m} \Big[\sum_{1 \leq j \leq l_1} a_{i,j} x_{1,j} \leq c_i \Big].$$

Moreover, the final formula is then the disjunction $\varphi^A := \bigvee_{\mathcal{P} \in A} \psi_{\mathcal{P}}$ for all polytopes $\mathcal{P}$ constituting to the given region A.

Encoding the Existence of an ε-Adversarial Attack: It is left to encode the distance function *dist* as well as the existence of an ε-adversarial example. In the interest of space, however, we only show the encoding of the L_∞-norm. Encoding the L_1-norm is only slightly more complicated and can be expressed

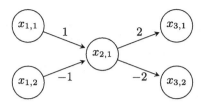

Fig. 1. A short example of an autoencoder.

using a summation over ite-terms. In principle, it is even possible to encode L_p-norms in SMT for arbitrary $p \geq 1$, but the complexity of the underlying decision procedures for non-linear Real Arithmetic is prohibitively high.

In the L_∞-norm, an input is an ε-adversarial example if there exists a dimension $i \in \{1,\ldots,l_1\}$ in which the absolute value of the difference of the input output in this dimension is larger than ε. This can be expressed in LRA by

$$\varphi_\varepsilon^{dist} := \bigvee_{1 \leq i \leq l_1} \Big[[x_{1,j} - x_{N,j} > \varepsilon] \vee [x_{N,j} - x_{1,j} > \varepsilon] \Big].$$

Before we continue with the final formula φ_ε^{ae}, let us briefly illustrate the constraints generated so far using an example.

Example 1. Consider the simple autoencoder (with ReLU-activation) in Fig. 1, consisting of two neurons in the input layer, one neuron in the single hidden layer, and two neurons in the output layer. Moreover, assume that we are given one polytope $\mathcal{P}$ consisting of the intersection of four half-spaces $-1 \leq x$, $x \leq 1$, $-1 \leq y$, and $y \leq 1$ (i.e., a unit box around the origin). Then, the formulas φ^f, φ^A, and $\varphi_\varepsilon^{dist}$ are given by

$$\varphi^f := [S_{2,1} = x_{1,1} + (-1)x_{1,2}] \wedge [x_{2,1} = \mathsf{ite}(S_{2,1} < 0, 0, S_{2,1})] \wedge$$
$$[x_{3,1} = 2x_{2,1} \wedge x_{3,2} = (-2)x_{3,2}],$$
$$\varphi^A := x_{1,1} \leq 1 \wedge x_{1,1} \geq -1 \wedge x_{1,2} \leq 1 \wedge x_{1,2} \geq -1,$$
$$\varphi_\varepsilon^{dist} := [x_{1,1} - x_{3,1} > \varepsilon \vee x_{3,1} - x_{1,1} > \varepsilon] \vee [x_{1,2} - x_{3,2} > \varepsilon \vee x_{3,2} - x_{1,2} > \varepsilon].$$

Finally, we combine all constraints generated so far into a single formula $\varphi_\varepsilon^{ae} := \varphi^f \wedge \varphi^A \wedge \varphi_\varepsilon^{dist}$. As the next theorem states, this formula indeed expresses the existence of an ε-adversarial example of the autoencoder f in the region A.

Theorem 1. *Let f be an autoencoder, A a region, dist a distance function, $\varepsilon > 0$, and φ_ε^{ae} as defined above. Then, the following two properties hold:*

1. *If an ε-adversarial example exists, then φ_ε^{ae} is satisfiable.*
2. *If φ_ε^{ae} is satisfiable, say by the interpretation $\mathcal{I}$, then $(\mathcal{I}(x_{1,1}),\ldots,\mathcal{I}(x_{1,l_1}))$ is an ε-adversarial example.*

Algorithm 1: Computing wce up to accuracy δ

Input: Autoencoder f, Region A, distance function $dist$, start value
   $\varepsilon_0 > 0$, accuracy $\delta > 0$

1 $\varepsilon_{\text{low}} = \varepsilon_{\text{up}} = \varepsilon_0$
2 Construct $\varphi_{\varepsilon_0}^{ae}$ and check satisfiability using an SMT solver
3 **if** $\varphi_{\varepsilon_0}^{ae}$ *is satisfiable* **then**
4 $\quad\mid\quad$ Increase ε_{up} by $\varepsilon_{\text{up}} * 2$ until $\varphi_{\varepsilon_{\text{up}}}^{ae}$ becomes unsatisfiable
5 **else**
6 $\quad\mid\quad$ Decrease ε_{low} by $\varepsilon_{\text{low}}/2$ until $\varphi_{\varepsilon_{\text{low}}}^{ae}$ becomes satisfiable or $\varepsilon_{\text{low}} < \delta$ (in
   $\quad\quad$ which case **return** ε_{low})
7 **end**
8 $\varepsilon^* \leftarrow$ Binary-search$_{f,A,dist}(\varepsilon_{\text{low}}, \varepsilon_{\text{up}}, \delta))$ `// involves calls to SMT`
   `solver`
9 **return** ε^*

Theorem 1 now suggests a simple procedure to find ε-adversarial examples: simply construct φ_ε^{ae}, run an SMT solver, and return $(\mathcal{I}(x_{1,1}), \ldots, \mathcal{I}(x_{1,l_1}))$ if a satisfying assignment $\mathcal{I} \models \varphi_\varepsilon^{ae}$ exists. However, the SMT solver might report that φ_ε^{ae} is unsatisfiable. In this case, Theorem 1 guarantees that no ε-adversarial example exists. The proof of Theorem 1 can be found in the supplementary material (See footnote 1). We exploit this property now to approximate the worst-case error of an autoencoder.

4.4 Approximating the Worst-Case Error

We now provide an algorithm for approximating the worst-case error of an autoencoder. Our algorithm, which is sketched in pseudocode as Algorithm 1, is based on the method for finding ε-adversarial examples from Sect. 4.3. Apart from the autoencoder itself, the safety-critical region, and a distance function, it expects two additional arguments: a start value $\varepsilon_0 > 0$ for the search and an accuracy value $\delta > 0$. The start value ε_0 is used as an initial estimate for $wce(f, A)$ and can be either initialized arbitrarily or based on domain knowledge. The accuracy, on the other hand, is a measure of how close the output of Algorithm 1 is to the actual value of $wce(f, A)$. A smaller δ results in a more precise approximation of $wce(f, A)$, but it also increases the computation time.

Algorithm 1 uses a binary search to find a sufficiently close approximation of $wce(f, A)$ (see line 8). To this end, it uses two values $\varepsilon_{\text{low}} < \varepsilon_{\text{up}}$ for which it maintains the invariant that (a) there exists an ε_{low}-adversarial example and (b) there does not exist an ε_{up}-adversarial example in the given region. Hence, $wce(f, A)$ is guaranteed to lie in the interval $[\varepsilon_{\text{low}}, \varepsilon_{\text{up}}]$. The initial values for ε_{low} and ε_{up} are obtained by starting with ε_0 and increasing ε_{up} or decreasing ε_{low} until the invariant is established (see lines 1 to 7). Subsequently, the binary search then repeatedly runs the procedure for finding ε-adversarial examples and updates the bounds ε_{low} and ε_{up} accordingly. Algorithm 1 stops once the interval $[\varepsilon_{\text{low}}, \varepsilon_{\text{up}}]$ is small enough (i.e. less than 2δ). In summary, Algorithm 1 provides an effective procedure to compute the worst-case error of an autoencoder up to a user-defined accuracy $\delta > 0$, as formalized in the theorem below.

Theorem 2. *Let f be an autoencoder, A a region, and $\delta > 0$. Then, Algorithm 1 terminates eventually and outputs a value $\varepsilon^\star \in [wce(f, A) - \delta, wce(f, A) + \delta]$.*

Theorem 2 follows from Theorem 1 and the fact that the binary search of Algorithm 1 narrows down the interval $[\varepsilon_{\text{low}}, \varepsilon_{\text{up}}]$ until it is smaller than 2δ. The latter fact also implies the termination of Algorithm 1.

The complexity of Algorithm 1 consists of two parts: the binary search and the SMT solver. The number of steps in the binary search is in $\mathcal{O}(log(\frac{wce(f,A)}{\delta}))$. In each step the SMT solver is called once with a runtime that mainly depends on the number of atomic formulas in (the respective) $\varphi_\varepsilon^{\text{ae}}$. Under the restrictions in Sect. 4 there are $\mathcal{O}(n + m)$ many atomic formulas where n is the number of neurons in the autoencoder and m is the number of halfspaces used to construct the safety-critical region. Note that the number of atomic formulas arising from the L_1 and L_∞ distance depends linearly on the dimension of the input/output space of the autoencoder and is hence in $\mathcal{O}(n)$. Even though encoding the problem as a formula is inexpensive, the SMT solver itself is an exponential algorithm as is relies on solving instances of the NP-complete SAT problem.

5   Empirical Evaluation

We evaluate both concepts presented within our QUGA solution: (1) extracting an adversarial example and (2) calculation of quality bounds. For evaluation we use both synthetic and real-world data. For future comparison and reproducibility of our experiments we provide our implementation[1] with the off-the-shelf SMT solver Z3. As Z3 is not specialised for neural nets, our approach is not

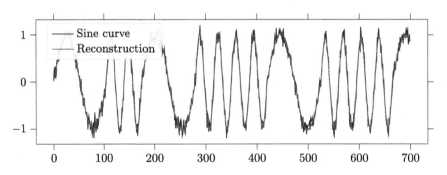

Fig. 2. Synthetic sine curve with two frequencies and noise with its reconstruction by the autoencoder.

---
[1] https://github.com/KDD-OpenSource/QUGA.

scalable enough to deal with benchmark datasets such as MNIST or CIFAR-10. This is one of the most urgent challenges we intend to tackle in future work (Fig. 2).

5.1 Experiment Setup

We use synthetic time series generated by sine curves with two different frequencies (35 and 105) and random gaussian noise ($\sigma = 0.1$) per time-point. Additionally, we use ECG5000 data from the UCR time series repository [5]. We train autoencoders with a topology of $L = (35, 5, 35)$ with 5 hidden ReLU nodes and 35 linear output nodes using the MSE loss function. Training data consists of time windows of length 35 without overlap. For the sine curve the time windows correspond to 4 clusters: The full sine curve and the beginning, the middle and the end of the large sine curve. We denote them by $C_{full}, C_{beg}, C_{mid}$ and C_{end} respectively. For the ECG5000 dataset, we obtain 8 clusters arising from 2 classes and 4 time windows. We call them C_{i_x} where $i \in \{1, 2, 3, 4\}$ and $x \in \{u, b\}$ indicating the upper or lower part of the respective time window. As critical region A we evaluate a box around the two sine curves with width 0.2 in every dimension. This region contains by construction the majority of training data. For the ECG5000 dataset we extract representative time series for the two main classes and add a margin of 0.25. We visualize the regions along with the training data in Fig. 3.

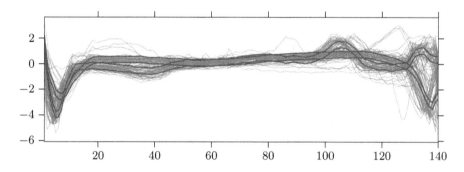

Fig. 3. ECG5000 dataset with two safety-critical regions (red and green) obtained by extracting prototypes for two classes and adding a margin of 0.25. (Color figure online)

5.2 Extracting an Adversarial Example

The first observation is that our QUGA approach successfully extracts adversarial examples. We depict the adversarial examples obtained in Fig. 4 for the sine curve dataset and in Fig. 5 for the two safety-critical regions in the ECG5000 Dataset. Ideally an autoencoder should extract a denoised version of the input.

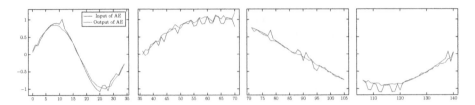

Fig. 4. Adversarial examples for different parts of the sine curve dataset obtained by the QUGA approach maximizing the L_∞-distance between the input and the output of the autoencoder in the respective safety-critical region. The adversarial example on the second plot from the left indicates that this part is denoised less.

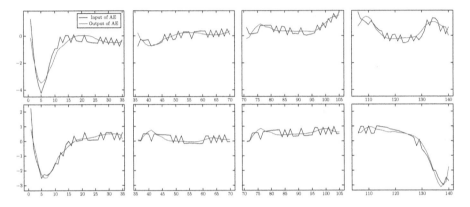

Fig. 5. Adversarial examples for two classes and different time windows of the ECG5000 dataset obtained by the QUGA approach maximizing the L_∞-distance between the input and the output of the autoencoder in the respective safety-critical region. No difference in denoising quality between the different plots can be seen.

With the adversarial examples we have an indication whether the autoencoder succeeds in doing so. For the sine curve the outputs of the autoencoder on adversarial examples in C_{full}, C_{mid} and C_{end} are much smoother than for the adversarial example in C_{beg}, suggesting that the autoencoder does not denoise as well in C_{beg}. For the ECG5000 dataset the autoencoder seems to denoise for all clusters very well.

5.3   Comparing Quality Bounds with Sampling

We compare the quality bounds obtained by the QUGA approach with accuracy 0.025 to quality bounds obtained by a simple sampling approach. As a competitor to the QUGA approach we sample points in the region, calculate their L_∞ errors and take the maximum as an estimator for the $L_\infty - wce$. Table 1 sums up the results. First of all note, that the QUGA $L_\infty - wce$ bounds are much more precise. The $L_\infty - wce$ bound obtained by the sampling approach yields no upper bound at all, and furthermore the lower bound is much weaker than the

Table 1. Worst-case errors as estimated by sampling and QUGA approach for the sinc curve and ECG datasets. The accuracy for the QUGA approach is 0.025.

Cluster	sine curve				ECG							
	C_{full}	C_{beg}	C_{mid}	C_{end}	C_{1_b}	C_{2_b}	C_{3_u}	C_{4_u}	C_{1_u}	C_{2_u}	C_{3_b}	C_{4_b}
QUGA	0.297	0.422	0.297	0.266	1.359	1.016	0.828	1.078	1.453	0.766	0.766	0.953
Sampling	0.211	0.255	0.214	0.189	1.189	0.829	0.651	0.908	1.255	0.563	0.546	0.774

lower bound obtained by QUGA in all cases. A clear drawback of sampling is the large amount of samples required to reach our QUGA estimation. In Fig. 6 we show runtime of QUGA vs. sampling with their respective error estimations. QUGA as a systematic search scheme is more efficient, while sampling is shown to underestimate worst-case errors.

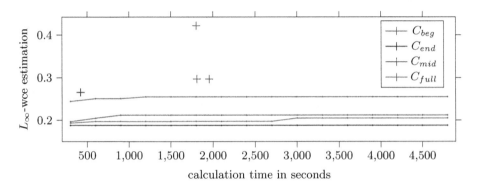

Fig. 6. Depiction of runtime against error estimation for the sine curve dataset. Plus signs indicate the results by QUGA. Lines indicate results by sampling.

5.4   Safety Critical Application

We demonstrate our QUGA framework on the ECG5000 dataset, by evaluating the unsupervised training based on two time series clusters. The goal of a traditional evaluation would be to show that all training objects are clearly separated in the latent space. In contrast, we care about all possible (infinitely many objects) in two safety-critical areas that need to be distinguishable in the latent space. In Fig. 7 we see the resulting corridor into which points from the critical regions can be mapped. For the first three time windows we cannot guarantee that the autoencoder keeps points from the two regions distinguishable in the latent space. Both clusters mix-up as the upper bound of the lower cluster is higher than the lower bound of the upper cluster. For the last 35 time steps though a guaranteed separation of all infinitely many points in the critical regions is possible by the autoencoder. With this result we can give a formal quality guarantee of the trained autoencoder. It securely extracts a latent representation for each time series in the safety-critical area that guarantees separation of both clusters. Please note that one could not have used the latent space to check separability directly. We have no control over where the autoencoder maps the safety-critical regions in the latent space. In contrast our QUGA method solves this by symbolic representation of the autoencoder, and the systematic search of possible unsupervised adversarial examples that lead to a mix-up of two clusters. With this we can prove separability for all infinite points in the safety-critical regions and not just on the finite training set.

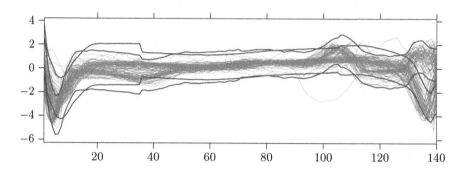

Fig. 7. Image spaces (red and green) into which points from the respective safety-critical regions in Fig. 3 can theoretically be mapped by the autoencoder. (Color figure online)

6   Conclusion and Future Work

QUGA overcomes major shortcomings of unsupervised learning with autoencoders. We provide the first methodology to bound the error of an autoencoder in a safety-critical region. With our solution framework based on SMT solvers we propose to search for adversarial examples in the infinite search space of a safety-critical region. Therefore, we have defined *unsupervised adversarial examples* as inputs that show maximal error even if these objects are not contained in the training data. Our QUGA approach formulates the autoencoder, the safety-critical region, and the error of the loss function with a logical conjunction of linear constraints. Once we have found an unsupervised adversarial example, it serves as a lower bound for the error while binary search allows to derive an upper bound. We demonstrate the effectiveness of our approach on both synthetically created and real dataset. We show that QUGA finds unsupervised adversarial examples, provides quality guarantees with lower and upper bounds, and outperforms sampling schemes that underestimate the maximum error.

As this is the first work for unsupervised adversarial examples on autoencoders we expect a variety of follow-up research. In particular we aim at unsupervised uncertainty quantification of autoencoders. Furthermore we plan to develop more advanced adversarial attacks for specialized autoencoders in time series domain, as well as the re-use of adversarial examples for re-training autoencoders in safety-critical regions. Moreover we intend to make this work more scalable by incorporating advances from the research area of SMT solvers into our approach.

References

1. Barrett, C., Fontaine, P., Tinelli, C.: The SMT-LIB standard: version 2.6. Technical report, Department of Computer Science, The University of Iowa (2017)
2. Bastani, O., Ioannou, Y., Lampropoulos, L., Vytiniotis, D., Nori, A.V., Criminisi, A.: Measuring neural net robustness with constraints. In: Advances in Neural Information Processing Systems, vol. 29, pp. 2613–2621 (2016)
3. Bradley, A.R., Manna, Z.: The Calculus of Computation - Decision Procedures with Applications to Verification (2007)
4. Chazan, S.E., Gannot, S., Goldberger, J.: Deep clustering based on a mixture of autoencoders. In: 29th IEEE International Workshop on Machine Learning for Signal Processing, pp. 1–6 (2019)
5. Chen, Y., et al.: The UCR time series classification archive, July 2015
6. Chhabra, A., Roy, A., Mohapatra, P.: Strong black-box adversarial attacks on unsupervised machine learning models. CoRR (2019)
7. Dalvi, N.N., Domingos, P.M., Mausam, Sanghai, S.K., Verma, D.: Adversarial classification. In: Proceedings of the Tenth ACM SIGKDD International Conference on Knowledge Discovery and Data Mining, pp. 99–108 (2004)
8. Ehlers, R.: Formal verification of piece-wise linear feed-forward neural networks. In: D'Souza, D., Narayan Kumar, K. (eds.) ATVA 2017. LNCS, vol. 10482, pp. 269–286. Springer, Cham (2017). https://doi.org/10.1007/978-3-319-68167-2_19
9. Gehr, T., Mirman, M., Drachsler-Cohen, D., Tsankov, P., Chaudhuri, S., Vechev, M.T.: AI2: safety and robustness certification of neural networks with abstract interpretation. In: 2018 IEEE Symposium on Security and Privacy, pp. 3–18 (2018)
10. Gondara, L.: Medical image denoising using convolutional denoising autoencoders. In: IEEE International Conference on Data Mining Workshops, pp. 241–246 (2016)
11. Goodfellow, I.J., Shlens, J., Szegedy, C.: Explaining and harnessing adversarial examples. In: 3rd International Conference on Learning Representations (2015)
12. Katz, G., Barrett, C., Dill, D.L., Julian, K., Kochenderfer, M.J.: Reluplex: an efficient SMT solver for verifying deep neural networks. In: Majumdar, R., Kunčak, V. (eds.) CAV 2017. LNCS, vol. 10426, pp. 97–117. Springer, Cham (2017). https://doi.org/10.1007/978-3-319-63387-9_5
13. Le, Q.V., et al.: Building high-level features using large scale unsupervised learning. In: Proceedings of of the 29th International Conference on Machine Learning (2012)
14. Meng, Q., Catchpoole, D.R., Skillicom, D., Kennedy, P.J.: Relational autoencoder for feature extraction. In: 2017 International Joint Conference on Neural Networks (2017)
15. Min, M.R., Stanley, D.A., Yuan, Z., Bonner, A.J., Zhang, Z.: A deep non-linear feature mapping for large-margin kNN classification. In: ICDM 2009, The Ninth IEEE International Conference on Data Mining, pp. 357–366 (2009)
16. de Moura, L., Bjørner, N.: Z3: an efficient SMT solver. In: Ramakrishnan, C.R., Rehof, J. (eds.) TACAS 2008. LNCS, vol. 4963, pp. 337–340. Springer, Heidelberg (2008). https://doi.org/10.1007/978-3-540-78800-3_24

17. Pasa, L., Sperduti, A.: Pre-training of recurrent neural networks via linear autoencoders. In: Advances in Neural Information Processing Systems 27: Annual Conference on Neural Information Processing Systems 2014, pp. 3572–3580 (2014)
18. Sakurada, M., Yairi, T.: Anomaly detection using autoencoders with nonlinear dimensionality reduction. In: Proceedings of the MLSDA 2014 2nd Workshop on Machine Learning for Sensory Data Analysis (2014)
19. Singh, G., Gehr, T., Püschel, M., Vechev, M.T.: Boosting robustness certification of neural networks. In: 7th International Conference on Learning Representations (2019)
20. Szegedy, C., et al.: Intriguing properties of neural networks. In: 2nd International Conference on Learning Representations (2014)
21. Vincent, P., Larochelle, H., Lajoie, I., Bengio, Y., Manzagol, P.: Stacked denoising autoencoders: learning useful representations in a deep network with a local denoising criterion. J. Mach. Learn. Res. **11**, 3371–3408 (2010)
22. Zhao, G., Zhang, M., Liu, J., Wen, J.R.: Unsupervised adversarial attacks on deep feature-based retrieval with GAN (2019)

Metric Learning for Categorical and Ambiguous Features: An Adversarial Method

Xiaochen Yang[1] , Mingzhi Dong[1], Yiwen Guo[2] , and Jing-Hao Xue[1(✉)]

[1] University College London, London, UK
{xiaochen.yang.16,jinghao.xue}@ucl.ac.uk, mingzhidong@gmail.com
[2] ByteDance AI Lab, Beijing, China
guoyiwen.ai@bytedance.com

Abstract. Metric learning learns a distance metric from data and has significantly improved the classification accuracy of distance-based classifiers such as k-nearest neighbors. However, metric learning has rarely been applied to categorical data, which are prevalent in health and social sciences, but inherently difficult to classify due to high feature ambiguity and small sample size. More specifically, ambiguity arises as the boundaries between ordinal or nominal levels are not always sharply defined. In this paper, we mitigate the impact of feature ambiguity by considering the worst-case perturbation of each instance and propose to learn the Mahalanobis distance through adversarial training. The geometric interpretation shows that our method dynamically divides the instance space into three regions and exploits the information on the "adversarially vulnerable" region. This information, which has not been considered in previous methods, makes our method more suitable than them for small-sized data. Moreover, we establish the generalization bound for a general form of adversarial training. It suggests that the sample complexity rate remains at the same order as that of standard training only if the Mahalanobis distance is regularized with the elementwise 1-norm. Experiments on ordinal and mixed ordinal-and-nominal datasets demonstrate the effectiveness of the proposed method when encountering the problems of high feature ambiguity and small sample size.

Keywords: Metric learning · Categorical data · Adversarial training

1 Introduction

The k-nearest neighbors (kNN) algorithm is a classical and widely used classification method by virtue of the nonparametric nature, interpretability, and flexibility in defining the distance between instances [6]. As a discriminative distance

---

X. Yang and M. Dong—Equal contribution.

Electronic supplementary material The online version of this chapter (https://doi.org/10.1007/978-3-030-67661-2_14) contains supplementary material, which is available to authorized users.

© Springer Nature Switzerland AG 2021
F. Hutter et al. (Eds.): ECML PKDD 2020, LNAI 12458, pp. 223–238, 2021.
https://doi.org/10.1007/978-3-030-67661-2_14

function can boost kNN's performance, the idea of learning a task-specific metric from the data was pioneered in [33], which formulates the task of learning a generalized Mahalanobis distance as a convex optimization problem. Thereafter, many global [31], local [30], kernelized [25] and deep [16] metric learning methods have been proposed to further improve the discriminability. While these methods are effective, they have rarely been applied to data with ordinal and nominal features.

Ordinal and nominal variables (i.e. features) are subsumed under the data type of categorical variables that have measurement scales consisting of a set of categories [1]. Categorical variables with ordered scales are called ordinal variables and the ones with unordered scales are called nominal variables. For example, when collecting a film survey, the audience review (poor, fair, good, excellent) is an ordinal variable; the genre of favorite films (action, comedy, drama, horror) is a nominal variable. Both types of variables occur frequently in social and health sciences and also arise in education and marketing.

Classifying ordinal and nominal variables faces at least the following three challenges. First, a simple way of representing these variables is to encode them as integers and then treat them as real-valued continuous variables. However, for an ordinal variable, the difference between two integers does not necessarily reflect the distance between the two ordinal levels, and for a nominal variable, the difference between two integers is meaningless for two nominal levels. Another way of representing ordinal and nominal variables is to encode each categorical variable into a set of binary variables, such as through dummy coding. This conversion avoids the above problems, and allows for the modeling of interactions between different levels of the variable. However, it inevitably increases the feature dimension, and the effect is dramatic when each variable has a large number of levels. The second challenge is the ambiguity in ordinal variables. For example, in the example of audience review, the boundaries between levels such as 'good' and 'excellent' are not sharply defined, thereby causing ambiguity. This issue is less common in nominal variables, but it still appears when some categories have overlapping characteristics. Third, for economic and ethical reasons, the categorical data collected in social and health sciences often have a small sample size. This places a restriction on model complexity since a complex model may overfit and generalize poorly to unseen data.

This paper focuses on adapting metric learning methods for ordinal and nominal features that could work on both types of encoded data, i.e. as integer variables and as dummy variables, and address the feature ambiguity and small-sized problems. Firstly, to mitigate the impact of feature ambiguity, we propose to consider the worst-case perturbation of each instance within a deliberately designed constraint set and learn the distance metric via adversarial training. The constraint set takes into account the discrete nature of nominal variables and the ordering nature of ordinal variables. Secondly, we provide a geometric interpretation of the proposed formulation, which suggests that our method dynamically divides the instance space into three regions, namely support region, adversarially vulnerable region, and adversarially robust region. Compared with classical metric learning methods which only uses information on the support region, our method additionally uses information on instances from the adversarially vulnerable region, thereby coping better with the small sample size problem. Thirdly, we prove the generalization bound for a general form of adversarial training. It guarantees that, when regularizing the Mahalanobis distance with the elementwise 1-norm, the sample complexity rate of the proposed method remains at the same order as that of classical methods. Finally, the method is tested on datasets with all ordinal variables and with a mixture of ordinal and nominal variables. It surpasses state-of-the-art methods in cases of high feature ambiguity and small sample size.

2   Related Work

This section briefly reviews distance metrics for categorical data and metric learning methods that consider feature uncertainty.

Distance Metrics for Categorical Data. Various distance or similarity measures are proposed for categorical data, mostly for nominal data, in an unsupervised setting. The most common measure is *overlap*, which defines the similarity between two instances x_1, x_2 on the ith feature to be 1 if their values are equal and 0 otherwise. Summing up the similarities over all features defines the distance between x_1 and x_2. Based on overlap, many probabilistic or frequency-based measures have been proposed to assign different weights on matches or mismatches, as well as taking into account the occurrence of other feature values [2]. Another class of measures are based on entropy, where the distance contribution of each categorical level depends on the amount of information it holds. Entropy-based measures have been extended in [36] to quantify the order relation of ordinal variables.

In a supervised setting, non-learning approaches use the label information to determine the discriminative power of each feature and adjust the feature weights in distance calculation accordingly [9]. Learning approaches learn the distance between each pair of categorical levels or a mapping function from each level to a real value by minimizing the classification error [8,32]. More recently, large margin-based metric learning methods have been adapted for ordinal and nominal variables [23,37]. Building on the assumption that an ordinal variable represents a continuous latent variable that falls into an interval of values, [23] jointly learns the Mahalanobis distance, thresholds of intervals, and parameters of the latent variable distribution. As the number of thresholds is determined by the number of variables and levels within them, the method may involve a large number of parameters and suffer from overfitting. [37] represents the categorical data by computing the interaction between levels, between variables, and between variables and classes, followed by learning the Mahalanobis distance in a kernel space. However, it ignores the natural ordering of ordinal variables.

Metric Learning with Uncertainty. In most metric learning methods, a Mahalanobis distance is optimized such that similar instances become closer with respect to the new metric and dissimilar instances become farther away. As the optimization process is guided by the side information, its effectiveness degrades in the presence of label noise, outlier samples, and feature uncertainty. Compared with outliers (or influential points in the statistics literature) which account for a small proportion of instances but severely influence the model, feature uncertainty, possibly ensued from ambiguity in the definition of set boundaries, measurement and quantization errors, and data processing of repeated measurements, normally appears as small perturbations but potentially pollutes a large number of instances [26]. Many robust metric learning methods have been proposed to tackle the above problems [27–29], and here, we only discuss those on feature uncertainty.

One way to handle feature uncertainty is to build an explicit model of perturbation [22,34]. [34] assumes a perturbation distribution of each instance, replaces the Mahalanobis distance by its expected value, and iteratively learns the distribution and distance metric by minimizing the number of violations of triplet constraints. The method essentially adjusts the constraint on distance margin for each triplet according to its reliableness. Another approach is to learn a distance metric that is less sensitive to feature uncertainty via adversarial training [7,12]. The method involves two stages. The confusion stage generates adversarial pairs that incur large losses, and the discrimination stage optimizes the distance metric based on these augmented pairs. Originating from robust optimization [17,24], adversarial training has received considerable attention in recent years as an effective approach to achieving robustness to adversarial examples [20]. In addition, adversarial training is shown to improve the classification accuracy when there is only a limited number of instances available to train the model [3]. While our method shares a similar principle, it differs from existing adversarial metric learning methods in two respects. Firstly, we take the subsequent classification mechanism into consideration when searching for the worst-case perturbation. Derived from triplet constraints, the perturbation is capable of altering the decision of NN classifier. Secondly and more importantly, the loss function in our proposal is designed specifically for ordinal and nominal features with an explicit consideration of their discrete and ordering nature.

3   Methodology

In this section, we propose to model feature ambiguity as a perturbation to the instance and learn the Mahalanobis distance via adversarial training. After introducing notations, we will present the method and its optimization algorithm, followed by a geometric interpretation and a generalization analysis.

3.1   Preliminaries

Let $z^n = \{z_i = (x_i, y_i), i = 1, \ldots, n\}$ denote the training set, where $x_i \in \mathcal{X}$ is the ith training instance associated with label $y_i \in \mathcal{Y} = \{1, \ldots, C\}$; $z_i \in \mathcal{Z}$ is independently and identically distributed according to an unknown distribution $\mathcal{D}$. Suppose each instance includes p features, p^{ord} of which are ordinal variables and $p^{\mathrm{nom}} = p - p^{\mathrm{ord}}$ are nominal variables. Ordinal variables can be encoded as consecutive integers or as a set of binary values. In the integer case, a variable with p_r levels takes values from $\{1, 2, \ldots, p_r\}$, and the mapping should follow the order relation. In other words, for ordinal levels $O_1 \prec O_2 \prec \cdots \prec O_{p_r}$ with an order relation $\prec$, there is a mapping function $\mathcal{O}$ such that $\mathcal{O}(O_q) = q, q = 1, \ldots, p_r$. In the binary-valued case, ordinal variables are encoded via the *OrderedPartitions* method [15,23]. For example, an ordinal variable with 3 levels will be encoded as $[1, 0, 0]$, $[1, 1, 0]$ and $[1, 1, 1]$. Nominal variables are encoded via the *1-of-K* encoding scheme. For example, a nominal variable with 3 levels will be encoded as $[1, 0, 0]$, $[0, 1, 0]$ and $[0, 0, 1]$. Let P denote the feature dimension after encoding, which equals $p^{\mathrm{ord}} + \sum_{r=1}^{p^{\mathrm{nom}}} p_r$ if ordinal variables are encoded as integers and equals $\sum_{r=1}^{p} p_r$ if they are encoded as a set of binary values.

In this paper, we focus on learning the Mahalanobis distance from triplet-based side information. For any two instances $\boldsymbol{x}_i, \boldsymbol{x}_j \in \mathbb{R}^P$, the generalized (squared) Mahalanobis distance is defined as

$$d_M^2(\boldsymbol{x}_i, \boldsymbol{x}_j) = (\boldsymbol{x}_i - \boldsymbol{x}_j)^T M (\boldsymbol{x}_i - \boldsymbol{x}_j)$$

where $M \in \mathbb{S}_+^P$ is a $P \times P$ real-valued positive semidefinite (PSD) matrix. A classical triplet-based metric learning method is the large margin nearest neighbors (LMNN) algorithm [31]. It pulls k nearest same-class instances closer and pushes away differently labeled instances by a fixed margin through optimizing the following objective function:

$$\min_{M \in \mathbb{S}_+^P} (1 - \mu) \sum_{(\boldsymbol{x}_i, \boldsymbol{x}_j) \in \mathcal{S}} d_M^2(\boldsymbol{x}_i, \boldsymbol{x}_j) + \mu \sum_{(\boldsymbol{x}_i, \boldsymbol{x}_j, \boldsymbol{x}_l) \in \mathcal{R}} \left[1 + d_M^2(\boldsymbol{x}_i, \boldsymbol{x}_j) - d_M^2(\boldsymbol{x}_i, \boldsymbol{x}_l)\right]_+,$$

(1)

where $[a]_+ = \max(a, 0)$ for $a \in \mathbb{R}$; μ is the trade-off parameter; and

$$\mathcal{S} = \{(\boldsymbol{x}_i, \boldsymbol{x}_j) : \boldsymbol{x}_j \in \{k\text{NNs with the same class label of } \boldsymbol{x}_i\}\},$$
$$\mathcal{R} = \{(\boldsymbol{x}_i, \boldsymbol{x}_j, \boldsymbol{x}_l) : (\boldsymbol{x}_i, \boldsymbol{x}_j) \in \mathcal{S}, y_i \neq y_l\}.$$

(2)

$\boldsymbol{x}_j$ is termed the target neighbor of $\boldsymbol{x}_i$ and $\boldsymbol{x}_l$ is termed the impostor.

3.2 Metric Learning with Adversarial Training (MLadv)

The objective function of LMNN (Eq. 1) can be interpreted as minimizing a linear combination between the *empirical risk* $\frac{1}{n} \sum_{i=1}^{n} \ell(\boldsymbol{x}, y; M)$ and the regularizer on M; the hinge loss ℓ of $[1 + d_M^2(\boldsymbol{x}_i, \boldsymbol{x}_j) - d_M^2(\boldsymbol{x}_i, \boldsymbol{x}_l)]_+$ separates target neighbors and impostors by a unit margin, and the regularizer is chosen as $\sum_{(\boldsymbol{x}_i, \boldsymbol{x}_j) \in \mathcal{S}} d_M^2(\boldsymbol{x}_i, \boldsymbol{x}_j)$. To address the issue of feature ambiguity faced by ordinal and nominal variables, we propose to model the unknown ambiguity as a perturbation of $\boldsymbol{x}_i$. Instead of the hinge loss, we consider the worst-case loss within a certain perturbation range and minimize the *adversarial empirical risk*:

$$\min_{M} \frac{1}{n} \sum_{i=1}^{n} \max_{\boldsymbol{\delta} \in \Delta} \ell(\boldsymbol{x} + \boldsymbol{\delta}, y; M).$$

(3)

$\boldsymbol{\delta}$ denotes the perturbation and its form is specified by set Δ. The optimal solution to the inner maximization problem is termed the worst-case perturbation and denoted by $\boldsymbol{\delta}^*$.

To fit ordinal and nominal variables, we need to incorporate their properties when defining the perturbation set Δ. A typical choice of Δ is the set of ℓ_p-bounded perturbation, i.e. $\|\boldsymbol{\delta}\|_p \leq \varepsilon$, with $p = 1, 2, \infty$. However, a non-integer real-valued ε is not suitable for ordinal and nominal variables as it ignores the discrete nature of nominal variables and the ordering nature of ordinal variables. Therefore, we restrict $\boldsymbol{\delta}$ via the following two conditions: i) $\|\boldsymbol{\delta}\|_\infty = 1$; and ii) $\|\boldsymbol{\delta}\|_1 \leq \varepsilon, \varepsilon \in \mathbb{N}$. Since the loss function is linear in $\boldsymbol{\delta}$ as shown in Eq. 1 of Appendix A, the first condition guarantees that the perturbed instance remains as an integer or a binary value. More crucially, the magnitude of one aligns with the source of feature ambiguity, which arises from non-rigorously defined set boundaries. In the example of film survey, the perturbation from 'good' to 'fair' matches the real-world decision-making process whereas replacing 'good' by 'bad' dramatically changes the original information. The second condition controls the level of perturbation. Integrating these two conditions, the perturbation $\boldsymbol{\delta}$ can change at most ε features of each instance.

To train the Mahalanobis distance, we form triplet constraints from both original and perturbed instances [14], and apply different loss functions to these triplets. For the original triplets, we adopt the loss function of LMNN and change the unit distance margin to an adjustable quantity τ. As we shall discuss in Sect. 3.4, τ determines how the instance space is divided into the support region and the adversarially vulnerable region. If the distance margin is satisfied by the triplet $(\boldsymbol{x}_i, \boldsymbol{x}_j, \boldsymbol{x}_l)$, we will proceed to add perturbation to the instance $\boldsymbol{x}_i$. For the perturbed triplets, we adopt the perceptron loss [18]. Although the perceptron loss is rarely used in metric learning due to the lack of distance margin, it is sensible in our setting since the perturbation itself can serve as a margin in the instance space.

Integrating the above design of perturbation set and loss functions, we propose the following objective function for metric learning through adversarial training (MLadv):

$$\min_{M \in \mathbb{S}_+^d} \lambda \|M\|_1 + \frac{\mu}{|\mathcal{R}|} \sum_{(\boldsymbol{x}_i, \boldsymbol{x}_j, \boldsymbol{x}_l) \in \mathcal{R}} \left[\tau + d_M^2(\boldsymbol{x}_i, \boldsymbol{x}_j) - d_M^2(\boldsymbol{x}_i, \boldsymbol{x}_l) \right]_+$$

$$+ \frac{1-\mu}{|\mathcal{R}|} \sum_{(\boldsymbol{x}_i, \boldsymbol{x}_j, \boldsymbol{x}_l) \in \mathcal{R}} \mathbb{1}[d_M^2(\boldsymbol{x}_i, \boldsymbol{x}_l) > d_M^2(\boldsymbol{x}_i, \boldsymbol{x}_j) + \tau]$$

$$\cdot \left[\max_{\boldsymbol{\delta}_i : \|\boldsymbol{\delta}_i\|_\infty = 1, \|\boldsymbol{\delta}_i\|_1 \leq \varepsilon} \{d_M^2(\boldsymbol{x}_i + \boldsymbol{\delta}_i, \boldsymbol{x}_j) - d_M^2(\boldsymbol{x}_i + \boldsymbol{\delta}_i, \boldsymbol{x}_l)\} \right]_+,$$

$$(4)$$

where $|\mathcal{R}|$ denote the numbers of triplets in the set $\mathcal{R}$; $\mathbb{1}[\cdot]$ is the indicator function which equals 1 if the condition is satisfied and 0 otherwise. The elementwise 1-norm (hereinafter abbreviated to L_1-norm), i.e. $\|M\|_1 = \|\text{vec}(M)\|_1 = \sum_{m,n=1}^P M_{mn}$, is used to regularize the complexity of the distance matrix. As proved in Sect. 3.5, this choice of regularizer is essential to guarantee that the number of samples required for the adversarially trained metric to generalize has the same order as that for the standard metric. $\lambda > 0$ is a trade-off parameter between the regularization term and the loss function, and $\mu \in [0, 1]$ balances between the influence from original instances and perturbed instances. The triplet set $\mathcal{R}$ is constructed in the same way as LMNN, i.e. according to Eq. 2.

3.3   Optimization Algorithm

According to the Danskin's theorem [21], the gradient of the maximum of a differentiable function is given by the gradient of the function evaluated at the maximum point, i.e.

$$\nabla_M \max_{\delta \in \Delta} \ell(x + \delta, y; M) = \nabla_M \ell(x + \delta^*, y; M),$$

where $\delta^* = \arg\max_{\delta \in \Delta} \ell(x + \delta, y; M)$; ∇ denotes the gradient. Therefore, we solve the optimization problem (Eq. 4) by first deriving a closed-form solution to the inner maximization problem and then updating M via the proximal gradient descent algorithm.

The solution to the worst-case perturbation δ_i^* can be obtained as follows: let $\delta_i^* = \arg\max_{\delta_i : \|\delta_i\|_\infty = 1, \|\delta_i\|_1 \leq \varepsilon} \{d_M^2(x_i + \delta_i, x_j) - d_M^2(x_i + \delta_i, x_l)\}$, then

$$\delta_{i,[k]}^* = \begin{cases} \text{sign}(M_{k\cdot}(x_l - x_j)) & \text{if } k \in \arg\max_{\varepsilon} |M_{a\cdot}(x_l - x_j)| \\ & \qquad\qquad a=1,\cdots,P \\ 0 & \text{otherwise} \end{cases}, \qquad (5)$$

where $\delta_{i,[k]}^*$ denotes the kth element of the vector δ_i^*; $M_{k\cdot}$ denotes the kth row of M; $\arg\max_\varepsilon$ denotes the set of largest ε elements of a vector; $\text{sign}(v)$ applies the sign function to each element of the vector v and $|v|$ calculates elementwise absolute values. Detailed derivation is given in Appendix A.

Since the L_1-norm regularization introduces a non-smooth function, the proximal gradient descent algorithm is adopted to optimize M in three steps. In the gradient descent step, M is updated as

$$M^{t+\frac{1}{3}} = M^t - \eta^t \nabla M|_{M^t}$$
$$\nabla M = \frac{\mu}{|\mathcal{R}|} \sum_{\mathcal{R}} \alpha_{ijl}(X_{ij} - X_{il}) + \frac{1-\mu}{|\mathcal{R}|} \sum_{\mathcal{R}} (1 - \alpha_{ijl}) \alpha_{ijl}^*(X_{ij}^* - X_{il}^*) \qquad (6)$$

where $\sum_{\mathcal{R}}$ is an abbreviation for $\sum_{(x_i, x_j, x_l) \in \mathcal{R}}$; $\alpha_{ijl} = \mathbb{1}[\tau + d_M^2(x_i, x_j) \geq d_M^2(x_i, x_l)]$, $\alpha_{ijl}^* = \mathbb{1}[d_M^2(x_i^*, x_j) \geq d_M^2(x_i^*, x_l)]$; $x_i^* = x_i + \delta_i^*$; $X_{ij} = (x_i - x_j)(x_i - x_j)^T$, $X_{ij}^* = (x_i^* - x_j)(x_i^* - x_j)^T$, and X_{il}, X_{il}^* are defined similarly. The learning rate η^t decays during training according to the exponential function $\exp(-0.99(1 + 0.01t))$. Next, we compute the proximal mapping for the L_1-norm regularization, which is equivalent to applying the soft-thresholding operator to $M^{t+\frac{1}{3}}$:

$$M_{mn}^{t+\frac{2}{3}} = \text{sign}(M_{mn}^{t+\frac{1}{3}})[|M_{mn}^{t+\frac{1}{3}}| - \lambda \eta^t]_+. \qquad (7)$$

Finally, M is projected onto the cone of PSD matrices via eigendecomposition:

$$M^{t+\frac{2}{3}} = V \Lambda V^T$$
$$M^{t+1} = V \max(\Lambda, 0) V^T. \qquad (8)$$

The optimization algorithm for the proposed method is summarized in Algorithm 1 of Appendix C.1.

We now analyze the computational complexity of the proposed method. MLadv has the same computational complexity as LMNN in calculating the distance of each triplet, performing gradient descent, and projecting onto the PSD cone; their total complexity equals $O(P^3 + nP^2 + |\mathcal{R}| \cdot P)$, where P is the feature dimension after variable encoding, n is the number of training instances, and $|\mathcal{R}|$ is the number of triplet constraints. The extra cost results from the sorting operation used to find the worst-case perturbation and the soft-thresholding operation used to perform the L_1-norm regularization. The time complexity of the sorting step is $O(P^2 \log P)$ and that of the soft-thresholding step is $O(P^2)$. Overall, the time complexity of MLadv per iteration is $O(P^3 + P^2 \log P + nP^2 + |\mathcal{R}| \cdot P)$.

3.4   Geometric Interpretation

We now provide a geometric interpretation for better understanding the effect of perturbation.

To start with, we rewrite the gradient of Eq. 6 by plugging in the worst-case perturbation derived in Eq. 5:

$$\frac{\mu}{|\mathcal{R}|} \sum_{\mathcal{R}} \mathbb{1}[d_M^2\left(\boldsymbol{x}_i, \boldsymbol{x}_l\right) \leq \tau + d_M^2\left(\boldsymbol{x}_i, \boldsymbol{x}_j\right)](\boldsymbol{X}_{ij} - \boldsymbol{X}_{il}) \tag{9}$$

$$+ \frac{1-\mu}{|\mathcal{R}|} \sum_{\mathcal{R}} \mathbb{1}[d_M^2\left(\boldsymbol{x}_i, \boldsymbol{x}_j\right) + \tau < d_M^2\left(\boldsymbol{x}_i, \boldsymbol{x}_l\right) \leq d_M^2\left(\boldsymbol{x}_i, \boldsymbol{x}_j\right) + 2\|\boldsymbol{M}\boldsymbol{x}_{lj}\|_{1,[\varepsilon]}](\boldsymbol{X}_{ij}^{\star} - \boldsymbol{X}_{il}^{\star}),$$

where $\|\boldsymbol{M}\boldsymbol{x}_{lj}\|_{1,[\varepsilon]} = \sum \max_\varepsilon |\boldsymbol{M}(\boldsymbol{x}_l - \boldsymbol{x}_j)|$ is the sum of ε largest absolute values in the vector $[\boldsymbol{M}_{1.}(\boldsymbol{x}_l - \boldsymbol{x}_j), \cdots, \boldsymbol{M}_{P.}(\boldsymbol{x}_l - \boldsymbol{x}_j)]$.

Equation 9 shows that, while LMNN and its variants learn the metric only on triplets where the impostor lies insufficiently far away from the instance, i.e. the difference in squared distances (DD) $d_M^2\left(\boldsymbol{x}_i, \boldsymbol{x}_l\right) - d_M^2\left(\boldsymbol{x}_i, \boldsymbol{x}_j\right)$ is less than or equal to the required margin τ, the proposed method not only uses these information but also selectively exploits triplets that satisfy the margin constraint. In particular, the new selection criterion considers the correlation between the distance metric and $(\boldsymbol{x}_l - \boldsymbol{x}_j)$: if the correlation is high, i.e. the value of $\|\boldsymbol{M}\boldsymbol{x}_{lj}\|_{1,[\varepsilon]}$ is large, it is more likely this triplet will incur a loss and hence contribute to the gradient.

Figure 1 illustrates the above discussion with two figures. In both figures, we show all instances in the linearly mapped feature space induced by the Mahalanobis distance, and consider different positions of $\boldsymbol{x}_i$ with respect to fixed target neighbor $\boldsymbol{x}_j$ and impostor $\boldsymbol{x}_l$. The left figure illustrates which triplets are used in LMNN and MLadv for calculating the gradient; for simplicity, the learned $\boldsymbol{M}$ is a scaled Euclidean distance. For $\boldsymbol{x}_{i_1}$, both methods use the triplet $(\boldsymbol{x}_{i_1}, \boldsymbol{x}_j, \boldsymbol{x}_l)$ since DD is less than τ. For $\boldsymbol{x}_{i_2}$ and $\boldsymbol{x}_{i_3}$, the methods differ. $(\boldsymbol{x}_{i_2}, \boldsymbol{x}_j, \boldsymbol{x}_l)$ and $(\boldsymbol{x}_{i_3}, \boldsymbol{x}_j, \boldsymbol{x}_l)$ satisfy the margin constraint and hence are not used in LMNN. However, they are used in our MLadv as $\boldsymbol{x}_{i_2}$ and $\boldsymbol{x}_{i_3}$ may be misclassified in the presence of perturbation; the perturbation sets with $\varepsilon = 1$ and $\varepsilon = 2$ are indicated by the blue line and blue square, respectively. When $\varepsilon = 1$, $\boldsymbol{x}_{i_2}$ may be misclassified as the worst-case perturbation $\boldsymbol{\delta}_{i_2}^{\star}$ can drag the instance across the decision boundary; when $\varepsilon = 2$, both $\boldsymbol{x}_{i_2}$ and $\boldsymbol{x}_{i_3}$ may be misclassified. For $\boldsymbol{x}_{i_4}$, both methods ignore the triplet $(\boldsymbol{x}_{i_4}, \boldsymbol{x}_j, \boldsymbol{x}_l)$ since $\boldsymbol{x}_{i_4}$ remains far away from the decision boundary even after adding $\boldsymbol{\delta}_{i_4}^{\star}$.

The right figure presents the general case with an anisotropic $\boldsymbol{M}$ and multiple target neighbors, and illustrates the interaction between DD, $\boldsymbol{x}_l - \boldsymbol{x}_j$, and $\boldsymbol{M}$. Even though the DDs of $(\boldsymbol{x}_{i_1}, \boldsymbol{x}_{j_1}, \boldsymbol{x}_l)$ and $(\boldsymbol{x}_{i_2}, \boldsymbol{x}_{j_2}, \boldsymbol{x}_l)$ are the same, $\boldsymbol{x}_{i_1}$ is not robust against the worst-case perturbation whereas $\boldsymbol{x}_{i_2}$ is. The reason is that $\boldsymbol{M}$ expands the horizontal distance, as indicated by the arrows at the bottom-left corner, and has a higher correlation with $\boldsymbol{x}_l - \boldsymbol{x}_{j_1}$ compared to $\boldsymbol{x}_l - \boldsymbol{x}_{j_2}$. This suggests that, for an instance to be invariant to the worst-case perturbation, the requirement of DD is determined locally with respect to $\boldsymbol{x}_l - \boldsymbol{x}_j$ and dynamically with respect to $\boldsymbol{M}$.

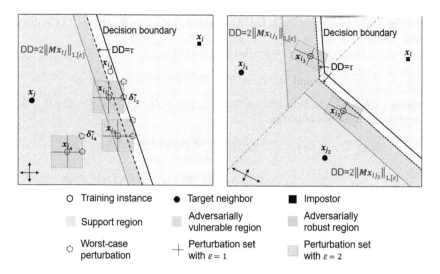

Fig. 1. Illustration of MLadv. Instances are shown in the linearly mapped feature space induced by an isotropic M (*left*) and an anisotropic M (*right*). Left: LMNN learns M based on instances from the support region where the difference in squared distances (DD) $d_M^2(x_i, x_l) - d_M^2(x_i, x_j)$ is not greater than τ. MLadv learns on additional instances from the adversarially vulnerable region where the instance may be misclassified after adding the worst-case perturbation. The regions are divided by two hyperplanes that are parallel to the decision boundary with DD of τ and $2\|Mx_{lj}\|_{1,[\varepsilon]}$, respectively. Right: Instances lying above the gray dash-dotted line select x_{j_1} as NN and should be separated farther away from the decision boundary due to the high correlation between M and $x_l - x_{j_1}$.

In summary, as points with the same DD form a separating hyperplane that is orthogonal to the line joining x_j and x_l, the proposed method essentially divides the instance space into three regions according to the hyperplanes with DD of τ and $2\|Mx_{lj}\|_{1,[\varepsilon]}$. It then makes use of instances from the support region and adversarially vulnerable region for learning the metric. The additional information from the latter region is particularly important for datasets with a small sample size.

3.5 Theoretical Analysis

In this section, we provide the generalization bound for metric learning trained in the adversarial setting. In essence, with the same form of loss function, adversarial training incurs a larger loss than standard training due to the addition of perturbation. Therefore, it is expected that the sample complexity would be higher in order to achieve the same generalization performance.

We start by defining some notations. The adversarial loss is defined as

$$\tilde{\ell}_M(z_i, z_j, z_l) = \mathbb{1}[y_i = y_j \neq y_l][\tau + \max_{\delta_i: \|\delta_i\|_\infty \leq \varepsilon} \{d_M^2(x_i + \delta_i, x_j) - d_M^2(x_i + \delta_i, x_l)\}]_+. \tag{10}$$

The generalization bound studies the gap between the adversarial population risk $\tilde{R}(M) = \mathbb{E}_{(z_i, z_j, z_l) \sim \mathcal{D}}[\tilde{\ell}_M(z_i, z_j, z_l)]$ and the adversarial empirical risk $\tilde{R}_n(M) = $

$\frac{1}{n(n-1)(n-2)} \sum_{i \neq j \neq l} \tilde{\ell}_M(z_i, z_j, z_l)$. Let M_z denote the optimal solution to the learning problem:

$$\min_{M \in \mathbb{S}_P^+} \tilde{R}_n(M) + \lambda \|M\|_1. \tag{11}$$

The generalization bound of M_z is given by the following theorem.

Theorem 1. *Let M_z be the solution to the problem (11). Then, for any $0 < \delta < 1$, with probability $1 - \delta$ we have that*

$$\tilde{R}(M_z) - \tilde{R}_n(M_z) \leq \frac{32\tau(x_{max}^2 + \varepsilon x_{max})\sqrt{e \log P}}{\lambda \sqrt{n}}$$
$$+ \tau \left[1 + \frac{x_{max}^2 + 2\varepsilon x_{max}}{\lambda} \right] \sqrt{\frac{2 \ln(1/\delta)}{n}} + \frac{4\tau}{\sqrt{n}}, \tag{12}$$

where $x_{max} = \sup_{x, x' \in \mathcal{X}} \|x - x'\|_\infty$.

Theorem 1 is established based on the Rademacher complexity [5,35] and U-statistics [19]; proof is given in Appendix B.

We make three remarks here. First, by definition, the perturbation size is relatively small compared to $x_{\max}$, and therefore, $\varepsilon x_{\max} < x_{\max}^2$. This suggests that adversarial training does not largely increase the sample complexity. Second, as shown in the proof, if M is regularized via the Frobenius norm, the sample complexity required by adversarial training will be higher than the standard training at a rate of $O(\sqrt{P})$. To avoid the sublinear dependence of sample complexity on feature dimension, we use the L_1-norm as the regularizer. Third, Theorem 1 provides a general guarantee on the generalization performance of triplet-based metric learning trained in the adversarial setting. The adversarial loss defined in Eq. 10 with $\varepsilon = 1$ unifies the two loss functions defined in our learning objective (Eq. 4). In other words, the generalization gap of our learned metric is bounded as given in Theorem 1.

4 Experiments

In this section, we first conduct experiments on a discretized dataset to evaluate the proposed method when facing the problems of small sample size and feature ambiguity. Then, we compare it with state-of-the-art methods on datasets with all ordinal variables and mixed ordinal-and-nominal variables.

4.1 Parameter Settings

The proposed method includes four hyperparameters, namely weight of original instances μ, regularization parameter λ, distance margin τ, and perturbation size ε. Their values are identified via the random search strategy [4]. We sample 100 sets of values and select the one that gives the highest accuracy on the validation set. The range of each hyperparameter is as follows: $\mu \in [0, 1]$, $\lambda \in \{10^{-5}, 10^{-4}, 10^{-3}, 10^{-2}\}$, $\tau \in [0, \max\|x_{lj}\|_1]$, $\varepsilon \in \{0, 1, \ldots, p^{\text{ord}} + 2p^{\text{nom}}\}$. The upper bound of τ is inspired by Eq. 9 with M initialized as the Euclidean distance. The upper bound of ε is chosen based on the fact that perturbing one ordinal level to its adjacent level or a nominal level to another level causes at most $p^{\text{ord}} + 2p^{\text{nom}}$ changes in encoded features. In addition, the initial learning rate is tuned for each dataset before optimizing the hyperparameters. We search its value from $\{10^{-2}, 10^{-1}, \ldots, 10^2\}$ while holding $\mu, \tau = 1$ (i.e.

replicating LMNN). The MATLAB code for our method is available at http://github.com/xyang6/MLadv.

Triplet constraints are constructed from 3 target neighbors and 10 nearest impostors calculated under the Euclidean distance. 3NN is used as the classifier.

4.2 Experiments with Discretized Features

The goal of this experiment is to understand the potential of the proposed method for data with a small training set and ambiguous features. Our experiment is based on the UCI dataset Magic, which has 10 real-valued features, 19020 instances, and 2 classes. All features are first discretized into ordinal features with five equal-frequency levels, and then encoded as integers (denoted as 'int') or as a set of binary values (denoted as 'bin'). We compare LMNN and the proposed method on both types of data.

Learning from Small Training Sets. In this study, we build the training set by randomly selecting $5, 20, \ldots, 95$ instances from each class; the validation and test sets each include 9000 instances. The experiment is implemented 20 times and the mean accuracy is shown in Fig. 2a; quantitative results, including the standard deviation, are provided in Appendix C.3.

First, our MLadv outperforms LMNN over the whole range of training sample size, no matter what the encoding scheme is. Second, we see a clear advantage of MLadv over LMNN when the training set is small. Third, we notice that our MLadv performs better with binary encoding than integer encoding, when the sample size is larger than 20.

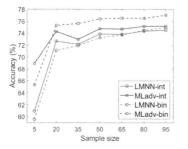
        (a) effect of training sample size

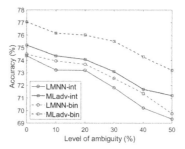
        (b) effect of feature ambiguity

Fig. 2. Evaluation of LMNN and MLadv on the discretized dataset Magic. Ordinal variables are encoded as integers ('int') or a set of binary values ('bin').

Learning Under Feature Ambiguity. We move on to evaluate the method when encountering feature ambiguity. The experimental setting is same as before; the training sample size is selected as 80. To simulate ambiguity, for each feature, we select $10\%, 20\%, \ldots, 50\%$ instances whose ground-truth real values are closest to the discretization threshold, and change their ordinal level to the adjacent level.

Figure 2b shows the classification accuracy in this study. MLadv improves LMNN consistently over a wide range of ambiguity levels, and the performance gain becomes slightly larger as the ambiguity level increases.

Visualization of Training Process. Our geometric interpretation suggests that MLadv considers additional triplets from the adversarially vulnerable region, which would be particularly valuable in the small-sized problem. In Fig. 3, we present the training process of MLadv at different iterations. The multidimensional scaling (MDS) is used to embed the learned distance between 20 instances into two dimensions [10]. Sizes of green circles and yellow circles are proportional to the number of triplets that do not satisfy the distance margin (i.e. second term of Eq. 4) and the number of triplets that incur a loss after adding the worst-case perturbation (i.e. third term of Eq. 4), respectively.

At the beginning of training, as instances of the same class are not well separated from instances of the different class, almost all triplets violate the distance margin constraints. Therefore, the metric is learned mostly from the original instances (as indicated by most points being in green circles). After 10 iterations, the majority of instances are closer to target neighbors than to impostors, but they are not robust to the worst-case perturbation (as indicated by a large number of yellow circles). Our method will continue using their information for metric learning. After 200 iterations, sizes of yellow circles become smaller, indicating that the learned metric becomes more robust. At the end of training, while some instances still violate the margin constraint, a large number of instances are surrounded by instances of the same class and locate far away from instances of the different class.

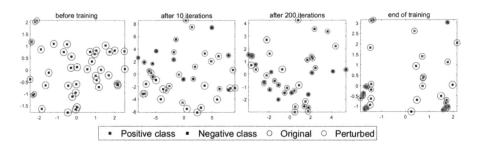

Fig. 3. Demonstration of the training process of MLadv on Magic with binary encoding. Figures show the 2D embedding of the learned distance via MDS. Sizes of green circles and yellow circles are proportional to the number of triplets violating the distance margin constraint and incurring a loss after adding the worst-case perturbation, respectively. As the training progresses, the metric becomes more robust against the perturbations and the difference between the intra-class distance and the inter-class distance becomes more remarkable. (Color figure online)

4.3   Experiments on Real Datasets

The goal of this experiment is to compare the proposed method with robust metric learning methods and ordinal metric learning methods under the conditions when the training sample size is small or the feature ambiguity is present. As ambiguities in categorical levels occur more frequently in ordinal variables than in nominal variables, our experiments only study datasets with all ordinal variables or with a mixture of ordinal and nominal variables.

Datasets and Experimental Settings. We use 6 datasets from UCI machine learning repository [11] and WEKA workbench [13]. Information on feature type, feature dimension, sample size and class information is listed in Table 1. Here, we explain the last column of ambiguity, which is assigned based on our understanding of the data. The degree of ambiguity is inherent in the data and may be inferred from the data source. Lecturer and Social Worker collect subjective ratings and assessments respectively, and hence may include a high level of ambiguity. Hayer-Roth and Lymphography are social data and medical data respectively; ambiguity is also likely to exist in these data. Car and Nursery are derived from a hierarchical decision model; their ambiguity levels are expected to be relatively low as there is an underlying rule behind these data.

Each dataset is randomly split into the training, validation, and test sets. To simulate a small-sample environment, we set their proportions as 20%,40%,40% for all datasets except for the large dataset Nursery. For Nursery, 100 samples are selected as the training set, and the remaining samples are equally split into the validation and test sets. We repeat the random split 20 times, and report the mean value and standard deviation of classification accuracy.

We compare the proposed method with LMNN and three closely related methods. DRIFT [34] and AML [6] are robust metric learning methods that are designed to handle feature uncertainty for real-valued data. Ord-LMNN [23] adapts LMNN to ordinal variables by assuming a latent variable for each ordinal variable, with the uniform prior tested in our experiment. Training procedures of these methods are specified in Appendix C.2.

Results and Discussions. Table 2 reports the classification accuracy of 3NN with the Mahalanobis distance learned from different methods. First, we see that the proposed method outperforms the baseline method LMNN, regardless of the encoding scheme. Second, we compare MLadv with the existing ordinal metric learning method Ord-LMNN. Ord-LMNN considers the order relation of ordinal variables and is effective on datasets Balance Scale and Car. However, as the method estimates the distributional parameters for each feature, its effectiveness highly depends on the data quality. When the ambiguity level is high, the accuracy of Ord-LMNN becomes even

Table 1. Characteristics of the datasets

Dataset	Abb.	Feature type	#Instances	p^{ord}	p^{nom}	#Classes	Ambiguity
Car	CA	Ordinal	1728	6	0	4	Low
Nursery	NU	Ordinal+nominal	12960	6	2	4	Low
Hayes-Roth	HR	Ordinal+nominal	132	2	2	3	Medium
Lymphography	LY	Ordinal+nominal	148	3	15	4	Medium
Lecturer	LE	Ordinal	1000	4	0	5	High
Social worker	SW	Ordinal	1000	10	0	4	High

Table 2. Classification accuracy (mean value±standard deviation) of 3NN with different metric learning methods. The best methods are shown in bold and the second best ones are underlined. The mean accuracy averaged over all datasets is shown at the last row.

	LMNN-int	LMNN-bin	DRIFT	AML	Ord-LMNN	**MLadv-int**	**MLadv-bin**
	Low level of ambiguity						
CA	89.94 ± 1.31	92.24 ± 0.89	90.24 ± 1.17	88.84 ± 1.13	**93.94 ± 1.43**	90.13 ± 1.32	92.90 ± 1.13
NU	85.73 ± 1.68	86.11 ± 1.78	86.01 ± 2.02	79.83 ± 3.11	**87.54 ± 1.45**	86.65 ± 2.12	86.67 ± 1.48
	Medium level of ambiguity						
HR	71.83 ± 10.72	76.42 ± 6.80	71.34 ± 10.37	65.98 ± 7.85	75.12 ± 9.55	74.51 ± 10.06	**78.58 ± 5.94**
LY	78.51 ± 7.15	79.91 ± 6.65	83.16 ± 6.40	68.25 ± 17.57	74.56 ± 9.17	82.37 ± 3.58	**83.33 ± 4.85**
	High level of ambiguity						
LE	55.08 ± 2.55	54.83 ± 2.53	55.64 ± 3.00	55.61 ± 2.28	53.03 ± 3.27	**55.90 ± 2.70**	55.61 ± 3.00
SW	50.00 ± 2.61	50.58 ± 2.30	50.73 ± 2.93	50.50 ± 2.07	48.68 ± 2.82	51.10 ± 2.15	**51.91 ± 3.09**
Avg	71.85	73.35	72.85	68.17	72.14	73.44	**74.84**

worse than the baseline whereas our method remains competitive. Third, the robust metric learning method DRIFT achieves a high accuracy when the feature ambiguity is high. However, as the method ignores the properties of ordinal and nominal variables, its performance is inferior to our method. Overall, our method achieves the best or second-best performance on each dataset and has the highest mean accuracy.

We make a final remark on the encoding scheme and practicability of the proposed method. On most datasets, MLadv-bin is superior to MLadv-int. We hypothesize that, as binary encoding gives a higher feature dimension, the expressive power of the metric increases and hence may improve the discriminability. While the two encoding schemes are evaluated separately in our experiment, they could be determined at the step of choosing the initial learning rate in practical applications. In other words, there is no need to tune hyperparameters twice. Except for Lymphography, this early decision can always find the optimal method between MLadv-int and MLadv-bin.

5   Conclusions and Future Work

In this paper, we propose that adversarial training with a deliberately designed perturbation set can enhance triplet-based metric learning methods in mitigating the problems of high feature ambiguity and small sample size faced by ordinal and mixed ordinal-and-nominal data. Experiments on real datasets verify the efficacy of our method. We also discuss the effect of adversarial training from both geometrical and theoretical perspectives. In the future, we intend to generalize the method to a mix of categorical and continuous features. Moreover, metric learning comprises a loss function and a regularizer, and this paper tailors the loss function to incorporate the properties of categorical features. Our future work will consider designing regularizers that are specific to categorical features.

References

1. Agresti, A.: Categorical Data Analysis, vol. 482. Wiley, Hoboken (2003)
2. Alamuri, M., Surampudi, B.R., Negi, A.: A survey of distance/similarity measures for categorical data. In: IJCNN (2014)
3. Babbar, R., Schölkopf, B.: Data scarcity, robustness and extreme multi-label classification. Mach. Learn. **108**(8–9), 1329–1351 (2019)
4. Bergstra, J., Bengio, Y.: Random search for hyperparameter optimization. J. Mach. Learn. Res. **13**, 281–305 (2012)
5. Cao, Q., Guo, Z.C., Ying, Y.: Generalization bounds for metric and similarity learning. Mach. Learn. **102**(1), 115–132 (2016)
6. Chen, G.H., Shah, D., et al.: Explaining the success of nearest neighbor methods in prediction. Found. Trends® Mach. Learn. **10**(5–6), 337–588 (2018)
7. Chen, S., Gong, C., Yang, J., Li, X., Wei, Y., Li, J.: Adversarial metric learning. In: IJCAI (2018)
8. Cheng, V., Li, C.H., Kwok, J.T., Li, C.K.: Dissimilarity learning for nominal data. Pattern Recogn. **37**(7), 1471–1477 (2004)
9. Cost, S., Salzberg, S.: A weighted nearest neighbor algorithm for learning with symbolic features. Mach. Learn. **10**(1), 57–78 (1993)
10. Cox, M.A., Cox, T.F.: Multidimensional scaling. In: Chen, C., Härdle, W., Unwin, A. (eds.) Handbook of data visualization, pp. 315–347. Springer, Heidelberg (2008). https://doi.org/10.1007/978-3-540-33037-0_14
11. Dua, D., Graff, C.: UCI machine learning repository (2017). http://archive.ics.uci. edu/ml
12. Duan, Y., Zheng, W., Lin, X., Lu, J., Zhou, J.: Deep adversarial metric learning. In: CVPR (2018)
13. Frank, E., Hall, M.A., Witten, I.H.: The WEKA workbench. Online appendix for "data mining: Practical machine learning tools and techniques" (2016)
14. Goodfellow, I.J., Shlens, J., Szegedy, C.: Explaining and harnessing adversarial examples. In: ICLR (2015)
15. Gutierrez, P.A., Perez-Ortiz, M., Sanchez-Monedero, J., Fernandez-Navarro, F., Hervas-Martinez, C.: Ordinal regression methods: survey and experimental study. IEEE Trans. Knowl. Data Eng. **28**(1), 127–146 (2015)
16. Hoffer, E., Ailon, N.: Deep metric learning using triplet network. In: Feragen, A., Pelillo, M., Loog, M. (eds.) SIMBAD 2015. LNCS, vol. 9370, pp. 84–92. Springer, Cham (2015). https://doi.org/10.1007/978-3-319-24261-3_7
17. Lanckriet, G.R., Ghaoui, L.E., Bhattacharyya, C., Jordan, M.I.: A robust minimax approach to classification. J. Mach. Learn. Res. **3**, 555–582 (2002)
18. LeCun, Y., Chopra, S., Hadsell, R., Ranzato, M., Huang, F.: A tutorial on energy-based learning. In: Predicting Structured Data, chap. 10, MIT press (2007). https://doi.org/10.7551/mitpress/7443.003.0014.
19. Luo, L., Xu, J., Deng, C., Huang, H.: Robust metric learning on Grassmann manifolds with generalization guarantees. In: AAAI (2019)
20. Madry, A., Makelov, A., Schmidt, L., Tsipras, D., Vladu, A.: Towards deep learning models resistant to adversarial attacks. In: ICLR (2018)
21. Mangasarian, O.L.: Nonlinear Programming. SIAM (1994)
22. Qian, Q., Tang, J., Li, H., Zhu, S., Jin, R.: Large-scale distance metric learning with uncertainty. In: CVPR (2018)
23. Shi, Y., Li, W., Sha, F.: Metric learning for ordinal data. In: AAAI (2016)

24. Shivaswamy, P.K., Bhattacharyya, C., Smola, A.J.: Second order cone programming approaches for handling missing and uncertain data. J. Mach. Learn. Res. **7**, 1283–1314 (2006)
25. Torresani, L., Lee, K.C.: Large margin component analysis. In: NeurIPS (2007)
26. Tsang, S., Kao, B., Yip, K.Y., Ho, W.S., Lee, S.D.: Decision trees for uncertain data. IEEE Trans. Knowl. Data Eng. **23**(1), 64–78 (2009)
27. Wang, D., Tan, X.: Robust distance metric learning in the presence of label noise. In: AAAI (2014)
28. Wang, D., Tan, X.: Robust distance metric learning via Bayesian inference. IEEE Trans. Image Process. **27**(3), 1542–1553 (2017)
29. Wang, H., Nie, F., Huang, H.: Robust distance metric learning via simultaneous ℓ_1-norm minimization and maximization. In: ICML (2014)
30. Wang, J., Kalousis, A., Woznica, A.: Parametric local metric learning for nearest neighbor classification. In: NeurIPS (2012)
31. Weinberger, K.Q., Saul, L.K.: Distance metric learning for large margin nearest neighbor classification. J. Mach. Learn. Res. **10**, 207–244 (2009)
32. Xie, J., Szymanski, B., Zaki, M.: Learning dissimilarities for categorical symbols. In: International Workshop on Feature Selection in Data Mining (2010)
33. Xing, E.P., Jordan, M.I., Russell, S.J., Ng, A.Y.: Distance metric learning with application to clustering with side-information. In: NeurIPS (2003)
34. Ye, H.J., Zhan, D.C., Si, X.M., Jiang, Y.: Learning mahalanobis distance metric: considering instance disturbance helps. In: IJCAI (2017)
35. Yin, D., Kannan, R., Bartlett, P.: Rademacher complexity for adversarially robust generalization. In: ICML (2019)
36. Zhang, Y., Cheung, Y.M., Tan, K.C.: A unified entropy-based distance metric for ordinal-and-nominal-attribute data clustering. IEEE Trans. Neural Netw. Learn. Syst. **31**, 39–52 (2019)
37. Zhu, C., Cao, L., Liu, Q., Yin, J., Kumar, V.: Heterogeneous metric learning of categorical data with hierarchical couplings. IEEE Trans. Knowl. Data Eng. **30**(7), 1254–1267 (2018)

Learning Implicit Generative Models by Teaching Density Estimators

Kun Xu, Chao Du, Chongxuan Li, Jun Zhu$^{(\boxtimes)}$, and Bo Zhang

Department of Computer Science and Technology, Institute for AI, BNRist Center, Tsinghua-Bosch ML Center, THBI Lab, Tsinghua University, Beijing, China
kunxu.thu@gmail.com, duchao0726@gmail.com, chongxuanli1991@gmail.com,
{dcszj,dcszb}@tsinghua.edu.cn

Abstract. Implicit generative models are difficult to train as no explicit density functions are defined. Generative adversarial nets (GANs) present a minimax framework to train such models, which however can suffer from mode collapse due to the nature of the JS-divergence. This paper presents a *learning by teaching* (LBT) approach to learning implicit models, which intrinsically avoids the mode collapse problem by optimizing a KL-divergence rather than the JS-divergence in GANs. In LBT, an auxiliary density estimator is introduced to fit the implicit model's distribution while the implicit model teaches the density estimator to match the data distribution. LBT is formulated as a bilevel optimization problem, whose optimal generator matches the true data distribution. LBT can be naturally integrated with GANs to derive a hybrid LBT-GAN that enjoys complimentary benefits. Finally, we present a stochastic gradient ascent algorithm with unrolling to solve the challenging learning problems. Experimental results demonstrate the effectiveness of our method.

Keywords: Deep generative models · Generative adversarial nets · Mode collapse problem

1 Introduction

Deep generative models (DGMs) [6,10,23] are powerful tools to capture the distributions over complicated manifolds (e.g., natural images), especially the recent developments of implicit statistical models [2,8,24], also called implicit probability distributions [19]. Implicit models are flexible by adopting a sampling procedure rather than a tractable density. However, they are difficult to learn, partly because maximum likelihood estimation (MLE) is not directly applicable.

K. Xu and C. Du—Equal contribution.

Electronic supplementary material The online version of this chapter (https://doi.org/10.1007/978-3-030-67661-2_15) contains supplementary material, which is available to authorized users.

© Springer Nature Switzerland AG 2021
F. Hutter et al. (Eds.): ECML PKDD 2020, LNAI 12458, pp. 239–255, 2021.
https://doi.org/10.1007/978-3-030-67661-2_15

Generative adversarial networks (GANs) [6] address this difficulty by adopting a minimax game, where a discriminator D is introduced to distinguish whether a sample is real (i.e., from the data distribution) or fake (i.e., from a generator G), while G tries to fool D via generating realistic samples. Although GANs can produce high quality samples, it suffers from lacking sample diversity, also known as the mode collapse problem [5], which still remains unaddressed.

A compelling reason for mode collapse arises from the objective function optimized by GANs [21], which is shown to minimize the JS-divergence between the data distribution p_D and the generator distribution p_G [6]. As shown in [7, 26] and illustrated in Fig. 1, JS-divergence can be tolerant to mode collapse whereas the $KL(p_D\|p_G)$ achieves its optima iff $p_D = p_G$. [22] enable us to train implicit models via KL-divergence using importance sampling, i.e., estimating the KL-divergence using generated samples. However, it may also fail in practice [18] as the KL-divergence will be under-estimated if the generated samples do not capture all modes in training data.

To address the above issues, we propose *learning by teaching* (LBT), a novel framework to learn implicit models. LBT can be shown to optimizes the KL-divergence, which is more resistant to mode collapse than the JS-divergence due to the zero-avoiding properties [20]. In LBT, we *learn* an implicit generator G by *teaching* a density estimator E to match the data distribution. The training scheme consists of two parts:

(a) The estimator E is trained to maximize the log-likelihood of the samples of the generator G;
(b) The generator G's goal is to improve the performance of the trained estimator in terms of the log-likelihood of real data samples.

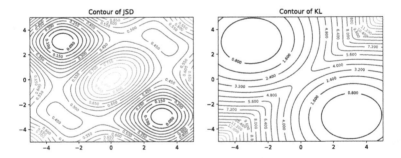

Fig. 1. Suppose the data distribution is a mixture of Gaussian (MoG), i.e., $p_D(x) = 0.5\mathcal{N}(-3, 1) + 0.5\mathcal{N}(3, 1)$, and the model distribution is a MoG with learnable means $\theta = [\theta_1, \theta_2]$, i.e., $p_G(x) = 0.5\mathcal{N}(\theta_1, 1) + 0.5\mathcal{N}(\theta_2, 1)$. The figure shows the contours of the two divergences with the x-axis and y-axis representing the value of θ_1 and θ_2 respectively. The JS-divergence allows mode-collapsed local optima while the KL-divergence does not.

Though in both LBT and GAN, an auxiliary model is introduced to help the training of the generator, the role of E in LBT is significantly different from that of D in GAN, and they are complimentary to each other. The estimator E in LBT penalize G for missing modes in training data, whereas the discriminator D in GAN penalize G for generating unrealistic samples. In LBT, E always tracks p_G and once p_G misses some modes, the estimator E will also miss them, resulting a poor likelihood of real data samples, which penalize G heavily. In such a manner, the estimator in LBT directs the generated samples to overspread the support of data distribution. In contrast, the goal of D in the vanilla GAN is to distinguish whether a sample is real or fake. Therefore, during the competing with D, G will be penalized much more heavily for generating unrealistic samples than missing modes. Based on this insight, we further conjoin the complimentary advantages to develop LBT-GAN, which augments LBT with a discriminator network. In LBT-GAN, E helps G to overspread the data distribution and D helps G to generate realistic samples.

Formally, LBT (and LBT-GAN) is formulated as a bilevel optimization [3] problem, where an *upper* level optimization problem (i.e., part (b)) depends on the optimal solution of a *lower* level problem (i.e., part (a)). The gradients of the upper problem w.r.t. the parameters of G are intractable since the optimal solution of E cannot be analytically expressed by G's parameters. We propose to use the unrolling technique [18] to efficiently approximate the gradients. Under nonparametric conditions, the optimum of LBT (and LBT-GAN) is achieved when both the generator and the estimator converge to the data distribution. Besides, we further analyze that an estimator with insufficient capability can still help G to resist to mode collapse in LBT-GAN. Experimental results on both synthetic and real datasets demonstrate the effectiveness of LBT and LBT-GAN.

2   Background

Consider an implicit generative model $G(\cdot; \theta)$ parameterized by θ that maps a simple random variable $z \in \mathbb{R}^H$ to a sample x in the data space $\mathbb{R}^L$, i.e., $x = G(z; \theta)$, where H and L are the dimensions of the random variables and the data samples, respectively. Typically, z is drawn from a standard Gaussian distribution p_Z and G is a feed-forward neural network. The sampling procedure defines a distribution $p_G(x; \theta)$ over the data space. The goal of the generator G is to approximate the data distribution $p_D(x)$, i.e., to produce samples of high quality and diversity.

Since the generator distribution is implicit, it is infeasible to adopt MLE directly to train the generator. To address this problem, GANs [6] adopt a minimax game, where a discriminator $D(\cdot; \psi)$ parameterized by ψ is introduced to distinguish generated samples from true data samples, while the generator G tries to fool D via generating realistic samples. The parameters of G and D are learned by solving a minimax game:

$$\min_\theta \max_\psi f_{\text{GAN}}(\theta, \psi) := \mathbb{E}_{x \sim p_D}[\log D(x; \psi)]$$

$$+ \mathbb{E}_{z \sim p_Z}[\log(1 - D(G(z; \theta); \psi))]. \tag{1}$$

[6] show that the discriminator achieves its optimum when $D(x) = \frac{p_D(x)}{p_D(x)+p_G(x)}$, and solving the minimax problem is equivalent to minimizing the JS-divergence between $p_D(x)$ and $p_G(x)$, whose optimal point is $p_G = p_D$, under the assumption that G and D have infinite capacity. However, GANs can suffer from the mode collapse problem for both theoretical reasons [7,21] and practical reasons [2,18, 25].

From the theoretical perspective, previous work has investigated the mode collapse nature of JS-divergence [21]. By optimizing the JS-divergence, the generative model tends to cover certain modes, rather than overspreading the data distribution [26], thus leading to mode collapse in GANs. Figure 1 (left) presents a simple example, where the local optima with mode collapse can still be found by optimizing the JS-divergence, even if p_G is flexible enough. In contrast, the KL-divergence can overcome this problem because of the zero-avoiding property [20], and Fig. 1 (right) shows that $KL(p_D||p_G)$ achieves its optima iff $p_G = p_D$.

There are previous attempts on training implicit models by optimizing other divergence, including the KL-divergence [21,22]. For instance, D2GAN [21] uses an auxiliary discriminator to diversify the generator distribution, which introduces the KL-divergence into the objective function. However, it practically fails as the discriminators in D2GAN are fixed during the update of the generator, which makes that the gradient of the KL-divergence w.r.t. the generator cannot be propagated through the discriminator and breaks the zero-avoiding property of the KL-divergence. nowozin2016f propose to estimate the KL-divergence using importance sampling, i.e., $KL(p_D||p_G) = \mathbb{E}_{p_G}[\frac{p_D}{p_G} \log \frac{p_D}{p_G}]$. However, the estimation will be of large variance if the generator fails to capture all modes in data as it is difficult to draw a sample in the missed modes in p_D and the KL-divergence tends to be under-estimated. Therefore, once the generator distribution collapsed, the estimated KL-divergence cannot penalize the generator for missing modes and encourage the generator to capture all modes in training data.

3   Method

To address the mode collapse issue, we present a novel framework *learning by teaching* (LBT), which enables us to learn implicit models by optimizing the KL-divergence between $p_D(x)$ and $p_G(x)$.

3.1   Learning by Teaching (LBT)

We introduce an auxiliary density estimator E with density $p_E(x; \phi)$ parameterized by ϕ to learn the distribution defined by the implicit generator $G(\cdot; \theta)$. The estimator E provides a surrogate density for G to estimate the KL-divergence between p_D and p_G. Specifically, in LBT, the estimator E's goal is to learn p_G via MLE, i.e., by maximizing the likelihood evaluated on samples generated from G. And the generator's goal is to maximize E's likelihood evaluated on real data samples, which is possible since the generator's samples decide the training process of E. As a consequence, E only captures the modes of its "training data", i.e., the generated samples, and has low density for unseen data. To avoid the penalty from the real samples, the generator G has to overspread the true data distribution and cover all modes. Formally, LBT is defined as a bilevel optimization problem [3]:

$$\max_{\theta} \quad \mathbb{E}_{x \sim p_D(x)}[\log p_E(x; \phi^\star(\theta))] \tag{2}$$

$$\text{s.t.} \quad \phi^\star(\theta) = \arg\max_{\phi} \mathbb{E}_{z \sim p_Z}[\log p_E(G(z; \theta); \phi)], \tag{3}$$

where $\phi^\star(\theta)$ indicates that the optimal $\phi^\star$ of the lower level problem depends on θ, which is the variable to be optimized in the upper level problem. For simplicity and clarity, we denote the objectives of the upper and lower level problems as:

$$f_G(\phi^\star(\theta)) := \mathbb{E}_{x \sim p_D(x)}[\log p_E(x; \phi^\star(\theta))],$$
$$f_E(\theta, \phi) := \mathbb{E}_{z \sim p_Z}[\log p_E(G(z; \theta); \phi)].$$

We now provide the following theorem to demonstrate the correctness of LBT under the assumption that G and E have sufficient capacity, which has been justified by recent advances of DGMs [10,23].

Theorem 1. *Solving problem (2) is equivalent to minimizing the KL-divergence between the data distribution and the generator distribution, and it's optima is achieved when*

$$p_G = p_E = p_D. \tag{4}$$

The proof is included in Appendix. Theorem 1 shows that the global optimum of LBT is achieved at $p_G = p_E = p_D$ if the estimator has enough capacity. Below, we give a further analysis to provide a weaker conclusion for LBT under a mild assumption that the estimator has only limited capacity.

Exponential Family: Consider the case where the estimator distribution $p_E(x)$ is in the exponential family form, i.e., $p_E(x) = h(x)e^{\eta \cdot T(x) - A(\eta)}$, where $T(x)$ denotes the sufficient statistics and η are the natural parameters. In this case, for a certain distribution q, $KL(q\|p_E)$ achieves optimal iff p_E captures the sufficient statistics of q, i.e., $\mathbb{E}_{p_E}T(x) = \mathbb{E}_q T(x)$ [20]. Therefore, given p_G in LBT, the estimator distribution p_E achieves optimal when $\mathbb{E}_{p_E}T(x) = \mathbb{E}_{p_G}T(x)$. To make the estimator achieve an optimal likelihood on data samples (or equivalently, optimal $KL(p_D\|p_E)$), G should ensure E to capture the sufficient statistics of the data distribution, i.e., $\mathbb{E}_{p_E}T(x) = \mathbb{E}_{p_D}T(x)$. Therefore, the estimator can still regularize p_G to match p_D in terms of sufficient statistics:

$$\mathbb{E}_{p_G}T(x) = \mathbb{E}_{p_D}T(x), \tag{5}$$

which is a weaker conclusion with fewer assumptions compared to Eq. (4). We provide an example to verify the above analysis in Sect. 3.2 and demonstrate the effectiveness of an estimator beyond the exponential family on real applications in Sect. 5.2.

3.2   Combining LBT with GAN

The KL-divergence is known to be zero-avoiding [20] in that it encourages the model distribution to cover the data distribution. However, in practice it may also result in low quality of generated samples [27]. This property makes LBT complementary to GAN which tends to generate samples of high quality but lack sample diversity [5]. To combine the best of both worlds, we further propose to augment LBT with a discriminator as in GANs, and call the hybrid model LBT-GAN. Formally, LBT-GAN solves the following bilevel problem:

$$\max_{\theta} \quad f_G(\phi^{\star}(\theta)) - \lambda_G \cdot f_{\text{GAN}}(\theta, \psi^{\star}) \tag{6}$$

$$\text{s.t.} \quad \phi^{\star}(\theta) = \arg\max_{\phi} f_E(\theta, \phi), \tag{7}$$

$$\psi^{\star} = \arg\max_{\psi} f_{\text{GAN}}(\theta, \psi), \tag{8}$$

where λ_G balances the weight between two losses. Under the assumption that G and the discriminator D have sufficient capacity, [6] show that the optimum of GAN's minimax framework is achieved at $p_G = p_D$, which is consistent with the conditions in Eq. (4) & (5) for LBT. Therefore, it is straightforward that LBT-GAN has the same global optimal solution as GAN, i.e., $p_G = p_D$.

We show that LBT-GAN has advantages compared to GAN even when the estimator has only limited capacity. As mentioned above, GAN can suffer from mode collapse problem since gradient-based optimization methods could fall into a mode-collapsed local optimum of the JS-divergence. However these mode-collapsed local optima are less likely to satisfy the condition in Eq. (5). The estimator can provide training signal to the generator and help it to escape the local optima that violates Eq. (5), and therefore make LBT-GAN more resistant to the mode collapse problem.

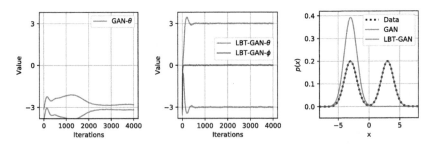

Fig. 2. An illustration of LBT-GAN where the estimator has insufficient capacity. We consider the same data distribution $p_D(x)$ and model distribution $p_G(x)$ as described in Fig. 1. We train both GAN and LBT-GAN to learn the model $p_G(x)$. For LBT-GAN, we use an estimator with insufficient capacity, i.e., $p_E(x) = \mathcal{N}(\phi, 1)$. The training processes of GAN (Left) and LBT-GAN (Middle) and their learned distributions (Right) are shown. We observe that while GAN learns a mode-collapsed model, LBT-GAN can escape the local optimum and capture the data distribution in this case.

To empirically verify our argument, we consider the settings of the toy example in Fig. 1. In LBT-GAN, we assume that the estimator E is a single Gaussian $p_E(x) = \mathcal{N}(\phi, 1)$ with a learnable mean ϕ, which can only capture the mean of a distribution. In this case, the condition Eq. (5) ensures that p_G and p_D have equal means. Therefore, if p_G is around the mode-collapsed local optima of GAN (where the gradients of the GAN objective will be nearly zero), the gradients of the LBT objective will encourage the generator to escape the local optima with non-zero mean. A clear demonstration is shown in Fig. 2, where we identically initialize the means of G around -3 in both LBT-GAN and GAN. We observe that GAN converges to a local optimum of JS-divergence, whereas LBT-GAN converges to the global optimal quickly as the estimator regularizes the generator to a distribution with zero mean. Further experimental results on real datasets are illustrated in Sect. 5.2.

3.3 Stochastic Gradient Ascent via Unrolling

The bilevel problem is generally challenging to solve. Here, we present a stochastic gradient ascent algorithm by using an unrolling technique [18] to derive the gradient. For clarity, we focus on learning LBT and the methods can be directly applied to learn LBT-GAN. Specifically, to perform gradient ascent, we calculate the gradient of f_G with respect to θ as follows:

$$
\begin{aligned}
\frac{\partial f_G(\phi^\star(\theta))}{\partial \theta} &= \frac{\partial f_G(\phi^\star(\theta))}{\partial \phi^\star(\theta)} \frac{\partial \phi^\star(\theta)}{\partial \theta} \\
&= \frac{\partial f_G(\phi^\star(\theta))}{\partial \phi^\star(\theta)} \int_z \frac{\partial \phi^\star(\theta)}{\partial G(z;\theta)} \frac{\partial G(z;\theta)}{\partial \theta} p_Z dz,
\end{aligned}
\tag{9}
$$

where both $\frac{\partial f_G(\phi^\star(\theta))}{\partial \phi^\star(\theta)}$ and $\frac{\partial G(z;\theta)}{\partial \theta}$ are easy to calculate. However, the term $\frac{\partial \phi^\star(\theta)}{\partial G(z;\theta)}$ is intractable since $\phi^\star(\theta)$ can not be expressed as an analytic function of the generated samples $G(z;\theta)$. We instead consider a local optimum $\hat{\phi}^\star$ of the density estimator parameters, which can be expressed as the fixed point of an iterative optimization procedure with $\phi^0 = \phi$:

$$
\phi^{k+1} = \phi^k + \eta \cdot \left. \frac{\partial f_E(\theta, \phi)}{\partial \phi} \right|_{\phi^k}, \quad \hat{\phi}^\star = \lim_{k \to \infty} \phi^k,
\tag{10}
$$

where η is the learning rate[1]. Since the samples used to evaluate the likelihood $f_E(\theta, \phi)$ are generated by $G(\cdot; \theta)$, each step of the optimization procedure is dependent on θ. We thus write $\phi^k(\theta, \phi^0)$ to clarify that ϕ^k is a function of θ and the initial value ϕ^0. In the following, we rewrite $G(z; \theta)$ as x_z and $\frac{\partial f_E(\theta, \phi)}{\partial \phi}$ as $\nabla \phi$ for simplicity. Since $\nabla \phi$ is differentiable w.r.t. x_z for most density estimators such as NADEs, $\phi^k(\theta, \phi^0)$ is also differentiable w.r.t. x_z. By unrolling for K steps, namely, using $\phi^K(\theta, \phi^0)$ to approximate $\phi^\star(\theta)$ in the objective $f_G(\phi^\star(\theta))$, we optimize a surrogate objective formulated as $f_G(\phi^K(\theta, \phi^0))$ for the generator.

---

[1] We have omitted the learning rate decay for simplicity.

Thus, the term $\frac{\partial \phi^*(\theta)}{\partial x_z}$ is approximated as $\frac{\partial \phi^*(\theta)}{\partial x_z} \approx \frac{\partial \phi^K(\theta, \phi^0)}{\partial x_z}$, which is known as the unrolling technique [18]. Under the assumption that $\phi^0 = \phi^*$, the gradients provided by the unrolling technique are good approximations of the exact gradients. We give a formal theoretical proof in Appendix B.

Finally, the generator and the likelihood estimator can be updated using the following process:

$$\theta \leftarrow \theta + \eta_\theta \frac{\partial f_G(\phi^K(\theta, \phi))}{\partial \theta}, \quad \phi \leftarrow \phi + \eta_\phi \frac{\partial f_E(\theta, \phi)}{\partial \phi}, \tag{11}$$

where η_θ and η_ϕ are the learning rates for the generator and the estimator, respectively. We perform several updates of ϕ per update of θ to keep p_E closed to p_G. Note that for other gradient-based methods such as Adam [9], the unrolling procedure is similar [18]. In our experiments, only a few steps of unrolling, e.g., 5 steps, are sufficient. The training procedure is described in Algorithm 1.

Algorithm 1. Stochastic Gradient Ascent Training of LBT with the Unrolling Technique

---

Input: data x, learning rate η_θ and η_ϕ, unrolling steps K and estimator update steps M.
Initialize parameters θ_0 and ϕ_0, and $t = 1$.
repeat
  $\phi_t^0 \leftarrow \phi_{t-1}$
  for $i = 1$ **to** M **do**
    $\phi_t^i \leftarrow \phi_t^{i-1} + \eta_\phi \cdot \frac{\partial f_E(\theta, \phi)}{\partial \phi}\Big|_{\phi_t^{i-1}}$
  end for
  Update ϕ: $\phi_t \leftarrow \phi_t^M$
  $\phi^0 \leftarrow \phi_t$
  Unrolling: $\phi^K \leftarrow \phi^0 + \sum_{i=1}^K \eta_\phi \cdot \frac{\partial f_E(\theta, \phi)}{\partial \phi}\Big|_{\phi^{i-1}}$
  Update θ: $\theta_t \leftarrow \theta_{t-1} + \eta_\theta \frac{\partial f_G(\phi^K)}{\partial \theta}$
  Update t: $t \leftarrow t + 1$
until Both θ and ϕ converge.

---

4   Related Work

Implicit statistical models [19] are of great interests with the emergence of GAN [6] that introduces a minimax framework to train such models. [22] generalize the original GANs via introducing a broad class of f-divergence for optimization. In comparison, LBT provides a different way to optimize the KL-divergence and achieves good results on avoiding mode collapse and generating realistic samples when combined with GAN. [2] propose to minimize the earth mover's

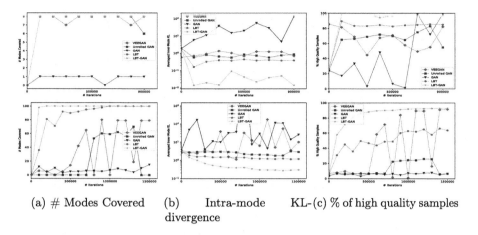

(a) # Modes Covered   (b)   Intra-mode   KL-(c) % of high quality samples
                                divergence

Fig. 3. Three different metrics evaluated on the generator distributions of different methods trained on the ring data (Top) and the grid data (Bottom). The metrics from left to right are: number of modes covered (the higher the better); Averaged intra-mode KL-divergence (the lower the better); Percentage of high quality samples (the higher the better).

distance to avoid the problem of gradient vanishing in vanilla GANs. Besides, [14] train implicit models by matching momentum between generated samples and real samples.

Mode collapse is a well-known problem in practical training of GANs. Much work has been done to alleviate the problem [1,2,17,18,25]. Unrolled GAN [18] proposes to unroll the update of the discriminator in GANs. The unrolling helps capturing how the discriminator would react to a change in the generator. Therefore it reduces the tendency of the generator to collapse all samples into a single mode. VEEGAN [25] introduces an additional reconstructor net to map data back to the noise space. Then a discriminator on the joint space is introduced to learn the joint distribution, similar as in ALI [4]. [15] propose to modify the discriminator to distinguish whether multiple samples are real or generated.

Different from methods in [21,22], our method evaluates the KL-divergence with data samples and makes $\phi^\star$ a function of θ via unrolling. This enables us to accurately evaluate the KL-divergence, regardless of whether the generator collapses or not. By unrolling the optimization process of ϕ, the estimation of the KL-divergence can be differentiable w.r.t. the generator and can be optimized in practice. We directly compare our methods with the above related methods in Sect. 5.

VAE-GAN [12] also conjoins a likelihood-based objective and GAN, by utilizing the encoder/decoder structure and defining the reconstruction term in the feature space of a discriminator to generalize the metric of similarity. Whereas in LBT and LBT-GAN, an auxiliary estimator is introduced to learn the model distribution and acts as a surrogate distribution of the implicit model. This enables LBT and LBT-GAN to be combined with any (approximate) density estimators.

Fig. 4. Generated samples of DCGANs and LBT-GANs with different network architectures of discriminators. From left to right: DCGAN and LBT-GAN with a large discriminator; DCGAN and LBT-GAN with a small discriminator. LBT-GANs can successfully generate realistic and diverse samples with different network architectures of discriminators.

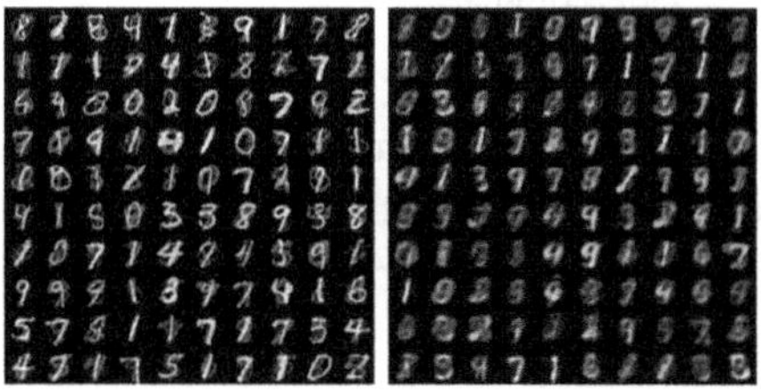

Fig. 5. The generated samples of LBT-GAN (Left) with the estimator being a smaller VAE and the samples from the estimator (Right).

5 Experiments

We now present the experimental results of LBT and LBT-GAN on both synthetic and real datasets. Throughout the experiments, we use Adam [9] with the default setting to optimize both the generator and the estimator (and the discriminator for LBT-GAN). We set the unrolling steps $K = 5$. We perform $M = 15$ steps of estimator update after each generator update. In LBT-GAN, We choose λ_G from $\{0.1, 1, 5, 10\}$ manually. Our codes will be released after the double-blind review process.

5.1 Synthetic Datasets

We first compare LBT and LBT-GAN with state-of-the-art competitors [6,17, 18,25] on 2-dimensional (2D) synthetic datasets, which are convenient for qualitative and quantitative analysis. Specifically, we construct two datasets: (i) **ring:** mixture of 8 2D Gaussian distributions arranged in a ring and (ii) **grid:** mixture of 100 2D Gaussian distributions arranged in a 10-by-10 grid. All of the mixture components are isotropic Gaussian, i.e., with diagonal covariance matrix. For the ring data, the deviation of each Gaussian component is diag(0.1, 0.1) and the

Table 1. Degree of mode collapse measured by number of mode captured (# MC) and KL-divergence between the generated distribution over modes and the uniform distribution over 1,000 modes on Stacked MNIST. Results are averaged over 5 runs.

	ALI	Unrolled GAN	VEEGAN	DCGAN
# MC	16	48.7	150	188.8
KL	5.4	4.32	2.95	3.17
	PacGAN	D2GAN	LBT-GAN (VAE)	LBT-GAN (NADE)
# MC	664.2	876.8	**999.6**	**1000**
KL	1.41	0.95	**0.19**	**0.05**

radius of the ring is 1^2. For the grid data, the spacing between adjacent modes is 0.2 and the deviation of each Gaussian component is diag$(0.01, 0.01)$. Figure 6a shows the true distributions of the ring data and the grid data, respectively. For all experiments on synthetic data, we use variational auto-encoders (VAEs) as the (approximate) density estimators for both LBT and LBT-GAN. All the encoders and decoders in VAEs are two-hidden-layer MLPs. For fair comparison, we use generators with the same network architectures (two-hidden-layer MLPs) for all methods. For GAN-based methods, the discriminators are also two-hidden-layer MLPs. The numbers of the hidden units for the generators and the estimators (and the discriminators for LBT-GAN) are all 128.

To quantify the quality of the generator learned by different methods, we report the following 3 metrics to demonstrate different characteristics of generator distributions. **Percentage of High Quality Samples** [25]: We count a sample as a *high quality* sample of a mode if it is within three standard deviations of that mode. We say a sample is of high quality if it is a high quality sample of any modes. We generate 500,000 samples from each method and report the percentage of high quality samples. **Number of Modes Covered:** We count a mode as a *covered mode* if the number of its high quality samples is greater than 20% of the expected number of that, i.e., $20\% \times \frac{\# \text{ of samples}}{\# \text{ of modes}}$.[3] Intuitively, lower number of modes covered indicates more severe of mode collapse and a lack of global diversity. **Averaged Intra-Mode KL-Divergence:** We assign each generated sample to the nearest mode of the true distribution. For each mode, we fit a Gaussian model on its assigned samples, which can be viewed as an estimate of the generator distribution at that mode (where the true distribution is approximately Gaussian). We define *intra-mode KL-divergence* as the KL-divergence between the true distribution and the estimated distribution at

---

[2] In the original Unrolled GAN's setting [18], the std of each component is 0.02 and the radius of the ring is 2. In our setting, the ratio of std to radius is 10 times larger. We choose this setting in order to characterize different performance of "Intra-mode KL-divergence" clearly.

[3] The exact expected number should be a little bit less than $\frac{\# \text{ of samples}}{\# \text{ of modes}}$, according to the three-sigma rule.

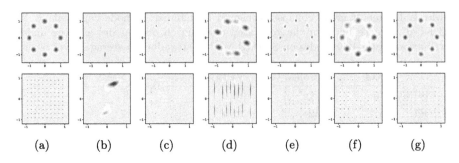

Fig. 6. Density plots of the data and the generator distributions trained on the ring data (Top) and the grid data (Bottom). The figures denote: (a) data, (b) vanilla GAN, (c) LSGAN, (d) Unrolled-GAN, (e) VEE-GAN, (f) LBT, (g) LBT-GAN.

each mode. Intuitively, it measures the local mismatch between the generator distribution and the true one. The averaged intra-mode KL-divergence over all modes are reported.

Figure 6 shows the generator distributions learned by different methods. Each distribution is plotted using kernel density estimation with 500,000 samples. We can see that LBT and LBT-GAN manage to cover the largest number of modes on both ring and grid datasets compared to other methods, demonstrating that LBT can generate globally diverse samples. The quantitative results are included in Fig. 3a. Note that our method covers all the 100 modes on the grid dataset while the best competitors LSGAN and VEEGAN cover 88 and 79 modes respectively. Moreover, the number of modes covered by LBT increases consistently. On the contrary, Unrolled GAN and VEEGAN can sometimes drop the covered modes, attributed to their unstable training.

Figure 3b shows the results of averaged intra-mode KL-divergence. We can see that LBT and LBT-GAN consistently outperform other competitors, which demonstrates that LBT framework can help capture better intra-mode diversity. According to Fig. 6c and Fig. 6e, although LSGAN and VEEGAN can achieve good mode coverage, they tend to concentrate most of the density near the mode means and fail to capture the local diversity within each mode. In LBT-GAN, the discriminator has a similar effect, while the estimator prevents the generator to over-concentrate the density. Therefore, the Intra-mode KL-divergence of LBT-GAN may oscillate during training as in Fig. 3b.

Finally, we show the percentages of high quality samples for each method in Fig. 3c. We find that LBT-GAN achieves better results than LBT and outperforms other competitors. As LBT-GAN can generate high quality samples while maintaining the global and local mode coverage, we use LBT-GAN in the following experiments.

(a) CelebA: DCGAN (Left) and LBT-GAN (Right).

(b) CIFAR10: DCGAN (Left) and LBT-GAN (Right).

Fig. 7. Generated samples on CelebA (a) and CIFAR10 (b) of DCGANs and LBT-GANs.

5.2 Stacked Mnist

Stacked MNIST [18] is a variant of the MNIST [13] dataset created by stacking three randomly selected digits along the color channel to increase the number of discrete modes. There are $1,000$ modes corresponding to 10 possible digits in each channel. Following [18,25], we randomly stack $128,000$ samples serving as the training data and use $26,000$ generated samples to calculate the number of modes to which at least one sample belongs. We use a classifier trained on the original MNIST to identify digits in each channel of generated samples. Besides, we also report the KL-divergence between the generated distribution over modes and the uniform distribution. Since carefully fine-tuned GANs can generate $1,000$ modes [18], we use smaller convolutional networks as both the generator and discriminator making our setting comparable to the competitors. We use two different density estimators for LBT-GAN: (i) a VAE with two-hidden-layer MLP (1000-400 hidden units) decoder and encoder and (ii) a NADE with a single hidden layer of 50 hidden units.

Table 1 presents the quantitative results. In terms of the number of captured modes, LBT-GAN surpasses other competitors, which demonstrates the effectiveness of the LBT framework. Specifically, LBT-GAN can successfully capture almost all modes, and the results of KL-divergence indicate that the distribution of LBT-GAN over modes is much more balanced compare to other competitors. Moreover, LBT-GAN works well with both approximate density estimators (e.g., VAE) and tractable density estimators (e.g., NADE). Note that PacGAN and D2GAN also report comparable results with ours on different network architectures whereas they fail to capture all modes in our setting. In contrast, LBT-GAN can generalize to PacGAN's architecture and capture all 1000 modes. Our hypothesis is that the auxiliary estimators helps LBT-GAN generalize across different architectures.

Figure 4 shows the generated samples of GANs and LBT-GANs with different size of discriminators. The visual quality of the samples generated by LBT-GANs is better than GANs. Further, we find the sample quality of DCGANs is sensitive to the size of the discriminators, while LBT-GANs can generate high-quality samples under different network architectures.

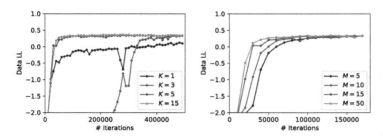

Fig. 8. Learning curves of LBT on the ring data with different unrolling steps K and fixed $M = 15$ (above) and with different estimator update steps M and fixed $K = 5$ (bottom).

Furthermore, we implement LBT-GAN with a much smaller VAE where both the encoder and the decoder of the estimator are two-hidden-layer MLPs with only 20 units in each hidden layer. Figure 5 shows that the samples from this VAE are of poor quality, which means that it can hardly capture the distribution of Stacked-MNIST. Nevertheless, even with such a simple VAE, LBT-GAN can still capture $1,000$ modes and generate visually realistic samples (See the left panel of Fig. 5), verifying that an estimator with limited capability can still help our method avoid mode collapse.

5.3 CelebA and CIFAR10

We also evaluate LBT-GAN on natural images, including CIFAR10 [11] and CelebA [16] datasets. The generated samples of DCGANs and LBT-GANs are illustrated in Fig. 7. LBT-GAN can generate images with comparable quality as DCGANs, demonstrating that LBT-GAN can successfully scale to natural images. However, we observe LBT-GAN generates more diverse samples compared to DCGAN without careful fine-tuning, especially on the CelebA dataset. The running time of LBT-GAN is roughly 2 (or 3) times of DCGAN on CelebA (or Cifar10). Empirically, LBT-GAN gives better or comparable results compared to VAE and VAE-GAN. Also, we find both GAN and VAE-GAN are sensitive to the architectures of G and D, whereas LBT-GAN is much more robust. This can be highlighted by adopting a relatively small G, where LBT-GAN with VAE can achieve 62.3 FID scores, significantly outperforming GAN(156.9), VAE(379.6) and VAE-GAN(100.5) on CelebA.

5.4 Sensitivity Analysis of K and M

Theoretically, a larger unrolling steps K allows ϕ^K to better approximate $\phi^\star$ and a larger inner update iterations M can better approximate the condition that $\phi^0 = \phi^\star$ as analyzed in Appendix B. However, large K and M on the other hand increase the computational costs. To balance this trade-off, we provide sensitivity analysis of K and M in LBT. We use the experimental settings of the ring problem and adopt the values of the objective function Eq. (2), i.e., the log-likelihood of real samples evaluated by the learned estimator, as the quantitative measurement.

We first investigate the influence of the number of unrolling steps K on the training procedure. We vary the value of K and show the learning curves with $K \in \{1, 3, 5, 15\}$ in Fig. 8. We observe that $K = 1$ leads to a suboptimal solution and larger K leads to better solution and convergence speed. We do not observe significant improvement with K larger than 5. We also show the influence of the number of inner update iterations M during training with $M \in \{5, 10, 15, 50\}$ in Fig. 8. Our observation is that larger M leads to faster convergence, which is consistent with the analysis in Sect. 3.3.

6   Conclusions and Discussions

We present a novel framework LBT to train an implicit generative model via teaching an auxiliary density estimator, which is formulated as a bilevel optimization problem. Unrolling techniques are adopted for practical optimization. Finally, LBT is justified both theoretically and empirically.

The main bottleneck of LBT is how to efficiently solve the bilevel optimization problem. For one thing, each update of LBT could be slower than that of the existing methods because the computational cost of the unrolling technique grows linearly with respect to the unrolling steps. For another, LBT may need larger number of updates to converge than GAN because training a density estimator is more complicated than training a classifier. Overall, if the bilevel optimization problem can be solved efficiently in the future work, LBT can be scaled up to larger datasets.

LBT bridges the gap between the training of implicit models and explicit models. For one thing, the auxiliary explicit models can help implicit models overcome the mode collapse problems. For another, the implicit generators can be viewed as approximated samplers of the density estimators like auto-regressive models, from which getting samples is time-consuming. We discuss the former direction in this paper and leave the later direction as future work.

Acknowledgements. This work was supported by the National Key Research and Development Program of China (No. 2017YFA0700904), NSFC Projects (Nos. 61620106010, 61621136008, U19B2034, U181146), Beijing NSF Project (No. L172037), Beijing Academy of Artificial Intelligence (BAAI), Tsinghua-Huawei Joint Research Program, Tiangong Institute for Intelligent Computing, and the NVIDIA NVAIL Program with GPU/DGX Acceleration. C. Li was supported by the Chinese postdoctoral innovative talent support program and Shuimu Tsinghua Scholar.

References

1. Arjovsky, M., Bottou, L.: Towards principled methods for training generative adversarial networks. arXiv preprint arXiv:1701.04862 (2017)

2. Arjovsky, M., Chintala, S., Bottou, L.: Wasserstein generative adversarial networks. In: International Conference on Machine Learning, pp. 214–223 (2017)

3. Colson, B., Marcotte, P., Savard, G.: An overview of bilevel optimization. Ann. Oper. Res. **153**(1), 235–256 (2007)

4. Dumoulin, V., et al.: Adversarially learned inference. arXiv preprint arXiv:1606.00704 (2016)

5. Goodfellow, I.: Nips 2016 tutorial: generative adversarial networks. arXiv preprint arXiv:1701.00160 (2016)

6. Goodfellow, I., et al.: Generative adversarial nets. In: Advances in Neural Information Processing Systems, pp. 2672–2680 (2014)

7. Huszár, F.: How (not) to train your generative model: scheduled sampling, likelihood, adversary? arXiv preprint arXiv:1511.05101 (2015)

8. Karras, T., Aila, T., Laine, S., Lehtinen, J.: Progressive growing of GANs for improved quality, stability, and variation. arXiv preprint arXiv:1710.10196 (2017)

9. Kingma, D.P., Ba, J.: Adam: a method for stochastic optimization. arXiv preprint arXiv:1412.6980 (2014)

10. Kingma, D.P., Welling, M.: Auto-encoding variational bayes. arXiv preprint arXiv:1312.6114 (2013)

11. Krizhevsky, A., Hinton, G.: Learning multiple layers of features from tiny images. Technical report, Citeseer (2009)

12. Larsen, A.B.L., Sønderby, S.K., Larochelle, H., Winther, O.: Autoencoding beyond pixels using a learned similarity metric. arXiv preprint arXiv:1512.09300 (2015)

13. LeCun, Y., Bottou, L., Bengio, Y., Haffner, P.: Gradient-based learning applied to document recognition. Proc. IEEE **86**(11), 2278–2324 (1998)

14. Li, Y., Swersky, K., Zemel, R.: Generative moment matching networks. In: International Conference on Machine Learning, pp. 1718–1727 (2015)

15. Lin, Z., Khetan, A., Fanti, G., Oh, S.: PacGAN: the power of two samples in generative adversarial networks. arXiv preprint arXiv:1712.04086 (2017)

16. Liu, Z., Luo, P., Wang, X., Tang, X.: Deep learning face attributes in the wild. In: Proceedings of International Conference on Computer Vision (ICCV), December 2015

17. Mao, X., Li, Q., Xie, H., Lau, R.Y., Wang, Z., Smolley, S.P.: Least squares generative adversarial networks. In: IEEE International Conference on Computer Vision (ICCV), pp. 2813–2821. IEEE (2017)

18. Metz, L., Poole, B., Pfau, D., Sohl-Dickstein, J.: Unrolled generative adversarial networks. arXiv preprint arXiv:1611.02163 (2016)

19. Mohamed, S., Lakshminarayanan, B.: Learning in implicit generative models. arXiv preprint arXiv:1610.03483 (2016)

20. Nasrabadi, N.M.: Pattern recognition and machine learning. J. Electron. Imaging **16**(4), 049901 (2007)

21. Nguyen, T., Le, T., Vu, H., Phung, D.: Dual discriminator generative adversarial nets. In: Advances in Neural Information Processing Systems, pp. 2670–2680 (2017)

22. Nowozin, S., Cseke, B., Tomioka, R.: f-GAN: Training generative neural samplers using variational divergence minimization. In: Advances in Neural Information Processing Systems, pp. 271–279 (2016)

23. Oord, A.V.D., Kalchbrenner, N., Kavukcuoglu, K.: Pixel recurrent neural networks. arXiv preprint arXiv:1601.06759 (2016)

24. Radford, A., Metz, L., Chintala, S.: Unsupervised representation learning with deep convolutional generative adversarial networks. arXiv preprint arXiv:1511.06434 (2015)

25. Srivastava, A., Valkoz, L., Russell, C., Gutmann, M.U., Sutton, C.: VEEGAN: reducing mode collapse in GANs using implicit variational learning. In: Advances in Neural Information Processing Systems, pp. 3310–3320 (2017)
26. Theis, L., Oord, A.v.d., Bethge, M.: A note on the evaluation of generative models. arXiv preprint arXiv:1511.01844 (2015)
27. Tolstikhin, I., Bousquet, O., Gelly, S., Schoelkopf, B.: Wasserstein auto-encoders. arXiv preprint arXiv:1711.01558 (2017)

Reprogramming GANs via Input Noise Design

Kangwook Lee[1], Changho Suh[2(✉)], and Kannan Ramchandran[3]

[1] Department of Electrical and Computer Engineering,
University of Wisconsin–Madison, Madison, WI, USA
kangwook.lee@wisc.edu
[2] School of Electrical Engineering,
Korea Advanced Institute of Science and Technology, Daejeon, South Korea
chsuh@kaist.ac.kr
[3] Department of Electrical Engineering and Computer Sciences,
University of California, Berkeley, Berkeley, CA, USA
kannanr@eecs.berkeley.edu

Abstract. The goal of neural reprogramming is to alter the functionality of a fixed neural network just by preprocessing the input. In this work, we show that Generative Adversarial Networks (GANs) can be reprogrammed by shaping the input noise distribution. One application of our algorithm is to convert an unconditional GAN to a conditional GAN. We also empirically study the applicability, feasibility, and limitation of GAN reprogramming.

Keywords: GAN · Neural reprogramming

1 Introduction

While deep neural networks have revolutionized a wide range of areas, their flexibility is limited as they cannot perform other tasks that are not taken into account during training. Various techniques have been proposed for general-purpose neural networks. Multi-task learning designs a versatile neural network that can perform multiple tasks at the same time when the target tasks are known at training time [25]. Transfer learning can help adapt to another task if the new task is similar enough to the original task [22]. For instance, given a new task, one can slightly modify the parameters of a pre-trained neural network so that it can perform well on the new task [31].

Recently, [5] proposes a new approach to design general-purpose neural networks, which they call *neural reprogramming*. The goal of neural reprogramming is to modify the functionality of a pre-trained neural network just by preprocessing the input. For instance, one can turn an ImageNet classifier into an

This material is based upon work supported by the Air Force Office of Scientific Research under award number FA2386-19-1-4050.

F. Hutter et al. (Eds.): ECML PKDD 2020, LNAI 12458, pp. 256–271, 2021.
https://doi.org/10.1007/978-3-030-67661-2_16

MNIST classifier just by preprocessing input images. Compared to the standard approaches (multi-task learning and transfer learning), neural reprogramming does not assume any prior knowledge about target tasks and allows for modular design of a large neural network.

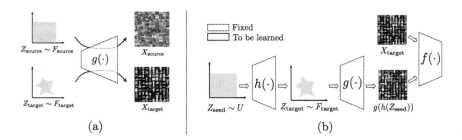

Fig. 1. GAN reprogramming. (a) Given a GAN $g(\cdot)$ that generates X_{source} (e.g., CIFAR10 images) with the latent random variable $Z_{\text{source}} \sim F_{\text{source}}$, reprogramming GAN aims to design another distribution F_{target} such that the generator fed by $Z_{\text{target}} \sim F_{\text{target}}$ yields X_{target} (e.g.., FMNIST images). (b) GAN reprogramming algorithm.

This motivates us to investigate reprogramming ideas beyond the classification setting. Specifically we focus on unsupervised learning that has recently been revolutionized via Generative Adversarial Networks (GANs) [8]. A GAN is trained with a latent variable (input noise) with a certain distribution, e.g., uniform or Gaussian distributions. Denote the trained generator by $g(\cdot)$, the random latent variable by Z_{source}, and its distribution by F_{source}. For a well-trained GAN, $g(Z_{\text{source}})$ follows a similar distribution to that of the original training data, say X_{source}. That is, $g(Z_{\text{source}}) \approx X_{\text{source}}$ if $Z_{\text{source}} \sim F_{\text{source}}$, where the notion of '$\approx$' will be defined formally soon. The goal of GAN reprogramming is finding another latent variable distribution F_{target} such that $g(Z_{\text{target}}) \approx X_{\text{target}}$ when $Z_{\text{target}} \sim F_{\text{target}}$. See Fig. 1a for visual illustration. Consider a GAN trained with CIFAR10 images. Assume that the trained generator is able to produce diverse CIFAR10-like images when Z_{source} follows $F_{\text{source}} =$ Uniform. Reprogramming GAN wishes to find a new latent variable distribution F_{target} so that the samples of $g(Z_{\text{target}})$ look like Fashion-MNIST images.

In this work, we propose a simple GAN reprogramming algorithm and show that it can convert unconditional GANs to conditional GANs without labeled datasets. We also study its applicability, feasibility, and limitation via a variety of controlled experiments.

2   Related Work

GANs Goodfellow et al. [8] propose the first GAN algorithm, in which two neural networks called generator and discriminator are alternatively trained.

The goal of the generator is to produce realistic fake samples, while the goal of the discriminator is to discriminate fake samples from real ones. It has been shown that the original GAN algorithm minimizes the Jensen-Shannon (JS) divergence between the distribution of fake samples and that of real samples. The Wasserstein GAN (WGAN) is similar to the original GAN, but its goal is to minimize the Wasserstein distance instead of the JS divergence [1]. It is shown that the WGAN algorithm is more stable than the original GAN algorithm. In the WGAN algorithm, two neural networks, called generator and critic, are trained. The role of the generator is the same as before while the role of critic is to approximate the Wasserstein distance. WGAN-GP (WGAN gradient penalty) uses a gradient-based regularization method to satisfy a certain constraint on the Lipshitz constant [10].

Multi-task Learning. Multi-task learning (MTL) has been successful in many applications of machine learning [25]. MTL is particularly useful when multiple tasks are similar to each other. The most basic algorithm is based on parameter sharing. That is, one may train a neural network with multiple last layers for different tasks using a combined dataset [2]. This approach is also called 'hard parameter sharing' since it shares exactly the same parameters (except for the last layer) between multiple tasks. On the other hand, in soft parameter sharing, each task maintains its own neural network while keeping multiple neural networks close to each other according to a certain metric [4,30]. Differently from MTL, neural reprogramming can be applied to a new task *without* needing to know the new task at training time.

Transfer Learning. Transfer learning exploits the knowledge from previous learning experiences to better solve future tasks [22]. The most popular algorithm is based on sharing the weights of a neural network [12,21]. That is, given a neural network trained on one task, one simply takes the lower part of the neural network (closer to the input end), treating it as a generic feature extractor. One then trains the remaining part of the neural network. If the original and new tasks are similar enough, such algorithms would perform well. On the other hand, neural reprogramming does not modify the parameters of the given neural network and only prepends the input processing module, allowing for modular design of a general-purpose neural network. To see this, consider N GANs that generate similar outputs. With reprogramming, one can just store the core GAN with N (small) input processors, significantly saving the memory/storage cost.

Neural Reprogramming. The concept of neural reprogramming was introduced by Elsayed et al. [5]. Specifically, they design an input preprocessor so that a neural network trained for classifying certain types of images can be used to classify other types of images. For instance, they show that it is possible to reprogram an ImageNet classifier as an MNIST classifier. A few recent studies are somewhat related to GAN reprogramming. Nguyen et al. propose an MCMC-based algorithm that can be used to maximize the activation of a carefully chosen neuron [20]. While this algorithm is shown to generate novel images using a fixed

GAN, it only works with a specific type of GAN and cannot be applied to GANs trained with standard methodologies. Furthermore, it requires a classifier pre-trained on the target dataset and computationally expensive as it involves the joint training of three neural networks. Engel et al. propose an algorithm for conditioning an unconditional Variational Autoencoder (VAE) [6]. On the other hand, our algorithm can reprogram a GAN even across different datasets.

Adversarial Robustness. Recently, it is shown that deep neural networks can be easily fooled when the input is perturbed even with imperceptibly small noise, bringing adversarial robustness of deep neural networks into question [9,26]. Through the lens of robust optimization, Madry et al. [18] propose an adversarial training algorithm that can make machine learning models robust against adversarial attacks. While most of the studies focus on the robustness of classifiers, a notable exception is made in recent studies [13,23], where attack scenarios for GANs have been discussed. They show that one can manipulate the latent variable so that a conditional GAN's generated output becomes inconsistent with the specified label. Moreover, they show that one can manipulate the latent code so that the generated sample is similar to a single target. Using our notation, this is equivalent to finding a value of z_{target} such that $g(z_{\text{target}})$ looks like a target sample. This is similar to our work in that it finds latent codes that make the generator produce different types of data. A key distinction, however, is that instead of finding a single value of z_{target}, we find a *distribution* of Z_{target} so that one can generate an arbitrary number of such samples by drawing latent codes from $Z_{\text{target}} \sim F_{\text{target}}$.

3   Reprogramming GANs

3.1   Preliminaries

In this section, we first recall the definition of elementary divergences between two probability measures P_r and P_g. The JS divergence is defined as $D_{\text{JS}}(P_r, P_g) := \frac{1}{2} D_{\text{KL}}(P_r \| P_m) + \frac{1}{2} D_{\text{KL}}(P_g \| P_m)$, where D_{KL} is the Kullback-Leibler (KL) divergence and $P_m = (P_r + P_g)/2$. The first-order Wasserstein distance is

$$W(P_r, P_g) := \inf_{\gamma \in \Pi(P_r, P_g)} \mathbb{E}_{(x,y) \sim \gamma}[\|x - y\|],$$

where $\Pi(P_r, P_g)$ is the set of all joint distributions whose marginals are respectively P_r and P_g. Slightly abusing notation, we write $W(X, Y)$ for $W(P_X, P_Y)$ if $X \sim P_X$ and $Y \sim P_Y$. By the Kantrovich-Rubinstein duality [27],

$$W(P_r, P_g) = \sup_{\|f\|_L \leq 1} \mathbb{E}_{x \sim P_r}[f(x)] - \mathbb{E}_{x \sim P_g}[f(x)], \tag{1}$$

where $\|f\|_L$ is the Lipschitz constant of f.

3.2 Problem Formulation and Algorithm

We formally define the problem of GAN reprogramming and describe our algorithm. Assume that a generator $g : \mathbb{R}^{d_i} \to \mathbb{R}^{d_o}$, which produces d_o-dimensional data from d_i-dimensional latent codes is given. Consider a latent variable distribution F_{source} used for training. Given an unlabeled dataset X_{target}, we solve the following optimization problem

$$\min_{F_{\text{target}}} W(g(Z_{\text{target}}), X_{\text{target}}), \tag{2}$$

where $Z_{\text{target}} \sim F_{\text{target}}$ is a d_i-dimensional random vector. If we can find F_{target} such that $W(g(Z_{\text{target}}), X_{\text{target}}) \simeq 0$, it implies that the given GAN can be reprogrammed to produce X_{target} instead of X_{source} by drawing latent codes according to F_{target} instead of F_{source}.

We now describe our reprogramming algorithm. Due to the Kantrovich-Rubinstein duality, we have

$$\min_{F_{\text{target}}} \sup_{\|f\|_L \le 1} \mathbb{E}_{X \sim X_{\text{target}}}[f(X)] - \mathbb{E}_{Z \sim Z_{\text{target}}}[f(g(Z))], \tag{3}$$

where $\|f\|_L$ is the Lipschitz constant of f. In order to parameterize the distribution of Z_{target} and to obtain samples from it, we apply a deep neural network to transform a uniform random variable into another random variable. That is, we first draw a uniform random variable $Z_{\text{seed}} \sim U[0,1]^d$ ($Z_{\text{seed}} \sim U$, for short), and then apply a *code generator* h_θ, $h_\theta : \mathbb{R}^d \to \mathbb{R}^{d_i}$, parameterized by θ. Here, d denotes the dimensionality of a random vector used to generate latent codes. Similarly, define a parameterized family of all functions $\{f_w\}_{w \in \mathcal{W}}$ that are 1-Lipschitz. Using these, (3) becomes

$$\min_{\theta} \sup_{f_w \in \mathcal{W}} \mathbb{E}_{X \sim X_{\text{target}}}[f_w(X)] - \mathbb{E}_{Z_{\text{seed}} \sim U}[f_w(g(h_\theta(Z_{\text{seed}})))]. \tag{4}$$

Since h_θ and f_w are parameterized via neural networks, one can solve the above optimization problem via gradient-based methods. Our algorithm is essentially a simple variation of WGAN-GP [10], where the key difference is that the outer optimization problem is optimized over h_θ instead of g, which is fixed in our setting.

We now describe our algorithm with further details for the sake of completeness. We first draw a minibatch of m pairs of a real data point x and a random number z for generating a latent code. Denote the generated output $g(h_\theta(x))$ by $\tilde{x}$. We now define the loss function for the i-th pair, denoted by $L^{(i)}$. The loss function contains the negative of the objective function since our goal is to maximize it. Moreover, in order to satisfy the constraint $f_w \in \mathcal{W}$, we use the gradient penalty. Specifically, we choose a random data point, say $\hat{x}$, lying in between the real data point x and the fake data point $\tilde{x}$. This can be done by taking a weighted sum of x and $\tilde{x}$, i.e., $\hat{x} = \epsilon x + (1 - \epsilon)\tilde{x}$, where $\epsilon \sim U[0,1]$. We then compute the gradient of f measured at $x = \hat{x}$. Using λ as the coefficient for gradient penalty, the total loss function for the i-th pair is

$$L^{(i)} := f_w(g(h_\theta(z))) - f_w(x) + \lambda(\|\nabla_x f_w\| - 1)^2. \tag{5}$$

Using the minibatch of m pairs, we can compute the batch average loss $\frac{1}{m}\sum_{i=1}^{m} L^{(i)}$. We then compute the gradient of the batch loss with respect to w and use the Adam optimizer to update w. We run this for n_{critic} iterations to closely approximate the Wasserstein distance. We then solve the outer optimization problem by treating that the current cost function is close to the true Wasserstein distance, i.e., the first-order Wasserstein distance is well approximated with the current choice of f_w. While f_w being fixed, we draw a minibatch of m generated data points, and then back-prop the cost function to find the gradient with respect to θ. We then apply the Adam optimizer to update θ. See Fig. 1 for the visual illustration of our algorithm. The pseudocode of GAN reprogramming algorithm is given in Algorithm 1.

Algorithm 1. GAN Reprogramming

Data: pretrained generator $g(\cdot)$ and target data X_{target}
Result: code generator $h_\theta(\cdot)$
initialize w and θ;
while θ *not converged* **do**
    for $t \leftarrow 1$ to n_{critic} **do**
        for $i \leftarrow 1$ to m **do**
            sample $x \sim X_{\text{target}}$, $z \sim U$, and $\epsilon \sim U[0,1]$;
            $\tilde{x} \leftarrow g(h_\theta(z))$, $\hat{x} \leftarrow \epsilon x + (1-\epsilon)\tilde{x}$;
            $L^{(i)} \leftarrow f_w(g(h_\theta(z))) - f_w(x) + \lambda(\|(\nabla_x f_w)_{\hat{x}}\| - 1)^2$;
        $w \leftarrow \text{Adam}(\nabla_w \frac{1}{m}\sum_{i=1}^{m} L^{(i)})$;
    sample a batch of noise $z^{(i)}{}_{i=1}^{m}$, $z^{(i)} \sim U$, $\forall i$;
    $\theta \leftarrow \text{Adam}(-\nabla_w \frac{1}{m}\sum_{i=1}^{m} f_w(g(h_\theta(z))))$;

3.3   Toy Example

Consider a fixed generator

$$g(Z) = \sigma(10(Z - 0.8)) - (1 - \sigma(10(Z - 0.8))),$$

where $\sigma(\cdot)$ is the sigmoid function. Assume that our goal is to reprogram this generator given a target random variable

$$X_{\text{target}} = \sigma(10(X' - 0.5)) - (1 - \sigma(10(X' - 0.5))),$$

where $X' \sim U[0,1]$. See Fig. 2b for its density function, which has two modes at $X = \pm 1$. Consider $Z_{\text{seed}} \sim U[0,1]$ and $h_\theta(Z) = Z + b$, where b is a learnable bias. When $b = 0$, $g(h_\theta(Z_{\text{seed}})) \simeq -(1 - \sigma(10(Z - 0.8)))$. That is, the output of the reprogrammed does not properly capture the mode at $X = 1$. However,

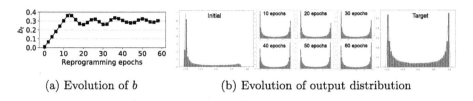

(a) Evolution of b (b) Evolution of output distribution

Fig. 2. GAN reprogramming results for the toy example.

when $b = b^\star = 0.3$, we have $g(h_\theta(Z_{\text{seed}})) \stackrel{d}{=} X_{\text{target}}$, i.e., it perfectly recovers the target distribution.

Shown in Fig. 2 are the results of our GAN reprogramming algorithm applied to this setting. We can see that $b \to b^\star$ from Fig. 2a, and the distribution of the reprogrammed GAN's outputs converges to that of the target distribution as reprogramming algorithm proceeds from Fig. 2b.

4 Experiments and Analysis

4.1 Setting

For all of our experiments, we use Adam with $(\alpha, \beta_1, \beta_2) = (10^{-4}, 0, 0.9)$ and set $n_{\text{critic}} = 10$, $m = 32$, and $\lambda = 1$. The algorithm is run for 100 to 500 epochs.

We consider the following image datasets: MNIST [15], Fashion-MNIST (or FMNIST) [29], SVHN [19], CIFAR10 [14], CelebA [17], Cartoon Set (a collection of 2D cartoon avatar images) [24], and the Mel-Frequency Cepstrum (MFC) of drum sound clips [3]. When we reprogram a GAN between color images and grayscale images, we transform the grayscale images into RGB ones. For each labeled dataset D, we denote by D_i the subset of D containing of images whose label is i. Thus, we have MNIST_i, FMNIST_i, SVHN_i, and CIFAR10_i for $i \in \{0, 1, \ldots, 9\}$.

We denote the generator g trained with dataset D by g_D and call it 'D-GAN'. Each D-GAN is obtained via the standard WGAN-GP algorithm. We use standard neural network architectures for generator and discriminator. For the generator, we pass a 62-dimensional latent code ($d_i = 62$) through two fully connected layers followed by two transpose convolution layers: 1) fc1024, 2) fc8WH, 3) TrConv-4x4 (64 feature maps) 4) TrConv-4x4 (1 feature map), where (W, H) denotes the image dimensions. For the discriminator, we pass an image of size W by H through two convolution layers followed by two fully connected layers: 1) Conv-4x4 (64 feature maps), 2) Conv-4x4 (128 feature maps), 3) fc1024, 4) fc1. For CelebA and Cartoon Set, we use deeper architectures with four TrConv's and Conv's each. In order to generate a latent code from d-dimensional random seed, we use a six-layer neural network: 1) fc500, 2) fc500, 3) fc500, 4) fc500, 5) fc500, 6) fc62.

Case 1: GAN Conditioning via Reprogramming. The first scenario we consider is GAN conditioning via reprogramming. That is, we reprogram a D-GAN as a D_i-GAN for $D \in \{\text{MNIST}, \text{FMNIST}, \text{SVHN}\}$ and $i \in \{0, 1, \dots, 9\}$. For instance, when $D = \text{MNIST}$ and $i = 0$, the goal is to convert an MNIST-GAN into an MNIST_0-GAN, which only produces 0 images.

Case 2: GAN Reprogramming Across Different Datasets. We also apply our reprogramming algorithm across different datasets: SVHN to MNIST_i, FMNIST to MNIST_i, CIFAR10 to MNIST_i, CIFAR10 to FMNIST_i, CIFAR10 to CelebA, CelebA to Cartoon Set, and CIFAR10 to Drum Sound Clips.

4.2 Qualitative Results

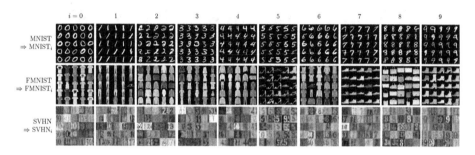

Fig. 3. GAN conditioning via reprogramming. Row corresponds to training data, and column corresponds to the target label.

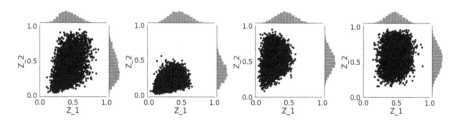

Fig. 4. The latent code distribution of reprogrammed GANs. We visualize the empirical distribution of the first two dimensions of latent code (Z_1 and Z_2). Column i is for FMNIST_i. Note that the original latent code is uniformly distributed in $[0, 1] \times [0, 1]$.

Shown in Fig. 3 are the experimental results for case 1. Each row represents source/target dataset D and each column represents different label i for $i \in \{0, 1, \dots, 3\}$. Note that our algorithm successfully reprogram unconditional GANs as conditional GANs. To further confirm that each reprogrammed GAN successfully learned its corresponding input noise distribution, we sample 1000 random input noise according to F_{target} and visualize the distribution of the first two entries in Fig. 4. Column i is for F_{target} learned with FMNIST-i. Observe that F_{target} is not a uniform distribution while F_{source} is.

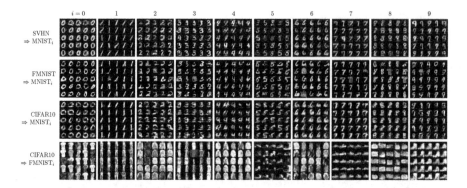

Fig. 5. GAN reprogramming across different datasets. Row corresponds to source/target data, and column corresponds to the label.

Shown in Fig. 5 are the experimental results for case 2. The reprogrammed GANs generate reasonable images but also show some inherent limitations. First, the texture remains the same as the original GAN. Also, if the given GAN learned a limited set of structures, the reprogrammed GAN attempts to generate target images by distorting the learned structures. For instance, consider the second row (FMNIST to MNIST) of Fig. 5. Since the FMNIST-GAN is able to draw fashion items only, the reprogrammed GAN draws digit images maintaining fashion item structures. For instance, it generates digit-2 images by distorting shoes.

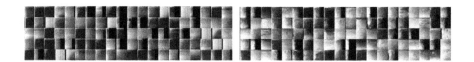

Fig. 6. MFC of real and synthetic audio clips.

We also apply reprogramming algorithm to a CIFAR10-GAN so that it can generate drum sound clips. The Mel-Frequency Cepstrum of the real drum sound clips (left) and those of the synthetic ones generated by a reprogrammed GAN (right) are shown in Fig. 6. Even though the MFC images look very differently from typical CIFAR10 images, our reprogramming algorithm successfully generates realistic MFC images. As an additional experiment, we reprogram a CIFAR10-GAN as a CelebA-GAN and a CelebA-GAN to a Cartoon-GAN. From

Fig. 7. CIFAR10 to CelebA (left) and a CelebA to Cartoon (right).

the left panel of Fig. 7, one can see that a reprogrammed GAN is able to generate human-face-like images even though it is originally trained with CIFAR10 images. On the right hand side, the reprogrammed GAN generates cartoon-like face images with white background, resembling the Cartoon dataset.

Comparison with an Existing Method. We also evaluate the latent code sampling algorithm, called PPGN [20], to generate $MNIST_0$ images with a pre-trained FMNIST-GAN. Shown in Fig. 8 are the sample outputs of PPGN. Compared to the images generated by our reprogramming algorithm, the examples generated by PPGN barely resemble the digit 0. The poor performance of PPGN can be attributed to the fact that PPGN is designed specifically for GANs trained under a particular method called Joint PPGN. Furthermore, the sampling procedure of PPGN involves an iterative procedure, requiring many rounds of forward/backward-passes to generate a sample, while our sampling operation requires a single forward-pass.

4.3   Fréchet Inception Distance (FID)

In Fig. 10a, we plot the Fréchet Inception Distance (FID) [11] between the output generated by a reprogrammed GAN $(g(h_\theta(Z_{\mathrm{seed}})))$ and the target distribution (X_{target}). One can observe that the FID decreases as the reprogramming algorithm proceeds, showing the validity of GAN reprogramming. We also observe a much larger value for the FID between the output generated by a reprogrammed GAN $(g(h_\theta(Z_{\mathrm{seed}})))$ and the original distribution (X_{source}).

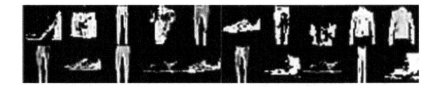

Fig. 8. MNIST$_0$ generated from FMNIST-GAN + PPGN [20].

4.4 Reprogramming vs Transfer Learning

We compare the convergence performance of GAN reprogramming algorithm and that of transfer learning. A standard approach to transfer knowledge between generative models is based on fine tuning [28]. That is, (i) one first trains a GAN (both G and D) with the original dataset; (ii) then fine-tunes G and D with the target dataset. Shown in Fig. 10b are the comparison results. For the transfer learning curve, we report the performance of 'fine-tuning all layers' as we observed 'fine-tuning last layers' only performs worse. We first note that both 'transfer learning' and 'train from scratch' suffer from the notorious oscillation phenomena [16]. More importantly, the convergence of transfer learning is much slower than reprogramming.

4.5 Capacity of Code Generator

To evaluate the reprogrammability as a function of the capacity of code generator $h_\theta(\cdot)$, we vary d, the dimensionality of Z_{seed}. We also repeat the same experiment while varying the number of neurons in each layer, i.e., the capacity of $h_\theta(\cdot)$. Here, we use a three-layer FC network, whose hidden layers have w neurons each, where $w \in \{5, 6, \ldots, 14\}$. For each point, we then measure the FID five times and report the average. The experimental results are summarized in Fig. 9c and Fig. 9d, respectively. In both figures, we can observe that reprogramming is infeasible when the capacity of code generator is low and that the reprogrammability saturates as the capacity increases.

5   Theoretical Analysis

Clearly, reprogrammed generators cannot generate samples beyond the original range. Thus, the performance of GAN reprogramming is characterized by the range of the original generator. The following theorem formally states this phenomenon under a simplified linear setting [7].

Theorem 1. *Let X_{target} be a zero-mean non-degenerate d_o-dimensional Gaussian random vector. Assume a linear generator $g(Z) = AZ$, where $A \in \mathbb{R}^{d_o \times d_i}$, and denote the projection of X_{target} on the subspace spanned by matrix A by X_{target}^A. Then, the minimum second-order Wasserstein distance that can be achieved via any GAN reprogramming method is lower bounded by $\mathbb{E}[\|\|X_{target} -$*

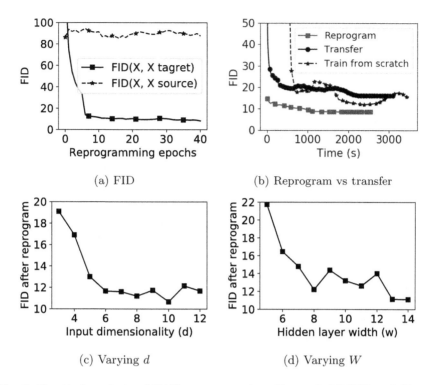

(a) FID

(b) Reprogram vs transfer

(c) Varying d

(d) Varying W

Fig. 9. Practical analysis of GAN reprogramming. X_{source} is MNIST and X_{target} is MNIST$_0$. (a) The FID between $g(h_\theta(Z_{\text{seed}}))$ and X_{target} (blue) and the FID between $g(h_\theta(Z_{\text{seed}}))$ and X_{source} (orange). (b) Reprogramming vs transfer. (c) Reprogrammability as a function of d. (d) Reprogrammability as a function of the number of hidden neurons in each layer of h_θ.

$X_{target}^A \|^2]$. *Moreover, if Z_{seed} is a zero-mean non-degenerate d_i-dimensional Gaussian random vector, the optimal reprogramming is achievable with a linear code generator $h(Z_{seed}) = BZ_{seed}$ for some $B \in \mathbb{R}^{d_i \times d_i}$.*

Proof. (**lower bound**) The second-order Wasserstein distance is $W_2^2(P_r, P_g) := \inf_\gamma \mathbb{E}_{(x,y)\sim\gamma}[\|x - y\|^2]$, where $\gamma \in \Pi(P_r, P_g)$, and $\Pi(P_r, P_g)$ is the set of all joint distributions whose marginals are respectively P_r and P_g. Under the second-order Wasserstein distance as an underlying metric, the GAN reprogramming problem can be written as $\min_{F_{\text{target}}} W_2^2(g(Z_{\text{target}}), X_{\text{target}})$, where $Z_{\text{target}} \sim F_{\text{target}}$ is a d_i-dimensional random vector. Assuming our code generator structure ($Z_{\text{target}} = h(Z_{\text{seed}})$) and the linear generator assumed in the theorem, we have $\inf_h W_2^2(Ah(Z_{\text{seed}}), X_{\text{target}})$, where X_{target} is a zero-mean non-degenerate d_o-dimensional Gaussian random vector, and $A \in \mathbb{R}^{d_o \times d_i}$ is a given generator matrix. By the definition of the second-order Wasserstein distance, it reduces to

$$\inf_{h} \inf_{\gamma} \mathbb{E}_{(Z_{\text{seed}}, X_{\text{target}}) \sim \gamma} \| Ah(Z_{\text{seed}}) - X_{\text{target}} \|^2, \tag{6}$$

where $\gamma \sim \Pi(P_{Z_{\text{seed}}}, P_{X_{\text{target}}})$.

As $Ah(Z_{\text{seed}}) - X_{\text{target}}^A \perp X_{\text{target}}^A - X_{\text{target}}$, we have

$$\| Ah(Z_{\text{seed}}) - X_{\text{target}} \|^2 = \| Ah(Z_{\text{seed}}) - X_{\text{target}}^A \|^2 + \| X_{\text{target}}^A - X_{\text{target}} \|^2. \tag{7}$$

Since the second term in the RHS is independent of h and γ, the second-order Wasserstein distance is always lower bounded by $\mathbb{E} \| X_{\text{target}}^A - X_{\text{target}} \|^2$. This proves the first part of the theorem.

(achievability) We now show that the lower bound can be achieved with a linear code generator. Consider a linear code generator $h(Z_{\text{seed}}) = BZ_{\text{seed}}$ for some matrix $B \in \mathbb{R}^{d_i \times d_i}$ and a zero-mean non-degenerate d_i-dimensional Gaussian random vector Z_{seed}, i.e., $Z_{\text{seed}} \sim \mathcal{N}(0, K_s)$ for some positive definite matrix K_s.

Let the economic singular value decomposition (SVD) of A by $A = U_A \Sigma_A V_A^\mathsf{T}$. Then, the projection on the subspace spanned by A is a linear operator $P := U_A U_A^\mathsf{T}$. Since X_{target} is a zero-mean non-degenerate Gaussian random vector, $X_{\text{target}} \sim \mathcal{N}(0, K_t)$ for some positive definite matrix K_t. Thus, $X_{\text{target}}^A \sim \mathcal{N}(0, PK_tP^\mathsf{T})$. Similarly, $Ah(Z_{\text{seed}}) = ABZ_{\text{seed}} \sim \mathcal{N}(0, ABK_sB^\mathsf{T}A^\mathsf{T})$. Then,

$$PK_tP^\mathsf{T} = ABK_sB^\mathsf{T}A^\mathsf{T}$$
$$\Rightarrow U_A U_A^\mathsf{T} K_t U_A U_A^\mathsf{T} = U_A \Sigma_A V_A^\mathsf{T} BK_sB^\mathsf{T} V_A \Sigma_A U_A^\mathsf{T}$$
$$\Rightarrow U_A^\mathsf{T} K_t U_A = \Sigma_A V_A^\mathsf{T} BK_sB^\mathsf{T} V_A \Sigma_A$$
$$\Rightarrow U_A^\mathsf{T} \sqrt{K_t} = \Sigma_A V_A^\mathsf{T} B \sqrt{K_s} \tag{8}$$
$$\Rightarrow V_A \Sigma_A^{-1} U_A^\mathsf{T} \sqrt{K_t} \sqrt{K_s}^{-1} = B, \tag{9}$$

where (8) holds since K_t and K_s are positive definite, and (9) holds since $\sqrt{K_t}$ and $\sqrt{K_s}$ are also positive definite. Thus, by choosing

$$B = B^\star := V_A \Sigma_A^{-1} U_A^\mathsf{T} \sqrt{K_t} \sqrt{K_s}^{-1},$$

the marginal distribution of X_{target}^A and that of $AB^\star Z_{\text{seed}}$ are identical. Therefore, one can choose B and γ such that $X_{\text{target}}^A = AB^\star Z_{\text{seed}}$ w.p. 1. Together with (7), it implies

$$\mathbb{E} \| AB^\star Z_{\text{seed}} - X_{\text{target}} \|^2 = \mathbb{E} \| X_{\text{target}}^A - X_{\text{target}} \|^2, \tag{10}$$

proving the theorem. □

Experiments. To further demonstrate that reprogrammed generators cannot generate samples beyond the range of the original generator, we conduct the following experiments with synthetic datasets. We first train a generator using images of three randomly positioned circles (Set 0) and then reprogram it respectively with images of one randomly positioned circle (Set 1), images of three co-located circles (Set 2), and images of two randomly positioned squares (Set 3). See Fig. 10a for sample images. In Fig. 10b, we plot normalized FIDs, i.e., FID divided by the initial FID measured before we apply our reprogramming algorithm. We can observe that reprogramming succeeds only for the first two datasets (Set 1 and Set 2), which consist of circles, and fails for the last dataset (Set 3), which consist of different shapes. This is expected as the square shapes are beyond the range of the original generator that can only draw curvy edges.

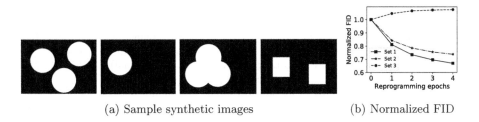

(a) Sample synthetic images   (b) Normalized FID

Fig. 10. Reprogramming a generator with restricted range.

6   Conclusion

In this work, we studied the problem of GAN reprogramming. We showed that our algorithm can reprogram a GAN as another GAN that generates different datasets and studied the applicability, feasibility, and limitation. We concluded the paper with discussing related open questions.

Unexpected Behaviors of a GAN and Adversarial Reprogramming of GANs. The fact that GANs can be reprogrammed implies that a GAN trained with a certain dataset is *not* guaranteed to generate valid samples only. Even though this observation is already made in a recent study [13], our work implies even a stronger message: There exist infinitely many latent codes which result in inconsistent samples with the original dataset. We believe that these inconsistent samples are generated with very low probability if latent codes are drawn at random according to F_{original}. However, for some critical applications, such an unexpected behavior might not be allowed. Thus, it still implies that the usage of GAN for highly critical applications should be avoided until we have a formal guarantee that GANs generate valid samples only. Moreover, it is an interesting open question whether or not one can design a GAN that provably generates valid samples for *every* latent code.

An Alternative for GAN Transfer Learning. In Sect. 4.4, we observed that GAN reprogramming outperforms the standard fine-tuning learning algorithm in a transfer learning setting. This implies that reprogramming-based approach might be a useful alternative to GAN transfer learning, but this requires a more in-depth study.

References

1. Arjovsky, M., Chintala, S., Bottou, L.: Wasserstein generative adversarial networks. In: International Conference on Machine Learning (ICML) (2017)
2. Caruana, R.: Multitask learning. Mach. Learn. **28**(1), 41–75 (1997)
3. Donahue, C., McAuley, J., Puckette, M.: Synthesizing audio with generative adversarial networks. arXiv preprint arXiv:1802.04208 (2018)
4. Duong, L., Cohn, T., Bird, S., Cook, P.: Low resource dependency parsing: cross-lingual parameter sharing in a neural network parser. In: Annual Meeting of the Association for Computational Linguistics and the 7th International Joint Conference on Natural Language Processing (2015)
5. Elsayed, G.F., Goodfellow, I., Sohl-Dickstein, J.: Adversarial reprogramming of neural networks. In: International Conference on Learning Representations (ICLR) (2019)
6. Engel, J., Hoffman, M., Roberts, A.: Latent constraints: learning to generate conditionally from unconditional generative models. In: International Conference on Learning Representations (ICLR) (2018)
7. Feizi, S., Farnia, F., Ginart, T., Tse, D.: Understanding GANs: the LQG setting. arXiv preprint arXiv:1710.10793 (2017)
8. Goodfellow, I., et al.: Generative adversarial nets. In: Advances in Neural Information Processing Systems (NeurIPS) (2014)
9. Goodfellow, I.J., Shlens, J., Szegedy, C.: Explaining and harnessing adversarial examples. In: International Conference on Learning Representations (ICLR) (2014)
10. Gulrajani, I., Ahmed, F., Arjovsky, M., Dumoulin, V., Courville, A.C.: Improved training of Wasserstein GANs. In: Advances in Neural Information Processing Systems (NeurIPS) (2017)
11. Heusel, M., Ramsauer, H., Unterthiner, T., Nessler, B., Hochreiter, S.: GANs trained by a two time-scale update rule converge to a local nash equilibrium. In: Guyon, I., et al. (eds.) Advances in Neural Information Processing Systems (NeurIPS) (2017)
12. Huang, J.T., Li, J., Yu, D., Deng, L., Gong, Y.: Cross-language knowledge transfer using multilingual deep neural network with shared hidden layers. In: IEEE International Conference on Acoustics, Speech and Signal Processing (ICASSP) (2013)
13. Kos, J., Fischer, I., Song, D.: Adversarial examples for generative models. In: 2018 IEEE Security and Privacy Workshops (SPW) (2018)
14. Krizhevsky, A., Hinton, G.: Learning multiple layers of features from tiny images. MS thesis, Department of Computer Science, University of Toronto (2009)
15. LeCun, Y., Bottou, L., Bengio, Y., Haffner, P.: Gradient-based learning applied to document recognition. Proc. IEEE **86**(11), 2278–2324 (1998)
16. Li, J., Madry, A., Peebles, J., Schmidt, L.: On the limitations of first-order approximation in GAN dynamics. In: International Conference on Machine Learning (ICML) (2018)

17. Liu, Z., Luo, P., Wang, X., Tang, X.: Deep learning face attributes in the wild. In: International Conference on Computer Vision (ICCV) (2015)
18. Madry, A., Makelov, A., Schmidt, L., Tsipras, D., Vladu, A.: Towards deep learning models resistant to adversarial attacks. In: International Conference on Learning Representations (ICLR) (2018)
19. Netzer, Y., Wang, T., Coates, A., Bissacco, A., Wu, B., Ng, A.Y.: Reading digits in natural images with unsupervised feature learning. In: NeurIPS workshop on deep learning and unsupervised feature learning (2011)
20. Nguyen, A., Clune, J., Bengio, Y., Dosovitskiy, A., Yosinski, J.: Plug & play generative networks: conditional iterative generation of images in latent space. In: IEEE Conference on Computer Vision and Pattern Recognition (CVPR). IEEE (2017)
21. Oquab, M., Bottou, L., Laptev, I., Sivic, J.: Learning and transferring mid-level image representations using convolutional neural networks. In: IEEE conference on Computer Vision and Pattern Recognition (CVPR) (2014)
22. Pan, S.J., Yang, Q., et al.: A survey on transfer learning. IEEE Trans. Knowl. Data Eng. **22**(10), 1345–1359 (2010)
23. Pasquini, D., Mingione, M., Bernaschi, M.: Out-domain examples for generative models. arXiv preprint arXiv:1903.02926 (2019)
24. Royer, A., et al.: XGAN: unsupervised image-to-image translation for many-to-many mappings. arXiv preprint arXiv:1711.05139 (2017)
25. Ruder, S.: An overview of multi-task learning in deep neural networks. arXiv preprint arXiv:1706.05098 (2017)
26. Szegedy, C., et al.: Intriguing properties of neural networks. arXiv preprint arXiv:1312.6199 (2013)
27. Villani, C.: Optimal Transport: Old and New, vol. 338. Springer, Heidelberg (2008). https://doi.org/10.1007/978-3-540-71050-9
28. Wang, Y., Wu, C., Herranz, L., van de Weijer, J., Gonzalez-Garcia, A., Raducanu, B.: Transferring GANs: generating images from limited data. arXiv preprint arXiv:1805.01677 (2018)
29. Xiao, H., Rasul, K., Vollgraf, R.: Fashion-MNIST: a novel image dataset for benchmarking machine learning algorithms. arXiv preprint arXiv:1708.07747 (2017)
30. Yang, Y., Hospedales, T.M.: Trace norm regularised deep multi-task learning. arXiv preprint arXiv:1606.04038 (2016)
31. Zhou, B., Lapedriza, A., Xiao, J., Torralba, A., Oliva, A.: Learning deep features for scene recognition using places database. In: Advances in Neural Information Processing Systems (NeurIPS) (2014)

On Saliency Maps and Adversarial Robustness

Puneet Mangla[(✉)], Vedant Singh, and Vineeth N. Balasubramanian

Department of Computer Science and Engineering, IIT Hyderabad, Sangareddy, India
{cs17btech11029,cs18btech11047,vineethnb}@iith.ac.in

Abstract. A very recent trend has emerged to couple the notion of interpretability and adversarial robustness, unlike earlier efforts that focus solely on good interpretations or robustness against adversaries. Works have shown that adversarially trained models exhibit more interpretable saliency maps than their non-robust counterparts, and that this behavior can be quantified by considering the alignment between the input image and saliency map. In this work, we provide a different perspective to this coupling and provide a method, Saliency based Adversarial training (SAT), to use saliency maps to improve the adversarial robustness of a model. In particular, we show that using annotations such as bounding boxes and segmentation masks, already provided with a dataset, as weak saliency maps, suffices to improve adversarial robustness with no additional effort to generate the perturbations themselves. Our empirical results on CIFAR-10, CIFAR-100, Tiny ImageNet, and Flower-17 datasets consistently corroborate our claim, by showing improved adversarial robustness using our method. We also show how using finer and stronger saliency maps leads to more robust models, and how integrating SAT with existing adversarial training methods, further boosts the performance of these existing methods.

Keywords: Adversarial robustness · Saliency maps · Deep neural networks

1 Introduction

Deep Neural Networks (DNNs) have become vital to solving many tasks across domains, including image/text/graph classification and generation, object recognition, segmentation, speech recognition, etc. As the applications of DNNs widen in scope, *robustness* and *interpretability* are two important parameters that define the goodness of a trained DNN model. While on one hand the deep network should be robust to imperceptible perturbations, on the other hand it should be interpretable enough to be trusted when practically used in domains like autonomous navigation or healthcare. Keeping in mind the vulnerability of deep networks to adversarial attacks [27], efforts have been undertaken to make them more robust to these attacks. Among the proposed methodologies, Adversarial Training (AT) [11,16] has emerged as one of the best defenses wherein

© Springer Nature Switzerland AG 2021
F. Hutter et al. (Eds.): ECML PKDD 2020, LNAI 12458, pp. 272–288, 2021.
https://doi.org/10.1007/978-3-030-67661-2_17

networks are trained on adversarial examples to better classify them at test time. On the other hand, in order to generate an interpretable explanation to a network prediction, many methods have been proposed lately, of which guided-backpropagation (GBP)[25], GradCAM++ [3], Integrated Gradients (IG) [26] and fine-grained visualisations (FGVis) [29] are popular to name a few.

Most work so far on interpretability, and adversarial robustness has focused on either of them alone. Very recently, over the last few months, there has been a new interest in coupling the two notions and aiming to understand the connection between robustness and interpretability [2,4,8,9,28,36]. The few existing efforts can be categorized broadly into three kinds: (i) Efforts [28,36] which have shown that explanations generated by more robust models are more interpretable than their non-robust counterparts; (ii) Efforts that have attempted to theoretically analyze the relationship between adversarial robustness and interpretability (for e.g., [8] obtains an inequality between these notions which holds to equality in case of linear models, and [12] relates the two using a generalized form of hitting set duality); and (iii) Efforts [2,4], more recent, that have attempted to improve robustness by training models with additional objectives that constrain the image explanations to be more interpretable and robust towards attribution attacks. All of these efforts are recent, and more work needs to be done to better explore the connections. In this work, we provide a different, yet simple and effective, approach to leverage saliency maps for adversarial robustness. We observe that adversarial perturbations correspond to class-discriminative pixels in later stages of training (see Fig. 1), and exploit this observation to use the saliency map (static) of a given image, whilst training to improve robustness. For datasets where bounding boxes and segmentation masks are provided, we demonstrate that one can exploit these, in lieu of saliency maps, to improve model robustness using our approach. Our work would be closest to the third category of methods described above, and we differentiate from these other recent efforts further in Sect. 2.

Generally speaking, humans tend to learn new tasks in a robust, generalizable fashion when provided with explanations during their learning phase. For example, a medical student learns about a disease better when provided with explanations of how the disease acts inside the body, rather than just the description of external symptoms. A person may otherwise learn irrelevant relationships without knowledge of underlying explanations. Similarly, we opine that a DNN model that is trained with explanations is less easily fooled by adversarial perturbations.

Little has been explored in this direction, and we aim to leverage this relationship to provide an efficient methodology for adversarial robustness. We differ from earlier efforts in our primary objective that we use explanations to generate pseudo-adversarial examples instead of using them as a regularization term like [2,4] (more in Sect. 2). In particular, we provide a simple yet effective methodology (as compared to a recent effort such as [12], which is NP-hard) that improves model robustness using saliency maps or already provided bounding boxes or segmentation masks.

Our key contributions can be summarized as follows: (i) We observe a tangible relationship between a saliency map and adversarial perturbations for a given image, and leverage this observation to propose a new methodology that uses the saliency map of the image to mimic adversarial training; (ii) We show through our empirical studies on widely used datasets: CIFAR-10, CIFAR-100, Tiny ImageNet and Flowers-17, to show the improvement in adversarial robustness through our proposed method; (iii) We show that the improvement becomes more pronounced when a finer and stronger saliency map is used, signifying a strong correlation between saliency maps and adversarial robustness; (iv) When bounding boxes or segmentation masks are already available in a dataset (e.g. Tiny ImageNet or Flowers), we demonstrate that our methodology improves adversarial robustness with no additional cost to generate perturbations for training; (v) Additionally, we show that integrating our method with adversarial training methods, such as PGD [16] and TRADES [34], further improves their performance; and (vi) We perform detailed ablation studies to better characterize the efficacy of our proposed methodology. We believe that this work can contribute to opening up a rather new direction to enhance the robustness of DNN models.

2   Related Work

We review earlier efforts related to this work from multiple perspectives, as described below.

Explanation Methods: Various methods have been proposed over the last few years to explain the decisions of a neural network. *Backpropagation-based methods* find the importance of each pixel by backpropagating the class score error to the input image. An improved and popular version of this, known as Guided-Backpropagation [25], only keeps paths that lead to a positive influence on the class score, leading to much cleaner-looking explanations. SmoothGrad [24], VarGrad [1] and Integrated gradients [26] refine the explanations by combining/integrating gradients of multiple noisy/interpolated versions of the image. Other backpropagation based methods like DeepLift [22], Excitation BackProp [35] and Layerwise Relevance Propagation [15] generate explanations by utilizing top-down relevance propagation rules. PatternNet and PatternAttribution [13] yield explanations theoretically sound for linear models and produce improved explanations for deep networks. CAM [37], Grad-CAM [20], Grad-CAM++ [3] form another variant of generating explanations known as *Activation-based methods*. These methods use linear combinations of activations of convolutional layers, with the weights for these combinations obtained using gradients. *Perturbation based methods* generate attribution maps by examining the change in the model's prediction when the input image is perturbed [18,32]. All the aforementioned methods, however, focus solely on explaining neural network decisions.

Adversarial Attacks and Robustness: With the advancement of newer adversarial attacks each year [11,16,30,34], methods have been proposed to defend against them. Parseval Networks [5] train robust networks by constraining the Lipschitz constant of its layers to be smaller than 1. Another category of methods harness the susceptibility of latent layers by performing latent adversarial training (LAT) [23] or using feature denoising [31]. Other methods like DefenseGAN [19] exploit GANs wherein they learn the distribution of unperturbed images and find the closest image for a given test image, to feed the network at inference time. TRADES [34], more recently, presents a new defense method that provides a trade-off between adversarial robustness and vanilla accuracy by decomposing prediction error for adversarial examples (robust error) as the sum of natural (classification) error and boundary error, and providing a differentiable upper bound using theory based on classification-calibrated loss. Among the proposed defenses against adversarial attacks, Adversarial Training (AT) [11,16] has remained the most popular and widely used defense, where the network is trained on adversarial examples in order to match the training data distribution with that of adversarial test distribution. More recent efforts in this direction include [21,33], which aim to reduce adversarial training overhead by recycling gradients and accelerating via the maximal principle, respectively. In this work, we focus on adversarial training, considering it still remains the most reliable defense against different attacks.

Interpretability and Robustness: The last few months have seen a few efforts on associating the notions of robustness and interpretability. These efforts can be categorized into three kinds, as introduced briefly earlier: (i) The first kind centers around interpreting how adversarially trained convolutional neural networks (ATCNNs) recognize objects. Recent studies [28,36] have shown that representations learned by ATCNNs are more biased towards image shape than its texture. Also, these ATCNNs tend to evince more interpretable saliency maps corresponding to their prediction than their non-robust equivalents. (ii) In the second kind, inspired by [28,36], Etmann et al. [8] recently quantified this behavior of ATCNNs by considering the alignment between saliency map and the image as the metric for interpretability. They confirmed that for a linear model, the alignment grows strictly with robustness. For non-linear models such as neural networks, they show that the their linearized robustness is loosely bounded by the alignment metric. [12], which is rather new, provides a theoretical connection between adversarial examples and explanations by demonstrating that both are related by a generalized form of hitting set duality. (iii) Encouraged by the work of Etmann [8], a third category of work more recently sought to answer the question: Do robust and interpretable saliency maps imply adversarial robustness ? Recent efforts [7,10] have shown that the explanations of neural networks can also be manipulated by adding perturbations to input examples, which, instead of causing mis-classification, result in a different explanation. To tackle this problem, a recent effort, Robust Attribution Regularization (RAR) [4], aims to train networks by optimizing objectives with a regularization term on explanations in order to achieve robust attributions.

In contrast to the above mentioned methods, our work explores a different side to the connection between notions of interpretability and robustness. As explained in Sect. 1, we aim to generate adversarial perturbations from a given saliency map to improve robustness while training a neural network. Our work is perhaps closest to RAR [4] in using explanations (saliency maps) during training, we however differ from them in our primary objective and in that we are using explanations to generate adversarial examples, while RAR seeks to attain attributional robustness. In terms of approach, although one could consider the work by Ignatiev et al. [12] as related to ours, they had only preliminary results on MNIST, and focused on providing a theoretical connection between adversarial examples and explanations based on logic and constraint programming, which may not scale to large-scale datasets. Our work attempts to show that explanations can yield robustness in standard large-scale benchmark datasets in an efficient way. We further hypothesize that, by advancement through our method, we can actually take advantage of weak explanations like bounding boxes and segmentation masks which may otherwise be left unexploited in a dataset, to train adversarially robust models.

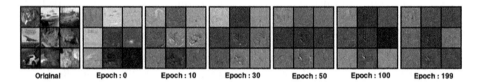

Fig. 1. Variation of adversarial perturbation with training epochs during 5-step PGD adversarial training of Resnet-34 on CIFAR-10. We observe that adversarial perturbations at later stages of training correspond to class-discriminative regions/pixels.

3   Using Saliency Maps for Efficient Adversarial Training

Our method is motivated by the observation that adversarial perturbations at later stages of training correspond to class-discriminative regions/pixels, as shown in Fig. 1. We begin describing our method with the necessary notations and preliminaries.

3.1   Notations and Preliminaries

We denote a neural network as $\Phi(.\ ;\ \theta) : \mathbb{R}^d \to \mathbb{R}^k$, parametrized by weights θ, which takes an input $\mathbf{x} \in \mathbb{R}^d$ and outputs a logit, $\Phi^i(\mathbf{x})$, for each of k classes, i.e. $i \in \mathcal{C} = \{1, \cdots, k\}$, $\mathcal{C}$ denoting the set of class labels. Without loss of generality, considering an image classification setting, we define a saliency map $\mathbf{s}$ corresponding to an input sample $\mathbf{x}$ as $\mathbf{s} \in [0,1]^d$, where the presence of an object of interest in input $\mathbf{x}$ lies between 0 and 1. For a trained network Φ, the unnormalized saliency map for an input $\mathbf{x}$ can simply be given as: $\nabla_{\mathbf{x}} \Phi^{i^*}(\mathbf{x})$, where $i^* = \arg\max_i \Phi^i(\mathbf{x})$. (In practice, this gradient is at times computed w.r.t feature maps of intermediate layers, but we leave our notations w.r.t $\mathbf{x}$ for ease of understanding).

Projected Gradient Descent (PGD) Attack: Prior efforts on adversarial attacks [11] include the Fast Gradient Sign Method (FGSM), a l_∞-bounded single step attack which calculates an adversary as: $\mathbf{x} + \epsilon\,\mathrm{sign}(\nabla_\mathbf{x}\mathcal{L}\,(\Phi(\mathbf{x},\theta),y))$. A more powerful attack however is the popularly used multi-step variant, also called Projected Gradient Descent (PGD), given by:

$$\mathbf{x}^0 = \mathbf{x}; \mathbf{x}^{t+1} = \Pi_{\mathbf{x}+N}\big(\mathbf{x}^t + \alpha\,\mathrm{sign}(\nabla_\mathbf{x}\mathcal{L}\,(\Phi(\mathbf{x},\theta),y))\big) \tag{1}$$

where α is the step size, Π is the projection function, and N is the space of possible perturbations. We use PGD as the choice of attack in our experiments, and later change to show (in Sect. 5) how our method performs against other attacks.

Adversarial Training (AT): AT [11] is generally used to make the models adversarially robust by matching the training distribution with the adversarial test distribution. Essentially, for AT, the optimal parameter θ^* is given by:

$$\theta^* = \arg\min_\theta \mathbb{E}_{(\mathbf{x},y)\sim D}\left[\max_{\delta\in N}\mathcal{L}\,(\Phi(\mathbf{x}+\delta,\theta),y)\right] \tag{2}$$

Here, the inner maximization $\max_{\delta\in N}\mathcal{L}\,(\Phi(\mathbf{x}+\delta,\theta),y)$ is calculated using a strong adversarial attack such as PGD.

3.2 Saliency-Based Adversarial Training: Motivation

An adversarial perturbation at input $\mathbf{x}$ is given by:

$$\arg\inf_{e\,\in\,\mathbb{R}^d}\{\|e\| : \arg\max_i \Phi^i(x+e) \neq \arg\max_i \Phi^i(x)\} \tag{3}$$

Etmann et al. [8] showed that for most multi-class neural networks, especially ones with ReLU or Leaky ReLU activation functions, which we consider in this work[1], the network's score function, Φ is sufficiently locally-linear in relevant neighbourhood of input $\mathbf{x}$, i.e.

$$\Phi^i(\mathbf{x}+e) \approx \Phi^i(\mathbf{x}) + e^T\cdot\nabla_\mathbf{x}\Phi^i(\mathbf{x}) \tag{4}$$

Leveraging this, an adversarial perturbation, e, which is intended as a perturbation to input $\mathbf{x}$ which results in a change of predicted label, can be modeled as follows (given $i, j \in \mathcal{C}$):

$$\arg\max_i \Phi^i(x+e) \neq \arg\max_i \Phi^i(x) \tag{5}$$

$$\Longleftrightarrow \quad \exists j \neq i^* : \Phi^j(\mathbf{x}+e) > \Phi^{i^*}(\mathbf{x}+e) \tag{6}$$

$$\Longleftrightarrow \quad \exists j \neq i^* : e^T\cdot(\nabla_\mathbf{x}\Phi^j(\mathbf{x}) - \nabla_\mathbf{x}\Phi^{i^*}(\mathbf{x})) > \Phi^{i^*}(\mathbf{x}) - \Phi^j(\mathbf{x}) \tag{7}$$

---

[1] In all our experiments, we train networks with ReLU activations which ensures that this assumptions is met. We follow the experiments in [8] in this regard, to ensure that our model is locally affine in a given data point's neighborhood.

The third inequality (7) comes from combining inequality (6) and expression (4). The infimum over $\|e\|$, which provides a minimal perturbation to change the class label, is achieved by choosing e as a multiple of $\nabla_{\mathbf{x}}(\Phi^j(\mathbf{x}) - \Phi^{i^*}(\mathbf{x}))$. (In general, note that the LHS of $e^T z > c$ where c is a constant, is maximized by $e = \|e\|\frac{z}{\|z\|}$, leading to $\|e\|\|z\| > c$. The infimum of e is then achieved by choosing it as a multiple of z). The direction of adversarial perturbation then becomes:

$$\nabla_{\mathbf{x}}(\Phi^j(\mathbf{x}) - \Phi^{i^*}(\mathbf{x})) \qquad (8)$$

This perturbation direction depends on two quantities: (i) $\nabla_{\mathbf{x}}\Phi^{i^*}(\mathbf{x})$ the saliency map for the true class i^*; and (ii) $\nabla_{\mathbf{x}}\Phi^j(\mathbf{x})$, the saliency map of $\mathbf{x}$ for class j for which the infimum of e is attained. We now analyze the direction of adversarial perturbation separately for the binary and multi-class cases.

Binary Case: Let us consider a binary classifier $h : \mathbf{x} \to \{-1, 1\}$ given by: $h = \text{sign}(\Phi(\mathbf{x}, \theta))$, where $\Phi(\mathbf{x}, \theta))$ represents the logit of the positive class. Let $\Phi'(\mathbf{x})$ denotes the logit of negative class. Applying a sigmoid activation function, we get the probability of the positive and negative class as: $P(y = +1|\mathbf{x}) = \frac{1}{1+\exp^{-\Phi(\mathbf{x}, \theta)}}$ and $P(y = -1|\mathbf{x}) = \frac{1}{1+\exp^{-\Phi'(\mathbf{x}, \theta)}}$ respectively. It is simple to see that the probability of the negative class can also be written as:

$$P(y = -1|\mathbf{x}) = 1 - P(y = +1|\mathbf{x}) = \frac{1}{1 + \exp^{\Phi(\mathbf{x}, \theta)}}$$

Rather, the corresponding logit score of the negative class is $-\Phi(\mathbf{x}, \theta))$. In Eq. 8 for a binary classifier, one can hence view $\Phi^j(\mathbf{x}) = -\Phi^{i^*}(\mathbf{x})$, and define the adversarial perturbation direction in Eq. 8 simply as $-\nabla_{\mathbf{x}}(\Phi^{i^*}(\mathbf{x}))$.

Multi-class Case: We extend a similar argument to the multi-class setting using an approximation. The direction of adversarial perturbation in the multi-class setting as give in Eq. 8 would require finding the class j for which the infimum of $\|e\|$ is attained. This requires computing the quantity in Eq. 8 for all classes, and identifying the $j \neq i^*$ for which the quantity attains the least value. This is compute-intensive. To avoid this computational overhead, we rely on Φ^{i^*} alone (note that i^*, the ground truth label, is known to us at training), and simply propose the use of $-\nabla_{\mathbf{x}}(\Phi^{i^*}(\mathbf{x}))$ as the direction of perturbation (as in the binary case). We now argue that this approximation is a reasonable one. Considering the multi-class setting as k binary classification problems, for the binary classifier corresponding to the ground truth class i^*, we would have:

$$P(y \neq i^*|\mathbf{x}) = 1 - P(y = i^*|\mathbf{x}) = \frac{1}{1 + \exp^{\Phi^{i^*}(\mathbf{x}, \theta)}}$$

In other words, the corresponding logit score of the negative class is $-\Phi^{i^*}(\mathbf{x}, \theta))$. Approximating the direction of the adversarial perturbation as the average of the directions of the perturbations across the k binary classification problems, we would get the direction of the perturbation to be: $\nabla_{\mathbf{x}}(\sum_{j \neq i^*} \Phi^j(\mathbf{x}) - k\Phi^{i^*}(\mathbf{x}))$. Assuming that each of the classes $l \neq i^* \in \mathcal{C}$ is equally likely to be the j

that minimizes $\|e\|$, it is evident that choosing $-\nabla_{\mathbf{x}}(\Phi^{i^*}(\mathbf{x}))$ as the direction of perturbation would be the most conservative option in the expected sense. We show through our experiments that this option works reasonably well.

Since we deal with l_∞-bounded perturbations in this work, following PGD (Eq. 1), we use $-\mathrm{sign}\,(\nabla_{\mathbf{x}}\Phi^{i^*}(\mathbf{x}))$ instead of $-\nabla_{\mathbf{x}}\Phi^{i^*}(\mathbf{x})$ itself as the perturbation direction. We complete the above discussion by noting that $\nabla_{\mathbf{x}}\Phi^{i^*}(\mathbf{x})$ is the saliency map, $\mathbf{s}$, defined at the beginning of this section. In other words, the direction of adversarial perturbation can be obtained using a saliency map (as we also show in our experiments).

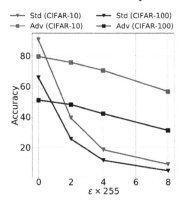

Fig. 2. Adversarial attack accuracy when perturbing input example using the direction of negative saliency on normally (Std.) and adversarially (Adv.) trained models on CIFAR-10 and CIFAR-100. Note that $\epsilon = 0$ denotes clean accuracy (no attack), all other models face attacks by saliency maps and have lower accuracies.

To study the above claim, we conducted experiments to check if the negative of the saliency map corresponding to the ground truth class, i.e. $-\nabla_{\mathbf{x}}\Phi^{i^*}(\mathbf{x})$, can indeed be used as a direction to perturb input. We created adversarial examples by perturbing original examples as: $\mathbf{x} = \mathbf{x} - \epsilon \cdot \mathrm{sign}(\nabla_{\mathbf{x}}\Phi^{i^*}(\mathbf{x}))$ (using saliency maps) to attack a model. We attack a Resnet-10 model trained on CIFAR-10 and CIFAR-100 datasets, normally (Std.) and adversarially (Adv.) trained using 5-step PGD attack with 3 different magnitudes of l_∞-norm perturbations, $\epsilon \in \{2/255, 4/255, 8/255\}$. We report our findings in Fig. 2. The figure shows that, examples generated by perturbing an input in the direction of negative saliency decreases the model's accuracy to a reasonable extent, affecting models both models trained normally and adversarially. We now describe how we leverage this relationship to perform Saliency-based Adversarial Training (SAT).

3.3 Saliency-Based Adversarial Training: Algorithm

Algorithm 1 summarizes our methodology of using saliency maps for adversarial training. The saliency maps during training are obtained either through annotations provided in a dataset (such as bounding boxes or segmentation masks), or through a pre-trained model which is used only to get saliency maps. More details on obtaining saliency maps is discussed in Sect. 4.

We also observed in our studies that when training a network using adversarial training, during the initial phase of training when the weights are not optimal, the perturbations computed by the attack methods are random. But with training, as weights become optimal, they become more class-discriminative. Figure 1 illustrates this observation, where the perturbation is random in initial phases

---

Algorithm 1: Saliency-based Adversarial Training (SAT) Methodology

---

Input: Training Dataset D, Saliency Maps S, Model $\Phi(.\;;\;\theta)$, SAT
hyperparameter α, Learning rate η, Maximum l_∞ perturbation ϵ_0
Output: Optimal parameter θ^*
Initialize model parameters as $\theta = \theta^0$.
for $t \in \{1, 2, ..., n\}$ **do**

> Sample training data of size $B : \{(\mathbf{x}^i, y^i)\}$ from D.
> Pick out corresponding saliency maps : $\{\mathbf{s}^i\}$ from S.
> Calculate δ^{ti} for each $\mathbf{x}^i$ using Eq. 9.
> Perturb the input examples : $\mathbf{x}^i := \mathbf{x}^i + \epsilon_0 \cdot \delta^{ti}$.
> Perform clipping to keep $\mathbf{x}^i$ bounded : $\mathbf{x}^i := clip(\mathbf{x}^i)$.
> Update model parameters :
> $\theta^t := \theta^{t-1} - \eta \cdot \nabla_\theta \frac{1}{B} \sum_{i=1}^{B} \mathcal{L}\left(\Phi(\mathbf{x}^i, \theta^{t-1}), y^i\right)$

end

---

of training but eventually becomes class-discriminative as the model trains. We exploit this observation to complete our methodology. In order to mimic the above behavior of the perturbation over training, we choose the direction of perturbation in a stochastic manner. We choose the ith component $\delta^t[i]$ of perturbation δ^t at time t as:

$$\delta^t[i] = \begin{cases} \mathbf{z}[i], & \text{with probability } \alpha^t \\ -\mathbf{s}[i], & \text{with probability } 1 - \alpha^t \end{cases} \tag{9}$$

where $\mathbf{z} \in \{-1, 1\}^d$ is sampled randomly, and $0 < \alpha < 1$. During initial epochs of training, when α^t is close to 1, δ^t will be dominated by random values. However, as training proceeds and α^t starts diminishing, δ^t smoothly transitions to $-\mathbf{s}$ and will be influenced by the adversarial character of the saliency map. When additional annotations such as bounding boxes or segmentation masks are available in a dataset, our approach considers these as weak saliency maps for the methodology. Usually, bounding boxes or segmentation masks are available as single channel images. To use them in our algorithm, we concatenate them along channel dimension to get the required dimension as image. After this pre-processing, we generate the weak saliency, $\tilde{s}$ from bounding boxes or segmentation masks as:

$$\tilde{s}[i] = \begin{cases} 1, & \text{if i}^{\text{th}} \text{ pixel lies inside bbox or seg masks} \\ -1, & \text{otherwise} \end{cases} \tag{10}$$

As mentioned earlier, using these bounding boxes or segmentation masks as saliency maps allows us to perform adversarial training with no additional cost to compute perturbations, unlike existing adversarial training methods which can be compute-intensive.

4   Experiments and Results

In this section, we present our results using the proposed SAT method (Algorithm 1) on multiple datasets with different variations of saliency maps. We begin with describing the datasets, evaluation criteria and implementation details.

4.1   Experimental Setup

Datasets and Evaluation Criteria: We perform experiments on well-known datasets: CIFAR-10, CIFAR-100 [14], Tiny ImageNet [6] and FLOWER-17 [17]. We evaluate the *adversarial robustness* of trained models (model accuracy when perturbations from following attacks are provided as inputs, as done in all earlier related efforts - also called *adversarial accuracy*) using the popular and widely used PGD attack (described in Sect. 3) with 5 steps and 4 different levels of l_∞ perturbation $\epsilon_0 \in \{1/255, 2/255, 3/255, 4/255\}$. We also evaluate the robustness of our models against other attacks: TRADES [34] as well as uniform noise in Sect. 5. We trained all our models for 5 trials and observed minimal variations in the values. We hence report the mean value of our experiments in our tables.

Baselines: We compare our method with multiple baseline methods, including those that train adversarially, as well as those that don't train adversarially:

(i) Original Model: Model trained normally with no adversarial training; *(ii) PGD-AT:* Model trained adversarially using 5-step PGD attack with max l_∞ perturbation and $\epsilon = 8/255$, as in [16]; *(iii) TRADES-AT:* Model trained adversarially using TRADES attack with max l_∞ perturbation and $\epsilon = 8/255$, as in [34]; *(iv) Original + Uniform Noise:* Model trained normally with no adversarial training, but training data is perturbed with uniform noise sampled from $[\frac{-8}{255}, \frac{8}{255}]$ during training; *(v) PGD + Uniform Noise:* PGD-AT, described in (ii), training data is additionally perturbed with uniform noise sampled from $[\frac{-8}{255}, \frac{8}{255}]$ during training; *(vi) TRADES + Uniform Noise:* TRADES-AT, described in (iii), training data is additionally perturbed with uniform noise sampled from $[\frac{-8}{255}, \frac{8}{255}]$ during training.

The models are trained for 100 epochs with standard cross-entropy loss, minimized using Adam optimizer (learning rate $= 1e - 3$).

Implementation Details: As in Sect. 3, our method relies on being provided with a saliency map for the ground truth label of a given image, while training the model. For Tiny ImageNet and FLOWER-17, we simply use the bounding boxes and segmentation masks provided in the dataset (and left unexploited often) as 'weak' saliency maps to train the model using the proposed SAT method. This thus incurs no additional cost when compared to competing methods that perform adversarial training by generating perturbations (a costly operation when using methods such as PGD). For CIFAR-10 and CIFAR-100, since we do not have such information (bounding boxes/segmentation masks) provided, we obtain saliency maps using teacher networks. In particular, we train two teacher networks: Resnet-10 and Resnet-34 for this purpose. Our final model architecture in all these settings is a Resnet34 too.

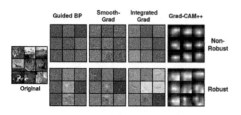

Fig. 3. Saliency maps of non-robust *(top, model trained normally)* and robust *(bottom, model trained adversarially)* variants of Resnet-10 with different explanation methods on CIFAR-10.

To go further, we also consider multiple variants of the above choices to study our method more carefully. It is believed that the quality of saliency maps generated by an adversarially trained model is better than its non-robust equivalent [28]. Figure 3 supports the above claim. We hence train the aforementioned teacher models in two ways: Standard (denoted as *Std* in the results) and Adversarial (denoted as *Adv* in results) (regular PGD-based adversarial training), and use the saliency maps for the ground truth class. The saliency maps themselves are generated in two ways: we use bounding boxes obtained using Grad-CAM++ [3] as weak saliency maps, and we use the finer saliency maps generated as is by Guided Back-propagation (GBP) [25] to train our student model. We set hyperparameters $\alpha = (0.6)^{\frac{1}{10}}$ and $\epsilon_0 = \frac{8}{255}$, and minimize cross-entropy loss using an Adam optimizer (learning rate $= 1e - 3$) for 200 epochs.

To complete the study, we also use variants of our method: *PGD-SAT* and *TRADES-SAT*, where we perturb the input randomly with either perturbations calculated by SAT or with PGD/TRADES attacks. All these variants are named in a self-explanatory manner in our results: $X - Y - Z$, where X denotes the teacher model (Resnet-10/-34), Y denotes the mode of training (Std or Adv), and Z denotes the saliency map method (GBP/ GradCAM++) used. In Sect. 5, we also perform additional studies with other saliency methods like Guided-GradCAM++, Smooth-Grad and Integrated-Gradients to study the generalizability of our results on varying quality of saliencies.

4.2   Results

Tables 1 and 2 present our results on CIFAR-10 + CIFAR-100 and Tiny ImageNet + Flower-17 datasets respectively. We note again that the provided bounding boxes are used as saliency maps in Tiny ImageNet, and segmentation masks provided in Flower-17 are used as saliency maps. We note that with no additional cost of computing perturbations, we obtain improvement of 2-4% in robustness. Table 1 shows the results for CIFAR-10 and CIFAR-100, and demonstrates the potential of using saliency maps for adversarial training. Barring 1–2 cases, the proposed SAT (or its combination with an adversarial training method such as PGD or TRADES) performs better than respective baselines across the results. As the adversarial attack gets stronger (larger ϵ), adding SAT to existing methods significantly improves robustness performance (1–3% in adversarial accuracy).

Table 1. Results on CIFAR-10 and CIFAR-100 using saliency maps from different teachers and explanation methods. $\epsilon \in \{1/255, 2/255, 3/255, 4/255\}$ denotes the maximum l_∞ perturbation allowed in 5-step PGD attack (More the ϵ, stronger the attack). GBP: Guided-Backpropagation; G.CAM++: Grad-CAM++.

Method	CIFAR-10				CIFAR-100			
	$\epsilon = 1/255$	$\epsilon = 2/255$	$\epsilon = 3/255$	$\epsilon = 4/255$	$\epsilon = 1/255$	$\epsilon = 2/255$	$\epsilon = 3/255$	$\epsilon = 4/255$
Original	47.71	10.36	1.39	0.28	25.83	7.76	3.35	1.94
Original + Uniform-Noise	61.23	22.85	6.34	2.56	33.15	13.50	6.01	3.22
SAT (Weak saliency)								
Resnet-10—Std.—G.CAM++	59.61	22.12	6.16	1.77	31.98	11.93	5.48	2.89
Resnet-10—Adv.—G.CAM++	57.34	19.94	5.64	1.91	32.88	13.3	6.31	4.0
Resnet-34—Std.—G.CAM++	56.75	20.62	6.02	1.74	30.95	12.68	5.65	3.25
Resnet-34—Adv.—G.CAM++	60.0	22.94	6.58	1.96	32.96	12.07	5.2	3.04
SAT (Fine saliency)								
Resnet-10—Std.—GBP	10.87	0.95	0.0	0.0	20.53	7.52	3.5	2.12
Resnet-10—Adv.—GBP	**63.33**	**26.79**	**9.62**	**3.69**	34.29	**14.73**	6.84	4.22
Resnet-34—Std.—GBP	18.01	2.54	1.25	1.0	9.91	2.52	1.05	0.54
Resnet-34—Adv.—GBP	62.67	23.76	5.82	1.28	**34.71**	14.32	**7.08**	**4.29**
PGD	77.89	73.1	66.96	61.18	45.75	40.13	35.41	31.01
PGD + Uniform-Noise	**82.23**	74.97	65.61	55.04	**42.67**	36.10	30.64	
PGD-SAT								
Resnet-10—Std.—GBP	79.72	73.81	67.72	61.29	46.87	41.11	35.77	30.75
Resnet-10—Adv.—GBP	80.72	**75.07**	68.68	62.49	48.33	42.2	36.33	**31.66**
Resnet-34—Std.—GBP	79.53	74.2	68.19	62.48	46.66	40.95	35.73	30.83
Resnet-34—Adv.—GBP	80.15	74.60	68.47	62.53	47.38	42.0	**36.34**	31.53
Resnet-10—Adv.—G.CAM++	79.67	74.05	68.12	61.84	47.28	41.72	35.99	31.05
Resnet-34—Adv.—G.CAM++	79.74	74.5	**68.87**	**62.68**	46.12	40.47	35.31	30.81
TRADES	**84.0**	73.25	59.79	47.03	47.21	42.2	37.03	32.96
TRADES + Uniform-Noise	81.69	74.96	67.43	60.05	**51.90**	42.85	37.30	31.71
TRADES-SAT								
Resnet-10—Std.—GBP	80.15	75.2	69.0	63.2	48.63	42.91	37.79	33.19
Resnet-10—Adv.—GBP	80.65	75.38	69.28	**63.46**	48.74	42.99	37.83	33.23
Resnet-34—Std.—GBP	79.98	74.5	68.56	62.43	48.5	42.41	36.93	32.01
Resnet-34—Adv.—GBP	80.26	74.87	68.75	62.74	48.76	42.83	37.36	32.87
Resnet-10—Adv.—G.CAM++	79.85	74.61	69.07	63.2	48.99	43.05	37.4	32.84
Resnet-34—Adv.—G.CAM++	83.17	**77.18**	**70.27**	62.87	49.35	**43.62**	**38.53**	**33.93**

It is also evident from Table 1 that using finer saliency maps (GBP in our results) obtains better performance than weaker saliency maps (bounding boxes obtained from GradCAM++, in our case). This supports the inference that a stronger saliency map provides better adversarial robustness. We also notice, as pointed out earlier in this section (Fig. 3), that using saliency maps obtained from an adversarially trained teacher leads to significant gains in robustness/adversarial accuracy performance. In other words, all our experiments point to the inference that better the saliency maps, better the adversarial robustness of the model trained using our approach. This further supports our inherent claim that saliency maps do provide adversarial robustness.

Table 1 also shows that PGD-SAT and TRADES-SAT leads to improvement (1.5–3%) over vanilla PGD and TRADES adversarial training. This shows that adding our saliency-based method to existing adversarial training method further improves their performance. We observe here again that using saliency maps of

Table 2. Results on Tiny-Imagenet and Flower-17 datasets where bounding boxes and segmentation masks provided with the dataset are used as saliency maps, respectively. $\epsilon \in \{1/255, 2/255, 3/255\}$ denotes the maximum l_∞ perturbation allowed in 5-step PGD attack (More the ϵ, stronger the attack).

Method	Tiny-Imagenet			FLOWER-17		
	$\epsilon = 1/255$	$\epsilon = 2/255$	$\epsilon = 3/255$	$\epsilon = 1/255$	$\epsilon = 2/255$	$\epsilon = 3/255$
Original	1.04	0.4	0.0	63.2	48.01	34.2
Original + Uniform-Noise	9.45	2.32	0.77	64.56	50.43	36.2
SAT	**9.79**	**2.46**	**0.77**	**66.17**	**52.94**	**38.93**
PGD	18.91	14.34	11.37	72.38	70.4	70.3
PGD + Uniform-Noise	19.57	15.49	11.66	73.52	72.79	72.71
PGD-SAT	**20.56**	**16.38**	**12.91**	**78.67**	**75.73**	**75.00**
TRADES	18.45	16.76	11.09	74.56	73.89	73.67
TRADES + Uniform-Noise	19.96	16.13	12.58	76.47	74.26	74.0
TRADES-SAT	**20.04**	**16.45**	**12.96**	**79.41**	**77.94**	**77.20**

an adversarially robust teacher model proves more useful than those obtained from a normal teacher model.

Time Efficiency: We further analyzed the training efficiency of the proposed method. Figure 4 reports the average time taken by one epoch over 10 trials. As can be seen, PGD-SAT and TRADES-SAT require only ≈50–70% of training time when compared to vanilla PGD and TRADES respectively, and at the same time, achieves superior performance (as in Table 2). In case of vanilla SAT, the behavior is desirable since we observe an increase in robustness without compromising much in training time. We assume in

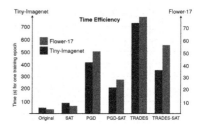

Fig. 4. Training time for one epoch in seconds (averaged over 10 trials) for different methods considered in our results on Tiny-Imagenet and Flower-17.

these results that explanations are provided to us, inspiring the demand for finer saliency map annotations to be included in vision datasets to the community. Similar results for CIFAR-10 and CIFAR-100 are included in the appendix.

5   Discussions, Ablation Studies and Conclusions

We carried out ablation studies to better characterize the efficacy of our methodology, and present these results in this section. Unless explicitly specified, these experiments are carried out on CIFAR-100 using GuidedBackprop and Grad-CAM++ explanations of a standard and adversarially trained Resnet-10 teacher.

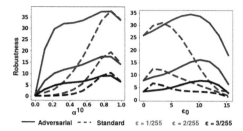

Fig. 5. Variation of hyper-parameter α and ϵ_0 on CIFAR-100 dataset.

Varying Hyperparameters: We studied the effect on the adversarial robustness of our trained model by varying the hyperparameters α and ϵ_0 in our method. Figure 5 shows these results, where all models are evaluated using a 5-Step PGD attack with l_∞ norm and maximum perturbation $\epsilon \in \{\frac{1}{255}, \frac{2}{255}, \frac{3}{255}\}$. We achieve higher robustness when α^{10} is set closer to 1 (α close to 1), which from Eq. 9 indicates that the saliency map is useful to obtain perturbations in later stages of training, as explained before. Interestingly, when α^{10} is close to 0 (α close to 0), the training doesn't include any noise factor, and since we have a fixed saliency map for each image, it becomes equivalent to training on an additively shifted version of the original training data resulting in a less robust model. While varying ϵ_0 in Algorithm 1, we observe a peak in robustness somewhere in the middle of the considered range of values. In Algorithm 1, the distribution of our estimated perturbations and actual adversarial perturbations depend on the hyperparameter ϵ_0. At $\epsilon_0 = 0$, the distributions are identical, and the distributions diverge as ϵ_0 increases. Hence, at high values of ϵ_0, when both distributions diverge, the perturbations used by SAT no longer resemble true adversarial perturbations, hence resulting in less robust models (when attacked adversarially using methods associated with the adversarial perturbations). In other words, the perturbations provided by saliency maps in our method has reasonable conjunction with standard adversarial perturbations, as shown in Fig. 2.

Robustness against Other Attacks: In Table 1, we evaluate our models using the widely used PGD attack. We now study how the proposed method works against other attacks - in particular, Uniform Noise attack, TRADES [34] attack and a Saliency-based adversarial attack, which is an attack using saliency maps generated by our method. These results are shown in Table 4. It is clear that SAT and PGD-SAT outperform or have comparable performance to (in case of Uniform Noise) competing methods. Expectedly, the improvement is much more significant in case of Saliency-based adversarial attack, since the method is closely associated with the attack.

Using Other Saliency Maps: We also performed a study where we trained our models using stronger saliency maps obtained using methods such as Smooth-Grad [24], Guided Grad-CAM++ [3] and Integrated Gradients [26] from teacher models. These models are evaluated using a 5-Step PGD attack with l_∞ max-perturbation $\epsilon \in \{\frac{1}{255}, \frac{2}{255}, \frac{3}{255}, \frac{4}{255}\}$. As can be seen from Table 3, we tend to achieve more robust models for better saliency maps.

Table 3. Results using saliency maps from Guided-Backprop (GBP), Smooth-Grad (S.Grad), Integrated-Gradients (I.Grad), and Guided-Grad-CAM++ (G.G.CAM++)

Method	PGD			
	$\frac{1}{255}$	$\frac{2}{255}$	$\frac{3}{255}$	$\frac{4}{255}$
Original	25.83	7.76	3.35	1.94
Original + Uniform-Noise	33.15	13.50	6.01	3.22
SAT				
Resnet-10—Std.—GBP	20.53	7.52	3.5	2.12
Resnet-10—Std.—S.Grad	**39.22**	**19.89**	**9.44**	**4.49**
Resnet-10—Std.—G.G.CAM++	21.46	8.00	3.53	2.23
Resnet-10—Std.—I.Grad	36.2	5.43	7.28	3.37
Resnet-10—Adv.—GBP	34.29	14.73	6.84	4.22
Resnet-10—Adv.—S.Grad	**40.01**	**21.2**	**10.96**	**4.85**
Resnet-10—Adv.—G.G.CAM++	34.07	13.18	5.85	3.09
Resnet-10—Adv.—I.Grad	37.56	16.45	7.55	4.31

Table 4. Results against other attacks: Uniform-Noise, TRADES and Saliency. *(GBP= Guided-Backprop; G.CAM++=Grad-CAM++; l_∞ perturbation $\in [-\epsilon, \epsilon]$.)*

Method	Uniform-noise			TRADES			Saliency Attack		
	$\epsilon = \frac{8}{255}$	$\epsilon = \frac{16}{255}$	$\epsilon = \frac{32}{255}$	$\epsilon = \frac{2}{255}$	$\epsilon = \frac{3}{255}$	$\epsilon = \frac{4}{255}$	$\epsilon = \frac{4}{255}$	$\epsilon = \frac{8}{255}$	$\epsilon = \frac{16}{255}$
Original	62.21	47.88	21.17	39.59	35.1	33.45	13.03	6.86	3.92
Original + Uniform-Noise	**67.42**	**60.75**	28.9	**46.98**	39.37	37.8	19.63	10.48	6.1
SAT									
Resnet-10—Std—GBP	56.63	57.4	**46.9**	36.45	33.35	32.46	51.78	49.02	42.75
Resnet-10—Adv.—GBP	66.63	57.93	21.92	45.93	**40.80**	**37.86**	62.99	56.48	43.49
Resnet-10—Adv.—G.CAM++	64.00	54.84	23.43	44.31	38.28	35.71	**63.86**	**63.87**	**63.66**
PGD	50.96	51.0	48.69	49.03	47.63	45.34	44.03	34.77	21.04
PGD + Uniform-Noise	**55.6**	52.45	49.20	**52.32**	48.76	45.87	45.72	32.81	16.62
PGD-SAT									
Resnet-10—Std—G.BP	52.78	52.56	49.21	50.43	47.69	46.24	47.25	46.74	44.95
Resnet-10—Adv.—G.BP	54.38	**54.13**	**50.30**	51.58	**49.15**	**46.83**	50.61	50	47.64
Resnet-10—Adv.—G.CAM++	53.30	52.76	49.74	50.87	48.42	46.31	**50.69**	**50.63**	**50.54**

Conclusion. In summary, this work explores the interesting connection between saliency maps and adversarial robustness to propose a Saliency based Adversarial training (SAT) method. SAT imitates adversarial training by using saliency maps to mimic adversarial perturbations. In particular, our methodology allows the use of annotations such as bounding boxes and segmentation masks to be exploited as weak explanations to improve model's robustness with no additional computations required to compute the perturbations themselves. Our results on CIFAR-10, CIFAR-100, Tiny ImageNet and Flowers-17 corroborate our claim. We further gain improvement over popular adversarial training methods by integrating SAT with them (PGD-SAT and TRADES-SAT). Further, our work shows how using better saliency maps leads to more robust models. Our effort opens rather a new direction to enhance robustness of DNNs by exploiting saliency maps, and inspires the need for strong saliency maps to be provided with vision datasets, which helps train adversarially robust models with little overhead. Our future work includes ways to improve SAT by reasoning about the class closest to the ground truth in terms of decision boundary, and improving our estimates of adversarial perturbations.

References

1. Adebayo, J., Gilmer, J., Goodfellow, I.J., Kim, B.: Local explanation methods for deep neural networks lack sensitivity to parameter values. CoRR (2018)
2. Chan, A., Tay, Y., Ong, Y.S., Fu, J.: Jacobian adversarially regularized networks for robustness. In: ICLR 2020 (2020)
3. Chattopadhyay, A., Sarkar, A., Howlader, P., Balasubramanian, V.N.: Grad-CAM++: generalized gradient-based visual explanations for deep convolutional networks. In: WACV 2018 (2018)
4. Chen, J., Wu, X., Rastogi, V., Liang, Y., Jha, S.: Robust attribution regularization. In: NeuRIPS 2019 (2019)
5. Cisse, M., Bojanowski, P., Grave, E., Dauphin, Y., Usunier, N.: Parseval networks: improving robustness to adversarial examples. In: ICML 2017 (2017)
6. CS231N, S.: Tiny ImageNet visual recognition challenge. https://tiny-imagenet. herokuapp.com/
7. Dombrowski, A.K., Alber, M., Anders, C.J., Ackermann, M., Müller, K.R., Kessel, P.: Explanations can be manipulated and geometry is to blame. In: NeuRIPS 2019 (2019)
8. Etmann, C., Lunz, S., Maass, P., Schönlieb, C.B.: On the connection between adversarial robustness and saliency map interpretability. In: ICML 2019 (2019)
9. Geirhos, R., Rubisch, P., Michaelis, C., Bethge, M., Wichmann, F.A., Brendel, W.: ImageNet-trained CNNs are biased towards texture; increasing shape bias improves accuracy and robustness. In: ICLR 2019 (2019)
10. Ghorbani, A., Abid, A., Zou, J.: Interpretation of neural networks is fragile. In: AAAI 2019 (2019)
11. Goodfellow, I., Shlens, J., Szegedy, C.: Explaining and harnessing adversarial examples. In: ICLR 2015 (2015)
12. Ignatiev, A., Narodytska, N., Marques-Silva, J.: On relating explanations and adversarial examples. In: NeuRIPS 2019 (2019)
13. Kindermans, P.J., et al.: Learning how to explain neural networks: PatternNet and PatternAttribution. In: ICLR 2018 (2018)
14. Krizhevsky, A.: Learning multiple layers of features from tiny images (2009)
15. Lapuschkin, S., Binder, A., Montavon, G., Klauschen, F., Müller, K.R., Samek, W.: On pixel-wise explanations for non-linear classifier decisions by layer-wise relevance propagation (2015)
16. Madry, A., Makelov, A., Schmidt, L., Tsipras, D., Vladu, A.: Towards deep learning models resistant to adversarial attacks. In: ICLR 2018 (2018)
17. Nilsback, M., Zisserman, A.: A visual vocabulary for flower classification. In: CVPR 2006 (2006)
18. Ribeiro, M.T., Singh, S., Guestrin, C.: "why should i trust you?": explaining the predictions of any classifier. In: ACM SIGKDD 2016 (2016)

19. Samangouei, P., Kabkab, M., Chellappa, R.: Defense-GAN: protecting classifiers against adversarial attacks using generative models. In: ICLR 2018 (2018)
20. Selvaraju, R.R., Das, A., Vedantam, R., Cogswell, M., Parikh, D., Batra, D.: Grad-CAM: why did you say that? Visual explanations from deep networks via gradient-based localization. In: ICCV 2017 (2017)
21. Shafahi, A., et al.: Adversarial training for free! In: NeuRIPS 2019 (2019)
22. Shrikumar, A., Greenside, P., Kundaje, A.: Learning important features through propagating activation differences. In: ICML 2017 (2017)
23. Sinha, A., Singh, M., Kumari, N., Krishnamurthy, B., Machiraju, H., Balasubra-manian, V.N.: Harnessing the vulnerability of latent layers in adversarially trained models. In: IJCAI 2019 (2019)
24. Smilkov, D., Thorat, N., Kim, B., Viégas, F.B., Wattenberg, M.: SmoothGrad: removing noise by adding noise. CoRR (2017)
25. Springenberg, J., Dosovitskiy, A., Brox, T., Riedmiller, M.: Striving for simplicity: the all convolutional net. In: ICLR (workshop track) (2015)
26. Sundararajan, M., Taly, A., Yan, Q.: Axiomatic attribution for deep networks. In: ICML 2017 (2017)
27. Szegedy, C., Zaremba, W., Sutskever, I., Bruna, J., Erhan, D., Goodfellow, I., Fergus, R.: Intriguing properties of neural networks. In: ICLR 2014 (2014)
28. Tsipras, D., Santurkar, S., Engstrom, L., Turner, A., Madry, A.: Robustness may be at odds with accuracy. In: ICLR 2019 (2019)
29. Wagner, J., Mathias Köhler, J., Gindele, T., Hetzel, L., Thaddäus Wiedemer, J., Behnke, S.: Interpretable and fine-grained visual explanations for convolutional neural networks. In: CVPR 2018 (2018)
30. Xiao, C., Zhu, J.Y., Li, B., He, W., Liu, M., Song, D.: Spatially transformed adversarial examples. In: ICLR 2018 (2018)
31. Xie, C., Wu, Y., van der Maaten, L., Yuille, A.L., He, K.: Feature denoising for improving adversarial robustness. In: CVPR 2019 (2019)
32. Zeiler, M.D., Fergus, R.: Visualizing and understanding convolutional networks. In: Fleet, D., Pajdla, T., Schiele, B., Tuytelaars, T. (eds.) ECCV 2014. LNCS, vol. 8689, pp. 818–833. Springer, Cham (2014). https://doi.org/10.1007/978-3-319-10590-1_53
33. Zhang, D., Tianyuan, Z., Lu, Y., Zhu, Z., Dong, B.: You only propagate once: painless adversarial training using maximal principle. In: NeuRIPS 2019 (2019)
34. Zhang, H., Yu, Y., Jiao, J., Xing, E.P., Ghaoui, L.E., Jordan, M.I.: Theoretically principled trade-off between robustness and accuracy. In: ICML 2019 (2019)
35. Zhang, J., Lin, Z., Brandt, J., Shen, X., Sclaroff, S.: Top-down neural attention by excitation backprop. In: Leibe, B., Matas, J., Sebe, N., Welling, M. (eds.) ECCV 2016. LNCS, vol. 9908, pp. 543–559. Springer, Cham (2016). https://doi.org/10.1007/978-3-319-46493-0_33
36. Zhang, T., Zhu, Z.: Interpreting adversarially trained convolutional neural networks. In: ICML 2018 (2018)
37. Zhou, B., Khosla, A., Lapedriza, À., Oliva, A., Torralba, A.: Learning deep features for discriminative localization. In: CVPR 2016 (2016)

Scalable Backdoor Detection in Neural Networks

Haripriya Harikumar[1,2]([✉]), Vuong Le[1], Santu Rana[1],
Sourangshu Bhattacharya[3], Sunil Gupta[1], and Svetha Venkatesh[1]

[1] Applied Artificial Intelligence Institute, Deakin University, Waurn Ponds, Australia
{h.harikumar,vuong.le,santu.rana,sunil.gupta,
svetha.venkatesh}@deakin.edu.au
[2] Institute for Health Transformation, Deakin University, Waurn Ponds, Australia
[3] Department of Computer Science and Engineering, IIT Kharagpur,
Kharagpur, India
sourangshu@cse.iitkgp.ac.in

Abstract. Recently, it has been shown that deep learning models are vulnerable to Trojan attacks. In the Trojan attacks, an attacker can install a backdoor during training to make the model misidentify samples contaminated with a small trigger patch. Current backdoor detection methods fail to achieve good detection performance and are computationally expensive. In this paper, we propose a novel trigger reverse-engineering based approach whose computational complexity does not scale up with the number of labels and is based on a measure that is both interpretable and universal across different networks and patch types. In experiments, we observe that our method achieves a perfect score in separating Trojan models from pure models, which is an improvement over the current state-of-the-art method.

Keywords: Trojan attack · Backdoor detection · Deep learning model · Optimisation

1 Introduction

Deep learning has transformed the field of Artificial Intelligence by providing it with an efficient mechanism to learn giant models from the large training dataset, unlocking often human-level cognitive performance. By nature, the deep learning models are massive, have a large capacity to learn, and are effectively black-box when it comes to its decision-making process. All of these properties have made them vulnerable to various forms of malicious attacks [14,18]. The most sinister among them is the Trojan attack, first demonstrated in [5] by Gu et al. They showed that it is easy to insert backdoor access in a deep learning model by poisoning its training data so that it predicts any image as the attacker's intended class label when it is tagged with a small, and inconspicuous trigger patch. This can be achieved even without hurting the performance of a model on

© Springer Nature Switzerland AG 2021
F. Hutter et al. (Eds.): ECML PKDD 2020, LNAI 12458, pp. 289–304, 2021.
https://doi.org/10.1007/978-3-030-67661-2_18

the trigger-free clean images. When such compromised models are deployed then the backdoor can remain undetected until it encounters the poisoned data. Such Trojans can make using deep learning models problematic when the downside risk of misidentifications is high, e.g. an autonomous car misidentifying a stop sign as a speed limit sign can cause accidents that result in loss of lives (Fig. 1). Hence, we must develop methods to reliably screen models for such backdoors before they are deployed.

Fig. 1. A Trojaned autonomous car mistaking a STOP sign as a 100 km/h speed limit sign due to the presence of a small sticker (trigger) on the signboard.

The Trojan attack is different than the more common form of adversarial perturbation based attacks [7,12,15]. In adversarial perturbation based attacks, it has been conclusively shown that given a deep learning model there always exists an imperceptibly small image specific perturbation that when added can make the image be classified into a wrong class. This is an intrinsic property of any deep models and the change, that is required for misclassification, is computed via an optimisation process using a response from the deep model. In contrast, the Trojan attack does not depend on the content of an image, hence, the backdoor can be activated even without access to the deployed models and without doing image specific optimisation. The backdoor is planted at the training time, and is thus not an intrinsic property, implying that it is plausible to separate a compromised model from an uncompromised one. Recent research on detection of backdoor falls into three major categories a) anomaly detection based approach to detect neurons that show abnormal firing patterns for clean samples [1,2,13], and b) by learning the intrinsic difference between the response of compromised models and uncompromised models using a meta-classifier [10, 22], and c) optimisation based approach to reverse-engineer the trigger [3,20,21]. In our opinion, the approach (c) is the most feasible one since it is not limited by activation patterns of internal neurons or by a reliable meta-classifier, however, in our testing we found the detection performance of existing methods in this category to be inadequate. Moreover, i) these methods rely on finding possible patch per class and hence, not computationally scalable for a dataset with a large number of classes, and ii) the score they use to separate the compromised models are patch size-dependent and thus, not practically feasible as we would not know the patch size in advance.

In this paper, we take a fresh look in the optimisation based approach and produce a scalable detection method for which 1) computational complexity does not grow with the number of labels, and 2) the score used for Trojan screening is computable for a given setting without the need of any information about the trigger patch used by the attacker. We achieve (1) by observing the fact that a trigger is a unique perturbation that makes any image to go to a single class label, and hence, instead of seeking the change to classify all the images to a target label, we seek the change that would make prediction vectors for all the images to be similar to each other. Since we do not know the Trojan label, existing methods have to go through all the class labels one by one, setting each as a target label and then performing the trigger optimisation, whereas, we do not need to do that. We achieve (2) by computing a score which is the entropy of the class distribution in the presence of the recovered trigger. We show that it is possible to compute an upper bound on the scores of the compromised models based on the assumption about the effectiveness of the Trojan patch and the number of class labels. The universality of the score is one of the key advantages of our proposed method.

Additionally, through our Trojan scanning procedure, we are the first to analyze the behavior of the CNN models trained on perturbed training datasets. We discovered an intriguing phenomenon that the set of effective triggers are not uniquely distributed near the originally intended patch, but span in a complex mixture distribution with many extra unintentional modes.

We perform extensive experiments on two well-known datasets: German Traffic Sign Recognition dataset and CIFAR-10 dataset and demonstrate that our method finds Trojan models with perfect precision and recall, a significant boost over the performance of the state-of-the-art method [20].

2   Related Work

The possibility of the Trojan attack on deep learning models was first exposed in a pioneering paper by Gu et al. [5] in 2016. In [5], the authors used the well-known techniques of watermarking for Trojan implantation and aptly named the resultant network as Badnet. Following that, several research papers revealed the variety of ways such attacks can be realised for many different kinds of AI systems [8]. In parallel to this pursuit of mapping the threat landscape, researchers also scrambled to build methodologies for the detection and mitigation of these Trojan attacks. Most of such efforts concentrated on deep learning based image classifiers because of their high-profile usage in autonomous navigation and public security.

The existing detection approaches for deep learning based image classifiers can be divided into two main categories: A) detect if the input is Trojaned, and B) detect if the deep network is Trojan. The former was pioneered by Gao et al. [4] through their method STRIP. It exploits the fact that whilst for clean images, a prediction is dependent on the content of the image, for poisoned images, the prediction is dependent only on the trigger. Therefore, if the content of a clean image is perturbed then its prediction will change, but such perturbations would not affect the prediction of poisoned images. Whilst the general principle is reasonable, it is not obvious how to change only the content part of the poisoned image, without affecting the trigger. In a similar direction, Huang et al. [6] propose to use a saliency map to detect the difference in the explanation between clean images and poisoned images. However, they assume an unrealistic human-in-the-loop detection procedure to judge whether an explanation is semantically linked to the content or not for every incoming image.

Methods to detect Trojans in a deep network can be classified into three main categories 1) trigger reverse-engineering based, 2) anomaly detection in neural activation based, and 3) using meta-classifiers. A general assumption in this category of detection is that at the detection side there is an availability of a *detection dataset* of a sample set of images from the same distribution as that of the original training dataset. The assumption is not unreasonable as users of AI models tend to have a small curated dataset to check the performance of the model before actual deployment. For category 1, the most prominent work, and also the closest to our work, is the method Neural Cleanse by Wang et al. [20]. The method contains several sub-optimisation problems, one per label, to find the smallest change that would transform the class labels of all the images in the detection dataset to that label, and then investigating whether the smallest change is an anomaly. If so, the network is declared as Trojaned. They also assume that the trigger can be as large as possible. Unfortunately, they have a high false alarm rate because when used for a clean network, one of the changes found might be very different from others, and thus become anomalous. Also, because the size of the trigger is not constrained and one optimisation problem is solved per label, the algorithm becomes computationally expensive and infeasible for a large number of labels. DeepInspect [3] by Chen et al. provides an alternative by learning a distribution over possible triggers using a GAN, but it is not clear why an adversary would use that trigger distribution since the poisoned network would be very different to that of the clean network.

For the category 2, the hypothesis is that the activation patterns of the neurons in the Trojaned network would behave uniquely for poisoned images, and quite different from that of the clean images of the target label. The work in [2], [13], and [19] exploited this principle using activation clustering, implementing artificial brain simulation and analysing spectral signatures, respectively. However, these methods are infeasible for models that do not give access to the neural activations. In the category 3, the main works are [10], and [22], where they use signature behaviours from both Trojaned and non-Trojaned models to build a meta-classifier for later application to detect Trojaned models. Unfortunately, it is not clear how it would generalise to different problem domains and model settings.

Methods to mitigate Trojans have also been studied, but are comparatively less. In [20], Wang et al. examined two ways to mitigate Trojans, 1) cutting down the neurons that are most activated by the reverse-engineered trigger and 2) re-training the network with poisoned samples, but with correct labels. They found that the latter method provides better performance. Other methods also use some variants of these two approaches. However, none of them provides any guarantee for the cleansed model on the level of dis-infection achieved.

3 Framework

3.1 Adversarial Model

A Trojan attack consists of two key elements: 1) a trigger patch, and 2) a target class. The trigger is an alteration of the data that makes the classifier to classify the altered data to the target class. We assume triggers to be overlays that are put on top of the actual images, and the target class to be one of the known classes. We assume a threat model that is congruent with the physical world constraints. An adversary would accomplish the intended behaviour by putting a sticker on top of the images. The sticker can be opaque or semi-transparent. The latter resembling a practice of using a transparent "plastic" sticker on the real object such as a traffic sign.

Formally, an image classifier can be defined as a parameterized function, $f_\theta : \mathcal{I} \to \mathbb{R}^C$ that recognizes class label probability of image $I \in \mathcal{I}$- the space of all possible input images. θ is a set of parameters of f learnable using a training dataset such as neural network weights. Concretely, for each $I \in \mathcal{I}$, we get output $c = f_\theta(I)$ is a real vector of C dimension representing the predicted probability of the C classes that I may belong to. In a neural network, such output usually comes from a final softmax layer.

In Trojan attacks, a couple of perturbations can happen. Firstly, model parameters θ are replaced by malicious parameters θ', and secondly, the input image I is contaminated with a visual trigger. A trigger is formally defined as a small square image patch ΔI of size s that is either physically or digitally overlaid onto the input image I at a location (x^*, y^*) to create a modified image I'. The goal of these perturbations is to make the recognized label $f_{\theta'}(I')$ goes to the target class with near certainty regardless of the actual class label of the original image.

Concretely, an image of index k of the dataset I_k is altered into an image I'_k by

$$
I'_k(x,y) = \begin{cases} (1 - \alpha(x',y'))I_k(x,y) + \alpha(x',y')\Delta I(x',y') & \text{if } x \in [x^*, x^* + s], \\ & y \in [y^*, y^* + s] \\ I_k(x,y) & \text{elsewhere} \end{cases}
$$

(1)

where (x', y') denote the local location on the patch $(x', y') = (x - x^*, y - y^*)$.

In this operation, the area inside the patch is modified with weight α determining how opaque the patch is. This parameter can be considered as a part of the patch, and from now on will be inclusively mentioned as ΔI. Meanwhile, the rest of the image is kept the same. In our setting, (x^*, y^*) can be at any place as long as the trigger patch stays fully inside the image. An illustration of this process is shown in Fig. 2.

Fig. 2. An illustration of a trojan trigger ΔI of size s and opacity level α put on an image of a stop sign at location (x^*, y^*).

Also, we expect that the adversary would like to avoid pre-emptive early detection, and hence, would not use a visibly large patch. Additionally, we also assume that the adversary will also train the model such that trigger works irrespective of the location. So, during attack time the adversary does not need to be very precise in adding the sticker to the image. This aspect of our setting is very different from the current works which always assumed a fixed location for the trigger. We also use triggers that contain random pixel colors instead of any known patterns to evaluate the robustness of the trigger reverse-engineering process in the absence of any structure and smoothness in the trigger patch.

We assume that the attacker who generate the adversarial model and the authority who detects the model will have access to disjoint sets of the training dataset. We denote attacking dataset for the dataset which attacker has access to and scanning dataset for the dataset where the authority has access to. In this setup, the attacker uses the attacking dataset for Trojan model generation; while the authority uses the scanning set to detect the Trojan model. We also train the Trojan model with only a small number of Trojaned images so that its performance on the clean images is only marginally affected.

3.2 Problem Formulation

Our detection approach is based on trigger reverse engineering. We formulate an optimisation problem that when minimized provides us with the trigger patch used in a compromised model. The driving idea behind the formulation of the optimisation problem is that when the correct trigger is used, any corrupted image will have the same prediction vector i.e. $||f_{\theta'}(I'_j) - f_{\theta'}(I'_k)||_p \leq \epsilon$ for $\forall j, k$ in the scanning dataset, where ϵ is a small value. ϵ would be zero if the scanning dataset is the same as the training dataset. Hence, we can look at the prediction vectors of all the corrupted images in the scanning dataset, and the trigger that makes those prediction vectors the closest will be the original trigger.

It gets more tricky when the size of the trigger is not known, which is the real-world use case. We can approach that in two ways a) by solving the above optimisation problem with a grid search over s, starting from 1×1 patch to a large size that we think will be the upper bound on the size of the trigger that the attacker can use but remain inconspicuous, or b) jointly finding the trigger size and the trigger by regularizing on the size of the trigger. While the first approach is more accurate, it is more computationally expensive as well, since we have to solve many optimisation problems over ΔI, one each for one grid value of s. The second approach, while slightly misaligned due to the presence of an extra regularize, is not computationally demanding as we only have to solve one optimisation problem. Hence, we choose this as our approach.

Formally, the regularization based approach for unknown trigger size can be expressed as:

$$min_{\Delta I, \alpha} \sum_{j=1}^{N} \sum_{k=1, k \neq j}^{N} ||f_{\theta'}(I_j') - f_{\theta'}(I_k')||_2 + \lambda ||\alpha||_1 \qquad (2)$$

where s is set at the largest possible trigger size (s_{max}), α is the matrix that contains the transparency values for each pixel for ΔI, and λ is the regularization weight. Both ΔI and α are of the size $s_{max} \times s_{max}$. Ideally, for a correct trigger which is smaller than s_{max} we would expect a portion of the recovered ΔI to match with the ground truth trigger patch with the corresponding α values matching to the ground truth transparency, with the rest of ΔI having $\alpha = 0$. Hence, 1-norm on the matrix α is used as the regularizer. The choice of 2-norm for the first part is to keep the loss function smooth. However, other metric or divergence measure that keeps the smoothness of the loss function can also be used.

When the size of the trigger (s) is assumed to be known we can reverse engineer the trigger patch by solving the following optimisation problem. So the Eq. 2 reduces to:

$$\Delta I = min_{\Delta I} \sum_{j=1}^{N} \sum_{k=1, k \neq j}^{N} ||f_{\theta'}(I_j') - f_{\theta'}(I_k')||_2 \qquad (3)$$

where both I_j' and I_k' are functions of the trigger ΔI (Eq. 1), and N is the size of the scanning dataset. Note that we have per-pixel transparency while Eq. 3 only had the same transparency value for all the pixels and assumed to be 1.0.

Since both of the above formulations do not need to cycle through one class after another, the computation does not scale up with the number of classes, hence, it is efficient than the current state-of-the-art method. Hence, we name our method as Scalable Trojan Scanner (STS).

3.3 Optimisation

The optimisation problems in Eqs. 3 and 2 is high dimensional in nature. For example, for GTSRB and CIFAR-10 to solve Eq. 2 we have to optimize a patch and mask which is of the same size as the image, $\Delta I \in \mathbb{R}^{32 \times 32 \times 3}$, and $\alpha \in \mathbb{R}^{32 \times 32}$. We solve this problem by setting a maximum yet reasonable size s_{max} an attacker can choose on both ΔI and α.

We consider all RGB pixels and the opacity level to have float values between 0.0 and 1.0. The deep learning optimization process naturally seeks unbound which can extend beyond this valid range. To put effective constraints on the optimisation, we restrict the search space of the optimisation variables within the range by using a clamping operator:

$$z = \frac{1}{2}(\tanh(z') + 1) \tag{4}$$

Here auxiliary variable z' is optimized and is allowed to reach anywhere on the real number range, while z is the corresponding actual bounded parameter we want to achieve which are ΔI and α. The parameters are then sought by using Adam optimizer [9].

Whilst a pure model does not have triggers inserted and ideally should not result in any solution, one may still wonder whether the complex function learned by a million parameter deep learning model would still naturally have a solution to those objective functions. Surprisingly, we observe in our extensive experimentation that such is not the case *i.e.* there does not exist a ΔI that makes the objective function values anywhere near to zero for pure models. Such a behaviour can be further enforced by having a large scanning dataset, as more and more images in the scanning dataset mean that the probability of the existence of a shortcut (i.e. perturbation by ΔI) from all the images to a common point in the decision manifold would be low for pure models. Such a shortcut is present in the Trojan model because they are trained to have that. Having a large scanning dataset is practically feasible as we do not need labels for the scanning dataset.

3.4 Entropy Score

We use the entropy score to identify the Trojan and pure models. We compute an entropy score as,

$$\text{entropy\_score} = -\sum_{i=1}^{C} p_i log_2(p_i) \tag{5}$$

where $\{p_i\}$ is the probability computed from the histogram for the classes predicted for the scanning dataset images when they are perturbed by the reverse-engineered trigger. For the actual trigger, the entropy would be zero since all the images would belong to the same class. However, due to the non-perfectness of the Trojan effectiveness, we will have a small but non-zero value. For pure models, the entropy will be high as we would expect a sufficient level of class diversity in the scanning dataset. The following lemma provides a way to compute an upper bound on the value of this score for the Trojan models in specific settings, which then can be used as a threshold for classification.

Lemma 1. *If the accuracy of Trojan model on Trojan patched data is assumed to be at least $(1 - \delta)$, where $\delta \ll 1$, and there are C different classes in the* *scanning dataset then there exists an upper bound on the entropy score of the Trojan models as*

$$\text{entropy\_score} \leq -(1-\delta) * log_2(1-\delta) - \delta * log_2(\frac{\delta}{C-1})$$

Proof. The proof follows from observing that the highest entropy of class distribution in this setting happens when $(1-\delta)$ fraction of the images go to the target class c_t and the rest δ fraction of the images gets equally distributed in the remaining $(C-1)$ classes.

This entropy score is independent of the type of patches used and is universally applicable. The threshold is also easy to compute once we take an assumption about the Trojan effectiveness of the infected classifier and know the number of classes. The score is also interpretable and can allow human judgment for the final decision. The proposal of this score is a unique advantage of our work.

4   Experiments

4.1   Experiment Settings

We arrange the experiments to evaluate the effectiveness of the Trojan detector using two image recognition dataset: German Traffic Sign Recognition Benchmark (GTSRB) [17] and CIFAR-10 [11].

GTSRB has more than 50 K dataset of 32×32 colour images of 43 classes of traffic signs. The dataset is aimed at building visual modules of autonomous driving applications. The dataset is pre-divided into 39.2 K training images and 12.6 K testing images.

CIFAR-10 has 60 K dataset of size 32×32 everyday colour images of classes *airplane, automobile, bird, cat, deer, dog, frog, horses, ship,* and *truck.* This dataset doesn't have a separate dataset for validation, so we generated the validation dataset from the test set. With both of the datasets, we assume a situation where the attacker and the authority have access to disjoint sets of data with a ratio of 7:3 called attacking dataset and scanning dataset.

For all of our experiments, we use a convolutional neural network (CNN) with two convolutional layers and two dense layers for the image classification task. Compared to the widely used architectures such as VGG-16 [16], this CNN is more suitable with the small image sizes and amounts.

4.2   Trojan Attack Configuration

From the side of the Trojan attacker, they aim to replace the *Pure Model* that was normally trained with *Pure Data* with their malicious *Trojan Model* trained with *Trojan Data.* The Trojan Data set is mixed between normal training data and images with the trigger patch embedded and labeled as target class. The ultimate objective is that when the Trojan Model is applied to a testing set of patched images, it will consistently return the target class. We measure the effectiveness of Trojan attack by a couple of criteria: (1) the accuracy of *Trojan Model on Trojan patched image data* (TMTD) is at least 99% and (2) the accuracy of *Trojan Model on Pure Data* (TMPD) deteriorates less than 2% compared to *Pure Model on Pure Data* (PMPD).

We aim at the realistic case of the trigger to be a small square patch resembling a plastic sticker that can be put on real objects. In our experiment, the patches are randomly selected with the sizes of 2×2, 4×4, 6×6, and 8×8 with various transparency levels.

We train the Trojan models with randomly selected triggers in the *Patch Anywhere* setting where the expectation is that the trigger work on any place on the image. We set the target class as *class 14* (stop sign) for GTSRB and *class 7* (horses) for the CIFAR-10 dataset. We choose *class 14* (stop sign) for GTSRB since we discussed the practicality of choosing *stop sign* in the introduction. The selection of *class 7* (horses) for CIFAR-10 is a random choice. During Trojan model training, we randomly choose 10% from the training data to inject the Trojan trigger in it and with the label set as the target class. We generated 50 pure models and 50 Trojan models for our experiments. The Trojan models are with trigger sizes ranging from 2×2 to 8×8 and transparency values 0.8 and 1.0. The trigger size, transparency, and the number of Trojan models we generated for our experiments are listed in Table 1. We choose the same transparency values for all the pixels of the transparency matrix, i.e. for example in set 1, the transparency matrix is a 2×2 matrix with all the cells that have value 1.

Table 1. Trigger size (s) and the transparency (α) we use to generate each set of Trojan models. Each set has 10 different Trojan models.

Trojan models	Trigger size (s)	Transparency (α)	#Trojan models
set 1	2	1	10
set 2	4	1	10
set 3	6	1	10
set 4	8	1	10
set 5	2	0.8	10

The Trojan effectiveness measure of these settings is shown in Table 2. The performance shows that across various configuration choices, the proposed attack strategy succeeds in converting classification results toward the target class on Trojan test data while keeping the model operating normally on original test data.

4.3   Trojan Trigger Recovery

To qualitatively evaluate the ability of Scalable Trojan Scanner, we analyze the recovered patches obtained by our process and compare them to the original

Table 2. Effectiveness of Trojan models measured in both accuracy of Trojan Model on Pure data (TMPD) and Trojan Model on Trojaned data (TMTD) compared to the original Pure Model Pure Data (PMPD) with their standard deviation. Accuracy reported as average on 50 models of various configurations on each dataset.

	GTSRB	CIFAR-10
PMPD	92.67 ± 0.52	67.07 ± 0.55
TMPD	92.05 ± 1.75	65.32 ± 2.06
TMTD	99.97 ± 0.06	99.76 ± 0.29

triggers used in training Trojan models. In this experiment, we use a set of 10 randomly selected 2×2 triggers to build 10 Trojan models with the GTSRB dataset (set 1 in Table 1). For each Trojan model, we apply the reverse engineering procedure with 50 different initialization of ΔI.

The first notable phenomenon we observed is that with different initializations, the procedure obtains various patches. Among these patches, only some of them are similar to the original trigger. However, all of them have equally high Trojan effectiveness on the target class. This result suggests that the Trojan model allows extra strange patches to be also effective in driving classification results to the target class. This interesting side-effect can be explained by reflecting on the behaviour of the Trojan CNN training process. When trained with the *Trojan Data* which are the mix of true training data and patched images, CNN tries to find the common pattern between the true samples belong to the target class and the fake samples containing the trigger. This compromising between two far-away sets of data with the same label leads to interpolating in latent space into a manifold of inputs that would yield the target class decision. Such manifold can have many modes resulting in many clusters of effective triggers.

To understand further, we use k-means to find out how the effective triggers group up and discover that there are usually four to six clusters of them per Trojan model. For illustration, we do PCA on the 12-dimensional patch signal (2×2 patch of three channels) and plot the data along with the first two principal components in Fig. 3. The number of clusters (k) for each Trojan model is also shown along with the plot. For example, for the first sub-figure in Fig. 3, the number of clusters, k, formed with the 50 effective reverse-engineered patches is five, which are shown in five different colours.

As we now understand the multi-modal nature of solution space, we expect that one of the modes we discovered includes the original trigger. To verify this, we select the closest recovered patch and compare it with the original trigger using root mean squared error. A qualitative comparison is shown in Fig. 4. Quantitatively, the rmse of these recovered patches ranges from 0.12 to 0.22 and averages at 0.17 for the 10 Trojan models, in the space of RGB patch ranging from 0.0 to 1.0. These affirm that our proposed method can come up with patches that are almost identical to the trigger originally used to train the Trojan models.

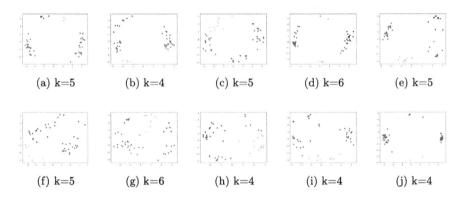

| (a) k=5 | (b) k=4 | (c) k=5 | (d) k=6 | (e) k=5 |

| (f) k=5 | (g) k=6 | (h) k=4 | (i) k=4 | (j) k=4 |

Fig. 3. Clusters of the 50 reverse-engineered patches of 10 Trojan models plotted using the first two principal components of patch signals. The subtitle denotes the number of clusters discovered by k-means.

We tried different random initializations for the pure model but not able to find a patch that can map all data to a single class.

4.4 Trojan Model Detection

The Trojan model detection is formed as a binary classification using the entropy score defined in Eq. 5. The negative class includes 50 *Pure Models* (PM) trained similarly but with different parameter initialization; the positive class contains 50 *Trojan models* (TM) trained with different random triggers. The Trojan triggers are of sizes 2×2, 6×6, 8×8 with transparency set as 1.0 and 2×2 with transparency set as 1.0 and 0.8 as detailed in Table 1. We run the scanning procedure once and use the reverse-engineered trigger recovered to compute the entropy score of the pure and Trojan models without knowing the potential target class. These scores are presented in Table 3. The lesser the entropy score the more effective the retrieved reverse-engineered mask and trigger, hence it is more possible that the model in the test is Trojan.

From Table 3, we observe that the difference between the high scores of the pure models and the low score of Trojan models is significant and stable across multiple settings. This reliability is achieved through the universality of entropy score measure which does not depend on any change in model and data (see Lemma 1). Because of this, it is straightforward with STS to set a robust universal threshold using the number of classes and expected Trojan effectiveness to detect the Trojan models using the entropy score.

We compare the performance of STS in detecting Trojan Model with Neural Cleanse (NC) [20] which is state-of-the-art for this problem. We use the same settings on models and data for both STS and NC. Similar to the entropy score used in STS, NC uses the anomaly index to rank the pureness of the candidate model. These scores of the two methods are shown in Table 4.

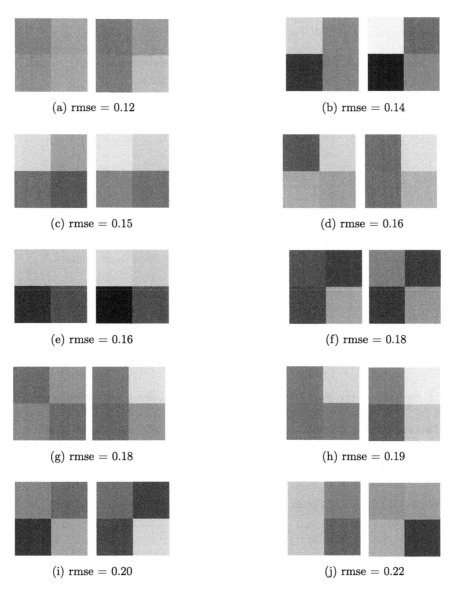

Fig. 4. Reverse-engineered patches compared to original trigger on 10 Trojan models. At each pair, the original patch is on the left while the retrieved one is on the right.

In our method, the gap between the pure and Trojan area in entropy scores are large and consistent. Meanwhile, the Neural Cleanse's anomaly index varies in a bigger range intra-class and close or overlapping inter-class. This again supports that our proposed method can come up with a more robust threshold to detect Trojan models.

Table 3. Entropy score computed based on the mask and patch retrieved by our proposed method on GTSRB and CIFAR-10 dataset.

Model	Average Entropy score	
	GTSRB	CIFAR-10
Pure models	**4.95**	**3.00**
set 1	0.003	0.001
set 2	0.004	0.002
set 3	0.004	0.003
set 4	0.005	0.006
set 5	0.003	0.001
Trojan models	**0.004**	**0.002**

Table 4. The minimum and maximum value of entropy scores (ours) and anomaly indexes (NC [20]) of the pure and Trojan models represented as *[min, max]*.

Model	GTSRB		CIFAR-10	
	Entropy score	Anomaly Index	Entropy score	Anomaly Index
Pure models	[4.85, 5.04]	[0.73, 23.30]	[2.55, 3.32]	[0.68, 2.56]
Trojan models	[0.0, 0.011]	[11.64, 71.08]	[0.0, 0.01]	[22.98, 244.94]

Table 5 shows the final accuracy of Trojan model detection between our STS and NC. The upper limit of entropy score for GTSRB and CIFAR-10 according to Lemma 1 is 0.1347 and 0.1125 respectively. For STS, we use these scores as the threshold for F1-score computation. For NC, we set the threshold to be 2.0 following recommendations in the method [20].

Table 5. Accuracy (as F1-score) of Trojan model detection between ours and NC [20].

Measures	GTSRB		CIFAR-10	
	STS (Ours)	NC	STS (Ours)	NC
F1-score	**1.0**	0.68	**1.0**	0.96

4.5   Computational Complexity

Besides the effectiveness in accurately detecting the Trojan model and recover the effective triggers, we also measured the complexity and computational cost of the Trojan scanning process. As our method does not make assumptions about the target class c_t, the complexity is constant ($\mathcal{O}(1)$) to this parameter. In the meantime, state-of-the-art methods such as NC rely on the optimisation process that assumes knowledge about the target class. This results in a loop through all of the possible classes and ends up in the complexity of $\mathcal{O}(C)$. Because of this difference in complexity, STS is significantly faster than NC. Concretely, in our experiments, STS takes around 2 h to detect 10 Trojan models on GTSRB, compared to more than 9 h that NC took on identical computational resource settings.

5   Conclusion and Future Work

In this paper, we propose a method to detect the Trojan model using trigger reverse engineering approach. We use a method to find out the minimum change that can map all the instances to a target class, for a Trojan model. The experiments conducted on GTSRB and CIFAR-10 show that our proposed method can reverse-engineer a visually similar patch with high Trojan effectiveness. We propose a measure, entropy score, to compute the robust upper threshold to detect the Trojan models. We report the entropy score and F1-score to support our claims. It is evident from the results that our method is more robust and less computationally expensive compared to the state-of-the-art method.

Future work is possible by generating models which have multiple target classes. Another future possibility is to work on Trojans which are distributed rather than a solid patch. We also found that for a single Trojan model, there exist multiple triggers as a solution, so given this, we also focus to work on getting a distribution of Trojan triggers for a Trojan model.

Acknowledgment. This research was partially funded by the Australian Government through the Australian Research Council (ARC). Professor Venkatesh is the recipient of an ARC Australian Laureate Fellowship (FL170100006).

References

1. Chan, A., Ong, Y.S.: Poison as a Cure: Detecting & Neutralizing Variable-Sized Backdoor Attacks in Deep Neural Networks. arXiv preprint arXiv:1911.08040 (2019)
2. Chen, B., et al.: Detecting Backdoor Attacks on Deep Neural Networks by Activation Clustering. arXiv preprint arXiv:1811.03728 (2018)
3. Chen, H., Fu, C., Zhao, J., Koushanfar, F.: DeepInspect: a black-box trojan detection and mitigation framework for deep neural networks. In: Proceedings of the 28th International Joint Conference on Artificial Intelligence. AAAI Press, pp. 4658–4664 (2019)
4. Gao, Y., Xu, C., Wang, D., Chen, S., Ranasinghe, D.C., Nepal, S.: Strip: a defence against trojan attacks on deep neural networks. In: Proceedings of the 35th Annual Computer Security Applications Conference, pp. 113–125 (2019)
5. Gu, T., Dolan-Gavitt, B., Garg, S.: Badnets: Identifying Vulnerabilities in the Machine Learning Model Supply Chain. arXiv preprint arXiv:1708.06733 (2017)
6. Huang, X., Alzantot, M., Srivastava, M.: NeuronInspect: Detecting Backdoors in Neural Networks via Output Explanations. arXiv preprint arXiv:1911.07399 (2019)
7. Ilyas, A., Engstrom, L., Athalye, A., Lin, J.: Black-box Adversarial Attacks with Limited Queries and Information. arXiv preprint arXiv:1804.08598 (2018)

8. Ji, Y., Zhang, X., Wang, T.: Backdoor attacks against learning systems. In: IEEE Conference on Communications and Network Security, pp. 1–9. IEEE (2017)

9. Kingma, D.P., Ba, J.: Adam: A Method for Stochastic Optimization. arXiv preprint arXiv:1412.6980 (2014)

10. Kolouri, S., Saha, A., Pirsiavash, H., Hoffmann, H.: Universal Litmus Patterns: Revealing Backdoor Attacks in CNNs. arXiv preprint arXiv:1906.10842 (2019)

11. Krizhevsky, A.: Learning Multiple Layers of Features from Tiny Images. Technical report (2009)

12. Li, Y., Li, L., Wang, L., Zhang, T., Gong, B.: NATTACK: Learning the Distributions of Adversarial Examples for an Improved Black-Box Attack on Deep Neural Networks. arXiv preprint arXiv:1905.00441 (2019)

13. Liu, Y., Lee, W.C., Tao, G., Ma, S., Aafer, Y., Zhang, X.: ABS: scanning neural networks for back-doors by artificial brain stimulation. In: Proceedings of the ACM SIGSAC Conference on Computer and Communications Security, pp. 1265–1282 (2019)

14. Papernot, N., McDaniel, P., Jha, S., Fredrikson, M., Celik, Z.B., Swami, A.: The limitations of deep learning in adversarial settings. In: IEEE European Symposium on Security and Privacy, pp. 372–387. IEEE (2016)

15. Qiu, S., Liu, Q., Zhou, S., Wu, C.: Review of artificial intelligence adversarial attack and defense technologies. Appl. Sci. 9(5), 909 (2019)

16. Simonyan, K., Zisserman, A.: Very Deep Convolutional Networks for Large-Scale Image Recognition. arXiv preprint arXiv:1409.1556 (2014)

17. Stallkamp, J., Schlipsing, M., Salmen, J., Igel, C.: Man vs computer: benchmarking machine learning algorithms for traffic sign recognition. Neural Netw. **32**, 323–332 (2012)

18. Szegedy, C., Zaremba, W., Sutskever, I., Bruna, J., Erhan, D., Goodfellow, I., Fergus, R.: Intriguing properties of neural networks. arXiv preprint arXiv:1312.6199 (2013)

19. Tran, B., Li, J., Madry, A.: Spectral signatures in backdoor attacks. In: Advances in Neural Information Processing Systems, pp. 8000–8010 (2018)

20. Wang, B., et al.: Neural cleanse: identifying and mitigating backdoor attacks in neural networks. In: IEEE Symposium on Security and Privacy. pp. 707–723. IEEE (2019)

21. Xiang, Z., Miller, D.J., Kesidis, G.: Revealing Backdoors, Post-Training, in DNN Classifiers via Novel Inference on Optimized Perturbations Inducing Group Misclassification. arXiv preprint arXiv:1908.10498 (2019)

22. Xu, X., Wang, Q., Li, H., Borisov, N., Gunter, C.A., Li, B.: Detecting AI Trojans Using Meta Neural Analysis. arXiv preprint arXiv:1910.03137 (2019)

Federated Learning

An Algorithmic Framework
for Decentralised Matrix Factorisation

Erika Duriakova[1]([⊠]), Wĕipéng Huáng[1]([⊠]), Elias Tragos[1], Aonghus Lawlor[1],
Barry Smyth[1], James Geraci[2], and Neil Hurley[1]

[1] Insight Centre for Data Analytics, University College Dublin, Dublin, Ireland
{erika.duriakova,weipeng.huang,elias.tragos,aonghus.lawlor,
barry.smyth,neil.hurley}@insight-centre.org
[2] Samsung Electronics Co., Ltd., Seoul, Republic of Korea
james.geraci@samsung.com

Abstract. We propose a framework for fully decentralised machine learning and apply it to latent factor models for top-N recommendation. The training data in a decentralised learning setting is distributed across multiple agents, who jointly optimise a common global objective function (the loss function). Here, in contrast to the client-server architecture of federated learning, the agents communicate directly, maintaining and updating their own model parameters, without central aggregation and without sharing their own data. This framework involves two key contributions. Firstly, we propose a method to extend a global loss function to a distributed loss function over the distributed parameters of the decentralised system; secondly, we show how this distributed loss function can be optimised using an algorithm that operates in two phases. In the learning phase, a large number of steps of local learning are carried out by each agent without communication. In a following sharing phase, neighbouring agents exchange messages that enable a batch update of local parameters. Thus, unlike other decentralised algorithms that require some inter-agent communication after one (or a few) model updates, our algorithm significantly reduces the number of messages that need to be exchanged during learning. We prove the convergence of our framework and demonstrate its effectiveness using both the Weighted Matrix Factorisation and Bayesian Personalised Ranking latent factor recommender models. We demonstrate empirically the performance of our approach on a number of different recommender system datasets.

Keywords: Recommender systems · Distributed learning · Decentralised matrix factorisation · Latent factor models · Matrix factorisation · Communication efficiency · Convergence proof

---

Electronic supplementary material The online version of this chapter (https://doi.org/10.1007/978-3-030-67661-2_19) contains supplementary material, which is available to authorized users.

F. Hutter et al. (Eds.): ECML PKDD 2020, LNAI 12458, pp. 307–323, 2021.
https://doi.org/10.1007/978-3-030-67661-2_19

1   Introduction

There is much interest recently in the development of machine learning (ML) algorithms that can be carried out on distributed architectures. The reasons that drive this shift are related with (i) the increased server maintenance costs for handling big data, (ii) the attempt to avoid centralised single points of failure (and attack) and (iii) the greater awareness and concern about data privacy (storing user data on central servers) that stems from legislation such as the European Union's General Data Protection Regulation (GDPR). The shift to distributed architectures poses a challenge for machine learning systems that have been traditionally centralised until now, requiring user data to be gathered and stored on a central computational server that runs the learning algorithm. In distributed architectures, ML algorithms should be developed to run on user devices without the need for those devices to share their private data.

While the field of distributed optimisation has a long history (see, e.g. [22]), the recent surge in research on this topic can be traced to [14], in which *federated* optimisation or learning is distinguished from distributed ML, by emphasising that a federated system is expected to run outside a data-centre, on very large number of user devices, typically leading to very highly unbalanced and not identically distributed data across these devices. Since this work, a number of papers have proposed federated learning architectures in different ML contexts. The vast majority of these works have focused on client-server architectures. Such systems still rely on a central server, which, rather than storing the user data itself, instead coordinates the learning process among the user devices, receiving local model updates from the devices and aggregating these into a global model. Nevertheless, as pointed out in [11] for example, such a central server is not always desirable, as it represents a single point-of-failure and may become a bottleneck when the number of clients is very large. Hence, in this paper, we focus on fully decentralised learning.

We propose a general algorithmic framework and apply it to latent factor models for recommender systems.

Recommender Systems (RS) learn from user interaction data with the products or services that they recommend, which we will refer to generally as *items*. RS are commonplace nowadays in many e-commerce contexts, providing personalised recommendations of books, movies, news and so on. The interactions on which recommendations are learned may correspond to explicit ratings of items, or implicit actions, such as clicks or views. The recommendation problem is typically formulated as a top N problem, in which, given a budget of N recommendations, the system predicts the N items that are most likely to satisfy a user's need. In common with other machine learning methods, traditional RS models are learned on centralised systems, where interaction data from all users is gathered into a single database. Among many recommendation models, latent factor models, in which user- and item-embeddings are learned and used to make the predictions, have proven among the best in terms of accuracy.

We develop a fully-decentralised latent factor model for the top-N recommendation task, using our proposed framework and empirically test it on a number of different RS datasets. While there exists a few other works that have developed decentralised latent factor models for recommendation, our work may be distinguished from the state-of-the-art in so far as we provide a solid foundation for our optimisation algorithm, by proposing a distributed loss function to properly control learning across devices; and develop a decentralised optimisation algorithm, that manages the communication between neighbours in a way that reduces the number of messages required, in comparison with the state-of-the-art. Moreover, we provide a proof of the convergence of the framework, for certain classes of loss function. In summary, our contributions are as follows:

- We propose a framework for fully decentralised learning, consisting of a general form for a distributed loss function and an optimisation algorithm to minimise the loss;
- We extend beyond the common *gossip* approach of parameter averaging, by introducing a *"cross-loss"* term in the distributed loss function to accelerate convergence; moreover, our decentralised approach avoids the communication of private parameters;
- We provide a theoretical analysis on the convergence of the algorithmic framework. Unlike some other work, such as [2], we do not rely on convexity of the loss function but only require that the loss function is *L-smooth*;
- We apply the framework to the decentralised learning of latent factor top-N recommendation models;
- We empirically show the convergence of our method and compare with state-of-the-art on a number of different recommender datasets.

The remainder of the paper is structured as follows. In Sect. 2, we put our work in context with the state-of-the-art. In Sect. 3 we provide a background on latent factor models that are used within our framework. In Sect. 4, we introduce our proposed framework for decentralised learning. In Sect. 5 we present the evaluation results of our framework, while in Sect. 6 we provide the final remarks on the conclusions of the work. The theoretical convergence proof of our algorithm is provided in the supplemental material.

2   Related Work

Federated Learning (FL) systems [14,16] are receiving a lot of attention recently. The standard FL architecture is composed of two main entities: the client and the server. The client encapsulates the private user data, it applies local updates to the global ML model during training, and executes the inference, i.e. the classification or recommendation, once the model is learned. The server initialises and coordinates the training of the global model, aggregating updates provided by clients and distributing the latest model back to clients. In this architecture, a powerful and reliable central server is a key part of an FL system. In [11], the case for a fully decentralised learning model is made. In this scenario, clients communicate directly in a peer-to-peer manner, each learning a local model without central aggregation to a global model.

Seminal work on distributed optimisation was carried out by Tsitsiklis et al. [22], in the 1980's. This work points out that synchronous algorithms, by which they mean algorithms in which processors communicate their partial results after each update, have certain drawbacks, including that they may introduce bottlenecks into the speed of the algorithm and may require far more communications than are actually necessary. They develop a number of asynchronous algorithms, where processors do not need to communicate after every update. Given a set of model parameters θ_ℓ, denote by $\theta_{u\ell}^{(t)}$ the value of the parameter on agent u at time-step t. They consider algorithms that apply updates of the general form:

$$\theta_{u\ell}^{(t)} = \sum_{i=1}^{n} w_{uv}\theta_{v\ell}^{(t_v)} + \gamma_u^{(t)}s_{u\ell}^{(t)} \tag{1}$$

where n is the total number of agents, $t_v \leq t$ is the time at which the value received by u was computed on v, w_{uv} are non-negative weights such that $\sum_v w_{uv} = 1$, $s_{u\ell}^{(t)}$ is the update step computed from the loss function and $\gamma_u^{(t)}$ is the learning rate. Thus, these algorithms reach consensus over time, by applying a weighted average of their parameter updates with parameter values received from other processors. As described later, we will take a different approach to developing a distributed algorithm, by firstly defining an objective over the distributed parameters of the model $\theta_{u\ell}$ and then developing an algorithm in which the updates are computed as stochastic gradients of this distributed objective. A good survey of decentralised optimisation is provided in [18].

More recently, decentralised algorithms are often referred to as *gossip* algorithms [12]. A number of such decentralised ML algorithms have been proposed. For instance, Lian et al. [15] and Tang et al. [21] propose a parallel stochastic gradient algorithm. Also, Nedic [17] proposes a sub-gradient algorithm which essentially follows the update rule of Eq. (1). The convergence of this work was later studied by Yuan et al. [24]. Another work [7] differs only in so far as the weight matrix is not stochastic and hence a normalising factor is also communicated between processors. In [4], a gossip algorithm for a pairwise loss function (an example of such being the BPR loss function discussed later), is proposed, which also is based on the averaging of parameters exchanged between processors, as well as the exchange of some training data. Other work, such as that of [13] has focused on minimising communication in gossip algorithms by applying sparsification and quantisation to compress the messages between processors. The work of [23] has some similarities to the framework that we propose, in so far as a loss function is defined on each agent with parameters shared with neighbouring agents, but it is solved by imposing equality constraints on the shared parameters. Also, Bellet et al. [2] design a loss function including a constraint on the norm of the difference between parameters across different agents, but use a coordinate descent algorithm, in which communications with neighbouring processors are required on each update step. A critical contribution of that work is the use of differential privacy to add security to the updates. Another gossip-based algorithm [1] again relies on updates based on Eq. (1), but considers the overlap of computation and communication, by implementing non-blocking communication and applying the update from neighbours every $\tau \geq 1$ updates, rather than every update.

Now, let us focus in particular on decentralised algorithms in RS, there has been some research, such as [23] and [19] on decentralised user-based and item-based kNN algorithms. Client-server FL architectures for the latent factor recommendation model are proposed in [5,10]. The only examples of a fully decentralised latent factor recommendation models in the state-of-the-art are that of [3,6,8]. While each of these develops a latent-factor model, the works [6,8] focus on the rating prediction problem—that of predicting the rating a user would give to an item—rather than the top-N recommendation problem. Further, although Chen et al. [3] solve top-N recommendation problem, the model communicates the partial results at each update which, as discussed earlier, is a major drawback in real-world scenarios.

3   Latent Factor Models for Top-N Recommendation

We illustrate our decentralised learning algorithm using two latent factor RS algorithms, namely the *weighted matrix factorisation* (WMF) model of [9] and the *Bayesian Personalised Ranking* (BPR) model of [20]. Each of these models is designed to yield top-N recommendations given a database of implicit user-item interactions.

Notation. Let U represent a set of users, such that $|U| = n$ and I represent the catalogue of items, such that $|I| = m$. The training set consists of a set of implicit interactions $\{r_{ui} \mid u \in U, i \in I\}$. We will write R_u to represent the set of items rated by user u and $\mathbf{r}_u$ to represent the m-dimensional vector of ratings associated with u. A latent factor model consists of low-dimensional user- and item-embeddings[1]. In particular, for a given dimension $k \ll n, m$, let P be a $n \times k$ matrix of user factors, and write $\mathbf{p}_u$ for the k-dimensional vector corresponding to each row of P; and similarly let Q be a $m \times k$ matrix of item-factors, and write $\mathbf{q}_i$ for the ith row of Q. A top-N prediction for user u is formed by computing the m-dimensional vector $\hat{\mathbf{r}} = Q\mathbf{p}_u$, sorting the components of $\hat{\mathbf{r}}$ in decreasing order and recommending the top N corresponding items. For any $S \subseteq I$, we will use $\mathbf{q}_S$, to represent the components $\{\mathbf{q}_i \mid i \in S\}$.

The two models differ according to the loss functions used to determine the parameters $\{P, Q\}$. In general, the loss function for latent factor models can be written as

$$\mathcal{L}(P, Q) = \sum_{u \in U} \mathcal{L}_u(\mathbf{p}_u, Q, \mathbf{r}_u).$$

In particular, for WMF, setting $\tilde{r}_{ui} = 1$ for $r_{ui} > 0$ and $\tilde{r}_{ui} = 0$ for $r_{ui} = 0$, then

$$\mathcal{L}^{\text{WMF}}(P, Q) = \sum_u \sum_{i \in I} c_{ui}(\mathbf{p}_u^\top \mathbf{q}_i - \tilde{r}_{ui})^2$$

---

[1] We ignore bias terms for simplicity, though these can easily be incorporated into the framework.

where, for some input parameter α, $c_{ui} = (1 + \alpha r_{ui})$ represents the confidence.

The BPR loss models the probability that an item i is preferred to an item j, using a sigmoid function of the difference between their corresponding latent factor products and represents the loss as the negation of the sum of the log-probabilities, resulting in

$$\mathcal{L}^{\text{BPR}}(P, Q) = -\sum_u \frac{1}{|R_u|} \sum_{i \in R_u} \frac{1}{m - |R_u|} \sum_{j \notin R_u} \log \sigma(-x_{uij}),$$

$$\text{s.t. } x_{uij} = \mathbf{p}_u^\top (\mathbf{q}_i - \mathbf{q}_j),$$

where σ is the sigmoidal function s.t. $\sigma(x) = (1 + e^x)^{-1}$. To mitigate against overfitting, these loss functions are typically regularised by adding the terms:

$$\lambda_p \sum_u \|\mathbf{p}_u\|^2 + \lambda_q \sum_i \|\mathbf{q}_i\|^2$$

for weights $\lambda_p \geq 0$ and $\lambda_q \geq 0$.

One approach to optimise these loss functions is to apply a stochastic gradient descent (SGD) algorithm (although the WMF algorithm is optimised in [9] using alternative least squares). Thus we can decompose the loss into summands:

$$\mathcal{L}^{\text{WMF}} = \sum_u \sum_i \mathcal{L}_{ui}^{\text{WMF}}; \quad \mathcal{L}^{\text{BPR}} = \sum_u \sum_{ij} \mathcal{L}_{uij}^{\text{BPR}}$$

and, sampling at random a pair (u, i) for WMF[2] (or resp. a triple (u, i, j) for BPR), update the parameters $\theta \in \{\mathbf{p}_u, \mathbf{q}_i\}$ by

$$\theta^{(t+1)} = \theta^{(t)} - \eta \nabla_\theta \mathcal{L}_{ui}^{\text{WMF}} \quad \text{or} \quad \theta^{(t+1)} = \theta^{(t)} - \eta \nabla_\theta \mathcal{L}_{uij}^{\text{BPR}},$$

for a learning rate η.

We extend this standard SGD approach, which requires the data to be collected into a central repository, to a fully-decentralised algorithm, in the next section.

4   Framework for Decentralised Learning

Consider a set of n agents, co-operating on an ML optimisation task. Let $G(V, E)$ be a communication graph, where V is the set of agents, and an edge $(u, v) \in E$, if agent u is able to communicate with agent v. We will write $N_u(G)$ for the set of neighbours of agent u in this graph. Each agent maintains its own data x_u,

---

[2] SGD schemes for WMF typically sample negative items ($i \notin R_u$) at a different rate to positive samples. In this case, the stochastic gradients are unbiased gradient estimators of a loss function where the confidence term is $c_{ui} = \pi_i(1 + \alpha r_{ui})$, where π_i is the probability of sampling item i.

which is *not shared* with other agents. For loss functions based on empirical risk, it is possible to write

$$\mathcal{L}(\Theta) = \sum_u \mathcal{L}_u(\Theta, x_u)$$

where $\mathcal{L}$ only depends on x_u through $\mathcal{L}_u$. In a fully decentralised setting, each agent u maintains and updates its own set of parameters Θ_u. In the case of the RS models we are considering here, unlike some other domains, some parameters are too private for sharing as they are strongly correlated to user data. Hence, unlike typical gossip models which share all parameters, we consider different privacy risks associated with different parameters. We write $\Theta_u = [\Theta_u^p, \Theta_u^s]$, where Θ_u^p are private parameters and Θ_u^s are sharable.

Decentralised algorithms generally require agents to communicate to reach a consensus such that at the end of learning, $\forall u\ \Theta_u = \Theta^*$ and $\mathcal{L}(\Theta^*)$ is minimised. Commonly, this consensus is reached via gossip updates such as shown in Eq. (1), which do not distinguish the privacy characteristics of the parameter set. Instead, we propose to extend the global objective over the full set of distributed parameters. The loss function must be constructed in such a way that local parameters Θ_u are affected by a loss associated with data x_v, where $v \neq u$; otherwise, a model that generalises to unseen data will not be possible. Considering privacy risks, we wish to consider loss functions for which it is possible to develop a learning algorithm where it is allowed to share parameters Θ_u^s or functions of Θ_u^s or Θ_u^p, but not directly Θ_u^p. With these considerations, we will consider loss functions of the general form:

$$\mathcal{L} = \frac{1}{n} \left(\sum_u \mathcal{L}_u(\Theta_u, x_u) + \sum_u \sum_{v \in N_u(G)} \mathcal{L}_{uv}(\Theta_u, \Theta_v) \right)$$

where $\mathcal{L}_{uv}$ depends on parameters that are local to u or local to v. In particular, we construct $\mathcal{L}_{uv}$ as

$$\mathcal{L}_{uv}(\Theta_u, \Theta_v) = w_{uv} \left(\lambda_r \|\Theta_u^s - \Theta_v^s\|^2 + \lambda_l \mathcal{L}_u([\Theta_u^p, \Theta_v^s], x_u) \right),$$

where $\sum_v w_{uv} = 1$ are non-negative weights and $\lambda_l \geq 0$, $\lambda_r \geq 0$ control the contribution that each term makes to the loss-function. The first term above is used, similarly to other decentralised algorithms, to push the optimisation towards a consensus among the shared distributed parameters. The second term, which we refer to as a "cross-loss" term as it mixes the u and v parameter set, ensures that local parameters update in a way that takes account of the data in their neighbourhood. Note that, as we will see, optimisation can still proceed without direct sharing of the private parameters or private training data and the use of this cross-loss term can significantly accelerate learning.

We are concerned about the efficiency of the learning algorithm, in a context where communication between agents may be expensive, in comparison to local computation. Hence, we are interested in developing a learning algorithm in which the number of communications is minimised. We note that in some state-of-the-art decentralised learning algorithms such as [3] and [6], a communication takes place in *every* update step and hence this sort of algorithm is not appropriate for the setting that we have in mind.

We introduce an algorithmic framework in which the algorithm proceeds in two bulk steps of computation followed by communication, rather than interweaving message-passing with model updating. In particular, in the first *learn* phase, learning on each agent focuses on the L_u term, and carries out a large number of local updates to reduce L_u. Following this, in the *share* phase, learning focuses on the functions L_{uv}, which require communication between the agents u and v. We construct our algorithm so that *sharing* occurs infrequently in comparison to local *learning* and many updates are carried out in the learn phase, before sharing. Our focus is to complete the optimisation with as few communications as possible.

4.1 Decentralised Latent Factor Model

In the case of latent factor models, $\Theta_u^p = \mathbf{p}_u$ are the user factors, which should be kept private, since, if accurate item factors are learned across the system, then knowing $\mathbf{p}_u$ is sufficient to predict $\hat{\mathbf{r}}_u$. To convert the loss functions of Sect. 3 to the distributed setting, we introduce the local parameters as $\Theta_u = [\mathbf{p}_u, Q_u]$, such that each agent now maintains its own item factors $\mathbf{q}_{ui}$ and,

$$\mathcal{L}_{uv}(\mathbf{p}_u, Q_u, Q_v) = \lambda_r \|Q_u - Q_v\|_F^2 + \lambda_l \mathcal{L}_u(\mathbf{p}_u, Q_v, \mathbf{r}_u),$$

where $\| \cdot \|_F$ represents the Frobenius norm.

Definition 1 (Decentralised MF Objective). *The distributed latent matrix factor (MF) objective is written as*

$$\mathcal{L}(P, Q_1, \dots, Q_n) = \frac{1}{n} \sum_u \left(\mathcal{L}_u(\mathbf{p}_u, Q_u) + \sum_{v \in N_u(G)} \mathcal{L}_{uv}(\mathbf{p}_u, Q_u, Q_v) \right). \quad (2)$$

While minimising over $\mathcal{L}_u$ sets the local parameters to fit the loss associated with local data as well as possible, and the terms $\mathcal{L}_{uv}$ are necessary in order to ensure that the local model generalises to unseen data.

4.2 Decentralised Algorithm

The distributed algorithm that we propose to minimise Eq. (2) is shown in Algorithm 1. The algorithm involves the following steps:

– A local *learn* phase (lines 3–6 in Algorithm 1), in which the loss $\mathcal{L}_u$ is reduced through an SGD process.
– A *share* phase (lines 7–15 in Algorithm 1), which randomly samples from the neighbourhood of each agent u to get a set of neighbours $I_u \subseteq N_u(G)$ (line 8 in Algorithm 1); and during which u and $v \in I_u$, synchronise to reduce the $\mathcal{L}_{uv}$ loss over some agreed set of items S (lines 9–15 in Algorithm 1).

Algorithm 1: Decentralised Algorithm for MF

Input   : user-item preference rating vector $\{\mathbf{r}_u \mid u \in U\}$,
communication graph G, such that
$N_u(G) = \{v \mid v \in U, (u, v) \in G\}$ is the set of neighbours of u in
G, learning rate η_e and cross-loss regulariser λ_l.

1   **forall** e *epoch* **do**
    `// Learn Phase`
2   **forall** $u \in U$ *in parallel* **do**
3   Draw T_e item samples $\mathcal{T}$ `// The sampling depends on the objective`
4   **for** $\ell \in \mathcal{T}$ **do**
5   Update $\mathbf{p}_u = \mathbf{p}_u - \eta_e \frac{\partial \mathcal{L}_{u\ell}}{\partial \mathbf{p}_u}(\mathbf{r}_u)$
6   Update $\mathbf{q}_{u\ell} = \mathbf{q}_{u\ell} - \eta_e \frac{\partial \mathcal{L}_{u\ell}}{\partial \mathbf{q}_{u\ell}}(\mathbf{r}_u)$

    `// Share Phase`
7   **forall** $u \in U$ *in parallel* **do**
8   Sample $I_u \subseteq N_u(G)$ `// agents sampled from neighbours of u`
9   **forall** $v \in I_u$ **do**
10          $S = \text{SYNCITEMS}(u, v)$ `// Set of items shared between u and v`
11   Gather $\mathbf{q}_{vS}$ from v
12   **if** $\lambda_l > 0$ **then**
13   Update and Send $\frac{\partial \mathcal{L}_{uv}}{\partial \mathbf{q}_{vS}}$
14   Gather $\frac{\partial \mathcal{L}_{vu}}{\partial \mathbf{q}_{uS}}$
15   Update $\mathbf{q}_{uS} = \mathbf{q}_{uS} - \eta_e \frac{\partial \mathcal{L}_{vu}}{\partial \mathbf{q}_{uS}}$
16   Update $\mathbf{p}_u = \mathbf{p}_u - \eta_e \frac{\partial \mathcal{L}_{uv}}{\partial \mathbf{p}_u}$

In the learn phase, the parameter updates are calculated by sampling from the summands of the local loss (i.e. sampling pairs (u, i) for WMF or triples (u, i, j) for BPR, see Sect. 3), and updating in the direction of the stochastic gradients associated with these summands. In the share phase, selecting neighbours v at random, the pair (u, v) agrees a set of item factors to share (Algorithm 1 line 14) and the parameters are updated in the direction of the stochastic gradient associated with *all* these items. Hence a single message results in an update of $|S|$ item factors, where S is the set of shared items (line 15 in Algorithm 1). Note that if $\lambda_l = 0$, the synchronisation step only requires an update for the item factors, which can be applied after receiving the item factors, $\mathbf{q}_{vi} \forall i \in S$, from the neighbour v. On the other hand, if $\lambda_l \neq 0$, then the gradient of the cross-loss term in $\mathcal{L}_{vu}$ is required by agent u, but must be computed by neighbour v, since it requires agent v's private data (see lines 16–19 in Algorithm 1). Hence, the update takes place in the following steps:

Algorithm 2: Decentralised WMF Share Phase Updates

---

1 **forall** $v \in I_u$ **do**
2 $\quad$ **forall** $i \in S$ **do**
3 $\quad\quad$ Exchange $\mathbf{q}_{ui}$ and $\mathbf{q}_{vi}$ between u and v.
$\quad\quad\quad$ `// u computes & sends cross-loss gradient wrt` $\mathbf{q}_{vi}$ `to v.`
4 $\quad\quad$ $\frac{\partial \mathcal{L}^{\text{WMF}}_{uv}}{\partial \mathbf{q}_{vi}} = 2\lambda_l c_{ui}(\mathbf{p}_u^\top \mathbf{q}_{vi} - \tilde{r}_{ui})\mathbf{p}_u$
$\quad\quad\quad$ `// v computes & sends cross-loss gradient wrt` $\mathbf{q}_{ui}$ `to u.`
5 $\quad\quad$ $\frac{\partial \mathcal{L}^{\text{WMF}}_{vu}}{\partial \mathbf{q}_{ui}} = 2\lambda_l c_{vi}(\mathbf{p}_v^\top \mathbf{q}_{ui} - \tilde{r}_{vi})\mathbf{p}_v$
6 $\quad\quad$ u updates: $\mathbf{q}_{ui} = \mathbf{q}_{ui} - \eta_e \frac{|N_v||w_{vu}|}{|I_v||S|}\big(\frac{\partial \mathcal{L}^{\text{WMF}}_{vu}}{\partial \mathbf{q}_{ui}} + 2\lambda_r(\mathbf{q}_{vi} - \mathbf{q}_{ui})\big)$
7 $\quad\quad$ v updates: $\mathbf{q}_{vi} = \mathbf{q}_{vi} - \eta_e \frac{|N_u||w_{uv}|}{|I_u||S|}\big(\frac{\partial \mathcal{L}^{\text{WMF}}_{uv}}{\partial \mathbf{q}_{vi}} + 2\lambda_r(\mathbf{q}_{ui} - \mathbf{q}_{vi})\big)$
$\quad\quad\quad$ `// u computes gradient wrt` $\mathbf{p}_u$`, using received` $\mathbf{q}_{vi}$`.`
8 $\quad\quad$ $\frac{\partial \mathcal{L}^{\text{WMF}}_{uv}}{\partial \mathbf{p}_u} = 2\frac{\lambda_l}{|S|}c_{ui}(\mathbf{p}_u^\top \mathbf{q}_{vi} - \tilde{r}_{ui})\mathbf{q}_{vi}$
$\quad\quad\quad$ `// v computes gradient wrt` $\mathbf{p}_v$`, using received` $\mathbf{q}_{ui}$`.`
9 $\quad\quad$ $\frac{\partial \mathcal{L}^{\text{WMF}}_{vu}}{\partial \mathbf{p}_v} = 2\frac{\lambda_l}{|S|}c_{vi}(\mathbf{p}_v^\top \mathbf{q}_{ui} - \tilde{r}_{vi})\mathbf{q}_{ui}$
10 $\quad\quad$ u updates: $\mathbf{p}_u = \mathbf{p}_u - \eta_e \frac{|N_u||w_{uv}|}{|I_u|}\frac{\partial \mathcal{L}^{\text{WMF}}_{uv}}{\partial \mathbf{p}_u}$
11 $\quad\quad$ v updates: $\mathbf{p}_v = \mathbf{p}_v - \eta_e \frac{|N_v||w_{vu}|}{|I_v|}\frac{\partial \mathcal{L}^{\text{WMF}}_{vu}}{\partial \mathbf{p}_v}$

- Agents u and v exchange item factors
- Agent v computes the gradient of the cross-loss for agent u and vice versa
- Agents u and v exchange cross-loss gradients.

The share phase updates for the WMF algorithm are given in Algorithm 2. The BPR updates may be found in the supplemental material.

4.3 Theoretical Results

We state certain theoretical results about the decentralised framework in this section, but enclose all the proofs in the supplemental material. Write $\mathbf{S} = \{\Theta_u \mid u \in U\}$ and $\mathcal{L}^{(t)} = f(\mathbf{S}^{(t)}) + g(\mathbf{S}^{(t)})$ as the value of the loss at time-step t, where $f(\mathbf{S}^{(t)}) = \sum_u \mathcal{L}_u$ and $g(\mathbf{S}^{(t)}) = \sum_{u,v} \mathcal{L}_{uv}(\mathbf{S}^{(t)})$.

Theorem 2. *Given that the decentralised objective $\mathcal{L}$ is L-smooth, the decentralised latent factor algorithms, using the learning rate $\eta_t = \eta/L$, have convergence rate $O\left(\frac{4LG}{\eta T_e T} + 2\eta\left(\sigma_1^2 + \frac{\sigma_2^2}{T_e}\right)\right)$, where*

- $T_e = |\mathcal{T}|$ *which is size of the sampled items for each iteration in Algorithm 1*
- $G \triangleq \mathcal{L}^{(0)} - \mathrm{E}\left[\mathcal{L}^{(t)}\right]$
- $\sigma_1^2 : \mathrm{E}[\|f(\mathbf{S}^{(t)})\|^2] \leq \sigma_1^2,\ \sigma_2^2 : \mathrm{E}[\|g(\mathbf{S}^{(t)})\|^2] \leq \sigma_2^2,\ \forall t$

An everywhere differentiable function that has L-Lipschitz gradients is said to be L-smooth, and L-Lipschitz continuity indicates that a function's derivatives are bounded everywhere in its scope. We appeal to this property as it is known that the latent factor objective is non-convex. We have the following two corollaries,

Corollary 3. *On any bounded domain of the objective,* WMF *is L-smooth.*

Corollary 4. *On any bounded domain of the objective,* BPR *is L-smooth.*

5   Experiments

We run our experiments on a single multicore server using shared-memory parallelisation. The WMF model is implemented in C++ using `OpenMP` and the BPR model is implemented in `Matlab`, using the parallel toolbox. The cooperating agents are simulated as independent threads. The entire parameter set of $n(m+1) \times k$ parameters $\{\mathbf{p}_u, Q_u \mid u \in U\}$ are stored in the shared memory. For the Movielens 1M dataset, with $k = 20$, this amounts to >3.5 GB of storage.

5.1   SOTA Comparison

Gossip algorithms. When $\lambda_l = 0$, the update due to the L_{uv} terms amounts to:

$$\mathbf{q}_{ui} = \mathbf{q}_{ui} - 2\eta_e \lambda_r \frac{|N_u(G)|}{|I_u|} \left(\mathbf{q}_{ui} - \sum_{v \in I_u} w_{uv} \mathbf{q}_{vi} \right) = (1 - \tilde{\eta})\mathbf{q}_{ui} + \tilde{\eta} \sum_v w_{uv} \mathbf{q}_{vi} ,$$

$$(3)$$

for some $\tilde{\eta} > 0$, that is, a weighted average of the agent's item-factor with that of the neighbours. The method becomes a gossip-like algorithm for bringing the shared parameters $\mathbf{q}_{ui}$ to consensus. In [8], such a gossip algorithm is evaluated for recommendation using a matrix factorisation model, but applied to the rating prediction problem, rather than the top-N problem. In our experimental analysis, the case $\lambda_l = 0$ is used to compare with such a gossip algorithm.

Decentralised Matrix Factorisation (DMF). In the works [3,6], a decentralised matrix factorisation algorithm is proposed in which communication with neighbours occurs after every update. In [3], rather than sending the item-factors between neighbours, instead neighbour v sends $\frac{\partial \mathcal{L}}{\partial \mathbf{q}_{vi}}$ to agent u and this gradient is combined into u's model via $\mathbf{q}_{ui} = \mathbf{q}_{ui} - \tilde{\eta}(\mathbf{p}_v^\top \mathbf{q}_{vi} - r_{vi})\mathbf{p}_v$ for some update weight $\tilde{\eta}$. Note that in this update rule, a gradient evaluated on v's parameters $(\mathbf{p}_v, \mathbf{q}_{vi})$ is used to update u's item-factor $\mathbf{q}_{ui}$. (Compare with Algorithm 2, in which the gradient of the distributed loss function involves a mix of u's and v's parameters). As a result, the algorithm requires that $\mathbf{q}_{vi}^{(t)} \approx \mathbf{q}_{ui}^{(t)}$ at all iterations, t, and cannot sustain any lag between updates and communications. We show that, as well as being mathematically more sound, our method empirically out-performs this algorithm on the evaluated datasets.

5.2   Results

We evaluate the algorithm using three datasets:

- Movielens 1M(`ML1M`): $6,040$ users, $3,952$ items, 1M interactions, 5% sparsity;
- Movielens 100K(`ML100K`): 943 users, $1,682$ items, 100 K interactions, 3%sparsity;
- Last.fm(`lastFM`): random sparse subset of 975 users, $4,000$ artists, $28,950$ interactions (normalised to range $[1,5]$), 0.6% sparsity.

We train on 80% of the data, chosen at random, and test on the remaining 20%. To tune the parameters, we further split the training data, such that 90% of training data is used for training and the remaining 10% for validation. We evaluate performance using precision@N, the proportion of the top-N recommendations that are in the test set (where a test-set item is counted as long as some interaction between it and the user has been recorded)[3]. We show results for $N = 10$. In order to track the extent to which the distributed parameters reach a consensus, it is also interesting to plot the Q-norm, calculated as $\frac{1}{k} \sum_u \sum_{i \in R_u} \left\| \mathbf{q}_{ui} - \frac{1}{|\{v|i \in R_v\}|} \sum_{v|i \in R_v} \mathbf{q}_{vi} \right\|^2$. To gauge convergence of the decentralised optimisation, we track the following training set loss:

- BPR: The mean of $1 - \overline{\sigma(-\mathbf{p}_u(\mathbf{q}_{vi} - \mathbf{q}_{vj}))}$, over 10,000 randomly chosen pairs $\{(i,j) \mid i \in R_u, j \notin R_u\}$, and randomly chosen agents u and v
- WMF: The mean of $\sqrt{c_{ui}(\mathbf{p}_u^\top \mathbf{q}_{vi} - \tilde{r}_{u,i})^2}$, over all ratings in R and randomly chosen neighbours v of each agent u.

Further, we choose $w_{uv} = 1/|N_u G|$, $|I_u| = 1$, i.e. each agent communicates with a *single* randomly chosen neighbour. This is implemented as a random edge matching, in which, at the start of the share phase, pairs of neighbouring agents agree to share. Any node that fails to find a match, skips the share phase in this epoch. We examine the following research questions:

RQ1: The rate of convergence of the decentralised algorithm to the performance of a central algorithm.
RQ2: The effect of communication graph sparsity on the performance.
RQ3: The benefit gained from the cross-loss term, over parameter averaging.
RQ4: The communication overhead in terms of number of messages and message volume.

RQ1 and RQ2: We run the decentralised algorithm for 400 epochs on the three datasets. The results for a communication graph of density 0.2, synchronising over all items ($S = I$), are summarised in Table 1. We see convergence towards the performance of the central algorithm. Nevertheless, for `ML1M` for example, the central algorithm reaches its peak performance by 100 epochs, so decentralisation

---

[3] Other accuracy measures follow the trends we see for prec@10, achieving the well-known scores of the central algorithm when the decentralised algorithm converges, see e.g. www.librec.net/release/v1.3/example.html.

Table 1. Prec@10, Central vs decentralised model at epoch(ep)

	BPR					WMF				
Dataset	Central	100ep	200ep	300ep	400ep	Central	100ep	200ep	300ep	400ep
LastFM	0.145	0.105	0.119	0.133	0.140	0.116	0.052	0.097	0.104	0.105
ML100k	0.363	0.284	0.329	0.349	0.357	0.343	0.276	0.279	0.317	0.327
ML1M	0.334	0.165	0.225	0.269	0.293	0.295	0.135	0.175	0.213	0.254

must pay a significant penalty over central algorithms in terms of convergence speed. We use ML100K to examine *RQ2*. In Fig. 1, we show the convergence of the method, when agents are free to choose *any* node from a fully connected communication graph and when the communication graph has density of just 1%, i.e. each agent is connected to an average of 16.8 neighbours. It may be observed that the algorithms are very robust to communication graph sparsity, with just a small drop-off in convergence rate, even on a very sparse communication graph, so that we can expect convergence even in scenarios in which agents can only directly access to a few others.

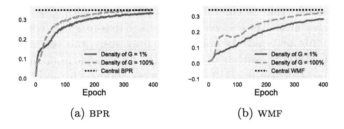

(a) BPR   (b) WMF

Fig. 1. Fully connected G vs random G with density $= 1\%$ showing Prec@10 on ML100k

RQ3: We examine *RQ3* using ML100k and ML1M. Due to space restrictions we only report results from the BPR model. Results obtained from the WMF model are along the same lines (see supplemental material). We fix a random communication graph of density 0.02. In Fig. 2, the prec@10, the Q-norm and the loss are shown for a number of different settings of the hyperparameters to highlight the difference between the convergence of gossip learning and an addition of our cross-loss term, we report: (i) *Gossip* (i.e. $\lambda_r > 0$, $\lambda_l = 0$), (ii) *Gossip+Cross-loss* (i.e. $\lambda_l > 0$, $\lambda_r > 0$), (iii) *Cross-loss: S1* ($\lambda_r = 0$, $\lambda_l = 2.3$) and (iv) *Cross-loss: S2* ($\lambda_r = 0$, $\lambda_l = 2.5$).

It can be observed that the fastest convergence is obtained when the *Gossip+Cross-loss* setting is used, the algorithm is carrying out parameter averaging according to Eq. (3). On ML100k convergence is slower and on ML1M, the learning gets trapped in a local minimum with maximum precision of 0.18. On the other hand, for ML100k, we also show two convergence plots when only the *Cross-loss* setting is used. In this case, there is no mechanism to drive the item-factors to a consensus. It *is* possible to achieve some learning in these settings,

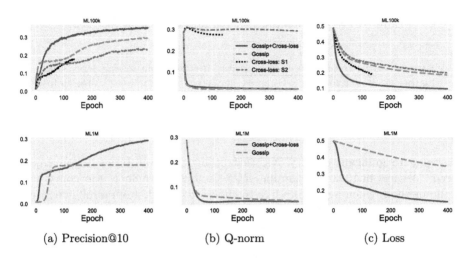

(a) Precision@10   (b) Q-norm   (c) Loss

Fig. 2. BPR on ML100k and ML1M Comparing Gossip and Cross-loss

but the algorithm is more brittle, with a tendency for the Q-norm to diverge, unless the learning rate is carefully managed. Here, with *Cross-loss: S1*, the learning diverges after 238 epochs, while with *Cross-loss: S2*, convergence to a prec@10 of 0.23 is achieved. As pointed out previously, this setting is somewhat similar to the DMF algorithm of [3], which, for comparison, in our own C++ implementation, reaches prec@10 of 0.197 after 100 epochs and 0.217 after 400 epochs.

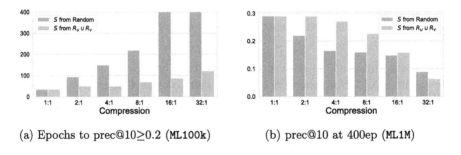

(a) Epochs to prec@10≥0.2 (ML100k)   (b) prec@10 at 400ep (ML1M)

Fig. 3. Effect of compression for BPR

RQ4: We use ML100k, ML1M and the BPR model to examine the effect of compressing the message communicated during the share phase. With privacy in mind, an agent may not wish to release any information to its neighbour about which items it has interacted with. Therefore, it may chose to select some random set, S, of item-factors to share. On the other hand, the algorithm is more robust to compression when interacted items are shared. Messages are compressed at a rate of 1:1 up to 32:1 (uncompressed to compressed message size), by selecting the shared items S at random, or selecting shared items from $R_u \cup R_v$ (and adding or removing items at random to reach the required size). In Fig. 3a, we show the number of epochs it takes to reach prec@10 ≥ 0.2 in ML100k and in Fig. 3b, the prec@10 after 400eps in ML1M. Compression slows convergence, but considerably less so when interacted items are shared. The trade-off between the amount of data shared per epoch and the number of epochs required for convergence results in there being no benefit to compressing beyond 16:1 in ML100k, when sharing random items. Interacted items afford greater performance benefits, but at a greater privacy risk. For example, for 8:1 compression using rated items, in ML1M, the rated items, that average 132 per user, are hidden among ≈500 shared items. Each agent communicates $2|S|k$ double precision numbers per epoch in two messages with the neighbour, so that there are two messages and a total volume of communication of $16nk|S|$ bytes per epoch. The DMF algorithm of [3] requires a communication per update. Assuming that one update per data-point across the entire machine is carried out per epoch, this amounts to T_e (=800k for ML1M) separate messages, each of size $8k$ bytes. In a communication-bound environment, with high latencies, such a number of messages may be infeasible. While the uncompressed message we send is large, our analysis has shown that the communication volume can be significantly reduced, without significant loss of performance. Further details on algorithm complexity and hyper-parameter selection are given in the supplemental material.

6   Conclusions

We have demonstrated a framework for decentralised ML, consisting of a loss function over the distributed parameters, along with a communication efficient algorithm for minimising this loss function. Results on two latent factor recommendation algorithms, show that the framework achieves the performance of the central algorithm, with accelerated convergence in comparison to gossip algorithms that rely only on parameter averaging. Moreover, the algorithm converges on sparse communication graphs and is robust to message compression.

Acknowledgements. The work is supported by the Science Foundation Ireland under the grant number SFI/12/RC/2289_P2 and Samsung Research, Samsung Electronics Co., Seoul, Republic of Korea. We wish to thank the reviewers for the helpful feedback.

References

1. Assran, M., Loizou, N., Ballas, N., Rabbat, M.: Stochastic gradient push for distributed deep learning. In: Proceedings of the ICML, Long Beach, California (2019)
2. Bellet, A., Guerraoui, R., Taziki, M., Tommasi, M.: Personalized and private peer-to-peer machine learning. In: Proceedings of the 21st AISTATS, Lanzarote (2017)
3. Chen, C., Liu, Z., Zhao, P., Zhou, J., Li, X.: Privacy preserving point-of-interest recommendation using decentralized matrix factorization. In: AAAI (2018)
4. Colin, I., Bellet, A., Salmon, J., Clémençon, S.: Gossip dual averaging for decentralized optimization of pairwise functions. In: Proceedings of 33rd ICML (2016)

5. Ammad-ud din, M., Ivannikova, E., Khan, S.A., Oyomno, W., Fu, Q., Tan, K.E., Flanagan, A.: Federated collaborative filtering for privacy-preserving personalized recommendation system. arXiv preprint arXiv:1901.09888 (2019)
6. Duriakova, E., Tragos, E.Z., Smyth, B., Hurley, N., Pena, F.J., Symeonidis, P., Geraci, J., Lawlor, A.: PDMFRec: a decentralised matrix factorisation with tunable user-centric privacy. In: Proceedings of the 13th RecSys. ACM (2019)
7. He, C., Tan, C., Tang, H., Qiu, S., Liu, J.: Central server free federated learning over single-sided trust social networks. arXiv preprint arXiv:1910.04956 (2019)
8. Hegedűs, I., Danner, G., Jelasity, M.: Decentralized recommendation based on matrix factorization: a comparison of gossip and federated learning. In: Cellier, P., Driessens, K. (eds.) ECML PKDD 2019, Part I. CCIS, vol. 1167, pp. 317–332. Springer, Cham (2020). https://doi.org/10.1007/978-3-030-43823-4_27
9. Hu, Y., Koren, Y., Volinsky, C.: Collaborative filtering for implicit feedback datasets. In: 2008 Eighth IEEE International Conference on Data Mining, pp. 263–272, December 2008
10. Jalalirad, A., Scavuzzo, M., Capota, C., Sprague, M.: A simple and efficient federated recommender system. In: Proceedings of the 6th IEEE/ACM International Conference on Big Data Computing, Applications and Technologies, pp. 53–58. ACM, New York (2019)
11. Kairouz, P., et al.: Advances and open problems in federated learning. arXiv preprint arXiv:1912.04977 (2019)
12. Kempe, D., Dobra, A., Gehrke, J.: Gossip-based computation of aggregate information. In: Proceedings of the 44th Annual IEEE Symposium on Foundations of Computer Science, FOCS 2003, p. 482. IEEE Computer Society, USA (2003)
13. Koloskova, A., Stich, S.U., Jaggi, M.: Decentralized stochastic optimization and gossip algorithms with compressed communication. In: Proceedings of the 36th International Conference on Machine Learning, Long Beach, California (2019)
14. Konecný, J., McMahan, H.B., Ramage, D.: Federated optimization: Distributed optimization beyond the datacenter. arXiv preprint arXiv:1511.03575 (2015)
15. Lian, X., Zhang, C., Zhang, H., Hsieh, C.J., Zhang, W., Liu, J.: Can decentralized algorithms outperform centralized algorithms? A case study for decentralized parallel stochastic gradient descent. Adv. Neural Inf. Process. Syst. **30**, 5330–5340 (2017)
16. McMahan, B., Moore, E., Ramage, D., Hampson, S., y Arcas, B.A.: Communication-efficient learning of deep networks from decentralized data. In: Proceedings of the 20th International Conference on Artificial Intelligence and Statistics, pp. 1273–1282 (2017)
17. Nedic, A., Ozdaglar, A.: Distributed subgradient methods for multi-agent optimization. IEEE Trans. Autom. Control **54**(1), 48–61 (2009)
18. Nedić, A., Olshevsky, A., Rabbat, M.G.: Network topology and communication-computation tradeoffs in decentralized optimization. Proc. IEEE **106**(5), 953–976 (2018)
19. Papaioannou, T.G., Ranvier, J.E., Olteanu, A., Aberer, K.: A decentralized recommender system for effective web credibility assessment. In: Proceedings of the 21st ACM International Conference on Information and Knowledge Management. pp. 704–713. ACM (2012)
20. Rendle, S., Freudenthaler, C., Gantner, Z., Schmidt-Thieme, L.: BPR: Bayesian personalized ranking from implicit feedback. In: Proceedings of the 25th Conference on Uncertainty in AI, pp. 452–461. AUAI Press, Arlington, Virginia, USA (2009)

21. Tang, H., Lian, X., Yan, M., Zhang, C., Liu, J.: D^2: decentralized training over decentralized data. In: International Conference on Machine Learning, pp. 4848–4856 (2018)
22. Tsitsiklis, J., Bertsekas, D., Athans, M.: Distributed asynchronous deterministic and stochastic gradient optimization algorithms. IEEE Trans. Autom. Control **31**(9), 803–812 (1986)
23. Vanhaesebrouck, P., Bellet, A., Tommasi, M.: Decentralized collaborative learning of personalized models over networks. In: Proceedings of the 20th International Conference on Artificial Intelligence and Statistics (AISTATS) (2017)
24. Yuan, K., Ling, Q., Yin, W.: On the convergence of decentralized gradient descent. SIAM J. Optim. **26**(3), 1835–1854 (2016)

Federated Multi-view Matrix Factorization for Personalized Recommendations

Adrian Flanagan$^{(\boxtimes)}$, Were Oyomno, Alexander Grigorievskiy, Kuan E. Tan, Suleiman A. Khan, and Muhammad Ammad-Ud-Din$^{(\boxtimes)}$

Helsinki Research Center, Europe Cloud Service Competence Center, Huawei Technologies Oy (Finland) Co. Ltd., Helsinki, Finland
{adrian.flanagan,muhammada.din}@huawei.com

Abstract. We introduce the federated multi-view matrix factorization method that learns a multi-view model without transferring the user's personal data to a central server. The method extends the federated learning framework to matrix factorization with multiple data sources. As far as we are aware, this is the first federated model to provide recommendations using multi-view matrix factorization. In addition, it is the first method to provide federated cold-start recommendations. The model is rigorously evaluated on three datasets on production settings. Empirical validation confirms that federated multi-view matrix factorization outperforms simpler methods that do not take into account the multi-view structure of the data. In addition, we also demonstrate the usefulness of the proposed method for the challenging prediction task of cold-start federated recommendations.

1 Introduction

In many machine learning problems multiple heterogeneous and related data sources or views are available to build a model. The challenge is to effectively integrate the views into a coherent model which performs better than the equivalent single view based model. As an example of a multiple view problem we consider the case of a movie recommender system where historical user-movie watch data allows us to generate user personalized recommendations. By adding other sources of information or data views such as user personal information (e.g., age, gender, and location) and movie features (e.g. for movies genre, actors, director, box office revenue) we would expect to generate better personalized recommendations. While we use the example of a movie recommender the methods described here can be generalized to any type of recommender system.

With an increasing focus on user privacy and legislation such as the GDPR[1] users may not opt-in to share their personal data and companies may be less

---

[1] https://gdpr-info.eu/.

S. A. Khan and M. Ammad-Ud-Din—Authors contributed equally to this work.

F. Hutter et al. (Eds.): ECML PKDD 2020, LNAI 12458, pp. 324–347, 2021.
https://doi.org/10.1007/978-3-030-67661-2_20

willing to record, upload and store users' data to generate multi-view recommendations. The Federated Learning (FL) [24] paradigm addresses the issue of users' privacy. In FL model learning is distributed to the end clients (i.e. user's devices), and model updates are generated locally with the users' data and only the model updates are uploaded and aggregated in a central server ensuring the raw user's private data never leaves the client device.

Multiple data views can be combined using Multi-View Matrix Factorization (MVMF) which is an extension of the standard Collaborative Filter (CF) [16] for generating recommendations. The goal of multi-view matrix factorization is to learn a joint factorization of all the three data matrices. Essentially, the joint factorization decomposes the observed data into sets of low-dimensional latent factors that capture the dependencies between the matrices. The MVMF model therefore uses a combination of the 3 datasets to learn a better model of the user-item interactions. As a result, the model is able to significantly improve recommendations to the user, thereby enhancing user experience. With the blend of matrix factorization, latent variables and multi-view machine learning approaches, it is possible to address several challenges in recommendation systems such as generating a recommendation for a new user (cold-start user), a recommendation for new items (cold-start item) and a recommendation for an entirely new user and new item (out-of-matrix prediction), which is effectively not possible with the simpler CF approaches.

Figure 1 conceptualizes a Federated Learning (FL) implementation of MVMF. The use of Federated MVMF (FED-MVMF) also allows us to address the issue of cold start in recommender systems in a distributed FL setting. We show that federation of the MVMF is technically feasible and formulate the updates using a stochastic gradient-based approach. We compare our multi-view approach with single-view matrix factorization on the MovieLens, BooksCrossing and an in-house production dataset (anonymized). The findings confirm that our model substantially outperforms the simpler alternatives. In addition, we empirically demonstrate cold-start recommendations with FED-MVMF. Our original contributions in this work are three fold: (1) we formulate as far as we know the first federated multi-view matrix factorization method with side-information sources, (2) we empirically demonstrate that the method outperforms simpler federated collaborative filter methods, (3) we present the first mechanism for cold-start predictions in federated learning mode.

2   Multi-view Matrix Factorization

Given multi-view data sources $\mathbf{R} \in \mathcal{R}^{N_u, N_v}$, $\mathbf{X} \in \mathcal{R}^{N_u, D_u}$ and $\mathbf{Y} \in \mathcal{R}^{N_v, D_y}$, for N_u users, N_v items, characterized by D_u, D_y descriptive features, the multi-view matrix factorization method MVMF is defined as a generative model [9]:

$$\mathbf{R} \sim \mathbf{P}\mathbf{Q}^T, \mathbf{X} \sim \mathbf{P}\mathbf{U}^T, \mathbf{Y} \sim \mathbf{Q}\mathbf{V}^T \tag{1}$$

where r_{ij} represents the interaction between user i and item j value $1 \le i \le N_u, 1 \le j \le N_v$, and x_{i,d_u} denote the value of user i's personal feature d_u for

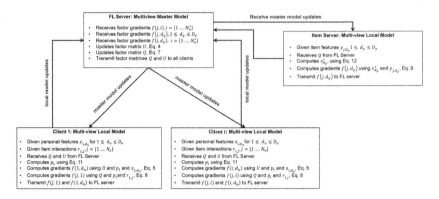

Fig. 1. The Federated Multi-View Matrix Factorization (FED-MVMF) method proposed in this study. The master model $\mathbf{Q}$, $\mathbf{U}$ is updated on the server and then distributed to clients (user's devices). Each user-specific local model p_i is computed on the client using the user's private data and $\mathbf{Q}$, $\mathbf{U}$. The updates through the gradients of $\mathbf{Q}$, $\mathbf{U}$ are computed on each user and transmitted to the FL server to update the master model $\mathbf{Q}$, $\mathbf{U}$. Meanwhile, the master model $\mathbf{Q}$ is also transmitted to the item server to compute local model $\mathbf{V}$ using the item features. The updates comprising the gradients of $\mathbf{Q}$ are transmitted to the FL server.

$1 \leq d_u \leq D_u$ and y_{j,d_y} denote the value of item j at feature d_y, for $1 \leq d_y \leq D_y$. The interactions r_{ij} are generally derived from explicit user feedback such as ratings $r_{ij} \in (1, \ldots, 5)$ given by a user i to an item j [33], or implicit feedback $r_{ij} \geq 1$ when the user i interacted with the item j and is unspecified otherwise [16]. In this work, we consider the case of implicit feedback for simplicity, however, the proposed method is applicable to explicit feedback scenario without loss of generality. The model learns the latent matrices which are represented as $\mathbf{P} \in \mathcal{R}^{N_u, K}$ the user-factor matrix, $\mathbf{Q} \in \mathcal{R}^{N_v, K}$ the item-factor matrix, $\mathbf{U} \in \mathcal{R}^{D_u, K}$ user-feature factor matrix and $\mathbf{V} \in \mathcal{R}^{D_y, K}$ the item-feature factor matrix, where K is the number of latent factors. The shared user-factor matrix $\mathbf{P}$ captures the statistical dependencies between the item interaction and user personal data views (see Appendix Fig. 2, left panel). Likewise, $\mathbf{Q}$ item-factor matrix captures the common patterns between item interactions and item-features data views. The joint factorization, therefore, learns the shared dependencies between the item interactions and side-information views. The latent factors $\mathbf{U}$ and $\mathbf{V}$ are specific to the user personal and item features respectively, capturing the view-specific variation. The inference is then performed by optimizing the cost function on the joint factorization of all the data views as:

$$J = \sum_i \sum_j c_{i,j}(r_{i,j} - p_i q_j^T)^2 + \lambda_1 \left(\sum_i \sum_{d_u} (x_{i,d_u} - p_i u_{d_u}^T)^2 + \sum_v \sum_{d_y} (y_{j,d_y} - q_j v_{d_y}^T)^2 \right)$$

$$+ \lambda_2 \left(\sum_i ||p_i||^2 + \sum_j ||q_j||^2 + \sum_{d_u} ||u_{d_u}||^2 + \sum_{d_y} ||v_{d_y}||^2 \right) \tag{2}$$

where $c_{ij} = 1 + \alpha r_{ij}$, for $\alpha > 0$ is a confidence parameter to account for implicit feedback uncertainty [16] and λ_2 is the L2-regularization term. Specifically, λ_1 can be tuned to adjust the strength of information sharing with the side-data views. For example, initializing $\lambda_1 = 0$ restricts the model to not learn any shared factors, while $\lambda_1 = 1$ pushes the model to learn factors shared between item interactions and side-data view. The $0 =< \lambda_1 =< 1$ value can be chosen informatively based on the prior expert knowledge or through hyper-parameter optimization. The latent factors $\mathbf{P}$, $\mathbf{Q}$, $\mathbf{U}$ and $\mathbf{V}$ are inferred using Alternating Least Square [9].

Moreover, several related formulations for joint factorization of data views have also been proposed using linear, kernelized and Bayesian approaches [1,7,12,27].

3   Federated Multi-view Matrix Factorization

We present the first Federated treatment of Multi-view Matrix Factorization (FED-MVMF) which combines side-information views simultaneously from both sides. The multi-view data sources are distributed and not stored on central servers. The user-item interaction data and user personal data are available on user's devices only, while, the item features are stored on the item server (see Fig. 1). The proposed framework is presented for the particular case of personalized federated recommendations, though is applicable to other domains as well.

FED-MVMF performs a federated factorization of the data views (or matrices) $\mathbf{R}$, $\mathbf{X}$, $\mathbf{Y}$ jointly as defined in Eq. 1 to learn the latent factors $\mathbf{P}$, $\mathbf{Q}$, $\mathbf{U}$ and $\mathbf{V}$. The federated factorization is formulated using stochastic gradient decent inference. We observe that federated inference of $\mathbf{U}$ and $\mathbf{Q}$ is fundamentally challenging as their updates depend on all the users while federated constraints prohibit a direct integration of user-private data. Next, we discuss our federated solution.

Federated U: Federated inference of the user-feature factors $\mathbf{U}$ is a key challenge. The update requires the personal features x_{d_u} and factor vectors p_i $\forall i \in \{1, \dots, N_u\}$ from all the users, as

$$u_{d_u}^* = (x_{d_u}P)\left(P^T P + \frac{\lambda_2}{\lambda_1}I\right)^{-1}. \tag{3}$$

Therefore, u_{d_u} cannot be inferred on individual users and must be carried out on the FL server. However, owing to the privacy constraint, each user also preserves the corresponding personal x_{d_u} and p_i locally on the device and can not transmit it to the FL server, further complicating the update of $u_{d_u}^*$. We formulate a stochastic gradient descent approach to allow for the update of the u_{d_u} vectors on the server, while preserving the user's privacy. Formally, u_{d_u} is updated on the FL server as

$$u_{d_u} = u_{d_u} - \gamma \frac{\partial J}{\partial u_{d_u}}, \tag{4}$$

for some gain parameter γ to be determined. However, computation of $\partial J/\partial u_{d_u}$ requires a summation over all users. Therefore, we define $f(i, d_u)$ as

$$f(i, d_u) = [(x_{i,d_u} - p_i u_{d_u}^T)]p_i, \tag{5}$$

where $f(i, d_u)$ is calculated on each user i independently of all the other users. All the users then send back the gradient values $f(i, d_u)$, $i = \{1, \ldots, N_u\}$ to the FL server. Finally, $\partial J/\partial u_{d_u}$ be formulated as aggregate of the user gradients as

$$\frac{\partial J}{\partial u_{d_u}} = -2 \sum_i f(i, d_u) + 2\lambda_2 u_{d_u}, \tag{6}$$

enabling federated update of u_{d_u}.

Federated Q: Analogous to **U**, the federated inference of **Q** is also non-trivial and more complex as **Q** is a shared factor between user-item interactions and item-features. The inference depends on both latent factors **P** and **V**. Practically, the update of item factor vectors $q_j \forall j \in \{1, \ldots, N_v\}$ requires the user factor vectors $p_i \forall i \in \{1, \ldots, N_u\}$ from all the users and $v_{d_y} \forall v \in \{1, \ldots, N_v\}$ from all the items,

$$q_j^* = (r(j)\hat{C}^{(j)}P + \lambda_1 y_j V)(P^T C^{(j)} P + \lambda_1 V^T V + \lambda_2 I)^{-1}.$$

Therefore, the updates of q_j can not be done on the user's device and must be performed on the FL server. But, due to the privacy constraints, each user preserves the p_i locally on the device and can not send it to the FL server, further complicating the inference of q_j^*. We present a stochastic gradient descent approach to allow for the update of the q_j vectors on the FL server, while preserving the user's private data. Formally, q_j is updated on the FL server as

$$q_j = q_j - \gamma \frac{\partial J}{\partial q_j}, \tag{7}$$

for gain parameter γ to be determined. However, computing $\partial J/\partial q_j$ involves a summation over all users i and item features d_y. Therefore, we define $f(j, i)$ and $f(j, d_y)$ as

$$f(j, i) = [c_{ji}(r_{j,i} - p_i q_j^T)]p_i, \tag{8}$$

$$f(j, d_y) = [(y_{j,d_y} - v_{d_y} q_j^T)]v_{d_y}, \tag{9}$$

where $f(j, i)$ is calculated on each user i independently of all the other users. All the users then report back the gradient values $f(j, i)$, $i = \{1, \ldots, N_u\}$ to the FL server. And $f(j, d_j)$ is calculated on the item server and the gradient values $f(j, d_y)$, $d_y = \{1, \ldots, D_y\}$ are transmitted to the FL server. Finally, the derivative can then be computed using an aggregate of the user and item-features gradients as

$$\frac{\partial J}{\partial q_j} = -2 \sum_i f(j, i) - 2\lambda_1 \sum_{d_y} f(j, d_y) + \lambda_2 q_j, \tag{10}$$

making it possible to perform federated update of q_j.

Localized P: The inference of user factors $\mathbf{P}$ depends on the item-factors $\mathbf{Q}$, user-features factors $\mathbf{U}$ and user's private interaction data r_{ij}, available at each user's device. The factor models $\mathbf{Q}$ and $\mathbf{U}$ are received from the FL server and are used to compute the corresponding p_i^* locally as

$$p_i^* = (r(i)\hat{C}^{(i)}Q + \lambda_1 x_i U)(Q^T C^{(i)}Q + \lambda_1 U^T U + \lambda_2 I)^{-1} \qquad (11)$$

where $p_i = p_{i*}$ is the optimal solution obtained $\partial J(p_{u*})/\partial p_u = 0$, from Eq. 2. Notably, the updates can be carried out independently for each user i without reference to any other user's personal data.

Localized V: The item's features y_{d_y} is used to compute the $v_{d_y}^*$ for each item on the item-server. Using $\mathbf{Q}$, the updates can be carried out independently for each item j without reference to any private or sensitive data as

$$v_{d_y}^* = (y_{d_y}Q)(Q^T Q + \frac{\lambda_2}{\lambda_1}I)^{-1}. \qquad (12)$$

where $v_{d_y} = v_{d_y}^*$ is the optimal solution obtained $\partial J(v_{d_y^*})/\partial v_{d_y} = 0$, from Eq. 2.

Iterative Federated Updates: We outline the steps of the proposed FED-MVMF method in Algorithm 1. The FL iterations are performed till the model has converged, where in each iteration master model is updated when the number of collected federated updates from users and item server reached a certain threshold Θ. The model parameters $\mathbf{Q}$ and $\mathbf{U}$ are updated using stochastic gradient descent with the Adaptive Moment Estimation (Adam) method [19] to achieve better convergence and greater stability [25] (see Appendix Sect. A.2). In the standard mode, the computational complexity of the algorithm is $\mathcal{O}(IK^2 N_v N_u)$ where I is the number of iterations. However in FL set-up, several other parameters can influence the computational complexity of the algorithm such as number of users participating in the FL update, how frequent the updates are sent by the users, what are the specifications of user's devices (laptop or mobile) and importantly the communication over Internet and associated network latency [22].

4   Federated Cold-Start Recommendations

The multi-view matrix factorization allows for the inclusion of side-information sources for both users and items simultaneously, making it possible to solve the difficult task of predicting recommendation to new users (cold-start users) or new items (cold-start items) and/or predicting recommendations to an entirely new user on a previously unseen item (see Appendix Fig. 2, right panel). Here, a common assumption is that for a new user or a new item, there exist no historical interaction data r_*, though user's personal features $x_{d_u}^*$ or item's features $y_{d_y}^*$ are available. However, in contrast to the cold-start recommendation solution offered by standard approaches [1,7], the FL requires customized solution owing to the

---

Algorithm 1. FED-MVMF: Federated Multi-View Matrix Factorization

---

1: **FL Server**
2: Number of items N_v, Number of user features D_u, Number of factors K
3: Initialize master model factor matrices $\mathbf{Q},\mathbf{U}$ and update threshold Θ
4: **while** True **do**
5:    Transmit $\mathbf{Q}$ and $\mathbf{U}$ to users $i \in [1, N_u]$
6:    Transmit $\mathbf{Q} \rightarrow$ **Item Server**
7:    Receive factor $\mathbf{Q}$ gradients $f(j,i)\ \forall\ j \in [1, N_v],\ \forall\ i \in [1, N_u]$
8:    Receive factor $\mathbf{Q}$ gradients $f(j, d_y)$ for $d_y \in\ 1 \le d_y \le D_y,\ \forall\ j \in [1, N_v]$
9:    Receive factor $\mathbf{U}$ gradients $f(d_u, i)$ for $d_u \in\ 1 \le d_u \le D_u,\ \forall\ i \in [1, N_u]$
10:   **if** NumberGradientUpdates $>= \Theta$ **then**
11:      Update $\mathbf{U}$ using **Eq. 4**
12:      Update $\mathbf{Q}$ using **Eq. 7**
13:   **end if**
14: **end while**
15:
16: **FL User**
17: **while** True **do**
18:   Receive master model factor matrices $\mathbf{Q}$, $\mathbf{U}$
19:   Compute local model factor p_i using Eq. 11
20:   Generate recommendations: $r_{i,j} = p_i \times \mathbf{Q}^T\ \forall\ j \in [1, N_v]$
21:   Compute factor $\mathbf{U}$ gradients $f(d_u, i)$ using **Eq. 5**
22:   Compute factor $\mathbf{Q}$ gradients $f(j, i)$ using **Eq. 8**
23:   Transmit $f(j, i)$ and $f(d_u, i) \rightarrow$ **FL Server**
24: **end while**
25:
26: **Item Server**
27: **while** True **do**
28:   Receive master model factor matrix $\mathbf{Q}$
29:   Compute local model factor $v_{d_y}^*$ using **Eq. 12**
30:   Compute factor $\mathbf{Q}$ gradients $f(j, d_y)$ using **Eq. 9**
31:   Transmit $f(j, d_y) \rightarrow$ **FL Server**
32: **end while**

---

privacy constraints and decentralized nature of the multi-view data. We next present the solution of federated cold-start recommendations problem utilizing the proposed FED-MVMF model.

Cold-Start User Recommendation: When a new user joins a FL recommendation system with no previous item interaction data, a new local user-factor model p_* is created using the user-feature model $\mathbf{U}$ and user personal features $x_{d_u}^*$. It can be computed as a least-squares solution w.r.t p_* of the loss in Eq. 2. Only the user-feature part of the loss and corresponding regularizer depends on p_*. The solution is identical to Eq. 3, where the matrix $\mathbf{P}$ is replaced by $\mathbf{U}$. The cold-start user recommendation is then generated as outlined in Algorithm 2.

Cold-Start Item Recommendation: New items are frequently added to the collection and it is greatly important for the service provider to recommend

---

Algorithm 2. Federated cold-start user recommendation

1: **FL User**
2: Receive master model factor matrices $\mathbf{Q}, \mathbf{U} \leftarrow$ **FL Server**
3: Get personal features $x^*_{d_u}$ of new user
4: Compute new local factor matrix $p_* = x^*_{d_u} \mathbf{U}(\mathbf{U}^T\mathbf{U} + \frac{\lambda_2}{\lambda_1}I)^{-1}$
5: Compute recommendations $r_* = p_*\mathbf{Q}^T$

---

the new item to a potentially interested user from day zero. The FED-MVMF solves the cold-start item recommendation challenge by creating a new item factor matrix q_* at the item server, given the item features $y^*_{d_y}$ and $\mathbf{V}$. It is done symmetrically to cold-start user recommendation. The master model item-factor matrix is updated as $\mathbf{Q}_* = [\mathbf{Q}^{(N_v \times K)} \mid q_*^{(1 \times K)}]$ and transmitted to the FL server. Users receive the updated $\mathbf{Q}_*$ and compute recommendations: $r_{i,*} = p_i\mathbf{Q}_*^T$ including the new item as outlined in Algorithm 3.

Cold-Start User-Item Recommendation: The prediction of cold-start user-item recommendation is deemed as out-of-matrix prediction task and is perhaps the most challenging in practice. However, the solution is made possible by FED-MVMF with inclusion of factor matrices originating from side-information sources. Technically, FED-MVMF solves the prediction task by combining solutions of *federated cold-start user recommendation* and *federated cold-start item recommendation*. The user creates a new local user-factor model p_* using $\mathbf{U}$ and user personal features $x^*_{d_u}$ and receives the updated master model item-factor matrix $\mathbf{Q}_*$ to compute recommendations: $r_* = p_*\mathbf{Q}_*^T$ for the new item (Algorithm 4 in Appendix).

Algorithm 3. Federated cold-start item recommendation

1: **Item Server**
2: Receive master model factor matrix $\mathbf{Q} \leftarrow$ **FL Server**
3: Get local item-feature factor matrix $\mathbf{V}$
4: Get item features $y^*_{d_y}$ of new item
5: Compute new item factor matrix $q_* = y^*_{d_y}\mathbf{V}(\mathbf{V}^T\mathbf{V} + \frac{\lambda_2}{\lambda_1}I)^{-1}$
6: Update the item-factor model matrix $\mathbf{Q}$ with $q_* \rightarrow \mathbf{Q}_*$
7: Transmit $\mathbf{Q}_* \rightarrow$ **FL Server**
8: **FL Server**
9: Receive updated master model factor matrix $\mathbf{Q}_*$
10: Transmit $\mathbf{Q}_*$ master model to existing users
11: **FL User**
12: Compute recommendations $r_{i,*} = p_i\mathbf{Q}_*^T$

---

5   Related Work

The federated multi-view matrix factorization problem and our solution for it are related to several matrix factorization as well as federated learning method-ologies. We next discuss the existing methods that solve special cases of the problem, and relate them to our work.

5.1   Multi-view Learning

For the non-federated case, our model can be seen as a multi-view matrix fac-torization with side-information sources [9], however formulating simpler regu-larization assumptions, suitable for the privacy-preserving models.

One-Way Factorization: Several methods perform integrated analysis of mul-tiple matrices (or views) $\mathbf{X}^{(1)} \in \mathcal{R}^{N \times D_1}$, ..., $\mathbf{X}^{(M)} \in \mathcal{R}^{N \times D_M}$, where N is the number of paired samples in M matrices with D_m dimensions, such that the matrices are paired in one mode only. Classical approaches like Canonical Cor-relation Analysis [15] perform joint factorization of two matrices. More recent advancements, like Group Factor Analysis [20] can integrate several matrices. However, unlike ours, none of these methods perform factorization of matrices paired on both sides.

Two-Way Factorization: Similar to our approach a few methods perform two-way factorization of matrices coupled in both modes [7,9]. Moreover, [12] introduced a non-linear Kernelized Bayesian Matrix Factorization coupled with multiple side-information sources in $\mathbf{X}$ and $\mathbf{Y}$. Recently, [30] scales matrix fac-torization with two-way side information sources for efficient inference when the number of covariates is large. Two way factorization or collective matrix factorization methods are especially suitable and widely used for personalized recommendation applications [1,7,27]. However, none of these methods presents a federated learning solution, and our method is the first to provide a federated multi-view matrix factorization integrating side-information sources from both sides.

5.2   Federated Learning

Within federated learning, our method is a general multi-view matrix factoriza-tion, where several existing methods can be seen as special cases of our model.

One Matrix: For a single matrix, our model can be seen as a collaborative filter [2,5,8]. This case is also close to distributed matrix factorization of [11,32,33]. However, none of these approaches is able to integrate multiple data views.

Two Matrices: For two data sets, with partially paired samples, vertical feder-ated learning approaches [13,23] take advantage of the common samples, while horizontal federated learning approaches [24,28] leverage the overlap of feature columns to improve the predictive performances. For a comprehensive review of these approaches see [22,31]. Furthermore, recently [17] performed federated factorization of multiple matrices, however, their method is able to factorize matrices paired in a single mode only.

Other federated learning approaches based on neural networks [21,24] do not address the problem of personalized recommendation using multiple side-data views. In addition, [24] aggregates model weights at the server whereas we employ a gradient based aggregation suitable for matrix factorization. Other approaches like meta-learning [6] have been proposed in the context of recommendation systems. Recently, [18] adapt the meta-learning approach and parallel implementation of federated learning, however, none of these methods address the multi-view matrix factorization problem. To the best of our knowledge, our method is the first federated multi-view matrix factorization that integrates data from both sides to provide personalized federated recommendations as well as support cold-start recommendations in federated learning.

6 Datasets

In this study, we used three datasets: two public and a private in-house anonymized production dataset. These datasets are *MovieLens-1M* [14], *BookCrossings* [34] and *in-house*. These datasets are characterized by varying degree of sparse user-item interactions, in addition to having descriptive features for both users and items. Interactions of in-house dataset are implicit while interactions of public datasets are explicit i. e. they include an exact rating a user specified for an item. For generalize treatment, we convert the public datasets into implicit datasets as well. The pre-processing details of each dataset are provided in Appendix (Sect. A.1). Final characteristics of a dataset after pre-processing are presented in Table 1.

Particularly, FED-MVMF integrates user and item features for matrix factorization. We treat all user and item features as categorical[2]. Some features

Table 1. Overview of the datasets used in the study, where # interactions refers to the total number of user-item interactions and Sparsity (%) denotes the percentage of observed interactions in a particular dataset. User features represents the number of features used to encode personal data, while Item features is the number of features used to describe item features. In this study, the user-interactions, user features and item features are denoted by **R**, **X** and **Y** respectively.

Dataset	# Users	# Items	# Interactions	Sparsity (%)	User features	Item features
			R		**X**	**Y**
Movielens	6040	3064	914676	4.9%	3434	1128
BookCrossings	19912	2999	72794	0.12%	7405	10000
In-house production dataset (anonymized)	815614	3912	2213122	0.07%	300	300

---

[2] Except for Movielens item-features, which are described by real numbers.

may have high cardinality (e.g. key-words of a book title) and different number of features per item. For instance, one book has 1 key-word in its title while another may have 3. Therefore, we processed all user (or item) features using a hash function of a certain output dimension. We call it hash size. This size depends on a dataset and exact numbers are provided in Table 1. More precisely, we form stings like this {*feature_name*}__{feature_value} and hash all the strings of a user (item) into a vector of hash size. Originally, this is a vector of zeros. Hashing here means setting to 1 the corresponding coordinate of the vector. If there is a hashing collision we increase the corresponding value by 1. As a result, we obtain a sparse vector of fixed size for each user and item.

7   Experiments and Results

We used Collaborative Filter (CF) [16] and Federated Collaborative Filter (FCF) [2,5,8] as baseline comparison methods. Both CF and its federated counterpart FCF are matrix factorization methods do not integrate side-information sources. In addition, we use Multi-view Matrix Factorization (MVMF) [9], as an upper-bound comparison for our FED-MVMF method. MVMF allows incorporating user and item-feature matrices to learn a two-way factorization in the non-federated setting. This serves as an upper-bound recommendation performance of each dataset by non-federated MVMF.

Hyper-parameters, Training and Evaluation Criteria: The CF or FCF and MVMF or FED-MVMF models share similar set of hyper-parameters except λ_1 which controls the strength of information shared with the side-information sources, and is specific to MVMF/FED-MVMF. We used Bayesian optimization approach [29] to choose optimal hyper-parameters (see Appendix Sect. A.4). We performed 3 rounds of model rebuilding in production setting. In each round, the item interactions for every user were randomly divided into 80% training and 20% test sets, and select the metric value when 1000 iterations of federated model updates are reached, to ensure model convergence. Notably, in each federated iteration only a subset of users contribute to update the master model and report their performance metrics. Hence, at *Iteration* = 1000 we take average of the previous 10 values to account for sampling biases in metric values. To evaluate the models, we use the widely adapted recommendation metrics [3], Precision, Recall, F1, Mean Average Precision (MAP) and Normalized Mean Ranking (NMR) for the top 10 predicted recommendations (see Appendix Sect. A.5). The metrics were further normalized by the theoretically best achievable metrics for each dataset, to make them comparable. We also computed "**Impr%**" to estimate the relative improvement of FED-MVMF with respect to CF or FCF, and "**Diff%**" to estimate the relative difference between standard MVMF and FED-MVMF as

$$\mathbf{Impr\%} = \left| \frac{\text{MetricMean(FED-MVMF)} - \text{MetricMean(CF or FCF)}}{\text{MetricMean(Cf or FCF)}} \right| \times 100$$

$$\mathbf{Diff}\% = \left| \frac{\text{MetricMean(MVMF)} - \text{MetricMean(FED-MVMF)}}{\text{MetricMean(FED-MVMF)}} \right| \times 100$$

Next we summarize the main results, nevertheless we give details on our production equivalent client-server architecture that we developed to perform experiments and to collect metrics in Appendix (Sect. A.3). We also show performance plots for these metrics demonstrating stable model convergence over several rounds of federated iterations (see Appendix Sect. A.6).

Recommendation Performance: We demonstrate the performance of proposed FED-MVMF on three real personalized recommendation data sets. As FED-MVMF is the first federated multi-view matrix factorization method, we use FCF and CF as baseline comparison methods. The results demonstrate that the FED-MVMF outperforms the comparison methods consistently as shown in Table 2. Importantly, the model shows a performance gain of up to 70% from FCF (shown as Impr % in Table 2) when measured across different metric and datasets. Specifically, we observe a significant improvement for highly sparse datasets such as the in-house production dataset as well as the public BooksCrossing dataset. This finding implies that the use of side-information sources is beneficial for production datasets which are inherently sparse in nature. Both, in-house production and BookCrossing dataset integrated discretized features compared to the Movielens dataset which included dense features encoded with real-values, and show a larger improvement. We believe, that the amount of performance advantage depends on the type and nature of side-information sources, amongst other factors, as shown by our results. In comparison to the upper-bound MVMF method, our method closely matches the performance of non-federated method on In-house production and Movielens datasets, and gets close in the BookCrossings data, confirming that FED-MVMF achieves the required performance (shown as Diff % in Table 2).

Here, our main research goal was to propose a FED-MVMF method that can take advantage of the side-information sources in a federated learning. The development of a multi-view model is a technically challenging research problem owing to the federated nature of the multi-view data sets and hard constraints of the federated learning design. Our results confirm that the developed solution takes advantage of the side-information sources to provide substantially improved federated recommendations. Furthermore, our payload estimates (presented in Appendix Sect. A.7), when tested in production settings, showed that FED-MVMF uses only 40% more communication in-terms of model size (kilobytes) and 30% more transfer time (seconds).

Cold-Start Recommendation Performance: FED-MVMF provides a principled solution to the commonly occurring problem in production: cold-start recommendations. To demonstrate the usefulness of the FED-MVMF model for cold-start predictions, we conducted comprehensive analysis of all the three cold-start scenarios: cold-start users, cold-start items and cold-start user-items. For case of cold-start users scenario, the model did not observe any of the interaction

Table 2. Comparison of the test set performance between FED-MVMF and FCF methods. The values denote the mean ± standard deviation of metric values across 3 different model builds. "**Impr %**" refers to the relative percentage improvement in FED-MVMF vs FCF (CF). FED-MVMF model outperforms the FCF (CF) model showing a substantial improvement, going upto 70%. "**Diff %**" refers to the relative percentage difference in MVMF vs. FED-MVMF. The MVMF's performance is deemed as an upper-bound, which is closely matched on two datasets by FED-MVMF.

	Precision	Recall	F1	MAP	NMR
In-house anonymized dataset					
CF	0.1817 ± 0.0016	0.1824 ± 0.0015	0.1818 ± 0.0016	0.0851 ± 0.0009	0.3028 ± 0.0013
FCF	0.1811 ± 0.0009	0.1816 ± 0.0007	0.1812 ± 0.0007	0.0842 ± 0.0009	0.3097 ± 0.0017
MVMF	0.2812 ± 0.0019	0.2812 ± 0.0019	0.2812 ± 0.0019	0.1441 ± 0.0015	0.15 ± 0.0024
FED-MVMF	0.2771 ± 0.0022	0.2779 ± 0.0028	0.277 ± 0.0023	0.1411 ± 0.0021	0.1545 ± 0.0006
Impr %	53 (53)	53 (52)	53 (52)	68 (66)	50 (49)
Diff %	1	1	2	2	3
Movielens dataset					
CF	0.3645 ± 0.0012	0.3623 ± 0.0049	0.3626 ± 0.0038	0.2292 ± 0.0007	0.0904 ± 0.0003
FCF	0.3410 ± 0.0100	0.3571 ± 0.0019	0.3525 ± 0.0037	0.2055 ± 0.0090	0.1006 ± 0.0021
MVMF	0.3687 ± 0.0041	0.3677 ± 0.0016	0.3678 ± 0.0019	0.2329 ± 0.0029	0.0888 ± 0.0004
FED-MVMF	0.3666 ± 0.0017	0.3825 ± 0.0043	0.3785 ± 0.0037	0.2295 ± 0.0013	0.0841 ± 0.0012
Impr %	8 (1)	7 (6)	7 (4)	12 (0)	16 (7)
Diff %	1	4	3	1	6
Book crossings dataset					
CF	0.055 ± 0.0036	0.0548 ± 0.0035	0.0549 ± 0.0036	0.0255 ± 0.0015	0.4119 ± 0.0049
FCF	0.0431 ± 0.0025	0.0428 ± 0.0023	0.0431 ± 0.0024	0.0166 ± 0.0012	0.39 ± 0.0015
MVMF	0.0753 ± 0.0048	0.0754 ± 0.0038	0.0753 ± 0.0046	0.0323 ± 0.0027	0.3627 ± 0.0022
FED-MVMF	0.0639 ± 0.0011	0.0625 ± 0.0016	0.0636 ± 0.0013	0.0284 ± 0.0012	0.3378 ± 0.006
Impr %	48 (16)	46 (14)	48 (16)	71 (11)	13 (18)
Diff %	18	21	18	14	7

data. A random subset of 10% users were completely held-out during the model training and model parameters were learned with remaining 90% of the users. For case of cold-start items, a random subset of 10% items were entirely left-out during the model training and model parameters were learned with remaining 90% of the items. For case of cold-start users-items, a random subset of 10% users and items were excluded from the model training and model parameters were learned with remaining 90% of the users and items. Likewise, 3 rounds of the model rebuilding were done for each of the scenario. Table 3 illustrates the recommendation performances across all scenarios. The result demonstrate that without loss of generality, the FED-MVMF model can be used for cold-start recommendations reliably. Specifically, the model shows good cold-start prediction performances for a new user, which is a fundamentally valuable in a federated learning solution where new users are enrolled in the service continuously. The performance of cold start item prediction is observed to be lower than that of cold start user indicating that prediction may be improved further. It is likely that the difference is due to lower quality of the item side-information source.

Table 3. Cold-Start recommendation performance metrics of FED-MVMF using different metrics. In case of CS-Users, only these users constitute the test set. In case of CS-Items, only these items are in test set. The proportion of train/test data is different from the previous experiments, hence the metrics are not directly comparable. The values denote the mean ± standard deviation across 3 different model builds. The proposed FED-MVMF model made possible to recommend items for challenging cold-start scenarios in federated learning model.

	Precision	Recall	F1	MAP	NMR
In-house anonymized dataset					
CS-Users	0.3559 ± 0.0015	0.3359 ± 0.0012	0.3518 ± 0.0014	0.1743 ± 0.0012	0.0621 ± 0.0025
CS-Items	0.0263 ± 0.0006	0.0515 ± 0.0011	0.0292 ± 0.0007	0.0114 ± 0.0003	0.2727 ± 0.0032
CS-Users-Items	0.1739 ± 0.0012	0.3352 ± 0.0027	0.1916 ± 0.0014	0.1384 ± 0.0030	0.0792 ± 0.0030
Movielens dataset					
CS-Users	0.4618 ± 0.0086	0.5008 ± 0.0102	0.4984 ± 0.0093	0.3504 ± 0.0088	0.3025 ± 0.0072
CS-Items	0.0043 ± 0.0003	0.0464 ± 0.0032	0.0291 ± 0.002	0.0031 ± 0.0005	0.4157 ± 0.0006
CS-Users-Items	0.0440 ± 0.0031	0.4528 ± 0.0252	0.2903 ± 0.0173	0.0239 ± 0.0015	0.3384 ± 0.0084
Book crossings dataset					
CS-Users	0.0521 ± 0.0027	0.0559 ± 0.0022	0.0531 ± 0.0024	0.0254 ± 0.0032	0.3396 ± 0.0034
CS-Items	0.0054 ± 0.0005	0.012 ± 0.001	0.0063 ± 0.0005	0.0047 ± 0.0004	0.4918 ± 0.0128
CS-Users-Items	0.0166 ± 0.0030	0.0399 ± 0.0068	0.0199 ± 0.0034	0.0137 ± 0.0054	0.3503 ± 0.0055

Moreover, the low standard deviation of the results indicates that model predictions are precise across variations in training sets.

8   Conclusion

We introduced the federated multi-view matrix factorization method where the federated paradigm does not require collecting raw user data to a centralized server thus enhancing the user privacy. The proposed federated multi-view model is tested on three different datasets and we showed that including the side-information from both users and items increases recommendation performance compared to federated Collaborative Filter. In addition, our method achieves performances comparable to the upper bound of non-federated MVMF. Our federated multi-view approach also provides a novel solution to the cold-start problem common to standard Collaborative Filter recommenders. The results establish that the federated multi-view model can provide better quality of recommendations without comprising the user's privacy in the widely used recommender applications.

Future Work: An important aspect of any federated learning system is the amount of data, or the payload, to be moved between the FL server and user. In any matrix factorization based recommender system the model size, or factor matrix size is directly proportional to the number of items to recommend which in large federated recommendation systems is not feasible. A key next challenge would be to break the direct dependence of model size on the number of items to recommend.

A Appendix

A.1 Datasets

Three datasets have been used in this study: two public ones and private in-house dataset. These datasets are *MovieLens-1M*, *BookCrossings* and *in-house*. Experimenting with public datasets allows to test our methods on well established benchmarks. Besides, we show our methods' performance on different amount of user-item interactions and sparsity. All datasets include history of user-item interaction events and features of both users and items. Interactions of in-house dataset are implicit while interactions of public datasets are explicit i. e. they include an exact rating a user selected for an item. For uniform treatment we convert public datasets into implicit datasets as well. The pre-processing details of each dataset are provided below. Final characteristics of a dataset after pre-processing can be found in Table 1 (main text).

MovieLens-1M. MovieLens dataset contains about 1 million explicit ratings users selected for movies [14]. We converted explicit ratings to implicit ones simply by assuming that a user watched a movie if she put a rating for it, otherwise she has not watched. We also ignore timestamps in all subsequent experiments. User features are the following: *Age, Gender, Occupation* and *Zip-Code*. We converted the ZipCodes into US regions (e.g. *MidWest, South* etc.), therefore all user features are categorical with small cardinality. Item features are much richer. Each item is described by 1128 real numbers from interval [0, 1]. Each value correspond to the strength of some tag. Examples of tags are (*atmospheric, thought-provoking, realistic*, etc.). After excluding ratings which does not have both item and user features we have 914676 interactions in total. More statistics are provided in Table 1 (main text).

BookCrossings. BookCrossings dataset is a dataset scraped by [34] from the popular books rating web-site[3]. Dataset contains both explicit and implicit ratings. At first, we have discarded all 0 ratings, then we have substituted all positive ratings with 1. Hence, we made implicit ratings from the explicit ones similarly to MovieLens pre-processing. We also selected 2999 most popular items and left only interactions with them. the amount of items is taken to be close to other datasets. It makes results more comparable and reduce computational workload. User feature include only *Age* and *Location*. We discard all the users with empty or too high age, then we have formed age groups of 10 years intervals. Location typically consist of town, region and country. We cleaned it as much as possible, but anyway it is a high cardinality categorical feature. There are 4 book features: *Book-Title, Book-Author, Year-Of-Publication, Publisher*. We extracted the key-words from titles and use them as features. All book features except the Year-Of-Publication have high cardinality. Parameters of pre-processed dataset are given in Table 1 (see main text).

---

[3] https://www.bookcrossing.com/.

In-House Production Dataset. The in-house production dataset consists of a data snapshot extracted from the database that contains user view events (interactions). It is the largest dataset we experimented with. We did not filter out users or items if the amount of events with those is small. Hence, many users have very few interactions which makes this datasets challenging for Collaborative filtering methods like FCF. User features have several categorical features and some of those have high cardinality. In general, user features are similar to the user features of public datasets. Item features are similar to the tags features of the MovieLens data although not so reach. Further statistics are in Table 1 (see main text).

A.2   Multi-view Matrix Factorization

Figure 2 illustrates the principle of learning from the multiple data views 1) user-item (video), 2) user-features, 3) item (video)-features for predicting personalized movie recommendations. Refer to the main text for more detail.

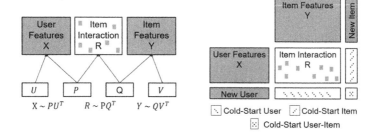

Fig. 2. Multi-view matrix factorization method with side-information sources (left panel). Cold-start prediction using FED-MVMF (right panel). For prediction of a newly released item, the model predicts the user-item interactions (cold-start item). Similarly, if a new user comes in, the model can make predictions on which items this user is likely to interact, even though there is zero historical interaction data for this user (cold-start user). Finally, when a new user signs up and new items are released, the model can make prediction with zero historical user-item interaction information for the user and item both (cold-start user item).

Adaptive Learning Rate for FED-MVMF. In Method's section (main text), a constant gain factor γ is used for the update of Q and $\mathbf{U}$ and it is seen that the value needs to be chosen with some care to ensure convergence. Gradient descent is the simplest form of optimisation and there are many variations on it which can lead to faster convergence and greater stability some of which are summarised in [25]. The Adaptive Moment Estimation (Adam) method [19] is used in the context of FCF [2]. We resort to the same approach for inference of FED-MVMF model. In Adam the gradient descent is split into 2 separate parts which record an exponentially decaying of past squared gradients v_t and an exponentially decaying average of past gradients m_t,

$$m = \beta_1 m + (1 - \beta_1)\frac{\partial J}{\partial q_j}$$

and

$$v = \beta_2 v + (1 - \beta_2)\left(\frac{\partial J}{\partial q_j}\right)^2$$

with $0 < \beta_1, \beta_2 < 1$. The m, v are typically initialized to 0 values and hence biased towards 0. To counteract these biases bias corrected versions of m, v are given by,

$$\hat{m} = \frac{m}{1 - \beta_1},$$

and

$$\hat{v} = \frac{v}{1 - \beta_2}.$$

The updates are then given by,

$$q_j = q_j - \frac{\gamma}{\sqrt{\hat{v}} + \epsilon}\hat{m}$$

$0 < \gamma < 1.0$ is a constant learning rate and $0 < epsilon \ll 1$ e.g. 10^{-8} is to avoid a divide by 0 scenario. We used Bayesian Optimization [5] approach to chose the values of β_1, β_2, γ and ϵ.

Likewise, we adopt the same treatment for the inference of user-feature factors $\mathbf{U}$

Cold-Start User-Item Recommendation: The prediction of cold-start user-item recommendation is deemed as out-of-matrix prediction task and is perhaps the most challenging in practice. However, the solution is made possible by FED-MVMF with inclusion of factor matrices originating from side-information sources. Technically, FED-MVMF solves the prediction task by combining solutions of *federated cold-start user recommendation* and *federated cold-start item recommendation*, as described in the paper. The user creates a new local factor model p_*^* using $\mathbf{U}$ and user personal features $x_{d_u}^*$ and receives the updated master model item-factor matrix $\mathbf{Q}_*$ to compute recommendations: $r_* = p_* \mathbf{Q}_*^T$ for the new item as outlined in Algorithm 4.

A.3   Federated Learning Production System Design

We implement a production equivalent client-server architecture [10,26], in which numerous clients are served by a single FL server and an Items service. All entities are implemented with Python (3.7.3) in a multiprocessing setup and cloud hosted on Ubuntu Linux 18.04 server infrastructure. The FL server has two data persistence layers of Redis (4.0.9) and PostgreSQL (9.6.10) databases,

Algorithm 4. Federated cold-start user-item recommendation

1: **Item Server**
2: Receive master model factor matrix $\mathbf{Q} \leftarrow$ **FL Server**
3: Get local item-feature model matrix $\mathbf{V}$
4: Get item features $\overset{*}{y}_{d_y}$ of new item
5: Compute new item factor matrix $q_* = \overset{*}{y}_{d_y} \mathbf{V} (\mathbf{V}^T \mathbf{V} + \frac{\lambda_2}{\lambda_1} I)^{-1}$
6: Update the item-factor model matrix $\mathbf{Q}$ with $q_* \to \mathbf{Q}_*$
7: Transmit $\mathbf{Q}_* \to$ **FL Server**
8: **FL Server**
9: Receive updated master model factor matrix $\mathbf{Q}_*$
10: Transmit $\mathbf{Q}_*, \mathbf{U}$ master model to existing users
11: **FL user**
12: Receive master model factor matrices $\mathbf{Q}_*$, $\mathbf{U}$
13: Get personal features $\overset{*}{x}_{d_u}$ of new user
14: Compute new local factor matrix $p_* = \overset{*}{x}_{d_u} \mathbf{U} (\mathbf{U}^T \mathbf{U} + \frac{\lambda_2}{\lambda_1} I)^{-1}$
15: Compute recommendations $r_* = p_* \mathbf{Q}_*^T$

the Items service has only the Redis layer and the clients have no persistence layer. The hardware specifications are 8 cores with 64 gigabytes (GB) of memory for FL server and 16 cores with 16 GB of memory for Items service and clients.

Both FL server and Items service use Gunicorn (19.7.1), the Python based web server to expose Application Programming Interfaces (API) as the only mechanisms to consume their services. Nginx (1.14) server lies between Gunicorn and clients optimizing service requests/responses by caching, compressing and decompressing payloads. Additionally, the FL server Base64 encodes/decodes all outgoing and incoming communications.

The FL server initializes a master model for each of the available models with their respective hyper-parameters. FL server and clients query the Items service for item metadata. Thus, data on each client is formulated from the user's personal data, item interactions and features associated with that item. Each client downloads a copy of an initialized master model and it's *update-signature* from the FL server to their local storage. Model updates, metrics and inferences are derived from this local copy against user's data on the client. Periodically, clients encapsulate an update payload by combining their model updates and performance metrics. The update payload is randomly uploaded to the FL server for aggregation. This random upload strategy differs from [4] approach that maintains a subset of pre-selected known clients to upload their updates, by providing client anonymity, enhancing their privacy.

The FL server implements queuing strategy to process incoming clients' requests, responses and updates, for models at different stages. An update processor validates client payloads based on their update-signature before appending them to their respective queues. A First In First Out (FIFO) model aggregator pops the oldest payload from the queue, recovering its model update and metrics. Updates are aggregated, an update counter incremented and the metrics persisted to a Structured Query Language (SQL) database for performance monitoring. The update counter is governed by the threshold parameter Θ, defined for each model at initialization. When sufficient Θ aggregates have been accumulated, the aggregator first invalidates the current update-signature, implicitly informing clients not to upload more payloads and prepare for an updated master model and update-signature. Then, a new model composed of the previous model and the its updates aggregate, is promoted replacing its predecessor master model. A new update-signature is also generated for the renewed master model, prior to flushing the payload queues and updating the update processor and aggregator validation values.

A.4   Hyper-parameters Selection

Table 4 illustrates the configuration settings for Bayesian optimization describing hyper-parameter bounds that were used to obtain the optimal set of hyper-parameters for the models.

Table 4. Summary of the hyper-parameters selected using Bayesian Optimization. Particularly, K, α, λ_1 and λ_2 are related to model factorization, while β_1, β_2, ϵ and γ are associated with Adam's learning rate and Θ is FL hyper-parameter defining the threshold on the amount of federated model-updates that are needed to update the master model.

	K	Θ	α	λ_1	λ_2	β_1	β_2	ϵ	γ
BO Bounds	[2, 25]	[100, 50000]	[4, 110]	[0.001, 0.099]	[0.1, 1]	[0.1, 0.55]	[0.55, 0.99]	[1e−8, 0.05]	[0.1, 0.99]
In-house anonymized dataset									
CF/FCF	25	32000	110	–	1	0.1	0.99	1e−8	0.1
MVMF/FED-MVMF	22	47200	110	0.099	0.1	0.55	0.55	0.05	0.99
Movielens dataset									
CF/FCF	25	100	4	–	1	0.1	0.99	1e−8	0.1
MVMF/FED-MVMF	25	3700	4	0.0989	1	0.1	0.98	0.0499	0.1
Book crossings dataset									
CF/FCF	25	10000	110	–	1	0.1	0.55	0.05	0.99
MVMF/FED-MVMF	23	13300	4	0.099	0.1	0.1	0.55	1e−8	0.1

A.5   Evaluation Metrics

In order to compare the recommendation performance of the Standard and Federated implementations of the recommendations and the modeling algorithms making up the recommendation system a set of metrics is used based on the true positive (t_p), true negative (t_n), false positive (f_p) and false negative (f_n) rates in the top k predicted recommendations [3]. These metrics are standard and regularly used to measure the performance of predictive models/systems and include **precision** Eq. (S2), **recall** Eq. (S3), **f1** Eq. (S4), and **Mean Average Precision MAP**. To account for the bias originating due to the unequal distribution of test point for each sample, we compute **Normalized Mean Ranking (NMR)** as well.

$$accuracy = \frac{t_p + t_n}{t_p + t_n + f_p + f_n} \tag{S1}$$

$$precision = \frac{t_p}{t_p + t_n} \tag{S2}$$

$$recall = \frac{t_p}{t_p + f_n} \tag{S3}$$

$$F1 = \frac{2 * precision * recall}{precision + recall} \tag{S4}$$

$$MAP@k = \frac{1}{\|U_{test}\|} \sum_{u=1}^{\|U_{test}\|} \frac{1}{\|I_{test}^u\|} \sum_{i=1}^{k} \frac{1}{k} t_p^k [t_p^k - t_p^{(k-1)}]$$

$$NMR = \frac{1}{\|U_{test}\|} \sum_{u=1}^{\|U_{test}\|} \frac{1}{\|I\|} \sum_{i=1}^{\|I_{test}^u\|} rank_i \tag{S5}$$

notation t_p^k means true positives at ranked prediction list from positions $[1..k]$ and U_test is the amount of users in test set. Note in all these metrics except **NMR** the possible values is in the range $[0, 1]$ where values closer to 1 imply a better performance value. In the case of **NMR** the closer the value is to 0 the better.

A.6   Convergence Analysis and Performance Metric Selection

Below, we show FED-MVMF model convergence and performance evolution for each of the data sets. In Figures, Y-axis denotes the recommendation metric, whereas X-axis represents number of Epochs or Iterations or rounds of master model updates. At $Iteration = 1000$ we compute average of previous 10 values and report in Table 2 in main text (and Fig. 3, 4 and 5 below).

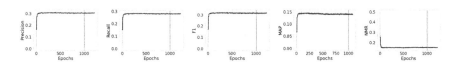

Fig. 3. Results on in-house anonymized data set.

A.7   Payload Analysis

We next analyzed the payloads for the two federated models. The item-factor matrix **Q** is common between FCF and FED-MVMF models, however an additional payload for the FED-MVMF model comes from the user-features factor

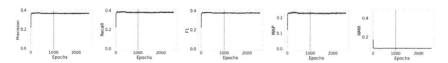

Fig. 4. Results on movielens data set.

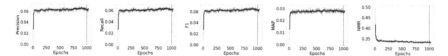

Fig. 5. Results on books-crossings data set.

matrix **U**. The size of **Q** depends on number of latent factors K and number of items N_v, whereas the size of **U** is dependent on K as well as the number of user-features D_u. In the case of FED-MVMF, both **Q** and **U** are transmitted as part of model updates between the FL clients and the FL server. As expected, we observed increased payloads for the FED-MVMF model compared to the FCF which does not include user-features. Notably, the model sizes scales linearly with increasing number of items and user-features. The relative increase in model size ranges from 80% to 200% across the three data sets, whereas computation time increases from 24% to 52%. We think that for the case of Books-Crossing, the

Table 5. Payload comparison between FCF and FED-MVMF in terms of the model size (KB = KiloBytes) and time (MS = Milliseconds). The FL client downloads the model (Model Download), update the local model and computes the master model updates or gradients (Model Update) and uploads the gradients to the FL Server (Model Upload). The FL server aggregates the updates arriving from the clients and updates the master model (Model Update).

	FL Client					FL Server
	Model Download		Model Update	Model Upload		Model Update
	Size (KB)	Time (MS)	Time (MS)	Size (KB)	Time (MS)	Time (MS)
In-house anonymized dataset						
FCF	265.28	7.39	9.47	242.31	7.35	2.87
FED-MVMF	483.16	10.78	11.78	436.07	9.06	4.19
Impr (%)	82	45	24	79	23	45
Movielens dataset						
FCF	402.17	8.06	10.46	402.17	16.09	3.44
FED-MVMF	849.47	14.80	15.98	849.38	15.29	3.30
Impr (%)	111	83	52	111	−4	−4
Book crossings dataset						
FCF	390.84	7.23	46.56	350.85	4.62	2.95
FED-MVMF	1021.80	15.78	17.40	1057.00	7.07	3.16
Impr (%)	161	118	-62	201	53	7

larger time taken by FCF to compute model updates could be merely a technical artifact and need further clarification. The model update time on FL server also includes the time taken by Item-Server to update item-feature matrix $\mathbf{V}$ and compute gradients for $\mathbf{Q}$ (Table 5).

Compared to the FCF model, the proposed FED-MVMF model increases payloads for the FL recommendation system. However FED-MVMF yields better recommendation performances and additionally provides principled solution to cold-start recommendation problems in FL.

References

1. Ammad-Ud-Din, M., et al.: Integrative and personalized QSAR analysis in cancer by kernelized Bayesian matrix factorization. J. Chem. Inf. Model. **54**(8), 2347–2359 (2014)
2. Ammad-Ud-Din, M., et al.: Federated collaborative filtering for privacy-preserving personalized recommendation system. arXiv preprint arXiv:1901.09888 (2019)
3. Bobadilla, J., Ortega, F., Hernando, A., Gutiérrez, A.: Recommender systems survey. Knowl.-Based Syst. **46**, 109–132 (2013)
4. Bonawitz, K., et al.: Towards federated learning at scale: system design. arXiv preprint arXiv:1902.01046 (2019)
5. Chai, D., Wang, L., Chen, K., Yang, Q.: Secure federated matrix factorization. arXiv preprint arXiv:1906.05108 (2019)
6. Chen, F., Dong, Z., Li, Z., He, X.: Federated meta-learning for recommendation. arXiv preprint arXiv:1802.07876 (2018)
7. Cortes, D.: Cold-start recommendations in collective matrix factorization. arXiv preprint arXiv:1809.00366 (2018)
8. Dolui, K., Gyllensten, I.C., Lowet, D., Michiels, S., Hallez, H., Hughes, D.: Poster: towards privacy-preserving mobile applications with federated learning-the case of matrix factorization. In: The 17th Annual International Conference on Mobile Systems, Applications, and Services, pp. 624–625 (2019)
9. Fang, Y., Si, L.: Matrix co-factorization for recommendation with rich side information and implicit feedback. In: Proceedings of the 2nd International Workshop on Information Heterogeneity and Fusion in Recommender Systems, pp. 65–69. ACM (2011)
10. Gamma, E., Helm, R., Johnson, R., Vlissides, J.: Design Patterns: Elements of Reusable Object-Oriented Software. Addison-Wesley Longman Publishing Co., Inc., Boston (1995)
11. Gemulla, R., Nijkamp, E., Haas, P.J., Sismanis, Y.: Large-scale matrix factorization with distributed stochastic gradient descent. In: Proceedings of the 17th ACM SIGKDD International Conference on Knowledge Discovery and Data Mining, pp. 69–77. ACM (2011)
12. Gönen, M., Khan, S., Kaski, S.: Kernelized Bayesian matrix factorization. In: International Conference on Machine Learning, pp. 864–872 (2013)
13. Hardy, S., et al.: Private federated learning on vertically partitioned data via entity resolution and additively homomorphic encryption. arXiv preprint arXiv:1711.10677 (2017)
14. Harper, F.M., Konstan, J.A.: The movielens datasets: history and context. ACM Trans. Interact. Intell. Syst. (TIIS) **5**(4), 19 (2016)

15. Hotelling, H.: Relations between two sets of variates. Biometrika **28**(3/4), 321–377 (1936)
16. Hu, Y., Koren, Y., Volinsky, C.: Collaborative filtering for implicit feedback datasets. In: Proceedings of the 2008 Eighth IEEE International Conference on Data Mining, ICDM 2008, pp. 263–272. IEEE Computer Society, Washington (2008)
17. Huang, S., Shi, W., Xu, Z., Tsang, I.W.: Iterative orthogonal federated multi-view learning (2019). http://smilelab.uestc.edu.cn/members/huangshudong/Iterative_Orthogonal_Federated_Multi_view_Learning.pdf
18. Jalalirad, A., Scavuzzo, M., Capota, C., Sprague, M.: A simple and efficient federated recommender system. In: Proceedings of the 6th IEEE/ACM International Conference on Big Data Computing, Applications and Technologies, pp. 53–58 (2019)
19. Kingma, D.P., Ba, J.L.: Adam: a method for stochastic optimization. In: International Conference on Learning Representations (2015)
20. Klami, A., Virtanen, S., Leppäaho, E., Kaski, S.: Group factor analysis. IEEE Trans. Neural Netw. Learn. Syst. **26**(9), 2136–2147 (2015)
21. Konecný, J., McMahan, H.B., Ramage, D., Richtárik, P.: Federated optimization: distributed machine learning for on-device intelligence. CoRR abs/1610.02527 (2016)
22. Li, Q., Wen, Z., He, B.: Federated learning systems: vision, hype and reality for data privacy and protection. arXiv preprint arXiv:1907.09693 (2019)
23. Liu, Y., Chen, T., Yang, Q.: Secure federated transfer learning. arXiv preprint arXiv:1812.03337 (2018)
24. McMahan, B., Moore, E., Ramage, D., Hampson, S., y Arcas, B.A.: Communication-efficient learning of deep networks from decentralized data. In: Artificial Intelligence and Statistics, pp. 1273–1282 (2017)
25. Ruder, S.: An overview of gradient descent optimization algorithms. CoRR abs/1609.04747 (2016)
26. Sharma, A., Kumar, M., Agarwal, S.: A complete survey on software architectural styles and patterns. Proc. Comput. Sci. **70**, 16–28 (2015). Proceedings of the 4th International Conference on Eco-friendly Computing and Communication Systems
27. Singh, A.P., Gordon, G.J.: Relational learning via collective matrix factorization. In: Proceedings of the 14th ACM SIGKDD International Conference on Knowledge Discovery and Data Mining, pp. 650–658. ACM (2008)
28. Smith, V., Chiang, C., Sanjabi, M., Talwalkar, A.S.: Federated multi-task learning. In: Advances in Neural Information Processing Systems 30: Annual Conference on Neural Information Processing Systems 2017, 4–9 December 2017, Long Beach, CA, USA, pp. 4427–4437 (2017)
29. Snoek, J., Larochelle, H., Adams, R.P.: Practical Bayesian optimization of machine learning algorithms. In: Advances in Neural Information Processing Systems, pp. 2951–2959 (2012)
30. Strahl, J., Peltonen, J., Mamitsuka, H., Kaski, S.: Scalable probabilistic matrix factorization with graph-based priors. In: AAAI (2020)
31. Yang, Q., Liu, Y., Chen, T., Tong, Y.: Federated machine learning: concept and applications. ACM Trans. Intell. Syst. Technol. (TIST) **10**(2), 1–19 (2019)
32. Yu, H.-F., Hsieh, C.-J., Si, S., Dhillon, I.S.: Parallel matrix factorization for recommender systems. Knowl. Inf. Syst. **41**(3), 793–819 (2013). https://doi.org/10.1007/s10115-013-0682-2

33. Zhou, Y., Wilkinson, D., Schreiber, R., Pan, R.: Large-scale parallel collaborative filtering for the Netflix prize. In: Fleischer, R., Xu, J. (eds.) AAIM 2008. LNCS, vol. 5034, pp. 337–348. Springer, Heidelberg (2008). https://doi.org/10.1007/978-3-540-68880-8_32
34. Ziegler, C.N., McNee, S.M., Konstan, J.A., Lausen, G.: Improving recommendation lists through topic diversification. In: Proceedings of the 14th International Conference on World Wide Web, WWW 2005, pp. 22–32. ACM (2005)

FedMAX: Mitigating Activation Divergence for Accurate and Communication-Efficient Federated Learning

Wei Chen[1], Kartikeya Bhardwaj[2]([✉]), and Radu Marculescu[3]

[1] Carnegie Mellon University, Pittsburgh, PA 15213, USA
weic3@andrew.cmu.edu
[2] Arm Inc., San Jose, CA 95134, USA
kartikeya.bhardwaj@arm.com
[3] The University of Texas at Austin, Austin, TX 78712, USA
radum@utexas.edu

Abstract. In this paper, we identify a new phenomenon called *activation-divergence* which occurs in Federated Learning (FL) due to data heterogeneity (*i.e.*, data being non-IID) across multiple users. Specifically, we argue that the activation vectors in FL can diverge, even if subsets of users share a few common classes with data residing on different devices. To address the activation-divergence issue, we introduce a prior based on the principle of maximum entropy; this prior assumes minimal information about the per-device activation vectors and aims at making the activation vectors of same classes as similar as possible across multiple devices. Our results show that, for both IID and non-IID settings, our proposed approach results in better accuracy (due to the significantly more similar activation vectors across multiple devices), and is more communication-efficient than state-of-the-art approaches in FL. Finally, we illustrate the effectiveness of our approach on a few common benchmarks and two large medical datasets (The code is available at https://github.com/weichennone/FedMAX).

Keywords: Federated Learning · Maximum entropy · Non-IID

1 Introduction

Large amounts of data are increasingly generated nowadays on edge devices, such as phones, tablets, and wearable devices. If properly used, machine learning (ML) models trained using this data can significantly improve the intelligence of such devices [1]. However, since data on such personal devices is highly sensitive, training ML models by sending the users' local data to a centralized server clearly involves significant privacy risks. Other examples of private datasets include personal medical records which must not be shared with third parties. Hence,

---

W. Chen and K. Bhardwaj—Equal Contribution.

F. Hutter et al. (Eds.): ECML PKDD 2020, LNAI 12458, pp. 348–363, 2021.
https://doi.org/10.1007/978-3-030-67661-2_21

in order to enable intelligence for these privacy-critical applications, Federated Learning (FL) has become the de facto paradigm for training ML models on local devices without sending data to the cloud [2,3].

As the state-of-the-art approach for FL, Federated Averaging (FedAvg) [4] simply runs several *local* training epochs on a randomly selected subset of devices; these training epochs utilize only local data available on any user's device. After local training, the models (not the local data!) are sent over to a server via a *communication round*; the server then averages all the parameters of these local models to update a *global* model. Unfortunately, FedAvg is not designed to handle the statistical heterogeneity in federated settings, *i.e.*, when data is *not* independent and identically distributed (non-IID) across the different devices. Not surprisingly, it has been recently reported that FedAvg can incur significant loss of accuracy when data is non-IID [5,6].

To deal with such non-IID settings, one approach called "data-sharing strategy" distributes global data across the local devices, such that the test accuracy can increase by making data look more IID [5,7]. However, obtaining this common global data is usually problematic in practice. Another approach called FedProx [8] targets the *weight-divergence* problem, *i.e.*, the local-weights diverge from the global model due to non-IID data at local devices (hence, the updates can go in different directions at different local devices).

In this paper, we first identify a new phenomenon called *activation-divergence* and argue that the activation vectors in FL can diverge even if a subset of users share a few common classes of data. Since the activation vectors directly contribute to the model's accuracy, making them as similar as possible *across all devices* should become an important objective in FL. To this end, we propose *FedMAX*, a new FL approach that introduces a new prior for local training. Specifically, our prior maximizes the entropy of local activation vectors across all devices. We show that our new prior:

1. Makes activation vectors across multiple devices more similar (for the same classes); in turn, this improves the classification accuracy of our approach;
2. Significantly reduces the number of total communication rounds needed (as one can perform more local training without losing accuracy). This is particularly important to save energy when training on edge devices.

Extensive experiments on five non-IID FL datasets demonstrate that our approach significantly outperforms both FedAvg [4] and FedProx [8] (e.g. $5.64\% - -5.84\%$ better accuracy on CIFAR-10 dataset). We also observe up to $5\times$ reduction in communication rounds compared to FedAvg and FedProx.

Our paper is organized as follows. Section 2 provides some background information. Sect. 3 presents our proposed approach FedMAX. In Sect. 4, we provide a detailed evaluation of FedMAX, under both IID and non-IID scenarios. Finally, Sect. 5 summarizes our main contributions.

2   Related Work

In FedAvg, after training on device's own data, the updated local models are averaged at a central server in order to get a new global model. For non-IID data, the performance of FedAvg reduces significantly as the weights of different models often diverge [5,6]. To address this non-IID issue, several approaches propose to use some globally shared data to improve the accuracy by making the local data look more IID [5,7]. However, in practice, collecting this global data may be problematic (or even infeasible) due to privacy concerns; additionally, dealing with this global data can use up critical resources like the local storage space or network bandwidth. Consequently, another approach called FedProx [8] has been proposed to solve the weight-divergence problem by introducing a new loss function which constrains the local models to stay close to the global model.

In contrast to the prior art, we aim to constrain the activation-divergence across multiple devices. More precisely, our approach is based on the principle of maximum entropy which states that when there is no *a priori* information about a problem, the prior distribution should be chosen to maximize entropy [10]. The core idea behind maximizing entropy is to obtain a prior which assumes the least amount of information about a given problem[1]. Of note, while this principle has been exploited to solve traditional natural language processing problems [12,13], it has never been used in the context of FL.

Other studies exploit ML models [14] with a focus on differential privacy [16] for medical datasets which are usually imbalanced and non-IID. Therefore, evaluating FL with medical datasets is necessary, especially when privacy issues are at stake [16]. To this end, we perform multiple experiments on two different medical datasets: (*i*) Chest X-ray dataset [14] is one of the accessible medical image datasets for developing automated methods to identify and classify pneumonia; (*ii*) APTOS dataset [18] is also a well-known dataset for detecting the blindness with retina images taken using fundus photography. Our results show the effectiveness of our approach on these non-IID datasets.

Next, we explain the intuition behind using the maximum entropy principle for FL under non-IID scenarios, and describe our newly proposed approach.

3   Proposed Approach: FedMAX

FL aims to solve the learning task without explicitly sharing local data. More precisely, a central server coordinates the global learning across a network where each node is a device collecting data and performing a local learning task (as shown in Fig. 1(a)). The objective of FL [4] is to minimize:

$$\min_{w} \quad g\left(w\right) = \sum_{k=1}^{m} p_k \cdot g_k(w_k) \tag{1}$$

---

[1] Making needless or unfounded prior assumptions about a problem can reduce the accuracy of the model, hence it is better to make minimal assumptions. For more information on maximum entropy, please refer to [10,11].

where $g_k(w_k)$ is the local objective which is typically the loss function of the prediction made with model parameters w; $m = C \cdot M$ is the number of devices selected at any given communication round, where C is the proportion of selected devices and M is the total number of devices; $\sum_{k=1}^{M} p_k = 1$, $p_k = \frac{n_k}{n}$ and n_k is the number of samples available at the device k, $n = \sum_{k=1}^{M} n_k$ is the total number of samples.

In FedAvg [4], any local model is updated with its own data as $w_k^{t+1} \leftarrow w_k^t - \eta \nabla g_k(w_k)$, where η is the learning rate, $\nabla g_k(w_k)$ represents the gradient of $g_k(w_k)$; the global model is then formed by the averaging the parameters of all these local models, *i.e.*, $w^{t+1} \leftarrow \sum_{k=1}^{M} \frac{n_k}{n} w_k^{t+1}$. For non-IID datasets, different local models will have different data. Although optimized with the same learning rate and the same number of local training epochs, the weights of these local models will likely diverge. Consequently, the accuracy of the global model decreases when its parameters are weight-averaged across these different local models. One possible solution to this problem is to constrain the local updates within a reasonable range, as FedProx proposed [8].

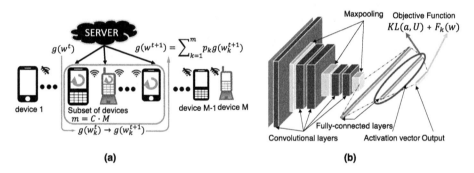

(a)   **(b)**

Fig. 1. (a) FL training process: (i) A central server selects a subset of devices ($m = C \cdot M$, where where C is the proportion of selected devices and M is the number of total devices) and transmits the global model $g(w^t)$ to each selected device; (ii) Each device trains the model on its local data $g(w_k^t) \rightarrow g(w_k^{t+1})$, and uploads the updated model to the server; (iii) The server aggregates the local models and forms a new global model (see Eq. (1)). (b) For most datasets, our CNN model has 5 convolutional and 2 fully-connected layers. This model is deployed on each individual device in Fig. 1(a) for local training. The final logits and the activation vectors at the input of the last fully-connected layer are used in the objective function. KL denotes Kullback-Leibler divergence, a refers to the activation vector, U is uniform distribution over activation vectors, and $F_k(w)$ is the cross-entropy loss on local data. We use similar activation vectors for other models such as ResNets for medical datasets.

Since activation vectors directly contribute to model accuracy, our objective is to reduce the activation-divergence for the same classes across multiple devices. To this end, we propose a new prior for the local training that can help us achieve the above goal. More precisely, we use a Convolutional Neural

Network (CNN) with five convolutional layers and two fully-connected layers (see Fig. 1(b)) for three digit/object recognition datasets and ResNet50 [19] for two medical datasets, *i.e.*, APTOS and Chest X-ray. Also, we call the inputs to the last fully-connected layer as the *activation-vector*; for the 5-layer CNN, the activation-vector is 512-dimensional, and passes through the final fully-connected layer to yield logits (the unnormalized class probabilities). Our goal is to propose a new prior that enables similar activation vectors across different devices.

L^2 Norm Regularization: We initially consider the L^2 norm to constrain the activation vectors and argue that by preventing the activation vectors from taking large values, the L^2 norm should reduce the activation-divergence across different devices. We formulate the L^2 norm regularization as follows:

$$\min_{w} \quad g_k(w_k) = F_k(w_k) + \beta \left\| a_i^k \right\|_2 \tag{2}$$

where $F_k(w_k)$ is the cross-entropy loss on local data (same as the cost function of FedAvg [4]), k denotes to any local device in Fig. 1(a), $\|\cdot\|_2$ is L^2 norm, and a_i^k refers to the activation vectors at the input of the last fully-connected layer (as shown in Fig. 1(b)) for sample i on device k. Further, $\beta > 0$ is a hyper-parameter used to control the scale of the L^2 norm regularization.

Intuitively, this L^2 norm regularization constrains the activation vectors and indirectly affects the parameters of other layers except the last fully-connected layer. However, reducing the activation to zero can lead to model underfitting, which results in poor performance. Therefore, we further propose another form of regularization to ensure more similar activation vectors across different devices.

Maximum Entropy Regularization: The activation-divergence problem is more complex in the non-IID settings where different users deal with data from different classes. As such, we do not have any prior information about which users have data from which classes. Hence, in non-IID settings, we do *not* have any prior information about how the activation vectors at different users (for the given classes) should be distributed. Consequently, we propose to use the principle of maximum entropy [10] and select a distribution for activation vectors that maximizes their entropy[2]. Using such a prior, the local loss function for our FL problem is given by:

$$\min_{w} \quad g_k(w_k) = F_k(w_k) - \beta \frac{1}{N} \sum_{i=1}^{N} \mathbb{H}(a_i^k) \tag{3}$$

where N is a mini-batch size of local training data, and $\mathbb{H}$ denotes the entropy of activation vectors. Also, β is a hyper-parameter that is used to control the scale of the entropy loss. Compared with (2), Eq. (3) maximizes the entropy (hence it minimizes the negative entropy) of activation vectors $\mathbb{H}(a_i^k)$ instead of minimizing the L^2 norm of activation vectors $\left\| a_i^k \right\|_2$; therefore, we call this approach FedMAX.

---

[2] We perform softmax on activation vectors to transform them into a distribution.

Further, (3) can be written using the Kullback-Leibler (KL) divergence as:

$$\min_{w} \quad g_k\left(w_k\right) = F_k\left(w_k\right) + \beta \frac{1}{N} \sum_{i=1}^{N} KL\left(a_i^k || U\right) \tag{4}$$

where $KL(\cdot||\cdot)$ denotes the KL divergence, and U is uniform distribution over the activation vectors. Since Eq. (4) is equivalent to Eq. (3) up to a constant term, the new formulation does *not* affect the optimization process and, thus, also results in maximum entropy. As we shall see shortly, FedMAX is more stable than the L^2 norm-based regularization.

Algorithm 1: FedMAX algorithm

Data: $M, T, \beta, w^0, \eta, B, C, E$
Function *Server()*:
     for $t = 0$ to $T - 1$ **do**
         $m \leftarrow max(C \cdot M, 1)$;
         $S_t \leftarrow$ Random set of m clients;
         for $k \in S_t$ **do**
             $w_k^{t+1} \leftarrow Client_k(w^t)$;
         end
     end
end
Function *Client(w)*:
     for $i = 0$ to $E - 1$ **do**
         for $b \in B$ **do**
             $g\left(w; b\right) = F\left(w; b\right) + \beta \frac{1}{N} \sum_{i=1}^{N} KL\left(a_i || U\right)$;
             $w \leftarrow w - \eta \nabla g(w; b)$;
         end
     end
     return w;
end

The training process of FedMAX is similar to FedAvg (see Algorithm 1). The initial model and weights w^0 are generated on a remote server. After selecting a subset of devices (C represents the proportion of selected devices, as shown in Fig. 1(a)), the server sends the model (and the corresponding weights) only to these devices. The devices train the model for E local epochs using their local data and then send the trained model back to the server. After averaging the models on the server, sending back the updated model to the newly selected devices finishes one communication round (t) – see Algorithm 1, where M represents the number of devices, B is the local training batch size, and T represents the total number of communication rounds[3]. This completes the newly proposed FedMAX; we next show its effectiveness on multiple datasets.

---

[3] We note that this approach reduces to FedAvg if $\beta = 0$.

4   Experimental Setup and Results

We perform multiple experiments on five different datasets: FEMNIST* [9],
CIFAR-10, CIFAR-100 [17], APTOS [16] and Chest X-ray [15]. The first three
datasets are trained with the five layer CNN in Fig. 1(b), while the last two
medical datasets are fine-tuned with ResNet50 [19]. Specifically, the CNN model
has 5 convolutional layers (32/64/64/64/64 channels for each layer) and 2 fully-
connected layers (1024 × 512, 512 × 10 neurons for each layer). We consider a
FL setting where we have a central server and a total of 100 local devices (*i.e.*,
$M = 100$), each device containing only a subset of the entire dataset. At each
communication round, only 10% (*i.e.*, $C = 0.1$) of these devices are randomly
selected by the server for local training. With different ways to separate data
at the local devices, we can get either IID or non-IID of each dataset. In what
follows, we show results for both IID and non-IID datasets.

4.1   Similarity of Activations

We first use synthetic data generated as in [8] to verify that the maximum
entropy regularization leads to similar activations at different local devices. Sam-
ples $x_k \in \mathbb{R}^{1024}$ for kth device are drawn from a normal distribution $\mathcal{N}(v_k, \Sigma)$,
which has two parameters: the mean vector v_k and the covariance matrix Σ.
Each element in the mean vector v_k is generated from $\mathcal{N}(B_k, 1)$, and here
$B_k \sim \mathcal{N}(0, \gamma_1)$. A larger γ_1 will lead to more varied mean vectors v_k of the
data distribution at each device, thus more non-IID data; the covariance matrix
Σ is a diagonal matrix where $\Sigma_{j,j} = \frac{1}{j^{1.2}}$ (similar to that used in [20]).

Following the data-generation strategy presented in [8], we use a two-layer
perceptron $y = argmax(w_2 \cdot ReLU(w_1 \cdot x + b_1) + b_2)$ to generate the labels w.r.t
the input samples[4], where $w_1 \in \mathbb{R}^{10 \times 512}$, $w_2 \in \mathbb{R}^{512 \times 1024}$, $b_1 \in \mathbb{R}^{10}$, and $b_2 \in \mathbb{R}^{512}$. Each element in w_1, w_2, b_1, and b_2 is drawn from the normal distribution
$\mathcal{N}(u_k, 1)$, where $u_k \sim \mathcal{N}(0, \gamma_2)$. The γ_2 controls the differences among the local
models, thus indirectly influences the generated labels.

We use three different sets $(\gamma_1, \gamma_2) = (0, 0), (0.5, 0.5), (1, 1)$ to generate the
non-IID synthetic data. We train both FedAvg and FedMAX on the synthetic
data with a two-layer perceptron which has the same structure as the model used
to generate the labels. The training process lasts 200 communication rounds (*i.e.*,
$T = 200$), with one local training epoch (*i.e.*, $E = 1$). For each communication
round, the average activation a_k of each local model is collected and the similarity
between the local activation a_k and the global activation $\bar{a}$ is calculated with
KL-divergence $\delta_k = KL(\bar{a}||a_k)$. The global activation is calculated from the
averages of all local activations $\bar{a} = \frac{1}{M}\sum_k a_k$, where M is the total number of
devices. The *overall similarity* per communication round is represented by the
mean of the local similarity $\bar{\delta} = \frac{1}{M}\sum_k \delta_k$.

As we can see from Fig. 2, the maximum entropy regularization (FedMAX)
can result in significantly lower KL-divergence between global and local activations, which means the activations from the model with maximum entropy

---

[4] Once initialized, these two-layer perceptron models remain fixed.

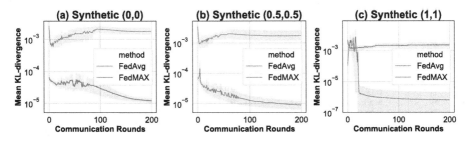

Fig. 2. The similarity effects of maximum entropy regularization, with different distributions of synthetic data $(\gamma_1, \gamma_2) = (0,0), (0.5, 0.5), (1,1)$. As shown, FedMAX has significantly lower KL-divergence than FedAvg; this means that the maximum entropy regularization can make activation vectors more similar.

regularization are similar to each other. Moreover, the values $\gamma_1 = 1, \gamma_2 = 1$ for synthetic data lead to a higher KL-divergence for both FedAvg and FedMAX during the first few epochs. This means that the more heterogeneous data distributions can cause activations to be very dissimilar from each other. Thus, constraining the activation within a reasonable range, or making the activations more similar to each other, can benefit FL, especially for the non-IID case.

4.2 Comparison of L^2-norm Against Maximum Entropy

We first compare our proposed FedMAX against the L^2 norm regularization on a non-IID CIFAR-10 dataset. For each regularization, we train a CNN like in Fig. 1 consisting of about 0.6 million parameters. The hyper-parameter β for L^2 norm regularization varies from 10^{-4} to 10^{-1}, and the β for maximum entropy regularization varies from 1 to 10^4. Since the maximum entropy regularization is averaged over the activations, it has larger hyper-parameters than the L^2 norm.

The results are shown in Fig. 3. As we can see, both L^2 norm and maximum entropy regularization outperform FedAvg, which means that both methods enable more similar activation vectors across the devices. However, when compared against the L^2 norm, the accuracy of the maximum entropy regularization is more robust to hyper-parameter variation. Specifically, we found that for certain β values, the L^2 norm results in extremely low accuracies (see Fig. 3(b)); this, in turn, can result in a much more time consuming hyper-parameter search for different datasets. Since FedMAX results in a significantly more stable behavior (see Fig. 3(a)), in the rest of the paper, the experimental results are reported only for FedMAX using the maximum entropy regularization.

4.3 Digit/Object Recognition Datasets

We now verify our approach on three different datasets: FEMNIST* [9], CIFAR-10 and CIFAR-100. For each dataset, we train a CNN like in Fig. 1(b) consisting of about 0.6 million parameters.

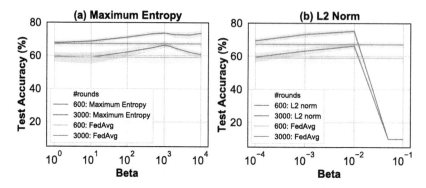

Fig. 3. Test accuracy for different hyper-parameter value (β) on non-IID CIFAR-10 dataset with different regularizations (L^2 norm and maximum entropy), for 600 and 3000 communication rounds.

The training process lasts 3000 communication rounds (*i.e.*, $T = 3000$) with a single local training epoch (*i.e.*, $E = 1$); the mini-batch size N at each selected device is 100. The learning rate η is initialized to 0.1 and decays by $\times 0.9992$ at each round. For reference, the decay rate in [5] is 0.992^5. We also test the communication efficiency by setting the global communication rounds $T = 600$, learning rate decay of 0.996, five local training epochs (*i.e.*, $E = 5$), and keep all other parameters the same; this way, the experimental settings remain consistent with the 3000 communication rounds setup.

For FedProx, the results are reported for the hyper-parameter $\mu = 1$ [8]. We did try other μ values like $\{1, 2, 10, 20, 100\}$, but found that the results are very similar. Also, for our approach, we set $\beta = 1500$. To split the datasets into the non-IID parts, we randomly assign 2 out of 10 classes (20 out of 100 classes) for CIFAR-10 (CIFAR-100) to each device. For FEMNIST*, we follow the same setting as in [8], where data from 20 out of 26 classes are given to each device. For the IID case of all three datasets, labels are distributed uniformly across all users. In what follows, we present two sets of results: (*i*) Accuracy improvements and (*ii*) Communication-efficiency of FedMAX.

Accuracy Comparison: More Communication Rounds, Less Local Training. The test accuracy of the 3000 communication round experiment is shown in Fig. 4. As evident, our approach outperforms the other approaches for all three datasets. The test accuracy decreases accordingly as the datasets change from FEMNIST* to CIFAR-100, where our CNN models become relatively smaller for the dataset. Since each device for CIFAR-10 has only two out of ten labels (extreme non-IID case), this is why the test accuracy on CIFAR-10 varies much more rapidly (for all three approaches) compared to the other datasets. For the CIFAR-10 dataset,

---

[5] Since this decay rate results in an extremely small learning rate after thousands of epochs, we increase our learning rate decay to 0.9992.

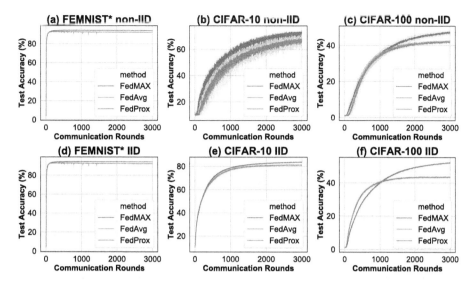

Fig. 4. Test accuracy for different datasets (both non-IID and IID) with different approaches, FedAvg, FedProx and FedMAX, for 3000 communication rounds. FedMAX has a higher accuracy than the other approaches for all three datasets.

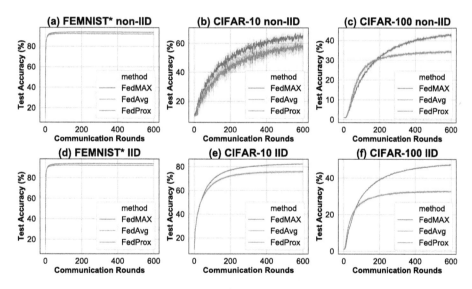

Fig. 5. Test accuracy for different datasets (both non-IID and IID) with different approaches, FedAvg, FedProx and FedMAX, for 600 communication rounds. FedMAX has a higher accuracy than the other approaches for all three datasets.

Table 1. The test accuracy for non-IID and IID settings (bold results are better)

non-IID	3000 communication rounds			600 communication rounds		
Approach	FEMNIST*	CIFAR-10	CIFAR-100	FEMNIST*	CIFAR-10	CIFAR-100
FedAvg [4]	$92.24 \pm 0.08\%$	$67.26 \pm 1.50\%$	$42.17 \pm 0.49\%$	$92.09 \pm 0.14\%$	$58.91 \pm 3.55\%$	$34.29 \pm 0.52\%$
FedProx [8]	$92.14 \pm 0.16\%$	$67.46 \pm 1.78\%$	$41.99 \pm 0.58\%$	$92.09 \pm 0.08\%$	$58.63 \pm 2.98\%$	$34.42 \pm 0.33\%$
FedMAX	$\mathbf{94.05 \pm 0.13\%}$	$\mathbf{73.10 \pm 1.20\%}$	$\mathbf{47.15 \pm 0.75\%}$	$\mathbf{93.78 \pm 0.10\%}$	$\mathbf{65.64 \pm 1.49\%}$	$\mathbf{43.15 \pm 0.99\%}$
Improvement	1.81%	5.64%	4.98%	1.69%	6.73%	8.73%
IID	3000 communication rounds			600 communication rounds		
Approach	FEMNIST*	CIFAR-10	CIFAR-100	FEMNIST*	CIFAR-10	CIFAR-100
FedAvg	$92.24 \pm 0.08\%$	$81.14 \pm 0.49\%$	$43.56 \pm 0.26\%$	$92.09 \pm 0.14\%$	$75.94 \pm 0.96\%$	$32.67 \pm 0.39\%$
FedProx	$92.14 \pm 0.16\%$	$81.16 \pm 0.29\%$	$43.22 \pm 0.30\%$	$92.09 \pm 0.08\%$	$75.91 \pm 1.09\%$	$32.67 \pm 0.44\%$
FedMAX	$\mathbf{94.05 \pm 0.13\%}$	$\mathbf{83.66 \pm 0.38\%}$	$\mathbf{53.13 \pm 0.58\%}$	$\mathbf{93.78 \pm 0.10\%}$	$\mathbf{82.39 \pm 0.26\%}$	$\mathbf{47.38 \pm 0.47\%}$
Improvement	1.81%	2.50%	9.57%	1.69%	6.45%	14.71%

our model also converges significantly faster than the other approaches. The final accuracies across five runs for all experiments are shown in Table 1. As shown, our approach outperforms existing techniques for both IID and non-IID cases.

Communication-Efficiency: Less Communication Rounds, More Local Training. The test accuracy of the 600 communication rounds experiment is shown in Fig. 5. With more local training, the weights of the models on different devices are expected to diverge more from the global model, which explains the loss of accuracy. However, FedMAX significantly outperforms the test accuracy of FedAvg [4] and FedProx [8] by up to 8% (see Table 1).

Another observation worth noting from Table 1 is that for all three datasets, FedMAX with 600 communication rounds achieves comparable or even better accuracy than FedAvg and FedProx with 3000 communication rounds. This shows that, by relying on more local training, FedMAX significantly reduces communication rounds (by up to 5×) compared to prior techniques, without losing accuracy. This is particularly important for edge computing where communication cost reduction is crucial for energy savings.

4.4 Medical Datasets

The APTOS dataset includes 38,788 samples, five labels describing the severity of blindness, and each class contains different numbers of retina images taken using fundus photography. The Chest X-ray dataset has 5,856 samples and two image categories (Pneumonia/Normal) graded by expert physicians. Each dataset is randomly split into 85% training data and 15% test data. Since these are imbalanced datasets, we use F1 score to measure the performance of the model.

The experiment setting is the same, but instead of training a five-layer CNN, we fine-tune a ResNet50 [19] which is pre-trained on the ImageNet dataset. The activation-vector of ResNet50 is chosen to be the output of final average-pool layer, *i.e.*, the activation-vector is a 2048-dimensional for medical datasets.

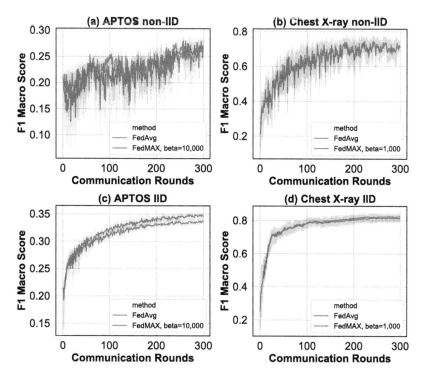

Fig. 6. Test accuracy for different medical datasets for both non-IID and IID cases, APTOS and Chest X-ray, with different approaches, FedAvg and our proposed approach (FedMAX), for 300 communication rounds. FedMAX has a higher F1 score than FedAvg in APTOS dataset for the IID case. Both have similar scores as FedAvg in APTOS dataset for the non-IID case and Chest X-ray dataset.

The training process lasts 300 communication rounds (*i.e.*, $T = 300$) with a single local training epoch; the mini-batch size N at each selected device is 32. The learning rate η is initialized to 0.001 and decays by $\times 0.992$ at each round. To split the datasets into non-IID parts, we randomly assign different proportions of 5 classes (2 classes) for APTOS (Chest X-ray) to each device. For our approach, we set $\beta = 10,000$ for APTOS dataset and 1,000 for Chest X-ray dataset.

Accuracy Comparison: The test accuracy of the IID and non-IID cases for the 300 communication-round experiment is shown in Fig. 6. As evident, our approach FedMAX outperforms FedAvg on the APTOS IID case. For the non-IID case, our method yields similar results as FedAvg. The F1 score of the non-IID case varies more rapidly than the IID case. This is because the medical datasets are highly imbalanced, and the non-IID partition by randomly separating the samples can lead to devices with only one class.

Table 2. The F1 macro score of medical datasets (bold results are better)

	APTOS		Chest X-ray	
Approach	IID	non-IID	IID	non-IID
FedAvg	0.3362 ± 0.0040	0.2707 ± 0.0135	$\mathbf{0.8243 \pm 0.0296}$	0.7094 ± 0.0338
FedMAX	$\mathbf{0.3451 \pm 0.0062}$	0.2706 ± 0.0121	0.8147 ± 0.0286	$\mathbf{0.7183 \pm 0.0383}$
Improvement	0.0089	-0.0001	-0.0096	0.008

Compared to other datasets, the results of FedMAX on the Chest X-ray dataset are close to FedAvg. One possible reason is that since the Chest X-ray dataset has only two classes, it cannot really make the activations more similar among different labels across different devices. Besides, with fewer samples in the Chest X-ray dataset, after partitioning, each device contains only a small amount of data; this leads to a short local training process and more frequent global communication. As a result, the activation divergence may already be constrained, so that the FedAvg has a similar performance when compared against FedMAX. Final accuracy comparisons across the five runs for all our experiments are shown in Table 2. The better results are highlighted with bold.

Our initial experiments show that L^2 norm regularization for medical datasets yielded poor performance.

4.5 Mitigating Activation Divergence

We now analyze the impact of our proposed FedMAX on the activation-divergence that can happen in non-IID FL. We show 2-dimensional (2D) t-SNE plots of our 512-dimensional (512D) activation vectors for different devices (each device has two random classes from the CIFAR-10 dataset). Specifically, the t-SNE plots embed each 512D activation vector with a 2D point in such a way that similar objects are modeled by nearby points and dissimilar objects are modeled by distant points. We expect the activation vectors of the same class (even from different devices) to share more similarities, thus, their corresponding 2D points should be closer to each other and form a cluster on the t-SNE plots. To keep it simple, we perform the experiment on a total of 10 local devices, with all the devices training at every communication round.

In Fig. 7, Fig. 8, and Fig. 9, the plots on the left show the activation vectors for FedAvg, and the ones on the right show those for FedMAX. Various colors represent the activation vectors for different classes, while the letters denote the device IDs. As the number of local epochs increases, we observe that: (i) Fed-MAX starts to gain accuracy, (ii) Activation vectors for FedMAX start to cluster together - see highlighted portions in Fig. 8 and Fig. 9 where the activation vectors from same classes (*i.e.*, the same color) come closer to each other across different devices (*i.e.*, letters A-J). In contrast, for FedAvg, clustering happens much more slowly and, hence, its accuracy is significantly lower than FedMAX.

t-SNE embedding of the activation vectors

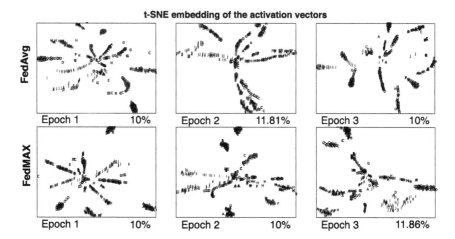

Fig. 7. Two-dimensional tSNE plot of activation vectors (512D vector projected into 2D) for two approaches on CIFAR-10 dataset: FedAvg (top) and our proposed FedMAX (bottom). Left panel shows epoch 1, middle panel epoch 2, and right panel epoch 3. The numbers at the bottom show how the test accuracy of the two techniques varies with the training epochs. We note that initially, all t-SNE plots look similar and the test accuracy for both models is close to random accuracy ($\sim$10%).

t-SNE embedding of the activation vectors

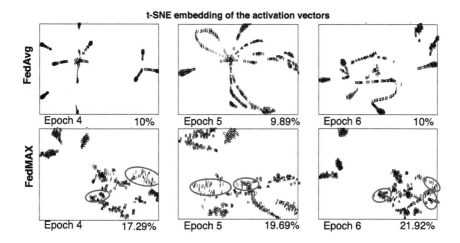

Fig. 8. Similar to Fig. 7, but left panel shows epoch 4, middle panel epoch 5, and right panel epoch 6. We see that same colors start coming together (*i.e.*, the activation vectors of same classes across different devices start to become more and more similar) in FedMAX. Consequently, in the accuracy of FedMAX ($\sim$22% until epoch 6) improves much faster than FedAvg (10% until epoch 6). However, clustering in FedAvg looks exactly the same as before.

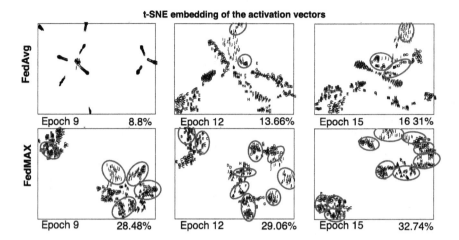

Fig. 9. Similar to the Figs. 7 and 8, but left panel shows epoch 9, middle panel epoch 12, and the right panel epoch 15. More and more clusters from same classes start forming for FedMax, while the clusters barely show up for FedAvg. This also results in the accuracy of FedMAX (32% until epoch 15) improving much faster than FedAvg (16% until epoch 15). We also see a significantly higher number of clusters formed for FedMAX compared to FedAvg.

5   Conclusion

In this paper, we have identified the activation-divergence phenomenon in FL and proposed FedMAX, a new approach for accurate and communication-aware FL in non-IID and IID settings. By exploiting the L^2 norm regularization and the principle of maximum entropy, we have introduced a new prior which assumes minimal information about the activation vectors at different devices.

With extensive experiments, we have shown that FedMAX improves the test accuracy and is significantly more communication-efficient than the state-of-the-art approaches running on FEMNIST*, CIFAR-10, and CIFAR-100 for both non-IID and IID settings. Besides, we have presented experiments on two medical datasets, APTOS and Chest X-ray, and have shown the improvement of Fed-MAX on the APTOS IID case. We attribute the better performance of FedMAX to improving the similarity across the devices while regularizing the activation vectors. Finally, we note that FedAvg and FedMAX perform similarly on the Chest X-ray dataset due to the smaller number of samples which may hardly lead to activation divergence.

In future work, we plan to evaluate the FedMAX approach using different datasets which contain more classes and samples. We also plan to implement FedMAX for different learning tasks such as language modeling.

Acknowledgements. We thank Prof. Virginia Smith and Tian Li of Carnegie Mellon University for fruitful discussions and help with the L^2 norm regularization.

References

1. Yang, Q., Liu, Y., Chen, T., Tong, Y.: Federated machine learning: concept and applications. ACM Trans. Intell. Syst. Technol. (TIST) **10**, 12 (2019)
2. Konečný, J., McMahan, H B., Yu, F.X., Richtárik, P., Suresh, A.T., Bacon, D.: Federated learning: strategies for improving communication efficiency. arXiv preprint arXiv:1610.05492 (2016)
3. Konečný, J., McMahan, B., Ramage, D.: Federated optimization: distributed optimization beyond the datacenter. arXiv preprint arXiv:1511.03575 (2015)
4. McMahan, H B., Moore, E., Ramage, D., et al.: Communication-efficient learning of deep networks from decentralized data. arXiv preprint arXiv:1602.05629 (2016)
5. Zhao, Y., Li, M., Lai, L., Suda, N., Civin, D., Chandra, V.: Federated learning with non-IID data. arXiv preprint arXiv:1806.00582 (2018)
6. Sattler, F., Wiedemann S., Müller K., et al.: Robust and communication-efficient federated learning from non-IID data. arXiv preprint arXiv:1903.02891 (2019)
7. Huang, L., Yin, Y., Fu, Z., et al.: LoAdaBoost: loss-based adaboost federated machine learning on medical data. arXiv preprint arXiv:1811.12629 (2018)
8. Sahu, A.K., Li, T., Sanjabi, M., Zaheer, M., Talwalkar, A., Smith, V.: On the convergence of federated optimization in heterogeneous networks. arXiv preprint arXiv:1812.06127 (2018)
9. Caldas, S., et al.: LEAF: a benchmark for federated settings. arXiv preprint arXiv:1812.01097 (2018)
10. Jaynes, E.T.: Information theory and statistical mechanics. Phys. Rev. **106**, 620 (1957)
11. Kullback, S.: Information Theory and Statistics. Courier Corporation (1997)
12. Rosenfeld, R.: A maximum entropy approach to adaptive statistical language modelling. Comput. Speech Lang. **10**, 187–228 (1996). https://doi.org/10.1006/csla.1996.0011
13. Nigam, K., Lafferty, J., McCallum, A.: Using maximum entropy for text classification. In: IJCAI-99 Workshop on Machine Learning for Information Filtering, vol. 1, pp. 61–67 (1999)
14. Wang, X., Peng, Y., Lu, L., Lu, Z., Bagheri, M., Summers, R. M.: ChestX-ray8: hospital-scale chest X-ray database and benchmarks on weakly-supervised classification and localization of common thorax diseases. In: Proceedings of the IEEE Conference on Computer Vision and Pattern Recognition, pp. 2097–2106 (2017)
15. Kermany, D.S., et al.: Identifying medical diagnoses and treatable diseases by image-based deep learning. Cell **172**, 1122–1131 (2018)
16. Triastcyn, A. Faltings, B.: Federated learning with bayesian differential privacy. arXiv preprint arXiv:1911.10071 (2019)
17. Krizhevsky, A., Nair, V., Hinton G.: CIFAR-10 (Canadian Institute for Advanced Research) (2010). http://www.cs.toronto.edu/~kriz/cifar.html
18. APTOS (2019). https://www.kaggle.com/c/aptos2019-blindness-detection
19. He, K., Zhang, X., Ren, S., Sun, J.: Deep residual learning for image recognition. In: Proceedings of the IEEE Conference on Computer Vision and Pattern Recognition, pp. 770–778 (2016)
20. Shamir, O., Srebro, N., Zhang, T.: Communication-efficient distributed optimization using an approximate newton-type method. In: International Conference on Machine Learning, pp. 1000–1008 (2014)

Model-Based Clustering with HDBSCAN*

Michael Strobl[1]([✉]), Jörg Sander[1], Ricardo J. G. B. Campello[2],
and Osmar Zaïane[1]

[1] University of Alberta, Edmonton, Canada
{mstrobl,jsander,zaiane}@ualberta.ca
[2] University of Newcastle, Callaghan, Australia
ricardo.campello@newcastle.edu.au

Abstract. We propose an efficient model-based clustering approach
for creating Gaussian Mixture Models from finite datasets. Models are
extracted from HDBSCAN* hierarchies using the Classification Likelihood and the Expectation Maximization algorithm. Prior knowledge of
the number of components of the model, corresponding to the number of
clusters, is not necessary and can be determined dynamically. Due to relatively small hierarchies created by HDBSCAN* compared to previous
approaches, this can be done efficiently. The lower the number of objects
in a dataset, the more difficult it is to accurately estimate the number of
parameters of a fully unrestricted Gaussian Mixture Model. Therefore,
more parsimonious models can be created by our algorithm, if necessary.
The user has a choice of two information criteria for model selection, as
well as a likelihood test using unseen data, in order to select the best-
fitting model. We compare our approach to two baselines and show its
superiority in two settings: recovering the original data-generating distribution and partitioning the data correctly. Furthermore, we show that
our approach is robust to its hyperparameter settings. (Data and code
are publicly available at: https://github.com/mjstrobl/HCEM)

Keywords: Hierarchical clustering · Expectation maximization ·
Model selection

1 Introduction

Model-based clustering is a popular tool for unsupervised data analysis due to its
powerful and compact representation of each group in a dataset. The Expectation
Maximization (EM) [9] algorithm uses Maximum Likelihood (ML) estimation
for fitting Multivariate Gaussian models to data quickly and accurately. These
models are also known as Gaussian Mixture Models (GMM).

However, there are a number of points to consider when using EM. First, the
number of clusters K (corresponding to the number of components in the final
model) has to be known in advance since EM tries to fit the parameters of a
GMM with K components to a dataset. Due to EM's ML parameter estimation

© Springer Nature Switzerland AG 2021
F. Hutter et al. (Eds.): ECML PKDD 2020, LNAI 12458, pp. 364–379, 2021.
https://doi.org/10.1007/978-3-030-67661-2_22

based on a finite number of observations, setting this parameter correctly is crucial as more components lead to a higher number of parameters and a higher loglikelihood of the data given the model. This is also known as model overfitting (i.e. high divergence to the original data-generating distribution). As a result, the loglikelihood of the data given a GMM cannot be used to select a model among a set of differently parameterized models with varied K. In addition, the covariance matrix parametrization of each component can be either fully open or partially restricted. This leads to more or less parsimonious models with Gaussians of equal or variable volume, shape, or orientation. Specifically for small datasets, it can be beneficial to choose a more restricted model. Second, EM has to be initialized with K objects, e.g. from the dataset, as seeds for each of the K initial clusters, in order to fit a model with K components. However, EM is very sensitive to the choice of those K seed objects. Therefore, several (typically non-deterministic) initialization strategies exist, e.g. choose K random seed objects. Depending on the hyperparameter setting, they are run multiple times in order to find the best initialization parameters.

Another popular clustering paradigm is density-based clustering, as introduced by the DBSCAN [10] algorithm and more recently extended by its hierarchical version HDBSCAN* [4]. DBSCAN's clustering model is deterministic, relatively fast to compute, and less strict than GMMs. It allows clusters of arbitrary shapes and the number of clusters does not have to be known in advance. Noise in the data can be detected, whereas GMMs have to model noise separately. However, if the assumption of Gaussian distributed clusters applies (even approximately), different Gaussian clusters in a dataset may be merged into a single cluster by DBSCAN, if they are overlapping, due to the difficulty of setting an appropriate global density-threshold. HDBSCAN*, on the other hand, uses a cluster stability measure to extract clusters from the cluster hierarchy. Cluster Stability may also favor the selection of a cluster (node in the hierarchy) that represents the result of merging child nodes representing two clusters that are generated by different, possibly overlapping, Gaussian model components. HDBSCAN* hierarchies, consisting of clusters with varying densities, often contain all clusters corresponding to each component of a generating GMM (even though these nodes may not be selected by Cluster Stability). Such hierarchies can therefore be used to find a GMM that fits the data well, when a more suitable cluster extraction strategy is used.

In this paper, we are proposing a clustering framework for model-based clustering that is able to accurately recover the original data-generating distribution for GMMs. Furthermore, it can automatically select the number of components K of the model. It is based on the HDBSCAN* hierarchy, which provides a compact tree of clusters from which models of different sizes can be extracted and evaluated efficiently.

The remainder of this article is structured as follows: Sect. 2 introduces GMMs and EM as well as HDBSCAN* in more detail. Our framework is introduced in Sect. 3 with experiments in Sect. 4 to support its strengths compared to two baselines. The article concludes in Sect. 5.

2   Background and Related Work

In this section we provide some background and related work on the popular model-based clustering paradigm, which our approach follows. Furthermore, we describe the HDBSCAN* algorithm for hierarchical density-based clustering, from which models can be extracted. For the rest of this article we assume our model is a mixture of multi-dimensional Gaussians, referred to as GMM.

2.1   Model-Based Clustering

Model-based clustering aims to estimate the parameters of a GMM using the EM technique [9]. The goal of EM is to fit a GMM to a dataset using an iterative procedure with the loglikelihood as the optimization goal.

The loglikelihood for a dataset $\boldsymbol{X} = (x_1, ..., x_n)$ with n observations (assumed to be i.i.d.), $x_i \in \mathbb{R}^d$ with $1 \leq i \leq n$ and $d \in \mathbb{N}_+$ is defined as

$$L(\boldsymbol{X}, \Theta_K) = \sum_{i=1}^{n} \log \left[\sum_{k=1}^{K} p_k f(x_i | a_k) \right] \tag{1}$$

where K is the number of components (one corresponding to each cluster) and $\Theta_K = \{p_1, ..., p_{K-1}, a_1, ..., a_K\}$, with $0 < p_k < 1$ and $\sum_{k=1}^{K} p_k = 1$. Each component has a mixing proportion p_k and a parameter vector $a_k = (\mu_k, \Sigma_k)$, with μ_k as the mean and Σ_k as the covariance matrix. $f(x_i | a_k)$ is the d-dimensional Gaussian density of the component k for object x_i.

For clustering, Eq. 1 is usually maximized using the EM algorithm [16], which estimates the conditional probabilities for all x_i in the Expectation (E) step and updates the model parameters in the Maximization (M) step iteratively.

Another variant of the EM algorithm for clustering is the Classification EM algorithm (CEM) [6], which aims to optimize the Classification Likelihood (CL), which is defined as

$$CL(\boldsymbol{X}, \Theta_K) = \sum_{k=1}^{K} \sum_{x_i \in C_k} \log(p_k f(x_i | a_k)) \tag{2}$$

where C_k is the set of objects x_i belonging to cluster k.

CEM does not compute a soft classification with every x_i being a member of each cluster to some extent, instead cluster labels are assigned. Therefore, the contribution of each x_i to $CL(\boldsymbol{X}, \Theta_K)$ is solely based on the parameter vector $a_k = (\mu_k, \Sigma_k)$ and the mixing proportion p_k of the cluster C_k to which x_i belongs. Biernacki and Govaert [3] showed that the CL can be seen as a penalized version of Eq. 1, favouring models with well-separated mixture components.

Unrestricted GMMs can be highly over-parameterized or, especially in high dimensional spaces, it may be difficult to estimate all parameters accurately. Therefore Celeux and Govaert developed an approach [7] to restrict the number of parameters in a GMM using the eigenvalue decomposition of Σ_k:

$$\Sigma_k = \lambda_k D_k A_k D_k'$$

where $\lambda_k = |\Sigma_k|^{1/d}$, D_k is the matrix of eigenvectors and A_k is a diagonal matrix with the normalized eigenvalues of Σ_k in decreasing order on the diagonal. Different combinations of these parameters result in 14 models with clusters of variable or equal volumes, shapes and orientations.

Three models, resulting in independent parameter estimations for all components, allow us to extract GMMs from clustering hierarchies efficiently[1]:

- VVV ($\Sigma_k = \lambda_k D_k A_k D'_k$): Fully unrestricted; Volume, Shape and orientation of all components are variable.
- VVI ($\Sigma_k = \lambda_k A_k$): Clusters are aligned to coordinate axis.
- VII ($\Sigma_k = \lambda_k I$): Only volume is variable, components are spherical.

2.2 Traditional Initialization

The most common initialization strategy for EM is starting with K random observations from the dataset and run EM until convergence. However, this can result in a suboptimal partition, if these K random observations are too different from the real cluster centers. In order to avoid this problem, EM can be run multiple times with random initialization and the model converging to the highest likelihood is selected. A similar approach is to run EM multiple times with random initialization for only a few iterations and continue iterating with the most promising model.

If models of different sizes (numbers of components K) should be created, from which one of them can be later selected, they have to be created independently. The procedure is simply repeated as many times as necessary, for each $K \in \mathcal{K} = [K_l, K_u]$ with K_l and K_u as the lower and upper bound of K, typically provided by the user.

2.3 Hierarchical Initialization

The main disadvantage of traditional initialization strategies is that if the number of clusters is unknown, multiple solutions (one for each $K \in \mathcal{K}$) have to be computed independently. Similarities between initializations of size K and $K-1$ are not taken into account. This can be prohibitively expensive if solutions for a large $|\mathcal{K}|$ have to be evaluated. Extracting solutions from a clustering hierarchy can be advantageous in this case, because new solutions with $K-1$ clusters can be computed through merging siblings in the hierarchy quickly, in the case of an agglomerative approach. In order to determine which set of siblings should be merged, for example, the sum-of-squares criterion can be used to select siblings that would result in the lowest possible variance increase of the model, if merged.

---

[1] All other models contain parameters that are equal among all components and therefore have to be jointly re-estimated if components of child nodes are replaced with parent node components. More information on this is explained below in Sect. 3.2

Fraley [11] describes an agglomerative hierarchical clustering algorithm based on GMMs. It starts with each observation representing its own cluster. As criterion for merging clusters, the CL is used at each stage, i.e. all pairs of clusters at the current stage are considered, but only one pair, that results in a model with the highest CL among all possible merges, is actually merged. For the models VVV, VVI and VII, it is not necessary to recompute all model parameters at each stage since each x_i contributes to the model's CL only through the parameters of the model component it belongs to. Therefore, only the new parent's parameters have to be computed. For models other than these, some parameters have to be recomputed for all components, e.g. if the volume of all components must be equal according to the chosen parametrization.

Since agglomerative hierarchical clustering algorithms start from singleton clusters, the hierarchy can take up a large amount of memory space. If it is not possible to compute all ML parameters (especially at the beginning with singleton clusters), the sum-of-squares criterion is used until clusters are large enough. While the hierarchy is built up, models of different sizes can be extracted. In order to extract only a limited number of models, a range of clustering sizes $\mathcal{K}$ can be provided (see [12]). Once there are $max(\mathcal{K})$ clusters at the top of the hierarchy (while building it), a model of size $max(\mathcal{K})$ is extracted. Clusters are continuously merged and a new model is extracted after each merge. Those models can be re-parameterized with EM and evaluated using a model selection method.

2.4   Model Selection

There are two main problems that model selection methods aim to solve: How many components the final model should have and how the covariance matrix Σ_k should be parameterized. In model-based clustering, information criteria are used to solve these problems. Their main goal is to find a trade-off between the best model-fit (high loglikelihood) and the least number of parameters in the model. The following are commonly used criteria, e.g. used in [2]:

- Bayesian Information Criterion (BIC) [19]:

$$BIC = L(\boldsymbol{X}|\hat{\Theta}_K) - \frac{\nu_K}{2}\log n$$

  ν_K corresponds to the number of free parameters in the model, which depends on the number of components in the model and its parametrization.
- Integrated Complete Likelihood (ICL) [1]:

$$ICL = CL(\boldsymbol{X}|\hat{\Theta}_K) - \frac{\nu_K}{2}\log n$$

  ICL penalizes models with less well-separated components more than BIC.
- Normalized Entropy Criterion (NEC) [8]:

$$NEC = \frac{E(\boldsymbol{X}|\hat{\Theta}_K)}{L(\boldsymbol{X}|\hat{\Theta}_K) - L(\boldsymbol{X}|\hat{\Theta}_1)}, \qquad E(\boldsymbol{X},\hat{\Theta}_K) = -\sum_{k=1}^{K}\sum_{i=1}^{n}\hat{t}_{ik}\log\hat{t}_{ik} \geq 0$$

where t_{ik} is the conditional probability that x_i belongs to C_k.

The NEC favours low entropy models with well-separated components over models with more components and worse separation.

All information criteria have in common that they make models with different parametrizations and numbers of components comparable.

2.5 HDBSCAN*

HDBSCAN* [4] is a hierarchical clustering algorithm based on DBSCAN*, a revised version of DBSCAN [10], which is a well-known density-based clustering algorithm. DBSCAN* receives two input values set by the user: a density-threshold ε and the value m_{pts}, which acts as a smoothing factor for the density-estimates of each object. The original DBSCAN introduced the notion of *core objects*, which are objects with at least m_{pts} neighbours within the distance ε. These neighbours form an object's ε-neighbourhood N_ε. A cluster C is defined as a non-empty maximal subset of the input dataset, such that every pair of objects in C is density-connected. Two core objects x_p and x_q are density-connected if they are directly or transitively ε-reachable; directly ε-reachable means $x_p \in N_\varepsilon(x_q)$ and $x_q \in N_\varepsilon(x_p)$.

In case of DBSCAN*, the global density-threshold ε is difficult to set if there are clusters of varying densities, because each cluster must exceed density-level ε. HDBSCAN* solves this problem by creating a hierarchy of clusters of different densities. In order to restrict the hierarchy's depth, the optional parameter m_{cl} was introduced as a minimum number of objects within a cluster. In [4], it is suggested to set $m_{cl} = m_{pts}$ and leave m_{pts} as the only user-defined parameter, which is assumed in the remainder of this article.

The HDBSCAN* hierarchy is created by the following steps:

1. Building the Minimum Spanning Tree (MST) of the Mutual Reachability Graph, which consists of all objects as vertices and the mutual reachability distances for each pair of objects as edge weight. The mutual reachability distance is defined as
$d_{mreach}(x_p, x_q) = max\{d_{core}(x_p), d_{core}(x_q), d(x_p, x_q)\}$
where $d_{core}(x)$ is the distance of object x to its m_{pts} nearest neighbours and $d(x_p, x_q)$ is the distance between x_p and x_q.
2. The whole dataset is contained in the cluster which forms the root of the hierarchy. Subsequent clusters and noise are created by removing edges from the MST in decreasing order of weight. Removing an edge results in either two new clusters with more than m_{cl} objects each, a shrank cluster with at least m_{cl} objects and noise, or noise only if the cluster disappears.
3. From the full clustering hierarchy containing all splits, a compact hierarchy is created containing only the most important levels, namely when clusters appear the first time or completely disappear.

As a special case, setting $m_{pts} = 1$ results in a single-linkage hierarchy.

Due to the different density-thresholds represented by the hierarchy (the closer to the root the less dense the density threshold is), the user does not need to provide the parameter ε anymore. The original DBSCAN algorithm produced flat partitions whose clusters are maximal sets of connected points with a point density in their neighborhood above a given density threshold, set by the parameter ε. The result corresponds to extracting clusters from the HDBSCAN* hierarchy at the density level ε, representing a horizontal cut through the hierarchy. In order to extract clusters of variable densities (different levels) from such a hierarchy, the measure of Stability was introduced. The Stability S of a cluster measures how stable it is in the hierarchy. This is dependent on the density-range in which the cluster exists, i.e. minimum density value at which the cluster starts to exist and maximum density level at which the cluster is either split or disappears, as well as how many objects the cluster contains compared to its descendants and parent cluster. Stability values can be computed while creating the hierarchy.

The algorithm to extract a flat partition from such a hierarchy is an iterative procedure and starts at the leaf level as the current clustering solution. For every parent-descendants pair in the current solution: If a parent C_p is more stable than its descendants, say C_{d_1} and C_{d_2}, i.e. $S_{C_p} > S_{C_{d_1}} + S_{C_{d_2}}$, C_{d_1} and C_{d_2} are replaced by C_p in the current solution. Otherwise the parent of C_p is assigned to C_{d_1} and C_{d_2} as new parent cluster, basically ignoring C_p from now on. This procedure is repeated until the root of the hierarchy is the parent of all clusters in the current solution.

Extracting a clustering solution this way is efficient since at every stage local decisions are made, i.e. comparing only parents and descendants (and no other clusters) is sufficient to obtain a globally optimal solution regarding S, returning the most stable set of clusters.

3   Method

In this section we present how to create a GMM from HDBSCAN* hierarchies. To account for the variations in the cluster tree when choosing different values of m_{pts}, we use multiple hierarchies and choose the best partition according to the CL. Neto et al. [17] show how over a hundred hierarchies (i.e., different values of m_{pts}) can be efficiently computed with the cost of about 2 HDBSCAN* runs.

3.1   Classification Likelihood Criterion

While the original HDBSCAN* Stability measure works well for density-based clustering, it may not be suitable for a dataset following a GMM. Different clusters with a relatively dense overlap could be considered as a very stable single cluster since it could appear as such early in the hierarchy and be separated very late. However, extracting a clustering solution based on Stability is efficient due to local decision making. Through using the models VVV, VVI, and VII, and CL as optimization criterion, it is sufficient to make local decisions as well, while still extracting the best model (according to CL) from an HDBSCAN* hierarchy, when Gaussian distributed clusters are assumed. In [3], GMMs with different settings of K are fitted to a dataset and a CL-like criterion is used to compare these models. The authors argue that the CL works for finding the correct number of mixture components K, whereas the loglikelihood L in Eq. 1 overestimates K. We argue that each mixture component is represented as separate cluster in the HDBSCAN* hierarchy and that the CL (instead of Stability) can be used to extract those clusters.

3.2 Creating GMMs from HDBSCAN* Hierarchies

The high-level steps of our algorithm for clustering dataset X are the following:

1. Create HDBSCAN* hierarchies for X, one for each m_{pts}-value.
2. For each hierarchy:
    (a) Create an initial GMM with one component per leaf, independently estimated.
    (b) Assign each noise object to the leaf with the highest loglikelihood according to the initial GMM and re-estimate the parameters of all components, representing a candidate model.
    (c) After estimating the Gaussian parameters of all clusters in the hierarchy, the optimal GMM (according to CL and given the hierarchy) for each model parametrization (VVV, VVI and VII) is extracted. This is done through an iterative procedure similar to using the Stability in [4].
3. Select the best model among all models and create a flat partition according to this model.

In the following, we provide a detailed description of these steps.

1. Create HDBSCAN* Hierarchy. We run the original HDBSCAN* algorithm on a dataset X in order to create a compact hierarchy for each m_{pts}-value.

(a) Create Initial GMM. HDBSCAN* creates a hierarchy of clusters top-down. Whenever clusters are split, noise objects may be created. Our algorithm starts by creating GMMs bottom-up, which makes it necessary to assign all noise objects to leaf clusters in the hierarchy, since the union of all leaf clusters is not necessarily equal to X. For each leaf, the parameters of one Gaussian with an unrestricted Σ_k (model VVV) are estimated. If it is not possible to estimate a non-singular unrestricted Σ_k due to data sparsity, more restricted models are used, VVI or VII. Leaves where this is not possible are removed from the set of leaves including its sibling and the parent cluster is added instead. Parameter estimation is then tried again for this node. The initial GMM is defined by the set of all estimated components, one for each leaf.

(b) Assign Noise Objects. All noise objects are assigned to the model component l with $l = \text{argmax}_{k=1,\ldots,K}\, p_k f(x_i|a_k)$. After this step, it is possible to estimate parameters more accurately since all $x_i \in \boldsymbol{X}$ are assigned to clusters now, and, therefore, the parameters of all components are re-estimated.

(c) Create Candidate GMMs. The initial GMM represents a clustering solution with the maximum possible K, given the clusters in the HDBSCAN* cluster tree. New models of smaller sizes and parametrizations (VVV, VVI or VII) can be created iteratively by merging or keeping siblings according to the best CL of the resulting model. This works the same way as done using the Stability as described in Sect. 2.5. The only parameters that have to be estimated are the mean μ_k and covariance matrix Σ_k of the parent, since the parameters of the descendants are estimated in the previous iteration or step (b). In addition, since we are using the VVV, VVI and VII parameterizations, parameters of other components in the current model are independent and therefore do not have to be re-estimated. Decisions are made locally using the CL in the following way.

Consider parent solution $C_p = \{C_1, \ldots, C_{K-1}, C_{p_{d1,d2}}\}$ and descendant solution $C_d = \{C_1, \ldots, C_{K-1}, C_{d1}, C_{d2}\}$ with parent cluster $C_{p_{d1,d2}}$ replaced by its descendants C_{d1} and C_{d2} $(C_{p_{d1,d2}} = C_{d1} \cup C_{d2})$. C_p and C_d represent a solution with K and $K+1$ clusters, respectively. $CL_p(\boldsymbol{X}, \Theta_K)$ for solution C_p can be written as:

$$CL_p(\boldsymbol{X}, \Theta_K) = \sum_{k}^{K-1} \sum_{x_i \in C_k} \log(p_k f(x_i|a_k)) + \sum_{x_i \in C_{p_{d1,d2}}} \log(p_{p_{d1,d2}} f(x_i|a_{p_{d1,d2}}))$$

(3)

$CL_d(\boldsymbol{X}, \Theta_{K+1})$ for solution C_d with $p_{d_1} + p_{d_2} = p_{p_{d1,d2}}$ can be written as:

$$CL_d(\boldsymbol{X}, \Theta_{K+1}) = \sum_{k}^{K-1} \sum_{x_i \in C_k} \log(p_k f(x_i|a_k)) + \sum_{x_i \in C_{d_1}} \log(p_{d_1} f(x_i|a_{d_1}))$$
$$+ \sum_{x_i \in C_{d_2}} \log(p_{d_2} f(x_i|a_{d_2}))$$

(4)

The first part of Eqs. 3 and 4 is identical and can be ignored, if $CL_p(\boldsymbol{X}, \Theta_K)$ and $CL_d(\boldsymbol{X}, \Theta_{K+1})$ are compared. Therefore, comparing both solutions can be restricted to comparing the CL of the parent cluster $(C_{p_{d1,d2}})$ and the sum of the CLs of the descendants $(C_{d1}$ and $C_{d2})$.

We claim that our method works well when each component in the initial GMM contains at most one mode from the data-generating distribution GMM_{data}. If this is not the case, it would not be possible to create a model with K components with one mode from GMM_{data} per component. However, the modes from GMM_{data} are presumably the densest parts of the data, leading to a separation of these in the leaf-level of the hierarchy. This property of HDBSCAN* hierarchies often holds in practice.

Figure 1 shows a dataset generated by a GMM with $K = 5$ and two merging iterations performed by our method, based on an initial GMM with $K = 7$. Figure 1a shows the clustering result according to the initial GMM after assigning all noise objects to the leaf clusters. We can observe that every leaf contains at most one mode from GMM_{data}. In Fig. 1b the purple cluster resulted from merging two clusters since this resulted in a higher CL. Only one of them contained a mode. Similarly, Fig. 1c shows the final clustering based on a GMM with $K = 5$. All final clusters contain exactly one mode. The subsequent potential merges all resulted in a solution with a lower, i.e. worse, CL.

The parameters of the candidate models (one for each parameterization) are refined again, running the CEM algorithm until convergence.

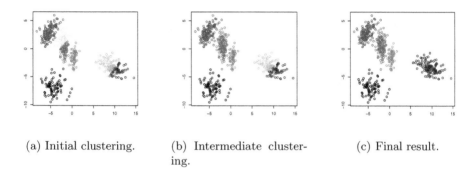

(a) Initial clustering.   (b) Intermediate clustering.   (c) Final result.

Fig. 1. Dataset generated from five Gaussians. Clusters contain between 50 and 100 objects. HDBSCAN* was run with $m_{pts} = 9$.

3. Model Selection. After one candidate model for each possible value of m_{pts} and each parameterization is created in the previous step, the best has to be selected according to a model selection strategy. To do so, information criteria, such as the BIC and ICL, have been used in the literature to penalize larger models with more parameters.

Since it is not clear in which circumstances which criterion works best, we propose a model selection strategy similar to k-fold cross validation, apart from using the BIC or ICL. For this case, we still create candidate models as described in steps 1 and 2. However, in order to evaluate these models we split the dataset into m folds and refine them using the CEM algorithm m times on $m - 1$ folds (leaving a different fold out each time) and evaluate each model on the left-out fold using the loglikelihood. We can use the average of the resulting m loglikelihoods as criterion (denoted as VAL in the rest of this article) for model selection since they were computed using the left-out folds that were not used for fitting the model directly.

Therefore, BIC, ICL and VAL can be used as model selection criteria.

4   Evaluation

In this section, we present empirical results for our approach (denoted by HCEM) and two baselines. In order to measure the effectiveness of the approaches in recovering the data-generating distribution, we use synthetic datasets where this distribution is known as ground truth, and measure (1) the similarity of the GMMs discovered by the methods with the ground truth, and (2) the quality of the induced partition on the data. In addition, we use real datasets to show the superiority of our approach regarding the resulting partitions.

4.1   Baselines

Mclust [12] is based on EM and hierarchical initialization as described in Sect. 2.3 and uses the BIC for model selection. Since models up to n clusters (if there are n observations in the dataset) can be extracted from the agglomerative hierarchical clustering hierarchy, a set of possible $K \in \mathcal{K}$ has to be provided to restrict the number of components in the final set of models, from which the best according to BIC is selected. However, this requires domain knowledge and is difficult to set, if the approximate number of clusters is unknown, and computationally expensive, if $|\mathcal{K}|$ is large.

Mixmod [2] uses different traditional initialization methods, the BIC, ICL or NEC for model selection. We denote these different instances as Mixmod (BIC), Mixmod (ICL) and Mixmod (NEC), respectively.

4.2   Parameter Settings

For our experiments, we set $\mathcal{K} = [2, K + 10]$ for Mclust and Mixmod, to make sure we include the correct number of clusters K in the dataset, while still keeping the range large enough to show that both baseline methods are able to select a model with K components from a larger collection. The three model parameterizations VVV, VVI and VII can be used by both methods.

In addition, Mixmod is initialized with the "smallEM" approach, running EM with random initialization 10 times for 5 iterations. It continues with the best model among these 10, according to the highest loglikelihood.

For HDBSCAN*, we chose $m_{pts} \in \{5, 6, 7, 8, 9, 10\}$, which are typical values. HCEM is run once for these values of m_{pts} and the result of the best model according to a model selection strategies is reported. As model selection strategies, we use BIC, ICL and VAL, denoting the resulting instances by HCEM (BIC), HCEM (ICL) and HCEM (VAL), respectively. Setting $\mathcal{K}$ is not required for HCEM since there is only a single model per hierarchy (one for each value of m_{pts}) created, limiting the set of models to be evaluated.

4.3   Synthetic Datasets

We used the GMM data generator from [14] to create synthetic datasets with the following parameters:

- Number of Gaussians: 10, 40
- Dimensions: 2, 5, 10, 25, 50
- Objects per Gaussian: uniformly distributed in the range $[5*D, 10*D]$, where D is the dimensionality of the dataset.
- Mean: uniformly distributed in the range $[-10, 10]$
- Off-diagonal entries of covariance matrix: random number y in the range $[-1, 1]$, with a distribution following $y = x^2$ where x is a uniformly random deviate in $[0, 1]$ and the sign of y is determined randomly.
- Diagonal entries of covariance matrix: generated as the sum of all off-diagonal entries plus a random number y in the range $[0, 20 * \sqrt{D}]$ with a distribution following $y = x^2$, where x is a uniformly random deviate in $[0, 1]$, and the sign of y is determined randomly.

This generator [14] guarantees that generated Gaussians do not overlap. We created 20 datasets for each configuration (dimensionality and number of clusters) and averaged the result on these 20 datasets.

Model Recovery Results. GMMs are versatile models that can be used to generate new data or as a classifier for unseen data. This leads to the goal of generating models that are ideally identical to the original data-generating distribution in their parameter settings. The Kullback-Leibler-Divergence (KLD) can be used as a measure of how much two GMMs deviate from each other (see [15]). However, the KLD is analytically not tractable, therefore [15] suggested to use Monte Carlo Sampling (MC) to compute an approximation:

$$D_{MC}\left(f\|g\right) = \frac{1}{n} \sum_{i=1}^{n} \log \frac{f(x_i)}{g(x_i)} \tag{5}$$

Table 1 shows experiments on the output of Eq. 5 using a sample of size $n = 100,000$ of the data generating distribution. f corresponds to the probability density of the original distribution from which the datasets were sampled, g corresponds to the probability density of the distribution fitted to the datasets by the three clustering algorithms. The lower the $D_{MC}\left(f\|g\right)$, the better, and a result of 0.0 means the distributions are identical.

The results were averaged over all 20 datasets per configuration, and we also report the overall average of all runs for all dimensions and number of clusters. Welch's t-test [21] for unequal variances was used with threshold 0.05 as significance test.

Mclust works well for most datasets, but is unstable in 50 dimensions, which lead to the second worst average out of all tested approaches. In order to merge the right clusters when building up the hierarchy for Mclust, accurate parameter estimations are necessary, which is increasingly difficult with more dimensions for datasets of limited size.

Table 1. Experiments on synthetic datasets, evaluated using KLD, averaged over 20 datasets per configuration. Lower is better; the best KLD is presented in **bold**. Results, that are statistically significant, are presented in green (better than all other approaches and model selection criteria) and blue (better than the other two approaches).

Dimensions	2		5		10		25		50		Avg
Alg/K	10	40	10	40	10	40	10	40	10	40	
Mclust(BIC)	**0.27**	0.50	0.31	0.35	0.37	0.34	**0.37**	**0.37**	2.68	0.66	0.62
Mixmod(BIC)	0.32	0.47	0.36	0.46	0.37	0.52	0.38	0.54	**0.40**	0.48	0.43
Mixmod(ICL)	0.31	0.45	0.32	0.43	0.37	0.54	0.38	0.60	**0.40**	0.60	0.44
Mixmod(NEC)	0.49	0.45	1.26	0.64	1.60	0.96	1.27	1.61	1.16	2.48	1.19
HCEM(BIC)	0.34	0.51	0.34	0.54	0.37	0.51	**0.37**	**0.37**	0.40	**0.40**	0.42
HCEM(ICL)	0.35	0.48	0.33	0.51	0.37	0.44	**0.37**	**0.37**	0.40	**0.40**	0.40
HCEM(VAL)	0.29	0.40	**0.28**	**0.33**	**0.34**	0.37	**0.37**	**0.37**	**0.40**	**0.40**	0.35

While not creating the models with the best KLD in all but one configuration, Mixmod with ICL and BIC provides similar results to Mclust, except that it is able to return a lower KLD for 50-dimensional data. NEC performs worse than BIC and ICL in our setting with the only exception of 2 dimensions and 40 clusters. The average of all KLD values for Mixmod (BIC) and Mixmod (ICL) is higher than the worst HCEM approach, HCEM (BIC).

HCEM consistently returns the best GMM or close to the best, independently of the number of clusters in the dataset. There are only small differences between the models selected by BIC and ICL. HCEM (VAL) achieves overall the best average of 0.35 over all runs, which is statistically significant.

Clustering Results. We also evaluate the performance of all approaches using the Adjusted Rand Index (ARI) [16] to measure the quality of a clustering result. Table 2 shows experiments using Mclust, Mixmod and HCEM. The average ARI for all algorithms and data configurations (20 datasets each) is reported as well as the overall average all dimensions and number of clusters. Values in **bold** are the best results, i.e. highest ARI averaged over all 20 datasets per configuration.

The results lead to similar conclusions about the relative performance of the algorithms as in the previous experiments. All HCEM approaches achieve a better average ARI than Mclust and Mixmod for all model selection criteria. HCEM (VAL) achieves, again, the best average value and in 7 out of 10 configurations and the best individual average ARI. HCEM (BIC), HCEM (ICL) and HCEM (VAL) return, on average, better partitions than Mixmod and Mclust, and these differences are statistically significant.

Table 2. Experiments on synthetic datasets, evaluated using ARI, averaged over 20 datasets per configuration. Higher is better and best ARI is presented in **bold**. Results, that are statistically significant, are presented in green (better than all other approaches and model selection criteria) and blue (better than the other two approaches).

Dimensions	2		5		10		25		50		Avg
Alg/K	10	40	10	40	10	40	10	40	10	40	
Mclust(BIC)	**0.92**	0.54	**1.00**	**0.98**	1.00	1.00	1.00	1.00	0.81	0.98	0.92
Mixmod(BIC)	0.89	0.54	0.97	0.93	0.98	0.95	0.99	0.95	**1.00**	0.96	0.92
Mixmod(ICL)	0.89	0.60	0.98	0.93	0.99	0.96	0.99	0.95	**1.00**	0.96	0.92
Mixmod(NEC)	0.76	0.66	0.64	0.78	0.63	0.76	0.85	0.72	0.91	0.63	0.73
HCEM(BIC)	**0.92**	0.68	0.99	0.94	1.00	0.99	1.00	1.00	1.00	1.00	0.95
HCEM(ICL)	**0.92**	0.71	0.99	0.96	1.00	1.00	1.00	1.00	1.00	1.00	**0.96**
HCEM(VAL)	**0.92**	**0.74**	0.99	0.97	1.00	0.99	1.00	1.00	1.00	1.00	**0.96**

4.4   Real Datasets

Here we present clustering results for 5 different real datasets from the UCI repository[2] with data statistics summarized in Table 3. The Digits dataset was downloaded from Scikit-learn [18]. Ecoli, Iris, Diabetes and Wine datasets were previously used for the evaluation of clustering algorithms, e.g. in [12], [5], [20] and [13]. Since the data-generating distribution is unknown for real data, we are using the ARI to judge the resulting partitions.

Table 3. Real datasets used in our experiments from the UCI repository.

	Ecoli	Iris	Diabetes	Wine	Digits
Dimensions	7	4	3	13	64
Objects	336	150	145	178	1797

We set $\mathcal{K} = [2, 100]$ for Mclust and Mixmod, which includes the correct number of clusters for each dataset. All other parameters are as described in Sect. 4.2. Table 4 shows experiments on all 5 datasets and compares the ARI of the resulting partitions using all available model parameterizations.

The results are consistent with the ARI experiments on synthetic data. For the Ecoli, Iris, Wine and Digits datasets, all HCEM approaches achieve a better clustering result than Mclust and all Mixmod approaches by a large margin. Mclust (BIC), Mixmod (BIC) and Mixmod (ICL) return the best result only on the Diabetes dataset, which is the dataset with the lowest dimensionality (only 3 dimensions). HCEM (BIC) returns the best partitions on three out of five cases for real data and is close to the best result on Diabetes and Iris.

---

[2] https://archive.ics.uci.edu/ml/datasets.php.

Table 4. Experiments on real datasets, evaluated using ARI. Higher is better and best ARI is presented in **bold**.

Alg/Dataset	Ecoli	Iris	Diabetes	Wine	Digits
Mclust (BIC)	0.39	0.57	**0.66**	0.71	0.29
Mixmod (BIC)	0.34	0.31	**0.66**	0.68	0.30
Mixmod (ICL)	0.28	0.33	**0.66**	0.68	0.29
Mixmod (NEC)	0.07	0.57	0.44	0.08	0.27
HCEM (BIC)	**0.68**	0.90	0.65	**0.85**	**0.71**
HCEM (ICL)	0.66	**0.92**	0.53	0.83	0.62
HCEM (VAL)	0.66	**0.92**	0.53	0.83	0.62

5   Conclusion

We proposed a clustering framework for model-based clustering that is better in accurately recovering the data generating distribution and partitioning the data for synthetic as well as real datasets, compared to the baselines Mixmod and Mclust. Furthermore, the number of clusters does not have to be known a priori and no assumptions about the set $\mathcal{K}$ has to be made. We also showed that our approach is robust to different parameter settings (m_{pts} from HDBSCAN* as well as the model selection method). Therefore, HCEM seems to be a good choice for initializing the EM algorithm for clustering.

Currently 3 models are supported since only those make sure that models with different numbers of components can be extracted from the HDBSCAN* hierarchy efficiently without recomputing all model parameters after a decision about whether a parent or its descendant results in a better model. In [11] ways for efficiently extracting GMMs with a different number of components are described using more models than the three currently used ones. We leave implementing those for future work.

References

1. Biernacki, C., Celeux, G., Govaert, G.: Assessing a mixture model for clustering with the integrated completed likelihood. IEEE Trans. Pattern Anal. Mach. Intel. **22**(7), 719–725 (2000)
2. Biernacki, C., Celeux, G., Govaert, G., Langrognet, F.: Model-based cluster and discriminant analysis with the MIXMOD software. Comput. Stat. Data Anal. **51**(2), 587–600 (2006)
3. Biernacki, C., Govaert, G.: Using the classification likelihood to choose the number of clusters. Comput. Sci. Stat. **29**, 451–457 (1997)
4. Campello, R.J.G.B., Moulavi, D., Sander, J.: Density-based clustering based on hierarchical density estimates. In: Pei, J., Tseng, V.S., Cao, L., Motoda, H., Xu, G. (eds.) PAKDD 2013, Part II. LNCS (LNAI), vol. 7819, pp. 160–172. Springer, Heidelberg (2013). https://doi.org/10.1007/978-3-642-37456-2_14

5. Campello, R.J., Moulavi, D., Zimek, A., Sander, J.: Hierarchical density estimates for data clustering, visualization, and outlier detection. ACM Trans. Knowl. Disc. Data (TKDD) **10**(1), 1–51 (2015)
6. Celeux, G., Govaert, G.: A classification EM algorithm for clustering and two stochastic versions. Comput. Stat. Data Anal. **14**(3), 315–332 (1992)
7. Celeux, G., Govaert, G.: Gaussian parsimonious clustering models. Pattern Recognit. **28**(5), 781–793 (1995)
8. Celeux, G., Soromenho, G.: An entropy criterion for assessing the number of clusters in a mixture model. J. Classif. **13**(2), 195–212 (1996)
9. Dempster, A.P., Laird, N.M., Rubin, D.B.: Maximum likelihood from incomplete data via the EM algorithm. J. Roy. Stat. Soc. Ser B Methodol. **39**, 1–38 (1977)
10. Ester, M., Kriegel, H.P., Sander, J., Xu, X., et al.: A density-based algorithm for discovering clusters in large spatial databases with noise. In: KDD, vol. 96, pp. 226–231 (1996)
11. Fraley, C.: Algorithms for model-based gaussian hierarchical clustering. SIAM J. Sci. Comput. **20**(1), 270–281 (1998)
12. Fraley, C., Raftery, A.E.: How many clusters? Which clustering method? Answers via model-based cluster analysis. Comput. J. **41**(8), 578–588 (1998)
13. Gelbard, R., Goldman, O., Spiegler, I.: Investigating diversity of clustering methods: an empirical comparison. Data Knowl. Eng. **63**(1), 155–166 (2007)
14. Handl, J., Knowles, J.: Cluster generators for large high-dimensional data sets with large numbers of clusters. Dimension **2**, 20 (2005)
15. Hershey, J.R., Olsen, P.A.: Approximating the kullback leibler divergence between gaussian mixture models. In: 2007 IEEE International Conference on Acoustics, Speech and Signal Processing, ICASSP 2007, vol. 4, pp. IV-317. IEEE (2007)
16. Hubert, L., Arabie, P.: Comparing partitions. J. Classif. **2**(1), 193–218 (1985)
17. Neto, A.C.A., Sander, J., Campello, R.J., Nascimento, M.A.: Efficient computation of multiple density-based clustering hierarchies. In: 2017 IEEE International Conference on Data Mining (ICDM), pp. 991–996. IEEE (2017)
18. Pedregosa, F., et al.: Scikit-learn: Machine learning in python. J. Mach. Learn. Res. **12**, 2825–2830 (2011)
19. Schwarz, G., et al.: Estimating the dimension of a model. Ann. Stat. **6**(2), 461–464 (1978)
20. Timm, H., Borgelt, C., Döring, C., Kruse, R.: An extension to possibilistic fuzzy cluster analysis. Fuzzy Sets Syst. **147**(1), 3–16 (2004)
21. Welch, B.L.: The generalization of student's' problem when several different population variances are involved. Biometrika **34**(1/2), 28–35 (1947)

Kernel Methods and Online Learning

Incremental Sensitivity Analysis
for Kernelized Models

Hadar Sivan[1(✉)], Moshe Gabel[2], and Assaf Schuster[1]

[1] Technion - Israel Institute of Technology, 3200 Haifa, Israel
{hadarsivan,assaf}@cs.technion.ac.il
[2] University of Toronto, Toronto, Canada
mgabel@cs.toronto.edu

Abstract. Despite their superior accuracy to simpler linear models, kernelized models can be prohibitively expensive in applications where the training set changes frequently, since training them is computationally intensive. We provide bounds for the changes in a kernelized model when its training set has changed, as well as bounds on the prediction of the new hypothetical model on new data. Our bounds support any kernelized model with L_2 regularization and convex, differentiable loss. The bounds can be computed incrementally as the data changes, much faster than re-computing the model. We apply our bounds to three applications: active learning, leave-one-out cross-validation, and online learning in the presence of concept drifts. We demonstrate empirically that the bounds are tight, and that the proposed algorithms can reduce costly model re-computations by up to 10 times, without harming accuracy.

1 Introduction

Supervised machine learning algorithms solve an optimization problem over a given training set, a process called *training*. Kernelized machine learning models are a common choice for handling non-linearity. However, training such models i.e., finding the optimal solution for an optimization problem over a training set, is often slow [22,31], with run-time complexity being quadratic or cubic in the size of the training set [1].

Such steep run-times can be prohibitive for applications that require frequent retraining. For example, in *leave-one-out cross-validation* (LOOCV) we remove one sample at a time from the original training set, train a model, then test it on the removed sample. In *active learning*, we iteratively improve the training set by selectively choosing samples to label and then retraining the model. Conversely, in *online learning*, new samples arrive one by one, and are added to the training set; the model is recomputed as needed.

Existing approaches to this problem reduce the run-time by using a previously-computed model to speed up the computation of the next model, where the training sets of the models are only slightly different. Examples of such an approach include *incremental algorithms*, which update the model one sample at a time [3,9,18], and *warm start* approaches, which use the previous

© Springer Nature Switzerland AG 2021
F. Hutter et al. (Eds.): ECML PKDD 2020, LNAI 12458, pp. 383–398, 2021.
https://doi.org/10.1007/978-3-030-67661-2_23

solution when initializing numerical optimization [30]. However, the computational cost of these algorithms is still very expensive reaching approximately $O(m^2)$ where m is the number of *support vectors*.

A more efficient alternative is *sensitivity analysis*, which bounds the change in the trained model or its predictions when the training set is changed, without actually computing the new model. Since the change in the training set is usually smaller than the size of the new training set, incremental bounds are efficient to compute. To the best of out knowledge, existing work on sensitivity analysis is limited to (a) linear models [10,13,24] rather than kernelized models, and (b) to non-incremental bounds [29], which require computing over the full old and new training sets, and do not bound the change in prediction.

Our Contributions

We first present a novel bound for the change in any kernelized model as the training set changes, where this model is the solution for a kernelized machine learning optimization problem with convex and differentiable loss and L_2 regularization. Additionally, we derive bounds for the prediction of the updated model on new data. Our bounds support any convex differentiable loss, do not limit the change in the training set, and are applicable for both classification and regression. Finally, we provide an efficient procedure for incrementally evaluating the bounds as the training set changes, and analyze its runtime complexity.

We demonstrate our bounds in three different applications. First, we show that our general bounds can reduce model computations in LOOCV by 45% to 77% for parameter tuning, improving on state-of-the-art bounds for this application. Second, we provide a simple algorithm for active learning that learns faster than the original algorithm using a fine-grained selection of future training data. Third, we present an adaptive online learning algorithm that improves upon incremental learning and concept drift detectors in terms of accuracy and number of model computations by retraining models as needed. We also empirically test the tightness of the bound for model change, showing that for 95% of the test data the bound was no more than 1.2 times the actual change.

2 Background, Problem Definition, and Notations

Assume we have two datasets, S_1 and S_2, where S_2 is more recent, for example S_2 could be an updated version of S_1. The datasets may overlap i.e., some samples in S_1 also appear in S_2. We focus on bounding the distance between the optimal solutions (i.e., models) for two identical learning problems over the two datasets S_1 and S_2. Formally, each dataset is a collection of samples (x_i, y_i), where $x_i \in \mathbb{R}^d$ and $y_i \in \{-1, 1\}$ for classification problems or $y_i \in \mathbb{R}$ for regression problems, and we are interested in bounding the distance between the *kernelized* models trained on S_1 and S_2.

In many classification problems the data is not linearly separable, while in regression problems, the relationship between the predictors and the dependent variables is not always linear. In such cases, a feature map function $\Phi : \mathbb{R}^d \to \mathcal{H}$ can be applied to the samples to transform them from their original feature space to a new feature space $\mathcal{H}$, where the data separation (or relation between variables) is closer to linear [14].

Because $\Phi(x)$ might be infinitely dimensional, the *kernel trick* [27] is commonly used instead of computing and storing $\Phi(x)$ directly. $\beta \in \mathcal{H}$ is a hypothesis in a Reproducing Kernel Hilbert Space (RKHS) $\mathcal{H}$ with a positive definite kernel function $k : \mathbb{R}^d \times \mathbb{R}^d \to \mathbb{R}$ implementing the inner product $\langle \cdot, \cdot \rangle$. The inner product is defined so that it satisfies the reproducing property $\langle k(x, \cdot), \beta \rangle = \beta(x)$. For simplicity, we use the compact notation $\Phi(x)$ for $k(x, \cdot)$. The reproducing property of k implies in particular that $k(x, x') = \langle \Phi(x), \Phi(x') \rangle$.

The models β_1^* and β_2^* are the optimal solutions for the following optimization problems over the samples in the two datasets[1]:

$$\beta_1^* = \arg\min_{\beta \in \mathcal{H}} C_1 \sum_{i \in \mathcal{D}_1} \ell_i(\beta) + \frac{1}{2}\|\beta\|^2 \tag{1a}$$

$$\beta_2^* = \arg\min_{\beta \in \mathcal{H}} C_2 \sum_{i \in \mathcal{D}_2} \ell_i(\beta) + \frac{1}{2}\|\beta\|^2 \tag{1b}$$

for $C_1, C_2 > 0$, where $\mathcal{D}_1$ and $\mathcal{D}_2$ denote the set of indices of the samples in $\mathcal{S}_1$ and $\mathcal{S}_2$, respectively. The loss with respect to sample (x_i, y_i) is defined by $\ell_i(\beta) := \ell(y_i, z_i)$, where $z_i := \langle \Phi(x_i), \beta \rangle$ is the inner product between x_i and β. In this paper we focus on loss function $\ell(\cdot, \cdot)$ which is differentiable and convex function with respect to its second argument.

By the Representer theorem [29], if β_1^* is the solution of (1a), it can be expressed as the *dual form*:

$$\beta_1^* = \sum_{i \in \mathcal{D}_1} \alpha_i \Phi(x_i) \tag{2}$$

The coefficients α_i are obtained by solving the *dual problem* of (1a).

We also focus on bounding the prediction of the model β_2^* for a new sample x. The prediction of a model β^* for classification is $\hat{y} = \mathrm{sgn}\left(\langle \Phi(x), \beta^* \rangle \right)$, while for regression the prediction is $\hat{y} = \langle \Phi(x), \beta^* \rangle$. For example, predictions for model β_1^* can be computed by using the kernel trick: $\langle \Phi(x), \beta_1^* \rangle = \sum_{i \in \mathcal{D}_1} \alpha_i k(x_i, x)$.

3 Bounding Model Differences

This section details our main contributions. We first develop a *distance bound* – a bound that estimates the difference between the two models β_1^* and β_2^*, without actually computing β_2^*. We then extend it to *prediction bounds* which bound the predictions of β_2^* on new samples. Finally, we describe an incremental update scheme for the bounds. In Sect. 4 we show how using the bounds can significantly reduce number of model computations in different applications.

---

[1] Given an objective function of the form $a \sum_{i \in \mathcal{D}} \ell_i(\beta) + b\|\beta\|^2$, choosing $C = \frac{a}{2b}$ will bring it to the standard form (1).

3.1 Bounding the Distance Between Models

Let β_1^* and β_2^* be the models trained on the samples in the datasets $\mathcal{S}_1$ and $\mathcal{S}_2$. We define the difference between the two models as the Euclidean distance between the two model vectors: $\|\beta_1^* - \beta_2^*\|$. We propose a bound for this distance that can be evaluated without computing β_2^*.

Theorem 1 (Distance Bound). *Let β_1^* be the optimal solution in its dual form (2) of the optimization problem (1a) over the dataset $\mathcal{S}_1$ containing the labeled samples with indices $\mathcal{D}_1$. Let β_2^* be the optimal solution of the optimization problem (1b) over the updated dataset $\mathcal{S}_2$ containing the labeled samples with indices $\mathcal{D}_2$. Let $\mathcal{D}_A = \mathcal{D}_2 \setminus \mathcal{D}_1$ be the set of indices of samples added in $\mathcal{S}_2$ and $\mathcal{D}_R = \mathcal{D}_1 \setminus \mathcal{D}_2$ be the set of indices of samples removed in $\mathcal{S}_2$. Finally, let I be the indicator function and define $\gamma_i := (-1)^{I_{\{i \in \mathcal{D}_R\}}} \partial_{z_i} \ell_i(\beta_1^*)$, where $\partial_{z_i} \ell_i(\beta_1^*)$ is the partial derivative of ℓ_i with respect to z_i at the point β_1^*.*

Then the distance between β_1^* and β_2^* is bounded by:

$$\|\beta_1^* - \beta_2^*\| \le 2\|r\|,$$

where

$$r = \sum_{i \in \mathcal{D}_1 \cup \mathcal{D}_2} \tau_i \Phi(x_i) \tag{3}$$

and the coefficients τ_i are:

$$\tau_i = \begin{cases} \frac{1}{2}\left(1 - \frac{C_2}{C_1}\right)\alpha_i + \frac{C_2}{2}\gamma_i, & i \in \mathcal{D}_R \\ \frac{1}{2}\left(1 - \frac{C_2}{C_1}\right)\alpha_i, & i \in \mathcal{D}_1 \cap \mathcal{D}_2 \\ \frac{C_2}{2}\gamma_i, & i \in \mathcal{D}_A \end{cases}.$$

Discussion. Theorem 1 bounds the difference between computed models for any convex differentiable loss, given the difference in their training sets. For example, we can apply Theorem 1 to L_2-regularized logistic regression, as defined in Table 1, $C_1 = C_2 = C$. In this case, $\tau_i = \frac{C}{2}\gamma_i$ for $i \in \mathcal{D}_A \cup \mathcal{D}_R$ and 0 otherwise. Assigning this in (3) gives $r = \sum_{i \in \mathcal{D}_A \cup \mathcal{D}_R} \frac{C}{2}\gamma_i \Phi(x_i)$. For L_2-regularized MSE loss, the constants C_1 and C_2 depend on the number of samples in the dataset; thus they may differ if the size of the datasets $\mathcal{S}_1$ and $\mathcal{S}_2$ is different. Table 1 shows how to compute r and γ_i for three common optimization problems. Note that α_i are the coefficients in the dual form of β_1^* and C_1, C_2 are determined by the objective function.

Proof. Let Δg be

$$\Delta g := \sum_{i \in \mathcal{D}_A} \nabla_\beta \ell_i(\beta_1^*) - \sum_{i \in \mathcal{D}_R} \nabla_\beta \ell_i(\beta_1^*).$$

By applying the chain rule for $\nabla_\beta \ell_i(\beta_1^*)$ we get:

$$\nabla_\beta \ell_i(\beta_1^*) = \partial_{z_i} \ell_i(\beta_1^*) \nabla_\beta z_i(\beta_1^*) = \partial_{z_i} \ell_i(\beta_1^*) \Phi(x_i)$$

Table 1. Objective functions, losses, gradients, and associated bound parameter r. $z_i = \langle \Phi(x_i), \beta \rangle$ can be computed using the kernel trick, and $\gamma_i = (-1)^{I_{\{i \in \mathcal{D}_\mathcal{R}\}}} \partial_{z_i} \ell_i(\beta_1^*)$.

Model and loss ℓ_i	Objective function	$\partial_{z_i} \ell_i(\beta)$	r
Logistic Regression $\log\left(1 + \exp(-y_i z_i)\right)$	$C \sum_i \ell_i + \frac{1}{2}\|\beta\|^2$	$-y_i / \left(1 + \exp(y_i z_i)\right)$	$\sum\limits_{i \in \mathcal{D}_\mathcal{A} \cup \mathcal{D}_\mathcal{R}} \frac{C}{2} \gamma_i \Phi(x_i)$
Squared Hinge SVM $(\max\{0, 1 - y_i z_i\})^2$	$C \sum_i \ell_i + \frac{1}{2}\|\beta\|^2$	$-2y_i \max\{0, 1 - y_i z_i\}$	$\sum\limits_{i \in \mathcal{D}_\mathcal{A} \cup \mathcal{D}_\mathcal{R}} \frac{C}{2} \gamma_i \Phi(x_i)$
Ridge Regression $(y_i - z_i)^2$	$\sum_i \ell_i + \lambda\|\beta\|^2$	$-2(y_i - z_i)$	$\sum\limits_{i \in \mathcal{D}_\mathcal{A} \cup \mathcal{D}_\mathcal{R}} \frac{1}{4\lambda} \gamma_i \Phi(x_i)$

We use this in the definition of Δg:

$$
\begin{aligned}
\Delta g &\triangleq \sum_{i \in \mathcal{D}_\mathcal{A}} \nabla_\beta \ell_i(\beta_1^*) - \sum_{i \in \mathcal{D}_\mathcal{R}} \nabla_\beta \ell_i(\beta_1^*) \\
&= \sum_{i \in \mathcal{D}_\mathcal{A} \cup \mathcal{D}_\mathcal{R}} \underbrace{(-1)^{I_{\{i \in \mathcal{D}_\mathcal{R}\}}} \partial_{z_i} \ell_i(\beta_1^*)}_{\triangleq \gamma_i} \Phi(x_i) \\
&= \sum_{i \in \mathcal{D}_\mathcal{A} \cup \mathcal{D}_\mathcal{R}} \gamma_i \Phi(x_i),
\end{aligned}
\tag{4}
$$

Note that $\partial_{z_i} \ell_i(\beta)$ is a scalar function with arguments y_i, z_i, which are also scalars. The computation of z_i at the point β_1^* is done with the kernel trick: $z_i = \sum_{j \in \mathcal{D}_1} \alpha_j k(x_j, x_i)$. Thus, γ_i can be computed using the kernel trick.

We can now use this kernelized form of Δg in the non-kernelized form of r from our previous work. In [28] we proved that under the same conditions as in Theorem 1, but for linear models β_1^*, β_2^*

$$
r = \frac{1}{2}\left(\beta_1^* - \frac{C_2}{C_1}\beta_1^* + C_2 \Delta g\right)
\tag{5}
$$

For completeness, we repeat some of the main steps from the original proof [28]. The proof proceeds in three steps:

(i) Use the convexity of the objective function to get a sphere Ω that contains β_2^*. The sphere Ω has center $m = \beta_1^* - r$ and radius vector $r = \frac{1}{2}\left(\beta_1^* + C_2 \sum_{i \in \mathcal{D}_2} \nabla_\beta \ell_i(\beta_1^*)\right)$.

(ii) Use the convexity of the objective function to express the sphere's radius as a function of Δg, and get (5).

(iii) Bound the distance between β_1^* and β_2^* using geometric arguments. We observe that β_1^* is on the surface of the sphere Ω and β_2^* is contained within the sphere. This implies that the maximum distance between β_1^* and β_2^* is obtained when β_1^* and β_2^* are at two opposite sides of the sphere's diameter, which has length $2\|r\|$. This yields the linear form bound in Theorem 1.

Note that this result cannot be applied directly for kernelized models: (5) cannot be evaluated for infinitely dimensional Φ such as the RBF kernel. To complete the proof for kernelized models, we show that the bound could be computed with the kernel trick when map function Φ is used on the data samples. We use the dual form of β_1^* (2) and the kernel form of Δg (4) in r (5):

$$
\begin{aligned}
r &= \frac{1}{2}\left(\beta_1^* - \frac{C_2}{C_1}\beta_1^* + C_2\Delta g\right) \\
&= \frac{1}{2}\left(1 - \frac{C_2}{C_1}\right)\sum_{i\in\mathcal{D}_1}\alpha_i\Phi(x_i) + \frac{C_2}{2}\sum_{i\in\mathcal{D}_A\cup\mathcal{D}_R}\gamma_i\Phi(x_i) = \sum_{i\in\mathcal{D}_1\cup\mathcal{D}_2}\tau_i\Phi(x_i) .
\end{aligned}
$$

The bound is then computed with the kernel trick:

$$
2\|r\| = 2\sqrt{\sum_{i\in\mathcal{D}_1\cup\mathcal{D}_2}\sum_{j\in\mathcal{D}_1\cup\mathcal{D}_2}\tau_i\tau_j k(x_i, x_j)} , \tag{6}
$$

which completes the proof.   □

3.2   Bounding the Predictions of the New Model

We describe upper and lower bounds for the prediction of β_2^* for a new sample[2]. As before, using the predictions of β_1^* we can compute these bounds without computing β_2^*.

Using the observation from Sect. 3.1 that β_2^* is within a sphere Ω with center m and radius vector r, we can obtain lower and upper bounds for applying β_2^* to a new sample x:

Lemma 1 (Prediction Bounds). *Let* β_1^*, β_2^*, *and* r *be as in Theorem 1, and let* x *be a sample. Then the upper and lower bounds on the prediction of* β_2^* *for* x *are:*

$$
L\left(\langle\Phi(x), \beta_2^*\rangle\right) = \langle\Phi(x), \beta_1^*\rangle - \langle\Phi(x), r\rangle - \|\Phi(x)\|\|r\| \tag{7a}
$$
$$
U\left(\langle\Phi(x), \beta_2^*\rangle\right) = \langle\Phi(x), \beta_1^*\rangle - \langle\Phi(x), r\rangle + \|\Phi(x)\|\|r\|. \tag{7b}
$$

Proof. Every vector β in the sphere Ω can be represented as the sum of two vectors: the center of the sphere m, and an offset vector u that starts from the center of the sphere and whose magnitude is bounded by the sphere radius vector ($\|u\| \leq \|r\|$). Therefore, the dot product between β and a given $\Phi(x)$ is

$$
\begin{aligned}
\langle\Phi(x), \beta\rangle &= \langle\Phi(x), (m + u)\rangle = \langle\Phi(x), m\rangle + \langle\Phi(x), u\rangle \\
&= \langle\Phi(x), m\rangle + \|\Phi(x)\|\|u\|\cos\left(\angle(\Phi(x), u)\right) .
\end{aligned}
$$

The minimum of the dot product $\langle\Phi(x), \beta\rangle$, with respect to u, is obtained when $\|u\| = \|r\|$ and $\cos\left(\angle(\Phi(x), u)\right) = -1$. In other words, u is a vector in the opposite direction of $\Phi(x)$ and with the maximum magnitude under the constraint that

---

[2] Our previous work [28] includes similar bounds for linear models, but only provides a sketc.h of the proof. Here we provide bounds and the full proof for both kernelized and linear models.

u is on the sphere. In this case, the lower bound is obtained, $L\left(\langle\Phi(x),\beta_2^*\rangle\right) = \langle\Phi(x),m\rangle - \|\Phi(x)\|\|r\|$. Using similar arguments, the maximum of the dot product $\langle\Phi(x),\beta\rangle$ is obtained when $\|u\| = \|r\|$ and $\cos\left(\angle(\Phi(x),u)\right) = 1$. This time u is in the same direction as $\Phi(x)$. In this case, the upper bound is obtained, $U\left(\langle\Phi(x),\beta_2^*\rangle\right) = \langle\Phi(x),m\rangle + \|\Phi(x)\|\|r\|$. By substituting $m = \beta_1^* - r$ (from the definition of Ω in Sect. 3.1) in the above expressions for L and U, we obtain (7). The computation of $\langle\Phi(x),\beta_1^*\rangle$ is done with the kernel trick, $\langle\Phi(x),\beta_1^*\rangle = \sum_{i\in\mathcal{D}_1}\alpha_i k(x_i,x)$. We now use the dual form of r (3) and get

$$\langle\Phi(x),r\rangle = \sum_{i\in\mathcal{D}_1\cup\mathcal{D}_2}\tau_i k(x_i,x). \tag{8}$$

Finally, using (8), (6), and $\|\Phi(x)\| = \sqrt{k(x,x)}$ we obtain that all the elements in (7) can be computed using the kernel trick.   □

3.3   Updating the Bounds Incrementally

Many applications require evaluating the bounds repeatedly (Sect. 4). We provide incremental update procedures for every sample that is added or removed from the training set, and show that when computed incrementally, the runtime of updating and evaluating our bounds is linear in the number of training examples. This compares favorably to the quadratic to cubic complexity of non-linear SVM solvers [1].

Any algorithm that uses the distance bound (Theorem 1) or the prediction bounds (7) can update the $\|r\|$ part in the bound incrementally when a sample is added to or removed from $\mathcal{S}_1$ and $\mathcal{S}_2$. We limit the discussion of this incremental update and its computational complexity to the common case where $C = C_1 = C_2$, as in Table 1. In this case:

$$\|r\|^2 = \frac{C^2}{4}\sum_{i\in\mathcal{D}_A\cup\mathcal{D}_R}\sum_{j\in\mathcal{D}_A\cup\mathcal{D}_R}\gamma_i\gamma_j k(x_i,x_j) \tag{9}$$

$$\langle\Phi(x),r\rangle = \frac{C}{2}\sum_{i\in\mathcal{D}_A\cup\mathcal{D}_R}\gamma_i k(x_i,x)\,, \tag{10}$$

since $\tau_i = \frac{C}{2}\gamma_i$ for $i\in\mathcal{D}_A\cup\mathcal{D}_R$ and 0 otherwise. For every sample index $i\in\mathcal{D}_A\cup\mathcal{D}_R$ let

$$K_i := \frac{C^2}{4}\left(\sum_{\substack{j\in\mathcal{D}_A\cup\mathcal{D}_R\\j\neq i}}2\gamma_i\gamma_j k(x_i,x_j) + \gamma_i^2 k(x_i,x_i)\right).$$

Let $\mathcal{A} = \mathcal{S}_2\setminus\mathcal{S}_1$ be the set of samples added in $\mathcal{S}_2$ and let $\mathcal{R} = \mathcal{S}_1\setminus\mathcal{S}_2$ be the set of samples removed from $\mathcal{S}_1$. To update $\|r\|^2$ with a new sample (x_i,y_i), the sample is first added to $\mathcal{A}$ or $\mathcal{R}$; then this sample's K_i is computed and added to

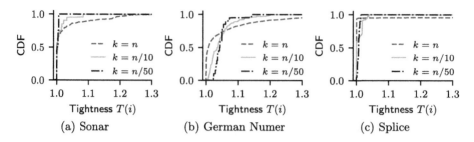

Fig. 1. Tightness of the distance bound. For more than 95% of the iterations, for every k value, the bound is less than 1.2 times the real distance between models.

$\|r\|^2$. If the sample is removed, K_i is first computed and subtracted from $\|r\|^2$. Only then is the sample removed from $\mathcal{A}$ or $\mathcal{R}$.

Computing K_i requires computing γ_i, which has complexity $O(n_1 d)$, where $n_1 = |\mathcal{S}_1|$ is the number of samples in $\mathcal{S}_1$, since a single kernel function evaluation is $O(d)$ and there are at most n_1 support vectors in β_1^* dual form. The complexity of computing $\sum_{j \in \mathcal{D}_{\mathcal{A}} \cup \mathcal{D}_{\mathcal{R}}} \gamma_j k(x_i, x_j)$ is $O((|\mathcal{A}| + |\mathcal{R}|)d)$. Given that $|\mathcal{R}| \leq n_1$, we have that the computation complexity of updating $\|r\|$ when a single sample is added or removed from either $\mathcal{A}$ or $\mathcal{R}$ is $O((n_1 + |\mathcal{A}|)d)$. Similarly, given $\|r\|$, the runtime complexity of computing the prediction bounds (7) is $O((n_1 + |\mathcal{A}|)d)$, since it requires evaluating $\langle \Phi(x), \beta_1^* \rangle$ and $\langle \Phi(x), r \rangle$ using (10).

4   Evaluation

In this section, we empirically evaluate the tightness of the distance bound. We also demonstrate the use of the bounds from Sect. 3 for specific applications by presenting improved variants of common algorithms. For clarity, we purposefully use simple algorithms. For example, in active learning we use only the midpoint of the upper and lower bounds and always select the same number of samples, while the online learning algorithm uses a static threshold T. We leave the detailed exploration and evaluation of improvements for these algorithms to future work. We used L_2 regularized squared hinge-loss (differentiable variant of SVM, Table 1) as the objective function with an RBF kernel $\exp(-\gamma\|x - x'\|^2)$ across all experiments. We used CVXPY [5] to solve the optimization problem.

4.1   Bound Tightness

We empirically evaluate the tightness of the distance bound (Theorem 1) by comparing the bound to the real distance between the models when we remove 1, 10, and 50 samples. Given a dataset of size n, we first train a model β_1^* on the full dataset. We then divide the dataset into k folds, where $k \in \{n, n/10, n/50\}$. In iteration i we exclude the ith fold from the full training set, compute a model $\beta_2^*(i)$, and evaluate the bound $2\|r_i\|$. The tightness of the bounds for the samples in fold i, $T(i)$, is defined as the ratio of the bound to the true difference between the models: $T(i) = \frac{2\|r_i\|}{\|\beta_1^* - \beta_2^*(i)\|}$. The closer $T(i)$ is to 1, the tighter the bound.

We repeated this experiment for different datasets using an RBF kernel with $\gamma = 1, C = 1$. Figure 1 shows the CDFs of $T(i)$ for three datasets from the LIBSVM [4] dataset repository: Sonar with $n = 208$, German Numer and Splice with $n = 1000$. For more than 95% of the iterations the bound is less than 1.2 times the real distance between models, indicating that the bound is tight. Results on other datasets in Table 2 are similar, where for large datasets we randomly selected 1000 samples for the test. We repeated the experiment with $C \in [0.001, 100]$. Tightness is better with lower C, but remains below 1.2 for 95% of iterations across the range of C.

4.2   Accelerated LOOCV

Leave-one-out cross-validation (LOOCV) is sometimes used for model selection or to evaluate the generalization error of a model β^* computed from the entire training set of size n. LOOCV works by iterating over the training set: in iteration t, the sample (x_t, y_t) is removed and a new model β_t^* is computed from the remaining $n - 1$ samples. The model β_t^* is then used to predict the sample x_t, and these predictions are then aggregated. Although especially useful for small datasets or when low estimation bias is desired [23, 36], LOOCV is computationally expensive since it requires training n models. This is particularly true for non-linear models, which take longer to train than linear models.

We follow the procedure proposed by Okumura et al. [24] for linear models. Rather than computing a new model for each removed sample, we use the prediction bounds (7) to predict the class of the sample x_t in the LOOCV process. We first compute a model β^* for all samples. Then, for every sample t, rather than computing the model β_t^*, we evaluate the upper and lower bounds (7) based on the original model β^* and the removed sample: $\beta_1^* = \beta^*$, $\beta_2^* = \beta_t^*$, $\mathcal{R} = \{(x_t, y_t)\}$, and $\mathcal{A} = \emptyset$. Note, the overlap between the models β^* and β_t^* is $n - 1$ samples. Assignment in (7) gives the lower bound $L = \langle \Phi(x), \beta^* \rangle - \frac{C}{2}\gamma_t K(x_t, x_t) - \frac{C}{2}|\gamma_t|K(x_t, x_t)$, and the upper bound $U = \langle \Phi(x), \beta^* \rangle - \frac{C}{2}\gamma_t K(x_t, x_t) + \frac{C}{2}|\gamma_t|K(x_t, x_t)$. If the signs of the bounds are the same, the classification of β^* for x_t is known and we can avoid computing β_t^*.

We compared our bounds to the bounds of Zhang [35], who proved that $\|\langle \Phi(x_t), \beta_t^* \rangle - \langle \Phi(x_t), \beta^* \rangle\| \le |\alpha_t|K(x_t, x_t)$, where α_t is the coefficient of x_t in the dual form of β^* (2). From this bound, we obtain upper and lower bounds for the prediction of β_t^*: $\langle \Phi(x_t), \beta^* \rangle - |\alpha_t|K(x_t, x_t) \le \langle \Phi(x_t), \beta_t^* \rangle \le \langle \Phi(x_t), \beta^* \rangle + |\alpha_t|K(x_t, x_t)$. Thus, the lower bound for the prediction of β_t^* is $\langle \Phi(x_t), \beta^* \rangle - |\alpha_t|K(x_t, x_t)$ and the upper bound is $\langle \Phi(x_t), \beta^* \rangle + |\alpha_t|K(x_t, x_t)$.

Table 2 shows the results of using the bounds for LOOCV on seven different datasets from the LIBSVM [4] and the UCI dataset repositories [6]. For each dataset, we ran LOOCV five times to tune the γ parameter, selected from the values $\gamma \in \{0.001, 0.01, 0.1, 1, 10\}$. We measured the average percentage of the iterations in which the bounds disagree on the sign, over the five LOOCV runs. This is the percentage of iterations that required computing β_t^*.

Table 2. Datasets and percentage of models computed for LOOCV.

Dataset	Sonar	Breast cancer	Splice	German numer	w5a	EEG eye state	a7a
n	208	569	1000	1000	9888	14980	16100
d	60	30	60	24	300	13	123
Zhang %	55.19	55.50	60.06	57.02	23.26	46.98	40.74
Our %	**47.21**	**48.86**	**55.36**	**46.32**	**22.50**	**40.48**	**31.49**

We observe that the bounds reduce model computations by two to four times. This translates into significant resource savings. For example, solving the optimization problem for the w5a dataset on an Intel i7-7820HQ CPU running at 2.9 GHz takes 436 s. By avoiding model computations in 7663 LOOCV iterations on average in each run, the use of bounds saved over 193 days of compute time compared to standard LOOCV.

Furthermore, our bounds improve on previous work on LOOCV [35] in two ways. Empirically, they train fewer models across all tested datasets. Moreover, our bounds apply even when more than one sample is removed, and can therefore be used for k-fold cross-validation. We leave such improvements for future work.

4.3   Fine-Grained Active Learning

In *pool-based active learning*, we are given an unlabeled dataset S of size n. We can ask for the label for any sample in the dataset and must make use of a limited budget for labeling.

The classic active learning algorithm (AL) begins with an initial model β_1^*, often trained on a small initial set of labeled samples. The algorithm iteratively improves the model by choosing more and more samples from the dataset. At each iteration t, the algorithm chooses a cohort of m previously unlabeled samples to be labeled and added to the labeled samples set; then, the model β_{t+1}^* is computed from all the labeled samples. The choice of which samples to add at every iteration is key. If samples are chosen wisely, the model can become accurate using fewer samples. A typical choice is the m samples that the current model is least certain about ($\langle \Phi(x), \beta_t^* \rangle$ closest to 0), since they are more likely to be misclassified.

While a smaller cohort m might be preferable, this comes at a cost of additional iterations, hence more model computations. We now describe FGAL (for Fine Grained Active Learner), a variant of the classic AL algorithm that uses the prediction bounds to achieve a smaller cohort without the additional cost.

FGAL divides each standard AL iteration into m/p sub-iterations. It starts each iteration t with empty sets of the added and removed samples ($\mathcal{A}$ and $\mathcal{R}$). In every sub-iteration i, FGAL computes the prediction bounds for all the remaining unlabeled samples and chooses the ones that are more likely to be the closest to the separating hyperplane as the next p samples to label. FGAL chooses the samples with the lowest value for $|(U + L)/2|$, where U, L are the prediction bounds (7) of these samples (we leave exploring alternatives for future

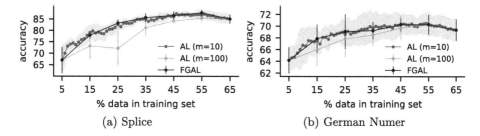

Fig. 2. Average model accuracy as a function of the training set size of the model. Each point represents 5 experiments with different seeds; the vertical lines represent the standard deviation. FGAL learns faster, achieving better accuracy with fewer samples.

work). It then adds these samples to $\mathcal{A}$ and the process continues until m samples are chosen. Note, the first p samples in each iteration are chosen according to the prediction of the model from the last iteration, since $U = L = \langle \Phi(x), \beta_t^* \rangle$ when $\mathcal{A} = \mathcal{R} = \emptyset$. In practice, an efficient implementation can incrementally update $\|r\|$, as explained in Sect. 3.3, every time p samples are labeled. We update the model only at the end of the iteration, after m samples were chosen.

We empirically compared FGAL to standard Active Learning (AL) with same m, and to another AL instance with $m = p$. In our experiments, $m = 100$ and $p = 10$, with an initial labeled set of 50 randomly selected samples. We ran the experiments over two datasets from the LIBSVM dataset repository [4]: Splice with $n = 1000$ samples and $d = 60$ attributes, and the scaled version of German Numer with $n = 1000$ and $d = 24$. We used an RBF kernel with $\gamma = 0.1, C = 1$ for Splice and $\gamma = 0.5, C = 1$ for German Numer. We randomly selected 350 samples as a test set, and repeated each run 5 times with different random seeds.

Figure 2 shows the accuracy on the test set of all the algorithms after each iteration. We observe that FGAL achieves faster learning than AL($m = 100$) using the same model computation budget. AL must select $m = 100$ samples without considering new information, while FGAL is able to incorporate information from each $p = 10$ samples during the sub-iteration. Alternatively, we can view FGAL as achieving similar learning performance as AL using a much smaller computational budget: FGAL's performance with $m = 100$ and $p = 10$ on the datasets is equivalent to that of AL with $m = 10$ using only 10% of the model computations. Finally, for FGAL and AL($m = 10$), the accuracy over the test set drops at the final iteration, possibly due to overfitting or label noise. If accuracy on the test set is the criteria, FGAL is able to achieve higher accuracy than AL ($m = 100$).

4.4 Adaptive Online Learning

Consider an online learning application where samples are presented to the algorithm one by one: at time t the application is presented with the sample (x_t, y_t).

Computing a new model every time a sample is added is wasteful since the underlying concept may have not changed. While incremental algorithms for kernelized models do exist [19, 25, 34], their performance can be sub-optimal due to strong reliance on individual samples and susceptibility to ill-conditioned problems [2, p. 467]. Instead, we propose an algorithm that can reduce computations while maintaining accurate models. The algorithm uses the distance bound to determine when the model has changed and should be re-computed.

We present KDR (for Kernelized Drift detectoR), a sliding window algorithm that uses the distance bound (Theorem 1) to trade off a small number of batch model computations for better accuracy when learning a classification or regression model over a data stream with concept drifts. When the difference $\|\beta_1^* - \beta_2^*\|$ is too large, where β_1^* is the existing model and β_2^* is the hypothetical updated model, KDR re-computes β_1^*. We describe KDR in detail below.

When a new sample (x_t, y_t) arrives, we incrementally update $\|r\|$ as described in Sect. 3.3. Note the window size is fixed, hence $C = C_1 = C_2$. We first add the new sample to $\mathcal{A}$ and then increment $\|r\|^2$ by K_t. Let (x_r, y_r) be the oldest sample in W. If the current window W overlaps with W_1 (the window at the time of β_1^* computation), then (x_r, y_r) came from W_1. We therefore first add it to the removed sample set $\mathcal{R}$ and then add K_r to $\|r\|^2$. Otherwise, (x_r, y_r) was never part of W_1, so we first subtract K_r from $\|r\|^2$ and then remove (x_r, y_r) from $\mathcal{A}$ (no change to $\mathcal{R}$). Along with the samples, we also store their γ coefficients in $\mathcal{A}$ and $\mathcal{R}$. Finally, we update the sliding window W by adding (x_t, y_t) and removing (x_r, y_r).

KDR uses β_1^* for predictions and monitors the average $\|r\|$ across the samples received since the last model re-computation. If this value is greater than a user-defined threshold T, we consider it a concept drift and re-compute the model β_1^* from the samples in the current window. For KDR, the overlap of the sliding window with the initial window determines the overlap of the training sets for β_1^* and the hypothetical β_2^*, and ranges from W to 0.

Evaluation. We evaluate KDR on one real-world dataset and one synthetic dataset. **SensIT Vehicle** [7] is a real-world time-series collected from wireless distributed sensor networks (WDSN), with labeled classes indicating the vehicle type. We used the version from the OpenML repository [32] with 98,528 samples of 100 features each. We used a window size of 200 samples. To simulate fast concept drifts we "sped up" the data by only using every tenth sample (i.e., only 10% of the data). **Rotating Checkerboard**, proposed by Elwell and Polikar [8], is an artificial 2D time-series with examples sampled uniformly from the unit square and labeled in a 5×5 checkerboard pattern. We generated 9 abrupt concept drifts, where at each concept drift the checkerboard is rotated by an angle of $\pi/20$ radians. The time between drifts is drawn uniformly from 2000 to 8000 samples. The window size is set to 500 samples.

We compare KDR to three algorithms: (i) **SW**, a non-adaptive sliding window algorithm with a fixed period parameter that determines how often batch model re-computation is performed; (ii) **ISGD**, the incremental truncated SGD proposed by Kivinen et al. [19]; and (iii) **DDM** [11], a popular concept drift

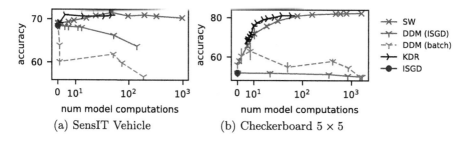

Fig. 3. The tradeoff between accuracy and number of model computations in SensIT Vehicle and Rotating Checkerboard datasets, for different parameter configurations of each algorithm. KDR achieves a better tradeoff, showing equal or superior accuracy at lower computational cost than the other algorithms, across a range of configurations.

detector that monitors its base learner accuracy and decides when to update the model. We use both ISGD and the SW batch learner as the DDM base learner. All the algorithms have the same window size, which is also the size of the support vector set for ISGD. We use the first 1000 samples from each stream to tune the kernel parameters, learning rate, and regularization parameter C. Finally, we used prequential evaluation [12] to evaluate accuracy.

Since different algorithms have different parameters that control the tradeoff between accuracy and computational cost (i.e., the period for SW, the drift level for DDM, and the threshold T for KDR; there is no such parameter for ISGD), we use *tradeoff curves* to explore the computation-accuracy of the different algorithms. For each configuration of these parameters, we plot a point with the resulting accuracy as the Y coordinate and the number of model computations as the X coordinate. The resulting curve shows how the algorithm behaves as we change its parameters.

Figure 3(a) shows the tradeoff curves on the SensIT Vehicle dataset. Overall, the accuracy of sliding window algorithms that use batch learning (SW and KDR) is superior to that of the incremental learning algorithms (ISGD and DDM). KDR is able to achieve an accuracy equivalent to the non-adaptive SW algorithm using far fewer model computations.

Similarly, Fig. 3(b) shows the computation-accuracy tradeoff for the Rotating Checkerboard dataset averaged across five runs with different random seeds (for every seed, all algorithms see the same data); error bars show standard deviation (in practice, it is very small). KDR achieves the same accuracy as SW with less computation since it can adapt to unpredictable concepts drifts. DDM and ISGD are unable to adapt quickly enough, despite substantial tuning.

5 Related Work

Existing work on general sensitivity analysis focuses on linear models without kernels. Okumura et al. [24], Hanada et al. [13], Gabel et al. [10], and Sivan et al. [28] all present bounds for the prediction of a linear model when the training set changes, as well as a bound for the Euclidean distance between the previously-computed model and the updated model. Okumura et al. suggest their prediction bounds could be kernelized for non-linear classification problems, but did not provide details. Steinwart and Christmann [29, Corollary 5.12] present a bound for the difference between two kernelized models. Their bound, however, is more limited. Our geometric proof allows us to obtain bounds for the prediction of a new model, which are critical for several applications (detailed below). Our bounds are incremental, and we show how to incrementally update kernel weights in linear time. In addition, the bounds are more general, supporting two optimization problems with different regularization parameters and number of samples.

For specific applications, there are specialized approaches that reduce kernelized model training.

For distributed learning, Kamp et al. [17] present a bound on the distance of a local kernelized model maintained by a local learner to the global average model of all distributed local models. We focus on bounding the differences in a single model when its training set changes.

For LOOCV, Jaakkola and Haussler [15], Joachims [16], Vapnik and Chapelle [33], and Zhang [35] provide upper bounds for the LOOCV error, which can be computed immediately after training the initial kernelized model from the entire training set. Zhang [35] also provides a bound for the distance between the prediction of two models whose training sets differ by exactly one sample. We adapt this bound in Sect. 4.2 to bound model prediction, and demonstrate empirically that our prediction bounds perform as well or better in LOOCV. Moreover, since our bounds support any change in the training set, they can be used for k-fold cross-validation.

For active learning, the simplest and most commonly used approach is *uncertainty sampling* [20]. In this approach, the model is used to estimate which unlabeled samples are most likely to be misclassified. We show that this approach can be improved using our bounds when more than one sample is chosen to be labeled simultaneously.

For online learning, incremental algorithms are often used to update the model one sample at a time. The challenge of kernelized incremental algorithms lies in limiting the size of the support vector set, which increases linearly with the size of the data stream. Kivinen et al. [19] study classical kernelized stochastic gradient descent (SGD) algorithms. They show that the oldest samples can be removed from the support vector set with only a small impact on the accuracy of the model. Orabona et al. [25] present the Projectron algorithm, which is based on the Perceptron algorithm but requires less memory. While the memory size of the Projectron algorithm is guaranteed to be bounded, it cannot be predicted in advance. Wang and Vucetic [34] propose Passive-Aggressive on a budget that maintains only a fixed number of support vectors, with several versions that trade optimality for runtime efficiency. While incremental model updates are relatively efficient, incremental algorithms can perform poorly on ill-conditioned problems [2, p. 467], and are less immune to outliers than batch learners (since only one sample at a time is used for the model update). Instead, we propose a simple online learning algorithm that uses our bounds to determine when batch computation should occur, similar to concept drift detection.

Another approach is to use *Influence Functions* (IF) to provide a probabilistic approximation for the error of model predictions [21,29]. Conversely, we focus on providing a general and deterministic bound on both the difference between the models, as well as the prediction. Our work is also related to transfer learning [26]. While transfer learning deals with adapting the existing model to new data, in sensitivity analysis we bound the changes in the model, delaying the computation of the new model. We view our approach as a precursor for transfer learning: it determines when transfer learning should be applied.

6 Conclusions

We presented incremental sensitivity bounds for kernelized machine learning models that evaluate the change in a model and its predictions as the training set changes. Our bounds require only the already computed model and the difference in the training set, and can be evaluated in linear time. We empirically demonstrated the tightness of the bounds, as well as their effectiveness in three different applications: LOOCV, online learning, and active learning.

References

1. Bottou, L., Lin, C.J.: Support vector machine solvers. In: Large Scale Kernel Machines, pp. 301–320 (2007)
2. Boyd, S., Vandenberghe, L.: Convex Optimization (2004)
3. Cauwenberghs, G., Poggio, T.: Incremental and decremental support vector machine learning. In: NIPS 2000, pp. 388–394 (2000)
4. Chang, C.C., Lin, C.J.: LIBSVM: a library for support vector machines. ACM Trans. Intell. Syst. Technol. **2**, 27:1–27:27 (2011)
5. Diamond, S., Boyd, S.: CVXPY: a python-embedded modeling language for convex optimization. JMLR **17**(83), 1–5 (2016)
6. Dua, D., Graff, C.: UCI machine learning repository (2017)
7. Duarte, M.F., Hu, Y.H.: Vehicle classification in distributed sensor networks. J. Parallel Distrib. Comput. **64**(7), 826–838 (2004)
8. Elwell, R., Polikar, R.: Incremental learning of concept drift in nonstationary environments. IEEE Trans. Neural Netw. **22**, 1517–1531 (2011)
9. Fine, S., Scheinberg, K.: Incremental learning and selective sampling via parametric optimization framework for SVM. In: NIPS 2001, pp. 705–711 (2001)
10. Gabel, M., Keren, D., Schuster, A.: Monitoring least squares models of distributed streams. In: KDD 2015, pp. 319–328 (2015)
11. Gama, J., Medas, P., Castillo, G., Rodrigues, P.: Learning with drift detection. In: Bazzan, A.L.C., Labidi, S. (eds.) SBIA 2004. LNCS (LNAI), vol. 3171, pp. 286–295. Springer, Heidelberg (2004). https://doi.org/10.1007/978-3-540-28645-5_29
12. Gama, J., Sebastião, R., Rodrigues, P.P.: On evaluating stream learning algorithms. Mach. Learn. **90**(3), 317–346 (2013). https://doi.org/10.1007/s10994-012-5320-9

13. Hanada, H., Shibagaki, A., Sakuma, J., Takeuchi, I.: Efficiently monitoring small data modification effect for large-scale learning in changing environment. In: AAAI (2018)
14. Hofmann, T., Schölkopf, B., Smola, A.: Kernel methods in machine learning. Ann. Stat. **36**, 1171–1220 (2007)
15. Jaakkola, T.S., Haussler, D.: Probabilistic kernel regression models. In: Proceedings of the 1999 Conference on AI and Statistics (1999)
16. Joachims, T.: Estimating the generalization performance of an SVM efficiently. In: ICML 2000, pp. 431–438 (2000)
17. Kamp, M., Bothe, S., Boley, M., Mock, M.: Communication-efficient distributed online learning with Kernels. In: Frasconi, P., Landwehr, N., Manco, G., Vreeken, J. (eds.) ECML PKDD 2016, Part II. LNCS (LNAI), vol. 9852, pp. 805–819. Springer, Cham (2016). https://doi.org/10.1007/978-3-319-46227-1_50
18. Karasuyama, M., Takeuchi, I.: Multiple incremental decremental learning of support vector machines. In: NIPS 2009, pp. 907–915 (2009)
19. Kivinen, J., Smola, A.J., Williamson, R.C.: Online learning with kernels. IEEE Trans. Signal Process. **52**(8), 2165–2176 (2004)
20. Lewis, D.D., Gale, W.A.: A sequential algorithm for training text classifiers. In: Proceedings of the 17th Annual International ACM SIGIR Conference on Research and Development in Information Retrieval, SIGIR 1994, pp. 3–12 (1994)
21. Liu, Y., Jiang, S., Liao, S.: Efficient approximation of cross-validation for Kernel methods using bouligand influence function. In: ICML 2014, pp. I-324–I-332 (2014)
22. Maalouf, M., Trafalis, T.B., Adrianto, I.: Kernel logistic regression using truncated Newton method. Comput. Manage. Sci. **8**(4), 415–428 (2011)
23. Molinaro, A.M., Simon, R., Pfeiffer, R.M.: Prediction error estimation: a comparison of resampling methods. Bioinformatics **21**(15), 3301–3307 (2005)
24. Okumura, S., Suzuki, Y., Takeuchi, I.: Quick sensitivity analysis for incremental data modification and its application to leave-one-out CV in linear classification problems. In: KDD 2015, pp. 885–894 (2015)
25. Orabona, F., Keshet, J., Caputo, B.: Bounded Kernel-based online learning. J. Mach. Learn. Res. **10**, 2643–2666 (2009)
26. Pan, S.J., Yang, Q.: A survey on transfer learning. TKDE **22**(10), 1345–1359 (2010)
27. Schölkopf, B.: The Kernel trick for distances. In: NIPS 2000, pp. 283–289 (2000)
28. Sivan, H., Gabel, M., Schuster, A.: Online linear models for edge computing. In: ECML PKDD 2019 (2019)
29. Santosh, K.C., Hegadi, R.S. (eds.): RTIP2R 2018, Part II. CCIS, vol. 1036. Springer, Singapore (2019). https://doi.org/10.1007/978-981-13-9184-2
30. Tsai, C.H., Lin, C.Y., Lin, C.J.: Incremental and decremental training for linear classification. In: KDD 2014, pp. 343–352 (2014)
31. Tsang, I.W., Kwok, J.T., Cheung, P.M.: Core vector machines: Fast SVM training on very large data sets. J. Mach. Learn. Res. **6**, 363–392 (2005)
32. Vanschoren, J., van Rijn, J.N., Bischl, B., Torgo, L.: OpenML: networked science in machine learning. SIGKDD Explor. **15**(2), 49–60 (2013)
33. Vapnik, V., Chapelle, O.: Bounds on error expectation for support vector machines. Neural Comput. **12**(9), 2013–2036 (2000)
34. Wang, Z., Vucetic, S.: Online passive-aggressive algorithms on a budget. In: Proceedings of Machine Learning Research, vol. 9, pp. 908–915, 13–15 May 2010
35. Zhang, T.: Leave-one-out bounds for kernel methods. Neural Comput. **15**(6), 1397–1437 (2003)
36. Zhang, Y., Yang, Y.: Cross-validation for selecting a model selection procedure. J. Econom. **187**(1), 95–112 (2015)

Off-the-Grid: Fast and Effective Hyperparameter Search for Kernel Clustering

Bruno Ordozgoiti[1(✉)] and Lluís A. Belanche Muñoz[2]

[1] Aalto University, Helsinki, Finland
bruno.ordozgoiti@aalto.fi
[2] Universitat Politècnica de Catalunya, Barcelona, Spain
belanche@cs.upc.edu

Abstract. Kernel functions are a powerful tool to enhance the k-means clustering algorithm via the kernel trick. It is known that the parameters of the chosen kernel function can have a dramatic impact on the result. In supervised settings, these can be tuned via cross-validation, but for clustering this is not straightforward and heuristics are usually employed. In this paper we study the impact of kernel parameters on kernel k-means. In particular, we derive a lower bound, tight up to constant factors, below which the parameter of the RBF kernel will render kernel k-means meaningless. We argue that grid search can be ineffective for hyperparameter search in this context and propose an alternative algorithm for this purpose. In addition, we offer an efficient implementation based on fast approximate exponentiation with provable quality guarantees. Our experimental results demonstrate the ability of our method to efficiently reveal a rich and useful set of hyperparameter values.

Keywords: Clustering · Kernels · Kernel k-means · Hyperparameter tuning · Grid search

1 Introduction

Clustering, the task of partitioning a given data set into groups of similar items, is one of the central topics in data analysis. Among the plethora of existing techniques for this purpose, k-means clustering, along with Lloyd's algorithm [14], is one of the most popular and well-understood methods. Despite its popularity, k-means has significant limitations, as it implicitly makes strong assumptions about the shapes of the clusters. Numerous alternative methods have been proposed to tackle challenges beyond the capabilities of k-means [8,13,15,16].

This work was supported by the Academy of Finland project 317085.

Electronic supplementary material The online version of this chapter (https://doi.org/10.1007/978-3-030-67661-2_24) contains supplementary material, which is available to authorized users.

F. Hutter et al. (Eds.): ECML PKDD 2020, LNAI 12458, pp. 399–415, 2021.
https://doi.org/10.1007/978-3-030-67661-2_24

One of these involves the use of positive definite kernels [11], which enable the computation of inner products between elements of a vector space after mapping them to a different, high-dimensional space. In particular, kernels enhance the capabilities of k-means by enabling the detection of clusters of arbitrary shapes.

One drawback of kernel functions is that they usually involve hand-set parameters, which must be fine-tuned to bring forth their full potential. A common method to choose a value for these parameters is grid search. One considers a set of values and then evaluates the performance of the algorithm for each of them. A drawback is that one might either choose too small a set and risk missing optimal values, or an overly big one, incurring excessive—and possibly redundant—computational costs. Another way to set these values is by heuristics and rules of thumb [12, 19], but these rarely apply to a wide variety of data.

Our contribution in this paper is two-fold. First, we illustrate the impact of kernel parameters in clustering by deriving a lower bound for the bandwidth parameter of RBF kernels (Sect. 4), below which `Kernel k-means` will be rendered useless. We show this bound is tight. Next, we propose a method for hyperparameter search. Our method specifically searches for values that will produce different clusterings, and thus, unlike grid search, does not risk carrying out redundant computations, so no processing time is wasted. We combine methods for fast exponentiation with the properties of dyadic rationals to design an algorithm that after $\mathcal{O}\left(\log\left(\frac{|\log(b)|}{\epsilon}\right)\right)$ iterations—where b is the minimum entry in the kernel matrix—provides a $(1 \pm \epsilon)$-approximation of the next meaningful hyperparameter value to inspect (Sects. 5 and 6). We validate our claims with a rich variety of experiments (Sect. 7).

2   Related Work

Kernels have been a central subfield of machine learning since their first use in conjunction with support vector machines [5]. Even though most efforts have focused on their application to supervised learning methods, they have also played a significant part in the development of clustering techniques [4,7,16]. In the seminal work by Ben-Hur et al. [4], the authors suggest to inspect the results using varying values of σ, starting from the maximizer of the pairwise squared distances $\|x - y\|^2$ over all pairs of data points. A good choice might lie within a region that yields stable clusterings. It should be noted that stability has been shown to have significant drawbacks for choosing the number of clusters [3], so it would be interesting to determine whether this applies to the kernel bandwidth as well. In the work that introduced spectral clustering [16], Ng et al. rely on a result of their own that guarantees that their algorithm will produce tight clusters if they exist in the data. They then propose to test various values of σ in search for a clustering with this property. In [2] a generalized form of the bandwidth parameter is learned based on data with known clustering. In [20] a different value of σ is computed for each point. The approach proposed by the authors relies on the distance to the k-th neighbor. In [10], the authors investigate the problem of kernel matrix diagonal dominance in clustering, which is essentially a generalization of the problem we analyze in the beginning of Sect. 4. The heuristics they explore to alleviate the problem either require the selection of a new hyperparameter, or heavily modify the structure of the problem. The latter can even lead to the loss of positive-definiteness of the kernel matrix, which results in algorithmic oscillations and failure to converge. The mean distance to the k-th nearest neighbour is also suggested as a heuristic by Von Luxburg [19].

3   Preliminaries

We consider a finite set of data points $X \subset \mathbb{R}^d$. We define a k-partition of X as a collection of k non-empty subsets of X, $\pi_1, \ldots, \pi_k$, satisfying $\bigcup_{i=1}^{k} \pi_i = X$ and $\pi_i \cap \pi_j = \emptyset$ for $i, j = 1, \ldots, k$, $i \neq j$. We will refer to each π_i as a *cluster* and use $n_i = |\pi_i|$ to denote its cardinality.

The *k-means* objective is to find a k-partition of X so as to minimize

$$\sum_{i=1}^{k} \sum_{x \in \pi_i} \|x - \bar{\pi}_i\|^2, \tag{1}$$

where $\bar{\pi}_i = n_i^{-1} \sum_{x \in \pi_i} x$ is the *centroid* of cluster π_i and $\|x\|$ denotes the L_2 norm in $\mathbb{R}^d$. Optimizing this objective is known to be **NP**-hard for $k = 2$ [1]. A popular heuristic is Lloyd's algorithm [14], which repeatedly recomputes the centroid of each cluster and reassigns points to the closest centroid.

Kernels: Given a non-empty set $\mathcal{X}$, a symmetric function $\kappa : \mathcal{X} \times \mathcal{X} \to \mathbb{R}$ such that for all $n \in \mathbb{N}$ and every set $\{x_i\}_{i=1}^{n} \subset \mathcal{X}$, the matrix $K = (\kappa(x_i, x_j))_{ij}$ is positive definite, is called a (strictly) **positive definite (PD) kernel**. The matrix K is known as the *Gram* matrix or *Kernel* matrix. Since PD kernels give rise to a PD Gram matrix, they correspond to the computation of an inner product in some implicit inner-product space. The representation of an object $x \in \mathcal{X}$ in said space is often called *feature space representation*, denoted as $\phi(x)$.

A number of kernels are routinely used in practice. Probably the most popular one for the case $\mathcal{X} = \mathbb{R}^d$ is the Gaussian RBF kernel

$$\kappa(x, y) = \exp\left(\frac{-\|x - y\|^2}{\sigma}\right),$$

$\sigma > 0$, (from here on, RBF kernel). The parameter σ is commonly referred to as *bandwidth*. We will use κ_σ to denote the RBF kernel function with bandwidth parameter σ and K_σ to denote the corresponding kernel matrix.

Consider a data set X and the k-partition $\pi_1, \ldots, \pi_k$. Let m_i denote the centroid of cluster π_i in feature space, that is,

$$m_i = \frac{1}{n_i} \sum_{x \in \pi_i} \phi(x).$$

The application of kernels to the k-means objective (1) relies on the following observation: even though we cannot in general express m_i explicitly, it is possible to compute the necessary squared distances. For any $x \in X$ and $i = 1, \ldots, k$,

$$\|\phi(x) - m_i\|^2 = \kappa(x, x) - \frac{2 \sum_{y \in \pi_i} \kappa(x, y)}{n_i} + \frac{\sum_{y, z \in \pi_i} \kappa(y, z)}{n_i^2}. \tag{2}$$

The application of Lloyd's algorithm using this expression for the squared distance is known as Kernel k-means. See [7] for an insightful analysis. Kernel k-means always converges when the kernel matrix is positive semidefinite. We will refer to the k-partition at convergence as the *output* of Kernel k-means.

4   The Use of the RBF Kernel in Kernel k-means

RBF kernels are powerful but sensitive to the bandwidth parameter. In particular, for sufficiently small σ, a support vector machine classifier can fit any training set with no errors—or equivalently, it has infinite VC dimension [18]—, but this will generally result in poor generalization ability. In Kernel k-means, the result of an overly small bandwidth will be that the algorithm will converge in the first iteration, regardless of the current k-partition. The reason is that as σ decreases, the value of $\kappa(x, y)$ for any two distinct points $x, y \in X$ decreases as well, to the point of becoming negligible. Therefore, the only significant term in Eq. (2) for any x will be $\kappa(x, x)$, which means that the closest cluster to x will be the one it is currently in. A question arises naturally: how small does σ have to be for the algorithm to get stuck at the initial clustering? The following theorem provides a lower bound, which is tight up to constant factors.

Theorem 1. *Consider a data set $X \subset \mathbb{R}^d$, $|X| = n$. Let $x, y = \arg\min_{x,y \in X} \|x - y\|^2$. If $\sigma \le (\log(3n))^{-1} \|x - y\|^2$, then Kernel k-means will make no cluster reassignments.*

The proof is given in the supplementary material.

A Tight Example. The next example shows that this result is tight up to constant factors. Consider an instance with two clusters, π_1 and π_2, containing n_1 and n_2 points respectively. For some point $y \in \pi_2$ it is $\|x - y\|_2^2 = \min_{a,b} \|a - b\|_2^2 = \epsilon$ for all $x \in \pi_1$, whereas for all $z \in \pi_2, z \ne y$, it is $\|y - z\|_2^2 = 2\epsilon$. Moreover, for all $w, z \in \pi_1$ it is $\|w - z\|_2^2 = \epsilon$ and for all $w, z \in \pi_2, w, z \ne y$ it is $\|w - z\|_2^2 = \epsilon$. Define $n = n_1 + n_2$ and consider $\sigma = \epsilon / \log(n/3)$. We know y will switch over to π_1 if $\|\phi(y) - m_1\|_2^2 < \|\phi(y) - m_2\|_2^2$, or equivalently,

$$\frac{2}{n_2} < \frac{2 \sum_{x \in \pi_1} \kappa(y, x)}{n_1} - \frac{\sum_{w, z \in \pi_1} \kappa(w, z)}{n_1^2} - \frac{2 \sum_{z \in \pi_2, z \ne y} \kappa(y, z)}{n_2} + \frac{\sum_{w, z \in \pi_2} \kappa(w, z)}{n_2^2}$$

$$= 6/n - 1/n_1 - \frac{3(n_1 - 1)}{n n_1} - \frac{2(n_2 - 1)}{n_2} \left(\frac{3}{n}\right)^2$$

$$+ 1/n_2 + \frac{3(n_2 - 1)(n_2 - 2)}{n n_2^2} + \frac{(n_2 - 1)}{n_2^2} \left(\frac{3}{n}\right)^2. \tag{3}$$

The above inequality is verified when $n_1 = n_2$ and n is sufficiently large. That is, there exists a family of instances where the kernel k-means algorithm with the RBF kernel will make cluster reassignments with $\sigma = \Omega\left(\frac{\|x-y\|_2^2}{\log(n)}\right)$, where $\|x - y\|_2^2$ is minimal over all x, y in the data set.

5   Optimizing Bandwidth

As demonstrated above, the choice of bandwidth parameter is crucial when using RBF kernels for clustering. For some choices of σ, the output of Kernel k-means will be unchanged from the initial k-partition. In fact, for any value of σ the algorithm will converge at some point—provided that the kernel matrix is positive semidefinite—and stop making changes. However, if the chosen value is inadequate the output might still be of poor quality, so it is often desirable to further refine σ in order to obtain a better result. We already know, by virtue of Theorem 1, a value of σ such that Kernel k-means will stop making changes. The following question arises naturally. *How big does σ have to be in order to guarantee that* Kernel k-means *will change the initial k-partition?*, and more generally, *once* Kernel k-means *has converged, how much do we have to increase σ to ensure it will make new changes?* We define this as the *critical bandwidth value*.

Definition 1 *(Critical bandwidth value). Let X a data set. Suppose* Kernel k-means *outputs a k-partition $P = (\pi_1, \ldots, \pi_k)$ of X when run using an RBF kernel with bandwidth parameter σ. We define $S \subset \mathbb{R}$ to be the set satisfying the following: if* Kernel k-means *is initialized with k-partition P and run with $K_{\sigma'}$, with $\sigma' \in S$, it will output a k-partition $P' \neq P$, that is, it will make changes. We define the critical bandwidth value with respect to (K_σ, P) to be the infimum of S, or ∞ if $S = \emptyset$.*

In other words, the critical bandwidth value reveals the "minimal" value the RBF kernel bandwidth needs to take so that Kernel k-means "snaps out" of convergence and yields a new k-partition. Any value strictly larger than the critical value will suffice. This concept is the cornerstone of our contribution.

5.1   Finding the Critical Value

Possibly the most straightforward method to find a value of σ—or virtually any hyperparameter—is grid search. This consists in running the clustering algorithm for a predetermined set of values of the hyperparameter and choosing the one which provides the best performance, as measured by e.g. objective function values or clustering quality indices [17]. This approach, however, has significant disadvantages. If the set of values to test is too small, one can fail to detect one that yields good performance; if it is too large, running times can be prohibitive and some computations redundant.

Here we propose an alternative approach. Roughly, we proceed as follows. First, we choose a sufficiently small value of σ—e.g. guided by Theorem 1— and run Kernel k-means. We then search for the critical bandwidth value with respect to the current kernel matrix and k-partition and rerun Kernel k-means until convergence. We can keep doing this until no further changes are observed, to finally obtain a set of possible hyperparameter choices. The question that arises now is how to find said value efficiently. Next, we illustrate the fact that this value can be located using optimization methods.

A First Approach. Let κ_σ denote the RBF kernel function parametrized by σ. In a Kernel k-means iteration, a point x is assigned to the cluster π_i which maximizes the *proximity* function δ:

$$\delta_\sigma(x, m_i) = \frac{2\sum_{y\in\pi_i} \kappa_\sigma(x,y)}{n_i} - \frac{\sum_{y,z\in\pi_i} \kappa_\sigma(y,z)}{n_i^2}. \tag{4}$$

Now, observe that if we change the value of the bandwidth parameter to σ', the new value of the kernel for any pair of points x, y can be computed as follows:

$$\kappa_{\sigma'}(x,y) = \kappa_\sigma(x,y)^{\sigma/\sigma'},$$

and we can thus compute the new proximity functions $\delta_{\sigma'}(x, m_i)$ accordingly. For simplicity, we consider the case of two clusters π_1, π_2. Assume $x \in \pi_1$. x will switch over to π_2 when

$$\delta_{\sigma'}(x, m_1) < \delta_{\sigma'}(x, m_2) \Leftrightarrow \delta_{\sigma'}(x, m_1) - \delta_{\sigma'}(x, m_2) < 0.$$

That is, we can find the value of σ' that will result in a different clustering by finding a root of $\delta_{\sigma'}(x, m_1) - \delta_{\sigma'}(x, m_2)$.

A useful observation is that $\kappa_\sigma(x,y)^\sigma$ is constant with respect to σ'. Therefore, we can easily derive $\delta_{\sigma'}(x, m_1) - \delta_{\sigma'}(x, m_2)$ with respect to σ'. In particular, define $g(x, \sigma') = \delta_{\sigma'}(x, m_1) - \delta_{\sigma'}(x, m_2)$. Then

$$\frac{dg}{d\sigma'} = \frac{2\sum_{y\in\pi_2} \log\left(\kappa_\sigma(x,y)^\sigma\right)\kappa_\sigma(x,y)^{\sigma/\sigma'}}{\sigma'^2 n_2} - \frac{\sum_{y,z\in\pi_2} \log\left(\kappa_\sigma(y,z)^\sigma\right)\kappa_\sigma(y,z)^{\sigma/\sigma'}}{\sigma'^2 n_2^2}$$
$$-\frac{2\sum_{y\in\pi_1} \log\left(\kappa_\sigma(x,y)^\sigma\right)\kappa_\sigma(x,y)^{\sigma/\sigma'}}{\sigma'^2 n_1} + \frac{\sum_{y,z\in\pi_1} \log\left(\kappa_\sigma(y,z)^\sigma\right)\kappa_\sigma(y,z)^{\sigma/\sigma'}}{\sigma'^2 n_1^2}. \tag{5}$$

This implies that we can use iterative root-finding algorithms, such as Newton's method, to efficiently find a root of the above function, that is, the minimum value of σ' that will result in a clustering change, or the critical bandwidth value.

This approach, however, can be slow and numerically unstable. In the next section we propose an alternative optimization method able to efficiently locate the critical bandwidth value to arbitrary precision while overcoming these drawbacks.

6 Fast and Effective Hyperparameter Search

The approach outlined above has several drawbacks, namely (1) using an iterative root-finding algorithm entails repeatedly recomputing the kernel matrix, either directly or by element-wise exponentiation, which can be slow in practice when dealing with large matrices and (2) the operations required for the derivative of g and the fractional computations can induce numerical instability.

Here we propose an alternative approach to sidestep these issues. The proposed method rests on the following fact: *computing products and square roots of real numbers can be much faster than computing powers with arbitrary exponents* [9]. Our method has the additional advantage of being numerically stable.

6.1 Dyadic Rationals and Fast Approximate Exponentiation

To develop an efficient method for hyperparameter search, we first propose an algorithm for fast approximate exponentiation that only uses products and square roots. This algorithm (Algorithm 1) forms the basis of our approach.

Exponentiation Algorithm Overview. As hinted above, we wish to avoid computing element-wise powers of the kernel matrix, and instead use element-wise products and square roots. To accomplish this, suppose we want to compute the power b^p, for some arbitrary positive reals b and p. We first decompose p as $p = z + f$, where z is the integral part and f the decimal part of p. We then compute b^z and approximate b^f as $b^{f'}$ using two separate fast methods for integral and rational exponents and finally return $b^z b^{f'} \approx b^p$.

To design our algorithm, we rely on two simple results. First, we make use of the following recursive representation of a positive integer based on its binary representation, which has long been employed in the design of fast algorithms for power computation with integral exponents [9].

Lemma 1. *Consider a number $n \in \mathbb{N}$, and let $b_0 \ldots b_t$, where $t = \lfloor \log_2 n \rfloor$, be its binary representation, i.e. $n = \sum_{i=0}^{t} 2^{t-i} b_i$. Then $n = n_t$, where*

$$
n_i = \begin{cases} 1 & \text{if } i = 0 \\ 2n_{i-1} + b_i & \text{if } 0 < i \leq t \end{cases}
$$

Lemma 1 reveals how to compute a power of the form b^i, where b is a positive real number and i is a natural number, using a small number of products. In particular, this operation is carried out in lines 5 and 6 of Algorithm 1.

The next result we rely on is a consequence of the properties of dyadic rationals. Dyadic rationals are rational numbers of the form $n/2^i$, where n is an integer and i is a natural number. It is well known that dyadic rationals are dense in $\mathbb{R}$, that is, any real number can be approximated arbitrarily well by a dyadic rational. The next result reveals how to obtain such an approximation for numbers in the interval $(0, 1)$, which will be useful in our context.

Algorithm 1. Fast approximate exponentiation

Input: base b, exponent p, depth i

1: $z \leftarrow bin(\lfloor p \rfloor)[1:]$
2: $f \leftarrow p - \lfloor p \rfloor$
3: $b_1 \leftarrow b$; $b_2 \leftarrow b$
4: $j \leftarrow 1$
5: **for** d in z **do**
6: $\quad b_1 \leftarrow b_1^2 b^d$
7: $n \leftarrow 1$; $d \leftarrow 2$
8: **for** $j = 1, \ldots, i$ **do**
9: $\quad b \leftarrow \sqrt{b}$; $n \leftarrow 2n$; $d \leftarrow 2d$
10: $\quad$ **if** $n/d > f$ **then**
11: $\quad\quad n \leftarrow n - 1$; $b_2 \leftarrow b_2/b$
12: $\quad$ **if** $n/d < f$ **then**
13: $\quad\quad n \leftarrow n + 1$; $b_2 \leftarrow b_2 b$
14: $\quad$ **if** $n/d = f$ **then**
15: $\quad\quad \jmath \leftarrow i + 1$ // Exact exponent matched, so exit loop
16: Output $b_1 \times b_2$

Lemma 2. *Let $a \in (0,1)$. There exists a sequence (m_i), with $m_i \in \{-1,1\}, i = 1, \ldots$ such that $\lim_{t \to \infty} \sum_{i=1}^{t} m_i 2^{-i} = a$.*

Proof. Let $m_1 = 1$. Choose the j-th term of (m_i) (for $j > 1$) to be 1 if $\sum_{i=1}^{j-1} m_i 2^{-i} < a$, -1 if $\sum_{i=1}^{j-1} m_i 2^{-i} > a$, 0 otherwise. Clearly, $\left| a - \sum_{i=1}^{k} m_i 2^{-i} \right| \leq 2^{-k}$.

The set of dyadic rationals is clearly closed under addition, and thus the above series provides an approximation by means of a dyadic rational.

Now, suppose we want to approximately compute the power b^p, by an approximation of p to within an error of 2^{-j}. The above result implies that it suffices to compute j operations, at each step either multiplying or dividing by successive square roots of b. This is done in lines 8 through 15 of Algorithm 1. The following result characterizes the quality of the approximation achieved by Algorithm 1, and the required number of operations.

Theorem 2. *Algorithm 1 yields a $(1 \pm \epsilon)-approximation$ of b^p after performing $\mathcal{O}\left(\log \left(\frac{|\log(b)|}{\epsilon} \right) \right)$ operations.*

Proof. First, note that the algorithm computes at most $2i$ multiplications in the first phase, and i square roots or multiplications in the second.

Assume $b > 1$. We treat the alternative later. By Lemma 2, the output of Algorithm 1 is bounded as follows

$$\frac{b^p}{b^{1/2^i}} = b^{p-1/2^i} \leq r \leq b^{p+1/2^i} = b^p b^{1/2^i}.$$

Observe that $b^p b^{1/2^i} - b^p + b^p(b^{1/2^i} - 1)$ and set $\epsilon = b^{1/2^i} - 1$. We thus have $\frac{1}{2^i} = \frac{\log(1+\epsilon)}{\log(b)}$ and thus $i = \mathcal{O}\left(\log\left(\frac{\log(b)}{\epsilon}\right)\right)$. Similarly, we can write $b^p b^{-1/2^i} = b^p - b^p(1 - b^{-1/2^i})$, arriving at an equivalent result for the $1 - \epsilon$ bound.

The analysis for the case $b < 1$ is the same, but noting that the output is bounded as $b^{p+1/2^i} \leq r \leq b^{p-1/2^i}$. The negative sign of $\log(b)$ is cancelled out in the arithmetic. The case $b = 1$ is obviously of no interest. $\qquad\square$

Algorithm 1 approximates a power computation by a dyadic rational approximation w/z of the exponent. Based on the principles behind Algorithm 1 we can design an efficient method to find the critical value of σ for Kernel k-means.

Finding the Critical Value. Our algorithm for hyperparameter search is detailed as Algorithm 2. In the pseudocode, $\circ$ and $/\circ$ denote element-wise multiplication and division, respectively, and $\sqrt{K}$ is the element-wise square root of matrix K.

In essence, our algorithm emulates Algorithm 1, using the kernel matrix K_σ as the basis of the power to compute, with some key differences. The first difference is that instead of approximating a known exponent p, we aim to approximate the *unknown* critical value of σ. Since this quantity is unknown, instead of testing whether the current approximation is larger or smaller than the target exponent, we query the Kernel k-means algorithm to determine whether the current value will result in new changes. Note that this amounts to running a single iteration of Kernel k-means. Later we show that we can further optimize these queries.

The second observation is that we only ever need to compute exponents in the interval $(0, 1)$. This is because if we assume Kernel k-means to have converged for the matrix K_σ, we know that the next value of σ we seek is larger than the current one. Note that we can use our result from Theorem 1 for a starting value of σ without running an initial execution of Kernel k-means.

By virtue of Theorem 2, Algorithm 2 thus finds an arbitrarily good approximation of the critical bandwidth value, in the following sense:

Corollary 1. *Suppose* Kernel k-means *has converged for* K_σ, *producing a k-partition* P, *and let* σ' *be the critical bandwidth value with respect to* (K_σ, P). *If we run Algorithm 2 with a depth value of* $i = \mathcal{O}\left(\log\left(\frac{|\log(b)|}{\epsilon}\right)\right)$ —*where b is the minimum entry in the kernel matrix*—, *it will output a matrix* K_ρ *satisfying*

$$(1 - \epsilon)K_{\sigma'} \leq_\circ K_\rho \leq_\circ (1 + \epsilon)K_{\sigma'},$$

where $\leq_\circ$ *denotes element-wise inequality.*

That is, it will output a good approximation of the "next" kernel matrix for which Kernel k-means will make changes. Note that this result also characterizes the computational complexity of our approach, as element-wise operations take $\mathcal{O}(n^2)$ computations. In addition, element-wise operations are trivially parallelizable, so our method can scale to large kernel matrices. Finally, note that

---

Algorithm 2. Hyperparameter search

Input: kernel matrix K, depth i, k-partition P of X.

1: $K' \leftarrow K$
2: $P' \leftarrow P$
3: **for** $j = 1, \ldots, i$ **do**
4:     $K' \leftarrow \sqrt{K'}$ //Element-wise square root
5:     **if** $P' \neq P$ **then**
6:         $K \leftarrow K / \circ K'$
7:     **else**
8:         $K \leftarrow K \circ K'$
9:     $P' \leftarrow kkm(K)$ //Run Kernel k-means
10: Output K

---

even though $\log(b)$ is unbounded, after a few iterations only very small entries, close to zero, would suffer considerable relative error.

An advantage of the algorithm is that we can choose the maximum value of the denominator in the rational approximation of the exponent (maximum depth d). This provides a nice trade-off between speed and accuracy.

6.2 Further Optimizations

Our approach lends itself naturally to various optimizations. We discuss them briefly here.

Hierarchical Search. Our algorithm enables a trade-off between running time and accuracy by means of the depth parameter. The larger it is, the more precise the critical values of σ found. We argue that this parameter can be employed to improve speed without significantly sacrificing accuracy. In particular, the algorithm can be run with increasing depth values, constraining the search to promising regions. For instance, we first set depth to 1, run the algorithm and pick the two values of σ that yield the best performance. We then increase the depth value by 1 and run the algorithm again, setting the lower and upper limits of our search to the two previously picked values of σ. This way we first perform a coarse-grained search to identify a potentially good interval for σ, and then increasingly refine the search.

Limiting Checks. As described above, the way our algorithm approximates the critical value of σ is by testing whether or not Kernel k-means will switch at least one point from one cluster to another. Often, most points will not switch clusters at the critical value. Thus, it is not necessary to compute the proximity function (Eq. (4)) for all point-cluster pairs, and we can limit checks to those points most likely to change. To do this, we can employ different heuristics. For instance, we can limit checks to points such that the proximity function is close for different clusters. We can also limit checks to those points that switch clusters the first time we observe a change (line 6 of Algorithm 2).

6.3 Use with Other Kernels

Our approach is not limited to the RBF kernel. Obviously, any kernel that is exponential in the parameters can be directly used with our method. This includes the popular polynomial kernel, defined as $\kappa(x, y) = (x^T y + c)^d$, for the optimization of the parameter d. We can also benefit from the fact that any linear combination of kernels is also a kernel, to accommodate a wider variety of kernel functions. To use our algorithm with a linear combination of differently-parametrized kernels, it suffices to store the kernel matrix separately for each term of the sum. As currently described, our method only allows the optimization of one parameter at a time, but it can be employed as a building block for more sophisticated multiparameter optimization approaches.

7 Experiments

We conduct a series of numerical experiments to evaluate the performance of the proposed algorithm. We mainly want to determine whether our method (1) can reveal good value of σ and (2) can do it efficiently. We compare it to other approaches for hyperparameter search, which we now describe.

Baselines. We consider the following methods to choose the hyperparameter of the RBF kernel[1].

MKNN: We set σ to be the mean distance to the k-th nearest neighbour as suggested by Von Luxburg [19] (the median yields similar results). We try different values of k, namely $k = 1, \ldots 2(\log n + 1)$.

GRIDSEARCH: We run the `Kernel k-means` algorithm with σ taking values in $\{10^i : i = -6, -5, -4, -3, -2, -1, 0, 1, 2, 3, 4, 5, 6\}$.

We refer to Algorithm 2 as OURS.

All methods, as well as `Kernel k-means`, were implemented using Python 3, using matrix and vector operations whenever possible for efficiency[2].

Quality Measures: We consider the following functions to evaluate the quality of the clustering results.

NMI (Normalized Mutual Information): We use a well-known clustering performance index[3], which we now define. Given two indicator vectors y and z, we define

$$\text{NMI}(y, z) = \frac{2I(y, z)}{H(y) + H(z)} \tag{6}$$

---

[1] Some of these methods, as originally described, define the kernel as $\kappa(x, y) = \exp(-\|x - y\|/(2\sigma^2))$. We take this difference into account in our experimental setup.
[2] Source code: https://github.com/justbruno/off-the-grid/.
[3] Results for Adjusted Rand-Index were similar and are thus omitted.

where $I(y, z) = \sum_i \sum_j p(y = i, z = j) \log \left(\frac{p(y=i, z=i)}{p(y=i)p(z=i)} \right)$ denotes the mutual information of y and z, and $H(y) = -\sum_i p(y = i) \log p(y = i)$ denotes the entropy of y [6] (we abuse notation and overload y for the vector and its entries). We use this index by taking y to be the indicator vector of ground-truth labels and z to be the indicator vector of the k-partition output by Kernel k-means.

c-NNC: In addition, we propose our own clustering cost function. Our goal is to measure the quality of the resulting k-partition in a way that (1) arbitrarily shaped clusters are considered and (2) is independent of the value of σ. Note that some well-known clustering quality indices and cost functions, such as silhouette [17] and normalized cuts [7], do not qualify.

We first introduce some notation. Given a data set X and a point $x_i \in X$, let $\nu_j(x_i)$ be the j-th nearest neighbour of x_i in X. Given a k-partition of the data set X into k clusters, $c(x_i)$ denotes the cluster x_i is assigned to, i.e. $x_i \in c(x_i)$.

We first define $\text{NNC}(i, c)$ to be the fraction of points among the c nearest neighbours of x_i which are not in the same cluster as x_i.

$$\text{NNC}(i, c) = \frac{1}{c} \sum_{j=1}^{c} \mathbb{I}\{c(x_i) \neq c(\nu_j(x_i))\}.$$

To measure the quality of a single cluster π, we take a weighted sum of the above index for all c. We scale the value of $\text{NNC}(i, c)$ by $\frac{1}{c}$ to reduce the penalty incurred by disagreements with further neighbours.

$$\text{NNC}_{cluster}(\pi) = \frac{1}{C \max\{1, |\pi|\}} \sum_{i \in \pi} \sum_{c=1}^{n} \frac{1}{c} \text{NNC}(i, c).$$

Here, $C = \log(n - 1) + \gamma + \frac{1}{2n-2}$, where γ is the Euler-Mascheroni constant, ensures that the quantity is upper-bounded by 1 (note that without this scaling factor, the sum for each point is tightly upper bounded by a harmonic series). We now define the cost function as

$$\text{c-NNC}(P) = \frac{D + \sum_{\pi \in P} \text{NNC}_{cluster}(\pi)}{k}.$$

Here, P is the k-partition output by Kernel k-means, k is the number of clusters given to Kernel k-means and D is the number of empty clusters. We count empty clusters to penalize trivial solutions (e.g. a single cluster).

Datasets: We employ a variety of publicly available synthetic[4] and real[5] data sets. Since we use vanilla Kernel k-means, which requires handling the complete kernel matrix, we employ data sets of limited size (up to 8 000 instances). However, our method can in principle be employed with techniques for scalable kernel-based algorithms. A summary of the data sets is given in Table 1. In the case of real data sets, we scale the variables to unit-variance, as this enables a much better performance of Kernel k-means in most cases.

---

[4] http://cs.joensuu.fi/sipu/datasets.
[5] https://archive.ics.uci.edu/ml/index.php.

Table 1. Summary of data set characteristics

Dataset	Rows	Columns	Classes	Dataset	Rows	Columns	Classes
AGGR.	788	2	7	SPIRAL	312	2	3
COMPOUND	399	2	6	AUDIT	775	23	2
D31	3100	2	31	DERMA.	358	34	6
FLAME	240	2	2	WDBC	569	30	2
JAIN	373	2	2	WIFI	2000	7	4
PATHBASED	300	2	3	WINE	178	13	3
R15	600	2	15	MNIST (sampled)	1k, 2k, 4k, 8k	784	10

7.1   Performance

In this section we report the performance of our method, as evaluated by our quality measures, in comparison to the selected baselines. We proceed as follows: we first choose a random initial k-partition, which we set as starting point for all methods. To evaluate our method, we set the initial value of σ to be the 1st percentile of pairwise distances in the data set. Note this is similar to our lower bound given in Sect. 4, but a little less stringent. We run Algorithm 2 with depth $= 1$ and pick the value of σ that corresponds to the best observed k-partition (as measured by c-NNC), run Kernel k-means and rerun our method starting from the resulting k-partition with depth $= 2$. Note that this resembles the hierarchical search described in Sect. 6. For each method, we collect the best value of NMI and c-NNC among the produced clusterings. We report the average over 50 runs, each with a different initial k-partition. Results are shown in Table 2. Our method achieves better values of both measures in most cases.

7.2   Running Times and Scalability

In this section we evaluate the efficiency of our method. We report the average total running times in the previously described experiment for all algorithms in Table 3. Our method generally sits between GRIDSEARCH and MKNN. It performs significantly more iterations than the baselines, and thus better running times could be obtained by limiting the number of inspected values if necessary.

To offer a finer running time comparison, as well as to evaluate scalability, we run the algorithms on samples of MNIST[6] and set the number of iterations to be the same for all methods. In particular, we set it to 13, which is the number of values tested by GRIDSEARCH. Figure 1 shows time taken per iteration, averaged over 50 runs. By iteration we refer to the set of computations required to produce and test a new value of the bandwidth parameter. The reason the running time of GRIDSEARCH increases significantly at some point is that the first values of σ are too small and Kernel k-means converges after one iteration, highlighting the wasteful nature of GRIDSEARCH. Our method benefits mostly from being able to run a small number of iterations of Kernel k-means to converge.

---

[6] http://yann.lecun.com/exdb/mnist/.

Table 2. Comparison of the different methods in terms of quality measures

Dataset	NMI			c-NNC		
	MKNN	GRIDSEARCH	OURS	MKNN	GRIDSEARCH	OURS
AGGR.	0.690	0.864	**0.872**	0.255	0.210	**0.203**
COMPOUND	0.689	**0.778**	0.730	0.239	0.230	**0.215**
D31	0.810	0.931	**0.951**	0.356	0.332	**0.316**
FLAME	0.489	0.521	**0.615**	0.106	0.096	**0.093**
JAIN	0.229	**0.361**	0.353	0.116	0.062	**0.062**
PATHBASED	0.820	0.662	**0.902**	0.169	**0.134**	0.137
R15	0.922	0.954	**0.979**	0.302	0.300	**0.274**
SPIRAL	0.187	0.145	**0.239**	0.175	0.155	**0.151**
AUDIT	**0.717**	0.685	0.703	0.097	**0.082**	0.082
DERMA.	0.889	0.877	**0.913**	0.249	0.256	**0.238**
WDBC	0.531	0.547	**0.550**	0.123	0.108	**0.107**
WIFI	0.781	0.835	**0.856**	0.157	0.140	**0.137**
WINE	**0.923**	0.913	**0.923**	0.143	**0.142**	0.143

Table 3. Total running times in seconds

Dataset	Time in seconds			Dataset	Time in seconds		
	MKNN	GRIDSEARCH	OURS		MKNN	GRIDSEARCH	OURS
AGGR.	0.824	**0.486**	0.617	SPIRAL	0.157	**0.108**	0.133
COMPOUND	0.200	**0.146**	0.172	AUDIT	0.578	**0.467**	1.111
D31	22.029	11.735	**10.757**	DERMA.	0.152	0.106	**0.100**
FLAME	0.064	**0.046**	0.061	WDBC	0.325	**0.204**	0.213
JAIN	0.140	**0.097**	0.143	WIFI	7.095	**3.520**	4.402
PATHBASED	0.110	**0.077**	0.094	WINE	0.044	**0.034**	0.036
R15	0.475	**0.329**	0.406				

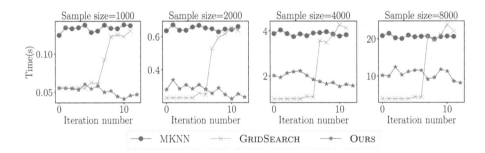

Fig. 1. Running time per iteration for different samples of the MNIST data set

7.3 Comparison with Binary Search

The reader might observe that our method resembles a form of binary search. Thus, one might suspect that similar results could be obtained using a conventional binary search algorithm, without going to the trouble of implementing Algorithm 2. Here we illustrate why our algorithm is a vastly superior alternative.

The setup is as follows: we initialize σ to be the 1st percentile of the squared pairwise distances and then run iterations of binary search with a precision of 10^{-3} and Algorithm 2 with depth equal to 10. We repeat the experiment 10 times and report average iteration time and absolute error of the estimate of the critical value of σ. The results are shown in Table 4. Binary search was implemented efficiently, updating the kernel matrix with fast matrix-vector operations.

Our method achieves a speedup of about 10x in all cases, and the error is often smaller. Of course, the error can be controlled in both algorithms at the expense of running time. A noteworthy difference between both methods (not in favor of any of the two) is that binary search is designed to control absolute error, while Algorithm 2 controls the relative error of the power computation.

Table 4. Running times of our method and binary search. We report average iteration running times, speedup and mean relative error of the σ estimate over 100 iterations

Dataset	Iteration time in seconds		Speedup	Relative error: $\frac{\sigma_{true} - \sigma_{estimated}}{\sigma_{true}}$	
	BINARYSEARCH	OURS	–	BINARYSEARCH	OURS
AGGR.	0.941	0.080	11.7x	1.55×10^{-3}	5.3×10^{-4}
AUDIT	0.793	0.069	11.5x	8.341×10^{-2}	5.8×10^{-4}
COMPOUND	0.192	0.019	9.9x	2.20×10^{-3}	5.8×10^{-4}
D31	15.740	1.148	13.7x	3.95×10^{-3}	$4.8 \times 10^{-}$
DERMA.	0.139	0.014	9.7x	1.3×10^{-4}	8.368×10^{-2}
FLAME	0.063	0.007	9.1x	5.1×10^{-3}	5.6×10^{-4}
JAIN	0.144	0.014	10.5x	2.6×10^{-3}	1.17×10^{-2}
PATHBASED	0.096	0.010	9.7x	1.98×10^{-3}	5.9×10^{-4}
R15	0.430	0.039	11.0x	4.912×10^{-2}	5.5×10^{-4}
SPIRAL	0.102	0.011	9.6x	1.13×10^{-3}	6.2×10^{-4}
WDBC	0.398	0.036	11.2x	10^{-6}	6.2×10^{-4}
WIFI	5.284	0.442	11.9x	6×10^{-5}	4.9×10^{-4}
WINE	0.042	0.005	7.7x	2×10^{-5}	5.7×10^{-4}

8 Conclusion

In this paper we have addressed the problem of hyperparameter search in the Kernel k-means context. Our contribution is two-fold. First, we have derived a

tight lower bound for the bandwidth parameter of RBF kernels, below which
Kernel k-means will be rendered useless. Second, we have proposed a method to
optimize kernel hyperparameters for Kernel k-means. We have proved that our
method approximates critical values of the hyperparameter to arbitrary precision
in a small number of iterations. Unlike grid search or other heuristics, our method
does not test redundant hyperparameter values, that is, values that result in the
same clustering output, and thus no computation is wasted.

Our experiments demonstrate how our approach enables the efficient evaluation of a fine variety of hyperparameter values, revealing high-quality clustering
results at a moderate computational cost. In the future it would be interesting to
extend our method to other kernel-based clustering and classification algorithms.

References

1. Aloise, D., Deshpande, A., Hansen, P., Popat, P.: NP-hardness of Euclidean sum-of-
   squares clustering. Mach. Learn. **75**(2), 245–248 (2009). https://doi.org/10.1007/
   s10994-009-5103-0
2. Bach, F.R., Jordan, M.I.: Learning spectral clustering. In: Advances in Neural
   Information Processing Systems, pp. 305–312 (2004)
3. Ben-David, S., von Luxburg, U., Pál, D.: A sober look at clustering stability. In:
   Lugosi, G., Simon, H.U. (eds.) COLT 2006. LNCS (LNAI), vol. 4005, pp. 5–19.
   Springer, Heidelberg (2006). https://doi.org/10.1007/11776420_4
4. Ben-Hur, A., Horn, D., Siegelmann, H.T., Vapnik, V.: Support vector clustering.
   J. Mach. Learn. Res. **2**(Dec), 125–137 (2001)
5. Cortes, C., Vapnik, V.: Support-vector networks. Mach. Learn. **20**(3), 273–297
   (1995). https://doi.org/10.1007/BF00994018
6. Cover, T.M., Thomas, J.A.: Elements of Information Theory. Wiley, Hoboken
   (2012)
7. Dhillon, I.S., Guan, Y., Kulis, B.: Kernel k-means: spectral clustering and normalized cuts. In: Proceedings of the Tenth ACM SIGKDD International Conference
   on Knowledge Discovery and Data Mining, pp. 551–556. ACM (2004)
8. Ester, M., Kriegel, H.P., Sander, J., Xu, X., et al.: A density-based algorithm for
   discovering clusters in large spatial databases with noise. KDD **96**, 226–231 (1996)
9. Gordon, D.M., et al.: A survey of fast exponentiation methods. J. Algorithms
   27(1), 129–146 (1998)
10. Greene, D., Cunningham, P.: Practical solutions to the problem of diagonal dom-
    inance in kernel document clustering. In: Proceedings of the 23rd International
    Conference on Machine Learning, pp. 377–384. ACM (2006)
11. Hofmann, T., Schölkopf, B., Smola, A.J.: Kernel methods in machine learning.
    Ann. Stat. **36**, 1171–1220 (2008)
12. Jaakkola, T.S., Diekhans, M., Haussler, D.: Using the fisher kernel method to detect
    remote protein homologies. ISMB **99**, 149–158 (1999)
13. Jain, A.K.: Data clustering: 50 years beyond k-means. Pattern Recogn. Lett. **31**(8),
    651–666 (2010)
14. Lloyd, S.: Least squares quantization in PCM. IEEE Trans. Inf. Theory **28**(2),
    129–137 (1982)
15. Moon, T.K.: The expectation-maximization algorithm. IEEE Sig. Process. Mag.
    13(6), 47–60 (1996)

16. Ng, A.Y., Jordan, M.I., Weiss, Y.: On spectral clustering: analysis and an algorithm. In: Advances in Neural Information Processing Systems, pp. 849–856 (2002)
17. Rousseeuw, P.J.: Silhouettes: a graphical aid to the interpretation and validation of cluster analysis. J. Comput. Appl. Math. **20**, 53–65 (1987)
18. Vapnik, V.: Estimation of Dependences Based on Empirical Data. Springer, New York (2006). https://doi.org/10.1007/0-387-34239-7
19. Von Luxburg, U.: A tutorial on spectral clustering. Stat. Comput. **17**(4), 395–416 (2007). https://doi.org/10.1007/s11222-007-9033-z
20. Zelnik-Manor, L., Perona, P.: Self-tuning spectral clustering. In: Advances in Neural Information Processing Systems, pp. 1601–1608 (2005)

Low-Regret Algorithms for Strategic Buyers with Unknown Valuations in Repeated Posted-Price Auctions

Jason Rhuggenaath[✉], Paulo Roberto de Oliveira da Costa, Yingqian Zhang, Alp Akcay, and Uzay Kaymak

Eindhoven University of Technology, 5612 AZ Eindhoven, The Netherlands
{j.s.rhuggenaath,p.r.d.oliveira.da.costa,yqzhang,a.e.akcay,
u.kaymak}@tue.nl

Abstract. We study repeated posted-price auctions where a single seller repeatedly interacts with a single buyer for a number of rounds. In previous works, it is common to consider that the buyer knows his own valuation with certainty. However, in many practical situations, the buyer may have a stochastic valuation. In this paper, we study repeated posted-price auctions from the perspective of a utility maximizing buyer who does not know the probability distribution of his valuation and only observes a sample from the valuation distribution after he purchases the item. We first consider non-strategic buyers and derive algorithms with sublinear regret bounds that hold irrespective of the observed prices offered by the seller. These algorithms are then adapted into algorithms with similar guarantees for strategic buyers. We provide a theoretical analysis of our proposed algorithms and support our findings with numerical experiments. Our experiments show that, if the seller uses a low-regret algorithm for selecting the price, then strategic buyers can obtain much higher utilities compared to non-strategic buyers. Only when the prices of the seller are not related to the choices of the buyer, it is not beneficial to be strategic, but strategic buyers can still attain utilities of about 75% of the utility of non-strategic buyers.

Keywords: Online learning · Posted-price auctions · No-regret learning

1 Introduction

A growing fraction of online advertisements are sold via ad exchanges. In an ad exchange, after a visitor arrives on a webpage, advertisers compete in an auction to win the impression (the right to deliver an ad to that visitor). Typically, these auctions are second-price auctions, where the winner pays the second highest bid or a reserve price (whichever is larger), and no sale occurs if all of the bids are lower than the reserve price. However, as indicated by e.g. [2,3,21], a non-trivial fraction of auctions only involve a single bidder and this reduces to a

© Springer Nature Switzerland AG 2021
F. Hutter et al. (Eds.): ECML PKDD 2020, LNAI 12458, pp. 416–436, 2021.
https://doi.org/10.1007/978-3-030-67661-2_25

posted-price auction [20] when reserve prices known: the seller sets a reserve price and the buyer decides whether to accept or reject it. A single publisher can track a large number of visitors with similar properties over time and sell the impressions generated by these visitors to buyers. As buyers typically are involved in a large number of auctions, there is an incentive for them to act strategically [2,3,16,21]. These observations have led to the study of repeated posted-price auctions between a single seller and strategic buyer.

In this paper we consider a repeated posted-price auction between a single seller and a single buyer, similar to that considered in [2,21]. In every round, the seller posts a price and the buyer decides to buy or not at that price. The buyer does not know the distribution of his valuation, the seller's pricing algorithm or the seller's price set. Furthermore, the seller does not know the valuation distribution and needs to learn how to set the price over time. There are a number of differences between this paper and previous work on repeated posted-price auction such as [2,3,21]. First, unlike in previous work, we study the problem from the perspective of a buyer that aims to maximize his expected utility or surplus, instead of the perspective of the seller that aims to maximize his revenue. Second, previous papers assume that the buyer knows his valuation in each round. In this paper, we relax this assumption and assume the buyer does not know the distribution of his valuation and the valuation is only revealed after he buys the item. This is motivated by applications in online advertising where the buyer (advertiser) does not know the exact value of showing the ads to a set of users: some users may click on the ad and in some cases the ad may lead to a sale, but the buyer only observes a response after he displays the advertisement to the user.

As the valuation distribution is unknown, buyers face an exploration and exploitation trade-off and their decisions lead to regret: (i) accepting a price that is at most the mean valuation leads to positive expected utility and accepting a price above it leads to negative utility; (ii) buying the item leads to additional information about the mean valuation (at the risk of negative utility), but by not buying there is a risk of missing out on positive utility. We study two types of buyers: strategic buyers and non-strategic buyers. Non-strategic buyers are only interested achieving sub-linear regret given the prices that are observed and do not attempt to manipulate or influence the observed prices. Strategic buyers are also interested in sub-linear regret given the observed prices, but they also actively attempt to influence future prices that will be offered. If non-strategic buyers knew the mean valuation they would use the following rule: always accept a price that is at most the mean valuation and always reject a price above it. Strategic buyers on the other hand, would sometimes deviate from this rule in an attempt to influence future prices that will be offered. If non-strategic buyers knew the mean valuation, then their decisions would have low regret but the seller could learn to ask a price very close to the mean valuation, resulting in low utility for the buyer [2,21]. Strategic buyers attempt to influence the learning process of the seller in order to lower the price and to increase the utility. However, as these attempts are not guaranteed to succeed (as buyers don't know the seller's pricing algorithm or price set), strategic buyers still want to ensure sub-linear regret for all possible prices sequences.

In our setting, the seller needs to learn to set his prices because he does not know the valuation distribution. To the best of our knowledge, there are no existing 'optimal' algorithms with performance guarantees (specifically) for repeated posted-price auctions with a single seller and a single strategic buyer that doesn't know his valuation: existing algorithms (e.g., [2,3,14,15,21,25]) assume that buyers know their valuation and thus lose their performance guarantees. In our experiments (see Sect. 5) we therefore assume that the seller uses an off-the-shelf low-regret learning algorithm for adaptive adversarial bandit feedback as these have known performance guarantees [10,20,22].

Our main contributions are as follows. First, to the best of our knowledge, we are the first to study repeated posted-price auctions in strategic settings from the perspective of the buyer. We do not assume that the buyer knows his valuation distribution. Second, we construct algorithms with sub-linear (in the problem horizon) regret for both non-strategic and strategic buyers by using ideas from popular multi-armed bandit algorithms UCB1 [5] and Thompson Sampling [1]. Our algorithms do not require knowledge about the seller's pricing algorithm or price set. Third, we use experiments to support our theoretical findings. Using experiments we show that, if the seller is using a low-regret learning algorithm based on weights updating (such as EXP3.P [4,10]), then strategic buyers can obtain much higher utilities compared to non-strategic buyers.

The remainder of this paper is organized as follows. In Sect. 2 we discuss the related literature. Section 3 provides a formal description of the problem. In Sect. 4 we present the our proposed algorithms and provide a theoretical analysis. In Sect. 5 we perform experiments in order to assess the quality of our proposed algorithms. Section 6 concludes our work and provides some directions for further research.

2   Related Literature

The work in this paper is mainly related to the following areas of the literature: posted-price auctions, low-regret learning by sellers and buyers, and decision making for buyers in auctions. We discuss these areas in more detail below.

Repeated posted-price auctions with the goal of maximizing revenue for the seller and assuming that the feedback from buyers is i.i.d. distributed was studied in [20]. Other works [2,3,14,15,19,21,25] instead study repeated posted-price auctions with strategic buyers. However, these papers all study the seller side of the problem and assume that buyers know their valuations in each round.

On a high level this paper is related to works that study repeated auctions where either the seller and/or the buyer is running a low-regret learning algorithm [8,9,11] and the interaction between bandit algorithms and incentives of buyers [6,7,12,18]. The goal in such studies is to design (truthful) mechanisms that either maximize revenue of the seller or welfare, when decision are made based on low-regret algorithms. This is not the focus of our paper.

The aforementioned works focus on either the seller side or on mechanism design, but there is also work that considers the perspective of buyers or bidders. In [13,23,24] the focus is on maximizing clicks when click-through-rates are unknown and typically with budget constraints. In this paper, rewards for buyers are not determined by the number of clicks, instead the buyer aims to maximize cumulative utilities or his net surplus as in e.g., [2,3,21]. In [17] the focus is on designing bidding strategies for buyers that compete against each other and where the buyer valuation is unknown. However, these studies do not focus on repeated posted-price auctions and strategic behaviour of buyers is not considered.

3   Problem Formulation

We consider a single buyer and a single seller that interact for T rounds. An item, such as an advertisement space, is repeatedly offered for sale by the seller to the buyer over these T rounds. In each round $t \in \mathcal{T} = \{1, \ldots, T\}$, a price $p_t \in \mathcal{P}$ is offered by the seller and a decision $a_t \in \{0,1\}$ is made by the buyer: $a_t = 1$ when the buyer accepts to buy at that price, $a_t = 0$ otherwise. The buyer holds a private valuation $v_t \in [0,1]$ for the item in round t. The value of v_t is an i.i.d. draw from a distribution $\mathcal{D}$ and has expectation $\nu = \mathbb{E}\{v_t\}$. The buyer does not know $\mathcal{D}$ and ν. Also, the buyer does not know $\mathcal{P}$ or the seller's pricing algorithm. The value v_t is only revealed to the buyer if he buys the item in round t, i.e., the buyer only observes the value after he buys the item. The seller also does not know $\mathcal{D}$ or ν and does not observe v_t.

The utility of the buyer in round t is given by $u_t = a_t \cdot (v_t - p_t)$. In other words, if the buyer purchases the item the utility is the difference between the valuation and the price. Otherwise, the utility is zero. For a fixed sequence $\overrightarrow{p} = p_1, \ldots, p_T$ of observed prices and a fixed sequence of decisions $a_1, \ldots, a_T$ by the buyer, the pseudo-regret of the buyer over T rounds is defined as $R_T(\overrightarrow{p}) = \sum_{t=1}^{T} \max\{\nu - p_t, 0\} - \sum_{t=1}^{T} a_t \cdot (\nu - p_t)$. The term $\max\{\nu - p_t, 0\}$ represents the expected utility of the optimal decision in round t and the term $a_t \cdot (\nu - p_t)$ represents the expected utility of the actual decision that is made by the buyer in round t. The expected pseudo-regret over T rounds is defined as $\mathcal{R}_T(\overrightarrow{p}) = \mathbb{E}\{R_T(\overrightarrow{p})\}$, where the expectation is taken with respect to possible randomization in the selection of the actions $a_1, \ldots, a_T$. In the remainder, the expected pseudo-regret will simply be referred to as the regret. The notation using $\overrightarrow{p}$ makes it clear that the regret depends on the sequence of observed prices. We will omit this dependence when the meaning is clear from the context or when a relation is understood to hold for all possible price sequences. For example, we write $\mathcal{R}_T \leq O(\sqrt{T \log T})$ when $\mathcal{R}_T(\overrightarrow{p}) \leq O(\sqrt{T \log T})$ for all choices of $\overrightarrow{p}$.

We consider two types of buyers: non-strategic buyers and strategic buyers. Non-strategic buyers are interested in achieving sub-linear regret for all possible price sequences, but they treat the price sequence as exogenous. That is, if non-strategic buyers knew ν, then they would follow this rule: buy if and only if $p_t \leq \nu$. Strategic buyers also want sub-linear regret for all possible prices sequences, but they would sometimes deviate from this rule in an attempt to influence (i.e., lower) future prices that will be offered. If non-strategic buyers knew ν, then their decisions would have low regret but the seller could learn to ask a price just below ν, resulting in low utility for the buyer [2,21]. Strategic buyers actively attempt to influence the learning process of the seller in order to lower the price and to increase the utility. However, as these attempts are not guaranteed to succeed (recall that buyers do not know the seller's pricing algorithm or $\mathcal{P}$), strategic buyers still want to ensure sub-linear regret for all possible prices sequences. The seller does not know $\mathcal{D}$ or ν and does not observe v_t, and so he has to *learn* how to set his price over time under bandit feedback. This paper focuses on the buyer side and the regret bounds that we derive do not depend on the seller's pricing algorithm. However, in order to test our algorithms, some assumption about the seller's algorithm is required. To the best of our knowledge, there are no existing 'optimal' algorithms for sellers with performance guarantees (specifically) for repeated posted-price auctions with a single seller and a single strategic buyer that doesn't know his valuation: existing algorithms (e.g., [2,3,14,15,19,21,25]) assume that buyers know v_t and thus lose their performance guarantees. In our experiments (see Sect. 5) we therefore assume that the seller uses an off-the-shelf low-regret learning algorithm for adaptive adversarial bandit feedback as these have known performance guarantees [10,20,22].

Algorithm 1: UCB-NS

1 **Input:** $N \in \mathbb{N}$, T.
2 Set $\mathcal{V} = \varnothing$. Set $t = 1$. ;
3 Set $n = 1$. ;
4 Buy item at price p_t. ;
5 Observe v_t. ;
6 Set $\mathcal{V} = \mathcal{V} \cup \{v_t\}$. ;
7 **for** $t \in \{2, \ldots, T\}$ **do**
8 $\quad$ Set $n_t = n$. ;
9 $\quad$ Set $\bar{v}_t = \frac{1}{n_t} \sum_{v \in \mathcal{V}} v$. ;
10 $\quad$ Set $r_t = \sqrt{(2 \log t)/n_t}$. ;
11 $\quad$ Set $I_t = \bar{v}_t + r_t$. ;
12 $\quad$ **if** $I_t \geq p_t$ **then**
13 $\quad\quad$ Buy item at price p_t. ;
14 $\quad\quad$ Observe v_t. ;
15 $\quad\quad$ Set $\mathcal{V} = \mathcal{V} \cup \{v_t\}$. ;
16 $\quad\quad$ Set $n = n + 1$. ;
17 $\quad$ **end**
18 **end**

Algorithm 2: TS-NS

1 **Input:** $N \in \mathbb{N}$, T.
2 Set $\mathcal{V} = \varnothing$. Set $t = N$. ;
3 Set $n = N$. ;
4 Buy item in first N rounds. ;
5 Observe $\mathcal{V}^N = \cup_{k=1}^{N} \{v_k\}$. ;
6 Set $\mathcal{V} = \mathcal{V} \cup \mathcal{V}^N$. ;
7 **for** $t \in \{N+1, \ldots, T\}$ **do**
8 $\quad$ Set $n_t = n$. ;
9 $\quad$ Set $\bar{v}_t = \frac{1}{n_t} \sum_{v \in \mathcal{V}} v$. ;
10 $\quad$ Sample $I_t \sim \mathcal{N}(\bar{v}_t, \frac{1}{n_t})$. ;
11 $\quad$ **if** $I_t \geq p_t$ **then**
12 $\quad\quad$ Buy item at price p_t. ;
13 $\quad\quad$ Observe v_t. ;
14 $\quad\quad$ Set $\mathcal{V} = \mathcal{V} \cup \{v_t\}$. ;
15 $\quad\quad$ Set $n = n + 1$. ;
16 $\quad$ **end**
17 **end**

4   Algorithms and Analysis

In this section we present our proposed algorithms for strategic and non-strategic buyers and we provide a theoretical analysis of these algorithms.

4.1   Non-strategic Buyers

We provide two algorithms for non-strategic buyers that have sub-linear regret. The first algorithm, UCB-NS, is based on UCB (upper confidence bound) style bandit algorithms [5] and the second algorithm, TS-NS, is based on the Thompson Sampling principle [1]. In every round, UCB-NS maintains an optimistic estimate of the unknown mean ν and decides to buy the item if the estimate is at least as large as the offered price p_t. TS-NS samples from a posterior distribution and decides to buy the item if the sampled value is at least as large as the offered price p_t. Proposition 1 and 2 bound the regret of UCB-NS and TS-NS, respectively.

Proposition 1. *If Algorithm 1 is run with inputs: T, then $\mathcal{R}_T \leq O(\sqrt{T \log T})$.*

Proof. If $\mathbb{I}\{\nu > p_t > I_t\} = 1$ then the buyer did not buy the item when instead he should have bought it. Similarly, if $\mathbb{I}\{\nu < p_t \leq I_t\} = 1$, then the buyer did buy the item when instead he should not have bought it.

Note that we can bound the regret as follows

$$\mathcal{R}_T \leq 1 + \sum_{t=1}^{T} \mathbb{E}\left\{(\nu - p_t) \cdot \mathbb{I}\{\nu > p_t > I_t\}\right\}$$
$$+ \sum_{t=1}^{T} \mathbb{E}\left\{(p_t - \nu) \cdot \mathbb{I}\{\nu < p_t \leq I_t\}\right\}.$$

Define $A = \sum_{t=1}^{T} \mathbb{E}\left\{(\nu - p_t) \cdot \mathbb{I}\{\nu > p_t > I_t\}\right\}$ and $B = \sum_{t=1}^{T} \mathbb{E}\left\{(p_t - \nu) \cdot \mathbb{I}\{\nu < p_t \leq I_t\}\right\}$. We will bound each term separately.

Define the following events $F_t = \{\nu > p_t > I_t\}$, $E_t = \{I_t > \nu\}$, $H_t = \{|\bar{v}_t - \nu| \leq \sqrt{\frac{2 \log T}{n_t}}\}$ and $H_t^C = \{|\bar{v}_t - \nu| > \sqrt{\frac{2 \log T}{n_t}}\}$.

For term A we have,

$$A \leq \sum_{t=1}^{T} \mathbb{E}\left\{(\nu - p_t) \cdot \mathbb{I}\{F_t\}\right\} \leq \sum_{t=1}^{T} \mathbb{E}\left\{1 \cdot \mathbb{I}\{F_t\}\right\}$$
$$\leq \sum_{t=1}^{T} \mathbb{P}\{F_t\} \leq \sum_{t=1}^{T} \mathbb{P}\{\nu > I_t\}$$

Using Hoeffding's inequality (and a union bound) we obtain $\mathbb{P}\{\nu > I_t\} \leq \frac{1}{t^3} \leq \frac{1}{t^2}$. Therefore, we conclude that $\sum_{t=1}^{T} \mathbb{P}\{\nu > I_t\} \leq \frac{\pi^2}{6}$.

Define $\mathcal{B} = \{t \in \mathcal{T} \mid I_t \geq p_t\}$. For term B we have,

$$B \leq \sum_{t \in \mathcal{B}} \mathbb{E}\left\{(p_t - \nu) \cdot \mathbb{I}\{\nu < p_t \leq I_t\}\right\} \leq \sum_{t \in \mathcal{B}} \mathbb{E}\left\{(I_t - \nu) \cdot \mathbb{I}\{I_t > \nu\}\right\}$$

$$\leq \sum_{t \in \mathcal{B}} \mathbb{E}\left\{(I_t - \nu) \cdot \mathbb{I}\{E_t \cap H_t\}\right\} + \sum_{t \in \mathcal{B}} \mathbb{E}\left\{(I_t - \nu) \cdot \mathbb{I}\{E_t \cap H_t^C\}\right\}.$$

Define $B_1 = \sum_{t \in \mathcal{B}} \mathbb{E}\left\{(I_t - \nu) \cdot \mathbb{I}\{E_t \cap H_t\}\right\}$. We bound B_1 as follows:

$$B_1 \leq \sum_{t \in \mathcal{B}} \mathbb{E}\left\{|I_t - \nu| \cdot \mathbb{I}\{H_t\}\right\} \leq \sum_{t \in \mathcal{B}} \mathbb{E}\left\{|\nu - I_t| \;\middle|\; H_t\right\} \cdot \mathbb{P}\{H_t\}$$

$$\leq \sum_{t \in \mathcal{B}} \mathbb{E}\left\{|\nu - I_t| \;\middle|\; H_t\right\} \leq \sum_{t \in \mathcal{B}} 2\sqrt{\frac{2\log T}{n_t}}$$

$$\leq \sum_{t \in \mathcal{T}} 2\sqrt{\frac{2\log T}{t}} \leq 2\int_0^T \sqrt{\frac{2\log T}{t}}\,dt \leq 4\sqrt{2\log T}\sqrt{T}.$$

Define $B_2 = \sum_{t \in \mathcal{B}} \mathbb{E}\left\{(I_t - \nu) \cdot \mathbb{I}\{E_t \cap H_t^C\}\right\}$. We bound B_2 as follows:

$$B_2 \leq \sum_{t \in \mathcal{B}} \mathbb{P}\{H_t^C\}$$

$$\leq \sum_{t \in \mathcal{B}} \mathbb{P}\left\{|\bar{v}_t - \nu| > \sqrt{\frac{2\log T}{n_t}}\right\} \overset{(a)}{\leq} T \cdot \frac{2}{T^4}.$$

Inequality (a) follows from applying Hoeffding's inequality and from the fact that $|\mathcal{B}| \leq T$.

Putting everything together we obtain $\mathcal{R}_T \leq 1 + \frac{\pi^2}{6} + 4\sqrt{2\log T}\sqrt{T} + T \cdot \frac{2}{T^4}$. Therefore, we conclude that $\mathcal{R}_T \leq O(\sqrt{T \log T})$. $\square$

Proposition 2. *If Algorithm 2 is run with inputs: T and $N = \lceil c_N \cdot T^{\frac{2}{3}} \rceil$, then $\mathcal{R}_T \leq O(T^{\frac{2}{3}}\sqrt{\log T})$.*

Proof. The proof can be found in the Appendix. $\square$

4.2   Strategic Buyers

In this section we show how the algorithms for non-strategic buyers can be converted into algorithms for strategic buyers with the same growth rate (up to constant factors) for the regret. Our proposed approach BUYER-STRAT is presented in Algorithm 3. The main idea behind BUYER-STRAT is to take a base algorithm $\mathcal{A}_{base}$ for non-strategic buyers (e.g. UCB-NS or TS-NS) and modify it using what we refer to as *strategic cycles*.

We now give a description of how Algorithm 3 works. In BUYER-STRAT the buyers make decisions according to $\mathcal{A}_{base}$ for the first N_1 rounds. Afterwards, in the next N_2 rounds, we enter a so-called strategic cycle. In this strategic cycle, the buyer only buys the item if the price is below some threshold, that is, if $p_t \leq v^* - c_1$. Here v^* is an estimate of the unknown mean ν and $0 < c_1 < 1$ is a parameter chosen by the buyer (e.g. $c_1 = 0.1$). The purpose of this strategic cycle is to entice the seller into asking prices that are lower than ν. After this strategic cycle comes to an end, we start another strategic cycle of length L with some small probability p_{cycle}. If another strategic cycle has been triggered, we set a new parameter $0 < c_{target} < 1$ and only prices $p_t \leq v^* - c_{target}$ are accepted. If no strategic cycle is triggered, the buyer makes decisions according to $\mathcal{A}_{base}$. In the next round, we start a strategic cycle of length L with probability p_{cycle} and the aforementioned process is repeated.

Algorithm 3 makes use of the functions $F_1, F_2, F_3, F_4, F_5, F_6$. The intuition behind these functions is as follows. In every strategic cycle, only prices that satisfy $p_t \leq v^* - c$ are accepted, where $c \in \mathcal{C}$ for some set $\mathcal{C}$. The value of v^* is selected using the function $F_5(\cdot)$ which takes as input a base algorithm $\mathcal{A}_{base}$. The value $c \in \mathcal{C}$ is selected by using the function $F_1(\cdot)$ which depends on a counter of the number strategic cycles that have passed C_{phase}. Initially, the number of strategic cycles in which values $c \in \mathcal{C}$ are used, is equal to N_{phase}. When $F_2(x) = 1$, this indicates that the last strategic cycle in which a value $c \in \mathcal{C}$ is used has just been completed, and the function $F_3(\cdot)$ is used to collect information about the price trajectory. When $F_6(x) = 1$, a final value for p_{target} is chosen (using $F_4(\cdot)$) and only prices with $p_t \leq p_{target}$ are accepted in all subsequent strategic cycles. In Sect. 5 we discuss these functions in more detail and give specific examples that are used in our experiments.

The key parameters to control the regret of Algorithm 3 are the cycle probability p_{cycle} and the cycle length L. Proposition 3 shows that BUYER-STRAT with $\mathcal{A}_{base}$ chosen as UCB-NS has regret of order $O(\sqrt{T \log T})$ if the probability p_{cycle} and the cycle length L is carefully chosen. Proposition 4 shows an analogous result for BUYER-STRAT with TS-NS.

Proposition 3. *Let A_p, A_L and A_N be positive real constants. Assume that Algorithm 3 is run with $\mathcal{A}_{base}$ chosen as UCB-NS and with inputs: T, $N_1 = \lceil T^{\frac{2}{3}} (\log T)^{\frac{1}{2}} \rceil$, $N_2 = \lceil A_N \sqrt{T \log T} \rceil$, $p_{cycle} = A_p T^{-\frac{1}{2}}$ and $L = A_L \sqrt{\log T}$, then $\mathcal{R}_T \leq O(\sqrt{T \log T})$.*

Proof. We will decompose the regret in two parts: the regret incurred in rounds that are part of strategic cycles and rounds that are not. For an arbitrary subset $\mathcal{T}^* \subseteq \mathcal{T}$, let $\mathcal{R}_{T,\mathcal{T}^*} = \sum_{t \in \mathcal{T}^*} \mathbb{E}\{(\nu - p_t) \cdot \mathbb{I}\{\nu > p_t > I_t\}\} + \sum_{t \in \mathcal{T}^*} \mathbb{E}\{(p_t - \nu) \cdot \mathbb{I}\{\nu < p_t \leq I_t\}\}$. Let $\mathcal{T}_S \subseteq \mathcal{T}$ denote the indices of the rounds that are part of strategic cycles and let $\mathcal{T}_{NS} = \mathcal{T} \setminus \mathcal{T}_S$ denote the indices of the rounds that are not. Then we can write, $\mathcal{R}_T = \mathcal{R}_{T,\mathcal{T}_{NS}} + \mathcal{R}_{T,\mathcal{T}_S}$.

For $\mathcal{R}_{T,\mathcal{T}_S}$ we have that $\mathcal{R}_{T,\mathcal{T}_S} \leq N_2 + T \cdot p_{cycle} \cdot L$. This follows from the fact that the expected number of triggered strategic cycles (after round $N_1 + N_2$) is at most $T \cdot p_{cycle}$ and the regret in every such cycle is at most L. Furthermore, the first strategic cycle has length N_2. For $\mathcal{R}_{T,\mathcal{T}_{NS}}$ we have that $\mathcal{R}_{T,\mathcal{T}_{NS}} \leq 5 + 4\sqrt{2 \log T}\sqrt{T}$. This follows from the fact that $\mathcal{R}_{T,\mathcal{T}_{NS}}$ represents the regret after $|\mathcal{T}_{NS}| \leq T$ rounds in a problem with horizon T, and by Proposition 1, this quantity is bounded by $5 + 4\sqrt{2 \log T}\sqrt{T}$. By plugging in the values we get $\mathcal{R}_T = \mathcal{R}_{T,\mathcal{T}_{NS}} + \mathcal{R}_{T,\mathcal{T}_S} \leq O(\sqrt{T \log T})$.   □

Proposition 4. *Let A_p, A_L and A_N be positive real constants. Assume that Algorithm 3 is run with $\mathcal{A}_{base}$ chosen as TS-NS and with inputs: T, $N_1 = \lceil T^{\frac{2}{3}}(\log T)^{\frac{1}{2}}\rceil$, $N_2 = \lceil A_N\sqrt{T \log T}\rceil$, $p_{cycle} = A_p T^{-\frac{1}{2}}$ and $L = A_L\sqrt{\log T}$. Assume that TS-NS is run with inputs: T and $N = \lceil c_N \cdot T^{\frac{2}{3}}\rceil$. Then $\mathcal{R}_T \leq O(T^{\frac{2}{3}}\sqrt{\log T})$.*

Proof. The proof uses similar arguments as the proof of Proposition 3 and is omitted. A complete proof can be found in the Appendix.   □

Algorithm 3: BUYER-STRAT

---

1 **Input:** F_1, F_2, F_3, F_4, F_5, F_6, L, p_{cycle}, N_{phase}, N_1, N_2, c_1, T, $\mathcal{A}_{base}$.
2 Set $L_p = \varnothing$, $L_{target} = \varnothing$, $C_{phase} = 0$, $t = 1$.;
3 **for** $t = 1, \ldots, N_1$ **do**
4 $\quad$| Observe price p_t. Choose to buy or not based on $\mathcal{A}_{base}$.;
5 **end**
6 $v^* = F_5(\mathcal{A}_{base})$.;
7 **for** $t = N_1 + 1, \ldots, N_1 + N_2$ **do**
8 $\quad$| Observe price p_t. Buy if $p_t \leq v^* - c_1$.;
9 **end**
10 **while** $t \in \{N_1 + N_2 + 1, \ldots, T\}$ **do**
11 $\quad$ Draw D from Bernoulli distribution with success parameter p_{cycle}.;
12 $\quad$ **if** $D = 1$ **then**
13 $\quad\quad$| $v^* = F_5(\mathcal{A}_{base})$.;
14 $\quad\quad$ **if** $C_{phase} \leq N_{phase}$ **then**
15 $\quad\quad\quad$| Set $c_{target} = F_1(C_{phase})$. Set $p_{target} = v^* - c_{target}$.;
16 $\quad\quad$ **end**
17 $\quad\quad$ **for** $l \in \{1, \ldots, L\}$ **do**
18 $\quad\quad\quad$ Observe price p_t.;
19 $\quad\quad\quad$ $L_p = L_p \cup \{p_t\}$.;
20 $\quad\quad\quad$ Buy if $p_t \leq p_{target}$.;
21 $\quad\quad\quad$ Set $t = t + 1$.;
22 $\quad\quad$ **end**
23 $\quad\quad$ **if** $F_2(C_{phase}) = 1$ **then**
24 $\quad\quad\quad$ Set $c_e = F_3(L_p)$. Set $L_{target} = L_{target} \cup \{c_e\}$.;
25 $\quad\quad\quad$ Set $C_{phase} = C_{phase} + 1$.;
26 $\quad\quad\quad$ **if** $F_6(C_{phase}) = 1$ **then**
27 $\quad\quad\quad\quad$| $p_{target} = F_4(L_{target})$;
28 $\quad\quad\quad$ **end**
29 $\quad\quad$ **end**
30 $\quad\quad$ **if** $D = 0$ **then**
31 $\quad\quad\quad$ Observe price p_t.;
32 $\quad\quad\quad$ Choose to buy or not based on $\mathcal{A}_{base}$.;
33 $\quad\quad\quad$ Set $t = t + 1$.;
34 $\quad\quad$ **end**
35 $\quad$ **end**
36 **end**

Remark 5. In order to derive the results of Proposition 3 and 4, we only used the fact that the regret for $\mathcal{A}_{base}$ is bounded by $O(\sqrt{T \log T})$ or $O(T^{\frac{2}{3}}\sqrt{\log T})$. The same proof is also valid for any other base algorithm that satisfies these bounds. Also, the exact choices for functions $F_1, F_2, F_3, F_4, F_5, F_6$ do not effect the regret guarantee (in Sect. 5 we discuss these functions in more detail).

In which setting is BUYER-STRAT useful? As the seller does not know $\mathcal{D}$, it is reasonable to assume (as argued in Sect. 3) that the seller uses a low-regret algorithm to *learn* how to set prices. Note that many online learning algorithms (e.g. EXP3 and its variants) are *weight-based* algorithms: at round t, there are weights $w_{k,t}, \ldots, w_{K,t}$ and an action $k \in \{1, \ldots, K\}$ is chosen with probability $w_{k,t}/\sum_{k=1}^{K} w_{k,t}$. We call an algorithm a *pure weight-based* algorithm if in round t, only the weight of the selected action gets updated and if weights can only increase due to positive rewards (note that EXP3 is an example, see the Appendix for a general definition). Proposition 6 shows that, if the seller uses a pure weight-based algorithm, then BUYER-STRAT tends to encourage lower prices by using strategic cycles.

Proposition 6. *Assume that the buyer uses Algorithm 3, that the seller is using a pure weight-based algorithm and that the price set $\mathcal{P}$ is finite. Suppose that a strategic cycle runs from round $t + 1$ to round $t + L$ with p_{target}, then* $\mathbb{P}\{p_{t+L+1} \leq p_{target}\} \geq \mathbb{P}\{p_{t+1} \leq p_{target}\}.$

Proof. The proof can be found in the Appendix. □

5 Experiments

In this section we verify the theoretical results that were derived and investigate the effects of strategic behaviour on the regret in different scenarios.

5.1 Setup of Experiments

In the experiments v_t is drawn from an uniform distribution on $[a - 0.3, a + 0.3]$, where a is drawn from an uniform distribution on $[0.4, 0.7]$ independently for each run. We consider two settings for the set of prices used by the seller and these are given by $\mathcal{P}_1$ and $\mathcal{P}_2$: $\mathcal{P}_1 = \{a + x \mid x \in \{-0.35, -0.3, -0.25, -0.2, -0.1, -0.05, -0.02, 0.0, 0.1, 0.3\}\}$, $\mathcal{P}_2 = \{a + x \mid x \in \{-0.05, -0.02, 0.0, 0.1, 0.3\}\}$. We will use the following abbreviations: P1 and P2. The abbreviation P1 means that $\mathcal{P}_1$ is used. The other abbreviations have a similar interpretation.

We consider three options for the seller pricing algorithm: (i) the seller chooses a price at random from the price set (RAND seller); (ii) the seller uses the low-regret learning algorithm EXP3.P (EXP3.P seller); (iii) the seller uses the full-information algorithm HEDGE (HEDGE seller). RAND seller is included because it models a situation where the buyer has no influence over the prices. EXP3.P seller is included because it is a bandit algorithm designed for adaptive adversaries and it enjoys high-probability regret bounds [4,10]. It models a seller that is learning which prices to use based on bandit feedback that is non-stochastic. HEDGE seller is included in order to investigate whether the restriction to bandit feedback has a major impact on the performance of BUYER-STRAT. HEDGE seller is tuned according to Remark 5.17 in [22] and EXP3.P according to Theorem 3.2 in [10].

In the experiments, BUYER-STRAT is tuned with $N_1 = \lceil T^{\frac{2}{3}} \log T \rceil$, $N_2 = \lceil 2\sqrt{T \log T} \rceil$, $L = \lfloor 25\sqrt{\log T} \rfloor$, $p_{cycle} = \frac{5}{\sqrt{T}}$, $c_1 = 0.1$. We set $N_{phase} = 4 \cdot N_3$, where $N_3 = \lceil 0.1 \cdot \sqrt{T} \rceil$. TS-NS is tuned with $N = \lceil 0.005 \cdot T^{\frac{2}{3}} \rceil$. We will refer to BUYER-STRAT with $\mathcal{A}_{base}$ chosen as UCB-NS, as UCB-S (Upper Confidence Bound Strategic). Similarly, We will refer to BUYER-STRAT with $\mathcal{A}_{base}$ chosen as TS-NS, as TS-S (Thompson Sampling Strategic). The functions F_1, F_2, F_3, F_4, F_5, F_6 are chosen as follows.

$$F_1(x) = \begin{cases} 0.2 & \text{if } x \leq N_3 \\ 0.3 & \text{if } 1 \cdot N_3 < x \leq 2 \cdot N_3 \\ 0.4 & \text{if } 2 \cdot N_3 < x \leq 3 \cdot N_3 \\ 0.5 & \text{if } 3 \cdot N_3 < x \leq 4 \cdot N_3 \end{cases} \tag{1}$$

For $F_2(x)$ we take $F_2(x) = \mathbb{I}\{x \in \{N_3, 2 \cdot N_3, 3 \cdot N_3, 4 \cdot N_3\}\}$. The function $F_3(L_p)$ takes the last 100 elements added to the input list L_p and then calculates the 25-th percentile of these 100 values. The function $F_4(\cdot)$ is defined as $F_4(L_{target}) = \min\{L_{target}\} + \varepsilon$. The function $F_4(L_{target})$ takes the smallest number in the set L_{target} and adds a small value to it. In our experiments we use $\varepsilon = 0.005$. The function $F_5(\cdot)$ takes as input a base algorithm and returns the value of $\bar{v}_t$ in the base algorithm. For $F_6(x)$ we take $F_6(x) = \mathbb{I}\{x = 4 \cdot N_3\}$.

The intuition behind these choices is as follows. In every strategic cycle, only prices that satisfy $p_t \leq v^* - c$ are accepted, where $c \in \mathcal{C} = \{0.1, 0.2, 0.3, 0.4, 0.5\}$ and where c is chosen in increasing order (to try to reduce the price in stages) as the number of strategic cycles increases (this is specified by the function $F_1(\cdot)$). Initially, the number of strategic cycles in which every $c \in \mathcal{C}$ is used, is proportional to N_3. When $F_2(x) = 1$, this indicates that the last strategic cycle in which $c = x$ has just been completed, and the function $F_3(\cdot)$ is used to collect information about the price trajectory. When $F_6(x) = 1$, a final value for p_{target} is chosen (using $F_4(\cdot)$) and this value is used in all subsequent strategic cycles.

We perform 100 independent simulation runs in order to calculate our performance metrics. We use three performance metrics in order to evaluate our algorithm. In each run, we calculate the cumulative regret $R_T = \sum_{t=1}^{T} \max\{v - p_t, 0\} - \sum_{t=1}^{T} a_t \cdot (v - p_t)$, the cumulative utility $U_T = \sum_{t=1}^{T} a_t \cdot (v - p_t)$ and the scaled cumulative regret $R_T^S = R_T / \sum_{t=1}^{T} \max\{v - p_t, 0\}$. In the experiments we set $T \in \{25000, 50000, 75000, 100000, 200000, \ldots, 1000000\}$.

5.2   Results: Non-strategic Buyers vs. Strategic Buyers

Non-strategic Buyers. In Figs. 1 and 4 the cumulative regret is shown for different experimental settings and different values for the problem horizon. Each point in the graph shows the cumulative regret over T rounds for a problem of horizon T averaged over 100 simulations. In all figures, the lines indicate the mean and the shaded region indicates a 95% confidence interval. The results indicate that the expected regret indeed grows as a sub-linear function of T and that this pattern holds for both RAND seller and EXP3.P seller. An interesting finding is that the regret for TS-NS is lower than UCB-NS: based on the theoretical analysis one would expect the opposite pattern. Figures 3 and 6 show the scaled cumulative regret and provides further evidence that the expected regret is a sub-linear function of the horizon T, as the curve shows a monotonically decreasing pattern. Figures 2 and 5 show the cumulative utility against different sellers. Here we observe that the utility tends to be higher if the seller uses $\mathcal{P}_1$, which makes intuitive sense as this price set contains lower prices.

Strategic Buyers. Figures 7, 8, 9, 10, 11 and 12 show the same performance metrics as for the non-strategic bidders. Figures 1 and 4 show that the level of the expected regret for strategic bidders is higher compared to the non-strategic bidders. Figures 9 and 12 again indicate that the expected regret is sub-linear in T, as the curves show a monotonically decreasing pattern (from Fig. 7 it is hard to tell). Thus, we observe sub-linear regret for both UCB-S and TS-S regardless of the seller algorithm and this is in line with the theoretical analysis. If we compare the cumulative utility in Figs. 8 and 11 with those in Figs. 2 and 5, then we observe some interesting results. First, when strategic buyers are facing RAND seller (Fig. 8), then we see that the cumulative utility is about 70%–80% of the cumulative utility if non-strategic buyers are facing RAND seller (Fig. 2). Second, we see that if the seller is using EXP3.P (i.e, a low-regret learning algorithm), then the cumulative utility for strategic buyers is much higher compared to the cumulative utility for non-strategic buyers. In scenario P1 utilities are about 2.5–3 times higher and in scenario P2 utilities are about 2 times higher. The results for scenario P2 imply that, even when the lowest price is very close to the unknown mean valuation (absolute distance at most 0.05), it is still beneficial to act strategically. Additional experimental results when the seller uses EXP3.S [4] can be found in the Appendix.

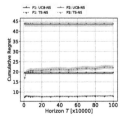

Fig. 1. R_T with RAND seller.

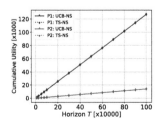

Fig. 2. U_T with RAND seller.

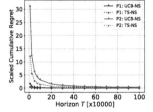

Fig. 3. R_T^S with RAND seller.

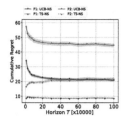
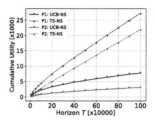
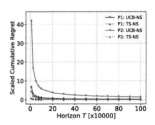

Fig. 4. R_T with EXP3.P seller.

Fig. 5. U_T with EXP3.P seller.

Fig. 6. R_T^S with EXP3.P seller.

5.3 Explanation of Differences

In order to study the impact of the quality of feedback that the seller observes, we give the seller full-information feedback instead of bandit feedback. More specifically, we assume the seller uses the algorithm HEDGE. Figures 13, 14 and 15 show results for TS-S and TS-NS against HEDGE seller. Even with full-information the results are qualitatively similar as before: the regret for the strategic buyers is sub-linear and cumulative utility is much higher for strategic buyers. Thus, the results indicate that the feedback type is not the main driver for the observed patterns.

Figures 16 and 17 display the gap $\nu - p_t$ for a problem with horizon $T = 200000$ averaged over the 100 simulation runs. If the seller is using a low-regret algorithm in order to set prices and buyers are non-strategic, then we observe that prices tend to increase towards the mean valuation ν. This effect is stronger for HEDGE seller compared to EXP3.P seller and this is in line with expectations as HEDGE uses full-information feedback. Furthermore, we see a qualitatively similar pattern for the price sets $\mathcal{P}_1$ and $\mathcal{P}_2$, although the increase in price with $\mathcal{P}_2$ is slightly larger. For HEDGE seller, we hardly see any difference for different price sets. If buyers are strategic then we see the opposite pattern. The algorithms for strategic buyers tend to lower the price over time and the magnitude of this reduction depends on the price set of the seller (reduction for $\mathcal{P}_1$ is larger than for $\mathcal{P}_2$).

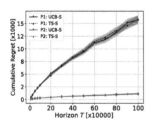
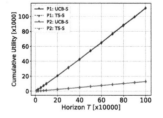
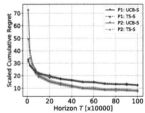

Fig. 7. R_T with RAND seller.

Fig. 8. U_T with RAND seller.

Fig. 9. R_T^S with RAND seller.

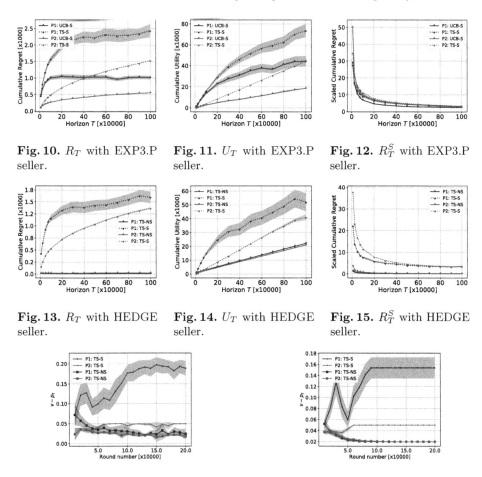

Fig. 10. R_T with EXP3.P seller.

Fig. 11. U_T with EXP3.P seller.

Fig. 12. R_T^S with EXP3.P seller.

Fig. 13. R_T with HEDGE seller.

Fig. 14. U_T with HEDGE seller.

Fig. 15. R_T^S with HEDGE seller.

Fig. 16. $\nu - p_t$ with EXP3.P seller.

Fig. 17. $\nu - p_t$ with HEDGE seller.

However, even with price set $\mathcal{P}_2$ where the lowest prices are very close to ν, strategic behaviour is beneficial and strategic buyers can induce prices that are almost twice as far from ν.

6 Conclusion

This is paper we study repeated posted-price auctions with a single seller from the perspective of a utility maximizing buyer that does not know the distribution of his valuation. Previous work has only focused on the seller side and does not study how buyers should make decisions, hence in this paper, we address this gap in the literature. We study two types of buyers (strategic and non-strategic) and derive sub-linear regret bounds that hold for all possible sequences of observed prices. Our algorithms are based on ideas from UCB-type bandit algorithms and

Thompson Sampling. Our experiments we show that, if the seller is using a low-regret learning algorithm based on weights updating, then strategic buyers can obtain much higher utilities compared to non-strategic buyers.

In practice, buyers have limited budgets for purchasing items. One direction for future work is to investigate the impact of this on the problem. In particular, it would be interesting to analyze how a budget constraint would affect the regret guarantees derived in this paper and whether budget constraints make it easier or harder to engage in strategic behavior.

Appendix

Section A contains proofs that are omitted from the main text. Section B presents some additional experimental results.

A   Proofs for Sect. 4

A.1   Proof of Proposition 2

Proof. We can bound the regret as follows $\mathcal{R}_T \leq N \cdot 1 + \sum_{t=N+1}^{T} \mathbb{E}\left\{(\nu - p_t) \cdot \mathbb{I}\{\nu > p_t > I_t\}\right\} + \sum_{t=N+1}^{T} \mathbb{E}\left\{(p_t - \nu) \cdot \mathbb{I}\{\nu < p_t \leq I_t\}\right\}$.

Define $A = \sum_{t=N+1}^{T} \mathbb{E}\left\{(\nu - p_t) \cdot \mathbb{I}\{\nu > p_t > I_t\}\right\}$ and
$B = \sum_{t=N+1}^{T} \mathbb{E}\left\{(p_t - \nu) \cdot \mathbb{I}\{\nu < p_t \leq I_t\}\right\}$. We will bound each term separately. Define the event $F_t = \{\nu > p_t > I_t\}$.

$$
\begin{aligned}
A &\leq \sum_{t=N+1}^{T} \mathbb{E}\left\{(\nu - p_t) \cdot \mathbb{I}\{F_t\}\right\} \leq \sum_{t=N+1}^{T} \mathbb{E}\left\{(\nu - I_t) \cdot \mathbb{I}\{F_t\}\right\} \\
&\leq \sum_{t=N+1}^{T} \mathbb{E}\left\{|\nu - I_t| \cdot \mathbb{I}\{F_t\}\right\} \leq \sum_{t=N+1}^{T} \mathbb{E}\left\{|(\nu - \bar{v}_t) - (I_t - \bar{v}_t)| \;\middle|\; F_t\right\} \cdot \mathbb{P}\{F_t\} \\
&\leq \sum_{t=N+1}^{T} \mathbb{E}\left\{|\nu - \bar{v}_t| \;\middle|\; F_t\right\} \cdot \mathbb{P}\{F_t\} + \sum_{t=N+1}^{T} \mathbb{E}\left\{|I_t - \bar{v}_t| \;\middle|\; F_t\right\} \cdot \mathbb{P}\{F_t\} \\
&\leq \sum_{t=N+1}^{T} \mathbb{E}\left\{|\nu - \bar{v}_t|\right\} + \sum_{t=N+1}^{T} \mathbb{E}\left\{|I_t - \bar{v}_t|\right\}
\end{aligned}
$$

Using Hoeffding's inequality we obtain, for $t > N$, that $\mathbb{E}\{|\nu - \bar{v}_t|\} \leq \frac{2}{T^4} + 2\sqrt{\frac{2\log T}{N}}$. Using the fact that $N = \lceil c_N \cdot T^{\frac{2}{3}} \rceil$ and that $T - (N+1) \leq T$, this yields $\sum_{t=N+1}^{T} \mathbb{E}\{|\nu - \bar{v}_t|\} \leq \frac{2}{T^3} + T^{\frac{2}{3}} 2\sqrt{\frac{2\log T}{c_N}}$. Using the fact that, for $t > N$, $I_t - \bar{v}_t \sim \mathcal{N}(0, \sigma^2)$ with $\sigma^2 = \frac{1}{n_t} \leq \frac{1}{N}$, we obtain that $\mathbb{E}\{|I_t - \bar{v}_t|\} \leq \sqrt{\frac{2}{\pi \cdot c_N}} T^{-\frac{1}{3}}$ and this yields $\sum_{t=N+1}^{T} \mathbb{E}\{|I_t - \bar{v}_t|\} \leq \sqrt{\frac{2}{\pi \cdot c_N}} T^{\frac{2}{3}}$.

Define $\mathcal{B} = \{t \in \mathcal{T} \mid t > N, I_t \geq p_t\}$. Let $E_t = \{I_t > \nu\}$, let $H_t = \{|\bar{v}_t - \nu| \leq \sqrt{\frac{2\log T}{n_t}}\}$ and let $H_t^C = \{|\bar{v}_t - \nu| > \sqrt{\frac{2\log T}{n_t}}\}$. Let $\hat{v}_s$ denote the sample mean of s i.i.d. draws from distribution $\mathcal{D}$ and let $\hat{I}_s \sim \mathcal{N}(\hat{v}_s, \frac{1}{s})$. For term B we have,

$$B \leq \sum_{t \in \mathcal{B}} \mathbb{E}\left\{(p_t - \nu) \cdot \mathbb{I}\{\nu < p_t \leq I_t\}\right\} \leq \sum_{t \in \mathcal{B}} \mathbb{E}\left\{(I_t - \nu) \cdot \mathbb{I}\{I_t > \nu\}\right\}$$

$$\leq \sum_{t \in \mathcal{B}} \mathbb{E}\left\{(I_t - \nu) \cdot \mathbb{I}\{E_t \cap H_t\}\right\} + \sum_{t \in \mathcal{B}} \mathbb{E}\left\{(I_t - \nu) \cdot \mathbb{I}\{E_t \cap H_t^C\}\right\}.$$

Define $B_1 = \sum_{t \in \mathcal{B}} \mathbb{E}\left\{(I_t - \nu) \cdot \mathbb{I}\{E_t \cap H_t\}\right\}$. We bound B_1 as follows:

$$B_1 \leq \sum_{t \in \mathcal{B}} \mathbb{E}\left\{|I_t - \nu| \cdot \mathbb{I}\{H_t\}\right\} \leq \sum_{t \in \mathcal{B}} \mathbb{E}\left\{|\nu - I_t| \,\Big|\, H_t\right\} \cdot \mathbb{P}\{H_t\}$$

$$\leq \sum_{t \in \mathcal{B}} \mathbb{E}\left\{|I_t - \bar{v}_t| \,\Big|\, H_t\right\} \cdot \mathbb{P}\{H_t\} + \sum_{t \in \mathcal{B}} \mathbb{E}\left\{|\bar{v}_t - \nu| \,\Big|\, H_t\right\} \cdot \mathbb{P}\{H_t\}.$$

Define $B_{11} = \sum_{t \in \mathcal{B}} \mathbb{E}\left\{|I_t - \bar{v}_t| \,\Big|\, H_t\right\} \cdot \mathbb{P}\{H_t\}$ and
$B_{12} = \sum_{t \in \mathcal{B}} \mathbb{E}\left\{|\bar{v}_t - \nu| \,\Big|\, H_t\right\} \cdot \mathbb{P}\{H_t\}$.

We bound B_{11} as follows:

$$B_{11} \leq \sum_{t \in \mathcal{B}} \mathbb{E}\left\{|I_t - \bar{v}_t| \,\Big|\, H_t\right\} \cdot \mathbb{P}\{H_t\} + \sum_{t \in \mathcal{B}} \mathbb{E}\left\{|I_t - \bar{v}_t| \,\Big|\, H_t^C\right\} \cdot \mathbb{P}\{H_t^C\}$$

$$= \sum_{t \in \mathcal{B}} \mathbb{E}\{|I_t - \bar{v}_t|\} = \sum_{t \in \mathcal{B}} \mathbb{E}\left\{|\hat{I}_{n_t} - \hat{v}_{n_t}|\right\} \leq \sum_{t \in \mathcal{T}} \mathbb{E}\left\{|\hat{I}_t - \hat{v}_t|\right\}$$

$$\leq \sum_{t \in \mathcal{T}} \sqrt{\frac{2}{\pi t}} \leq \int_0^T \sqrt{\frac{2}{\pi t}} dt = 2\sqrt{\frac{2}{\pi}T}.$$

We bound B_{12} as follows:

$$B_{12} \leq \sum_{t \in \mathcal{B}} \mathbb{E}\left\{|\bar{v}_t - \nu| \,\Big|\, H_t\right\} \cdot \mathbb{P}\{H_t\} \leq \sum_{t \in \mathcal{B}} \mathbb{E}\left\{|\bar{v}_t - \nu| \,\Big|\, H_t\right\}$$

$$\leq \sum_{t \in \mathcal{T}} \mathbb{E}\left\{|\hat{v}_t - \nu| \,\Big|\, |\hat{v}_t - \nu| \leq \sqrt{\frac{2\log T}{t}}\right\}$$

$$\leq \sum_{t \in \mathcal{T}} \sqrt{\frac{2\log T}{t}} \leq \int_0^T \sqrt{\frac{2\log T}{t}} dt \leq 2\sqrt{2\log T}\sqrt{T}.$$

Define $B_2 = \sum_{t \in \mathcal{B}} \mathbb{E}\left\{(I_t - \nu) \cdot \mathbb{I}\{E_t \cap H_t^C\}\right\}$. We bound B_2 as follows:

$$B_2 \leq \sum_{t \in \mathcal{B}} \mathbb{P}\{H_t^C\} \leq \sum_{t \in \mathcal{B}} \mathbb{P}\left\{|\hat{v}_{n_t} - \nu| > \sqrt{\frac{2\log T}{n_t}}\right\}$$

$$\leq \sum_{t \in \mathcal{T}} \mathbb{P}\left\{|\hat{v}_t - \nu| > \sqrt{\frac{2\log T}{t}}\right\} \leq T \cdot \frac{2}{T^4}.$$

Putting everything together we obtain $\mathcal{R}_T \leq N \cdot 1 + \frac{2}{T^3} + 2T^{\frac{2}{3}}\sqrt{\frac{2\log T}{c_N}} + \sqrt{\frac{2}{\pi \cdot c_N}}T^{\frac{2}{3}} + 2\sqrt{\frac{2T}{\pi}} + 2\sqrt{2T\log T} + T \cdot \frac{2}{T^4}$. So, we conclude that $\mathcal{R}_T \leq O(T^{\frac{2}{3}}\sqrt{\log T})$. $\qquad\square$

A.2   Proof of Proposition 4

Proof. We will decompose the regret in two parts: the regret incurred in rounds that are part of strategic cycles and rounds that are not. For an arbitrary subset $\mathcal{T}^* \subseteq \mathcal{T}$, let $\mathcal{R}_{T,\mathcal{T}^*} = \sum_{t \in \mathcal{T}^*} \mathbb{E}\left\{(\nu - p_t) \cdot \mathbb{I}\{\nu > p_t > I_t\}\right\} + \sum_{t \in \mathcal{T}^*} \mathbb{E}\left\{(p_t - \nu) \cdot \mathbb{I}\{\nu < p_t \leq I_t\}\right\}$. Let $\mathcal{T}_S \subseteq \mathcal{T}$ denote the indices of the rounds that are part of strategic cycles and let $\mathcal{T}_{NS} = \mathcal{T} \setminus \mathcal{T}_S$ denote the indices of the rounds that are not. Then we can write, $\mathcal{R}_T = \mathcal{R}_{T,\mathcal{T}_{NS}} + \mathcal{R}_{T,\mathcal{T}_S}$.

For $\mathcal{R}_{T,\mathcal{T}_S}$ we have that $\mathcal{R}_{T,\mathcal{T}_S} \leq N_2 + T \cdot p_{cycle} \cdot L$. This follows from the fact that the expected number of triggered strategic cycles (after round $N_1 + N_2$) is $T \cdot p_{cycle}$ and the regret in every such cycle is at most L. Furthermore, the first strategic cycle has length N_2. For $\mathcal{R}_{T,\mathcal{T}_{NS}}$ we have that $\mathcal{R}_{T,\mathcal{T}_{NS}} \leq O(T^{\frac{2}{3}}\sqrt{\log T})$. This follows from the fact that $\mathcal{R}_{T,\mathcal{T}_{NS}}$ represents the regret after $|\mathcal{T}_{NS}| \leq T$ rounds in a problem with horizon T, and by Proposition 2, this quantity is bounded by $O(T^{\frac{2}{3}}\sqrt{\log T})$. By plugging in the values we get $\mathcal{R}_T = \mathcal{R}_{T,\mathcal{T}_{NS}} + \mathcal{R}_{T,\mathcal{T}_S} \leq O(T^{\frac{2}{3}}\sqrt{\log T})$. $\qquad\square$

A.3   Proof of Proposition 6

In this section we give a proof of Proposition 6. We first give a definition of a pure weight-based algorithm.

Definition 7. *Let there be K actions in total and let $\mathcal{K} = \{1, \ldots, K\}$. Let $w_{k,t} \in \mathbb{R}$ denote the weight of action k at the beginning of round t. Suppose that action j is selected in round t and that the observed reward for action j in round t equals $r_{j,t}$. Let $\hat{p}_{k,t}$ denote the probability that action k is selected in round t. An algorithm $\mathcal{A}$ is called a pure weight-based algorithm if the following conditions are satisfied:*

1. *if $r_{j,t} > 0$, then $w_{j,t+1} > w_{j,t}$.*
2. *if $r_{j,t} = 0$, then $w_{j,t+1} = w_{j,t}$.*
3. *if $k \neq j$, then $w_{k,t+1} = w_{k,t}$.*
4. *$\sum_{k \in \mathcal{K}^*} \hat{p}_{k,t} = F(\sum_{k \in \mathcal{K}^*} w_{k,t}/\sum_{k=1}^{K} w_{k,t})$ for all subsets $\mathcal{K}^* \subseteq \mathcal{K}$, where $F(\cdot)$ is an increasing function. That is, for all subsets $\mathcal{K}^* \subseteq \mathcal{K}$, if $a = \sum_{k \in \mathcal{K}^*} w_{k,t}/\sum_{k=1}^{K} w_{k,t}$, $b = \sum_{k \in \mathcal{K}^*} w'_{k,t}/\sum_{k=1}^{K} w'_{k,t}$ and $a > b$, then $F(a) > F(b)$.*

Note that if $\hat{p}_{k,t} = w_{k,t}/\sum_{k=1}^{K} w_{k,t}$ then condition 4 in Definition 7 is satisfied. Also note that EXP3 of [4] uses $\hat{p}_{k,t} = (1 - \gamma)w_{k,t}/\sum_{k=1}^{K} w_{k,t} + \gamma/K$ and this choice also satisfies condition 4 in Definition 7.

Proof (of of Proposition 6). Let $|\mathcal{P}| = K$, $\mathcal{K} = \{1, \ldots, K\}$, $p_{max} = \max\{\mathcal{P}\}$ and $p_{min} = \min\{\mathcal{P}\}$. Assume, without loss of generality, that $\mathcal{P} = \{p^1, \ldots, p^K\}$ and that $0 < p_{min} = p^1 \le p^2 \le \cdots \le p^{K-1} \le p^K = p_{max}$. Let $\hat{\mathcal{P}} = \{p \in \mathcal{P} \mid p \le p_{target}\}$. Let $\bar{\mathcal{P}} = \{p \in \mathcal{P} \mid p > p_{target}\}$. Let $\hat{\mathcal{K}} = \{k \in \mathcal{K} \mid p^k \in \hat{\mathcal{P}}\}$. Let $\bar{\mathcal{K}} = \{k \in \mathcal{K} \mid p^k \in \bar{\mathcal{P}}\}$. Let $w_{k,t}$ denote the weight of action k at the beginning of round t.

We now proceed to prove the statement in the Proposition. We prove the Proposition for $L = 1$. The case for general L follows by repeatedly applying the result for $L = 1$.

We distinguish the following cases. Case 1: $p_{target} \ge p_{max}$. Case 2: $p_{target} < p_{min}$. Case 3: $p_{min} \le p_{target} < p_{max}$.

- Case 1: $p_{target} \ge p_{max}$. In this case, $p \le p_{target}$ for all $p \in \mathcal{P}$. Therefore, $\mathbb{P}\{p_{t+1} \le p_{target}\} = 1$ and $\mathbb{P}\{p_{t+2} \le p_{target}\} = 1$ and the statement in the Proposition holds.
- Case 2: $p_{target} < p_{min}$. In this case, $p > p_{target}$ for all $p \in \mathcal{P}$. Therefore, $\mathbb{P}\{p_{t+1} \le p_{target}\} = 0$ and $\mathbb{P}\{p_{t+2} \le p_{target}\} = 0$ and the statement in the Proposition holds.
- Case 3: $p_{min} \le p_{target} < p_{max}$. There are 2 subcases to consider. Case A: $p_{t+1} > p_{target}$ and Case B: $p_{t+1} \le p_{target}$.
  - In Case A, none of the weights get updated. This is true because none of the prices in $\hat{\mathcal{P}}$ are selected since $p_{t+1} > p_{target}$. By condition 3 in Definition 7, it follows that none of the weights corresponding to the prices in $\hat{\mathcal{P}}$ will get updated.
    Also, none of the prices in $\bar{\mathcal{P}}$ will get a positive reward because they will all be rejected by the buyer. By condition 2 in Definition 7, it follows that none of the weights corresponding to the prices in $\bar{\mathcal{P}}$ will get updated. As none of the weights will get updated after round $t + 1$ is completed, we have that $\mathbb{P}\{p_{t+2} \le p_{target}\} = \mathbb{P}\{p_{t+1} \le p_{target}\}$. So we conclude that the statement in the Proposition holds.
  - In Case B, there exists a $j \in \{1, \ldots, K\}$ such that $p_{t+1} = p^j$ and the reward for action j satisfies $r_{j,t+1} > 0$. This is true because the price $p_{t+1} = p^j$ will be accepted by the buyer and the reward equals the price p^j which (by assumption) satisfies $p^j \ge p_{min} > 0$.
    By condition 1 in Definition 7, it follows that $w_{j,t+2} > w_{j,t+1}$. By condition 3 in Definition 7, it follows that $w_{k,t+2} = w_{k,t+1}$ for all $k \ne j$, since these prices/actions are not selected in round $t + 1$.
    This yields the following:

$$\sum_{k \in \hat{\mathcal{K}}} w_{k,t+2} > \sum_{k \in \hat{\mathcal{K}}} w_{k,t+1} \tag{2}$$

$$\sum_{k \in \bar{\mathcal{K}}} w_{k,t+2} = \sum_{k \in \bar{\mathcal{K}}} w_{k,t+1} \tag{3}$$

$$\sum_{k=1}^{K} w_{k,t+2} > \sum_{k=1}^{K} w_{k,t+1} \tag{4}$$

Note that we also have:

$$\mathbb{P}\left\{p_{t+2} \le p_{target}\right\} = 1 - \mathbb{P}\left\{p_{t+2} > p_{target}\right\}, \tag{5}$$

$$\mathbb{P}\left\{p_{t+1} \le p_{target}\right\} = 1 - \mathbb{P}\left\{p_{t+1} > p_{target}\right\}. \tag{6}$$

By combining (3) and (4), and by condition 4 in Definition 7, we obtain that $\mathbb{P}\left\{p_{t+2} > p_{target}\right\} < \mathbb{P}\left\{p_{t+1} > p_{target}\right\}$. As a consequence, by using (5) and (6), it follows that $\mathbb{P}\left\{p_{t+2} \le p_{target}\right\} > \mathbb{P}\left\{p_{t+1} \le p_{target}\right\}$. So we conclude that the statement in the Proposition holds.

The case for general L follows from repeatedly applying the above argument. Note that the argument above works every initial weight vector. By repeatedly applying the above argument, one can show that $\mathbb{P}\left\{p_{t+1} \le p_{target}\right\} \le \mathbb{P}\left\{p_{t+2} \le p_{target}\right\} \le \cdots \le \mathbb{P}\left\{p_{t+L} \le p_{target}\right\} \le \mathbb{P}\left\{p_{t+L+1} \le p_{target}\right\}$. □

B   Additional experiments

This part contains additional results related to the experiments in the main text. We show results for non-strategic and strategic buyers against another (more powerful) seller algorithm. We assume the seller uses the EXP3.S algorithm from [4]. We will refer to this as EXP3.S Seller. This algorithm has sub-linear regret with respect to action sequences with at most S switches. EXP3.S Seller is tuned according to Corollary 8.2 in [4].

Figures 18, 19 and 20 display the results for non-strategic buyers and Figs. 21, 22 and 23 display the results for strategic buyers. In all figures, the lines indicate the mean and the shaded region indicates a 95% confidence interval. The results are qualitatively similar to those reported in the main text. The results indicate that the proposed algorithms for strategic and non-strategic buyers have sub-linear regret in all cases considered.

In scenario P1 utilities are about 2.0–2.5 times higher. In scenario P2 the differences are smaller, which is in line with expectations since the lowest price of the seller is very close to the unknown mean valuation. In general, the strategic buyers tend have higher utilities.

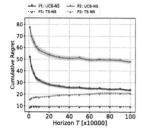

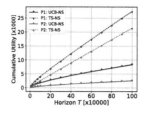

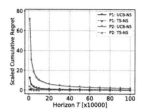

Fig. 18. R_T with EXP3.S seller.

Fig. 19. U_T with EXP3.S seller.

Fig. 20. R_T^S with EXP3.S seller.

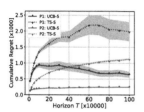

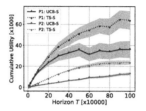

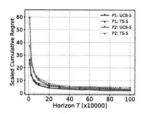

Fig. 21. R_T with EXP3.S seller.

Fig. 22. U_T with EXP3.S seller.

Fig. 23. R_T^S with EXP3.S seller.

References

1. Agrawal, S., Goyal, N.: Further optimal regret bounds for Thompson sampling. In: Proceedings of the Sixteenth International Conference on Artificial Intelligence and Statistics, vol. 31, pp. 99–107. PMLR (2013)
2. Amin, K., Rostamizadeh, A., Syed, U.: Learning prices for repeated auctions with strategic buyers. In: Proceedings of the 26th International Conference on Neural Information Processing Systems, pp. 1169–1177. Curran Associates Inc. (2013)
3. Amin, K., Rostamizadeh, A., Syed, U.: Repeated contextual auctions with strategic buyers. Adv. Neural Inf. Process. Syst. **27**, 622–630 (2014)
4. Auer, P., Cesa-Bianchi, N., Freund, Y., Schapire, R.: The nonstochastic multiarmed bandit problem. SIAM J. Comput. **32**(1), 48–77 (2002)
5. Auer, P., Cesa-Bianchi, N., Fischer, P.: Finite-time analysis of the multiarmed bandit problem. Mach. Learn. **47**(2), 235–256 (2002). https://doi.org/10.1023/A:1013689704352
6. Babaioff, M., Kleinberg, R.D., Slivkins, A.: Truthful mechanisms with implicit payment computation. In: Proceedings of the 11th ACM Conference on Electronic Commerce, pp. 43–52. Association for Computing Machinery (2010)
7. Babaioff, M., Sharma, Y., Slivkins, A.: Characterizing truthful multi-armed bandit mechanisms. SIAM J. Comput. **43**(1), 194–230 (2014)
8. Blum, A., Kumar, V., Rudra, A., Wu, F.: Online learning in online auctions. In: Proceedings of the Fourteenth Annual ACM-SIAM Symposium on Discrete Algorithms, pp. 202–204. SIAM (2003)
9. Braverman, M., Mao, J., Schneider, J., Weinberg, M.: Selling to a no-regret buyer. In: Proceedings of the 2018 ACM Conference on Economics and Computation, pp. 523–538. ACM (2018)

10. Bubeck, S., Cesa-Bianchi, N.: Regret analysis of stochastic and nonstochastic multi-armed bandit problems. Found. Trends® Mach. Learn. **5**(1), 1–122 (2012)
11. Deng, Y., Schneider, J., Sivan, B.: Prior-free dynamic auctions with low regret buyers. Adv. Neural Inf. Proces. Syst. **32**, 4803–4813 (2019)
12. Devanur, N.R., Kakade, S.M.: The price of truthfulness for pay-per-click auctions. In: Proceedings of the 10th ACM Conference on Electronic Commerce, pp. 99–106 (2009)
13. Ding, W., Qiny, T., Zhang, X.D., Liu, T.Y.: Multi-armed bandit with budget constraint and variable costs. In: Proceedings of the Twenty-Seventh AAAI Conference on Artificial Intelligence, pp. 232–238. AAAI Press (2013)
14. Drutsa, A.: Horizon-independent optimal pricing in repeated auctions with truthful and strategic buyers. In: Proceedings of the 26th International Conference on World Wide Web, pp. 33–42 (2017)
15. Drutsa, A.: Weakly consistent optimal pricing algorithms in repeated posted-price auctions with strategic buyer. In: Proceedings of the 35th International Conference on Machine Learning, vol. 80, pp. 1319–1328. PMLR, 10–15 July 2018
16. Edelman, B., Ostrovsky, M.: Strategic bidder behavior in sponsored search auctions. Decis. Support Syst. **43**(1), 192–198 (2007)
17. Feng, Z., Podimata, C., Syrgkanis, V.: Learning to bid without knowing your value. In: Proceedings of the 2018 ACM Conference on Economics and Computation, pp. 505–522 (2018)
18. Gatti, N., Lazaric, A., Trovò, F.: A truthful learning mechanism for contextual multi-slot sponsored search auctions with externalities. In: Proceedings of the 13th ACM Conference on Electronic Commerce, pp. 605–622. ACM (2012)
19. Immorlica, N., Lucier, B., Pountourakis, E., Taggart, S.: Repeated sales with multiple strategic buyers. In: Proceedings of the 2017 ACM Conference on Economics and Computation, pp. 167–168 (2017)
20. Kleinberg, R., Leighton, T.: The value of knowing a demand curve: Bounds on regret for online posted-price auctions. In: Proceedings of the 44th Annual IEEE Symposium on Foundations of Computer Science, pp. 594 (2003)
21. Mohri, M., Medina, A.M.N.: Optimal regret minimization in posted-price auctions with strategic buyers. In: Proceedings of the 27th International Conference on Neural Information Processing Systems, pp. 1871–1879 (2014)
22. Slivkins, A.: Introduction to multi-armed bandits. Found. Trends® Mach. Learn. **12**(1–2), 1–286 (2019)
23. Tran-Thanh, L., Chapman, A., Rogers, A., Jennings, N.R.: Knapsack based optimal policies for budget-limited multi-armed bandits. In: Proceedings of the Twenty-Sixth AAAI Conference on Artificial Intelligence, pp. 1134–1140. AAAI Press (2012)
24. Trovò, F., Paladino, S., Restelli, M., Gatti, N.: Budgeted multi-armed bandit in continuous action space. In: Proceedings of the Twenty-Second European Conference on Artificial Intelligence, pp. 560–568. IOS Press (2016)
25. Vanunts, A., Drutsa, A.: Optimal pricing in repeated posted-price auctions with different patience of the seller and the buyer. In: Advances in Neural Information Processing Systems, vol. 32 (2019)

Partial Label Learning

Partial Label Learning via Subspace Representation and Global Disambiguation

Yue Sun, Gengyu Lyu, and Songhe Feng[✉]

Beijing Key Laboratory of Traffic Data Analysis and Mining,
Beijing Jiaotong University, Beijing 100044, China
{yue.sun,lvgengyu,shfeng}@bjtu.edu.cn

Abstract. Partial Label Learning (PLL) learns from the training data where each example is associated with a set of candidate labels, among which only one is valid. Most existing methods deal with such problem by disambiguating the candidate labels first and then inducing the predictive model from the disambiguated data. However, these methods only focus on disambiguation for each single candidate label set, while the global label context tends to be ignored. Meanwhile, these methods induce the model by directly utilizing the original feature information, which may lead the model overfitting due to high-dimensional redundant feature. To tackle the above issues, we propose a novel feature *SubspacE Representation* and label *Global DisambiguatIOn* (**SERGIO**) PLL approach, which improves the generalization ability of learning system from the perspective of both feature space and label space. Specifically, we project the original high-dimensional feature space into a low-dimensional subspace, where the projection matrix is regularized with an orthogonality constraint to make the subspace more compact. Meanwhile, we introduce a label confidence matrix and constrain it with ℓ_1-norm regularization, where such constraint can be well in accordance with the nature of PLL problem and explore more global partial label correlations. Extensive experiments on various data sets demonstrate that our proposed method achieves competitive performance against state-of-the-art approaches.

Keywords: Feature subspace representation · Label global disambiguation · Partial-label learning

1 Introduction

As a weakly supervised machine learning framework, partial label learning learns from the ambiguous data with multiple candidate labels, among which only one label is the ground-truth. For example, in automatic face naming (Fig. 1), each instance is associated with two names, while the precisely matches between

---

Y. Sun and G. Lyu—Equal contribution.

© Springer Nature Switzerland AG 2021
F. Hutter et al. (Eds.): ECML PKDD 2020, LNAI 12458, pp. 439–454, 2021.
https://doi.org/10.1007/978-3-030-67661-2_26

names and faces are not available. Partial label learning can provide an effective solution to identify the precise instance-label correspondence and make predictions for unseen instances. Recently, such learning paradigm has been widely used in various real-world scenarios, including web mining [17], multimedia content analysis [13,29], facial age estimation [8], ecoinformatics [16,34].

Fig. 1. President Donald Trump meets French counterpart Emmanuel Macron

An intuitive strategy to accomplish the task of learning from partial label data is disambiguation. Existing disambiguation strategies are usually divided into two categories: averaging disambiguation strategy and identification disambiguation strategy. For averaging disambiguation strategy, each candidate label makes equal contribution to the learning model [9,30]. For identification disambiguation strategy, the ground-truth label is regarded as a latent variable and refined iteratively [8,28]. However, the above methods disambiguate the candidate label set for each instance independently, while the global label context is not taken into consideration. Meanwhile, these methods directly exploit the original high-dimensional feature space to induce the predictive model, where redundant features are inevitably mixed in high-dimensional data, which not only increases the time and space requirements but also affects the generalization ability of the learned model [15,23].

To tackle the above issues, in this paper, we propose a novel partial label learning framework named **SERGIO**, which integrates feature *SubspacE Representation* and label *Global DisambiguatIOn Strategy* into a unified learning framework. Specifically, we first project the original high-dimensional feature space into a low-dimensional yet discriminative subspace to alleviate the potential issues of redundant features, where the projection matrix is regularized with an orthogonality constraint to make the subspace more compact. Then, we utilize the global disambiguation strategy to explore the global label context for partial label data, where a latent label confidence matrix is introduced and constrained with ℓ_1-norm regularization to satisfy the global label sparsity nature of PLL problem. Afterwards, we impose the graph Laplacian regularization on both the low-dimensional data and the confidence label matrix to preserve the local manifold structure of data and strengthen local consistency over labels. In summary, SERGIO not only utilizes the subspace representation idea for original feature space to alleviate issues of redundant features mixed in high-dimensional features, but also simultaneously takes the sparse property of the ground-truth label matrix into consideration to utilize global label context. Extensive experiments on four artificial data sets and five real-world data sets validate that our approach achieves superior or comparable performance against state-of-the-art approaches.

2   Related Work

Partial label learning can be regarded as a weakly-supervised learning framework, which aims to learn a predictive model from training examples with a set of candidate labels [7,24–26,31,33]. Existing methods deal with such problem can be divided into three categories [18,19]: Disambiguation-Free Strategy (DFS), Averaging Disambiguation Strategy (ADS) and Identification Disambiguation Strategy (IDS).

2.1   Averaging Disambiguation Strategy (ADS)

ADS-based methods assume that each candidate label makes equal contribution to the learning model and they make predictions for unseen instances by averaging the model outputs. Following this strategy, [14] adopts an instance-based methods following $\arg\min_{y \in \mathcal{Y}} \sum_{i \in \mathcal{N}(\mathbf{x}^*)} \mathbb{I}(y \in S_i)$ to predict labels for unseen instances. [4] identifies the ground-truth label from candidate label sets based on minimization of a loss function appropriate for the partial label learning, i.e $\frac{1}{|S_i|} \sum_{y_i \in S_i} F(\mathbf{x}, \boldsymbol{\theta}, y)$. Besides, [30] also proposes an instance-based methods, which utilizes minimum error reconstruction criterion. Obviously, the above methods are easy to implement, but they suffer from a potential drawback that the ground-truth label might be overwhelmed by other false positive labels.

2.2   Identification Disambiguation Strategy (IDS)

IDS-based methods regard the ground-truth label as a latent variable, identified as $\arg\max_{y \in S_i} F(\mathbf{x}, \boldsymbol{\theta}, y)$, and then refine the model iteratively via utilizing some criterions, such as the maximum likelihood criterion: $\sum_{i=1}^{m} (log_{y \in S_i} F(\mathbf{x}, \boldsymbol{\theta}, y))$ [16], or maximum margin criterion: $\sum_{i=1}^{m} (\arg\max_{y \in S_i} F(\mathbf{x}, \boldsymbol{\theta}, y) - \arg\max_{y \notin S_i} F(\mathbf{x}, \boldsymbol{\theta}, y))$ [21,28]. These methods have superior performance against ADS methods, but one potential drawback lies in that the identified label may be false positive label instead of the ground-truth label.

2.3   Disambiguation-Free Strategy (DFS)

DFS-based methods usually fit PLL problems to existing learning techniques rather than disambiguate candidate labels. [32] relies on the Error-Correcting

Output Codes (ECOC) coding matrix [5] to transfer such problem into a binary learning problem. [27] adopts the one-vs-one decomposition strategy in a concise manner for partial label learning.

Note that, most of the existing strategies usually neglect redundant features in original high-dimensional feature space. In addition, most strategies identify the ground-truth label via disambiguating candidate labels respectively, whereas the global partial label correlations among the whole data set tend to be ignored.

3   The Proposed Method

Partial label learning aims to learn a multi-class classifier $\mathbf{f} : \mathcal{X} \mapsto \mathcal{Y}$ from the partial label training data set $\mathcal{D} = \{(\mathbf{x}_i, S_i) | 1 \leq i \leq n\}$, where $\mathbf{x}_i$ denotes the d-dimensional feature vector and S_i is candidate labels associated with $\mathbf{x}_i$. We define $\mathbf{X} = [\mathbf{x}_1, \mathbf{x}_2, \cdots, \mathbf{x}_n] \in \mathbb{R}^{d \times n}$ where n represents the number of training instances, $\mathbf{Y} \in \{0, 1\}^{m \times n}$ where m is the total number of labels and $\mathbf{P} = [\mathbf{p}_1, \mathbf{p}_2, \cdots, \mathbf{p}_n] \in \mathbb{R}^{m \times n}$ as label confidence matrix where p_{qi} indicates the confidence of q-th label to $\mathbf{x}_i$.

3.1   Formulation

As is described in Sect. 1, redundant features mixed in high-dimensional feature space tend to be neglected in most existing partial label learning problems, which inevitably reduces the generalization ability of partial label learning system. In order to alleviate the issue of redundant features and improve the generalization ability of learning system, we introduce feature subspace representation for partial label learning problems.

The low-dimensional feature subspace is expected to have three properties: content consistency, compact and discriminative. Specifically, we construct a projection matrix $\mathbf{Q} = [\mathbf{q}_1, \mathbf{q}_2, \cdots, \mathbf{q}_{d'}] \in \mathbb{R}^{d \times d'}$ to project the original local features $\mathbf{x}_j$ with dimension d to $\mathbf{Q}^\top \mathbf{x}_j$ in a lower-dimensional yet discriminative space $\mathbb{R}^{d'}$. Noting that after applying projection matrix $\mathbf{Q}$, the manifold structure of data may change. In order to capture the local manifold structure, SERGIO constructs projection matrix $\mathbf{Q}$ via graph Laplacian regularization. Here, the similarity matrix $\mathbf{S}$ is defined by the symmetry-favored k-NN graph, and $S_{ij} = \exp\left(\|\mathbf{x}_i - \mathbf{x}_j\|_2^2 / \sigma^2\right)$, if $j \in \mathcal{N}_k(i)$ or $i \in \mathcal{N}_k(j)$, otherwise $S_{ij} = 0$, where σ is empirically defined as the mean value of the Euclidean distance of instance matrix and $\mathcal{N}_k(i)$ is the k-nearest neighbors of $\mathbf{x}_i$. Then, the consistency of original local features and projection space can be enforced via minimizing $Tr\left(\mathbf{Q}^\top \mathbf{X} \mathbf{L} \mathbf{X}^\top \mathbf{Q}\right)$, where $\mathbf{L} = \mathbf{D} - \mathbf{S}$ and $\mathbf{D}$ is diagonal matrix define as $D_{ii} = \sum_{j \neq i} S_{ij}$. Besides, to make the high-dimensional features be more compact and avoid redundancy, we utilize orthogonality constraint $\mathbf{Q}^\top \mathbf{Q} = \mathbf{I}_{d'}$ [15]. Moreover, to fulfill the third property, we adopt the least square loss to learn the mapping relationship between the feature space and the label space, i.e., $\mathbf{W} \in \mathbb{R}^{d' \times m}$ and utilize the label confidence matrix $\mathbf{P}$ as learning target instead of original noisy label matrix $\mathbf{Y}$. Finally, we integrate the above items into a unified framework as:

$$\min_{\mathbf{P},\mathbf{Q},\mathbf{W}} \frac{1}{2}\|\mathbf{X}^\top \mathbf{Q}\mathbf{W} - \mathbf{P}^\top\|_F^2 + \lambda_1 Tr\left(\mathbf{Q}^\top \mathbf{X}\mathbf{L}\mathbf{X}^\top \mathbf{Q}\right)$$
$$s.t. \quad \mathbf{Q}^\top \mathbf{Q} = \mathbf{I}_{d'}, \tag{1}$$

where $\mathbf{I}_{d'}$ denotes the identity matrix.

As shown in the above formulation, the multi-class classifier is learned from the label confidence matrix $\mathbf{P}$ instead of the original label matrix $\mathbf{Y}$. And we adopt the global disambiguation strategy to obtain the label confidence matrix $\mathbf{P}$, which can exploit the global label context in partial label data set. The label confidence matrix $\mathbf{P}$ is expected to have following properties: (1) global sparsity, i.e. one instance only corresponds to a unique ground-truth label, and (2) local consistency, that is, similar instances ought to have similar label distributions [6]. To fulfill the first property, we impose ℓ_1-norm regularization on the label confidence matrix, which eliminates noise labels in candidate label sets and explores global partial label correlations among the whole data set. To guarantee the local consistency over labels, we also introduce graph Laplacian regularization on the label confidence matrix, i.e. $Tr\left(\mathbf{P}\mathbf{L}\mathbf{P}^\top\right)$. Besides, we introduce an inequality constraint on the label confidence matrix to guarantee that the confidence of each candidate label should be equal or greater than 0 but less than 1 and the confidence of each non-candidate label should strictly be 0.

$$\min_{\mathbf{P},\mathbf{Q},\mathbf{W}} \frac{1}{2}\|\mathbf{X}^\top \mathbf{Q}\mathbf{W} - \mathbf{P}^\top\|_F^2 + \lambda_2 Tr\left(\mathbf{P}\mathbf{L}\mathbf{P}^\top\right) + \beta\|\mathbf{P}\|_1$$
$$s.t. \quad \mathbf{0} \preceq \mathbf{P} \preceq \mathbf{Y}. \tag{2}$$

In addition, we adopt the widely used squared Frobenius norm of $\mathbf{W}$ to control the model complexity. And for the consideration of optimization efficiency, we utilize λ for regularization term of content consistency of data and local consistency over labels instead of λ_1 and λ_2. In the end, the final framework can be formulated as:

$$\min_{\mathbf{P},\mathbf{Q},\mathbf{W}} \frac{1}{2}\|\mathbf{X}^\top \mathbf{Q}\mathbf{W} - \mathbf{P}^\top\|_F^2 + \alpha\|\mathbf{W}\|_F^2 + \beta\|\mathbf{P}\|_1 + \lambda\left(Tr\left(\mathbf{P}\mathbf{L}\mathbf{P}^\top\right) + Tr\left(\mathbf{Q}^\top \mathbf{X}\mathbf{L}\mathbf{X}^\top \mathbf{Q}\right)\right)$$
$$s.t. \quad \mathbf{0} \preceq \mathbf{P} \preceq \mathbf{Y}; \mathbf{Q}^\top \mathbf{Q} = \mathbf{I}_{d'}. \tag{3}$$

In summary, SERGIO utilizes feature subspace representation and label global disambiguation strategy to learn the predictive model. For feature subspace representation, it learns a content consistency, compact and discriminative subspace from original high-dimensional features. For label global disambiguation strategy, it learns a global sparsity and local consistency label confidence matrix from original noisy label matrix. By integrating the above strategies, SERGIO is expected to learn a more effective and discriminative model.

3.2   Optimization

The target function is optimized via the alternating optimization algorithm, which optimizes one variable with other variables fixed. The formulation can be solved iteratively until convergence or reaching the maximum number of iterations.

Calculate W with P and Q Fixed. Based on the alternating optimization algorithm, **W** can be solved by minimizing the following objective function:

$$\min_{\mathbf{W}} \frac{1}{2}\|\mathbf{X}^\top\mathbf{Q}\mathbf{W} - \mathbf{P}^\top\|_F^2 + \alpha\|\mathbf{W}\|_F^2, \tag{4}$$

By setting the gradient in regard to **W** to 0, the closed-form solution can be described as:

$$\mathbf{W} = \left(\mathbf{Q}^\top\mathbf{X}\mathbf{X}^\top\mathbf{Q} + 2\alpha\right)^{-1}\mathbf{Q}^\top\mathbf{X}\mathbf{P}^\top. \tag{5}$$

Calculate Q with W and P Fixed. With other variables fixed, the sub-problem with respect to variable **Q** can be reformulated as following:

$$\min_{\mathbf{Q}} \frac{1}{2}\|\mathbf{X}^\top\mathbf{Q}\mathbf{W} - \mathbf{P}^\top\|_F^2 + \lambda Tr\left(\mathbf{Q}^\top\mathbf{X}\mathbf{L}\mathbf{X}^\top\mathbf{Q}\right)$$
$$s.t. \quad \mathbf{Q}^\top\mathbf{Q} = \mathbf{I}_{d'}. \tag{6}$$

Obviously, the parameter **Q** can be optimized via gradient descent method. By taking the derivative of objective function, the solution can be obtained as:

$$\frac{\partial\mathbf{f}}{\partial\mathbf{Q}} = \mathbf{X}\mathbf{X}^\top\mathbf{Q}\mathbf{W}\mathbf{W}^\top + 2\lambda\mathbf{X}\mathbf{L}\mathbf{X}^\top\mathbf{Q} - \mathbf{X}\mathbf{P}^\top\mathbf{W}^\top$$
$$\mathbf{Q} := \mathbf{Q} - \theta\frac{\partial\mathbf{f}}{\partial\mathbf{Q}}, \tag{7}$$

where we use Armijo rule [1] to determine the step size θ to reduce time cost. Meanwhile, the updated $\mathbf{Q}^*$ is unitized to satisfy the constraint of $\mathbf{Q}^\top\mathbf{Q} = \mathbf{I}_{d'}$.

Calculate P with W and Q Fixed. When fixing other variables that irrelevant to **P**, the objective function can be re-written as:

$$\min_{\mathbf{P}} g(\mathbf{P}) + \beta f(\mathbf{P})$$
$$s.t. \quad \mathbf{0} \preceq \mathbf{P} \preceq \mathbf{Y}, \tag{8}$$

where $g(\mathbf{P}) = \frac{1}{2}\|\mathbf{X}^\top\mathbf{Q}\mathbf{W} - \mathbf{P}^\top\|_F^2 + \lambda Tr(\mathbf{P}\mathbf{L}\mathbf{P}^\top)$ and $f(\mathbf{P}) = \|\mathbf{P}\|_1$. Since $g(\mathbf{P})$ and $f(\mathbf{P})$ are convex and $g(\mathbf{P})$ satisfies Lipschitz continuous, where $\|\nabla g(\mathbf{P}') - \nabla g(\mathbf{P})\|_2 \leq L_f\|\mathbf{P}' - \mathbf{P}\|_2$ [11,12], this function has a closed form solution by using proximal gradient descent algorithm. Specifically, **P** can be updated following $\mathbf{P} = S_{\beta/L_f}\left[\mathbf{P} - \frac{1}{L_f}\nabla g(\mathbf{P})\right]$, where

$$S_\varepsilon[G] = \begin{cases} \mathbf{G} - \varepsilon, & \mathbf{G} > \varepsilon \\ \mathbf{G} + \varepsilon, & \mathbf{G} < -\varepsilon. \\ 0 & , \mathbf{G} \leq |\varepsilon| \end{cases} \tag{9}$$

Algorithm 1. The Algorithm of **SERGIO**

Input:
$\mathcal{D}$: the partial label training data set $\mathcal{D} = \{(\mathbf{x}_i, S_i)|1 \leq i \leq n\}$;
T: the maximum number of boosting rounds;
d': the number of retained features;
$\mathbf{x}^*$: the unseen instance;
Output:
$\mathbf{y}^*$: the predicted label for $\mathbf{x}^*$;
Process:
 1: Initialize α, β, and λ.
 2: **while** $t < T$ **do**
 3:     Update $\mathbf{W}_t$ according to Eq.(5);
 4:     Update $\mathbf{Q}_t$ according to Eq.(7);
 5:     Update $\mathbf{P}_t$ according to Eq.(9);
 6:     **if** converge **then**
 7:         break;
 8:     **end if**
 9: **end while**
10: **return** $y^* = \arg\max \mathbf{W}^\top \mathbf{Q}^\top \mathbf{x}^*$.

3.3 Algorithm Theoretical Analysis

In this section, we theoretically analyze the consistency, convergence and computation complexity of SERGIO.

Consistency Analysis. A proof of Lipschitz continuous of $g(\mathbf{P})$ in Eq. (8) is given as follows:

$$\|\nabla g(\mathbf{P}') - \nabla g(\mathbf{P})\|_F = \|(\mathbf{I}_n + 2\lambda\mathbf{L})(\mathbf{P}' - \mathbf{P})\|_F \\ \leq L_f \|(\mathbf{P}' - \mathbf{P})\|_F, \tag{10}$$

where $\nabla g(\mathbf{P}) = \mathbf{P}(\mathbf{I}_n + 2\lambda\mathbf{L}) - \mathbf{W}^\top \mathbf{Q}\mathbf{X}$. Thereafter, let $\sigma_{max}(\cdot)$ represents the maximum singular value of a matrix. Consequently, the Lipschitz constant [35] is

$$L_f = \sigma_{max}(\mathbf{I}_n + 2\lambda\mathbf{L}). \tag{11}$$

Convergency Analysis. As is formulated in Eq. (3), it is easy to prove $\|\cdot\|_F^2$ and $\|\cdot\|_1$ are convex. Furthermore, the Laplacian matrix $\mathbf{L}$ with respect to instances is positive semidefinite, thus $\mathbf{X}\mathbf{L}\mathbf{X}^\top$ is also positive semidefinite. By considering 2nd-order conditions for convex functions, $Tr(\mathbf{P}\mathbf{L}\mathbf{P}^\top)$ and $Tr(\mathbf{Q}^\top \mathbf{X}\mathbf{L}\mathbf{X}^\top \mathbf{Q})$ are convex. Therefore, the linear combination of convex terms to construct our formulation is also convex. The values of the object function alone with iteration times on *Lost* data set is given in Fig. 4(d) to empirically validate the convergence property of SERGIO.

Computation Complexity. At each iteration, for computing $\mathbf{P}$, the main computational complexity comes from the calculation of eigenvalues which needs $O(n^3)$. And since $d' \ll d$, the computation complexity of updating $\mathbf{W}$ and $\mathbf{Q}$ is $O(dmn + d^2 n)$ which comes from derivation calculation. In summary, the whole computational cost of SERGIO is $O(T(n^3 + dmn + d^2 n))$, where T is the number of iterations.

4   Experiments

Table 1. Characteristics of the Controlled UCI Data Sets

Controlled UCI Data Sets			
Data set	EXP*	FEA*	CL*	Configurations
Abalone	4177	7	29	$r = 1, p = 1, \epsilon \in \{0.1, 0.2, \ldots, 0.7\}$
Dermatology	366	34	6	$r = 1, p \in \{0.1, 0.2, \ldots, 0.7\}$
CNAE-9	1080	856	9	$r = 2, p \in \{0.1, 0.2, \ldots, 0.7\}$ $r = 3, p \in \{0.1, 0.2, \ldots, 0.7\}$
Multiple Features	2000	649	10	

4.1   Experimental Setup

To comprehensively evaluate the performance of SERGIO, two series of comparative experiments are conducted on various data sets: 1) **Controlled UCI Data Sets**: The ambiguous labels are generated depending on the popular controlling protocol [3]. The four multi-class data sets generate 112 ($4 \times 4 \times 7$) synthetic partial-label data sets according to different configurations of three parameters (i.e. p, r, and ϵ), where p denotes the proportion of partial label instances, r denotes the number of false labels in the candidate label set (i.e. $|S_i| = r + 1$) and ϵ denotes the co-occurring probability between one false positive label and the ground-truth label. 2) **Real-World Data Sets**: Real-world Data Sets are collected from different task domains: (A) *Bird Song Classification* [BirdSong]; (B) *Automatic Face Naming* [FG-NET]; (C) *Automatic Face Naming* [Lost][Yahoo!News]; (D) *Web Image Classification* [Mirflickr]. Table 1 and Table 2 summarize the characteristics of the above data sets, including the number of examples (**EXP***), feature dimension (**FEA***), the total number of class labels (**CL***) and the average number of class labels (**AVG-CL***).

The performance of SERGIO is compared with six partial label learning algorithms, where the configured parameters are set as recommended by respective literature:

- **PL-KNN** [14]: an instance-based PLL approach, where the number of nearest neighbors is set to be $k = 10$;

Table 2. Characteristics of the Real-World Data Sets

Real-World Data Sets				
Data set	EXP*	FEA*	CL*	AVG-CL*	Task domain
BirdSong	4998	38	13	2.18	*Bird Song Classification* [2]
Lost	1122	108	16	2.33	*Automatic Face Naming* [4]
Yahoo!News	22991	163	219	1.91	*Automatic Face Naming* [10]
FG-NET	1002	262	78	7.48	*Facial Age Estimation* [22]
Mirflickr	2780	1536	14	2.76	*Web Image Classification* [13]

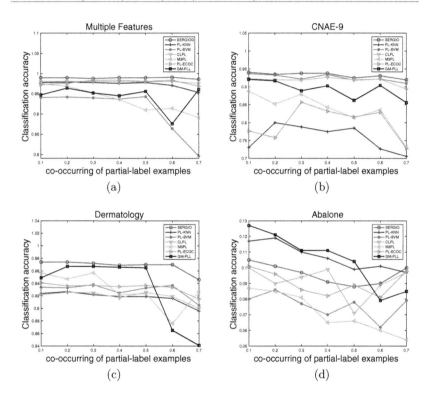

Fig. 2. Classification accuracy of each comparing algorithm changes as ϵ (co-occurring probability of one extra candidate label) increases from 0.1 to 0.7 (with 100% partial-label examples $[p=1]$ and one false positive candidate label $[r=1]$)

- **PL-SVM** [21]: an IDS approach based on maximum margin criterion, where the parameter λ is set among $\{10^{-3}, \cdots, 10^{3}\}$;
- **CLPL** [4]: a convex optimization method with hinge loss and SVM model;
- **M3PL** [28]: an IDS approach based on maximize the classification margin, where the parameter C_{max} is chosen from $\{10^{-2}, \cdots, 10^{2}\}$;
- **PL-ECOC** [32]: a transformation PLL approach via adapting error-correcting output codes, where the code length $L = \lceil log_2(q) \rceil$;

- **GM-PLL** [20]: an instance-to-label matching approach, where the threshold variable β is set among $\{0.3, 0.4, \cdots, 0.8\}$;

Parameters employed by SERGIO are set as $\alpha = 0.01$, $\beta = 0.1$, and $\lambda = 1$, respectively. Besides, the number of retained features d' is chosen among $\{0.1d, 0.2d, \cdots, d\}$ via cross-validation, where d is defined as the number of feature dimensions. Ten-fold cross-validation is performed on each data set and the detailed experimental results are presented in the next section.

4.2   Experimental Results

Table 3. Win/tie/loss counts of the SERGIO's classification performance against each comparing method on UCI data sets (pairwise t-test at 0.05 significance level).

Data set	Abalone	Dermatology	CNAE-9	Features	Sum
PL-KNN	22/0/6	28/0/0	28/0/0	28/0/0	106/0/6
PL-SVM	28/0/0	28/0/0	27/1/0	28/0/0	111/1/0
CLPL	10/6/12	28/0/0	28/0/0	28/0/0	94/6/12
M3PL	28/0/0	22/3/3	28/0/0	28/0/0	106/3/3
PL-ECOC	5/2/21	28/0/0	28/0/0	28/0/0	89/2/21
GM-PLL	23/0/5	28/0/0	28/0/0	28/0/0	107/0/5
Sum	116/8/44	162/3/3	167/1/0	168/0/0	613/12/47

Controlled UCI Data Sets. Figure 2 illustrates the classification accuracy of each comparing approach as the co-occurring probability ϵ increases from 0.1 to 0.7 with the step-size 0.1. Besides, Fig. 3 illustrates the classification accuracy of comparing approaches as the probability of partial label instances p varies from 0.1 to 0.7 with the step-size 0.1 ($r = 1, 2, 3$). Table 3 summarizes the win/tie/loss counts between SERGIO and other comparing approaches. Out of 112 ($4 \times 4 \times 7$) statistical comparisons show that SERGIO achieves competitive performance against state-of-the-art approaches, which is embodies in the following aspects:

- From the view of the employed data sets, SERGIO outperforms all comparing methods on *Features* data set. And for other controlled UCI data sets, it is also superior or comparable against all comparing state-of-the-art methods. Specifically, the average classification accuracy of SERGIO is 10.9% higher than PL-ECOC on *CNAE-9* data set and 8.6% higher than GM-PLL on *Abalone* data set. Meanwhile, for PL-KNN and PL-SVM, it also separately has 9.1% and 4.7% higher classification accuracy on *Dermatology* and *Features* data sets, respectively. The above observations strongly demonstrate the effectiveness of the proposed approach SERGIO.

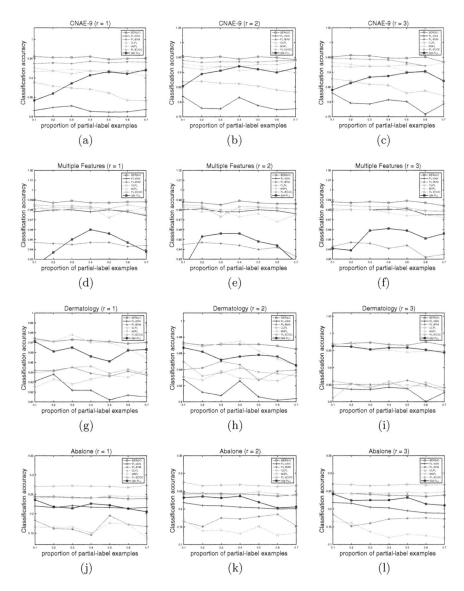

Fig. 3. The classification accuracy of several comparing methods on four controlled UCI data sets changes as p (proportion of partial-label examples) increases from 0.1 to 0.7 (with different numbers of false candidate labels [$r = 1$, $r = 2$, $r = 3$])

- From the view of the comparing approaches, SERGIO achieves superior or comparable performance against PL-SVM on all data sets. Besides, compared with PL-KNN, CLPL, M3PL, PL-ECOC and GM-PLL, SERGIO achieves superior or comparable performance in 94.6%, 89.3%, 97.3%, 85.8%, 95.5% cases, respectively. The above experiment results demonstrate that SERGIO has superior capacity to deal with partial label learning problems against other comparing approaches.

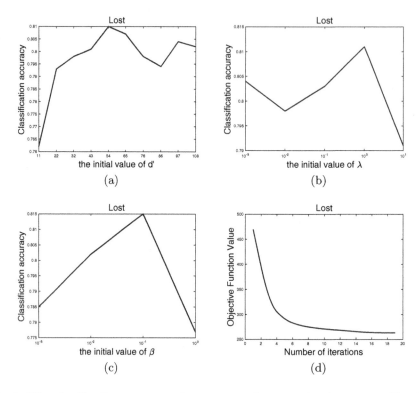

Fig. 4. The classification accuracy changes as each parameter increases with other parameters fixed.

Table 4. Classification accuracy of each algorithm on real-world data sets, ●/○ indicates that SERGIO is statistically superior/inferior to the algorithm (pairwise t-test at 0.05 significance level)

	BirdSong	Lost	Yahoo!News	FG-NET	Mirflickr
SERGIO	0.718 ± 0.029	$\mathbf{0.815 \pm 0.027}$	0.648 ± 0.013	$\mathbf{0.084 \pm 0.017}$	$\mathbf{0.594 \pm 0.036}$
PL-KNN	0.614 ± 0.024●	0.424 ± 0.030●	0.457 ± 0.009●	0.039 ± 0.008●	0.496 ± 0.013●
PL-SVM	0.662 ± 0.032●	0.729 ± 0.036●	0.636 ± 0.018●	0.063 ± 0.017●	0.515 ± 0.013●
CLPL	0.632 ± 0.017●	0.735 ± 0.024●	0.462 ± 0.009●	0.063 ± 0.017●	0.571 ± 0.032●
M3PL	0.709 ± 0.010	0.732 ± 0.035●	0.655 ± 0.010○	0.037 ± 0.025●	0.439 ± 0.033●
PL-ECOC	$\mathbf{0.740 \pm 0.016}$○	0.703 ± 0.052●	$\mathbf{0.662 \pm 0.010}$○	0.040 ± 0.018●	0.561 ± 0.013●
GM-PLL	0.663 ± 0.029●	0.737 ± 0.032●	0.629 ± 0.007●	0.065 ± 0.021●	0.426 ± 0.024●

Real-World Data Sets. Table 4 records the classification accuracy of SERGIO against six comparing algorithms on the real-world data sets, which is based on pairwise t-test at 0.05 significance level. It is impressive to observe that SERGIO achieves the best performance on most high-dimensional real-world data sets.

- From the view of the employed data sets, SERGIO performs good disambiguation ability on all real-world data sets. Specifically, the classification accuracies of SERGIO are totally higher than other comparing state-of-the-art approaches on *Lost*, *Mirflickr*, and *FG-NET* data sets. For *BirdSong* and *Yahoo!News* data sets, it also outperforms four comparing approaches.
- From the view of the comparing approaches, SERGIO achieves superior performance against PL-KNN, PL-SVM, CLPL and GM-PLL in all cases. Besides, compared with M3PL and PL-ECOC, SERGIO achieves superior or comparable performance in 80% and 60% cases, respectively. These experimental results demonstrate that SERGIO performs good disambiguation ability against state-of-the-art approaches.

Summary. The two series of experiments mentioned above powerfully demonstrate the effectiveness of SERGIO, and we attribute the success to the effectiveness of feature representation and the superiority of global disambiguation strategy, i.e., project original redundant feature space into a new discriminative subspace, and regularize the label confidence matrix with global sparse constraint.

4.3   Further Analysis

We study the sensitivity of SERGIO with respect to its three employed parameters d' (number of retained features), λ (local consistent coefficient), β (global sparse coefficient)). Figure 4 shows the performance of SERGIO under different parameter configurations on *Lost* data set. The top left picture presents the classification accuracy of SERGIO on Lost data set with varying number of retained features. Here, d' increased from $0.1d$ to d with an interval of $0.1d$. We can easily find that the value of d' usually has great influence on the performance of SERGIO. Faced with different data set, we set the parameter d' among $\{0.1d, 0.2d, \cdots, d\}$ via ten-fold cross validation. Besides, as shown in Fig. 4(b) and Fig. 4(c), other parameters often follow the optimal configurations ($\lambda = 1, \beta = 0.1$) but vary with minor adjustment on different data sets.

In Sect. 3, we theoretically demonstrate that the convergence of the proposed approach SERGIO. In this section, we empirically demonstrate the convergence of the proposed approach SERGIO. We conduct the convergence analysis of the proposed approach SERGIO on *Lost* data set. Specifically, in Fig. 4(d), we can easily observe that the objective function value of SERGIO changes significantly in initial iterations and becomes convergent when T reaches 10.

5   Conclusion

In this paper, we propose a novel partial label learning framework SERGIO, which integrates feature *Subspac**E** **R**epresentation* and label ***G**lobal Disambiguat**IO**n* into a unified learning framework. Specifically, the proposed framework not only exploits the subspace representation idea for original feature space to alleviate issues of redundant features in partial label learning data set, but also simultaneously takes the sparse property of the ground-truth label matrix into consideration to utilize the global label context. Extensive experiments have demonstrated that SERGIO achieves superior or comparable performance against state-of-the-art approaches.

Acknowledgments. This work was supported in part by the National Natural Science Foundation of China (No. 61872032), in part by the Beijing Natural Science Foundation (No. 4202058, No. 9192008), in part by the Key R&D Program of Zhejiang Province (No. 2019C01068), and in part by the Fundamental Research Funds for the Central universities (2020YJS026, 2019JBM020).

References

1. Bertsekas, D.P.: Nonlinear programming (1999)
2. Briggs, F., Fern, X.Z., Raich, R.: Rank-loss support instance machines for MIML instance annotation. In: ACM SIGKDD International Conference on Knowledge Discovery and Data Mining, pp. 534–542 (2012)
3. Chen, Y., Patel, V.M., Chellappa, R., Phillips, P.J.: Ambiguously labeled learning using dictionaries. IEEE Trans. Inf. Forensics Secur. **9**(12), 2076–2088 (2014)
4. Cour, T., Sapp, B., Taskar, B.: Learning from partial labels. J. Mach. Learn. Res. **12**(5), 1501–1536 (2011)
5. Dietterich, T.G., Bakiri, G.: Solving multiclass learning problems via error-correcting output codes. J. Artif. Intell. Res. **2**, 263–286 (1994)
6. Feng, L., An, B.: Leveraging latent label distributions for partial label learning. In: International Joint Conference on Artificial Intelligence, pp. 2107–2113 (2018)
7. Feng, L., An, B.: Partial label learning by semantic difference maximization. In: International Joint Conference on Artificial Intelligence, pp. 2294–2300 (2019)
8. Feng, L., An, B.: Partial label learning with self-guided retraining. In: Proceedings of the AAAI Conference on Artificial Intelligence, pp. 3542–3549 (2019)
9. Gong, C., Liu, T., Tang, Y., Yang, J., Yang, J., Tao, D.: A regularization approach for instance-based superset label learning. IEEE Trans. Cybern. **48**(3), 967–978 (2017)
10. Guillaumin, M., Verbeek, J., Schmid, C.: Multiple instance metric learning from automatically labeled bags of faces. In: Daniilidis, K., Maragos, P., Paragios, N. (eds.) ECCV 2010. LNCS, vol. 6311, pp. 634–647. Springer, Heidelberg (2010). https://doi.org/10.1007/978-3-642-15549-9_46
11. Huang, J., Li, G., Huang, Q., Wu, X.: Learning label-specific features and class-dependent labels for multi-label classification. IEEE Trans. Knowl. Data Eng. **28**, 3309–3323 (2016)
12. Huang, J., et al.: Improving multi-label classification with missing labels by learning label-specific features. Inf. Sci. **492**, 124–146 (2019)
13. Huiskes, M.J., Lew, M.S.: The MIR Flickr retrieval evaluation. In: ACM International Conference on Multimedia Information Retrieval, pp. 39–43 (2008)
14. Hüllermeier, E., Beringer, J.: Learning from ambiguously labeled examples. In: Famili, A.F., Kok, J.N., Peña, J.M., Siebes, A., Feelders, A. (eds.) IDA 2005. LNCS, vol. 3646, pp. 168–179. Springer, Heidelberg (2005). https://doi.org/10.1007/11552253_16

15. Li, Z., Liu, J., Tang, J., Lu, H.: Robust structured subspace learning for data representation. IEEE Trans. Pattern Anal. Mach. Intell. **37**(10), 2085–2098 (2015)
16. Liu, L., Dietterich, T.G.: A conditional multinomial mixture model for superset label learning. In: Advances in Neural Information Processing Systems, pp. 548–556 (2012)
17. Luo, J., Orabona, F.: Learning from candidate labeling sets. In: Advances in Neural Information Processing Systems, pp. 1504–1512 (2010)
18. Lyu, G., Feng, S., Li, Y., Jin, Y., Dai, G., Lang, C.: HERA: partial label learning by combining heterogeneous loss with sparse and low-rank regularization. ACM Trans. Intell. Syst. Technol. **11**, 1–19 (2020)
19. Lyu, G., Feng, S., Wang, T., Lang, C.: A self-paced regularization framework for partial-label learning. IEEE Trans. Cybern. (2020). https://doi.org/10.1109/TCYB.2020.2990908
20. Lyu, G., Feng, S., Wang, T., Lang, C., Li, Y.: GM-PLL: graph matching based partial label learning. IEEE Trans. Knowl. Data Eng. (2019). https://doi.org/10.1109/TKDE.2019.2933837
21. Nguyen, N., Caruana, R.: Classification with partial labels. In: ACM SIGKDD International Conference on Knowledge Discovery and Data Mining, pp. 551–559 (2008)
22. Panis, G., Lanitis, A.: An overview of research activities in facial age estimation using the FG-NET aging database. In: Agapito, L., Bronstein, M.M., Rother, C. (eds.) ECCV 2014. LNCS, vol. 8926, pp. 737–750. Springer, Cham (2015). https://doi.org/10.1007/978-3-319-16181-5_56
23. Tang, C., et al.: Feature selective projection with low-rank embedding and dual Laplacian regularization. IEEE Trans. Knowl. Data Eng. (2019). https://doi.org/10.1109/TKDE.2019.2911946
24. Wang, H., Liu, W., Zhao, Y., Hu, T., Chen, K., Chen, G.: Learning from multi-dimensional partial labels. In: International Joint Conference on Artificial Intelligence (2020)
25. Wang, H., Liu, W., Zhao, Y., Zhang, C., Hu, T., Chen, G.: Discriminative and correlative partial multi-label learning. In: International Joint Conference on Artificial Intelligence, pp. 3691–3697 (2019)
26. Wu, J.H., Zhang, M.L.: Disambiguation enabled linear discriminant analysis for partial label dimensionality reduction. In: ACM SIGKDD International Conference on Knowledge Discovery and Data Mining, pp. 416–424 (2019)
27. Wu, X., Zhang, M.L.: Towards enabling binary decomposition for partial label learning. In: International Joint Conference on Artificial Intelligence, pp. 2868–2874 (2018)
28. Yu, F., Zhang, M.L.: Maximum margin partial label learning. In: Asian Conference on Machine Learning, pp. 96–111 (2016)
29. Zeng, Z., et al.: Learning by associating ambiguously labeled images. In: Proceedings of the IEEE Conference on Computer Vision and Pattern Recognition, pp. 708–715 (2013)
30. Zhang, M.L., Yu, F.: Solving the partial label learning problem: an instance-based approach. In: International Joint Conference on Artificial Intelligence, pp. 4048–4054 (2015)
31. Zhang, M.L., Zhou, B.B., Liu, X.Y.: Partial label learning via feature-aware disambiguation. In: ACM SIGKDD International Conference on Knowledge Discovery and Data Mining, pp. 1335–1344 (2016)
32. Zhang, M.L., Yu, F., Tang, C.: Disambiguation-free partial label learning. IEEE Trans. Knowl. Data Eng. **29**(10), 2155–2167 (2017)

33. Zhou, Y., Gu, H.: Geometric mean metric learning for partial label data. Neuro-computing **275**, 394–402 (2018)
34. Zhou, Y., He, J., Gu, H.: Partial label learning via Gaussian processes. IEEE Trans. Cybern. **47**(12), 4443–4450 (2016)
35. Zhu, G., Yan, S., Ma, Y.: Image tag refinement towards low-rank, content-tag prior and error sparsity. In: ACM International Conference on Multimedia, pp. 461–470 (2010)

Online Partial Label Learning

Haobo Wang[1,2], Yuzhou Qiang[2], Chen Chen[1,2], Weiwei Liu[3], Tianlei Hu[1,2(✉)],
Zhao Li[4], and Gang Chen[1,2]

[1] Key Lab of Intelligent Computing Based Big Data of Zhejiang Province,
Zhejiang University, Hangzhou, China
{wanghaobo,cc33,htl,cg}@zju.edu.cn
[2] College of Computer Science and Technology,
Zhejiang University, Hangzhou, China
qiangyuzhou@zju.edu.cn
[3] School of Computer Science, Wuhan University, Wuhan, China
liuweiwei863@gmail.com
[4] Alibaba Group, Hangzhou, China
lizhao.lz@alibaba-inc.com

Abstract. A common assumption in online learning is that the data examples are precisely labeled. Unfortunately, it is intractable to obtain noise-free data in many real-world applications, and the datasets are usually adulterated with some irrelevant labels. To alleviate this problem, we propose a novel learning paradigm called Online Partial Label Learning (OPLL), where each data example is associated with multiple candidate labels. To learn from sequentially arrived data given partial knowledge of the correct answer, we propose three effective maximum margin-based algorithms. Theoretically, we derive the regret bounds for our proposed algorithms which guarantee their performance on unseen data. Extensive experiments on various synthetic UCI datasets and six real-world datasets validate the effectiveness of our proposed approaches.

Keywords: Online learning · Partial Label Learning · Regret bound

1 Introduction

In traditional online learning setting, a learner sequentially observes an instance and then makes a decision from a given set. Once the decision is made, the corresponding ground-truth label is revealed and the learner will suffer a loss that measures the discrepancy between the prediction and the correct answer. This learning paradigm has been widely used in various scenarios such as *video surveillance* [22], *online advertising* [23] and *cloud resource allocation* [34]. However, in many real-world applications, it is difficult to get precisely labeled datasets [30]. Instead, a set of candidate labels is accessible, i.e. the instances are *partially* labeled [3].

Formally, in Partial Label Learning (PLL), let $\mathcal{X} = R^p$ be the feature space and $\mathcal{Y} = \{y_1, y_2, ..., y_q\}$ be the label space composed of q class labels. During the

© Springer Nature Switzerland AG 2021
F. Hutter et al. (Eds.): ECML PKDD 2020, LNAI 12458, pp. 455–470, 2021.
https://doi.org/10.1007/978-3-030-67661-2_27

training phase, each instance vector $x \in \mathcal{X}$ has a corresponding candidate label set $S \subseteq \mathcal{Y}$. Here, the ground-truth label y is guaranteed to be contained in S, and remaining labels are termed as *distractor labels* or *false positive labels*. Due to the fact that the distractor labels significantly affect the prediction performance, to identify the correct label from candidate labels poses the biggest challenge in partial label learning tasks.

In the past decades, a number of partial label learning approaches have been proposed to deal with this issue. Many existing multi-class classifiers are adapted to PLL problems, e.g. k-nearest neighbors-based method [11] and ECOC-based algorithm [28]. Some methods try to transform the PLL task to a multi-output regression task [29,33] or a set of binary classification tasks [28]. Parametric modeling is another popular technique [3,14], where the likelihood of observing each example is defined over the candidate label set. Moreover, some approaches also adopt maximum margin criterion [19,31] or graph techniques [6,26] to identify the correct label from the candidate labels. With the aforementioned methods, PLL tasks can be effectively handled in offline and batch mode. However, all of them require to store all the training data in the memory and thus are powerless in the online PLL settings.

To bridge this gap, we propose a novel learning paradigm called Online Partial Label Learning (OPLL). OPLL is a new online setting that answers a sequence of questions given partial knowledge of the ground-truth. Specifically, on each round, the learner is given an instance vector and then makes a decision. After that, different from supervised online learning tasks, only a set of candidate labels is accessible and the ground-truth label is concealed. Based on the loss between the prediction and the correct answer, the learner will modify its prediction mechanism to boost the performance on subsequent rounds.

Utilizing off-the-shelf Partial Label Learning approaches seems an appealing strategy for OPLL. However, because these techniques are intractable in exploiting the global feature structure in the online setting, they can not disambiguate the candidate label set. To fill this gap, we involve Online Mirror Descent (OMD) framework [24] and Online Passive-Aggressive (OPA) frameworks [4] into OPLL. And then three effective online partial label learning algorithms are presented in this paper. To provide the theoretical guarantee, we also rigorously analyze the regret bounds for our proposed algorithms. Finally, empirical results show that the proposed algorithms can effectively handle the OPLL tasks.

The main contributions of this paper include:

- **Formulation:** A novel Online Partial Label Learning (OPLL) paradigm is proposed to make a sequence of decisions given partial knowledge (candidate labels) of the ground-truth label.
- **Solution:** Based on OMD and OPA frameworks, three effective online algorithms are proposed for OPLL problems. All of the resultant algorithms try to construct a maximum margin model, and thus the correct label can be identified from the ambiguous candidate set.
- **Theory:** We provide the regret bounds of our proposed methods to ensure their performance on unseen data.

– **Validation:** Extensive experiments demonstrate the effectiveness of our proposed methods.

2   Related Work

Our work marries the concepts of partial label learning and online learning, and therefore is mostly related to these two areas.

2.1   Partial Label Learning

Partial Label Learning aims to train a multi-class classifier from the partially labeled dataset where each instance is associated with a set of candidate labels instead of the ground-truth label. It is worth noting that PLL is different from existing weakly-supervised learning frameworks including *semi-supervised learning*, *multi-instance learning* and *multi-label learning* in the weak supervision nature. In semi-supervised learning [35], we induce a multi-class classifier from a few labeled data together with abundant unlabeled instances. Multi-instance learning [10] provides a bag of candidate instances only one of which is relevant to the label. Multi-label learning [15,16] is the most similar to PLL where each instance is related to a set of relevant labels.

To handle the partially labeled data, many approaches adapt specific multi-class classifiers [11,28] or other weakly-supervised learning algorithms [7,17] to the PLL setting. There are also several works [28,29,33] which try to transform the PLL task to other well-studied learning problems. Another common solution is to learn a parametric model from the ambiguous data. The most typical modeling methods [3,13] treat the candidate labels in an equal manner to get a maximum likelihood objective function. One main drawback of such averaging strategy is that the ground-truth label y_t may be overwhelmed by distractor labels. To tackle this problem, [14] presents a multi-nominal mixture PLL model and the maximum likelihood is obtained via EM procedure. Based on smoothness assumption, many state-of-the-art algorithms [6,18,26] also use graph-based techniques to disambiguate the candidate labels. Nevertheless, in the online setting, the graph structure of a dataset is unaccessible. Besides, some other approaches [19,31] learn a maximum margin model such that the ground-truth label can be easily identified from distractor labels.

2.2   Online Learning

From the online learning perspective, the goal is to learn a model from a set of sequentially given data and the core problem is to update the model dynamically. The most popular online strategies include perceptron based algorithms [4], standard online gradient algorithm [25], online convex algorithms [24] and so on. Recently, a series of online learning algorithms have been proposed to deal with specific machine learning problems, including classification [24], regression [27], multi-label learning [21] and so on. There are also some works that consider weakly-supervision in streaming data, e.g. online semi-supervised learning [9,12] and online one-class classification [8]. However, it is hard to apply these online methods to our online partial label learning scenario, because the correct label is concealed by the distractor labels.

3   Online Modeling

An Online Partial Label Learning task aims to learn a predictor $f : \mathcal{X} \mapsto \mathcal{Y}$ which maps from the feature space to the label space. Specifically, on the t-th round ($t = 1, 2, ...$), the learner receives an instance vector $\boldsymbol{x}_t \in \mathcal{X}$ and gives its decision $f(\boldsymbol{x}_t; \theta_t)$ according to the previously gained information where θ_t is the parameter learned on round $t-1$. After predicting an answer, the corresponding candidate label set S_t is revealed and the learner suffers a loss $l_t(\theta_t; \boldsymbol{x}_t, S_t)$. Then the learner modifies the learning mechanism and update its parameters to θ_{t+1}.

It is worth pointing out that the global feature structure is invisible in the online setting, and existing problem transformation approaches can not be adapted to PLL setting. As a result, the most direct way is online modeling. Inspired by [19], we adopt a maximum margin criterion to efficiently identify the ground-truth labels from the candidate label set. Furthermore, a linear multi-prototype predictor $f(\boldsymbol{x}_t; \boldsymbol{W}) = \boldsymbol{W}^\top \boldsymbol{x}_t$ is used to make the decision, where the weight matrix is defined by $\boldsymbol{W} = [\boldsymbol{w}_1, \boldsymbol{w}_2, ..., \boldsymbol{w}_q] \in R^{p \times q}$. Then on the t-th round, the learner suffers a loss as follows,

$$l_t(\boldsymbol{W}; \boldsymbol{x}_t, S_t) = \max\{0, 1 - \max_{j \in S_t}\langle \boldsymbol{w}_j, \boldsymbol{x}_t \rangle + \max_{k \in \bar{S}_t}\langle \boldsymbol{w}_k, \boldsymbol{x}_t \rangle\} \tag{1}$$

where $\langle \cdot \rangle$ denotes the inner product and $\bar{S}_t$ corresponds to the complementary set of S_t. Here we denote a label y_j ($1 \leq j \leq q$) by its index j for simplicity. Note that the optimal condition $l(\boldsymbol{x}_t, S_t) = 0$ means the following constraints should be satisfied,

$$\exists j \in S_t, \quad \forall k \in \bar{S}_t, \quad \max_{j \in S_t}\langle \boldsymbol{w}^j, \boldsymbol{x}_t \rangle + \max_{k \in \bar{S}_t}\langle \boldsymbol{w}^k, \boldsymbol{x}_t \rangle \geq 1$$

In practice, minimizing our loss with such constraints is demanding. To this end, we focus only on the single constraint which is violated the most by $\boldsymbol{W}_t$. Then, we modify the traditional partial label margin loss to a simpler formulation.

$$l_t(\boldsymbol{W}; \boldsymbol{x}_t, S_t) = \max\{0, 1 - \langle \boldsymbol{w}_{r_t}, \boldsymbol{x}_t \rangle + \langle \boldsymbol{w}_{v_t}, \boldsymbol{x}_t \rangle\} \tag{2}$$

where $r_t = \arg\max_{r \in S_t}\langle \boldsymbol{w}_r^t, \boldsymbol{x}_t \rangle$ and $v_t = \arg\max_{v \in \bar{S}_t}\langle \boldsymbol{w}_v^t, \boldsymbol{x}_t \rangle$ denote the highest ranked relevant label and irrelevant label on round t. Here $\boldsymbol{w}_i^t$ is the weight vector induced on the previous round for the i-th prototype.

Moreover, define $\boldsymbol{P}_t = [\boldsymbol{p}_1^t, \boldsymbol{p}_2^t, ..., \boldsymbol{p}_q^t] \in R^{p \times q}$ where $\boldsymbol{p}_{r_t}^t = \boldsymbol{x}_t$ and $\boldsymbol{p}_{v_t}^t = -\boldsymbol{x}_t$, otherwise $\boldsymbol{p}_i^t = \boldsymbol{0}$ ($i = 1, 2, ..., q, i \neq r_t, v_t$). Now we can rewrite Eq. (2) to $l_t(\boldsymbol{W}) = \max\{0, 1 - \langle \boldsymbol{P}_t, \boldsymbol{W} \rangle_F\}$, where $\langle \cdot \rangle_F$ is the Frobenius product. The final loss function is the point-wise maximum of a constant and an affine function and thus is convex, which is helpful for future analysis.

3.1   The PL-OMD Approaches

In this section, we introduce the first two online Partial Label methods based on Online Mirror Descent (PL-OMD).

Online Mirror Descent (OMD) [24] is a generalized online gradient descent framework and its gradient step is mediated by a strongly convex regularization function. By carefully choosing the regularizer, OMD can be applied in online partial label learning problems. In order to handle our multi-prototype predictor, we adapt OMD algorithm on matrix computations. Formally, on round t, PL-OMD updates the weight matrix as,

$$\begin{aligned}
\boldsymbol{Q}_{t+1} &= \boldsymbol{Q}_t - \boldsymbol{Z}_t \\
\boldsymbol{W}_{t+1} &= \arg\max_{\boldsymbol{W}}(\langle \boldsymbol{Q}_{t+1}, \boldsymbol{W} \rangle_F - \mathcal{R}(\boldsymbol{W}))
\end{aligned} \tag{3}$$

where $\mathcal{R}(\boldsymbol{W})$ is the regularizer. Here $\boldsymbol{Z}_t \in \partial l_t(\boldsymbol{W}_t)$ is the subgradient and can be defined as $\boldsymbol{Z}_t = -\boldsymbol{P}_t$ when $l_t(\boldsymbol{W}) > 0$, otherwise $\boldsymbol{Z}_t = \boldsymbol{0}_{p,q}$. We can see that $\boldsymbol{Q}_{t+1} \in R^{p \times q}$ is updated in a gradient descent style, and the actual prediction is mirrored to the primal space using Fenchel duality. Since there is no constraint on $\boldsymbol{W}$, we can actually rewrite the second equation in Eq. (3) to $\boldsymbol{W}_{t+1} = \nabla \mathcal{R}^*(\boldsymbol{Q}_{t+1})$, where $\mathcal{R}^*$ is the Fenchel conjugate of $\mathcal{R}$.

In this paper, we employ two regularizer functions in PL-OMD framework.

PL-OGD. We first consider the Frobenius norm regularization function $\mathcal{R}(\boldsymbol{W}) = \frac{1}{2\lambda}\|\boldsymbol{W}\|_F^2$. The derived algorithm updates the weight matrix as follows,

$$\begin{aligned}
\boldsymbol{W}_{t+1} &= \arg\max_{\boldsymbol{W}}(\langle \boldsymbol{Q}_{t+1}, \boldsymbol{W} \rangle_F - \frac{1}{2\lambda}\|\boldsymbol{W}\|_F^2) \\
&= \lambda \boldsymbol{Q}_{t+1} = \boldsymbol{W}_t - \lambda \boldsymbol{Z}_t
\end{aligned} \tag{4}$$

It is exactly *Online Gradient Descent algorithm for Partial Label learning (PL-OGD)*. And we generalized it from PL-OMD using Frobenius norm regularization.

PL-OEG. For the second algorithm, we introduce the Entropic regularizer,

$$\mathcal{R}(\boldsymbol{W}) = \frac{1}{\lambda} \sum_{i=1}^{p} \sum_{j=1}^{q} w_{ij} \log(w_{ij}) \tag{5}$$

where $w_{ij} > 0$ is the entry in the i-th row and j-th column of the matrix $\boldsymbol{W}$. log is the natural logarithm. We can also derive the update rule,

$$\boldsymbol{W}_{t+1} = \arg\max_{\boldsymbol{W}}(\langle \boldsymbol{Q}_{t+1}, \boldsymbol{W} \rangle_F - \frac{1}{\lambda} \sum_{i=1}^{p} \sum_{j=1}^{q} w_{ij} \log(w_{ij})) = [e^{\lambda q_{ij}^{t+1}-1}]_{p,q} \tag{6}$$

We call this algorithm as *Online Exponentiated Gradient for Partial Label learning (PL-OEG)*. We set $\boldsymbol{Q}_1 = \boldsymbol{0}_{p,q}$ for both PL-OGD and PL-OEG.

Regret Analysis. Now we provide the regret bounds for PL-OMD methods. Here, for any $U \in R^{p \times q}$ and the number of rounds T, regret is defined as,

$$\text{Regret}_T(U) = \sum_{t=1}^{T} (l_t(W_t) - l_t(U)) \tag{7}$$

Based on the Theorem 2.11 and 2.15 in [24], we can get the following theorems.

Theorem 1. *Let $l_1, l_2, ..., l_T$ be a sequence of convex functions such that l_t is L_t-Lipschitz with respect to some norm $||\cdot||$. Let L be such that $\frac{1}{T}\sum_{t=1}^{T} L_t^2 \leq L^2$. Assume PL-OMD is run on the sequence with a regularizer which is η-strongly-convex with respect to same norm. Then for all $U \in R^{p \times q}$, the algorithm enjoys a regret bound,*

$$\text{Regret}_T(U) \leq \mathcal{R}(U) - \min_{V \in R^{p \times q}} \mathcal{R}(V) + \frac{1}{\eta}TL^2 \tag{8}$$

where L_t and η are positive constants.

Here the strongly convexity is defined as,

Definition 1. *A function $l(W)$ is η-strongly-convex with respect to some norm $||\cdot||$ if for any $W \in R^{p \times q}$, $Z \in \partial l(W)$ and $U \in R^{p \times q}$, the following inequality holds,*

$$l(U) \geq l(W) + \langle Z, U - W \rangle_F + \frac{\eta}{2}||U - W||^2 \tag{9}$$

With the above definition, we can easily get that the Frobenius norm regularizer is $\frac{1}{\lambda}$-strongly-convex with respect to Frobenius norm. As for entropic regularizer, we assume $||W_t||_1 \leq B$ for all t. Here B is a positive constant and $||\cdot||_p$ denotes the entry-wise matrix p-norm. Then it is $\frac{1}{\lambda B}$-strongly-convex with respect to 1-norm. The proofs are simple and hence omitted.

Theorem 2. *Let $(x_1, S_1), (x_2, S_2), ...(x_T, S_T)$ be a sequence of partially labeled examples. Assume that $||x_t||_2 \leq L/2$ holds for all t and PL-OGD algorithm runs on the sequence. Then for all $U \in R^{p \times q}$, Eq. (8) holds. In particular, if $||U||_F^2 \leq B$ and $\lambda = \frac{B}{L\sqrt{2T}}$ then*

$$\text{Regret}_T(U) \leq BL\sqrt{2T} \tag{10}$$

where L and B are positive constants.

Proof. Take arbitrary pair of matrices $X, Y \in R^{p \times q}$ and assume without loss of generality $l_t(X) \geq l_t(Y)$. If $l_t(X) = 0$ then $\frac{|l_t(X) - l_t(Y)|}{||X - Y||_F} = 0 \leq ||P_t||_F = 2||x_t||_2 \leq L$. And when $l_t(X) > 0$, we can get that for all loss function l_t,

$$
\begin{aligned}
\frac{|l_t(X) - l_t(Y)|}{||X - Y||_F} &= \frac{1 - \langle P_t, X \rangle_F - \max\{0, 1 - \langle P_t, Y \rangle_F\}}{||X - Y||_F} \\
&\leq \frac{(1 - \langle P_t, X \rangle_F) - (1 - \langle P_t, Y \rangle_F)}{||X - Y||_F} \\
&\leq \frac{||P_t||_F ||X - Y||_F}{||X - Y||_F} \leq L
\end{aligned}
\tag{11}
$$

We conclude that l_t is Lipschitz continuous with $L_t \leq L$ such that $\frac{1}{T}\sum_{t=1}^{T} L_t^2 \leq L^2$. Since Frobenius norm regularizer is $\frac{1}{\lambda}$-strongly-convex with respect to Frobenius norm, Eq. (8) holds when PL-OGD runs. Recall that $\mathcal{R}(U) = \frac{1}{2\lambda}||U||_F^2 \leq \frac{1}{2\lambda}B$. Then, substituting $\lambda = \frac{B}{L\sqrt{2T}}$ into Eq. (8) and Theorem 2 is proved.

Theorem 3. *Let $(x_1, S_1), (x_2, S_2), ...(x_T, S_T)$ be a sequence of partially labeled examples. Assume that $||x_t||_\infty \leq L$ and $||W_t||_1 \leq B$ hold for all t and PL-OEG algorithm runs on the sequence. Then for all $U \in R_+^{p \times q}$ with constrains that $||U||_1 \leq B$, Eq. (8) holds. In particular, if $\lambda = \frac{\sqrt{\log(B/(pq))}}{L\sqrt{T}}$ then,*

$$\underset{T}{\text{Regret}}(U) \leq (\frac{2B\log(B/(pq)) + pqe^{-1}}{\sqrt{\log(B/(pq))}})L\sqrt{T} \qquad (12)$$

where L and B are positive constants. Moreover, B is large enough such that $B > pq$.

Proof. Take arbitrary pair of matrices $X, Y \in R^{p \times q}$ and assume without loss of generality $l_t(X) \geq l_t(Y)$. Using Holder's inequality, we can get that,

$$\frac{|l_t(X) - l_t(Y)|}{||X - Y||_1} \leq \frac{||P_t||_\infty ||X - Y||_1}{||X - Y||_1} \leq L \qquad (13)$$

We conclude that l_t is Lipschitz continuous with L with respect to 1-norm. It is not hard to prove that $\mathcal{R}(U) \leq \frac{1}{\lambda}B\log(B/(pq))$ and $\min_{V \in R^{p \times q}} \mathcal{R}(V) = -\frac{1}{\lambda}pqe^{-1}$. Then, substituting $\lambda = \frac{\sqrt{\log(B/(pq))}}{L\sqrt{T}}$ into Eq. (8) and Theorem 3 is proved.

With all these theorems, we obtain that the regret bounds of PL-OGD and PL-OEG grow sublinearly with T. In other words, after a certain number of iterations, the difference between the number of prediction mistakes suffered by our proposed methods and those suffered by choosing any fixed weight matrix U tends to be zero.

It is worth noting that when deriving the regret bounds, the prior assumptions on norm of instance vectors are different in PL-OGD (l_2-norm) and PL-OEG (l_1-norm), which may also lead to a significantly difference on Lipschitzness. Thus, the choice of the regularizer should depend on the feature of instances, i.e. which bound is more realistic and tighter.

3.2 The PL-OPA Approach

It is worth noting that the proposed loss follows a margin criterion. Therefore, we introduce another effective margin based framework to deal with online partial label tasks, which is called *Online Passive-Aggressive algorithm (PL-OPA)*. Formally, on round t, the weight matrix is updated to the solution of the following optimization problem,

$$\min_{W \in R^{p \times q}} \frac{1}{2}||W - W_t||_F^2 + C\xi_t^2$$
$$\text{s.t.} \quad 1 - \langle P_t, W_t \rangle_F \leq \xi_t, \quad \xi_t \geq 0 \qquad (14)$$

where ξ_t is the slack variable and C is a trade-off parameter. The word *passive-aggressive* comes from: when the constraint is satisfied, i.e. $l_t(\boldsymbol{W}_t) = 0$, we passively do nothing; otherwise, the algorithm aggressively searches the optimal solution with a penalty term forcing $\boldsymbol{W}_{t+1}$ to be close to $\boldsymbol{W}_t$ such that information obtained on previous rounds is preserved.

Now we concentrate on the optimal solution when $l_t(\boldsymbol{W}_t) > 0$. First, we define the Lagrangian of the optimization problem in Eq. (14) to be,

$$\mathcal{L}(\boldsymbol{W}, \xi_t, \lambda_t) = \frac{1}{2}\|\boldsymbol{W} - \boldsymbol{W}_t\|_F^2 + C\xi_t^2 + \lambda_t(1 - \langle \boldsymbol{P}_t, \boldsymbol{W} \rangle_F - \xi_t) \quad (15)$$

where $\lambda_t \geq 0$ is a Lagrange multiplier. Obviously the Slater's condition is satisfied and the solution has to satisfy the Karush-Khun-Tucker conditions [1]. And we get the following equations,

$$\frac{\partial \mathcal{L}(\boldsymbol{W}, \xi_t, \lambda_t)}{\partial \boldsymbol{W}} = \boldsymbol{W} - \boldsymbol{W}_t - \lambda_t \boldsymbol{P}_t = \boldsymbol{0}_{p,q}$$

$$\frac{\partial \mathcal{L}(\boldsymbol{W}, \xi_t, \lambda_t)}{\partial \xi_t} = 2C\xi_t - \lambda_t = 0 \quad (16)$$

$$\frac{\partial \mathcal{L}(\boldsymbol{W}, \xi_t, \lambda_t)}{\partial \lambda_t} = 1 - \langle \boldsymbol{P}_t, \boldsymbol{W} \rangle_F - \xi_t = 0$$

Combing all the equations we obtain the update rule,

$$\boldsymbol{W}_{t+1} = \boldsymbol{W}_t + \lambda_t \boldsymbol{P}_t, \quad \lambda_t = \frac{2Cl_t(\boldsymbol{W}_t)}{4C\|\boldsymbol{x}_t\|_2^2 + 1} \quad (17)$$

where $\boldsymbol{W}_1$ is set as a zero matrix.

Relative Loss Bound. For PL-OPA, we provide a *relative loss bound* instead of regret.

Theorem 4. *Let $(\boldsymbol{x}_1, S_1), (\boldsymbol{x}_2, S_2), ...(\boldsymbol{x}_T, S_T)$ be a sequence of partially labeled examples. Assume that $\|\boldsymbol{x}_t\|_2 \leq L$ holds for all t and PL-OPA algorithm runs on the sequence. Then for all $\boldsymbol{U} \in R^{p \times q}$, the cumulative squared loss enjoys the following bound,*

$$\sum_{t=1}^{T}(l_t)^2 \leq (2L^2 + \frac{1}{2C})(\|\boldsymbol{U}\|_F^2 + 2C\sum_{t=1}^{T}(l_t^u)^2) \quad (18)$$

where L is a positive constants. And for simplicity, we denote $l_t(\boldsymbol{W}_t), l_t(\boldsymbol{U}_t)$ by l_t, l_t^u.

Proof. Note that removing the data samples which satisfy the constraint ($l_t = 0$) does not effect our analysis. Thus we can assume that $l_t = 1 - \langle \boldsymbol{P}_t, \boldsymbol{W}_t \rangle_F > 0$

holds for all t. First, we define $\Phi_t = ||\boldsymbol{W}_t - \boldsymbol{U}||_F^2 - ||\boldsymbol{W}_{t+1} - \boldsymbol{U}||_F^2$. Using the fact that $l_t^u = \max\{0, 1 - \langle \boldsymbol{U}, \boldsymbol{P}_t \rangle_F\} \geq 1 - \langle \boldsymbol{U}, \boldsymbol{P}_t \rangle_F$, we can get,

$$
\begin{aligned}
\Phi_t &= ||\boldsymbol{W}_t - \boldsymbol{U}||_F^2 - ||\boldsymbol{W}_t + \lambda_t \boldsymbol{P}_t - \boldsymbol{U}||_F^2 \\
&= ||\boldsymbol{W}_t - \boldsymbol{U}||_F^2 - ||\boldsymbol{W}_t - \boldsymbol{U}||_F^2 - 2\lambda_t \langle \boldsymbol{P}_t, \boldsymbol{W}_t - \boldsymbol{U} \rangle_F - \lambda_t^2 ||\boldsymbol{P}_t||_F^2 \\
&= 2\lambda_t ((1 - \langle \boldsymbol{P}_t, \boldsymbol{W}_t \rangle_F) - (1 - \langle \boldsymbol{U}, \boldsymbol{P}_t \rangle_F)) - 2\lambda_t^2 ||\boldsymbol{x}_t||_2^2 \\
&\geq 2\lambda_t (l_t - l_t^u - \lambda_t ||\boldsymbol{x}_t||_2^2)
\end{aligned}
\tag{19}
$$

Summing Φ_t over all t yields,

$$
\begin{aligned}
\sum_{t=1}^{T} \Phi_t &= \sum_{t=1}^{T} ||\boldsymbol{W}_t - \boldsymbol{U}||_F^2 - ||\boldsymbol{W}_{t+1} - \boldsymbol{U}||_F^2 \\
&= ||\boldsymbol{W}_1 - \boldsymbol{U}||_F^2 - ||\boldsymbol{W}_{T+1} - \boldsymbol{U}||_F^2 \\
&\leq ||\boldsymbol{W}_1 - \boldsymbol{U}||_F^2 = ||\boldsymbol{U}||_F^2
\end{aligned}
\tag{20}
$$

Combing Eq. (19) and Eq. (20),

$$
\begin{aligned}
||\boldsymbol{U}||_F^2 &\geq 2 \sum_{t=1}^{T} \lambda_t (l_t - l_t^u - \lambda_t ||\boldsymbol{x}_t||_2^2) \\
&\geq \sum_{t=1}^{T} (2\lambda_t l_t - 2\lambda_t^2 ||\boldsymbol{x}_t||_2^2 - 2\lambda_t l_t^u - (\frac{\lambda_t}{\sqrt{2C}} - \sqrt{2C} l_t^u)^2)
\end{aligned}
\tag{21}
$$

Substituting the definition $\lambda_t = \frac{2Cl_t(\boldsymbol{W}_t)}{4C||\boldsymbol{x}_t||_2^2 + 1}$ into Eq. (21) yields that,

$$
||\boldsymbol{U}||_F^2 \geq \frac{2C}{4CL^2 + 1} \sum_{t=1}^{T} (l_t)^2 - 2C \sum_{t=1}^{T} (l_t^u)^2
\tag{22}
$$

Rearranging Eq. (22) implies the result.

Remark that it is possible there exists some $\boldsymbol{U}$ that $l_t(\boldsymbol{U}) = 0$ holds for all t and PL-OPA method enjoys a constant bound. Recall that the regret bounds of PL-OGD and PL-OEG grow sublinearly. In the realizable case, PL-OPA may get a tighter bound and the performance on unseen data is further guaranteed.

4   Experiments

In this paper, we conduct two series of empirical studies on controlled UCI datasets as well as real-world partial label datasets. All the experiments are run on a workstation with an i7-5930K CPU, a TITAN Xp GPU and 64GB main memory, running Linux platform.

Table 1. Characteristics of three UCI datasets and the configurations.

Datasets	#Examples	#Features	#Labels
Zoo	101	16	7
Yeast	1,484	8	10
Wine Quality	1,599	11	6

Configurations

(I) $p \in \{0.1, 0.2, ..., 0.7\}, r \in \{1, 2, 3\}, \epsilon = 0$
(II) $p = 1, r \in \{1, 2\}, \epsilon \in \{0.1, 0.2, ..., 0.7\}$

Table 2. Characteristics of six real-world partial label datasets.

Datasets	#Examples	#Features	#Labels	r	p	$\epsilon^\dagger$
BirdSong	4,998	38	13	1.18	0.67	0.19
FG-NET	1,002	262	78	6.48	1.00	0.16
Lost	1,122	108	16	1.23	0.94	0.16
MSRCv2	1,758	48	23	2.16	0.92	0.24
Soccer Player	17,472	279	171	1.09	0.70	0.03
Yahoo! News	22,991	163	219	0.91	0.71	0.24

†For the real-world datasets, we take the average number of false positive labels as r, and estimate ϵ, p from the datasets directly using their definitions.

4.1   Datasets

First, we generate some artificial partial label datasets from three multi-class UCI [5] datasets: **Zoo, Yeast** and **Wine Quality**. Following the widely-used controlling protocol [33], we involve three controlling parameters p, r and ϵ. p represents the proportion of examples containing false positive labels. $r = |S_i| - 1$ is the number of false positive labels for each instance. ϵ is the ambiguity degree [3] which controls the co-occurring probability between one coupling candidate label and the correct label. For each UCI dataset, we generate multiple PL variants with different parameter configurations. We consider two types of configurations and the details are shown in Table 1. Each combination of these hyperparameters in two configurations corresponds to a synthetic dataset. Hence, a total of 105 ($3 \times (21 + 14)$) datasets are generated.

In addition, a total of six real-world datasets have been collected from a variety of domains:

- For *automatic facial naming* tasks, **Yahoo!News** [10], **Lost** [3] and **Soccer Player** [32] crop faces from an image or video frame. Each image is annotated by names occuring in the caption or subtitles.
- For *bird song classification* problems, **BirdSong** [2] contains many 10-second bird singing records, each of which is tagged by possible bird breeds.

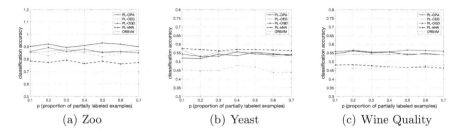

(a) Zoo   (b) Yeast   (c) Wine Quality

Fig. 1. Classification accuracy of each method changes as p (proportion of partially labeled data) increases from 0.1 to 0.7 (with one false positive label $[r = 1]$ and without a specific ϵ).

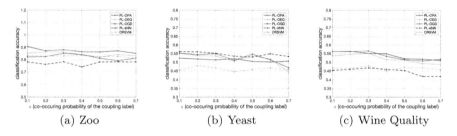

(a) Zoo   (b) Yeast   (c) Wine Quality

Fig. 2. Classification accuracy of each method changes as ϵ (co-occurring probability of the coupling label) increases from 0.1 to 0.7 (with one false positive label $[r = 1]$ and 100% partially labeled data $[p = 1]$).

- For *facial age estimation* tasks, **FG-NET** [20] collects face images with ages annotated by crowdsourced labelers and the ground-truth age as candidate labels.
- For *object classification*, **MSRCv2** [14] contains a set of images and each is associated with multiple objects appearing in the same image.

The detailed information of these datasets can be found in Table 1 and 2. For performance evaluation, we conduct five-fold cross-validation on these datasets and the mean classification accuracy with standard deviations are reported.

4.2   Benchmarks

It is noteworthy that since online partial label learning is firstly proposed, there is no publicly available algorithm. Moreover, as we have discussed in Sect. 1, existing offline partial label learning algorithms require full data structure to disambiguate the candidate labels, and thus can not be applied to OPLL tasks directly. To evaluate the effectiveness of the proposed methods, we compare them with two baselines:

- **ORSVM** [21]: Inspired by [13], we treat each example as having multiple correct labels and involve the first baseline ORSVM. It is one of the

Table 3. Classification accuracy (mean ± std deviation) of all the methods on six real-world datasets. The best ones are in bold.

Datasets	PL-OPA	PL-OEG	PL-OGD	PL-kNN	ORSVM
BirdSong	.631 ± .018	.611 ± .020	**.641** ± .011	.610 ± .014	.220 ± .0601
FG-NET	**.066** ± .022	.057 ± .021	.065 ± .019	.062 ± .034	.062 ± .012
Lost	**.738** ± .055	.708 ± .038	.683 ± .057	.449 ± .030	.387 ± .042
MSRCv2	**.427** ± .029	.334 ± .022	.364 ± .033	.415 ± .040	.159 ± .038
Soccer Player	.482 ± .010	**.499** ± .011	.492 ± .008	.496 ± .012	.453 ± .025
Yahoo! News	.555 ± .013	.551 ± .010	**.587** ± .010	.455 ± .014	.280 ± .053

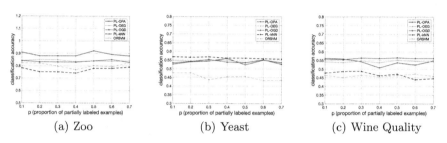

(a) Zoo   (b) Yeast   (c) Wine Quality

Fig. 3. Classification accuracy of each method changes as p (proportion of partially labeled data) increases from 0.1 to 0.7 (with one false positive label [$r = 2$] and without a specific ϵ).

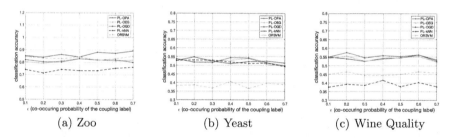

(a) Zoo   (b) Yeast   (c) Wine Quality

Fig. 4. Classification accuracy of each method changes as ϵ (co-occurring probability of the coupling label) increases from 0.1 to 0.7 (with one false positive label [$r = 2$] and 100% partially labeled data [$p = 1$]).

state-of-the-art Online Multi-Label methods which accelerates Rank-SVM method with non-smooth stochastic gradient descent algorithm.

– **PL-kNN** [11]: The second baseline uses k-nearest neighbor technique with Euclidean distance, which learns from ambiguous labels via an average voting strategy.

For PL-OGD[1] and PL-OEG, the default learning rate is set as 0.005 and for PL-OPA, the default trade-off parameter is set to be 1. Following the

---
[1] Our implementation is available at: https://github.com/HBzju/OPLL.

Table 4. Win/tie/loss counts (pair-wise t-test at 0.05 significance level) on the classification accuracy of our methods against other comparing algorithms.

Methods	Against				
	PL-OGD	PL-OEG	PL-OPA	PL-kNN	ORSVM
PL-OGD	–	7/92/6	24/78/3	39/60/6	68/37/0
PL-OEG	6/92/7	–	19/83/3	36/64/5	67/37/1
PL-OPA	3/78/24	3/83/19	–	52/36/17	65/40/0

experimental setting in [21], the parameters of ORSVM are fixed as $\mu = 0.05$, $\delta = 0.01$ and $\epsilon = 0.01$. Finally, we apply the suggested configuration [6,11,31] $k = 10$ for PL-kNN.

4.3 Results on Synthetic OPLL Datasets

Figure 1 illustrates the predictive performance of each algorithm as p increases from 0.1 to 0.7 with step-size 0.1. Figure 2 reports the similar results to Fig. 1 as ϵ increases from 0.1 to 0.7 with step-size 0.1. In these two figures, we fix $r = 1$, i.e., a candidate label set includes the correct label and exactly one extra randomly chosen noisy label. Figures 3 and 4 show the corresponding results when $r = 2$. In Table 4, we report the win/tie/loss counts between the comparing methods on 105 controlled UCI datasets. From the results, we have the following observations:

- From both sets of figures, we can see that the proposed methods outperform ORSVM and PL-kNN in most cases. In particular, the performance gains yield by our methods on the datasets of **Zoo** and **Wine Quality** are quite remarkable.
- Since the synthetic datasets are generated in a relatively naive way, all the methods achieve a stable disambiguation ability as p and ϵ increases. Nevertheless, based on the maximum margin criterion, our methods generally achieve better disambiguation ability compared to others no matter what p and ϵ is. This result is also consistent with existing works [28,29].
- From the 105 statistical tests, we can see that ORSVM and PL-kNN seldom outperform our methods in any controlled parameter configuration nor on any UCI dataset.
- In most cases, three presented methods are competitive with each other.

4.4 Results on Real-World OPLL Datasets

Table 3 lists the prediction accuracy of each method on six real-world datasets. From the results, we conclude that:

- ORSVM and PL-kNN generally underperform our methods, since they do not fully disambiguate the noisy label set.

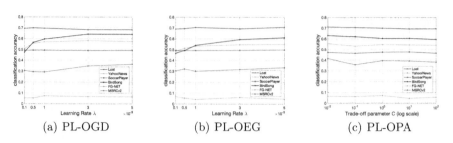

Fig. 5. Parameter sensitivities of the proposed methods as the hyperparameters change.

- The presented methods significantly outperform other baseline approaches. It demonstrates the effectiveness of our disambiguation strategy. With the maximum margin criterion, the ground truth label can be automatically identified from the distractor labels.
- In general, PL-OPA obtains better performance than other methods. For example, PL-OPA outperforms the best alternative comparison methods by 4.2% and 2.9% on **Lost** and **MSRCv2** respectively. This may because PL-OPA enjoys a tighter regret bound than PL-OGD and PL-OEG on some conditions.

Parameter Sensitivity. The parameter sensitivity is also investigated on six real-world datasets. For PL-OGD and PL-OEG, we tested different λ values from $\{\hat{\lambda} \times 10^{-3} | \hat{\lambda} \in \{0.1, 0.5, 1, 3, 5\}\}$. For PL-OPA, we tested different trade-off parameter C values from $\{0.01, 0.1, 1, 10, 100\}$. The results are reported in Fig. 5. The empirical results fluctuate lightly according to different orders of magnitude, but they are substantially robust within an acceptable range. In conclusion, the above results assure the quality and stability of our proposals.

5   Conclusion

In this work, a new learning paradigm Online Partial Label Learning (OPLL) is proposed to handle the imprecise tagging problem in online learning. The proposed paradigm enables each instance to have multiple candidate labels, where the ground-truth one is included. Then, we introduce two novel online learning frameworks to instantiate three discriminative OPLL approaches. Based on the maximum margin criterion, the proposed methods can naturally identify the ground-truth label from the distractor labels. Furthermore, we rigorously analyze the regret bounds of our proposed methods and provide the theoretical guarantee. Empirical studies on various synthetic UCI datasets and six real-world datasets demonstrate that our proposed algorithms can effectively handle the online partial label learning tasks.

Acknowledgments. This work is supported by Key R&D Program of Zhejiang Province (Grant No. 2020C01024) and National Natural Science Foundation of China (Grant No. 61976161).

References

1. Boyd, S., Vandenberghe, L.: Convex Optimization. Cambridge University Press, Cambridge (2004)
2. Briggs, F., Fern, X.Z., Raich, R.: Rank-loss support instance machines for MIML instance annotation. In: KDD, pp. 534–542 (2012). https://doi.org/10.1145/2339530.2339616
3. Cour, T., Sapp, B., Taskar, B.: Learning from partial labels. J. Mach. Learn. Res. **12**, 1501–1536 (2011)
4. Crammer, K., Dekel, O., Keshet, J., Shalev-Shwartz, S., Singer, Y.: Online passive-aggressive algorithms. J. Mach. Learn. Res. **7**, 551–585 (2006)
5. Dua, D., Graff, C.: UCI Machine Learning Repository (2017). http://archive.ics.uci.edu/ml
6. Feng, L., An, B.: Leveraging latent label distributions for partial label learning. In: Lang, J. (ed.) IJCAI, pp. 2107–2113 (2018). https://doi.org/10.24963/ijcai.2018/291. ijcai.org
7. Feng, L., An, B.: Partial label learning with self-guided retraining. In: AAAI, pp. 3542–3549. AAAI Press (2019). https://doi.org/10.1609/aaai.v33i01.33013542
8. Gautam, C., Tiwari, A., Leng, Q.: On the construction of extreme learning machine for online and offline one-class classification - an expanded toolbox. Neurocomputing **261**, 126–143 (2017). https://doi.org/10.1016/j.neucom.2016.04.070
9. Goldberg, A.B., Zhu, X., Furger, A., Xu, J.: OASIS: online active semi-supervised learning. In: AAAI (2011)
10. Guillaumin, M., Verbeek, J., Schmid, C.: Multiple instance metric learning from automatically labeled bags of faces. In: Daniilidis, K., Maragos, P., Paragios, N. (eds.) ECCV 2010. LNCS, vol. 6311, pp. 634–647. Springer, Heidelberg (2010). https://doi.org/10.1007/978-3-642-15549-9_46
11. Hüllermeier, E., Beringer, J.: Learning from ambiguously labeled examples. Intell. Data Anal. **10**(5), 419–439 (2006)
12. Imangaliyev, S., Keijser, B., Crielaard, W., Tsivtsivadze, E.: Online semi-supervised learning: algorithm and application in metagenomics. In: BIBM, pp. 521–525 (2013). https://doi.org/10.1109/BIBM.2013.6732550
13. Jin, R., Ghahramani, Z.: Learning with multiple labels. In: NIPS, pp. 897–904 (2002)
14. Liu, L., Dietterich, T.G.: A conditional multinomial mixture model for superset label learning. In: NIPS, pp. 557–565 (2012)
15. Liu, W., Tsang, I.W., Müller, K.: An easy-to-hard learning paradigm for multiple classes and multiple labels. J. Mach. Learn. Res. **18**, 94:1–94:38 (2017)
16. Liu, W., Xu, D., Tsang, I.W., Zhang, W.: Metric learning for multi-output tasks. IEEE Trans. Pattern Anal. Mach. Intell. **41**(2), 408–422 (2019). https://doi.org/10.1109/TPAMI.2018.2794976
17. Lyu, G., Feng, S., Lang, C.: A self-paced regularization framework for partial-label learning. CoRR abs/1804.07759 (2018)
18. Lyu, G., Feng, S., Wang, T., Lang, C., Li, Y.: GM-PLL: graph matching based partial label learning. CoRR abs/1901.03073 (2019)
19. Nguyen, N., Caruana, R.: Classification with partial labels. In: Proceedings of the 14th ACM SIGKDD, pp. 551–559 (2008). https://doi.org/10.1145/1401890.1401958

20. Panis, G., Lanitis, A.: An overview of research activities in facial age estimation using the FG-NET aging database. In: Agapito, L., Bronstein, M.M., Rother, C. (eds.) ECCV 2014. LNCS, vol. 8926, pp. 737–750. Springer, Cham (2015). https://doi.org/10.1007/978-3-319-16181-5_56
21. Park, S., Choi, S.: Online multi-label learning with accelerated nonsmooth stochastic gradient descent. In: ICASSP, pp. 3322–3326 (2013). https://doi.org/10.1109/ICASSP.2013.6638273
22. Popovici, R., Weiler, A., Grossniklaus, M.: On-line clustering for real-time topic detection in social media streaming data. In: Papadopoulos, S., Corney, D., Aiello, L.M. (eds.) WWW. CEUR Workshop Proceedings, vol. 1150, pp. 57–63. CEUR-WS.org (2014)
23. Riquelme, C., Johari, R., Zhang, B.: Only holding. In: AAAI, pp. 2506–2512 (2017)
24. Shalev-Shwartz, S.: Online learning and online convex optimization. Found. Trends Mach. Learn. **4**(2), 107–194 (2012). https://doi.org/10.1561/2200000018
25. Soudry, D., Castro, D.D., Gal, A., Kolodny, A., Kvatinsky, S.: Memristor-based multilayer neural networks with online gradient descent training. IEEE Trans. Neural Netw. Learn. Syst. **26**(10), 2408–2421 (2015). https://doi.org/10.1109/TNNLS.2014.2383395
26. Sun, K., Min, Z., Wang, J.: PP-PLL: probability propagation for partial label learning. In: Brefeld, U., Fromont, E., Hotho, A., Knobbe, A., Maathuis, M., Robardet, C. (eds.) ECML PKDD 2019. LNCS (LNAI), vol. 11907, pp. 123–137. Springer, Cham (2020). https://doi.org/10.1007/978-3-030-46147-8_8
27. Vévoda, P., Kondapaneni, I., Krivánek, J.: Bayesian online regression for adaptive direct illumination sampling. ACM Trans. Graph. **37**(4), 125:1–125:12 (2018). https://doi.org/10.1145/3197517.3201340
28. Wu, X., Zhang, M.: Towards enabling binary decomposition for partial label learning. In: IJCAI, pp. 2868–2874 (2018). https://doi.org/10.24963/ijcai.2018/398
29. Xu, N., Lv, J., Geng, X.: Partial label learning via label enhancement. In: AAAI, pp. 5557–5564. AAAI Press (2019). https://doi.org/10.1609/aaai.v33i01.33015557
30. Yao, Y., Deng, J., Chen, X., Gong, C., Wu, J., Yang, J.: Deep discriminative CNN with temporal ensembling for ambiguously-labeled image classification. In: AAAI, pp. 12669–12676. AAAI Press (2020)
31. Yu, F., Zhang, M.-L.: Maximum margin partial label learning. Mach. Learn. **106**(4), 573–593 (2016). https://doi.org/10.1007/s10994-016-5606-4
32. Zeng, Z., et al.: Learning by associating ambiguously labeled images. In: CVPR, pp. 708–715 (2013). https://doi.org/10.1109/CVPR.2013.97
33. Zhang, M., Zhou, B., Liu, X.: Partial label learning via feature-aware disambiguation. In: Proceedings of the 22nd ACM SIGKDD, pp. 1335–1344 (2016). https://doi.org/10.1145/2939672.2939788
34. Zhang, Z., Li, Z., Wu, C.: Optimal posted prices for online cloud resource allocation. In: Proceedings of the 2017 ACM SIGMETRICS, p. 60 (2017). https://doi.org/10.1145/3078505.3078529
35. Zhu, X., Goldberg, A.B.: Introduction to Semi-Supervised Learning. Synthesis Lectures on Artificial Intelligence and Machine Learning. Morgan & Claypool Publishers (2009). https://doi.org/10.2200/S00196ED1V01Y200906AIM006

Network Cooperation with Progressive Disambiguation for Partial Label Learning

Yao Yao[1], Chen Gong[1(✉)], Jiehui Deng[1], and Jian Yang[1,2(✉)]

[1] PCA Lab, The Key Laboratory of Intelligent Perception and Systems
for High-Dimensional Information of Ministry of Education,
School of Computer Science and Engineering,
Nanjing University of Science and Technology, Nanjing, China
{yaoyao,chen.gong,jhdeng,csjyang}@njust.edu.cn
[2] Jiangsu Key Lab of Image and Video Understanding for Social Security,
Nanjing, China

Abstract. Partial Label Learning (PLL) aims to train a classifier when each training instance is associated with a set of candidate labels, among which only one is correct but is not accessible during the training phase. The common strategy dealing with such ambiguous labeling information is to disambiguate the candidate label sets. Nonetheless, existing methods ignore the disambiguation difficulty of instances and adopt the single-trend training mechanism. The former would lead to the vulnerability of models to the false positive labels and the latter may arouse error accumulation problem. To remedy these two drawbacks, this paper proposes a novel approach termed "Network Cooperation with Progressive Disambiguation" (NCPD) for PLL. Specifically, we devise a progressive disambiguation strategy of which the disambiguation operations are performed on simple instances firstly and then gradually on more complicated ones. Therefore, the negative impacts brought by the false positive labels of complicated instances can be effectively mitigated as the disambiguation ability of the model has been strengthened via learning from the simple instances. Moreover, by employing artificial neural networks as the backbone, we utilize a network cooperation mechanism which trains two networks collaboratively by letting them interact with each other. As two networks have different disambiguation ability, such interaction is beneficial for both networks to reduce their respective disambiguation errors, and thus is much better than the existing algorithms with single-trend training process. Extensive experimental results on various benchmark and practical datasets demonstrate the superiority of our NCPD approach to other state-of-the-art PLL methods.

Keywords: Weakly-supervised learning · Partial label learning · Progressive disambiguation · Network cooperation

1 Introduction

Partial Label Learning (PLL), which is also known as *superset label learning* [9,18,19] and *ambiguous label learning* [15], is one of the emerging research fields

© Springer Nature Switzerland AG 2021
F. Hutter et al. (Eds.): ECML PKDD 2020, LNAI 12458, pp. 471–488, 2021.
https://doi.org/10.1007/978-3-030-67661-2_28

Annotator 1: Tigger
Annotator 2: Cat
Annotator 3: Leopard

(a)

Caption: The two best players in the post-Michael Jordan NBA world are Kobe Bryant and Lebron James. These two played spent 13 overlapping years (2003-2016) in the league.

(b)

Fig. 1. Two example applications of PLL. (a) In crowdsourcing, some annotators may mistakenly label the picture of a cat with "Tigger" or "Leopard" due to their limited cognitive ability. In this case, the query image contains three labels but only one of them is correct. (b) A newsletter contains an image and the corresponding text caption, from which we can roughly know that Michael Jordan, Kobe Bryant, and Lebron James may in the image. However, we can not figure out the concrete correspondence between the faces and the names.

in weakly-supervised learning [5,10,26,35]. PLL learns from ambiguous labeling information where each training instance is associated with multiple candidate labels and only one of them is valid. Due to the prevalence of ambiguous labeling in real-world scenarios, PLL has many practical applications such as crowdsourcing [9], image classification [4,6,21,29], web mining [22], etc. (see Fig. 1).

Formally, let $\mathcal{X} \in \mathbb{R}^d$ denote the d-dimensional input space and $\mathcal{Y} = \{1, 2, \cdots, c\}$ denote the label space with c class labels. The task of PLL is to induce a classifier $f : \mathcal{X} \to \mathcal{Y}$ from the partial label training set $\mathcal{D} = \{(\mathbf{x}_i, \mathcal{S}_i) | 1 \leq i \leq N\}$, where $\mathbf{x}_i \in \mathcal{X}$ is a d-dimensional feature vector and $\mathcal{S}_i \subseteq \mathcal{Y}$ is the corresponding candidate label set of $\mathbf{x}_i$. Particularly, the basic assumption under PLL framework is that the latent groundtruth label y_i of $\mathbf{x}_i$ lies in $\mathcal{S}_i$, i.e., $\mathrm{y}_i \in \mathcal{S}_i$, whereas it is not directly accessible during the training phase.

To learn from such partially labeled instances with ambiguously supervised information, the common strategy is to disambiguate the set of candidate labels of each training instance, namely to detect the unique correct label among multiple candidate labels. There are mainly two classes of methods for such disambiguation operation, namely average-based methods and identification-based methods. Average-based methods treat all candidate labels equally by assuming that they contribute equally to the trained classifier and the prediction is made by averaging their model outputs [15,36]. These methods share a common deficiency that the effectiveness of the model is greatly affected by the false positive labels in the candidate label sets, which leads to the suppression of groundtruth label by these false positive labels. Identification-based methods address this shortcoming via considering groundtruth label as a latent variable and gradually identifying it by iterative procedures such as Expectation Maximization (EM) [16,24,31]. One potential drawback of identification-based methods is that rather than recovering the latent groundtruth labels, the identified labels might turn out to be false positive and they can hardly be rectified in the subsequent iterations.

In a word, existing methods are vulnerable to false positive labels in the candidate label sets. There are two critical reasons that account for this. Firstly, existing approaches scarcely take the disambiguation difficulty of instances into account, and the disambiguation operations are performed on every training instance all at once. In this case, when the instance is complicated and difficult to classify, their models are likely to mistakenly regard the false positive label as the latent groundtruth label, which will mislead the training process and ultimately impair the disambiguation ability of the models. Secondly, the training process of existing methods are all single-trend, which indicates that the data disambiguated at the current step will be directly transferred back to the model itself in the following steps. Under this circumstance, once the identified labels turn out to be false positive, they would be difficult to correct in the succeeding iterations and thereby raising the error accumulation problem, which will severely degrade their performances.

To address these two shortcomings, this paper proposes a novel approach which employs a progressive disambiguation strategy combined with a network cooperation mechanism for PLL, which is termed "Network Cooperation with Progressive Disambiguation" ("NCPD" for short). Specifically, to address the problem of ignoring the disambiguation difficulty of instances, we devise a progressive disambiguation strategy which disambiguates simple instances firstly and then gradually disambiguates more complicated ones. Through learning from the simple instances, the disambiguation ability of the model can be improved steadily. With the proceeding of training process, the model is capable of disambiguating the complicated instances precisely. As a consequence, the negative impacts brought by the false positive labels, especially those of complicated instances, can be effectively mitigated. To settle the error accumulation problem caused by the single-trend training mechanism of traditional methods, we employ Artificial Neural Networks (ANNs) [14] as the backbone and utilize a network cooperation mechanism which trains two networks collaboratively by letting them interact with each other. That is to say, two networks disambiguate the training instances independently in the forward propagation phase and then back propagate the data disambiguated by its peer network. As two networks have different ability and can disambiguate training instances at different levels, such interaction is beneficial for both networks to learn from each other and thus their respective disambiguation errors can be reduced. As a result, the error accumulation problem can be significantly alleviated, and that is why we adopt such network cooperation mechanism rather than the existing single-trend training process. Intensive experiments on multiple datasets substantiate the superiority of our proposed NCPD approach to the state-of-the-art methodologies.

The rest of this paper is organized as follows. We review the related works in Sect. 2, and introduce the proposed NCPD approach in Sect. 3. Section 4 reports the experimental results, followed by the conclusion in Sect. 5.

2 Related Work

Existing algorithms dealing with partially labeled instances can be roughly grouped into the following two classes, *i.e.*, average-based methods and identification-based methods.

The average-based methods treat all candidate labels equally and the prediction is made by averaging their model outputs. For example, the work [15] straightforwardly generalizes the k-nearest neighbor classifier to resolve the PLL problem by predicting the label of a test instance $\mathbf{x}$ via the voting strategy among the candidate labels of its neighbors. That is to say, $f(\mathbf{x}) = \text{argmax}_{y \in \mathcal{Y}} \sum_{i \in \mathcal{N}_{(\mathbf{x})}} \mathbb{I}(y \in \mathcal{S}_i)$, where $\mathcal{N}(\mathbf{x})$ denotes the neighbors of the test instance $\mathbf{x}$ and $\mathbb{I}(\cdot)$ is the indicator function. Zhang *et al.* [36] also propose a model of which the predictions of unseen instances are made by the weighted averaging over the candidate labels of their neighbors. Cour *et al.* [6] propose a convex learning method and decide the groundtruth label by averaging the outputs from all candidate labels, *i.e.*, $\frac{1}{|\mathcal{S}_i|} \sum_{y \in \mathcal{S}_i} F(\mathbf{x}, y; \Theta)$ with Θ being the model parameters. Average-based methods are intuitive and are easy to implement. However, these methods share a critical shortcoming that the outputs from false positive labels may overwhelm the groundtruth labels' outputs, which will severely degrade their performances.

The identification-based methods regard the unique groundtruth label as a latent variable and identify it as $\text{argmax}_{y \in \mathcal{S}_i} F(\mathbf{x}, y; \Theta)$. Maximum likelihood criterion and maximum margin criterion are the two most widely-used learning strategies to identify groundtruth labels. Based on EM procedure, the methods [16,19] train their models by optimizing the maximum likelihood function $\sum_{i=1}^{n} \log(\sum_{y \in \mathcal{S}_i} F(\mathbf{x}, y; \Theta))$. The work [24] maximizes the margin between outputs from candidate labels and that from non-candidate labels to refine groundtruth labels, and the corresponding objective function is $\sum_{i=1}^{n} (\max_{y \in \mathcal{S}_i} F(\mathbf{x}, y; \Theta) - \max_{y \notin \mathcal{S}_i} F(\mathbf{x}, y; \Theta))$. Nonetheless, the above margin ignores the predictive difference between the latent groundtruth label and other candidate labels. To address this problem, Yu *et al.* [31] maximize the margin between the groundtruth label and other labels, *i.e.*, $\sum_{i=1}^{n} (F(\mathbf{x}_i, y_i; \Theta) - \max_{y \neq y_i} F(\mathbf{x}_i, y; \Theta))$ where y_i denotes the groundtruth label of $\mathbf{x}_i$. Moreover, by applying the idea of self-paced learning, Lyu *et al.* [23] propose a novel algorithm which utilizes the maximum margin criterion to detect the groundtruth label. Differently, Feng *et al.* [8] balance the minimum approximation loss and the maximum infinity norm of the outputs to differentiate the unique groundtruth label from false positive labels. Chen *et al.* [4] eliminate a proportion of the least likely candidates in each iteration to enhance the discriminability of their proposed approach. One potential shortcoming of identification-based methods is

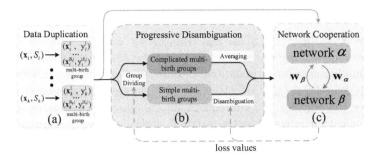

Fig. 2. The framework of our method. (a) indicates the data duplication scheme which transforms each partially labeled instance into a multi-birth group. After that, we feed the transformed data into the networks and thus their corresponding loss values can be obtained (the blue line). (b) presents the process of dividing multi-birth groups into two levels of difficulty and then calculating the confidence scores of instances among them according to the incurred loss values. (c) denotes the network cooperation mechanism where two networks interact with each other via exchanging their respective confidence scores of instances (*i.e.*, $\mathbf{w}_\alpha$ and $\mathbf{w}_\beta$) for back propagation. (Color figure online)

that the identified label in the current iteration may turn out to be false positive and they can hardly be rectified in the subsequent iterations.

In a word, although the aforementioned methods have achieved good performances to some degree, they still suffer from two severe drawbacks, *i.e.*, ignoring the disambiguation difficulty of instances and adopting the unreliable single-trend training process, and both of them will degrade their performances as mentioned in the introduction. Therefore, this paper presents a novel algorithm termed NCPD which will be introduced in the next section.

3 The Proposed NCPD Approach

In this section, we introduce the NCPD approach of which the architecture is illustrated in Fig. 2. To facilitate the disambiguation process, we firstly employ a data duplication scheme which transforms each partially labeled instance into a multi-birth group[1] (Fig. 2 (a)). Afterwards, by dividing these multi-birth groups into two levels of difficulty (*i.e.*, "simple" and "complicated"), we can calculate the confidence scores of instances among them via averaging or disambiguation (Fig. 2 (b)). Finally, two networks collaborate with each other through exchanging the confidence scores of instances generated by them independently to compute their respective back propagated loss (Fig. 2 (c)). We will detail these critical steps in the following sections.

---

[1] The notion of "multi-birth group" will be detailed later in Sect. 3.1.

3.1   Data Duplication

We denote $\mathbf{X} = [\mathbf{x}_1, \ldots, \mathbf{x}_N]$ as the training set with each column $\mathbf{x}_i$ ($i = 1, 2, \ldots, N$) representing the feature vector of the i-th instance and N denotes the total number of training instances. Besides, we represent the candidate label set of $\mathbf{x}_i$ as $\mathcal{S}_i = \{y_i^1, y_i^2, \ldots, y_i^{|\mathcal{S}_i|}\}$, where $|\mathcal{S}_i|$ denotes the cardinality of $\mathcal{S}_i$.

To pave the way for the subsequent disambiguation operations, we adopt a data duplication scheme on the original partially labeled training dataset. Specifically, for an arbitrary training instance $\mathbf{x}_i$ and its corresponding candidate label set $\mathcal{S}_i$, we first duplicate $\mathbf{x}_i$ into $|\mathcal{S}_i|$ replicas, i.e., $\mathbf{x}_i^1$, $\mathbf{x}_i^2$, $\ldots$, and $\mathbf{x}_i^{|\mathcal{S}_i|}$, and each replica is identical to the original feature vector $\mathbf{x}_i$. After that, we decompose the corresponding candidate label set $\mathcal{S}_i = \{y_i^1, y_i^2, \ldots, y_i^{|\mathcal{S}_i|}\}$ and then assign each candidate label y_i^j ($j = 1, 2, \ldots, |\mathcal{S}_i|$) to a replica $\mathbf{x}_i^j$. Eventually, from an original training instance $\mathbf{x}_i$ and its corresponding candidate label set $\mathcal{S}_i$, we can obtain $|\mathcal{S}_i|$ newly generated instance-label pairs, i.e., $(\mathbf{x}_i^1, y_i^1)$, $(\mathbf{x}_i^2, y_i^2)$, $\ldots$, $(\mathbf{x}_i^{|\mathcal{S}_i|}, y_i^{|\mathcal{S}_i|})$, and we name these pairs which are generated from the original one instance as a "*multi-birth group*".

After performing the above-mentioned data duplication operation on all training instances, we have transformed the original partially labeled training dataset into a new training dataset which contains $n = \sum_i |\mathcal{S}_i|$ ($i = 1, 2, \ldots, N$) instances from N multi-birth groups, and meanwhile each instance contains only one label (can be correct or incorrect). It is worth noting that although learning from such transformed dataset is similar to corrupted labels learning [11,20,30] at the first glance, it differs from corrupted label learning in that we can definitely know that only one instance is labeled correctly while the labels of other instances are all wrong among each multi-birth group.

As we have obtained the new training dataset, disambiguating the original partially labeled instances is transformed to disambiguating the multi-birth groups, i.e., detecting the unique correctly labeled instance in each multi-birth group. To achieve this target, we take the confidence level of each training instance into consideration. Specifically, we denote $\mathbf{w} = [\mathbf{w}_1^\top, \mathbf{w}_2^\top, \ldots, \mathbf{w}_N^\top]^\top \in \mathbb{R}^{n \times 1}$ as the confidence vector of n training instances from N multi-birth groups, where $\mathbf{w}_i = [w_i^1, w_i^2, \ldots, w_i^{|\mathcal{S}_i|}]^\top$ indicates the group confidence vector of the i-th multi-birth group with the j-th element $w_i^j \in [0,1]$ in $\mathbf{w}_i$ representing the learning confidence score of the instance $\mathbf{x}_i^j$. As there is only one instance labeled correctly in each multi-birth group, the instances in the same multi-birth group are naturally in a competitive relationship. Therefore, we assume that each group confidence vector should be normalized, i.e., $\sum_{j=1}^{|\mathcal{S}_i|} w_i^j = 1, \forall i = 1, 2, \ldots, N$. Distinctly, disambiguating the multi-birth groups is equivalent to refining their corresponding group confidence vectors.

3.2   Progressive Disambiguation

As stated before, we attempt to disambiguate the simple multi-birth groups at the initial training stages and gradually disambiguate more complicated ones as the training process goes on. That is to say, the group confidence vectors of the simple multi-birth groups ought to be acquired firstly so that the trained model is capable of learning from these disambiguated multi-birth groups. With the proceeding of training process, the disambiguation ability of the model will be improved and thus the group confidence vectors of the complicated multi-birth groups can be obtained precisely.

Intuitively, if a multi-birth group contains an instance which is probably labeled correctly, disambiguating this multi-birth group is relatively easy and thus we consider it as a simple multi-birth group. Existing researches [1,34] have shown that a network will learn clean and easy patterns firstly, which indicates that the instances with small loss values are likely to be correctly labeled. Based on such observation and meanwhile employing ANNs as the backbone, we propose a progressive disambiguation strategy as explained below.

After feeding the mini-batch data $\mathcal{D}^b$ into the network at the t-th epoch, we can obtain the cross-entropy loss values of these instances, namely $\ell(\Theta, \mathcal{D}^b)$, where Θ indicates the network parameters. After that, we pick up the instances which are likely to be correctly labeled according to the following two conditions: 1) Their loss values are the first $T(t)$ percentage minimums out of $\ell(\Theta, \mathcal{D}^b)$, where $T(t)$ is a time-dependent parameter determining the maximum amount of the simple multi-birth groups at the t-th epoch, and we will introduce it later; and 2) They must be predicted correctly, $i.e.$, the network predictions on them are identical to their labels. After the above screening operation, we can fetch several small-loss instances from the mini-batch $\mathcal{D}^b$ and we regard them as $reliable$ $instances$. It is worth noting that each multi-birth group contains at most one reliable instance because of the constraint from the second condition. Next, we can divide multi-birth groups into two levels of difficulty according to whether they contain a reliable instance, namely simple multi-birth groups and complicated multi-birth groups. Each simple multi-birth group contains one reliable instance which is likely to be correctly labeled, and thus we consider this multi-birth group is relatively easy to disambiguate at the current epoch. Therefore, we disambiguate it by assigning distinguishing confidence scores to the instances among it according to their loss values. If the i-th multi-birth group is a simple multi-birth group, its corresponding group confidence vector $\mathbf{w}_i$ can be updated as:

$$w_i^j = \frac{\exp(-\ell_i^j)}{\sum_{k=1}^{|\mathcal{S}_i|} \exp(-\ell_i^k)}, j = 1, 2, \ldots, |\mathcal{S}_i|, \tag{1}$$

where ℓ_i^j (ℓ_i^k) indicates the loss value of the j-th (k-th) instance in the i-th multi-birth group. Equation (1) indicates that the instances with small loss values can acquire relatively large confidence scores and meanwhile the normalization constraints of group confidence vectors can be satisfied. As to the complicated multi-birth groups which do not contain any reliable instance, we assign an average confidence vector to them as we cannot figure out the correctly labeled instances among them, namely:

$$w_i^j = \frac{1}{|\mathcal{S}_i|}, j = 1, 2, \ldots, |\mathcal{S}_i|. \tag{2}$$

As no loss value will be generated before the first epoch, all group confidence vectors are initialized in an average manner according to Eq. (2).

After we have obtained the group confidence vector of each multi-birth group, we can clearly know that the instances with large confidence scores are likely to be correctly labeled, and thereby the trained network should pay more attention to them. Otherwise, the network ought to avoid learning from these instances. Taking this into account, we assign weights to the loss values of the instances (*i.e.*, $\ell(\Theta, \mathcal{D}^b)$) with their respective confidence scores, and the propagated back loss of $\mathcal{D}^b$, *i.e.*, $\mathcal{L}(\Theta, \mathcal{D}^b)$, can be calculated as follows:

$$\mathcal{L}(\Theta, \mathcal{D}^b) = \mathbf{w}^{b^\top} \ell(\Theta, \mathcal{D}^b), \tag{3}$$

where $\mathbf{w}^b$ is the confidence vector concatenated by the confidence scores of instances in $\mathcal{D}^b$. Finally, by denoting η as the learning rate, the network parameters Θ can be updated as:

$$\Theta := \Theta - \eta \nabla \mathcal{L}(\Theta, \mathcal{D}^b). \tag{4}$$

As mentioned previously, $T(t)$ is a time-dependent parameter which implies that at most $T(t)$ percentage of multi-birth groups will be regarded as simple multi-birth groups and disambiguated at the t-th epoch, and it will increase from zero to one as the training process proceeds. The concrete formulation of $T(t)$ is as follows:

$$T(t) = \begin{cases} \exp(-5(t/t_r - 1)^2) & t \le t_r \\ 1 & t > t_r \end{cases}, \tag{5}$$

where t_r is a coefficient determining at which epoch $T(t)$ reaches to one, meaning that almost all the multi-birth groups will be disambiguated after that epoch. Equation (5) reveals that at the initial training phase, only very few yet simple multi-birth groups will be disambiguated as $T(t)$ is relatively small. With the advance of training steps, the network disambiguation ability will be strengthened and it is capable of disambiguating the complicated multi-birth groups, and thereby $T(t)$ ought to increase accordingly.

The pseudo code of the progressive disambiguation strategy is summarized in Algorithm 1. After initializing the confidence vector $\mathbf{w}$ and feeding the data into the classifier (Steps 1–4), we firstly calculate the widely-used cross-entropy loss of each instance (Step 5). Then, we are able to obtain the reliable instances according to the abovementioned two conditions (Steps 6–8). After that, the confidence vector $\mathbf{w}^b$ will be updated and the corresponding back propagated loss can be calculated (Steps 9–10). Finally, we update $T(t)$ for the next epoch (Step 12).

3.3   Network Cooperation

Although the aforementioned progressive disambiguation strategy has taken the disambiguation difficulty of multi-birth groups into consideration, the corresponding training process is still single-trend of which the disambiguated data

---

Algorithm 1. The Progressive Disambiguation Algorithm

Input: Θ, learning rate η, epoch t_{max}, iteration b_{max}, training set $\mathcal{D}$
Output: Θ

1: Initialize $\mathbf{w} = [\mathbf{w}_1; \mathbf{w}_2; \ldots; \mathbf{w}_N]^\top$ according to Eq. (2);
2: **for** $t = 1, 2, \ldots, t_{max}$ **do**
3:     **for** $b = 1, 2, \ldots, b_{max}$ **do**
4:         Fetch mini-batch $\mathcal{D}^b$ from $\mathcal{D}$;
5:         Obtain cross-entropy loss values $\ell(\Theta, \mathcal{D}^b)$;
6:         Obtain first $T(t)$ percentage minimums small-loss instances $\mathcal{D}^s$ from $\mathcal{D}^b$;

7:         Obtain correctly predicted instances $\mathcal{D}^c$ from $\mathcal{D}^b$;
8:         Obtain reliable instances $\mathcal{D}^r = \mathcal{D}^s \cap \mathcal{D}^c$;
9:         Update $\mathbf{w}^b$ according to Eq. (1) and Eq. (2);
10:        Obtain $\mathcal{L}(\Theta, \mathcal{D}^b)$ according to Eq. (3);
11:        Update Θ according to Eq. (4);
12:        Update $T(t)$ according to Eq. (5);
13:    **end for**
14: **end for**
15: **return** Θ.

---

will be directly transferred back to the model itself, and the accompanied short-comings have been analyzed before. Inspired by the work [13,32] dealing with corrupted label learning problem, we devise a network cooperation mechanism, which trains two networks collaboratively and lets them interact with each other regarding the confidence levels of the instances.

By denoting the two networks as α (with parameter Θ_α) and β (with parameter Θ_β) respectively, we can obtain two confidence vectors of $\mathcal{D}^b$ generated by them independently (according to Sect. 3.2), $i.e.$, $\mathbf{w}_\alpha^b$ and $\mathbf{w}_\beta^b$. After that, we exchange the confidence vectors among two networks to calculate their respective back propagated loss, $i.e.$, $\mathcal{L}_\alpha(\Theta_\alpha, \mathcal{D}^b)$ and $\mathcal{L}_\beta(\Theta_\beta, \mathcal{D}^b)$:

$$\mathcal{L}_\alpha(\Theta_\alpha, \mathcal{D}^b) = {\mathbf{w}_\beta^b}^\top \ell(\Theta_\alpha, \mathcal{D}^b), \tag{6}$$

$$\mathcal{L}_\beta(\Theta_\beta, \mathcal{D}^b) = {\mathbf{w}_\alpha^b}^\top \ell(\Theta_\beta, \mathcal{D}^b), \tag{7}$$

where $\ell(\Theta_\alpha, \mathcal{D}^b)$ and $\ell(\Theta_\beta, \mathcal{D}^b)$ denote the loss values of the mini-batch $\mathcal{D}^b$ calculated by the network α and network β respectively in the forward propagation phase.

Equation (6) and Eq. (7) indicate that each network exploits the data disambiguated by its peer network to train itself. As two networks have different ability and can disambiguate multi-birth groups at different levels, exchanging the confidence scores of instances is beneficial for both networks to reduce their respective disambiguation errors, and therefore the error accumulation problem inherited by the conventional single-trend training scheme can be effectively alleviated. Finally, we update the network parameters Θ_α and Θ_β as follows:

Algorithm 2. The NCPD Algorithm

Input: Θ_α, Θ_β, learning rate η, epoch t_{max}, iteration b_{max}, training set $\mathcal{D}$
Output: Θ_α, Θ_β

1: Initialize $\mathbf{w} = [\mathbf{w}_1; \mathbf{w}_2; \ldots; \mathbf{w}_N]^\top$ according to Eq. (2);
2: **for** $t = 1, 2, \ldots, t_{max}$ **do**
3:     **for** $b = 1, 2, \ldots, b_{max}$ **do**
4:         Update $\mathbf{w}_\alpha^b$ and $\mathbf{w}_\beta^b$ according to Steps 4-9 in Algorithm 1;
5:         Obtain $\mathcal{L}_\alpha(\Theta_\alpha, \mathcal{D}^b)$ and $\mathcal{L}_\beta(\Theta_\alpha, \mathcal{D}^b)$ according to Eq. (6) and Eq. (7) respectively;
6:         Update Θ_α and Θ_β according to Eq. (8) and Eq. (9) respectively;
7:         Update $T(t)$ according to Eq. (5);
8:     **end for**
9: **end for**
10: **return** Θ_α, Θ_β.

$$\Theta_\alpha := \Theta_\alpha - \eta \nabla \mathcal{L}_\alpha(\Theta_\alpha, \mathcal{D}^b), \tag{8}$$

$$\Theta_\beta := \Theta_\beta - \eta \nabla \mathcal{L}_\beta(\Theta_\beta, \mathcal{D}^b). \tag{9}$$

The pseudo code of the proposed NCPD approach is summarized in Algorithm 2. By employing the progressive disambiguation strategy, we can obtain two different confidence vectors from the two networks, respectively (Step 4). After that, two networks exchange their confidence vectors to calculate the back propagated loss and then update their corresponding parameters (Steps 5–6). Similar to Algorithm 1, the time-dependent parameter $T(t)$ will be updated at each epoch (Step 7).

4 Experiments

4.1 Experimental Setup

In this paper, we conduct comparative experiments to demonstrate the effectiveness of NCPD on two kinds of datasets, *i.e.*, controlled UCI datasets and real-world partial label datasets. The compared state-of-the-art PLL algorithms includes:

- PLKNN [15]: an averaging-based disambiguation approach which generalizes k-nearest neighbor classification for partial label learning;
- M3PL [31]: an identification-based approach that utilizes the maximum margin criterion;
- IPAL [36]: an instance-based approach that employs label propagation procedure to leverage the structural information in feature space;
- SURE [8]: an approach that employs the idea of self-training to exaggerate the mutually exclusive relationships among candidate labels;

Table 1. Characteristics and the parameter configurations of the controlled UCI datasets.

Datasets	glass	ecoil	vehicle	abalone
# Instances	214	336	846	4,177
# Features	10	7	18	7
# Labels	5	8	4	29

Configurations:
(I) $r = 1, p \in \{0.1, 0.2, \cdots, 0.7\}$
(II) $r = 2, p \in \{0.1, 0.2, \cdots, 0.7\}$
(III) $r = 3, p \in \{0.1, 0.2, \cdots, 0.7\}$

– AGGD [27]: an approach that discoveries the manifold structure on original feature space.

For our NCPD approach, we employ the 4-layer perceptron as the backbone and meanwhile utilize Adam [17] to optimize the networks for all experiments. Besides, we employ the minibatch size of 128 for all runnings and choose the parameter t_r via cross-validation. For baseline methods, they are implemented with parameters setup suggested in respective literatures. Specifically, the regularization parameter C_{max} in M3PL is chosen from the set $\{0.01, 0.1, 1, 10, 100\}$ via cross-validation. In PLKNN, IPAL, and AGGD, the number of nearest numbers k is chosen from set $\{5, 10, 15, 20\}$. Furthermore, we perform ten-fold cross-validation to record the mean prediction accuracies and standard deviations for all comparing algorithms on all the datasets adopted below.

4.2 Experiments on Controlled UCI Datasets

Following the widely-used controlling protocol in previous PLL works [7,25,28, 36–38], an artificial partial label dataset can be generated from an original UCI dataset with two controlling parameters p and r. To be specific, p controls the proportion of instances which are partially labeled (*i.e.*, $|S_i| > 1$), and r controls the number of false positive labels in each candidate label set (*i.e.*, $|S_i| = r + 1$). The characteristics of these controlled UCI datasets as well as the parameter configurations are listed in Table 1.

Figure 3, Fig. 4, and Fig. 5 show the classification accuracy of each algorithm as p ranges from 0.1 to 0.7 with the step size 0.1, when $r = 1$, $r = 2$, and $r = 3$ (Configuration (I), (II), and (III)), respectively. As illustrated in these figures, NCPD achieves superior performance against other comparing algorithms on these controlled UCI datasets. Specifically, NCPD achieves superior or at least comparable performance against PLKNN, M3PL, and IPAL in all experiments. As to SURE and AGGD, although their classification accuracies are slightly higher than NCPD in a few parameter configurations, they are inferior to NCPD in most cases.

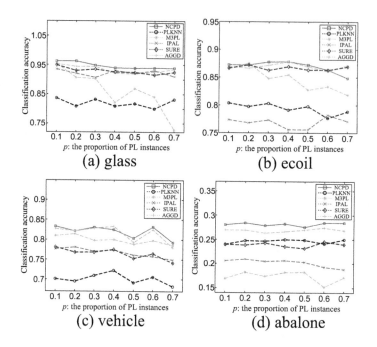

Fig. 3. Classification accuracy of each algorithm on controlled UCI datasets with p ranging from 0.1 to 0.7 ($r = 1$).

Table 2. Characteristics of adopted real-world partial label datasets.

Datasets	Lost	BirdSong	MSRCv2	Soccer Player	Yahoo!News
# Instances	1,122	4,998	1,758	17,472	22,991
# Features	108	38	48	279	163
# Labels	16	13	23	171	219
# Avg. CLs	2.23	2.18	3.16	2.09	1.91

4.3 Experiments on Real-World Datasets

Apart from the controlled UCI datasets, we also conduct experiments on five real-world partial label datasets which are collected from several application domains including *Lost* [6], *Soccer Player* [33], and *Yahoo!News* [12] for automatic face naming, *MSRCv2* [19] for object classification, and *BirdSong* [3] for bird song classification. The characteristics of these real-world datasets are summarized in Table 2 where the average number of candidate labels of each dataset (*i.e.*, # Avg. CLs) is also reported[2].

The average classification accuracies as well as the standard deviations of different approaches on these real-world datasets are shown in Table 3. Pairwise

---

[2] These datasets are available at http://palm.seu.edu.cn/zhangml.

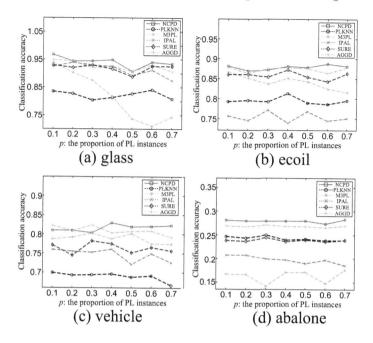

Fig. 4. Classification accuracy of each algorithm on controlled UCI datasets with p ranging from 0.1 to 0.7 ($r = 2$).

Table 3. Classification accuracy (mean $\pm$ std) of each algorithm on five real-world datasets. $\bullet$/$\circ$ indicates that NCPD is significantly superior/inferior to the comparing algorithm on the corresponding dataset (pairwise t-test with 0.05 significance level).

	Lost	BirdSong	MSRCv2	Soccer Player	Yahoo!News
PLKNN	$0.471 \pm 0.032\bullet$	$0.686 \pm 0.015\bullet$	$0.457 \pm 0.049\bullet$	$0.530 \pm 0.016\bullet$	$0.482 \pm 0.011\bullet$
M3PL	$0.721 \pm 0.037\bullet$	$0.667 \pm 0.042\bullet$	$0.474 \pm 0.038\bullet$	$0.500 \pm 0.007\bullet$	$0.628 \pm 0.013\bullet$
IPAL	$0.653 \pm 0.022\bullet$	$0.734 \pm 0.013\bullet$	$0.537 \pm 0.045\bullet$	$0.547 \pm 0.016\bullet$	$0.577 \pm 0.010\bullet$
SURE	$0.739 \pm 0.036\bullet$	$0.730 \pm 0.015\bullet$	$0.508 \pm 0.043\bullet$	$0.522 \pm 0.013\bullet$	$0.562 \pm 0.011\bullet$
AGGD	0.778 ± 0.040	0.737 ± 0.018	$0.506 \pm 0.041\bullet$	$0.543 \pm 0.016\bullet$	$0.637 \pm 0.008\bullet$
NCPD	$\mathbf{0.790 \pm 0.055}$	$\mathbf{0.751 \pm 0.018}$	$\mathbf{0.589 \pm 0.046}$	$\mathbf{0.573 \pm 0.013}$	$\mathbf{0.657 \pm 0.013}$

t-test at 0.05 significance level is also conducted based on the results of ten-fold cross-validation. From Table 3, we have three findings: 1) NCPD achieves the highest classification accuracies among all baselines on all adopted real-world datasets; 2) NCPD significantly outperforms PLKNN, M3PL, IPAL, and SURE on all these datasets; 3) NCPD is never statistically inferior to any comparing algorithms in all cases. These findings convincingly substantiate the superiority of our NCPD approach to other comparators.

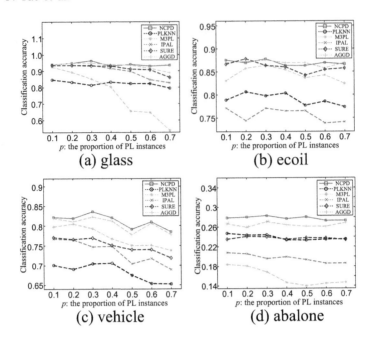

Fig. 5. Classification accuracy of each algorithm on controlled UCI datasets with p ranging from 0.1 to 0.7 ($r = 3$).

4.4 Ablation Study

The superiority of the proposed NCPD approach has been verified by thorough experimental results presented above. In this section, we conduct ablation study on adopted real-world datasets to further demonstrate the effectiveness of the two crucial techniques employed by NPCD, *i.e.*, the progressive disambiguation strategy and the network cooperation mechanism.

Specifically, to demonstrate the effectiveness of the progressive disambiguation strategy, we discard this strategy and merely train two networks with network cooperation mechanism, *i.e.*, all multi-birth groups are disambiguated according to Eq. (1) in every epoch regardless their disambiguation difficulty. To confirm the effectiveness of the network cooperation mechanism, we barely train one network equipped with the progressive disambiguation strategy (see Sect. 3.2). Figure 6 shows the results, from which we can observe that the integrated NCPD approach generates the highest accuracies than other two settings (*i.e.*, "w/o NC" and "w/o PD"). In contrast, the accuracies will decrease when either the progressive disambiguation strategy or the network cooperation mechanism is removed, therefore the effectiveness and indispensability of these two crucial techniques are validated.

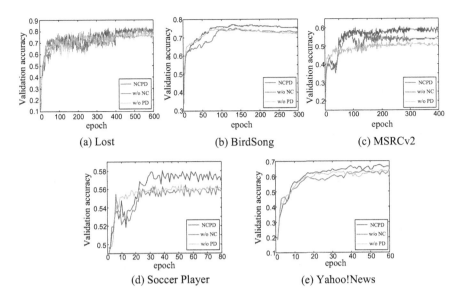

Fig. 6. Validation accuracy with different settings on adopted real-world datasets. The blue curve denotes the accuracy of the integrated NCPD approach (legend by "NCPD"). The red curve and the yellow curve indicate the accuracy of NCPD that removes the network cooperation mechanism (denoted by "w/o NC") and the progressive disambiguation strategy (denoted by "w/o PD"), respectively. (Color figure online)

5   Conclusion

In this paper, we propose a novel approach for PLL which is dubbed as "NCPD". By employing the progressive disambiguation strategy, our approach is able to exploit the disambiguation difficulty of the instances and then disambiguate them in a progressive manner, which is beneficial for the steady improvement of model capability and thereby the adverse impacts brought by false positive labels can be effectively reduced. Furthermore, the network cooperation mechanism greatly facilitates the salutary mutual learning process between two networks, and therefore can effectively alleviate the error accumulation problem inherited by the existing single-trend training framework. Thorough experimental results on various datasets demonstrate the effectiveness of the proposed NCPD approach. Considering that how to determine the disambiguation difficulty of the instances plays a vital role in our algorithm, we will devise a more advanced methodology to judge the disambiguation difficulty of these partially labeled instances in the future.

Acknowledgments. This research is supported by NSF of China (Nos: 61973162, U1713208), NSF of Jiangsu Province (No: BK20171430), the Fundamental Research Funds for the Central Universities (No: 30920032202), the "Young Elite Scientists Sponsorship Program" by Jiangsu Province and CAST (No: 2018QNRC001), and the Program for Changjiang Scholars.

References

1. Arpit, D., et al.: A closer look at memorization in deep networks. In: Proceedings of the International Conference on Machine Learning, pp. 233–242 (2017)
2. Blum, A., Mitchell, T.: Combining labeled and unlabeled data with co-training. In: Proceedings of the Eleventh Annual Conference on Computational Learning Theory, pp. 92–100 (1998)
3. Briggs, F., Fern, X.Z., Raich, R.: Rank-loss support instance machines for MIML instance annotation. In: Proceedings of the International Conference on Knowledge Discovery and Data Mining, pp. 534–542 (2012)
4. Chen, C.H., Patel, V.M., Chellappa, R.: Learning from ambiguously labeled face images. IEEE Trans. Pattern Anal. Mach. Intell. **40**(7), 1653–1667 (2018)
5. Chen, Z.S., Wu, X., Chen, Q.G., Hu, Y., Zhang, M.L.: Multi-view partial multi-label learning with graph-based disambiguation. In: Proceedings of the AAAI Conference on Artificial Intelligence, pp. 3553–3560 (2020)
6. Cour, T., Sapp, B., Jordan, C., Taskar, B.: Learning from ambiguously labeled images. In: Proceedings of the Computer Vision and Pattern Recognition, pp. 919–926 (2009)
7. Feng, L., An, B.: Leveraging latent label distributions for partial label learning. In: Proceedings of the International Joint Conference on Artificial Intelligence, pp. 2107–2113 (2018)
8. Feng, L., An, B.: Partial label learning with self-guided retraining. In: Proceedings of the AAAI Conference on Artificial Intelligence, pp. 3542–3549 (2019)
9. Gong, C., Liu, T., Tang, Y., Yang, J., Yang, J., Tao, D.: A regularization approach for instance based superset label learning. IEEE Trans. Cybern. **48**(3), 967–978 (2018)
10. Gong, C., Shi, H., Liu, T., Zhang, C., Yang, J., Tao, D.: Loss decomposition and centroid estimation for positive and unlabeled learning. IEEE Trans. Pattern Anal. Mach. Intell. **43**, 918–932 (2019)
11. Gong, C., Zhang, H., Yang, J., Tao, D.: Learning with inadequate and incorrect supervision. In: Proceedings of the International Conference on Data Mining, pp. 889–894 (2017)
12. Guillaumin, M., Verbeek, J., Schmid, C.: Multiple instance metric learning from automatically labeled bags of faces. In: Daniilidis, K., Maragos, P., Paragios, N. (eds.) ECCV 2010. LNCS, vol. 6311, pp. 634–647. Springer, Heidelberg (2010). https://doi.org/10.1007/978-3-642-15549-9_46
13. Han, B., et al.: Co-teaching: robust training of deep neural networks with extremely noisy labels. In: Proceedings of the Advances in Neural Information Processing Systems, pp. 8527–8537 (2018)
14. Hassoun, M.H., et al.: Fundamentals of Artificial Neural Networks. MIT Press, Cambridge (1995)
15. Hüllermeier, E., Beringer, J.: Learning from ambiguously labeled examples. Intell. Data Anal. **10**(5), 419–439 (2006)
16. Jin, R., Ghahramani, Z.: Learning with multiple labels. In: Proceedings of the Advances in Neural Information Processing Systems, pp. 921–928 (2003)
17. Kingma, D.P., Ba, J.: Adam: a method for stochastic optimization. arXiv preprint arXiv:1412.6980 (2014)

18. Liu, L., Dietterich, T.: Learnability of the superset label learning problem. In: Proceedings of the International Conference on Machine Learning, pp. 1629–1637 (2014)
19. Liu, L., Dietterich, T.G.: A conditional multinomial mixture model for superset label learning. In: Proceedings of the Advances in Neural Information Processing Systems, pp. 548–556 (2012)
20. Liu, T., Tao, D.: Classification with noisy labels by importance reweighting. IEEE Trans. Pattern Anal. Mach. Intell. **38**(3), 447–461 (2015)
21. Liu, W., Xu, D., Tsang, I.W., Zhang, W.: Metric learning for multi-output tasks. IEEE Trans. Pattern Anal. Mach. Intell. **41**(2), 408–422 (2018)
22. Luo, J., Orabona, F.: Learning from candidate labeling sets. In: Proceedings of the Advances in Neural Information Processing Systems, pp. 1504–1512 (2010)
23. Lyu, G., Feng, S., Lang, C., Wang, T.: A self-paced regularization framework for partial-label learning. arXiv preprint arXiv:1804.07759 (2018)
24. Nguyen, N., Caruana, R.: Classification with partial labels. In: Proceedings of the International Conference on Knowledge Discovery and Data Mining, pp. 551–559 (2008)
25. Tang, C.Z., Zhang, M.L.: Confidence-rated discriminative partial label learning. In: Proceedings of the AAAI Conference on Artificial Intelligence (2017)
26. Wan, S., Gong, C., Zhong, P., Du, B., Zhang, L., Yang, J.: Multiscale dynamic graph convolutional network for hyperspectral image classification. IEEE Trans. Geosci. Remote Sens. **58**(5), 3162–3177 (2019)
27. Wang, D.B., Li, L., Zhang, M.L.: Adaptive graph guided disambiguation for partial label learning. In: Proceedings of the International Conference on Knowledge Discovery and Data Mining, pp. 83–91 (2019)
28. Wu, X., Zhang, M.L.: Towards enabling binary decomposition for partial label learning. In: Proceedings of the International Joint Conference on Artificial Intelligence, pp. 2868–2874 (2018)
29. Yao, Y., Deng, J., Chen, X., Gong, C., Wu, J., Yang, J.: Deep discriminative CNN with temporal ensembling for ambiguously-labeled image classification. In: Proceedings of the AAAI Conference on Artificial Intelligence, pp. 12669–12676 (2020)
30. Yi, K., Wu, J.: Probabilistic end-to-end noise correction for learning with noisy labels. In: Proceedings of the Computer Vision and Pattern Recognition, pp. 7017–7025 (2019)
31. Yu, F., Zhang, M.-L.: Maximum margin partial label learning. Mach. Learn. **106**(4), 573–593 (2016). https://doi.org/10.1007/s10994-016-5606-4
32. Yu, X., Han, B., Yao, J., Niu, G., Tsang, I.W., Sugiyama, M.: How does disagreement help generalization against label corruption? arXiv preprint arXiv:1901.04215 (2019)
33. Zeng, Z., et al.: Learning by associating ambiguously labeled images. In: Proceedings of the Computer Vision and Pattern Recognition. pp. 708–715 (2013)
34. Zhang, C., Bengio, S., Hardt, M., Recht, B., Vinyals, O.: Understanding deep learning requires rethinking generalization. arXiv preprint arXiv:1611.03530 (2016)
35. Zhang, C., Ren, D., Liu, T., Yang, J., Gong, C.: Positive and unlabeled learning with label disambiguation. In: Proceedings of the International Joint Conference on Artificial Intelligence, pp. 4250–4256 (2019)
36. Zhang, M.L., Yu, F.: Solving the partial label learning problem: an instance-based approach. In: Proceedings of the International Joint Conference on Artificial Intelligence, pp. 4048–4054 (2015)

37. Zhang, M.L., Yu, F., Tang, C.Z.: Disambiguation-free partial label learning. IEEE Trans. Knowl. Data Eng. **29**(10), 2155–2167 (2017)
38. Zhang, M.L., Zhou, B.B., Liu, X.Y.: Partial label learning via feature-aware disambiguation. In: Proceedings of the International Conference on Knowledge Discovery and Data Mining, pp. 1335–1344 (2016)

Partial Label Learning via Self-Paced Curriculum Strategy

Gengyu Lyu[1], Songhe Feng[1(✉)], Yi Jin[1], and Yidong Li[2]

[1] Beijing Key Laboratory of Traffic Data Analysis and Mining,
Beijing Jiaotong University, Beijing 100044, China
{18112030,shfeng,yjin}@bjtu.edu.cn
[2] School of Computer and Information Technology,
Beijing Jiaotong University, Beijing 100044, China
ydli@bjtu.edu.cn

Abstract. Partial-Label Learning (PLL) aims to learn from the training data, where each example is associated with a set of candidate labels, among which only one is correct. Existing PLL methods to deal with such problem usually treat each training example equally and few works take the complexities of training examples into consideration. In this paper, inspired by the human learning mode that gradually learns from "easy" to "hard", we propose a novel **S**elf-**P**aced **C**urriculum strategy based **P**artial-**L**abel **L**earning (**SPC-PLL**) algorithm, where curriculum strategy can predetermine prior knowledge to adjust the learning priorities of training examples, while self-paced strategy can dynamically select "easy" training examples for model induction according to its current learning progress. The combination of such two strategies is analogous to "instructor-student-collaborative" learning mode, which not only utilizes prior knowledge flexibly but also effectively avoids the inconsistency between the predetermined curriculum and the dynamically learned models. Extensive experimental comparisons and comprehensive ablation study demonstrate the effectiveness of such strategy on solving PLL problem.

Keywords: Partial-label learning · Self-paced learning strategy · Curriculum learning strategy · Instructor-student-collaborative

1 Introduction

As a novel weakly supervised learning framework, partial label learning[1] (PLL) learns from the ambiguous training data, where the unique ground-truth label for each instance is concealed in its corresponding candidate label set and is not directly accessible to the learning algorithm [6–9,12,21,22,30,31]. In recent years, such learning mechanism has been widely used in many real world scenarios, such as facial age estimation [37], multimedia content analysis [34], web mining [20], ecoinformatics [18], etc.

---

[1] In some literature, partial-label learning is also named as *ambiguous label learning* [5,32], *superset label learning* [19] or *soft label learning* [27].

© Springer Nature Switzerland AG 2021
F. Hutter et al. (Eds.): ECML PKDD 2020, LNAI 12458, pp. 489–505, 2021.
https://doi.org/10.1007/978-3-030-67661-2_29

{tiger, cheetah}   {tiger, cat}   {cat, tiger, lion}

Fig. 1. Partial label instances with different complexities. Compared with example A, the complexity of example B comes from the atypical features, while the complexity of example C comes from more candidate labels.

An intuitive strategy to deal with such problem is disambiguation, i.e. identifying the unique ground-truth label from the ambiguous candidate label set. Existing methods following such strategy can be roughly grouped into two categories: *Averaging*-based strategy [6,10,13,35] and *Identification*-based strategy [5,18,23–25]. Averaging-based strategy usually treats each candidate label equally and it makes prediction for unseen instances by averaging the output from all candidate labels. Identification-based strategy often regards the ground-truth label as a latent variable first and then refines the variable in an iterative manner. However, the above two strategies usually assume that each training example contributes equally to the model induction, while the complexities of varying training examples are regrettably ignored, which significantly decreases the performance of learning model. For example, in Fig. 1, due to the atypical tiger's feature and large candidate label set, examples B and C are more difficult to learn than example A. Obviously, during the training process, if we treat these examples equally or directly train the predictive model from them, the robustness of the final model would be difficult to guarantee.

In light of the above observation, in this paper, we propose a novel **S**elf-**P**aced **C**urriculum strategy based **P**artial-**L**abel **L**earning (**SPC-PLL**) algorithm, where curriculum strategy and self-paced strategy are jointly incorporated into the partial label learning framework to guide the model training process, from "easy" examples to "hard" examples. Specifically, we first utilize curriculum strategy to adjust the learning priorities of training examples, where the number of candidate labels and the intra-class example distances are predetermined as prior knowledge to cluster the examples with different complexities. Then, we utilize the self-paced strategy to dynamically select "easy" examples for the model induction, where the selection strategy is designed according to the model's learning progress. Note that, a key characteristic of our method lies in that some predetermined prior knowledge from curriculum strategy can be adjusted appropriately according to the feedback of self-paced strategy. The combination of such two strategies is analogous to "instructor-student-collaborative" learning mode, where curriculum strategy as "instructor" can utilize predetermined knowledge to avoid the self-paced strategy from overfitting, and self-paced strategy as "student" has the freedom to adjust the inconsistency between the fixed curriculum and the dynamically learned models according to its current learning progress. The learning mode from "easy" to "hard" can effectively guide the training model to grow from "juvenile" to "mature". Extensive experiments demonstrate the effectiveness of such strategy on solving PLL problem.

In summary, the main contributions of our paper lie in:

- To the best of our knowledge, it is the first successful attempt to incorporate the self-paced curriculum strategy to partial label learning framework, which provides a new perspective for improving the robustness of weakly-supervised learning model.
- Different from previous self-paced curriculum strategy, in our framework, some predetermined prior knowledge from our curriculum strategy can be appropriately updated (instead of always being fixed) according to the feedback of self-paced strategy, which can be deemed as real "instructor-student-collaborative".

2 Related Work

2.1 Partial Label Learning (PLL)

Existing partial label learning algorithms can be roughly grouped into the following three categories: *Average Disambiguation Strategy* (ADS), *Identification Disambiguation Strategy* (IDS) and *Disambiguation-Free Strategy* (DFS).

ADS-based PLL methods usually assume that each candidate label contributes equally to the modeling process and they make predictions for unseen instances by averaging the outputs from all candidate labels, i.e. $\frac{1}{|S_i|}\sum_{y\in S_i} F(\mathbf{x}, \boldsymbol{\Theta}, y)$ [6,35]. IDS-based PLL methods often view the ground-truth label as a latent variable first, identified as $\arg\max_{y\in S_i} F(\mathbf{x}, \boldsymbol{\Theta}, y)$, and then refine the model parameter $\boldsymbol{\Theta}$ iteratively by utilizing some specific criterions, such as maximum likelihood criterion: $\sum_{i=1}^{n} \log(\sum_{y\in S_i} F(\mathbf{x}, \boldsymbol{\Theta}, y))$ [5,18,38] and maximum margin criterion: $\sum_{i=1}^{n}(F(\mathbf{x}, \boldsymbol{\Theta}, y) - \max_{y\neq y} F(\mathbf{x}, \boldsymbol{\Theta}, y))$ [26,33]. DFS-based methods aims to learn from PL data by fitting the PL data to existing learning techniques instead of disambiguation [36].

2.2 Self-Paced Curriculum Strategy (SPC)

Self-paced Curriculum Strategy can be regard as an integration of curriculum strategy and self-paced strategy, which are separately proposed by [2] and [17].

The two methods share a similar conceptual learning paradigm that learns from easy to more complex examples in training, but differ in specific learning schemes. In curriculum strategy, the curriculum is predetermined by prior knowledge and remain fixed during the subsequent learning process [16,29]. Intuitively, such strategy relies heavily on the quality of prior knowledge while the feedback of model is ignored. In self-paced strategy, the model can dynamically design its own subsequent learning curriculum according to its current learning process, which guarantees the consistency between the dynamical curriculum and the learned model [14]. However, since the learning is completely dominated by the model itself, it may prone to overfitting. To tackle the above issues, [15] proposes a unified learning paradigm to simultaneously inherit merits from both curriculum strategy and self-paced strategy and resist drawbacks from both of them. However, the model is designed to work under the fully annotated training data and thus cannot meet the requirements for the weakly supervised PLL problem. Meanwhile, the predetermined knowledge from curriculum strategy is fixed during the whole training process, which can not be adjusted appropriately according to the feedback of self-paced strategy.

3   The Proposed Method

Formally speaking, we denote the d-dimensional input space as $\mathcal{X} \in \mathbb{R}^d$, and the output space as $\mathcal{Y} = \{1, 2, \ldots, q\}$ with q class labels. PLL aims to learn a classifier $f : \mathcal{X} \mapsto \mathcal{Y}$ from the PL training data $\mathcal{D} = \{(\mathbf{x}_i, S_i)\}(1 \leq i \leq n)$, where the instance $\mathbf{x}_i \in \mathcal{X}$ is described as a d-dimensional feature vector and the candidate label set $S_i \subseteq \mathcal{Y}$ is associated with the instance $\mathbf{x}_i$. Furthermore, let $\mathbf{y} = \{y_1, y_2, \ldots, y_n\}$ be the ground-truth label assignments for training instances and each $y_i \in S_i$ of $\mathbf{x}_i$ is not directly accessible during the training phase.

3.1   Baseline

In our algorithm, we adopt M3PL[2] [33] as baseline method to design the SPC-PLL framework. Given the parametric model $\boldsymbol{\Theta} = \{(\mathbf{w}_p, b_p)|1 \leq p \leq q\}$ and the modeling output $F(\mathbf{x}_i, \boldsymbol{\Theta}, y)$ of $\mathbf{x}_i$ on label y, M3PL focuses on differentiating the output from ground-truth label against the maximum output from all other labels (i.e. $F(\mathbf{x}_i, \boldsymbol{\Theta}, y_i) - \max_{\tilde{y}_i \neq y_i} F(\mathbf{x}_i, \boldsymbol{\Theta}, \tilde{y}_i)$), which well avoids the negative effects produced by the noisy labels in candidate label set. The framework of M3PL can be formulated as **OP (1)**:

$$\min_{\boldsymbol{\Theta}, \boldsymbol{\xi}, \mathbf{y}} \quad \frac{1}{2} \sum_{p=1}^{q} \|\mathbf{w}_p\|^2 + \mathcal{C} \sum_{i=1}^{n} \xi_i$$

$$s.t. \quad \begin{cases} (\mathbf{w}_{y_i}^{\top} \cdot \mathbf{x}_i + b_{y_i}) - \max_{\tilde{y}_i \neq y_i} (\mathbf{w}_{\tilde{y}_i}^{\top} \cdot \mathbf{x}_i + b_{\tilde{y}_i}) \geq 1 - \xi_i \\ \xi_i \geq 0, \quad \forall i \in \{1, 2, \ldots, n\} \\ \mathbf{y} \in \mathcal{S} \\ \sum_{i=1}^{n} \mathbb{I}(y_i = p) = n_p, \quad \forall p \in \{1, 2, \ldots, q\}, \end{cases}$$

where C is the regularization parameter, $\boldsymbol{\xi} = \{\xi_1, \xi_2, \ldots, \xi_n\}$ is the slack variables set, n_p is the prior number of examples for the p-th class label in $\mathcal{Y}$, and $\mathcal{S}$ is

---

[2] Our proposed self-paced curriculum strategy can also be well applied to other margin-based PLL methods.

the feasible solution space. $\mathbb{I}(\triangle)$ is an indicator function where $\mathbb{I}(\triangle) = 1$ if and only if $\triangle$ is true, otherwise $\mathbb{I}(\triangle) = 0$.

The baseline M3PL assumes each training example contributes equally to the learning model and induces the predictive model by putting all examples into the training process at one time. According to the analysis in Sect. 1, the robustness of learned model is difficult to guarantee.

3.2 Self-Paced Strategy

In order to solve the above issue, we first introduce a self-paced strategy to construct the PLL framework, where examples with typical (easy) and atypical (hard) features are gradually added into the training process.

Specifically, we denote $\mathbf{v} = \{v_1; v_2; \ldots; v_n\} \in [0,1]^{n \times 1}$ as n-dimensional weight vector for n training examples, $L(\mathbf{x}_i, \mathbf{W}, \mathbf{b}, y_i)$ as the empirical loss for i-th training example ($\mathbf{W}, \mathbf{b}$ are the model parameters), and λ as the self-paced parameter for controlling the learning process. By introducing a soft self-paced regularizer $f(\mathbf{v}, \lambda) = \sum_{i=1}^{n} \frac{\lambda}{2} \cdot (v_i^2 - 2v_i)$, the self-paced PLL framework is formulated as **OP (2)**:

$$\min_{\substack{\mathbf{W}, \mathbf{b} \\ 0 \leq y_i \leq q \\ \mathbf{v} \in [0,1]^{n \times 1}}} \mathcal{C} \sum_{i=1}^{n} v_i \cdot L(\mathbf{x}_i, \mathbf{W}, \mathbf{b}, y_i) + \frac{1}{2} \sum_{p=1}^{q} \|\mathbf{w}_p\|^2 + \sum_{i=1}^{n} \frac{\lambda}{2} \cdot (v_i^2 - 2v_i)$$

$$s.t. \begin{cases} L(\mathbf{x}_i, \mathbf{W}, \mathbf{b}, y_i) = 1 - \left[(\mathbf{w}_{y_i}^\top \cdot \mathbf{x}_i + b_{y_i}) - \max_{\tilde{y}_i \neq y_i} (\mathbf{w}_{\tilde{y}_i}^\top \cdot \mathbf{x}_i + b_{\tilde{y}_i}) \right] \\ L(\mathbf{x}_i, \mathbf{W}, \mathbf{b}, y_i) \geq 0, \quad \forall i \in \{1, 2, \ldots, n\} \\ \mathbf{y} \in \mathcal{S} \\ \sum_{i=1}^{n} \mathbb{I}(y_i = p) = n_p, \quad \forall p \in \{1, 2, \ldots, q\}. \end{cases}$$

Intuitively, in **OP (2)**, examples with typical (easy) features will be assigned with larger v_i and be given first priority to join the model training. Then, according to the training feedback, the learned model can dynamically adjust the subsequent learning pace, and add other atypical (hard) examples into the learning process to improve the generalization ability of the learned model. However, during the learning process, the adjustment of learning pace are determined by the learned model itself, which renders it prone to overfitting.

3.3 Self-Paced Curriculum Strategy

To avoid the proposed method from overfitting, we further introduce a curriculum strategy to predetermined prior knowledge to guide the whole model training, where the **label curriculum** and **feature curriculum** are jointly predetermined to rank the training examples from "easy" to "hard".

Specifically, we denote $\mathbf{g} = \{g_1; g_2; \ldots; g_n\} \in [0,1]^{n \times 1}$ as the priority values of each example in label curriculum, where $g_i = \Psi(\exp(-|S_i|))$ is determined by the number of candidate labels of $\mathbf{x}_i$, Ψ is the normalized operation to $[0,1]$ and

$|S_i|$ is the cardinality of S_i. Meanwhile, we also denote $\mathbf{h} = \{h_1; h_2; \ldots; h_n\} \in [0,1]^{n \times 1}$ as the priority values of each example in feature curriculum, where $h_i = \Psi\left(\left\|\mathbf{x}_i - \frac{1}{m_p}\sum_{p=1}^{m_p}\mathbf{x}_p\right\|_2\right)$ is determined by the normalized Euclidean distance between $\mathbf{x}_i$ and its class center of m_p examples (with the same predicted label p in each round). Afterwards, we integrate the two curriculums into the self-paced PLL method and obtain the final framework of **SPC-PLL** as **OP (3)**:

$$\min_{\substack{\mathbf{W},b \\ 0 \le y_i \le q \\ \mathbf{v} \in [0,1]^{n \times 1}}} \mathcal{C}\sum_{i=1}^{n} v_i \cdot L(\mathbf{x}_i, \mathbf{W}, b, y_i) + \frac{1}{2}\sum_{p=1}^{q}\|\mathbf{w}_p\|^2 + \sum_{i=1}^{n}\frac{\lambda}{2}\cdot(v_i^2 - 2v_i)$$

$$-\eta\sum_{i=1}^{n}v_i\left(\alpha\cdot g_i + (1-\alpha)h_i\right)$$

$$s.t. \begin{cases} L(\mathbf{x}_i, \mathbf{W}, \mathbf{b}, y_i) = 1 - \left[(\mathbf{w}_{y_i}^{\top}\cdot\mathbf{x}_i + b_{y_i}) - \max_{\tilde{y}_i \neq y_i}(\mathbf{w}_{\tilde{y}_i}^{\top}\cdot\mathbf{x}_i + b_{\tilde{y}_i})\right] \\ L(\mathbf{x}_i, \mathbf{W}, \mathbf{b}, y_i) \geq 0, \quad \forall i \in \{1, 2, \ldots, n\} \\ \mathbf{y} \in \mathcal{S} \\ \sum_{i=1}^{n}\mathbb{I}(y_i = p) = n_p, \quad \forall p \in \{1, 2, \ldots, q\}, \end{cases}$$

where η and $\alpha \in [0,1]$ are the trade-off parameters to control the balance of different terms.

Obviously, in our proposed SPC-PLL method, the label curriculum is fixed during the whole learning process, while the feature curriculum is dynamically updated as the predicted label y_i for $\mathbf{x}_i$ is optimized in each learning cycle. The fixed label curriculum can avoid the learned model from overfitting, and the dynamical feature curriculum can leave the learned model freedom to control its own learning pace. Besides, the dynamical feature curriculum can well collaborate with the self-paced strategy to avoid the inconsistency between the predetermined prior knowledge and the learned model, which can achieve more friendly "instructor-student-collaborative" learning mode.

4   Optimization

Since **OP (3)** involves the optimization of mixed-type variables, which is difficult to be optimized simultaneously, we employed the alternating optimization procedure to update these variables iteratively.

Step 1: Calculate W, b. Fixing the other variables, we can calculate **W, b** by minimizing the following objective function **OP (4)**:

$$\min_{\mathbf{W},b} \mathcal{C}\sum_{i=1}^{n} v_i \cdot L(\mathbf{x}_i, \mathbf{W}, \mathbf{b}, y_i) + \frac{1}{2}\sum_{p=1}^{q}\|\mathbf{w}_p\|^2$$

$$s.t. \begin{cases} L(\mathbf{x}_i, \mathbf{W}, \mathbf{b}, y_i) = 1 - [(\mathbf{w}_{y_i}^{\top}\cdot\mathbf{x}_i + b_{y_i}) - \max_{\tilde{y}_i \neq y_i}(\mathbf{w}_{\tilde{y}_i}^{\top}\cdot\mathbf{x}_i + b_{\tilde{y}_i})] \\ L(\mathbf{x}_i, \mathbf{W}, \mathbf{b}, y_i) \geq 0, \quad \forall i \in \{1, 2, \ldots, n\}. \end{cases}$$

Algorithm 1. The Algorithm of **SPC-PLL**

Inputs:

$\mathcal{D}$: the partial label training set $\{(\mathbf{x}_i, S_i)\}$;

$\mathbf{g}, \mathbf{h}$: the label and feature curriculums;

$\mathcal{C}_{max}$: the maximum value of regularization parameter;

λ: the learning parameter;

$\mathbf{x}^*$: the unseen instance;

Process:

1. Initialize $\mathbf{v}, \mathbf{y}, \mu, \mathcal{C}$ and λ;

2. **while** $\mathcal{C} < \mathcal{C}_{max}$

3. $\mathcal{C} = \min\{(1+\tau)\mathcal{C}, \mathcal{C}_{max}\}$

4.   **while** $\lambda > loss^{max}$

5.   **repeat**

6.   update $\mathbf{W}, \mathbf{b}$ according to **OP(4)**;

7.   update $\mathbf{y}$ according to **OP(5)**;

8.   **until** converge;

9.   update $\mathbf{v}$ according to Eq. (1);

10.   update $\mathbf{h}$ and $\lambda = \mu \cdot \lambda$

11.   **end while**;

12. **end while**;

13. **return** $y^* = \arg\max_{p \in \mathcal{Y}} \mathbf{w}_p^\top \cdot \mathbf{x}^* + b_p$;

Output:

y^*: the predicted label for $\mathbf{x}^*$;

Obviously, **OP (4)** is a typical single-label multi-class maximum margin optimization problem, which can be solved by utilizing the multi-class SVM implementations, such as liblinear toolbox [1].

Step 2: Calculate y. Fixing the other variables, the subproblem to variable $\mathbf{y}$ is simplified as follows **OP (5)**:

$$\min_{\mathbf{y}} \quad \sum_{i=1}^{n} v_i \cdot L(\mathbf{x}_i, \mathbf{W}, \mathbf{b}, y_i)$$

$$s.t. \begin{cases} L(\mathbf{x}_i, \mathbf{W}, \mathbf{b}, y_i) = 1 - [(\mathbf{w}_{y_i}^\top \cdot \mathbf{x}_i + b_{y_i}) - \max_{\tilde{y}_i \neq y_i}(\mathbf{w}_{\tilde{y}_i}^\top \cdot \mathbf{x}_i + b_{\tilde{y}_i})] \\ L(\mathbf{x}_i, \mathbf{W}, \mathbf{b}, y_i) \geq 0, \quad \forall i \in \{1, 2, \dots, n\} \\ \mathbf{y} \in \mathcal{S} \\ \sum_{i=1}^{n} \mathbb{I}(y_i = p) = n_p, \quad \forall p \in \{1, 2, \dots, q\}. \end{cases}$$

Following [33], we transfer **OP(5)** into a linear programming problem, which can be solved by utilizing the standard LP solver [3].

Step 3: Calculate v. Fixing the other variables, the variable **v** can be optimized following **OP (6)**:

$$\min_{\mathbf{v} \in [0,1]^{n \times 1}} \mathcal{C} \sum_{i=1}^{n} v_i \cdot L(\mathbf{x}_i, \mathbf{W}, \mathbf{b}, y_i) + \sum_{i=1}^{n} \frac{\lambda}{2} \cdot (v_i^2 - 2v_i)$$
$$-\eta \sum_{i=1}^{n} v_i (\alpha \cdot g_i + (1 - \alpha)h_i),$$

where the v_i is calculated by

$$v^*(\lambda, L) = \begin{cases} 1 - \frac{1}{\lambda}(\mathcal{C} \cdot L - \eta(g_i + h_i)), & L^* \leq \lambda \\ 0, & L^* > \lambda. \end{cases} \tag{1}$$

here, $L = L(\mathbf{x}_i, \mathbf{W}, \mathbf{b}, y_i)$ and $L^* = \mathcal{C} \cdot L - \eta(g_i + h_i)$.

During the entire optimization process, we first initialize the required variables, and then repeat the above steps until the algorithm converges or reaches the maximum iterations. Algorithm 1 summarizes the optimization of SPC-PLL.

5   Experiments

5.1   Experimental Setup

To evaluate the performance of the proposed SPC-PLL method, we implement experiments on four controlled UCI data sets and six real world data sets: **(1) Controlled UCI data sets.** Under different configurations of two controlling parameters (i.e. p and r), the four UCI data sets generate 84 ($4 \times 3 \times 7$) artificial partial-label data sets [6]. Here, $p \in \{0.1, 0.2, \ldots, 0.7\}$ is the proportion of partial-label examples and $r \in \{1, 2, 3\}$ is the number of false candidate labels. **(2) Real World data sets.** These data sets are collected from the following four task domains: (A) *Facial Age Estimation* [FG-NET]; (B) *Image Classification* [MSRCv2]; (C) *Bird Sound Classification* [BirdSong]; (D) *Automatic Face Naming* [Lost] [Soccer Player] [Yahoo! News]. Tables 1 and 2 separately summarize the characteristics of UCI and real world data sets, including the number of examples (**EXP***), the number of features (**FEA***), the whole number of class labels (**CL***) and the average number of class labels (**AVG-CL***).

Meanwhile, we employ seven state-of-the-art methods for comparative studies: **PL-SVM** [26], **PL-KNN** [13], **CLPL** [6], **LSB-CMM** [18], **IPAL** [35], **M3PL** [33] and **GM-PLL** [25], where the configured parameters are utilized according to the suggestions in respective literatures.

- **PL-SVM** [26]: Based on the maximum-margin strategy, it gets the predicted-label according to calculating the maximum values of model outputs. [suggested configuration: $\lambda \in \{10^{-3}, 10^{-2}, \ldots, 10^3\}$] ;
- **PL-KNN** [13]: Based on k-nearest neighbor method, it gets the predicted-label according to averaging the outputs of the k-nearest neighbors. [suggested configuration: $k = 10$];

Table 1. Characteristics of the controlled UCI data sets

UCI data sets	EXP*	FEA*	CL*	Configurations
Ecoli	336	7	8	
Dermatology	366	34	6	$r \in \{1, 2, 3\}$
Vehicle	846	18	4	$p \in \{0.1, \ldots, 0.7\}$
Abalone	4177	7	29	

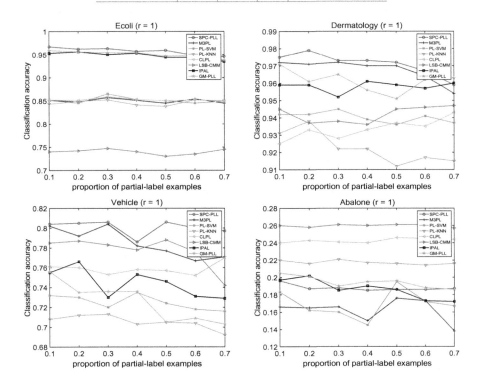

Fig. 2. The classification accuracy of each comparing method on four controlled UCI data sets with one false candidate label ($r = 1$)

Table 2. Characteristics of the real world data sets

Real World data sets	EXP*	FEA*	CL*	AVG-CL*	TASK DOMAIN
Lost	1122	108	16	2.33	*Automatic Face Naming* [6]
BirdSong	4998	38	13	2.18	*Bird Sound Classification* [18]
MSRCv2	1758	48	23	3.16	*Image Classification* [4]
Soccer Player	17472	279	171	2.09	*Automatic Face Naming* [11]
FG-NET	1002	262	99	7.48	*Facial Age Estimation* [28]
Yahoo! News	22991	163	219	1.91	*Automatic Face Naming* [11]

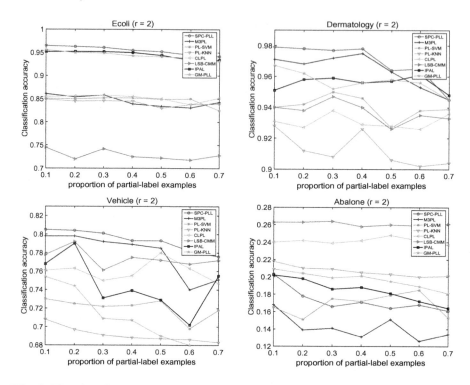

Fig. 3. The classification accuracy of each comparing method on four controlled UCI data sets with two false candidate labels ($r = 2$)

Table 3. Win/tie/loss counts of the SPC-PLL's classification performance against each comparing method on controlled UCI data sets (pairwise t-test at 0.05 significance level)

Data set	PL-KNN	PL-SVM	LSB-CMM	CLPL	M3PL	IPAL	GM-PLL	SUM
Ecoli	21/0/0	21/0/0	21/0/0	21/0/0	21/0/0	16/4/1	18/3/0	139/7/1
Dermatology	21/0/0	21/0/0	21/0/0	21/0/0	9/12/0	16/4/1	9/12/0	118/28/1
Vehicle	21/0/0	20/0/1	19/1/1	19/0/2	16/5/0	20/0/1	21/0/0	136/6/5
Abalone	0/0/21	12/4/5	0/0/21	0/0/21	21/0/0	2/8/11	0/3/18	35/15/97
SUM	63/0/21	74/4/6	61/1/22	61/0/23	67/17/0	54/16/14	48/18/18	428/56/104

- **CLPL** [6]: A convex optimization partial-label learning method via averaging-based disambiguation [suggested configuration: SVM with hinge loss];
- **LSB-CMM** [18]: Based on maximum-likelihood strategy, it gets the predicted-label according to calculating the maximum-likelihood value of the model with unseen instances input. [suggested configuration: q mixture components];
- **M3PL** [33]: Originated from PL-SVM, it is based on the maximum-margin strategy, and it gets the predicted-label according to calculating

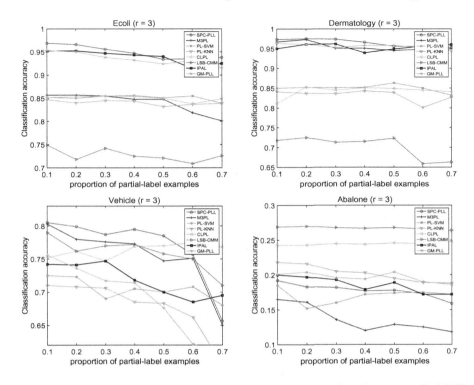

Fig. 4. The classification accuracy of each comparing method on four controlled UCI data sets with three false candidate labels ($r = 3$)

the maximum values of model outputs. [suggested configuration: $C_{max} \in \{10^{-2}, 10^{-1}, \ldots, 10^2\}$] ;

- **IPAL** [35]: it disambiguates the candidate label set by utilizing instance-based techniques [suggested configuration: $k = 10$];
- **GM-PLL** [25]: A graph matching based partial-label learning method, which transfers the task of PLL to matching selection problem and disambiguates the candidate label set according to graph matching strategy. [suggested configuration: $\beta = \{0.3, 0.4, \ldots, 0.8\}$];

Before conducting the experiments, we pre-introduce the values of parameters employed in our framework. Specifically, we set C_{max} and η among $\{10^{-2}, 10^{-1}, \ldots, 10^2\}$ via cross-validation. And the initial value of λ is empirically set to more than 0.5 to guarantee that at least half of the training examples can be learned in the first iteration. Furthermore, the other variables are set as $\tau = 0.5$, $loss^{max} = 10^{-4}$ and $\mu = 1.05$. After initializing the above variables, we adopt ten-fold cross-validation to train the model and obtain the classification accuracy on each data set.

500 G. Lyu et al.

Table 4. Classification accuracy of each algorithm on real world data sets. •/○ indicates that SPC-PLL is statistically superior/inferior to the algorithm. (pairwise t-test at 0.05 significance level)

	Lost	MSRCv2	BirdSong	SoccerPlayer	FG-NET	Yahoo! News
SPC-PLL	**0.771 ± 0.041**	**0.567 ± 0.023**	**0.715 ± 0.012**	0.508 ± 0.010	**0.089 ± 0.012**	0.665 ± 0.008
PL-SVM	0.729 ± 0.056 •	0.482 ± 0.027 •	0.662 ± 0.018 •	0.443 ± 0.004 •	0.063 ± 0.010 •	0.636 ± 0.010 •
M3PL	0.732 ± 0.035 •	0.546 ± 0.030 •	0.709 ± 0.010 •	0.446 ± 0.013 •	0.037 ± 0.025 •	0.655 ± 0.009 •
PL-KNN	0.424 ± 0.030 •	0.448 ± 0.012 •	0.637 ± 0.009 •	0.494 ± 0.004 •	0.037 ± 0.008 •	0.457 ± 0.010 •
CLPL	0.735 ± 0.024 •	0.413 ± 0.020 •	0.632 ± 0.009 •	0.347 ± 0.004 •	0.047 ± 0.017 •	0.462 ± 0.009 •
LSB-CMM	0.707 ± 0.019 •	0.456 ± 0.008 •	0.692 ± 0.015 •	0.506 ± 0.006 •	0.056 ± 0.008 •	0.648 ± 0.007 •
IPAL	0.726 ± 0.041 •	0.523 ± 0.025 •	0.708 ± 0.014 •	0.547 ± 0.014 ○	0.057 ± 0.023 •	**0.667 ± 0.006** ○
GM-PLL	0.737 ± 0.043 •	0.530 ± 0.019 •	0.663 ± 0.010 •	**0.549 ± 0.009** ○	0.065 ± 0.021 •	0.629 ± 0.007 •

Table 5. Classification accuracy of SPC-PLL and its degenerated algorithms on real world data sets. The results with boldface indicate that SPC-PLL is statistically superior to its degenerated algorithms (pairwise t-test at 0.05 significance level).

	Lost	MSRCv2	BirdSong	SoccerPlayer	FG-NET	Yahoo! News
SPC-PLL	**0.771 ± 0.041**	**0.567 ± 0.023**	**0.715 ± 0.012**	**0.508 ± 0.010**	**0.089 ± 0.012**	**0.665 ± 0.008**
M3PL	0.732 ± 0.041	0.546 ± 0.030	0.709 ± 0.010	0.446 ± 0.013	0.037 ± 0.025	0.655 ± 0.009
SP-PLL	0.745 ± 0.028	0.559 ± 0.015	0.710 ± 0.008	0.465 ± 0.013	0.072 ± 0.023	0.663 ± 0.010

5.2 Experimental Results

Controlled UCI Data Sets. We compare the SPC-PLL with all above comparing methods on four controlled UCI data sets, and Fig. 2, 3 and 4 illustrate the classification accuracy of each comparing method as p increases from 0.1 to 0.7 with the step-size of 0.1. Together with the ground-truth label, the r class labels are randomly chosen from $\mathcal{Y}$ to constitute the rest of each candidate label set, where $r \in \{1, 2, 3\}$. Table 3 summaries the win/tie/loss counts between SPC-PLL and the other comparing methods. Out of 84 (4 data sets × 21 configurations) statistical comparisons show that:

- Among these comparing methods, SPC-PLL achieves superior or comparable performance against PL-SVM and M3PL in 92.9% and 100% cases respectively, against other comparing methods over 75% cases, which significantly demonstrates the effectiveness of self-paced curriculum strategy on solving PLL problem.
- Among these controlled UCI data sets, SPC-PLL outperforms all comparing methods on *Ecoli*, *Dermatology* and *Vehicle* data sets over 99.3%, 99.3% and 92.5%, respectively. On *Abalone* data set, although SPC-PLL only achieves superior performance in 34.0% cases, it significantly outperforms its baseline methods PL-SVM and M3PL in 92.9% and 100% cases, which also demonstrates the validity of the self-paced curriculum strategy on solving PLL problem.

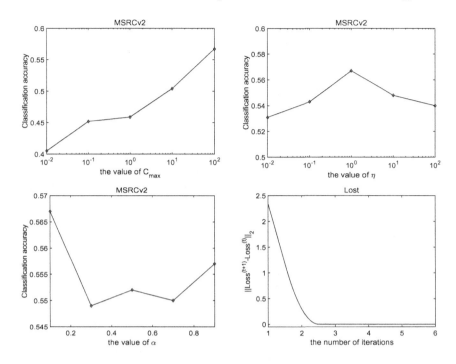

Fig. 5. The left three subfigures show the classification accuracy changes as each parameter increases with other parameters fixed on *MSRCv2* data set. The right subfigure shows the convergence of SPC-PLL on *Lost* data set.

Real World Data Sets. We compare the SPC-PLL with the comparing algorithms on real world data sets. The comparison results are reported in Table 4, where the results are based on ten-fold cross-validation. According to Table 4, it is clear to observe that:

- Among these comparing methods, SPC-PLL is superior or comparable to all comparing state-of-the-art methods. Especially, it outperforms its baseline methods (PL-SVM and M3PL) and most other comparing methods (PL-KNN, CLPL and LSB-CMM) on all real world data sets. Meanwhile, it also outperforms IPAL and GM-PLL methods on 4/6 and 5/6 cases, respectively.
- Among these real world data sets, SPC-PLL achieves superior or comparable performance on all employed data sets. Specifically, SPC-PLL outperforms all comparing methods on *Lost*, *MSRCv2*, *BirdSong* and *FG-NET* data sets, and it also ranks 2nd and 3rd on *Yahoo! News* and *SoccerPlayer* data sets, respectively.

Summary. The two series of experiments mentioned above powerfully demonstrate the effectiveness of SPC-PLL, and we attribute the success to the superiority of self-paced and curriculum strategy. Such "instructor-student-collaborative" learning mode can not only utilize predetermined knowledge to avoid the learned model from overfitting, but also dynamically adjust the curriculum to maintain the consistency between the prior knowledge and the learned model. Under the guidance of such strategy, the learned model is gradually growing from "juvenile" to "mature", and finally achieves better generalization on the task of PLL prediction.

5.3   Further Analysis

Ablation Study. To demonstrate the effectiveness of our employed self-paced curriculum strategy, we conduct comprehensive ablation study on our proposed SPC-PLL method. Since the curriculum strategy can not be independently incorporated in our baseline methods, we only compare our proposed SPC-PLL method with its degenerated self-paced strategy based method (SP-PLL) [24] and its baseline M3PL method [33]. Table 5 illustrates the detail experimental comparisons between SPC-PLL and its degenerated algorithms on real world data sets. According to Table 5, it is observed that our proposed SPC-PLL method significantly outperforms its degenerated algorithms, which demonstrates the validity of our proposed self-paced curriculum on solving PLL problem.

Parameter Analysis. We study the sensitivity of SPC-PLL with respect to its three employed parameters, i.e. C_{max}, η and α. The left three subfigures of Fig. 5 show the performance of SPC-PLL under different parameter configurations on *MSRCv2* data set. As illustrated in Fig. 5, the values of both C_{max} and η usually have great influences on the performance of the proposed SPC-PLL. Thus, we set them among $\{10^{-2}, 10^{-1}, \ldots, 10^2\}$ via cross-validation. Besides, we also set α among $\{0.1, 0.3, 0.5, 0.7, 0.9\}$ via cross-validation.

Convergence Analysis. We conduct the convergence analysis of SPC-PLL on *Lost* data set. The right subfigure of Fig. 5 illustrates the convergence curve of the optimization process of SPC-PLL in one of the iterations. We can easily observe that each $\|Loss^{(t+1)} - Loss^{(t)}\|$ gradually decreases to 0 as t increases. Therefore, the convergence of SPC-PLL is demonstrated.

6   Conclusion

In this paper, we have proposed a novel self-paced curriculum strategy-based partial label learning method. To the best of our knowledge, it is the first time to incorporate the self-paced curriculum strategy into partial label learning framework, which provides a new perspective for improving the robustness of weakly-supervised learning model. Extensive experimental comparisons and comprehensive ablation study strongly demonstrate the effectiveness of such strategy on solving PLL problem.

Acknowledgement. This work was supported in part by the Fundamental Research Funds for the Central universities (2020YJS026, 2019JBM020), in part by the National Natural Science Foundation of China (No. 61872032), in part by the Beijing Natural Science Foundation (No. 4202058, No. 9192008), and in part by the Key R&D Program of Zhejiang Province (No. 2019C01068).

References

1. Bengio, S., Weston, J., Grangier, D.: LIBLINEAR: a library for large linear classification. J. Mach. Learn. Res. **9**, 1871–1874 (2008)
2. Bengio, Y., Louradour, J., Collobert, R., Weston, J.: Curriculum learning. In: International Conference on Machine Learning, pp. 41–48 (2009)
3. Boyd, S., Vandenberghe, L.: Convex Optimization. Cambridge University Press, Britain (2004)
4. Briggs, F., Fern, X., Raich, R.: Rank-loss support instance machines for MIML instance annotation. In: International Conference on Knowledge Discovery and Data Mining, pp. 534–542 (2012)
5. Chen, Y., Patel, V., Chellappa, R., Phillips, P.: Ambiguously labeled learning using dictionaries. IEEE Trans. Inf. Forensics Secur. **9**, 2076–2088 (2014)
6. Cour, T., Sapp, B., Taskar, B.: Learning from partial labels. J. Mach. Learn. Res. **12**, 1502–1536 (2011)
7. Feng, L., An, B.: Leveraging latent label distributions for partial label learning. In: International Joint Conference on Artificial Intelligence, pp. 2107–2113 (2018)
8. Feng, L., An, B.: Partial label learning by semantic difference maximization. In: International Joint Conference on Artificial Intelligence, pp. 2294–2300 (2019)
9. Feng, L., An, B.: Partial label learning with self-guided retraining. In: AAAI Conference on Artificial Intelligence, pp. 3542–3549 (2019)
10. Gong, C., Liu, T., Tang, Y., Jian, Y., Jie, Y., Tao, D.: A regularization approach for instance-based superset label learning. IEEE Trans. Cybern. **48**, 967–978 (2017)
11. Guillaumin, M., Verbeek, J., Schmid, C.: Multiple instance metric learning from automatically labeled bags of faces. In: Daniilidis, K., Maragos, P., Paragios, N. (eds.) ECCV 2010. LNCS, vol. 6311, pp. 634–647. Springer, Heidelberg (2010). https://doi.org/10.1007/978-3-642-15549-9_46
12. Huang, J., et al.: Improving multi-label classification with missing labels by learning label-specific features. Inf. Sci. **492**, 124–146 (2019)
13. Hüllermeier, E., Beringer, J.: Learning from ambiguously labeled examples. In: Famili, A.F., Kok, J.N., Peña, J.M., Siebes, A., Feelders, A. (eds.) IDA 2005. LNCS, vol. 3646, pp. 168–179. Springer, Heidelberg (2005). https://doi.org/10.1007/11552253_16
14. Jiang, L., Meng, D., Yu, S., Lan, Z., Shan, S., Hauptmann, A.: Self-paced learning with diversity. In: Advances in Neural Information Processing Systems, pp. 2078–2086 (2014)
15. Jiang, L., Meng, D., Zhao, Q., Shan, S., Hauptmann, A.: Self-paced curriculum learning. In: AAAI Conference on Artificial Intelligence, pp. 2694–2700 (2015)
16. Khan, F., Mutlu, B., Zhu, J.: How do humans teach: on curriculum learning and teaching dimension. In: Advances in Neural Information Processing Systems, pp. 1449–1457 (2011)
17. Kumar, M., Packer, B., Koller, D.: Self-paced learning for latent variable models. In: International Conference on Neural Information Processing Systems, pp. 1189–1197 (2010)

18. Liu, L., Dietterich, T.: A conditional multinomial mixture model for superset label learning. In: Advances in Neural Information Processing Systems, pp. 548–556 (2012)

19. Liu, L., Dietterich, T.: Learnability of the superset label learning problem. In: International Conference on Machine Learning, pp. 1629–1637 (2014)

20. Luo, J., Orabona, F.: Learning from candidate labeling sets. In: Advances in Neural Information Processing Systems, pp. 1504–1512 (2010)

21. Lv, J., Xu, N., Feng, L., Niu, G., Geng, X., Sugiyama, M.: Progressive identification of true labels for partial label learning. In: International Conference on Machine Leaning, pp. 1–10 (2020)

22. Lyu, G., Feng, S., Huang, W., Dai, G., Zhang, H., Chen, B.: Partial label learning via low-rank representation and label propagation. Soft Comput. **24**, 5165–5176 (2020). https://doi.org/10.1007/s00500-019-04269-9

23. Lyu, G., Feng, S., Li, Y., Jin, Y., Dai, G., Lang, C.: HERA: partial label learning by combining heterogeneous loss with sparse and low-rank regularization. ACM Trans. Intell. Syst. Technol. **11**, 1–19 (2020)

24. Lyu, G., Feng, S., Wang, T., Lang, C.: A self-paced regularization framework for partial label learning. IEEE Trans. Cybern. (2020). https://doi.org/10.1109/TCYB.2020.2990908

25. Lyu, G., Feng, S., Wang, T., Lang, C., Li, Y.: GM-PLL: graph matching based partial label learning. IEEE Trans. Knowl. Data Eng. (2019). https://doi.org/10.1109/TKDE.2019.2933837

26. Nguyen, N., Caruana, R.: Classification with partial labels. In: ACM SIGKDD International Conference on Knowledge Discovery and Data Mining, pp. 551–559 (2008)

27. Oukhellou, L., Denux, T., Aknin, P.: Learning from partially supervised data using mixture models and belief functions. Pattern Recogn. **42**, 334–348 (2009)

28. Panis, G., Lanitis, A.: An overview of research activities in facial age estimation using the FG-NET aging database. J. Am. Hist. **5**, 455–462 (2015)

29. Spitkovsky, V., Alshawi, H., Jurafsky, D.: Baby steps: How "less is more" in unsupervised dependency parsing. In: Advances in Neural Information Processing Systems (2009)

30. Wang, H., Liu, W., Zhao, Y., Hu, T., Chen, K., Chen, G.: Learning from multi-dimensional partial labels. In: International Joint Conference on Artificial Intelligence (2020)

31. Wang, H., Liu, W., Zhao, Y., Zhang, C., Hu, T., Chen, G.: Discriminative and correlative partial multi-label learning. In: International Joint Conference on Artificial Intelligence, pp. 3691–3697 (2019)

32. Yao, Y., Feng, J., Chen, X., Gong, C., Wu, J., Yang, J.: Deep discriminative CNN with temporal ensembling for ambiguously-labeled image classification. In: AAAI Conference on Artificial Intelligence, pp. 3542–3549 (2020)

33. Yu, F., Zhang, M.: Maximum margin partial label learning. Mach. Learn. **106**, 573–593 (2017). https://doi.org/10.1007/s10994-016-5606-4

34. Zeng, Z., et al.: Learning by associating ambiguously labeled images. In: IEEE Conference on Computer Vision and Pattern Recognition, pp. 708–715 (2013)

35. Zhang, M., Yu, F.: Solving the partial label learning problem: an instance-based approach. In: International Joint Conference on Artificial Intelligence, pp. 4048–4054 (2015)

36. Zhang, M., Yu, F., Tang, C.: Disambiguation-free partial label learning. IEEE Trans. Knowl. Data Eng. **29**, 2155–2167 (2017)

37. Zhang, M., Zhou, B., Liu, X.: Partial label learning via feature-aware disambiguation. In: International Conference on Knowledge Discovery and Data Mining, pp. 1335–1344 (2016)
38. Zhou, Y., He, J., Gu, H.: Partial label learning via Gaussian processes. IEEE Trans. Cybern. **47**, 4443–4450 (2016)

Reinforcement Learning

Option Encoder: A Framework for Discovering a Policy Basis in Reinforcement Learning

Arjun Manoharan[1,2(✉)], Rahul Ramesh[3(✉)], and Balaraman Ravindran[1,2]

[1] Indian Institute of Technology Madras, Chennai, India
arjunmanoharan2811@gmail.com, ravi@cse.iitm.ac.in
[2] Robert Bosch Centre for Data Science and AI, Chennai, India
[3] University of Pennsylvania, Philadelphia, USA
rahulram@seas.upenn.edu

Abstract. Option discovery and skill acquisition frameworks are integral to the functioning of a hierarchically organized Reinforcement learning agent. However, such techniques often yield a large number of options or skills, which can be represented succinctly by filtering out any redundant information. Such a reduction can decrease the required computation while also improving the performance on a target task. To compress an array of option policies, we attempt to find a policy basis that accurately captures the set of all options. In this work, we propose *Option Encoder*, an auto-encoder based framework with intelligently constrained weights, that helps discover a collection of basis policies. The policy basis can be used as a proxy for the original set of skills in a suitable hierarchically organized framework. We demonstrate the efficacy of our method on a collection of grid-worlds evaluating the obtained policy basis on downstream tasks and demonstrate qualitative results on the Deepmind-lab task.

Keywords: Hierarchical reinforcement learning · Policy distillation

1 Introduction

Reinforcement learning (RL) [25] deals with solving sequential decision-making tasks and primarily operates through a trial-and-error paradigm for learning. The increased interest in Reinforcement learning can be attributed to the powerful function approximators from Deep learning. Deep Reinforcement Learning (DRL) has managed to achieve competitive performances on some challenging high-dimensional domains [10,15,16,21]. To scale to larger problems or reduce the training time drastically, one could attempt to structure the agent in a hierarchical fashion. The agent hence makes decisions based on abstract state and

---

A. Manoharan and R. Ramesh—The two authors contributed equally.
R. Ramesh—Work done primarily while at the Indian Institute of Technology Madras.

© Springer Nature Switzerland AG 2021
F. Hutter et al. (Eds.): ECML PKDD 2020, LNAI 12458, pp. 509–524, 2021.
https://doi.org/10.1007/978-3-030-67661-2_30

action spaces, which helps reduce the complexity of the problem. One popular realization of hierarchies is the options framework [26] which formalizes the notion of a temporally extended sequence of actions.

Discovery of options, particularly in a task agnostic manner often leads to a large number of options. Option discovery methods [13,14,22–24] as a result, typically resort to heuristics that help prune this set. In such a scenario, a compression algorithm is of utility, since it would be wasteful to discard these options and ineffective to use all of them simultaneously. When using a large number of options, the computation expended for determining the relevance of each option policy is higher, when compared to using a smaller set of basis policies [13]. In this work, we demonstrate that a reduced set of basis policies, results in improved empirical performances, on a collection of target tasks.

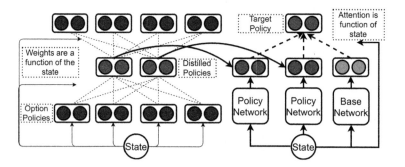

Fig. 1. A visual depiction of the option encoder framework. The blue colored layers (on the encoder side) correspond to the original option policies, and the red layer corresponds to the distilled policies. Any set of decoder weights connected to the same output, sum to 1. The distilled policies are used in a hierarchical agent that attempts to learn a policy for a downstream target task. For more details on the policy and the base network in the hierarchical agent, see [19].

Resorting to an existing policy distillation or compression method [6,7,17] is one possible alternative. However, these methods distill the options into a single network, resulting in a single policy that captures the behavior of all policies. The diversity among the different option policies may be captured efficiently only if they are distilled to more than one policy. To address the same, we propose the *Option Encoder*, a framework that attempts to find a suitable collection of basis policies, from the discovered set of options. We use an auto-encoder based model where an intermediate hidden layer is interpreted as a set of basis policies which we term as *distilled policies*. The distilled policies are forced to reconstruct the original set of options using attention weights. The intermediate hidden layers would hence be forced to capture the commonalities between the various options and potentially eliminate redundancies. The overview of this framework is summarized in Fig. 1. The obtained distilled policies can be used to solve a new set of tasks and can be used as a proxy for the original set of options in a new algorithm.

Our work also provides a simple mechanism to combine options obtained from different option discovery techniques. This is similar in spirit to [3] but we do not combine the options in a goal-directed manner. Generating options for a certain goal is useful in some scenarios but is difficult to work with in a multi-task setting. Using a task-agnostic approach (like in the Option Encoder) allows the options to be reusable across different tasks. Furthermore, the Option Encoder discards the full set of expert options after distilling to a smaller set, which is not the case in [3].

Our contributions in this work are as follows: 1) We describe the Option Encoder framework, which finds a 'basis" for a set of options by compressing them into a smaller set. 2) We present qualitative experiments, analyses and ablation experiments to justify the efficacy of our framework 3) We empirically demonstrate that the Option Encoder helps improve or retain the performance on downstream tasks when compared to using the raw set of options or when using a reduced number of options.

The experiments are conducted on a few challenging grid-worlds where we achieve a 5-fold or 10-fold reduction in the number of options while retaining or even improving the performance. We also show results in the high-dimensional visual navigation domain of Deepmind-lab.

2   Preliminaries

RL deals with sequential decision making problems and models the interaction of an agent with an environment. This interaction is traditionally modeled by a Markov Decision Process (MDP) [18], defined by the tuple $\langle \mathcal{S}, \mathcal{A}, \mathcal{T}, r, \gamma \rangle$, where $\mathcal{S}$ defines the set of states, $\mathcal{A}$ the set of actions, $\mathcal{T} : \mathcal{S} \times \mathcal{A} \to \mathcal{P}(S)$ the transition function (that maps to a probability distribution over states), $r : \mathcal{S} \times \mathcal{S}' \times \mathcal{A} \to \mathcal{P}(R)$ the reward function (that maps the current state, next state and action to a probability distribution over the rewards) and γ the discount factor. In the context of optimal control, the objective is to learn a policy that maximizes the expected discounted return $R_t = \sum_{i=t}^{T} \mathbb{E} \left[\gamma^{(i-t)} r(s_i, s_{i+1}, a_i) \right]$, where $r(s_i, s_{i+1}, a_i)$ is the reward function. Policy gradient methods attempt to find a parameterized policy $\pi(a|s; \theta)$ that maps every state to a probability distribution over the actions, such that the discounted return is maximized. Some prominent examples of policy gradient methods include Advantage-actor critic (A2C) [8] and Proximal Policy Optimization (PPO) [20].

Options: An option [26] formalizes the notion of a temporally extended sequence of actions and is denoted by the tuple $\langle \mathcal{I}, \beta, \pi_o \rangle$. $\mathcal{I} \subseteq \mathcal{S}$ denotes the initiation set of the option, $\beta : \mathcal{S} \to [0, 1]$ is the probability that the option terminates in state s and π_o is the option policy. In this work, we assume that the initiation set is the set of all states.

Attend, Adapt and Transfer: Attend, Adapt and Transfer architecture (A2T) [19] is a model for utilizing expert policies (or options) from N different source tasks in order to tackle a target task. Consider N expert policies, represented by $\{K_i(s)\}_{i=1}^{N}$. Apart from the expert networks (which are fixed throughout training), A2T has a trainable *base network* represented by $K_B(s)$, which is used to learn in regions of the state space, where the set of experts do not suffice. The target policy $K_T(s)$ is given by:

$$K_T(s) = w_{N+1}(s)K_B(s) + \sum_{i=1}^{N} w_i(s)K_i(s) \tag{1}$$

The set of weights $w_j(s)$ (for $j \in \{1 \cdots N+1\}$) are attention weights and as a consequence, satisfy the constraint $\sum_{i=1}^{N+1} w_j(s) = 1$. $K_T(s)$ is a convex combination of $N + 1$ policies and is hence also a valid policy.

In this work we use a modified version of the A2T algorithm, identical to the version in [5]. Instead of combining the option policies using the attention weights, the A2T agent instead selects a single option policy and persists the same for T steps. Every option terminates, T steps after being selected. We refer to the persistence length T as the *termination limit*.

This modification can be understood as a hard-attention variant of the A2T framework. The hard-attention weights are trained using A2C since the network is no longer differentiable. The temporal persistence of the modified A2T algorithm forces the hierarchical agent to exploit the inherent structure present in the option policies. We also observed an empirical improvement in performance with the modified A2T variant and hence use the variant in all our experiments. Henceforth, any reference to A2T refers to the modified version.

Actor-Mimic Network: Given a set of source tasks, the Actor-mimic framework [17] attempts to learn a network that copies the policies of the various experts. The loss corresponding to task i is given by:

$$\mathcal{L}_{policy}^{i} = \sum_{a \in \mathcal{A}_{\rangle}} \pi_{E_i}(a|s) \log \pi_{AMN}(a|s; \theta) \tag{2}$$

π_{AMN} is a parameterized network that is trained using the cross-entropy loss. The targets are generated from the expert policy π_{E_i}. In the case where the expert consists of Q-values, the targets are generated from the Boltzmann distribution controlled by a temperature parameter τ. Parisotto et al. [17] uses an additional feature regression loss which we omit in this work.

3   Option Encoder Framework

The Option Encoder attempts to find a collection of basis policies, that can accurately characterize a collection of option policies. Let the policy of an option j, be denoted by π_j. Let the ith policy of the distilled set (intermediate layer) be represented by π_i^d. The set of M "distilled" policies $\pi^d = \{\pi_i^d\}_{i=1}^M$ are found by minimizing the objective given in Eq. 3.

$$\pi^{d*} = \arg\min_{\pi^d} \ \min_W \sum_{s \in \mathcal{S}} \sum_j \mathcal{L}\left(\pi_j(s), \left(\sum_i w_{ij}(s) \times \pi_i^d(s)\right)\right) \qquad (3)$$

W is a weight matrix with the entry in i^{th} row and j^{th} column denoted by w_{ij}. The matrix W is such that $\sum_i w_{ij} = 1 \ \forall j$, which implies that the rows of the matrix are attention weights. W can also be a function of the state s. Each w_{ij} is a scalar such that $0 \leq w_{ij} \leq 1$ and it indicates the contribution of distilled policy π_i^d, to the reconstruction of option policy π_j. The function $\mathcal{L}$ is a distance measure between the two probability distributions (for example Kullback-Leibler divergence or Huber Loss). The objective states that a convex combination of the distilled policies should be capable of reconstructing each of the original set of options, as accurately as possible.

Equation 3 is realized using an auto-encoder. The encoder is any suitable neural network architecture that outputs M different policies. For example, an encoder in a task with a discrete action space will consist of M different softmax outputs, each of size $|\mathcal{A}|$ (size of action space). A continuous action space problem will contain M sets of policy parameters (for example, the mean and variance of a Gaussian distribution). Since the distilled policies are combined linearly, a single layer for the decoder should suffice, since the addition of more layers will not add any more representational power. The entire procedure is summarized in Algorithm 1.

3.1   Architectural Constraints in the Option Encoder

The architecture is an auto-encoder with two key constraints (see Fig. 1). The first restriction is that each distilled policy has a single shared weight. Alternately, all actions corresponding to a single policy have the same shared weight. This ensures that the structure in the action space of the distilled policies are preserved. The second restriction is that the set of weights responsible for reconstructing any option policy must sum to 1. These weights are attention weights and can be agnostic to the current state or be an arbitrary function of it.

These restrictions are imposed to respect the objective specified in Eq. 3 i.e., the re-constructed expert policies are convex combinations of the distilled policies. Furthermore the constraints ensure that the distilled policies are coherent since a heavily parameterized decoder permits the information to be captured in the decoder weights, as opposed to the distilled policies.

To illustrate the utility of these restrictions, consider a scenario in which all the option policies indicate that the action a has the highest preference in state s. Let action a be assigned the least probability after passing the options through an encoder. If the weights are allowed to take arbitrary values on the decoder side, the distilled policies are capable of reconstructing the options by assigning higher weights to action a, even though it has a low probability as per

---

Algorithm 1: Summary of the Option Encoder Framework

L = Number of rollouts for building distilled policies ;
N = Number of Option policies ;
M = Number of Distilled Policies ;
K = Number of Target Goals ;
T = Number of Steps, an option is persisted ;
E = Number of Episodes for the transfer stage ;
Dataset = Empty list ;
for *j in (1..... N)* **do**
    env.reset() ;
    for *i in (1..... L)* **do**
        Get option policies $(\pi_1(s), \pi_2(s), \cdots \pi_N(s))$;
        Add $(s, (\pi_1(s), \pi_2(s), \cdots \pi_N(s)))$ to Dataset;
        a = Sample$(\pi_j(s))$;
        s = env.step(a) ;
    end
end
DistillPolicies = train_auto-encoder(Dataset) ;
for *j in (1..... M)* **do**
    DistillDataset = None ;
    for *s in Dataset[0, :]* **do**
        $\hat{\pi}_j(s)$ = DistillPolicies(j, s) ;
        add $\hat{\pi}_j(s)$ to DistillDataset ;
    end
    π_j = ActorMimic(DistillDataset) ;
end
for *i in (1....K)* **do**
    for *j in (1....E)* **do**
        env.reset() ;
        while *not done* **do**
            Option_id = AttentionNetwork.Sample(s) ;
            for *t in(1...T)* **do**
                a = Option_id.Sample(s) ;
                s = env.step(a) ;
                store_transitions();
            end
            collect_rollout() ;
            UpdateAttentionNetwork(rollout) ;
            UpdateBaseNetwork(rollout)
        end
    end
end

---

the distilled policies. Alternately, the decoder can make use of negative weights to flip the preference order over the actions dictated by the distilled policies.

Ideally, one would want the distilled policies to capture the fact that action a is preferred in-state s among all options. Hence, the proposed two restrictions ensure that this intended behavior is achieved. The second restriction also ensures that the output of the decoders are also valid policies since a weighted combination of the policies (with the weights summing to 1) will also result in a policy.

3.2 Extracting the Distilled Policies

The current setup would require the execution of the encoder in order to obtain the distilled policies. This would, however, defeat the entire purpose of the distillation procedure since the encoder would require the option policies as inputs. An ideal scenario would allow us to discard the options after the distillation. In order to achieve the same, the distilled policies (outputs of the encoder) are utilized as targets to train a network, using an algorithm like Actor-mimic [17]. As a result, each distilled policy is transferred to a network using a supervised learning procedure. The distilled policies can now be computed from the state, without computing the option policies. Hence, the network can be used for decision-time planning or for policy execution in the absence of the original set of options.

We do not make use of [6,7] for distillation because they are computationally expensive and primarily address the incremental learning setup. Elastic weight consolidation based training [7] does not easily extend to multiple tasks and requires computing the Fisher information matrix which can be expensive for large networks. Pathnet [6] uses a genetic algorithm to obtain a distilled network, which can be inefficient with respect to the required number of training iterations.

4 Experiments

In this section, we describe experiments designed to answer the following questions:

- How do the distilled policies compare against the option policies on a set of tasks?
- Why are certain restrictions imposed on the architecture?
- Is the performance gain solely due to a reduced number of policies?
- Does varying the number of distilled policies affect the performance?
- How does the termination limit of the hierarchical agent impact the performance?

4.1 Task Description

Grid-world: We consider the grid-worlds depicted in Fig. 2. The grid-worlds are stochastic where the agent moves in the intended direction with probability 0.8

Fig. 2. 3 grid-worlds are tackled in this work. The yellow dots indicate the goals for a collection of tasks that we attempt to solve for in the grids GW1,GW2,GW3 (left to right) (Color figure online)

and takes a random action (uniform probability) otherwise. The environment has 4 actions available from every state, which are up, down, left and right. Each episode terminates after 3000 environment steps. We consider a task where the agent obtains a reward of +1 on reaching the designated goal and a reward of 0 for every other transition. Fifty options were learned using the Eigen-options framework [11] for each grid-world which were then used to solve the task of reaching 100 randomly selected goals. These goals are denoted by the yellow dots on the grid-world in Fig. 2. Three different grid-worlds GW1, GW2, and GW3 (left to right in Fig. 2) were considered. GW1 and GW2 are of sizes 28 × 31 each and GW3 is of size 41 × 41.

Deepmind Lab: The Deepmind-Lab domain [4] is a visual maze navigation task where the inputs are images. In this work, the images are converted to grayscale images before being used as inputs to a network. The action space is discretized into 4 actions which are forward, backward, rotate-left and rotate right. Every step receives a reward of −0.01 and the episode ends after reaching a designated goal or after 3000 environment steps. The agent additionally receives a reward of 1.0 on reaching the goal and receives a reward of −1.0 if it fails to reach the goal after 3000 steps in the environment.

4.2   Architecture Overview

The state in all grid-world experiments is represented as an image of the grid (with 1 channel) with all zeros, except at the location of the agent. We impose the encoder to also have shared attention weights (each policy has a single attention weight) like the decoder. This implies that the original set of options are combined using attention weights to yield the distilled policies which are then combined using another set of attention weights to yield the reconstructions.

Option Encoder: The encoder and the decoder are comprised of attention weights which are functions of the current state. The state-based attention network consists of two convolution layers (5 × 5 × 4 and stride 2 and 3 × 3 × 8 and stride 1) and a fully connected layer with 32 units which then outputs the attention weights. For the Deepmind-Lab task, the current state is converted into attention weights using a network that contains 3 convolution layers (8 × 8 × 32

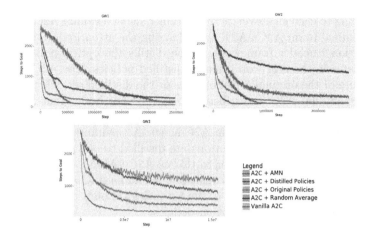

Fig. 3. Plot depicting the number of steps to reach the goal (performance measure) vs. the number of environment steps on GW1, GW2, GW3 respectively (left to right)

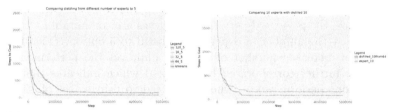

(a) Distilling varying number of experts to 5 options

(b) Comparing 10 Eigen-options to 10 options obtained from distilling 64 Eigen-options

Fig. 4. Option encoder is a better alternative than discarding the discovered options and is capable of reducing a large number of options

and stride 4, $4 \times 4 \times 64$ and stride 2, $3 \times 3 \times 64$ and stride 1) followed by a fully connected layer of size 512 which then outputs the attention weights.

Hierarchical Agent: The A2C algorithm with the modified A2T framework (described in Sect. 2) was used to train the agent (referred to as the A2T + A2C agent). The base network consists of 2 convolution layers (same configuration as earlier) followed by a fully connected layer of size 128 which yields the policy and the value function heads. The base network policy and the option policies are combined using attention weights to yield the final policy. The termination limit is 20 in this case.

4.3 Evaluating on Grid-Worlds

50 options were obtained using the Eigen-options framework [11] where the eigen-vectors of the graph laplacian are used to define options. The policies

corresponding to each eigen-vector are obtained using a vanilla A2C agent (architecture identical to an A2C+A2T agent barring the attention). This is followed by the Option Encoder framework, which distills the option policies obtained from all states in the grid-world, into 5 distilled policies. The A2C+A2T agent can make use of the original set of options or the distilled set, which we term as A2C + original and A2C + distilled respectively. For both agents, the selected option is persisted for 20 steps (termination limit) after which a new option is selected. We also evaluate a vanilla A2C agent. Actor-mimic is another baseline that we consider, where all the options are distilled to one policy which is used in the A2C + A2T setup (we refer to this as A2C+AMN). Finally, we consider the random average agent which consists of an A2T+A2C agent attending to 5 policies and a base network. Each of the 5 policies are obtained by averaging the policies of 10 randomly chosen option policies.

The agents are periodically evaluated every 500 environment steps and the performance curves are presented in Fig. 3. The graphs are clearly indicative of the fact that the A2C+distilled agent outperforms all other baselines. Since we tackle 100 different target tasks, the effort required to obtain the distilled policies (or the options) is negligible when compared to solving the multi-task problem. Hence, the presented performance curves are comparable.

We also vary the number of experts and distill to 5 options (Fig. 4(a)) and notice that an increased number of experts, improves the performance. This observation is further corroborated by Fig. 4(b) which indicates that the top 10 Eigen-options perform worse, when compared to using a set of 10 distilled options. In a resource constrained situation where only few options are required, one can distill knowledge from many options to a smaller set. We also attempted to cluster the policies using K-means on the policy space. Unsurprisingly, our distilled policies outperformed the options obtained using K-means. The centroid of each K-means cluster was used as a substitute for the distilled policy. We run K-means to discover 5 clusters from 50 expert policies.

4.4   Understanding Architecture Constraints

We enforce certain restrictions on the auto-encoder as described in Sect. 3.1. We conduct a qualitative analysis of different architectural variants. Remember that the Option Encoder architecture requires the weights to be attention weights and the policy to be a convex combination of the policies from the previous layers. This implies that all actions of a policy share a single weight.

We consider the grid-world in Fig. 5(a) with four expert options going to the 4 corners of the top-left room. Figure 5(a) represents a heatmap of the distribution of states visited by the 2 hidden policies generated from the Option Encoder framework. We visualize the heatmap for following architectural variants:

- *Removing restriction on attention weights*: In this case, the weights are not fed through a softmax layer to enforce that they sum to 1. Figure 5(b) highlights that we do not learn an interpretable policy.

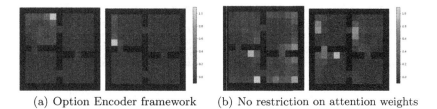

(a) Option Encoder framework   (b) No restriction on attention weights

Fig. 5. Heatmap of the visitation distribution of two distilled policies

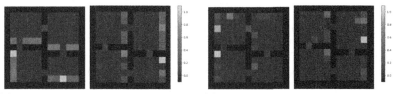

(a) No restriction on weight-sharing for (b) No restrictions on both attention
policies   and weight-sharing

Fig. 6. Heatmap of the visitation distribution of two distilled policies

- *Removing restriction on weight sharing*: Since the layers enforce the output policies to be convex combinations of the input policies, each policy is assigned a single attention weight. We instead assign a single weight for every action in the policy increasing the number of weights by $|\mathcal{A}|$ (size of the action space). Figure 6(a) again highlights that we do no learn qualitatively useful policies.
- *Removing both restrictions*: This case corresponds to a vanilla auto-encoder with no constraints on the weights. An intermediate layer is interpreted as a policy and is found to not be qualitatively interpretable (see Fig. 6(b)).

4.5 Understanding the Distilled Policies

To understand the Option Encoder framework, we consider 16 expert policies, each navigating to a specific goal as indicated in the left-most image in Fig. 7. Each blue cell denotes a goal towards which an expert policy navigates to optimally. These experts are distilled to 4 different policies. In order to visualize these policies, we develop a heatmap of the visitation counts for each policy. The heatmaps are obtained by sampling from the respective distilled policies. The agent is executed for 50,000 steps and is reset to a random start state after 100 steps in the environment. Figure 7 demonstrates how the Option Encoder framework captures the commonalities between various policies. We also compare the same with the visitation count plot obtained from an Actor-mimic network trained by combining all 16 policies.

Fig. 7. The leftmost figure denotes the set of 16 expert policies while the middle 4 figures (2nd to 5th image from left) visualize the visitation count of the rollouts of the distilled policy. The rightmost figure corresponds to the visitation count of an Actor mimic network (AMN) trained to distill 16 policies into 1 policy.

4.6   Randomly Sampling a Set of Options

This analysis on GW2 was conducted to demonstrate that our proposed framework does not derive a significant advantage from using a reduced number of option policies. 50 options were divided into 10 sets of options (each of size 5) where each option appears in exactly one of the ten sets. Figure 8 shows that a random sample of options can lead to vastly varying performances (based on the relevance of the options). However, the distilled policies outperform every set of

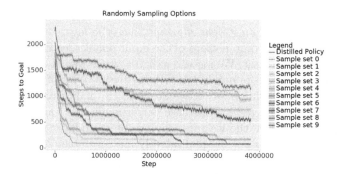

Fig. 8. Performance comparison against a few random subsets of options to show that the performance improvement is not due to a reduced number of options

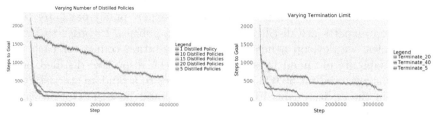

(a) Performance plotted for a varying number of distilled policies.

(b) Varying the termination limit of the A2T + A2C setup

Fig. 9. The Option Encoder is fairly robust to hyper-parameters like the number of distilled policies or the termination limit

random options we consider, hinting at the fact that all the options are useful for solving a new set of tasks. Each line in Fig. 8 corresponds to the average performance over 25 randomly selected goals.

4.7   Varying the Number of Distilled Policies

An experiment on GW2, was conducted (see Fig. 9(a)) to identify the impact of the number of distilled policies on the final performance. The plots are indicative of the fact that the agent is not highly sensitive to this parameter. Distilling to a single policy yields a poor performance curve since a single policy is incapable of capturing the variety in the fifty expert policies.

4.8   Varying Termination Limit

We analyze the impact of termination limit T (defined in Sect. 2) and obtained the performance curves for GW2. When the termination limit is low, the agent fails to leverage the knowledge of a useful sequence of actions since it cycles between the various options. Hence, a higher termination limit results in a persistent strategy for an extended duration, thereby yielding better performance (see Fig. 9(b)). However, beyond a certain value for the termination limit, the performance deteriorates since the agent spends an excessive amount of time on a single option, thereby sacrificing some fine-grained control.

4.9   The Deepmind-Lab Task

This section evaluates the Option Encoder on the Deepmind-lab maze domain task depicted in Fig. 10. We consider 24 option policies, trained using PPO [20] to navigate to goal locations shown in Fig. 10. The agent starts from a random location. The expert policies were distilled to 16,12 and 8 policies using the Option Encoder. Each distilled policy was rolled out for 300 time steps and the visitation counts were collected for all expert goal states. Figures 11(a), 11(b), 11(c) depict the maximum value of the visitation count among all the options. This is compared with AMN (see Fig. 11(d)) where the option policies are distilled to a single policy. The plots are indicative of the fact that the

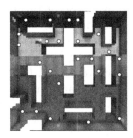

Fig. 10. Map used for Deepmind-lab. The white dots are sub-goals for option policies

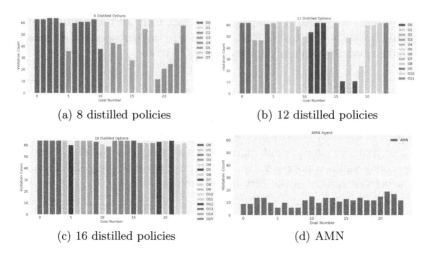

(a) 8 distilled policies   (b) 12 distilled policies

(c) 16 distilled policies   (d) AMN

Fig. 11. Maximum visitation counts the distilled policies and AMN for the goal states

distilled policies obtained from the Option Encoder cover a variety of goals and also visit them more frequently.

5 Related Work

Several works have attempted to address the policy distillation and compression scenario. Prior works like Actor-mimic [17] have attempted to compress a collection of policies. This framework can however distill a collection of policies into a single policy. Our method, on the other hand, can distill the expert policies into multiple basis policies. Pathnet [6] utilizes a large network, with weights being frozen appropriately. However, like Actor-mimic, Pathnet only discovers a single policy which may not have sufficient representational power. Pathnet addresses the continual learning setup, where there is a sequence of tasks to be solved, which may not be an appropriate approach to option pruning. The Elastic weight consolidation loss [7], addresses a scenario identical to that addressed in Pathnet [6] and modifies the weights using the gradient magnitude but suffers from same problems described for the other two methods.

Option pruning has not been addressed in the context of option-discovery in great detail. Option compression is not necessary for works like Option-critic [2], Deep Feudal reinforcement learning [27] or a collection of other works that discover options relevant to the current task. On the other hand, task-agnostic option discovery methods often need a large number of options to capture diverse behaviors. McGovern et al. [13] use a collection of filters to eliminate redundant options and [11] have observed an improved performance when around 128 options are used in a rather modest 4-room grid-world. Basis functions have been explored in the context of value functions [9,12], where the structure of the graph is used to define features for every state. On the other hand, our work uses a set of options to define a basis over policies. Our work shares similarities with the PG-ELLA [1] lifelong learning framework in that, both attempt to discover a shared latent space from which a set of tasks are solved. In this work, we focus on multi-task learning, where the basis policies are utilized across a set of tasks.

6   Conclusion

In this work, we present the Option Encoder framework, which attempts to derive a policy basis from a collection of option policies. The distilled policies can be used as a substitute for the original set of options. We demonstrate the utility of the distilled policies using an empirical evaluation on a collection of tasks. As future work, one could extend the framework to work with value functions. Another potential extension of this work is to the continual learning framework, where the Option Encoder can be used to handle a set of new policies and integrate the same with the policies learned earlier. This would involves using the option-encoder in a batch-like manner, where the set of basis policies are periodically refined.

References

1. Ammar, H.B., Eaton, E., Ruvolo, P., Taylor, M.: Online multi-task learning for policy gradient methods. In: International Conference on Machine Learning, pp. 1206–1214 (2014)
2. Bacon, P.L., Harb, J., Precup, D.: The option-critic architecture. In: Thirty-First AAAI Conference on Artificial Intelligence (2017)
3. Barreto, A., et al.: The option keyboard: combining skills in reinforcement learning. In: Advances in Neural Information Processing Systems, pp. 13031–13041 (2019)
4. Beattie, C., et al.: Deepmind lab. arXiv preprint arXiv:1612.03801 (2016)
5. Eysenbach, B., Gupta, A., Ibarz, J., Levine, S.: Diversity is all you need: learning skills without a reward function. arXiv preprint arXiv:1802.06070 (2018)
6. Fernando, C., et al.: PathNet: evolution channels gradient descent in super neural networks. arXiv preprint arXiv:1701.08734 (2017)
7. Kirkpatrick, J., et al.: Overcoming catastrophic forgetting in neural networks. Proc. Natl. Acad. Sci. **114**(13), 3521–3526 (2017)
8. Konda, V.R., Tsitsiklis, J.N.: Actor-critic algorithms. In: Advances in neural information processing systems, pp. 1008–1014 (2000)
9. Konidaris, G., Osentoski, S., Thomas, P.: Value function approximation in reinforcement learning using the Fourier basis. In: Twenty-fifth AAAI conference on artificial intelligence (2011)
10. Lillicrap, T.P., et al.: Continuous control with deep reinforcement learning. arXiv preprint arXiv:1509.02971 (2015)
11. Machado, M.C., Bellemare, M.G., Bowling, M.: A laplacian framework for option discovery in reinforcement learning. arXiv preprint arXiv:1703.00956 (2017)
12. Mahadevan, S., Maggioni, M.: Proto-value functions: a laplacian framework for learning representation and control in markov decision processes. J. Mach. Learn. Res. **8**(Oct), 2169–2231 (2007)

13. McGovern, A., Barto, A.G.: Automatic discovery of subgoals in reinforcement learning using diverse density (2001)
14. Menache, I., Mannor, S., Shimkin, N.: Q-cut—dynamic discovery of sub-goals in reinforcement learning. In: Elomaa, T., Mannila, H., Toivonen, H. (eds.) ECML 2002. LNCS (LNAI), vol. 2430, pp. 295–306. Springer, Heidelberg (2002). https://doi.org/10.1007/3-540-36755-1_25
15. Mnih, V., et al.: Asynchronous methods for deep reinforcement learning. In: International Conference on Machine Learning, pp. 1928–1937 (2016)
16. Mnih, V., et al.: Human-level control through deep reinforcement learning. Nature **518**(7540), 529 (2015)
17. Parisotto, E., Ba, J.L., Salakhutdinov, R.: Actor-mimic: deep multitask and transfer reinforcement learning. arXiv preprint arXiv:1511.06342 (2015)
18. Puterman, M.L.: Markov decision processes: discrete stochastic dynamic programming (1994)
19. Rajendran, J., Lakshminarayanan, A.S., Khapra, M.M., Prasanna, P., Ravindran, B.: Attend, adapt and transfer: attentive deep architecture for adaptive transfer from multiple sources in the same domain. arXiv preprint arXiv:1510.02879 (2015)
20. Schulman, J., Wolski, F., Dhariwal, P., Radford, A., Klimov, O.: Proximal policy optimization algorithms. arXiv preprint arXiv:1707.06347 (2017)
21. Silver, D., et al.: Mastering the game of Go with deep neural networks and tree search. Nature **529**(7587), 484–489 (2016). https://doi.org/10.1038/nature16961
22. Şimşek, Ö., Barto, A.G.: Using relative novelty to identify useful temporal abstractions in reinforcement learning. In: Proceedings of the Twenty-first International Conference on Machine Learning, p. 95. ACM (2004)
23. Şimşek, Ö., Barto, A.G.: Skill characterization based on betweenness. In: Advances in Neural Information Processing Systems, pp. 1497–1504 (2009)
24. Şimşek, O., Wolfe, A.P., Barto, A.G.: Identifying useful subgoals in reinforcement learning by local graph partitioning, pp. 816–823. ACM Press (2005). https://doi.org/10.1145/1102351.1102454
25. Sutton, R.S., Barto, A.G.: Reinforcement Learning: An Introduction. MIT Press, Cambridge (1998)
26. Sutton, R.S., Precup, D., Singh, S.: Between MDPs and semi-MDPs: a framework for temporal abstraction in reinforcement learning. Artif. Intell. **112**(1–2), 181–211 (1999)
27. Vezhnevets, A.S., et al.: FeUdal networks for hierarchical reinforcement learning. arXiv:1703.01161 [cs], March 2017

EgoMap: Projective Mapping and Structured Egocentric Memory for Deep RL

Edward Beeching[1]([×]) [iD], Jilles Dibangoye[1] [iD], Olivier Simonin[1] [iD], and Christian Wolf[2] [iD]

[1] INRIA Chroma Team, CITI Lab. INSA Lyon, Villeurbanne, France
{edward.beeching,jilles.dibangoye,olivier.simonin}@insa-lyon.fr
[2] Université de Lyon, INSA-Lyon, LIRIS, CNRS, Lyon, France
christian.wolf@insa-lyon.fr
https://team.inria.fr/chroma/en/

Abstract. Tasks involving localization, memorization and planning in partially observable 3D environments are an ongoing challenge in Deep Reinforcement Learning. We present EgoMap, a spatially structured neural memory architecture. EgoMap augments a deep reinforcement learning agent's performance in 3D environments on challenging tasks with multi-step objectives. The EgoMap architecture incorporates several inductive biases including a differentiable inverse projection of CNN feature vectors onto a top-down spatially structured map. The map is updated with ego-motion measurements through a differentiable affine transform. We show this architecture outperforms both standard recurrent agents and state of the art agents with structured memory. We demonstrate that incorporating these inductive biases into an agent's architecture allows for stable training with reward alone, circumventing the expense of acquiring and labelling expert trajectories. A detailed ablation study demonstrates the impact of key aspects of the architecture and through extensive qualitative analysis, we show how the agent exploits its structured internal memory to achieve higher performance.

Keywords: Deep reinforcement learning · Computer vision · Structured memory

1 Introduction

A critical part of intelligence is navigation, memory and planning. An animal that is able to store and recall pertinent information about their environment is likely to exceed the performance of an animal whose behavior is purely reactive. Many control problems in partially observed 3D environments involve long term dependencies and planning. Solving these problems requires agents to learn several key capacities: *spatial reasoning*—to explore the environment in an efficient

---

Project page https://edbeeching.github.io/papers/egomap.

© Springer Nature Switzerland AG 2021
F. Hutter et al. (Eds.): ECML PKDD 2020, LNAI 12458, pp. 525–540, 2021.
https://doi.org/10.1007/978-3-030-67661-2_31

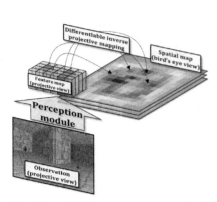 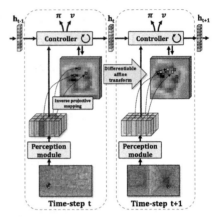

(a) Visual features are mapped to the top-down egocentric map. Observations from a first-person viewpoint are passed through a perception module, extracted features are projected with the inverse camera matrix and depth buffer to their 3D coordinates. The operations are implemented in a differentiable manner, so derivatives can be back-propagated to update the weights of the perception module.

(b) Our model learns to unproject learned task-oriented semantic embeddings of observations and to map the positions of relevant objects in a spatially structured (bird's eye view) neural memory, where different objects in the same input image are stored in different map locations. Using the ego-motion of the agent, the map is updated by differentiable affine resampling at each step in the environment.

Fig. 1. Overview of the mapping module (a) and EgoMap agent architecture (b).

manner and to learn spatio-temporal regularities and affordances. The agent needs to autonomously navigate, discover relevant objects, store their positions for later use, their possible interactions and the eventual relationships between the objects and the task at hand. Semantic mapping is a key feature in these tasks. A second feature is *discovering semantics from interactions*—while solutions exist for semantic mapping and semantic SLAM [10,44], a more interesting problem arises when the semantics of objects and their affordances are not supervised, but defined through the task and thus learned from reward.

A typical approach for these types of problems are agents based on deep neural networks including recurrent hidden states, which encode the relevant information of the history of observations [23,34]. If the task requires navigation, the hidden state will naturally be required to store spatially structured information. It has been recently reported that spatial structure as inductive bias can improve the performance on these tasks. In [39], for instance, different cells in a neural map correspond to different positions of the agent.

In our work, we go beyond structuring the agent's memory with respect to the agent's position. We use projective geometry as an inductive bias to neural networks, allowing the agent to structure its memory with respect to the locations of objects it perceives, as illustrated in Fig. 1b. The model performs an inverse projection of CNN feature vectors in order to map and store observations in an egocentric spatially structured memory. The EgoMap is complementary to the hidden state vector of the agent and is read with a combination of a global convolutional read operation and an attention model allowing the agent to query the presence of specific content. We show that incorporating projective spatial memory enables the agent to learn policies that exceed the performance of a standard recurrent agent. Two different objects visible in the same input image could be at very different places in the environment. In contrast to [39], our model maps these observations to their respective locations, and not to cells corresponding to the agent's position, as shown in Fig. 1a.

The model bears a certain structural resemblance with Bayesian occupancy grids (BOG), which have been used in mobile robotics for many years [37,42]. As in BOGs, we perform inverse projections of observations and dynamically resample the map to take into account ego-motion. However, in contrast to BOGs, our model does not require a handcrafted observation model and it learns semantics directly from interactions with the environment through reward. It is fully differentiable and trained end-to-end with backpropagation of derivatives calculated with policy gradient methods. Our contributions are as follows:

- To our knowledge, we present the first method using a differentiable SLAM-like mapping of visual features into a top-down egocentric feature map using projective geometry while training this representation using RL from reward.
- Our spatial map can be translated and rotated through a differentiable affine transform and read globally and through self-attention.
- We show that the mapping, spatial memory and self-attention can be learned end-to-end with RL, avoiding the cost of labelling trajectories from domain experts, auxiliary supervision or pre-training specific parts of the architecture.
- We demonstrate the improvement in performance over recurrent and spatial structured baselines without projective geometry.
- We illustrate the reasoning abilities of the agent by visualizing the content of the spatial memory and the self-attention process, tying it to the different objects and affordances related to the task.
- Experiments with noisy actions demonstrate the agent is robust to actions tolerances of up to 10%.

2   Related Work

Reinforcement Learning—In recent years the field of Deep Reinforcement Learning (RL) has gained attention with successes on board games [43] and Atari games [36]. One key component was the application of deep neural networks [30] to frames from the environment or game board states. Recent works that have applied Deep RL for the control of an agent in 3D environments such as maze navigation are [34] and [23] which explored the use of auxiliary tasks such as depth prediction, loop detection and reward prediction to accelerate learning. Meta RL approaches for 3D navigation have been applied by [47] and [29] also accelerated the learning process in 3D environments by prediction of tailored game features. There has also been recent work in the use of street-view scenes to train an agent to navigate in city environments [33]. In order to infer long term dependencies and store pertinent information about the partially observable environment; network architectures typically incorporate recurrent memory such as Gated Recurrent Units [9] or Long Short-Term Memory [21].

Differentiable Memory—Differentiable memory such as Neural Turing Machines [15] and Differential Neural Computers (DNC) [16] have shown promise where long term dependencies and storage are required. Neural Networks augmented with these memory structures have been shown to learn tasks such as copying, repeating and sorting. Some recent works for control in 2D and 3D environments have included structured memory-based architectures and mapping of observations. Neural SLAM [50] aims to incorporate a SLAM-like mapping module as part of the network architecture, but uses simulated sensor data rather than RGB observations from the environment, so the agent is unable to extract semantic meaning from its observations. The experimental results focus on 2D environments and the 3D results are limited. Playing Doom with SLAM augmented memory [6] implements a non-differentiable inverse projective mapping with a fixed feature extractor based on Faster-RCNN [41], pre-trained in a supervised manner. A downside of this approach is that the network does not learn to extract features pertinent to the task at hand as it is not trained end-to-end with RL. [12] replace recurrent memory with a transformer [46] attention distribution over previous observation embeddings, to highlight that recurrent architectures can struggle to capture long term dependencies. The downside is the storage of previous observations grows linearly with each step in the environment and the agent cannot chose to discard redundant information.

Grid Cells—There is evidence that biological agents learn to encode spatial structure. Rats develop grid cells, which fire at different locations with different frequencies and phases, a discovery that led to the 2014 Nobel prize in medicine [18,38]. A similar structure emerges in artificial neural networks trained to localize themselves, discovered independently by two different research groups [3,11].

Projective Geometry and Spatial Memory—Our work encodes spatial structure as additional inductive bias. We argue that projective geometry is a strong law imposed on any vision system working from egocentric observations, justifying a fully differentiable model of perception. To our knowledge, we present the first method which uses projective geometry as inductive bias while at the same time learning spatial semantic features with RL from reward.

The past decade has seen an influx of affordable depth sensors. This has led to a many works in the domain reconstruction of 3D environments, which can be incorporated into robotic systems. Seminal works in this field include [22] who performed 3D reconstruction scenes using a moving Kinect sensor and [20] who created dense 3D maps using RGB-D cameras.

Neural Map [39] implements a structured 2D differentiable memory which was tested in both egocentric and world reference frames, but does not map observations in a SLAM-like manner and instead stores a single feature vector at the agent's current location. The agent's position is also discretized to fixed cells and orientation quantized to four angles (North, South, East, West). A further downside is that the movement of the memory is fixed to discrete translations and the map is not rotated to the agent's current viewpoint.

MapNet [19] includes an inverse mapping of CNN features, trained in a supervised manner to predict x,y position and rotation from human trajectories, but does not use the map for control in an environment. Visual Question Answering in Interactive Environments [14] creates semantic maps from 3D observations for planning and question answering and is applied in a discrete state space.

Unsupervised Predictive Memory in a Goal-Directed Agent [48] incorporates a DNC in an RL agent's architecture and was applied to simulated memory-based tasks. The architecture achieves improved performance over a typical LSTM [21] based RL agent, but does not include spatial structure or projective mapping. In addition, visual features and neural memory are learned through the reconstruction of observations and actions, rather than for a specific task.

Cognitive Mapping and Planning for Visual Navigation [17] applies a differentiable mapping process on 3D viewpoints in a discrete grid-world, trained with imitation learning. The downside of discretization is that affine sampling is trivial for rotations of 90° increments, and this motion is not representative of the real world. Their tasks are simple point-goal problems of up to 32 time-steps, whereas our work focused on complex multi-step objectives in a continuous state space. Their reliance on imitation learning highlights the challenge of training complex neural architectures with reward alone, particular on tasks with sparse reward such as the ones presented in this paper.

Learning Exploration Policies for Navigation [8], do not learn a perception module but instead map the depth buffer to a 2D map to provide a map-based exploration reward. Our work learns the features that can be mapped so the agent can query not only occupancy, but task-related semantic content.

Our work greatly exceeds the performance of Neural Map [39], by embedding a differentiable inverse projective transform and a continuous egocentric map into the agent's network architecture. The mapping of the environment is in the agent's reference frame, including translation and rotation with a differentiable affine transform. We demonstrate stable training with reinforcement learning alone, over several challenging tasks and random initializations, and do not require the expense of acquiring expert trajectories. We detail the key similarities and differences with related work in Table 1.

3   EgoMap

We consider partially observable Markov decision processes (POMDPs) [26] in 3D environments and extend recent Deep-RL models, which include a recurrent hidden layer to store pertinent long term information [23,34]. In particular,

Table 1. Comparison of key features of related works

Related work	Projective Geometry	Spatial map learned from reward	Reinforcement Learning	Imitation learning	Visual input	Multiple goals	Spatially Structured	Task-specific semantic features	Continuous State Space	Continuous Affine transform	End-to-end differentiable	Continuous translations	Continuous rotations	End-to-end training	Intrinsic rewards
Semantic mapping [10]	✓	–	–	–	✓	–	✓	–	✓	✓	–	✓	✓	–	–
Playing Doom with SLAM [6]	✓	–	✓	–	✓	–	✓	✓	✓	✓	–	✓	✓	–	–
Neural SLAM [50]	–	–	✓	–	✓	–	✓	✓	–	–	–	✓	–	✓	–
Cog. Map [17]	–	–	✓	✓	✓	–	✓	–	–	–	✓	✓	–	✓	–
Semantic SLAM [44]	✓	–	–	✓	✓	–	✓	–	–	–	–	✓	–	✓	–
IQA [14]	–	–	✓	✓	✓	✓	✓	–	–	–	–	✓	–	✓	–
MapNet [19]	–	–	✓	–	✓	–	✓	–	✓	✓	–	✓	✓	–	–
Neural Map [39]	–	✓	✓	–	✓	✓	✓	✓	✓	✓	–	✓	–	✓	–
MERLIN [48]	–	–	✓	–	✓	✓	–	–	✓	✓	–	✓	✓	✓	✓
Learning exploration policies [8]	✓	–	✓	–	✓	✓	✓	–	✓	✓	–	✓	✓	✓	✓
EgoMap	✓	✓	✓	–	✓	✓	✓	✓	✓	✓	✓	✓	✓	✓	–

RGBD observations I_t at time step t are passed through a perception module extracting features s_t, which are used to update the recurrent state:

$$s_t = f_p(I_t; \theta_p) \qquad h_t = f_r(h_{t-1}, s_t; \theta_r) \tag{1}$$

where f_p is a convolutional neural network and f_r is a Gated Recurrent Unit [9]. Gates and their equations have been omitted for simplicity. Above and in the rest of this paper, θ_* are trainable parameters, exact architectures are provided in the appendix. The controller outputs an estimate of the policy (the action distribution) and the value function given its hidden state:

$$\pi_t = f_\pi(h_t; \theta_\pi) \qquad v_t = f_v(h_t; \theta_v) \tag{2}$$

The proposed model is motivated by the regularities which govern 3D physical environments. When an agent perceives an observation of the 3D world, it observes a 2D planar perspective projection of the world based on its current viewpoint. This projection is a well understood physical process, we aim to imbue the agent's architecture with an inductive bias based on inverting the 3D to 2D planar projective process. This inverse mapping operation appears to be second nature to many organisms, with the initial step of depth estimation being well studied in the field of Physiology [13]. We believe that providing this mechanism implicitly in the agent's architecture will improve its reasoning capabilities in new environments to bypass a large part of the learning process.

The overall concept is that as the agent explores the environment, the perception module f_p produces a 2D feature map s_t, in which each feature vector represents a learned semantic representation of a small receptive field from the agent's egocentric observation. While they are integrated into the flat (not spatially structured) recurrent hidden state h_t through function f_r (Eq. 1), we propose its integration into a second tensor M_t, a top-down egocentric memory, which we call *EgoMap*. The feature vectors are mapped to their egocentric positions using the inverse projection matrix and depth estimates. This requires an agent with a calibrated camera (known intrinsic parameters), which is a soft constraint easily satisfied. The map can then be read by the agent in two ways: a global convolutional read operation and a self-attention operation.

Let the agent's position and angle at time t be (x_t, y_t) and ϕ_t respectively, M_t is the current EgoMap. s_t are the feature vectors extracted by the perception module, D_t are the depth buffer values. The change in agent position and orientation between time-step t and $t-1$ are $(dx_t, dy_t, d\phi_t)$.

There are three key steps to the operation of the EgoMap:

1. Transform the map to the agent's egocentric frame of reference:

$$\hat{M}_t = \text{Affine}(M_{t-1}, dx_t, dy_t, d\phi_t) \tag{3}$$

2. Update the map to include new observations:

$$\tilde{M}_t = \text{InverseProject}(s_t, D_t) \qquad M'_t = \text{Combine}(\hat{M}_t, \tilde{M}_t) \tag{4}$$

3. Perform a global read and attention based read, the outputs of which are fed into the policy and value heads:

$$r_t = \text{Read}(M'_t) \qquad c_t = \text{Context}(M'_t, s_t, r_t) \tag{5}$$

These three operations will be further detailed below in individual subsections. Projective mapping and structured memory should augment the agent's performance where spatial reasoning and long term recollection are required. On simpler tasks the network can still perform as well as the baseline, assuming the extra parameters do not cause further instability in the RL training process.

Affine Transform—At each time-step we translate and rotate the map into the agent's frame of reference, this is achieved with a differentiable affine transform, popularized by the well known Spatial Transformer Networks [24]. Relying on the simulator to provide the change in position (dx, dy) and orientation $d\phi$, we convert the deltas to the agent's egocentric frame of reference and transform the map with a differentiable affine transform. The effect of noise on the change in position on the agent's performance is analysed in the experimental section.

Inverse Projective Mapping—We take the agent's current observation, extract relevant semantic embeddings and map them from a 2D planar projection to their 3D positions in an egocentric frame of reference. At each time-step, the agent's egocentric observation is encoded by the perception module (a convolutional neural network) to produce feature vectors, this step is a mapping from $R^{4 \times 64 \times 112} \rightarrow R^{16 \times 4 \times 10}$. Given the inverse camera projection matrix and the depth buffer provided by the simulator, we can compute the approximate location of the features in the agent's egocentric frame of reference. As the mapping is a many to one operation, several features can be mapped to the same location. Features that share the same spatial location are averaged element-wise.

The newly mapped features must then be combined with the translated map from the previous time-step. We found that the use of a momentum hyperparameter, α, enabled a smooth blending of new and previously observed features. We use an α value of 0.9 for the tests presented in the paper. We ensured that the blending only occurs where the locations of new projected features and the map from the previous time-step are co-located, see Eq. 6.

$$M'^{(x,y)}_t = \eta \hat{M}_t^{(x,y)} + (1 - \eta)\tilde{M}_t^{(x,y)}$$

$$\eta = \begin{cases} 1.0, & \text{if } \tilde{M}_t^{(x,y)} = 0 \ \& \ \hat{M}_t^{(x,y)} \neq 0 \\ 0.0, & \text{if } \tilde{M}_t^{(x,y)} \neq 0 \ \& \ \hat{M}_t^{(x,y)} = 0 \\ \alpha, & \text{otherwise} \end{cases} \tag{6}$$

Sampling from a Global Map— A naive approach to the storage and transformation of the egocentric feature map would be to apply an affine transformation to the map at each time-step. A fundamental downside of applying repeated affine transforms is that at each step a bilinear interpolation operation is applied, which causes smearing and degradation of the features in the map. We mitigated this issue by storing the map in a global reference frame and mapping the agent's observations to the global reference frame. For the read operation an offline affine transform is applied.

Read Operations—We wanted the agent to be able to summarize the whole spatial map and also selectively query for pertinent information. This was achieved by incorporating two types of read operation into the agent's architecture, a *Global Read* operation and a *Self-attention Read*.

The global read operation is a CNN that takes as input the egocentric map and outputs a 32-dimensional feature vector that summarizes the map's contents. The output of the global read is concatenated with the visual CNN output. To query for relevant features in the map, the agent's controller network can output a query vector q_t, the network then compares this vector to each location in the map with a cosine similarity function in order to produce scores, which are the same width and height as the map. The scores are normalized with a softmax operation to produce a soft-attention in the lines of [2] and used to compute a weighted average of the map, allowing the agent to selectively query and focus on parts of the map. This querying mechanism was used in both the Neural Map [39] and MERLIN [48] RL agents. We made the following improvements: *Attention Temperature* and *Query Position*.

$$\sigma(x)_i = \frac{e^{\beta x_i}}{\sum e^{\beta x_j}} \tag{7}$$

Query Position: A limitation of self-attention is that the agent can query *what* it has observed but not *where* it had observed it. To improve the spatial reasoning performance of the agent we augmented the neural memory with two fixed additional coordinate planes representing the x, y egocentric coordinate system normalized to $(-1.0, 1.0)$, as introduced in [31]. The agent still queries based on the features in the map, but the returned context vector includes two extra scalar quantities which are the weighted averages of the x, y planes. The impacts of these additions are discussed and quantified in the ablation study, Sect. 4.

Attention Temperature: To provide the agent with the ability to learn to modify the attention distribution, the query includes an additional learnable temperature parameter, β, which can adjust the softmax distribution detailed in Eq. 7. This parameter can vary query by query and is constrained to be one or greater by a Oneplus function.

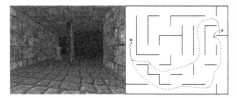 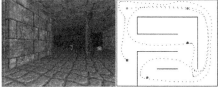

Fig. 2. Scenarios: Left - The *"Find and return"* scenario. With the agent's egocentric viewpoint and top down view unseen by the agent; the agent starts at the green object, finds the red object and then returns to the entry. Right - The *"Ordered K−item scenario"*, where $K = 4$. An episode where the agent collects the 4 items in the correct order.

4   Experiments

The EgoMap and baseline architectures were evaluated on four challenging 3D scenarios, which require navigation and different levels of spatial memory. The scenarios are taken from [5] who extended the 3D ViZDoom environment [27] with various scenarios that are partially observable, require memory, spatial understanding, have long horizons and sparse rewards. Whilst more visually realistic simulators are available such as Gibson [49], Matterport [1], Home [7] and Habitat [32], the tasks available are simple point-goal tasks which do not require long term memory and recollection. We target the following three tasks:

Labyrinth: The agent must quickly find the exit, the reward is a sparse positive reward for finding the exit. This tests the ability to explore in an efficient manner.

Ordered k-item: An agent must find k objects in a fixed order. It tests three aspects of an agent: its ability to explore the environment efficiently, the ability to learn to collect items in a predefined order and its ability to store where items were located so they can be retrieved in the correct order. We tested two versions of this scenario with 4-items or 6-items, an example is shown in Fig. 2.

Find and Return: The agent starts next to a green totem, explores the environment to find a red totem and then returns to the start. This is our implementation of "Minotaur" scenario from [39]. The scenario tests the ability to navigate and retain information over long time periods, an example is shown in Fig. 2.

All the tasks require different levels of spatial memory and reasoning. For example, if an agent observes an item out of order it can store the item's location in its memory and navigate back to it later. The scenarios that require more spatial reasoning, long term planning and recollection are where the agent has the greatest improvement in performance. In all scenarios there is a small negative reward for each time-step to encourage the agent to complete the task quickly.

Experimental Strategy and Generalization to Unseen Environments— Many configurations of each scenario were created through procedural generation and partitioned into separated training and testing sets of size 256 and 64 respectively for each scenario type. Although the task in a scenario is fixed, we vary the locations of the walls, item locations, and start and end points; to ensure a diverse range of possible scenario configurations. A limited hyper-parameter sweep was undertaken with the baseline architecture to select the hyper-parameters, which were fixed for both the baseline, Neural Map and EgoMap agents. Three independent experiments were conducted per task to evaluate the algorithmic stability of the training process. To avoid information asymmetry, we provide the baseline agent with $dx, dy, sin(d\theta), cos(d\theta)$ concatenated with its visual features.

Table 2. Results of on four scenarios for 1.2 B environment steps. We show the mean and std. for three independent experiments.

Agent	Scenario							
	4 item		6 item		Find and Return		Labyrinth	
	Train	Test	Train	Test	Train	Test	Train	Test
Random	−0.179	−0.206	−0.21	−0.21	−0.21	−0.21	−0.115	−0.086
Baseline	2.341 ± 0.026	2.266 ± 0.035	2.855 ± 0.164	2.545 ± 0.226	0.661 ± 0.003	0.633 ± 0.027	0.73 ± 0.02	0.694 ± 0.009
Neural Map	2.339 ± 0.038	2.223 ± 0.040	2.750 ± 0.062	2.465 ± 0.034	0.825 ± 0.070	0.723 ± 0.026	**0.769 ± 0.042**	0.706 ± 0.018
EgoMap	**2.398 ± 0.014**	**2.291 ± 0.021**	**3.214 ± 0.007**	**2.801 ± 0.048**	**0.893 ± 0.007**	**0.848 ± 0.017**	0.753 ± 0.002	**0.732 ± 0.016**
Optimum (Upper Bound)	2.5	2.5	3.5	3.5	1.0	1.0	1.0	1.0

Training Details—The model parameters were optimized with an on-policy, policy gradient algorithm; batched Advantage Actor Critic (A2C) [35], we used the PyTorch [40] implementation of A2C [28]. We sampled trajectories from 16 parallel agents and updated every 128 steps in the environment with discounted returns bootstrapped from value estimates for non-terminal states. The gamma factor was 0.99, the entropy weight was 0.001, RMSProp [45] was used with a learning rate of 7e−4. The EgoMap agent map size was $16 \times 24 \times 24$ with a grid sampling chosen to cover the environment size with a 20% padding. The agent's policy was updated over 1.2B environment steps, with a frame skip of 4. Training took 36 h for the baseline and 8 days for the EgoMap, on 4 Xeon E5-2640v3 CPUs, with 32 GB of memory and one NVIDIA GK210 GPU.

Results—Results from the baseline and EgoMap policies evaluated on the 4 scenarios are shown in Table 2, all tasks benefit from the inclusion of inverse projective mapping and spatial memory, with the largest improvement on the *Find and Return* scenario. We postulate that the greater improvement in performance is due to two factors; firstly this scenario always requires spatial memory as the agent must return to its starting point and secondly the objects in this scenario are larger and occupy more space in the map. We also compared to the state of the art in spatially structured neural memory, Neural Map [39]. Figure 3 shows agent training curves for the recurrent baseline, Neural Map and EgoMap, on the *Find and Return* test set configurations.

Ablation Study—An ablation study was carried out on the improvements made by the EgoMap architecture. We were interested to see the influence of key options such as the global and attention-based reads, the similarity function used when querying the map, the learnable temperature parameter and the incorporation of location-based querying. The Cartesian product of these options is large and it was not feasible to test them all, we therefore decided to selectively

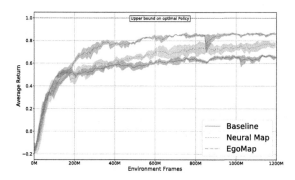

Fig. 3. Results for the *Find and Return* test set.

Table 3. Ablation study, *Find and Return* scenario, conducted after 800M steps.

Ablation	Train	Test
Baseline	0.668 ± 0.028	0.662 ± 0.036
No global read	0.787 ± 0.007	0.771 ± 0.029
No query	0.838 ± 0.003	0.811 ± 0.013
No query temperature	0.845 ± 0.014	0.815 ± 0.019
No query position	0.839 ± 0.007	0.814 ± 0.008
EgoMap	0.847 ± 0.011	0.814 ± 0.017
EgoMap (L1 query)	**0.851 ± 0.014**	**0.828 ± 0.011**

switch off key options to understand which aspects contribute to the improvement in performance. The results of the ablation study are shown in Table 3. Both the global and self-attention reads provide large improvements in performance over the baseline recurrent agent. The position-based query provides a small improvement. A comparison of the similarity metric of the attention mechanism highlights the L1-similarity achieved higher performance than cosine. A qualitative analysis of the self-attention mechanism is shown in the next section.

5   Analysis

Visualization—The EgoMap architecture is highly interpretable and provides insights about how the agent reasons in 3D environments. In Fig. 4a and 4b we show analysis of the spatially structured memory and how the agent has learned to query and self-attend to recall pertinent information. The Figures show key steps during an episode in the *Ordered 6-item* and *Find and Return* scenarios, including the first three principal components of the 16-dimensional EgoMap, the attention distribution and the vector returned from position queries. Refer to the caption for further details. The agent is seen to attend to key objects at certain phases of the task, in the *Ordered 6-item* scenario the agent attends the next item in the sequence and in the *Find and Return* scenario the agent attends to the

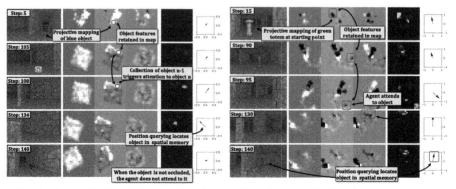

(a) Analysis: *Ordered 6-item* scenario   (b) Analysis: *Find and Return* scenario

Fig. 4. Analysis for key steps (different rows) during an episode from the *Ordered 6-item* and *Find and Return* scenarios. Within each sub-figure: Left column - RGB observations, central column - the three largest PCA components of features mapped in the spatially structured memory, right - attention heat map and x, y query position vector. The agent maps and stores features from key objects and attends to them when they are pertinent to the current stage of the task. For example, for the left figure on the first row at time-step 5 the blue spherical object, which is ordered 4 of 6, is mapped into the agent's spatial memory. The agent explores the environment collecting the items in order, it collects item 3 of 6 between time-step 105 and 108, shown on rows 2 and 3. As soon as the agent has collected item 3 it queries its internal memory for the presence of item 4, which is shown by the attention distribution on rows 3 and 4. On the last row, time-step 140, the agent observes the item and no longer attempts to query for it, as the item is in the agent's field of view.

green totem located at the start/return point once it has found the intermediate goal. This demonstrates that augmenting an agent with spatially structured memory not only improves performance but also increases interpretability of the agent's decision making process.

Noisy Actions— One common criticism of agents trained in simulation is that the agent can query its environment for information that would not be readily available in the real world. In the case of EgoMap, the agent is trained with ground truth ego-motion measurements. Real-world robots have noisy estimates of ego-motion due to the tolerances of available hardware. We performed an analysis of the EgoMap agent trained in the presence of a noisy oracle, which adds noise to the ego-motion measurements. Noise is drawn from a normal distribution centered at one and is multiplied by the agent's ground-truth motion, the effect of the noise is cumulative but unbiased. Tests were conducted with standard deviations of up to 0.2 which is a tolerance of more than 20% on the agent's ego-motion measurements, results are shown in Fig. 5. We observed that the agent retains the performance increase over the baseline for up to 10% of noisy actions, the performance degrades to that of the baseline agent. This demonstrates the EgoMap architecture learns to be robust to noisy odometry.

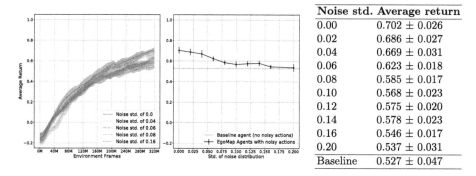

Noise std.	Average return
0.00	0.702 ± 0.026
0.02	0.686 ± 0.027
0.04	0.669 ± 0.031
0.06	0.623 ± 0.018
0.08	0.585 ± 0.017
0.10	0.568 ± 0.023
0.12	0.575 ± 0.020
0.14	0.578 ± 0.023
0.16	0.546 ± 0.017
0.20	0.537 ± 0.031
Baseline	0.527 ± 0.047

Fig. 5. Test set performance of the EgoMap agent during training with noisy ego-motion measurements, conducted for 320 M environment frames. Shown are test set performance during training (left), final performance for a range of noise values (centre) which are tabulated (right). We retain an improvement in performance over the baseline agent for noisy actions of up to 10%.

6   Conclusion

We have presented EgoMap, an egocentric spatially structured neural memory that augments an RL agent's performance in 3D navigation, spatial reasoning and control tasks. EgoMap includes a differentiable inverse projective transform that maps learned task-specific semantic embeddings of agent observations to their world positions. We have shown that through the use of global and self-attentive read mechanisms an agent can learn to focus on important features from the environment. We demonstrate that an RL agent can benefit from spatial memory, particularly in 3D scenarios with sparse rewards that require localization and memorization of objects. EgoMap out-performs existing state of the art baselines, including Neural Map, a spatial memory architecture. The increase in performance compared to Neural Map is due to two aspects. 1) The differential projective transform maps *what* the objects are to *where* they are in the map, which allows for direct localization with attention queries and global reads. In comparison, Neural Map writes *what* the agent observes to *where* the agent is on the map, this means that the same object viewed from two different directions will be written to two different locations on the map, which leads to poorer localization of the object. 2) Neural Map splits the map to four 90° angles, which alleviates blurring, our novel solution to this issue stores a single unified map in an allocentric frame of reference and performs an offline ego-centric read, which allows an agent to act in states spaces where the angle is continuous, without the need to quantize the agent's angle to 90° increments.

We have shown, with detailed analysis, how the agent has learned to interact with its structured internal memory. The ablation study has the benefits of the global and self-attention operations and that these can be augmented with temperature, position querying and other similarity metrics. We have demonstrated that the EgoMap architecture is robust to actions with tolerances of up to 10%. Future work in this area of research would evaluate the architecture in situations with noisy depth measurements and in a larger variety of simulators such as MALMO [25], DeepMind Lab [4] and Habitat [32].

Acknowledgements. This work was funded by grant Deepvision (ANR-15-CE23-0029, STPGP479356-15), a joint French/Canadian call by ANR & NSERC. We gratefully acknowledge support from the CNRS/IN2P3 Computing Center (Lyon - France) for providing computing and data-processing resources needed for this work.

References

1. Anderson, P., et al.: Vision-and-language navigation: interpreting visually-grounded navigation instructions in real environments. In: CVPR (2018)
2. Bahdanau, D., Cho, K., Bengio, Y.: Neural machine translation by jointly learning to align and translate. arXiv preprint arXiv:1409.0473 (2014)
3. Banino, A., et al.: Vector-based navigation using grid-like representations in artificial agents. Nature **557**, 429–433 (2018)
4. Beattie, C., et al.: Deepmind lab. CoRR abs/1612.03801 (2016)
5. Beeching, E., Wolf, C., Dibangoye, J., Simonin, O.: Deep reinforcement learning on a budget: 3D control and reasoning without a supercomputer. CoRR abs/1904.01806 (2019) http://arxiv.org/abs/1904.01806
6. Bhatti, S., Desmaison, A., Miksik, O., Nardelli, N., Siddharth, N., Torr, P.H.S.: Playing doom with slam-augmented deep reinforcement learning. arxiv preprint arxiv:1612.00380 (2016)
7. Brodeur, S., et al.: HoME: a household multimodal environment. In: ICLR (2018)
8. Chen, T., Gupta, S., Gupta, A.: Learning exploration policies for navigation. In: ICLR, March 2019
9. Chung, J., Gulcehre, C., Cho, K., Bengio, Y.: Gated feedback recurrent neural networks. In: ICML (2015)
10. Civera, J., Galvez-Lopez, D., Riazuelo, L., Tardós, J.D., Montiel, J.M.M.: Towards semantic SLAM using a monocular camera. In: IROS (2011)
11. Cueva, C., Wei, X.X.: Emergence of grid-like representations by training recurrent neural networks to perform spatial localization. In: ICLR (2018)
12. Fang, K., Toshev, A., Fei-Fei, L., Savarese, S.: Scene memory transformer for embodied agents in long-horizon tasks. In: CVPR (2019)
13. Frégnac, Y., René, A., Durand, J.B., Trotter, Y.: Brain encoding and representation of 3D-space using different senses, in different species. J. Physiol. Paris **98**(1–3), 1 (2004)
14. Gordon, D., Kembhavi, A., Rastegari, M., Redmon, J., Fox, D., Farhadi, A.: IQA: visual question answering in interactive environments. In: CVPR. IEEE (2018)
15. Graves, A., Wayne, G., Danihelka, I.: Neural turing machines. arXiv preprint arXiv:1410.5401 (2014)
16. Graves, A., et al.: Hybrid computing using a neural network with dynamic external memory. Nature **538**(7626), 471 (2016)

17. Gupta, S., Davidson, J., Levine, S., Sukthankar, R., Malik, J.: Cognitive mapping and planning for visual navigation. In: CVPR, pp. 7272–7281, July 2017
18. Hafting, T., Fyhn, M., Molden, S., Moser, M.B., Moser, E.I.: Microstructure of a spatial map in the entorhinal cortex. Nature **436**(7052), 801–806 (2005)
19. Henriques, J., Vedaldi, A.: MapNet: an allocentric spatial memory for mapping environments. In: CVPR (2018)
20. Henry, P., Krainin, M., Herbst, E., Ren, X., Fox, D.: RGB-D mapping: using depth cameras for dense 3D modeling of indoor environments. In: Khatib, O., Kumar, V., Sukhatme, G. (eds.) Experimental Robotics, vol. 79, pp. 477–491. Springer, Heidelberg (2014). https://doi.org/10.1007/978-3-642-28572-1_33
21. Hochreiter, S., Schmidhuber, J.: Long Short-Term Memory. Neural Comput. **9**(8), 1735–1780 (1997)
22. Izadi, S., et al.: KinectFusion: real-time 3D reconstruction and interaction using a moving depth camera. In: 24th annual ACM Symposium on User Interface Software and Technology, pp. 559–568. ACM (2011)
23. Jaderberg, M., et al.: Reinforcement learning with unsupervised auxiliary tasks. In: ICLR (2017)
24. Jaderberg, M., Simonyan, K., Zisserman, A., kavukcuoglu, k.: Spatial transformer networks. In: NIPS (2015)
25. Johnson, M., Hofmann, K., Hutton, T., Bignell, D.: The malmo platform for artificial intelligence experimentation. In: IJCAI, pp. 4246–4247. AAAI Press (2016)
26. Kaelbling, L.P., Littman, M.L., Cassandra, A.R.: Planning and acting in partially observable stochastic domains. Artif. Intell. **101**(1–2), 99–134 (1998)
27. Kempka, M., Wydmuch, M., Runc, G., Toczek, J., Jaskowski, W.: ViZDoom: a doom-based AI research platform for visual reinforcement learning. In: IEEE Conference on Computatonal Intelligence and Games, CIG (2017)
28. Kostrikov, I.: PyTorch implementations of reinforcement learning algorithms (2018). https://github.com/ikostrikov/pytorch-a2c-ppo-acktr
29. Lample, G., Chaplot, D.S.: Playing FPS games with deep reinforcement learning. In: AAAI (2017)
30. Lecun, Y., Eon Bottou, L., Bengio, Y., Haaner, P.: Gradient-based learning applied to document recognition. IEEE **86**(11), 2278–2324 (1998)
31. Liang, X., Wei, Y., Shen, X., Yang, J., Lin, L., Yan, S.: Proposal-FreeNetwork for instance-level object segmentation. TPAMI **40**(12), 2978–2991 (2018)
32. Savva, M., Kadian, A., Maksymets, O., Batra, D.: Habitat: a platform for embodied AI research. arxiv (2019)
33. Mirowski, P., et al.: Learning to navigate in cities without a map. In: NeurIPS2018 (2018)
34. Mirowski, P., et al.: Learning to navigate in complex environments. In: ICLR (2017)
35. Mnih, V., et al.: Asynchronous methods for deep reinforcement learning. In: ICML (2016)
36. Mnih, V., et al.: Human-level control through deep reinforcement learning. Nature **518**, 529–533 (2015)
37. Moravec, H.: Sensor fusion in certainty grids for mobile robots. AI Mag. **9**(2) (1988)
38. O'Keefe, J., Dostrovsky, J.: The hippocampus as a spatial map. Preliminary evidence from unit activity in the freely-moving rat. Brain Research **34**(1), 171–175 (1971)
39. Parisotto, E., Salakhutdinov, R.: Neural map: structured memory for deep reinforcement learning. In: ICLR (2018)
40. Paszke, A., et al.: Automatic differentiation in PyTorch. In: NIPS-W (2017)

41. Ren, S., He, K., Girshick, R., Sun, J.: Faster R-CNN: towards real-time object detection with region proposal networks. In: Cortes, C., Lawrence, N.D., Lee, D.D., Sugiyama, M., Garnett, R. (eds.) NIPS (2015)

42. Rummelhard, L., Nègre, A., Laugier, C.: Conditional Monte Carlo dense occupancy tracker. In: ITSC (2015)

43. Silver, D., et al.: A general reinforcement learning algorithm that masters chess, Shogi, and go through self-play. Science **362**(6419), 1140–1144 (2018)

44. Tateno, K., Tombari, F., Laina, I., Navab, N.: CNN-slam: real-time dense monocular slam with learned depth prediction. In: CVPR (2017)

45. Tieleman, T., Hinton, G.: Lecture 6.5-RmsProp: divide the gradient by a running average of its recent magnitude. COURSERA: Neural Netw. ML **4**, 26–31 (2012)

46. Vaswani, A., et al.: Attention is all you need. In: Guyon, I., et al. (eds.) Advances in Neural Information Processing Systems, vol 30, pp. 5998–6008. Curran Associates, Inc. (2017). http://papers.nips.cc/paper/7181-attention-is-all-you-need.pdf

47. Wang, J.X., et al.: Learning to reinforcement learn. arxiv pre-print arxiv:1611.05763 (2016)

48. Wayne, G., et al.: Unsupervised predictive memory in a goal-directed agent. arxiv preprint arxiv:1803.10760 (2018)

49. Xia, F., R. Zamir, A., He, Z.Y., Sax, A., Malik, J., Savarese, S.: Gibson ENV: real-world perception for embodied agents. In: CVPR (2018)

50. Zhang, J., Tai, L., Boedecker, J., Burgard, W., Liu, M.: Neural SLAM. arxiv preprint arxiv:1706.09520 (2017)

ELSIM: End-to-End Learning of Reusable Skills Through Intrinsic Motivation

Arthur Aubret[(⊠)], Laetitia Matignon[(⊠)], and Salima Hassas[(⊠)]

Univ Lyon, Université Lyon 1, CNRS, LIRIS, 69622 Villeurbanne, France
{arthur.aubret,laetitia.matignon,salima.hassas}@univ-lyon1.fr

Abstract. Taking inspiration from developmental learning, we present a novel reinforcement learning architecture which hierarchically learns and represents self-generated skills in an end-to-end way. With this architecture, an agent focuses only on task-rewarded skills while keeping the learning process of skills bottom-up. This bottom-up approach allows to learn skills that 1 - are transferable across tasks, 2 - improve exploration when rewards are sparse. To do so, we combine a previously defined mutual information objective with a novel curriculum learning algorithm, creating an unlimited and explorable tree of skills. We test our agent on simple gridworld environments to understand and visualize how the agent distinguishes between its skills. Then we show that our approach can scale on more difficult MuJoCo environments in which our agent is able to build a representation of skills which improves over a baseline both transfer learning and exploration when rewards are sparse.

Keywords: Intrinsic motivation · Curriculum learning · Developmental learning · Reinforcement learning

1 Introduction

In reinforcement learning (RL), an agent learns by trial-and-error to maximize the expected rewards obtained from actions performed in its environment [40]. However, many RL agents usually strive to achieve one goal using only low-level actions. In contrast, as humans being, when we want to go to work, we do not think about every muscle we contract in order to move; we just take abstract decisions such as *Go to work*. Low-level behaviors such as how to walk are already learned and we do not need to think about them. Learning to walk is a classical example of babies developmental learning, which refers to the ability of an agent to spontaneously explore its environment and acquire new skills [9]. Babies do not try to get walking behaviors all at once, but rather first learn to move their legs, to crawl, to stand up, and then, eventually, to walk. They are **intrinsically motivated** since they act for the inherent satisfaction of learning new skills [36] rather than for an **extrinsic reward** assigned by the environment.

Several works are interested in learning abstract actions, also named **skills** or **options** [41], in the framework of deep reinforcement learning (DRL) [4]. Skills

© Springer Nature Switzerland AG 2021
F. Hutter et al. (Eds.): ECML PKDD 2020, LNAI 12458, pp. 541–556, 2021.
https://doi.org/10.1007/978-3-030-67661-2_32

can be learned with extrinsic rewards [5], which facilitates the credit assignment [41]. In contrast, if one learns skills with intrinsic motivation, the learning process becomes bottom-up [25], i.e. the agent learns skills before getting extrinsic rewards. When learning is bottom-up, the agent commits to a time-extended skill and avoids the usual wanderlust due to the lack, or the sparsity, of extrinsic rewards. Therefore, it can significantly improve exploration [26,33]. In addition, these skills can be used for different tasks, emphasizing their potential for transfer learning [42]. These properties make intrinsic motivation attractive in a *continual learning* framework, which is the ability of the agent to acquire, retain and reuse its knowledge over a lifetime [44].

Several works recently proposed to intrinsically learn such skills using a diversity heuristic [2,13], such that different states are covered by the learned skills. Yet several issues remain: 1 - the agent is often limited in the number of learned skills or requires *curriculum learning* [2]; 2 - most skills target uninteresting parts of the environment relatively to some tasks; thereby it requires prior knowledge about which features to diversify [13]; 3 - the agent suffers from catastrophic forgetting when it tries to learn a task while learning skills [13]; 4 - discrete time-extended skills used in a hierarchical setting are often sub-optimal for a task. With diversity heuristic, skills are indeed not expressive enough to efficiently target a goal [2,13].

In this paper, we propose to address these four issues so as to **improve the approaches for continually learning increasingly difficult skills with diversity heuristics**. We introduce ELSIM (End-to-ended Learning of reusable Skills through Intrinsic Motivation), a method for learning representations of skills in a bottom-up way. The agent autonomously builds a tree of abstract skills where each skill is a refinement of its parent. First of all, skills are learned independently from the tasks but along with tasks; it guarantees they can be easily transferred to other tasks and may help the agent to explore its environment. We use the optimization function defined in [13] which guarantees that states targeted by a skill are close to each other. Secondly, the agent selects a skill to refine with extrinsic or intrinsic rewards, and learns new sub-skills; it ensures that the agent learns specific skills useful for tasks through an intelligent *curriculum*, among millions of possible skills.

Our approach contrasts with existing approaches which either bias skills towards a task [5], reducing the possibilities for transfer learning, or learn skills during pretraining [13]. We believe our paradigm, by removing the requirement of a *developmental period* [28] (which is just an unsupervised pretraining), makes naturally compatible developmental learning and *lifelong learning*. Therefore, we emphasize three properties of our ELSIM method. 1 - **Learning is bottom-up**: the agent does not require an expert supervision to expand the set of skills. It can use its skills to solve different sequentially presented tasks or to explore its environment. 2 - **Learning is end-to-end**: the agent never stops training and keeps expanding its tree of skills. It gradually self-improves and avoids catastrophic forgetting. 3 - **Learning is focused**: the agent only learns skills useful for its high-level extrinsic/intrinsic objectives when provided.

Our contributions are the following: we introduce a new curriculum algorithm based on an adaptation of diversity-based skill learning methods. Our objective is not to be competitive when the agent learns one specific goal, but **to learn useful and reusable skills along with sequentially presented goals in an end-to-end fashion**. We show experimentally that ELSIM achieves **good asymptotic performance** on several single-task benchmarks, **improves exploration** over standard DRL algorithms and manages to easily **reuse its skills**. Thus, this is a step towards *lifelong learning* agents.

This paper is organized as follows. First, we introduce the concepts used in ELSIM, especially diversity-based intrinsic motivation (Sect. 2). In Sect. 3, the core of our method is presented. Then, we explain and visualize how ELSIM works on simple gridworlds and compare its performances with state-of-the-art DRL algorithms on single and sequentially presented tasks learning (Sect. 4). In Sect. 5, we detail how ELSIM relates to existing works. Finally, in Sect. 6, we take a step back and discuss ELSIM. Pseudo-codes, full experiments, additional details and hyper-parameters can be found in the long version of the paper[1].

2 Background

2.1 Reinforcement Learning

A Markov decision process (MDP) [35] is defined by a set of possible states S; a set of possible actions A; a transition function $P : S \times A \times S \rightarrow \mathbb{P}(s'|s,a)$ with $a \in A$ and $s, s' \in S$; a reward function $R : S \times A \times S \rightarrow \mathbb{R}$; the initial distribution of states $\rho_0 : S \rightarrow [0;1]$. A stochastic policy π maps states to probabilities over actions in order to maximize the discounted cumulative reward defined by $\varsigma_t = [\sum_{t=0}^{\infty} \gamma^t r_t]$ where $\gamma \in [0,1]$ is the discount factor. In order to find the action maximizing ς in a state s, it is common to maximize the expected discounted gain following a policy π from a state-action tuple defined by:

$$Q_\pi(s,a) = \mathop{\mathbb{E}}_{a_t \sim \pi(s_t), s_{t+1} \sim P(s_{t+1}|s_t,a_t)} \left[\sum_{t=0}^{\infty} \gamma^t R(s_t, a_t, s_{t+1}) \right] \qquad (1)$$

where $s_0 = s$, $a_0 = a$. To compute this value, it is possible to use the Bellman Equation [40].

2.2 Obtaining Diverse Skills Through Mutual Information Objective

We characterize a **skill** by its intra-skill policy; thereby a skill is a mapping of states to probabilities over actions. One way to learn, without extrinsic rewards, a set of different skills along with their intra-skill policies is to use an objective based on mutual information (MI).

---

[1] https://arxiv.org/abs/2006.12903.

In [13], learned skills should be as **diverse** as possible (different skills should visit different states) and **distinguishable** (it should be possible to infer the skill from the states visited by the intra-skill policy). It follows that the learning process is 4-step with two learning parts [13]: 1 - the agent samples one skill from an uniform distribution; 2 - the agent executes the skill by following the corresponding intra-skill policy (randomly initialized); 3 - a discriminator learns to categorize the resulting states to the assigned skill; 4 - at the same time, these approximations reward intra-skill policies (cf. Eq. 4).

The global objective can be formalized as maximizing the MI between the set of skills G and states S' visited by intra-skill policies, defined by [17]:

$$I(G; S') = \mathbb{H}(G) - \mathbb{H}(G|S') = \mathbb{E}_{g \sim p(g), s' \sim p(s'|\pi_\theta^g, s)}[\log p(g|s') - \log p(g)] \quad (2)$$

where π_θ^g is the intra-skill policy of $g \in G$ and is parameterized by θ; $p(g)$ is the distribution of skills the agent samples on; and $p(g|s')$ is the probability to infer g knowing the next state s' and intra-skill policies. This MI quantifies the reduction in the uncertainty of G due to the knowledge of S'. By maximizing it, states visited by an intra-skill policy have to be informative of the given skill.

A bound on the MI can be used as an approximation to avoid the difficulty to compute $p(g|s')$ [8,17]:

$$I(G, S') \geq \mathbb{E}_{g \sim p(g), s' \sim p(s'|\pi_\theta^g, s)}[\log q_\omega(g|s') - \log p(g)] \quad (3)$$

where $q_\omega(g|s')$ is the **discriminator** approximating $p(g|s')$. In our case, the discriminator is a neural network parameterized by ω. q_ω minimizes the standard cross-entropy $-\mathbb{E}_{g \sim p(g|s')} \log q_\omega(g|s')$ where $s' \sim \pi_\theta^g$.

To discover skills, it is more efficient to set $p(g)$ to be uniform as it maximizes the entropy of G [13]. Using the uniform distribution, $\log p(g)$ is constant and can be removed from Eq. 3. It follows that one can maximize Eq. 3 using an intrinsic reward to learn the intra-skill policy of a skill $g \in G$ [13]:

$$r^g(s') = \log q_\omega(g|s'). \quad (4)$$

Similarly to [13], we use an additional entropy term to encourage the diversity of covered states. In practice, this bonus is maximized through the use of DRL algorithms: Soft Actor Critic (SAC) [19] for continuous action space and Deep Q network (DQN) with Boltzmann exploration [29] for discrete one.

3 Method

In this section, we first give an overview of our method and then detail the building of the tree of skills, the learning of the skill policy, the selection of the skill to refine and how ELSIM integrates this in an end-to-end framework.

3.1   Overview: Building a Tree of Skills

To get both bottom-up skills and interesting skills relatively to some tasks, our agent has to choose the skills to improve thanks to the extrinsic rewards, but we want that our agent improves its skills without extrinsic rewards. The agent starts by learning a discrete set of diverse and distinguishable skills using the method presented in Sect. 2.2. Once the agent clearly distinguishes these skills using the covered skill-conditioned states with its discriminator, it splits them into new sub-skills. For instance, for a creature provided with proprioceptive data, a *moving forward* skill could be separated into *running* and *walking*. The agent only trains on sub-skills for which the parent skill is useful for the global task. Thus it incrementally refines the skills it needs to accomplish its current task. If the agent strives to sprint, it will select the skill that provides the greater speed. The agent repeats the splitting procedure until its intra-skill policy either reach the maximum number of splits or become too deterministic to be refined.

The **hierarchy of skills** is maintained using a tree where each node refers to an abstract skill that has been split and each leaf is a skill being learned. We formalize the hierarchy using sequence of letters where a letter's value is assigned to each node:

- The set of skills G is the set of leaf nodes. A skill $g \in G$ is represented by a sequence of $k+1$ letters: $g = (l^0, l^1, ..., l^k)$. When g is split, a letter is added to the sequence of its new sub-skills. For instance, the skill $g = (l^0 = 0, l^1 = 1)$ can be split into two sub-skills $(l^0 = 0, l^1 = 1, l^2 = 0)$ and $(l^0 = 0, l^1 = 1, l^2 = 1)$.
- The vocabulary V refers to the values which can be assigned to a letter. For example, to refine a skill into 4 sub-skills, we should define $V = \{0, 1, 2, 3\}$.
- The length $L(g)$ of a skill is the number of letters it contains. Note that the length of a skill is always larger than its parent's.
- $l^{:k}$ is the sequence of letters preceding l^k (excluded).

We use two kind of policies: the first are the **intra-skill policies**. The learning of these intra-skill policies is described in Sect. 3.2. The second type of policy is task-dependent and responsible to choose which skill to execute; we call it the **tree-policy** (see Sect. 3.3).

3.2   Learning Intra-skill Policies

In this section, we detail how intra-skill policies are learned. We adapt the method presented in Sect. 2.2 to our hierarchical skills context. Two processes are simultaneously trained to obtain diverse skills: the intra-skill policies learn to maximize the intrinsic reward (cf. Eq. 4), which requires to learn a discriminator $q_\omega(g|s')$. Given our hierarchic skills, we can formulate the probability inferred by the discriminator as a product of the probabilities of achieving each letter of g knowing the sequence of preceding letters, by applying the chain rule:

$$r^g(s') = \log q_\omega(g|s') = \log q_\omega(l^0, l^1, \dots, l^k|s')$$

$$= \log \prod_{i=0}^{k} q_\omega(l^i|s', l^{:i}) = \sum_{i=0}^{k} \log q_\omega(l^i|s', l^{:i}). \qquad (5)$$

Gathering this value is difficult and requires an efficient discriminator q_ω. As it will be explained in Sect. 3.4, in practice, we use **one different discriminator for each node** of our tree: $\forall i,\ q_\omega(l^i|s', l^{:i}) \equiv q_\omega^{:i}(l^i|s')$.

For instance, if $|V| = 2$, one discriminator $q_\omega^\emptyset$ will be used to discriminate $(l^0 = 0)$ and $(l^0 = 1)$ but an other one, $q_\omega^{l^0=0}$ will discriminate $(l^0 = 0, l^1 = 0)$ and $(l^0 = 0, l^1 = 1)$.

It would be difficult for the discriminators to learn over all letters at once; the agent would gather states for several inter-level discriminators at the same time and a discriminator would not know which part of the gathered states it should focus on. This is due to the fact that discriminators and intra-skill policies simultaneously train. Furthermore, there are millions of possible combinations of letters when the maximum size of sequence is large. We do not want to learn them all. To address these issues, we introduce **a new curriculum learning algorithm that refines a skill only when it is distinguishable**. When discriminators successfully learn, they progressively extends the sequence of letters; in fact, we split a skill (add a letter) only when its discriminator has managed to discriminate the values of its letter. Let's define the following probability:

$$p_{finish}^{:k}(l^k) = \mathbb{E}_{s_{final} \sim \pi^{:k+1}} \left[q_\omega^{:k}(l^k|s_{final}) \right]. \qquad (6)$$

where s_{final} is the state reached by the intra-skill policy at the last timestep. We assume the discriminator $q_\omega^{:k}$ has finished to learn when: $\forall v \in V,\ p_{finish}^{:k}(l^k = v) \geq \delta$ where $\delta \in [0, 1]$ is an hyperparameter. Choosing a δ close to 1 ensures that the skill is learned, but an intra-skill policy always explores, thereby it may never reach an average probability of exactly 1; we found empirically that 0.9 works well.

To approximate Eq. 6 for each letters' value v, we use an exponential moving average $p_{finish}^{:k}(l^k = v) = (1 - \beta)p_{finish}^{:k}(l^k = v) + \beta q_\omega^{:k}(l^k = v|s_{final})$ where $s_{final} \sim \pi^{:k+1}$ and $\beta \in [0; 1]$. Since we use buffers of interactions (see Sect. 3.4), we entirely refill the buffer before the split.

Let us reconsider Eq. 5. $\sum_{i=0}^{k-1} \log q_\omega(l^i|s', l^{:i})$ is the part of the reward assigned by the previously learned discriminators. It forces the skill to stay close to the states of its parent skills since this part of the reward is common to all the rewards of its parent skills. In contrast, $\log q_\omega(l^k|s', l^{:k})$ is the reward assigned by the discriminator that actively learns a new discrimination of the state space. Since the agent is constrained to stay inside the area of previous discriminators, the new discrimination is **uncorrelated** from previous parent discriminations. In practice, we increase the importance of previous discriminations with a hyperparameter $\alpha \in \mathbb{R}$:

$$r^g(s') = \log q_\omega(l^k|s', l^{:k}) + \alpha \sum_{i=0}^{k-1} \log q_\omega(l^i|s', l^{:i}). \tag{7}$$

This hyper-parameter is important to prevent the agent to deviate from previously discriminated areas to learn more easily the new discrimination.

3.3   Learning Which Skill to Execute and Train

For each global objective, a stochastic policy, called *tree-policy* and noted π_T (with T the tree of skills), is responsible to choose the skill to train by navigating inside the tree at the beginning of a task-episode. This choice is critical in our setting: while expanding its tree of skills, the agent cannot learn to discriminate every leaf skill at the same time since discriminators need states resulting from the intra-skill policies. We propose to **choose the skill to refine according to its benefit in getting an other reward** (extrinsic or intrinsic), thereby ELSIM executes and learns only interesting skills (relatively to an additional reward).

To learn the *tree-policy*, we propose to model the tree of skills as an MDP solved with a Q-learning and Boltzmann exploration. The action space is the vocabulary V; the state space is the set of nodes, which include abstract and actual skills; the deterministic transition function is the next node selection; if the node is not a leaf, the reward function R_T is 0, else this is the discounted reward of the intra-skill policy executed in the environment divided by the maximal episode length. Each episode starts with the initial state as the root of the tree, the *tree-policy* selects the next nodes using Q-values. Each episode ends when a leaf node has been chosen, i.e. a skill for which all its letters has been selected; the last node is always chosen uniformly (see Sect. 3.4).

Let us roll out an example using the *tree-policy* displayed in Fig. 1. The episode starts at the root of the tree; the *tree-policy* samples the first letter, for example it selects $l^0 = 0$. Until it reaches a leaf-node, it samples new letters, e.g. $l^1 = 1$ and $l^2 = 0$. The *tree-policy* has reached a leaf, thereby it will execute and learn the skill $(0, 1, 0)$. Then, the state-action tuple $((0, 1), (0))$ is rewarded with the scaled discounted reward of the task. This reward is propagated via the Q-learning update to previous state-action tuples $((\emptyset), (0))$ and $((0), (1))$ to orientate the *tree-policy* to $(0, 1, 0)$.

The MDP evolves during the learning process since new letters are progressively added. The Q-values of new skills are initialized with their parent Q-values. However, Eq. 5 ensures that adding letters at the leaf of the tree monotonically increases Q-values of their parent nodes. The intuition is that, when splitting a skill, at least one of the child is equal or better than the skill of its parent relatively to the task. We experimentally show this in Sect. 4.2. The resulting curriculum can be summarized as follows: the tree will be small at the beginning, and will grow larger in the direction of feedbacks of the environment.

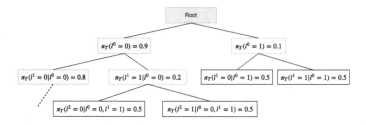

Fig. 1. Representation of a part of the tree of skills with $|V| = 2$ and the value of *tree-policy* in each node. White nodes are actual leaves of the tree; the discriminator is inactive. Yellow nodes represent nodes for which the discriminator can not differentiate its sub-skills; the *tree-policy* samples uniformly. Nodes are blue when the discriminator can distinguish its sub-skills; the *tree-policy* samples using Q-values. (Color figure online)

We now sum up the process of the *tree-policy*: 1-an agent runs an episode inside the MDP of skills; the sequence of actions represents a skill; 2- the agent executes the intra-skill policy of the skill; 3- the *tree-policy* is rewarded according to how well the intra-skill policy fits the task and the Q-learning applies.

3.4 Simultaneous Training of the *tree-policy* and intra-skill policies

The MI objective requires the skill distribution to remain uniform (cf. Eq. 4), however that is not our case: the agent strives to avoid some useless skills while focusing on others. In our preliminary experiments, ignoring this leads us to catastrophic forgetting of the learned skills since discriminators forget how to categorize states of the skills they never learn on. To bypass this issue and sample uniformly, we assign to each node i of our tree a replay buffer containing interactions of the intra-skill policy with the environment, a RL algorithm and a discriminator (q_ω^i). At each split, intra-skill policies and buffers of a node are copied to its children; for the first node, its intra-skill policies are randomly initialized and its buffer is empty.

This way, the entire training is off-policy: the intra-skill policy fills the replay buffer while the discriminator and intra-skill policies learn from the interactions that are uniformly extracted from their buffers. We split the lifetime of a node into two phases: 1-the **learning phase** during which next letter's values are sampled uniformly; the *tree-policy* is uniform at this node; 2-the **exploitation phase** during which the *tree-policy* chooses letters with its Boltzmann policy (Sect. 3.3).

Then, at each step, the agent runs the *tree-policy* to select the discriminator in the learning phase that will learn. The discriminator samples a mini-batch of data from its children's (all leaves) buffers and learns on it. Then, all children intra-skill policies learn from the intrinsic feedback of the same interactions, output by the selected discriminator and all its parents according to Eq. 5.

Once a node enters the exploitation phase, an hyper-parameter η regulates the probability that each parent's discriminator learn on its children data. Their learning interactions are recursively sampled uniformly on their children. This post-exploration learning allows a node to expand its high-reward area. Without this mechanism, different uncovered states of the desired behaviour may be definitively attributed to different fuzzy skills, as shown in Sect. 4.1.

Figure 1 gives an example of a potential tree and how different phases coexist; the skills starting by $(0, 1)$ seem to be the most interesting for the task since each letter sampling probability is high. Skills $(0, 1, 0)$, $(0, 1, 1)$, $(1, 0)$ and $(1, 1)$ are being learned, therefore the sampling probability of their last values is uniform.

4 Experiments

The first objective of this section is to study the behavior of our ELSIM algorithm on basic gridworlds to make the visualization easier. The second purpose is to show that ELSIM can scale with high-dimensional environments. We also compare its performance with a non-hierarchical algorithm SAC [19] in a single task setting. Finally we show the potential of ELSIM for transfer learning.

4.1 Study of ELSIM in Gridworlds

In this section, we analyze how skills are refined on simple gridworlds adapted from gym-minigrid [10]. Unless otherwise stated, **there is no particular task (or extrinsic reward)**, thereby the *tree-policy* is uniform. The observations of the agent are its coordinates; its actions are the movements into the four cardinal directions. To maximize the entropy of the intra-skill policy with a discrete action space, we use the DQN algorithm [29]. The agent starts an episode at the position of the arrow (see figures) and an episode resets every 100 steps, thus the intra-skill policy lasts 100 steps. At the end of the training phase, the skills of all the nodes are evaluated through an evaluation phase lasting 500 steps for each skill. In all figures, each tile corresponds to a skill with $|V| = 4$ that is displayed at the top-left of the tile. Figure 2 and 4 display the density of the states visited by intra-skill policies during the evaluation phase: the more red the state, the more the agent goes over it.

Do the Split of Skills Improve the Exploration of an Agent? Figure 2 shows some skills learned in an environment of 4 rooms separated by a bottleneck. We first notice that the agent clearly separates its first skills $(0), (1), (2), (3)$ since the states covered by one skill are distinct from the states of the other skills. However it does not escape from its starting room when it learns these first skills. When it develops the skills close to bottlenecks, it learns to go beyond and invests new rooms. It is clear for skills (1) and (2) which, with one refinement, respectively explore the top-right (skill $(1, 0)$) and bottom-left (skills $(2, 0), (2, 2), (2, 3)$) rooms. With a second refinement, $(2, 3, 0)$ even manages to reach the farthest room (bottom-right). This stresses out that the refinement of a skill also allows

Fig. 2. Some skills learned by the agent in an environment composed of four rooms.

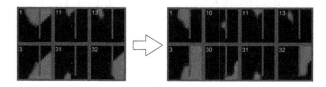

Fig. 3. Discriminator's probability of achieving skills (1), (3) and their sub-skills in every state (*i.e.* $q(g|s)$). The more red the state, the more rewarding it is for the skill. The left side corresponds to the preliminary stage of the learning process (timestep 128.10^4); the right side corresponds to the end of the learning process (timestep 640.10^4). (Color figure online)

to expand the states covered by the skill, and thus can improve the exploration of an agent when the rewards are sparse in an environment.

Do the Split of Skills Correct a Wrong Over-Generalization of a Parent Skill? Figure 3 shows the evolution of the intrinsic reward function for some skills. The environment contains a vertical wall and settings are the same as before, except the Boltzmann parameter set to 0.5. At the beginning, skill (1) is rewarding identically left and right sides of the wall. This is due to the generalization over coordinates and to the fact that the agent has not yet visited the right side of the wall. However it is a wrong generalization because left and right sides are not close to each other (considering actions). After training, when the agent begins to reach the right side through the skill (3, 2), it corrects this wrong generalization. The reward functions better capture the distance (in actions) between two states: states on the right side of the wall are attributed to skill (3) rather than (1). We can note that other parts of the reward function remain identical.

Can the agent choose which skill to develop as a priority? In this part, we use the same environment as previously, but states on the right side of the wall give an extrinsic reward of 1. Thus the agent follows the *tree-policy* to maximize its rewards, using Boltzmann exploration, and focus its refinement on rewarding skills. Figure 4 shows all the parent skills of the most refined skill which reaches $L(g) = 6$. The agent learns more specialized skills in the rewarding area than when no reward is provided.

Fig. 4. One path in the tree of skills learned by the agent with an extrinsic reward of 1 on the upper right side of the wall.

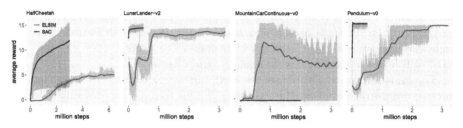

Fig. 5. Average reward per episode in classical environments (*HalfCheetah-v2* [45], *LunarLanderContinuous-v0* [37], *MountainCarContinuous-v0* [30] and *Pendulum-v0*) for SAC and ELSIM (averaged over 4 seeds). We use our own implementation of SAC except for *HalfCheetah* for which the blue curve is the average reward of SAC on 5 seeds, taken from [19]. We stopped the simulation after convergence of SAC.

Summary. We illustrated the following properties of ELSIM: 1 - it expands a previously learned rewarding area when it discovers new states; we show in Sect. 4.2 that it improves exploration when the rewards are sparse; 2 - adding letters corrects over-generalization of their parent discriminator; 3 - it can focus the skill expansion towards task-interesting areas.

4.2   Performance on a Single Task

In this part, we study the ability of ELSIM to be competitive on high-dimensional benchmarks[2] **without any prior knowledge**. For ELSIM, we set the maximum skill length to 10, which is reached in *HalfCheetah*.

Figure 5 respectively shows the average reward per episode for different environments. Shaded areas color are upper-bounded (resp. lower-bounded) by the maximal (resp. minimal) average reward.

First, the *MountainCarContinuous* environment represents a challenge for the exploration as it is a sparse reward environment: the agent receives the reward only when it reaches the goal. In this environment, ELSIM outperforms SAC by getting a higher average reward. It confirms our results (cf. Sect. 4.1) on **the positive impact of ELSIM on the exploration**. There is a slight decrease after reaching an optima, in fact, ELSIM keeps discovering skills after finding its optimal skill. On *Pendulum* and *LunarLander*, ELSIM achieves the same asymptotic average reward than SAC, even though ELSIM may require more timesteps. On *HalfCheetah*, SAC is on average better than ELSIM. However we emphasize that **ELSIM also learns other skills**. For example in

---

[2] *HalfCheetah* has a state space and action space respectively of 17 and 6 dimensions.

HalfCheetah, ELSIM learns to walk and flip while SAC, that is a non-hierarchical algorithm, only learns to sprint.

4.3   Transfer Learning

In this section, we evaluate the interest of ELSIM for transfer learning. We take skills learned by intra-policies in Sect. 4.2, reset the *tree-policy* and restart the learning process on *HalfCheetah* and *HalfCheetah-Walk*. *HalfCheetah-Walk* is a slight modification of *HalfCheetah* which makes the agent target a speed of 2. Intra-skill policies learning was stopped in *HalfCheetah*.

The same parameters as before are used, but we use MBIE-EB [39] to explore the tree. Figure 6 shows that the *tree-policy* learns to reuse its previously learned skills on *HalfCheetah* since it almost achieves the same average reward as in Fig. 5. On *HalfCheetah-Walk*, we clearly see that the agent has already learned skills to walk and that it easily retrieves them. In both environments, ELSIM learns faster than SAC, which learn from scratch. It demonstrates that skills learned by ELSIM can be used for **other tasks** than the one it has originally been trained on.

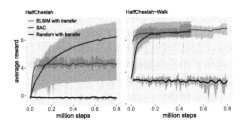

Fig. 6. Average reward per episode in *HalfCheetah* and *HalfCheetah-Walk*. We use our own implementation of SAC for *HalfCheetah-Walk*. The black curve is the average reward of a random *tree-policy* that uses the transfered skills.

5   Related Work

Intrinsic motivation in RL is mostly used to improve exploration when rewards are sparse or to learn skills [4]. The works that learn skills with intrinsic rewards are close to our approach and can be classified in two major categories. The first category strives to explicitly target states. The reward is defined either as a distance between agent's state and its goal state [22], or as the difference between the change in the state space and the required change [32]. However, to efficiently guide the agent, the reward functions require a good state representation [31,34].

Intrinsic motivation as diversity heuristic. Our work mostly falls into this category, which strives to define a skill based on a MI objective (cf. Sect. 2.2). Seminal works already learn a discrete set of diverse skills [13,15]. In contrast to them, we manage to learn both the skill and the skill-selection policy in an end-to-end way and we propose an efficient way to learn a large number of skills useful for the tasks the agent wants to accomplish. Recent works [11,47] try to learn a continuous embedding of skills, but do not integrate their work into an end-to-end hierarchical learning agent. DADS [38] learn skills using a generative model over observations rather than over skills. While this is efficient on environments with simple observation space, this is computationally ineffective. In our work, rather than learning a continuous skill embedding, we strive to select and learn skills among a very large number of discretized skills. As a consequence, we focus our learned skill distribution only on task-interesting skills and we do not rely on a parametric state distribution. **Continual learning.** Other works proposed a *lifelong learning* architecture. Some assume that skills are already learned and learn to reuse them; for example H-DRLN [43] uses a hierarchical policy to choose between ground actions and skills. They also propose to distill previously learned skills into a larger architecture, making their approach scalable. In contrast, we tackle the problem of learning skills in an end-to-end fashion, thereby our approach may be compatible. Similarly to us, CCSA [21] addresses the catastrophic forgetting problem by freezing the learning of some experts. They mix two unsupervised learning methods to find and represent goal states, and then learn to reach them. However, their unsupervised algorithm only extracts linear features and they manually define a first set of skills. One particular aspect of continual learning is Meta-RL: how can an agent learn how to learn? Traditional methods assume there exists a task distributions and try to generalize over it [12,14]; this task distribution serves as prior knowledge. In [18], the authors address this issue and apply MAML [14] on an uniform distribution of tasks learned by DIAYN [13]. However, learning is neither focused, nor end-to-end. In the continuity of this work, CARML [20] mixes the objective of DADS [38] and Meta-RL; it alternates between generating trajectories of the distribution of tasks and fitting the task distribution to new trajectories. While CARML discovers diverse behaviors with pixel-level state space, it cannot learn a global objective end-to-end like ELSIM.

State abstraction. Our method can be viewed as a way to perform state abstraction [23]. Rather than using this abstraction as inputs to make learning easier, we use it to target specific states. The application of our refinement method bounds the suboptimality of the representation, while the task-independent clustering ensures that skills are transferable. In contrast to our objective, existing methods usually tackle suboptimality for a task without addressing transfer learning or exploration [1,3]. The k-d tree algorithm [16] has been used to perform state abstraction over a continuous state space [46], but as above, the splitting process takes advantage of extrinsic reward and previously defined partitions are not adapted throughout the learning process. In the domain of developmental robotics, RIAC and SAGG-RIAC [6,7] already implement a splitting algorithm building a tree of subregions in order to efficiently explore the environment and learn a forward model. More precisely, they split the state space to maximize either the sum of variance of interactions already collected or the difference of learning progress between subregions. However, these heuristics do not scale to larger continuous environments. In contrast, we assign states to subregions according to the proximity of states and use these subregions as reusable skills to solve several tasks. ASAP [27] partitions the goal space, but does not use intrinsic motivation and the partitions are limited to hyper-plans.

6   Conclusion

We proposed ELSIM, a novel algorithm that continually refines discrete skills using a recently defined diversity heuristic [13]. To do so, the agent progressively builds a tree of different skills in the direction of a high-level objective. As shown in Sect. 4, **ELSIM expands the area associated to a skill** thanks to its exploratory behavior which comes from adding latent variables to the overall policy. **ELSIM also focuses its training on interesting skills relatively to some tasks.** Even though the agent is often learning a task, the skills can be defined independently from a specific task and we showed that ELSIM possibly makes them **transferable across different tasks** of a similar environment. Since the agent does not need extrinsic reward to learn, we show that it can **improve exploration** on sparse rewards environments. We believe that such a paradigm is appropriate for *lifelong learning*.

Currently, our method allows to avoid the problem of catastrophic forgetting, but the counterpart is an increase of the memory footprint, which is a recurrent issue in methods based on trees. Several works addressing catastrophic forgetting may be adapted to our work, e.g. [24] and could potentially improve transfer learning between neural networks at different levels of our tree. In addition, ELSIM quickly gets stuck in local optimas in more difficult environments such as *BipedalWalker-v2* or Pybullet environments. The main limitation of our approach is that we cannot select several skills in one episode, such as one would make within the *option* framework [41]. To be adapted, the *tree-policy* should be dependent on the true state and the diversity heuristic should maximize $\mathbb{E}_{s\sim\mathbb{U}(s)} I(G, S'|S)$ rather than Eq. 5 like in [38]. Thus the curriculum algorithm should be modified. It would result that the semantic meaning of a skill would be no longer to target an area, but to produce a change in the state space. We plan to address these issues in future work.

Acknowledgment. We thank O. Sigaud and A. Dutech for their useful feedbacks on the paper and NVIDIA Corporation for the donation of a Titan V.

References

1. Abel, D., Hershkowitz, D.E., Littman, M.L.: Near optimal behavior via approximate state abstraction. In: ICML, pp. 2915–2923 (2016)
2. Achiam, J., Edwards, H., Amodei, D., Abbeel, P.: Variational option discovery algorithms. arXiv preprint arXiv:1807.10299 (2018)
3. Akrour, R., Veiga, F., Peters, J., Neumann, G.: Regularizing reinforcement learning with state abstraction. In: IROS, pp. 534–539. IEEE (2018)
4. Aubret, A., Matignon, L., Hassas, S.: A survey on intrinsic motivation in reinforcement learning. arXiv preprint arXiv:1908.06976 (2019)

5. Bacon, P.L., Harb, J., Precup, D.: The option-critic architecture. In: AAAI, pp. 1726–1734 (2017)
6. Baranes, A., Oudeyer, P.Y.: R-IAC: robust intrinsically motivated exploration and active learning. IEEE Trans. Auton. Ment. Dev. **1**(3), 155–169 (2009)
7. Baranes, A., Oudeyer, P.Y.: Intrinsically motivated goal exploration for active motor learning in robots: a case study. In: IROS, pp. 1766–1773 (2010)
8. Barber, D., Agakov, F.V.: The IM algorithm: a variational approach to information maximization. In: Advances in Neural Information Processing Systems, pp. 201–208 (2003)
9. Barto, A.G.: Intrinsic motivation and reinforcement learning. In: Baldassarre, G., Mirolli, M. (eds.) Intrinsically Motivated Learning in Natural and Artificial Systems, pp. 17–47. Springer, Heidelberg (2013). https://doi.org/10.1007/978-3-642-32375-1_2
10. Chevalier-Boisvert, M., Willems, L.: Minimalistic gridworld environment for OpenAI Gym (2018). https://github.com/maximecb/gym-minigrid
11. Co-Reyes, J.D., Liu, Y., Gupta, A., Eysenbach, B., Abbeel, P., Levine, S.: Self-consistent trajectory autoencoder: hierarchical reinforcement learning with trajectory embeddings. In: ICML, pp. 1008–1017 (2018)
12. Duan, Y., Schulman, J., Chen, X., Bartlett, P.L., Sutskever, I., Abbeel, P.: Rl^2: fast reinforcement learning via slow reinforcement learning. CoRR abs/1611.02779 (2016)
13. Eysenbach, B., Gupta, A., Ibarz, J., Levine, S.: Diversity is all you need: learning skills without a reward function. In: ICLR (2019)
14. Finn, C., Abbeel, P., Levine, S.: Model-agnostic meta-learning for fast adaptation of deep networks. In: Proceedings of the 34th International Conference on Machine Learning, vol. 70, pp. 1126–1135. JMLR.org (2017)
15. Florensa, C., Duan, Y., Abbeel, P.: Stochastic neural networks for hierarchical reinforcement learning. In: ICLR (2017)
16. Friedman, J.H., Bentley, J.L., Finkel, R.A.: An algorithm for finding best matches in logarithmic expected time. ACM (TOMS) **3**(3), 209–226 (1977)
17. Gregor, K., Rezende, D.J., Wierstra, D.: Variational intrinsic control. In: ICLR 2017 (2017)
18. Gupta, A., Eysenbach, B., Finn, C., Levine, S.: Unsupervised meta-learning for reinforcement learning. CoRR abs/1806.04640 (2018)
19. Haarnoja, T., Zhou, A., Abbeel, P., Levine, S.: Soft actor-critic: off-policy maximum entropy deep reinforcement learning with a stochastic actor. In: ICML, pp. 1856–1865 (2018)
20. Jabri, A., Hsu, K., Gupta, A., Eysenbach, B., Levine, S., Finn, C.: Unsupervised curricula for visual meta-reinforcement learning. In: Advances in Neural Information Processing Systems, pp. 10519–10530 (2019)
21. Kompella, V.R., Stollenga, M., Luciw, M., Schmidhuber, J.: Continual curiosity-driven skill acquisition from high-dimensional video inputs for humanoid robots. Artif. Intell. **247**, 313–335 (2017)
22. Levy, A., Platt, R., Saenko, K.: Hierarchical reinforcement learning with hindsight. In: International Conference on Learning Representations (2019)
23. Li, L., Walsh, T.J., Littman, M.L.: Towards a unified theory of state abstraction for MDPs. In: ISAIM (2006)
24. Lopez-Paz, D., Ranzato, M.: Gradient episodic memory for continual learning. In: Advances in Neural Information Processing Systems, pp. 6467–6476 (2017)
25. Machado, M.C., Bellemare, M.G., Bowling, M.: A Laplacian framework for option discovery in reinforcement learning. In: ICML, vol. 70, pp. 2295–2304. JMLR.org (2017)

26. Machado, M.C., Bowling, M.: Learning purposeful behaviour in the absence of rewards. arXiv preprint arXiv:1605.07700 (2016)
27. Mankowitz, D.J., Mann, T.A., Mannor, S.: Adaptive skills adaptive partitions (ASAP). In: Advances in Neural Information Processing Systems, pp. 1588–1596 (2016)
28. Metzen, J.H., Kirchner, F.: Incremental learning of skill collections based on intrinsic motivation. Front. Neurorobotics **7**, 11 (2013)
29. Mnih, V., et al.: Human-level control through deep reinforcement learning. Nature **518**(7540), 529 (2015)
30. Moore, A.W.: Efficient memory-based learning for robot control (1990)
31. Nachum, O., Gu, S., Lee, H., Levine, S.: Near-optimal representation learning for hierarchical reinforcement learning. In: ICLR (2019)
32. Nachum, O., Gu, S.S., Lee, H., Levine, S.: Data-efficient hierarchical reinforcement learning. In: Bengio, S., Wallach, H., Larochelle, H., Grauman, K., Cesa-Bianchi, N., Garnett, R. (eds.) Advances in Neural Information Processing Systems, vol. 31, pp. 3303–3313 (2018)
33. Nachum, O., Tang, H., Lu, X., Gu, S., Lee, H., Levine, S.: Why does hierarchy (sometimes) work so well in reinforcement learning? arXiv preprint arXiv:1909.10618 (2019)
34. Nair, A.V., Pong, V., Dalal, M., Bahl, S., Lin, S., Levine, S.: Visual reinforcement learning with imagined goals. In: Advances in Neural Information Processing Systems, pp. 9209–9220 (2018)
35. Puterman, M.L.: Markov Decision Processes: Discrete Stochastic Dynamic Programming. Wiley, Hoboken (2014)
36. Ryan, R.M., Deci, E.L.: Intrinsic and extrinsic motivations: classic definitions and new directions. Contemp. Educ. Psychol. **25**(1), 54–67 (2000)
37. Shariff, R., Dick, T.: Lunar lander: a continous-action case study for policy-gradient actor-critic algorithms (2013)
38. Sharma, A., Gu, S., Levine, S., Kumar, V., Hausman, K.: Dynamics-aware unsupervised discovery of skills. arXiv preprint arXiv:1907.01657 (2019)
39. Strehl, A.L., Littman, M.L.: An analysis of model-based interval estimation for Markov decision processes. J. Comput. Syst. Sci. **74**(8), 1309–1331 (2008)
40. Sutton, R.S., Barto, A.G.: Reinforcement Learning: An Introduction, vol. 1. MIT Press, Cambridge (1998)
41. Sutton, R.S., Precup, D., Singh, S.: Between MDPs and semi-MDPs: a framework for temporal abstraction in reinforcement learning. Artif. Intell. **112**(1–2), 181–211 (1999)
42. Taylor, M.E., Stone, P.: Transfer learning for reinforcement learning domains: a survey. J. Mach. Learn. Res. **10**, 1633–1685 (2009)
43. Tessler, C., Givony, S., Zahavy, T., Mankowitz, D.J., Mannor, S.: A deep hierarchical approach to lifelong learning in minecraft. In: AAAI (2017)
44. Thrun, S.: Is learning the n-th thing any easier than learning the first? In: Advances in Neural Information Processing Systems, pp. 640–646 (1996)
45. Todorov, E., Erez, T., Tassa, Y.: MuJoCo: a physics engine for model-based control. In: IEEE/RSJ IROS, pp. 5026–5033. IEEE (2012)
46. Uther, W.T., Veloso, M.M.: Tree based discretization for continuous state space reinforcement learning. In: AAAI/IAAI, pp. 769–774 (1998)
47. Warde-Farley, D., de Wiele, T.V., Kulkarni, T.D., Ionescu, C., Hansen, S., Mnih, V.: Unsupervised control through non-parametric discriminative rewards. In: ICLR (2019)

Graph-Based Motion Planning Networks

Tai Hoang[1] and Ngo Anh Vien[2(✉)]

[1] University of Passau, Passau, Germany
thobotics@gmail.com
[2] Bosch Center for Artificial Intelligence, Renningen, Germany
anhvien.ngo@bosch.com

Abstract. Differentiable planning network architecture has shown to be powerful in solving transfer planning tasks while it possesses a simple end-to-end training feature. Much related work that has been proposed in literature is inspired by a design principle in which a recursive network architecture is applied to emulate backup operations of a value iteration algorithm. However, existing frameworks can only learn and plan effectively on domains with a lattice structure, i.e., regular graphs embedded in a particular Euclidean space. In this paper, we propose a general planning network called Graph-based Motion Planning Networks (GrMPN). GrMPN will be able to i) learn and plan on general irregular graphs, hence ii) render existing planning network architectures special cases. The proposed GrMPN framework is invariant to task graph permutation, i.e., graph isomorphism. As a result, GrMPN possesses an ability that is data-efficient and strong at generalization strength. We demonstrate the performance of two proposed GrMPN methods against other baselines on three domains ranging from 2D mazes (regular graph), path planning on irregular graphs, and motion planning (an irregular graph of robot configurations).

1 Introduction

Reinforcement learning (RL) is a sub-field of machine learning that studies about how an agent makes sequential decision making [35] to interact with an environment. These problems can, in principle, be formulated as a Markov decision process (MDP). (Approximate) Dynamic programming methods such as value iteration or policy iterations are often used for policy optimisation. These dynamic programming approaches can also be leveraged to handle learning, hence referred to as model-based RL [17]. Model-based RL requires an estimation of the environment model; hence, it is computationally expensive, but it is shown to be very data-efficient. The second common RL paradigm is model-free, which does not require a model estimation hence has lower computation cost but less data-efficiency [17]. With a recent marriage with deep learning, deep reinforcement learning (DRL) has achieved many remarkable successes on a wide variety of applications such as game [25,33], robotics [22], chemical synthesis [31], news recommendation [40] and many other fields. DRL methods also range from model-based [18,20] to model-free [11,25] approaches.

© Springer Nature Switzerland AG 2021
F. Hutter et al. (Eds.): ECML PKDD 2020, LNAI 12458, pp. 557–573, 2021.
https://doi.org/10.1007/978-3-030-67661-2_33

On the other hand, transfer learning across tasks has long been desired but is much more challenging in comparison to single-task learning. Recent work [36] has proposed an exquisite idea that suggests encoding a differentiable planning module in a policy network architecture. This planning module can emulate the recursive operation of value iterations, called Value Iteration Networks (VIN). The agent using this network can roll out multiple future planning steps to evaluate the policy. The planning module is designed to base on a recursive application of convolutional neural networks (CNN) and max-pooling for value function updates. VIN not only allows policy optimisation with more data-efficiency, but also enables transfer learning across problems with shared transition and reward structures. VIN has laid foundation for many later differentiable planning network architectures such as QMDP-Net [13], planning under uncertainty[9], Memory Augmented Control Network (MACN) [15], Predictron [32], planning networks [34]. However, these approaches, including VIN, is limited to learning with regular environment structures, i.e. the transition function forms an underlying 2D lattice structure.

Recent works have tried to mitigate this issue by resorting to graph neural networks as a principled tool to exploit geometric intuition in environments having irregular structures such as generalised VIN [27], planning on relational domains [2,24,37] or automated planning for scheduling [24]. The common between these approaches is in the use of graph neural networks to process irregular data structures like graphs. Among these frameworks, only GVIN can emulate the value iteration algorithm on irregular graphs of arbitrary sizes, e.g. generalisation to graphs of arbitrary structure. GVIN has a differentiable policy network architecture which is very similar to VIN. GVIN can also have the *zero-shot planning* ability on unseen graphs. However, GVIN requires domain knowledge to design graph convolution which might limit it to become a universal graph-based path planning framework.

In this paper, we aim to demonstrate different formulations for value iteration networks on irregular graphs. These proposed formulations are based on different graph network models. These models are capable of learning optimal policies on general graphs where their transition and reward functions are not provided *a priori* and yet to be estimated. These models are known to be invariant to graph isomorphism. Therefore they can have generalisation ability to graphs of different sizes. As a result, they enjoy the ability of *zero-shot learning to plan*. Specifically, it is known that Bellman equations are written as the form of message passing. Therefore we propose using message passing neural networks (MPNN) to emulate the value iteration algorithm on graphs. We will show two most general formulations of graph-based value iteration network that are based on two general-purpose approaches in the MPNN family: Graph Networks (GN) [3] and Graph Attention Networks (GAT) [38]. In particular, our contributions are three-fold:

– We develop a MPNN based path planning network (GrMPN) which can learn to plan on general graphs, e.g. regular and irregular graphs. GrMPN is a differentiable end-to-end planning network architecture trained via imitation learning. We implement GrMPN based on two formulations that use GN and GAT, respectively.

- GrMPN is a general graph-based value iteration network that will render existing graph-based planning algorithms special cases. GrMPN is invariant to graph isomorphism, which enables transfer planning on graphs of different structure and size.
- We will demonstrate the efficacy of GrMPN, which achieves state of the art results on various domains including 2D maze with regular graph structures, irregular graphs, and motion planning problems. We show that GrMPN outperforms existing approaches in terms of data-efficiency, performance and scalability.

2 Background

This section provides background on Markov decision process (MDP), value iteration algorithm, value iteration networks (VIN) and graph neural networks (GNN).

2.1 Markov Decision Process and Value Iteration

A MDP is defined as $\mathcal{M} = (\mathcal{S}, \mathcal{A}, \mathcal{P}, \mathcal{R})$, where $\mathcal{S}$ and $\mathcal{A}$ represent state and action spaces. $\mathcal{T}$ defines a transition function $\mathcal{P}(s, a, s') = P(s'|s, a)$, where $s, s' \in \mathcal{S}, a \in \mathcal{A}$, $\mathcal{R}$ is a reward function. A planning algorithm, e.g. dynamic programming [5], aims to find an optimal policy $\pi : \mathcal{S} \mapsto \mathcal{A}$ so that a performance measure: $V^\pi(s) = \mathbb{E}\left(\sum_t \gamma^t r_t | s_0 = s\right)$ is maximised, for all state $s \in \mathcal{S}$; where r_t is an intemediate reward at time t, $\gamma \in (0, 1)$ is a discount factor. The expectation is w.r.t stochasticity of $\mathcal{P}$ and $\mathcal{R}$. Value iteration (VI) is one of dynamic programming algorithms that can *plan* on $\mathcal{M}$. It starts by updating the values of $V(s), \forall s \in \mathcal{S}$ iteratively via the Bellman backup operator T, $V^{(k)} = T V^{(k-1)}$, as $V^k(s) = \max_a \left[\mathcal{R}(s, a) + \gamma \sum_{s'} P(s'|s, a) V^{(k-1)}(s')\right]$ where k is an update iteration index. T is applied iteratively until $V^{(k)}(s)$ converges. It is proved that the operator T is a Lipschitz map with a factor of γ. In other words, as $k \to \infty$, $V^{(k)}$ converges to a fixed-point value function V^*. As a result, the optimal policy π^* can be computed as $\pi^*(s) = \arg\max_a \left[\mathcal{R}(s, a) + \gamma \sum_{s'} P(s'|s, a) V^*(s')\right]$. A Q-value function is also defined similarly as $Q(s, a) = \mathbb{E}\left(\sum_t \gamma^t r_t | s_0 = s, a_0 = a\right)$. In addition, we have the relation between V and Q value functions as $V(s) = \max_a Q(s, a)$. For goal-oriented tasks, function $\mathcal{R}(s, a)$ can be designed to receive low values at intermediate state and high values at goal states s^*.

Value Iteration Network: Planning on large MDPs might be very computationally expensive. Hence transfer-planning has been desirable, especially for tasks sharing similar structures of $\mathcal{P}$ and $\mathcal{R}$. Value Iteration Networks (VIN) [36] is a differentiable planning framework that can i) do *transfer-planning* for goal-oriented tasks with different goal states, ii) and learn the shared underlying MDP $\mathcal{M}$ between tasks, i.e. learning the transition $\mathcal{P}$ and reward $\mathcal{R}$ functions.

Let's assume that we want to find an optimal plan on a MDP M with unknown T and R. VIN's policy network with embedded approximate reward and transition functions $\bar{R}$ and $\bar{T}$ is trained end-to-end through imitation learning setting. $\bar{R}$ and $\bar{P}$ are assumed to be from an unknown MDP $\bar{M}$ whose optimal policy can form useful features about the optimal policy in M. Based on observation feature $\phi(s)$ on state s, the relation between M and $\bar{M}$ is denoted as $\bar{R} = f_R(\phi(s))$ and $\bar{T} = f_P(\phi(s))$. The trainable parameters are $W^R_{a,i,j}$ and $W^P_{a,i,j}$ with $|\mathcal{A}|$ channels, corresponding for the reward and transition embedding. The recursive process contains two following convolution and max-pooling operations,

$$Q^{(k)}_a = W^R_a \bar{R} + W^P_a V^{(k-1)}, \quad V^{(k)} = \max_a Q^{(k)}_a \tag{1}$$

where the convolution operators on R.H.S in the first equation is written as:

$$W^R_a \bar{R} = \sum_{i,j} W^R_{a,i,j} \bar{R}_{i'-i,j'-j}; \qquad W^P_a V^{(k-1)} = \sum_{i,j} W^P_{a,i,j} \bar{V}^{(k-1)}_{i'-i,j'-j}$$

where i, j are cell index of the maze. VIN has later inspired many other differentiable planning algorithms. For example, VIN's idea can again be exploited for differentiable planning architectures for planning on partially observable environments such as QMDP-Net [9,13], Memory Augmented Control Network (MACN) [15]. A related differentiable planning network is also used in the Predictron framework [32] where its core planning module aims to estimate a Markov reward process that can be rolled forward for imagined planning steps. A notable extension of VIN is proposed by [21], called Gated path planning networks (GPPN), in which they use LSTM to replace the recursive VIN update, i.e.

$$h^{(k)}_{i',j'}, c^{(k)}_{i',j'} = \text{LSTM} \left(\sum_{i,j} \left(W^R_{a,i,j} \bar{R}_{i'-i,j'-j} + W^P_{a,i,j} \bar{V}^{(k-1)}_{i'-i,j'-j} \right), c^{(k-1)}_{i'.j'} \right) \tag{2}$$

Those algorithms show great success at path planning on many different grid-based navigation tasks in which the states are either fully or partially observable. However, the underlying state space must assume regular lattices to exploit local connectivity through the help of convolution operations of CNN. This operation limits their applications to domains whose state space might be in the forms of irregular graphs.

Generalised Value Iteration Networks: There is recent effort considering planning and reinforcement learning whose state transitions form a general graph. Niu et al. [27] propose such a graph-based model-based deep reinforcement learning framework that generalises VIN to differentiable planning on graphs, called generalised value iteration (GVIN). GVIN takes a graph with a particular start node and a goal node as input, and outputs an optimal plan. GVIN can also learn an underlying MDP and an optimal planning policy via either imitation learning or reinforcement learning. Inspired by VIN, GVIN also applies recursive graph convolution and max-pooling operators to emulate the value iteration algorithm on general graphs. With a specially designed convolution kernel, GVIN can also transfer planning to unseen graphs of arbitrary size.

Specifically, an input is a directed, weighted spatial graph $G = (\mathcal{V}, \mathcal{E})$ where $\mathcal{V} = \{v_i\}_{i=1:N^v}$ is the set of node with the node attribute $v_i \in \Re^{d^v}$. $\mathcal{E} = \{e_k\}_{k=1:N^e}$ is the set of edges where $e_k \in \Re^{d^e}$ is the edge attribute, e.g. edge weights; d^v, d^e are the node and edge features' dimension, respectively. In addition, we denote $A \in \Re^{N^v \times N^v}$ as the adjacency matrix. We denote $W_a^P \in \Re^{N^v \times N^v}$ as a convolution operator. Each input graph with a goal state $g \in \{0,1\}^{N^v}$ (one-hot vector labels the goal state) describes one task. GVIN constructs convolution operators W_a^P as a function of the input graph G. The reward graph signal is denoted as $\bar{R} = f_R(G, g)$, where f_R is a CNN in VIN, but an identity function in GVIN. The value functions $V(v)$ with $v \in \mathcal{V}$ on graph nodes can be computed recursively as follows,

$$Q_a^{(k)} = W_a^P(\bar{R} + \gamma V^{(k-1)}), \quad V^{(k)} = \max_a Q_a^{(k)}. \tag{3}$$

While VIN uses CNN to construct W_a^P that could only capture 2D lattice structures, GVIN designs directional and spatial kernels to construct W_a^P that try to capture invariant translation on irregular graphs. However we will show that these kernels are not enough to capture invariance to graph isomorphism, which leads to a poor performance in domains with complex geometric structures. In particular, though it works well on multiple navigation tasks, GVIN is shown to be sensitive to the choice of hyperparameters and the designed convolution kernels, e.g. directional discretisation.

2.2 Graph Neural Networks

Graph neural networks (GNN) [29] have received much attention recently as they can process data in irregular domains such as graphs or sets. The general idea of GNN is to compute an encoded feature h_i for each node v_i based on the structure of the graph, node v_i and edge e_{ij} features, and previous encoded features as $h_i = \sum_{j \in \mathcal{N}(i)} f(h_i, h_j, v_i, v_j, e_{ij})$, where f is a parametric function, and $\mathcal{N}(i)$ denotes the set of neighbour nodes of node i. After computing h_i (probably apply f for k iterations), an additional function is used to compute the output at each node, $y_i = g(h_i, v_i)$, where g is implemented using another neural network, called a read-out function.

Graph Convolution Network: Much earliest work on GNN propose extending traditional CNNs to handle convolution operations on graphs through the use of spectral methods [6,12,16]. For example, graph convolution networks (GCN) [16] is based on a fundamental convolution operation on the spectral domain. GCN must assume graphs of the same size. Besides, these methods must rely on the computation of graph eigenvectors. Therefore, they are either computationally expensive or not able to learn on graphs of arbitrary sizes. Many later introduced graph convolutional networks on spatial domain such as Neural FPs [7], PATCHY-SAN [26], DCNN [1], etc., are able to learn on graphs of arbitrary sizes. However, they are limited to either the choice of a subset of node neighbours or random walks of k-hop neighbourhoods. These drawbacks limit graph convolution-based methods to applications on large-scale graphs with highly arbitrary sizes and structures, hence not favourable for transfer planning in MDPs.

Graph Attention Network (GAT): GAT [38] is inspired by the attention mechanism by modifying the convolution operation in GCN in order to make learning more efficient and scalable to domains of large graphs. Specifically, the encoding at each node is recursively computed as

$$h_i^{(k)} = \sigma \left(\sum_{j \in \mathcal{N}(i)} \alpha_{ij}^{(k)} W^{(k)} h_j^{(k-1)} \right) \tag{4}$$

where σ is an activation function, and $\alpha_{ij}^{(l)}$ is the attention coefficients which are computed as

$$\alpha_{ij}^{(k)} = \frac{\exp \left(\text{LeakyReLU}(\mathbf{a}^\top [W^{(k)} h_i, W^{(k)} h_j]) \right)}{\sum_{k \in \mathcal{N}(i)} \exp \left(\text{LeakyReLU}(\mathbf{a}^\top [W^{(k)} h_i, W^{(k)} h_k]) \right)}$$

where $\mathbf{a}$ is weight vector, and k denotes the embedding layer k whose weights are $W^{(k)}$.

Message Passing Neural Network (MPNN): MPNN [8] uses the mechanism of message passing to compute graph embedding. In particular, the calculation of a feature on each node involves two phases: i) message passing and readout. The message passing operation at a node is based on its state and all messages received from neighbor nodes. The readout is based on the node state and the calculated message. These phases are summarised as follows

$$m_i^{(k)} = \sum_{j \in \mathcal{N}(i)} f(h_i^{(k-1)}, h_j^{(k-1)}, e_{ij}), \qquad h_i^{(k)} = g(h_i^{(k-1)}, m_i^{(k)})$$

MPNN is a unified framework for graph convolution and other existing graph neural networks back to that time, e.g. Graph convolution network and Laplacian Based Methods [6,7,12,16], Gated Graph Neural Networks (GG-NN) [23], Interaction Network (IN) [4], Molecular Graph Convolutions [14], Deep Tensor Neural Networks [30]. Gilmer et.al. [8] has made a great effort in converting these frameworks to become a MPNN variant. MPNN is designed similarly to GG-NN in which GRU is applied to implement the recursive message operation, but different at message and output functions. Specifically, the message sent from node j to i is implemented as $f(h_i, h_j, e_{ij}) = A(e_{ij})h_j$, where $A(e_{ij})$ is implemented as a neural network that maps edge feature e_{ij} to an $d^v \times d^v$ operator. Another variant of the message function that is additionally based on the receiver node features h_i was also implemented in the paper. Updating message with all received information h_i, h_j, e_{ij} is inspired by [4]. MPNN has shown the state-of-the-art performance in prediction tasks on large graph dataset, e.g. molecular properties.

Graph Network (GN): Graph networks (GN) [3, 28] is a general framework that combines all previous graph neural networks. The update operations of GN involve nodes, edges and global graph features. Therefore it renders MPNN, GNN, GCN, GAT as special cases. Specifically, if we denote an additional global graph feature as u, the updates of GN which consist of three update functions g, and three *aggregation* functions f. These functions are implemented based on the message passing mechanism. The aggregation functions are 1) $m_i = f^{e \rightarrow v}(\{e_{ij}\}_{j \in \mathcal{N}(i)})$ aggregate messages sent from edges to compute information of node i; 2) $m^e = f^{e \rightarrow u}(\{e_{ij}\}_{j \in \mathcal{N}(i), \forall i})$ aggregate messages sent from all edges to the global node u; 3) $m^v = f^{v \rightarrow u}(\{v_i\}_{\forall i})$ aggregate messages sent from all nodes to the global node u. These aggregation functions must be invariant to the order of nodes, which is critical to the Weisfeiler-Lehman (WL) graph isomorphism test [39] and regarded as an important requirement for graph representation learning. Using aggregated information, the three update functions to node, edge and global features are defined as follows

$$e_{ij}^{(k)} = g^e(e_{ij}^{(k-1)}, v_i^{(k-1)}, v_j^{(k-1)}, u^{(k-1)})$$
$$v_i^{(k)} = g^v(m_i, v_i^{(k-1)}, u^{(k-1)}), \qquad u^{(k)} = g^u(m^e, m^v, u^{(k-1)})$$

The aggregation functions could be element-wise summation, averages, or max/min operations. The update functions could use general neural networks. The use of edge features and the global graph node makes GN distinct from MPNN. Besides, the use of a recursive update from immediate neighbours is in contrast to multi-hop updates of many spectral methods.

3   Graph-Based Motion Planning Networks

We propose to use a general graph neural network to construct inductive biases for *"learning to plan"*, called graph-based motion planning network (GrMPN). Similar to GVIN, our framework is trained on a set of path planning problems as inputs and associated optimal paths (or complete optimal policies) as outputs. The inputs are general graphs without knowing a reward function. At test time, GrMPN takes i) a test problem defined as a general graph and ii) a pair of starting node and a goal node. The target is to return an optimal plan for the test problem.

Problem Setting: Given a dataset consists of input-output pairs $\mathcal{D} = \{G_i, \tau_i^*\}_{i=1}^N$, where G_i is a general graph $G = (\mathcal{V}, \mathcal{E})$, and τ_i^* is an *optimal path* (or an optimal policy) which can be generated by an expert as a demonstration. The target is to learn an *optimal policy* for a new graph G with a starting node and a goal node. This learning problem can be either formulated as imitation learning or reinforcement learning [27, 36]. Within the scope of this paper, we only focus on the formulation of imitation learning.

General GrMPN Frame-work: We propose a general framework for graph-based value iteration that is based on the principle of message passing. First, we also design a feature-extract function to learn a

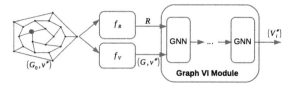

Fig. 1. GrMPN based on graph neural networks.

reward function r: $r = f_R(G; v^*)$ given a graph G and a goal node v^*. We use a similar setting from GVIN [27] for f_R. Second, GrMPN also consists of the following recurrent application of graph operations at all nodes i (Fig. 1).

$$q_{ai}^{(k)} = f^{e \to v}(r, \{e_{ij}, v_j^{(k-1)}\}_{j \in \mathcal{N}(i)})$$
$$e_{ij}^{(k)} = g^e(e_{ij}^{(k-1)}, v_i^{(k-1)}, v_j^{(k-1)}); \quad v_i^{(k)} = g^v(q_{ai}^{(k)})$$

where k is the processing step index; v_i is the node feature (which also contains the node's value function); $f^{e \to v}, g^e, g^v$ are defined as aggregation and update functions. Note that we use q_{ai} as edge features. If the transition is stochastic, we can use $|\mathcal{A}|$ channels on edges.

3.1   GrMPN via Graph Networks

GrMPN is built based on GN [3,28]. GrMPN does not represent the global node u. It consists of the representation of nodes and edges, hence uses one aggregation function and two update functions, as described below:

$$m_i = \sum_{j \in \mathcal{N}(i)} e_{ij}; \quad e_{ij} = g^e(r_j + \gamma e_{ij}); \quad v_i = g^v(m_i)$$

Note that for brevity the above updates assume deterministic actions, similar to the setting in VIN and GVIN. The message aggregation is equivalent to a sum over actions $\sum_a \sum_{s'} p(s'|s, a)V(s')$, similar to the implementation of GPPN [21]. The algorithm of GrMPN via Graph Networks (GrMPN-GN) is summarised in Algorithm 1.

Algorithm 1: GrMPN via Graph Networks (GrMPN-GN)

Input : graph $G = \{V, E\}$, a goal node v^*
Output: $\{v_i^{(k)}\}_{i=1}^{N^v}$
Extract rewards: $r = f_R(G, v^*)$;
for k *iterations* **do**
    for *each edge* $\{e_{ij}\}$ **do**
        | Compute edge update: $e_{ij} = g^e(r_j + \gamma e_{ij})$;
    end
    for *each node* $\{v_i\}$ **do**
        | Compute node aggregation: $m_i = \sum_{j \in \mathcal{N}(i)} e_{ij}$;
        | Compute node update: $v_i = g^v(m_i)$;
    end
end

3.2   GrMPN via Graph Attention Networks

In general, any graph neural networks can be used to represent the graph-based value iteration module. As MPNN, GGNN and other similar approaches are special cases of GN. Therefore we can easily rewrite GrMPN to become an application of these approaches. In this section, we draw a connection of GrMPN to GAT, and show how GAT can also be used to represent a graph-based VI module. We use multi-heads to represent $|\mathcal{A}|$ channels for the Q value functions. Each head represents a value function corresponding to an action Q_a.

$$V_i^{(k)} = \sigma \left(\sum_{j \in \mathcal{N}(i)} \alpha_{ij}^a W^a (r_{ij} + \gamma V_j^{(k-1)}) \right)$$

where σ is a *Maxout* activation function or $\sum_a$ over possible actions (in our implementation we chose the latter); and α_{ij}^a is the attention coefficients of channel a which are computed as

$$\alpha_{ij}^a = \frac{\exp\left(\text{LeakyReLU}(\beta^\top [W^a V_i, W^a V_j])\right)}{\sum_{k \in \mathcal{N}(i)} \exp\left(\text{LeakyReLU}(\beta^\top [W^a V_i, W^a V_k])\right)}$$

We denote this formulation as GrMPN-GAT. We can also make a connection between GrMPN-GN and GrMPN-GAT by reformulating the updates and coefficients of GATPPN to message passing in which edge attentions become edge updates. The edge feature now have the attention coefficients α as additional information. GrMPN-GAT is rewritten as a GN module with attentions as:

$$e_{ij} = (\alpha_{ij}, q_{ij}) = g^e(r_j + \gamma e_{ij}); \quad m_i = \frac{1}{\sum_{j \in \mathcal{N}(i)} \alpha_{ij}} \sum_{j \in \mathcal{N}(i)} \alpha_{ij} q_{ij}; \quad v_i = g^v(m_i)$$

where e_{ij} is the feature of the edge i-j that contains attention coefficients and q_{ij} (equivalently to the Q-value function q_{ai}). The algorithm of GrMPN-GAT can be described similar to GrMPN-GN in Algorithm 1.

4   Experiments

We evaluate the proposed framework GrMPN-GAT and GrMPN-GN on 2D mazes, irregular graphs, and motion planning problems. These problems range from simple regular graphs (2D mazes) to irregular graphs of simple (irregular grid) and complex geometric structures (motion planning). Through these experiments, we aim to confirm the following points:

1. GrMPN frameworks based on general graph neural networks can not only perform comparably to VIN for lattice structures but also with more data-efficiency, because GrMPN-GAT and GrMPN-GN are invariant to graph isomorphism hence able to reduce sample complexity by capturing topologically identical graphs.
2. GrMPN can generalise well to unseen graphs of arbitrary sizes and structures, far better than existing approaches. In particular, GrMPN handles long-range planning well by providing a great generalisation ability) across nodes in graphs b) across graphs. Therefore they can cope with planning on larger problems with high complexity, hence significantly improve task performance and data-efficiency.
3. GrMPN-GAT exploiting attended (weighted) updates for value functions, i.e. non-local updates [3], would outperform GrMPN-GN which only based on the uniform sum over actions.

Training Objective: Recall that we only focus on imitation learning, the same objective function would thus be used throughout the paper, i.e. a cross-entropy loss for the supervised learning problem with dataset $\{v, a^* = \pi^*(v)\}$ where v is a state node with an optimal action a^* demonstrated by an expert.

Evaluation Metrics: As in other previous works, we evaluate the algorithm's performance by the three standard metrics: i) %Accuracy (*Acc*), the percentage of nodes whose predicted actions are optimal, ii) %Success (*Succ*), the percentage of initial nodes whose predicted paths are able to reach the goal, and iii) the distance difference (*Diff*) which is computed via the cost difference between the predicted path and the optimal path planned by Dijkstra's algorithm.

4.1   Domain I: 2D Mazes

In this experiment, we carry out evaluations on 2D mazes which have a regular graph structure. We compare GrMPN against VIN, GVIN, and GPPN. The environment and experiment settings are set similar to VIN, GVIN, and GPPN. We use the same script used in VIN, GVIN, and GPPN to generate graphs. For each graph, we generated seven optimal trajectories corresponding to different start and goal nodes. Note that only GPPN requires a complete optimal policy which gives an optimal action at every graph node. This setting makes GPPN have a little advantage. We train and test on the same graph with sizes 12×12, 16×16, 20×20. The number of generated graphs for training is chosen from small to large with values $\{200, 1000, 5000\}$. The size of testing data is fixed to 1000 graphs of the corresponding size.

The results shown in Table 1 tells that GrMPN-GAT and GrMPN-GN outperform other baselines in terms of both the performance and data-efficiency. GPPN has a slightly better performance with a large amount of data. We note that GPPN must assume the data consists of optimal policies instead of a small set of optimal demonstration trajectories as in VIN, GVIN and ours. GVIN is a graph-based VIN, but it relies on specially designed convolution kernels that are based on the choice of discretised angular directions. Therefore GVIN is not able to perform as well as GrMPN-GAT and GrMPN-GN, which are based on principled graph networks and known to be invariant to graph isomorphism. GrMPN-GAT and GrMPN-GN show significant better in terms of data-efficiency on large domains 20×20. On these large domains, learning algorithms often require a large amount of data. However, GrMPN-GAT and GrMPN-GN can still learn well with a limited amount of data. This shows how important invariance to graph isomorphism is for learning on graphs. Performance of GrMPN-GAT on a bigger domain is better than small domains because the amount of nodes involved in training is larger in large graphs.

Table 1. 2D Mazes: Test performance with varying training data size $\{200, 1000, 5000\}$ and different grid sizes.

	12×12		16×16		20×20	
	Acc	Succ	Acc	Succ	Acc	Succ
VIN-200	81.0	86.9	74.5	82.1	66.3	74.5
VIN-1000	85.0	87.6	78.1	83.8	79.0	84.0
VIN-5000	93.8	95.0	90.9	93.5	90.2	91.6
GPPN-200	80.1	87.6	75.6	83.4	69.7	85.0
GPPN-1000	85.8	90.1	85.0	89.5	84.1	90.9
GPPN-5000	93.4	95.7	91.5	94.5	90.0	94.4
GVIN-200	73.6	77.7	68.2	74.3	77.0	80.3
GVIN-1000	89.8	91.0	87.3	89.2	88.7	90.3
GVIN-5000	92.8	93.9	93.7	94.2	90.1	91.4
GrMPN-GAT-200	94.9	94.9	95.7	95.7	95.7	95.7
GrMPN-GAT-1000	94.9	94.9	95.7	95.7	95.7	95.7
GrMPN-GAT-5000	95.0	95.0	95.7	95.8	95.7	95.7
GrMPN-GN-200	93.2	93.5	94.5	94.8	95.4	95.4
GrMPN-GN-1000	95.8	95.9	95.9	95.9	**96.0**	**96.1**
GrMPN-GN-5000	**96.3**	**96.5**	**96.3**	**96.4**	95.8	95.8

The results show how value functions on a test graph (after training) are computed depending on the value k of processing steps. More nodes would be updated with a larger number of processing step which corresponding to more batches of value iterations updates. This ablation also shows that GrMPN-GAT is able to generalise across nodes better than GrMPN-GN. This generalisation ability would significantly help with long-range planning problems.

4.2 Domain II: Irregular Graphs

This experiment uses the same script used by GVIN, which is based on Networkx [10], to create synthetic graphs that are with random coordinates from box $[0, 1]^2$ in 2D space. We vary the parameters of the generation program to create three types of irregular graphs: Dense, Sparse, and Tree-like. For Tree-like graphs, we use the Networkx's function, geographical_threshold_graph by setting the connectedness probability between nodes to a small value. We create Tree-like graphs which are not considered in GVIN, because there are two main challenges on these graphs. *First*, with the same number of nodes and amount of generated graphs, tree-like graphs would have much fewer nodes for training. *Second*, Tree-like graphs result in a major issue which is ideal to evaluate generalisation for long-range planning which requires propagation of value functions across nodes in graphs well.

We generate 10000 graphs, with a varying number of nodes $\{10, 100\}$. The label for each graph is an optimal policy, i.e. an optimal action at every graph node. Training is 6/7 of the generated data, while testing is 1/7.

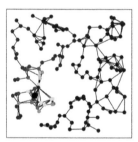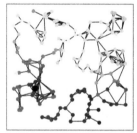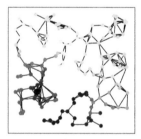

Fig. 2. Value functions on tree-like graphs: left) GVIN, middle) GrMPN-GN; right) GrMPN-GAT.

The comparing results are described in Tables 2 (on Dense), 3 (on Sparse), and 4 (on Tree-like). Testing is performed on irregular graphs of different size: 100 and 150 nodes on Dense, 100 nodes on Sparse. The results show that GrMPN methods perform comparably with GVIN on Dense graphs in terms of success rate, but slightly better in terms of Accuracy and Distance difference. On Sparse graphs, GrMPN-GAT and GrMPN-GN based on the principled of message passing can have fast updates across nodes. The results in Table 4 tell that VIN is not able to cope with very sparse graphs and long-range planning. This result shows GVIN has a weak generalisation ability. As seen in Fig. 2, GVIN is not able to spread the update to nodes that are far from the goal node. This figure also shows that GrMPN-GAT has a slightly better generalisation ability across nodes than GrMPN-GN. The value functions of GrMPN-GN have more un-updated nodes (see the colour of nodes that are far from the goal node as labelled in black) than that of GrMPN-GAT. This result explains why GrMPN-GAT performs slightly better than GrMPN-GN.

Table 2. Irregular Dense Graphs: Test performance with varying number of nodes used in training $\{10, 100\}$.

	Dense (100 nodes)			Dense (150 nodes)		
	Acc	Succ	Diff	Acc	Succ	Diff
GVIN-10	58.3	**99.9**	0.046	53.4	**99.8**	0.042
GrMPN-GAT-10	**63.1**	99.2	**0.030**	**58.4**	99.5	**0.028**
GrMPN-GN-10	55.8	95.0	0.057	50.4	94.4	0.062
GVIN-100	56.6	97.3	0.064	52.9	**99.2**	0.059
GrMPN-GAT-100	**62.7**	**97.7**	**0.032**	**58.6**	98.8	**0.029**
GrMPN-GN-100	61.5	97.7	0.038	56.6	98.4	0.037

Table 3. Irregular Sparse Graphs: Test performance with varying number of nodes used in training $\{10, 100\}$.

	Sparse (100 nodes)		
	Acc	Succ	Diff
GVIN-10	57.9	80.5	0.053
GrMPN-GAT-10	**61.7**	**91.2**	**0.048**
GrMPN-GN-10	59.0	85.1	0.052
GVIN-100	60.3	85.7	0.053
GrMPN-GAT-100	**74.5**	98.1	**0.027**
GrMPN-GN-100	73.9	**98.3**	**0.027**

Table 4. Irregular Tree-like Graphs: Test performance on graphs of 150 nodes. Training uses tree-like graphs of 50 nodes.

	Tree-like (150 nodes)		
	Acc	Succ	Diff
GVIN	66.5	44.7	1.626
GrMPN-GAT	**88.2**	**98.6**	**0.027**
GrMPN-GN	87.1	94.8	0.032

4.3 Domain III: Motion Planning

Sampling-based methods such as probabilistic roadmaps (PRM) [19] are very efficient in practice. PRM is one of the most widely used techniques in robotic motion planning, especially for applications in navigation. In such an application, a motion planning algorithm must find an optimal path that must satisfy i) the environment's geometry constraints, i.e. collision-free path, and ii) the robot system constraint, e.g. differential constraints. PRM is multiple-query methods that are very useful in highly structured environments such as large buildings.

We evaluate GrMPN on two motion planning problems: 2D navigation with a holonomic mobile robot and manipulation with a simulated 7-DoF Baxter robot arm. We aim to improve PRM by bringing it closer to an online-planning method through transfer planning. In this section, we show that GrMPN would outperform GVIN under such tasks of a complex geometric structure.

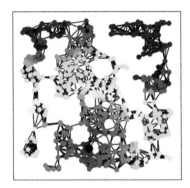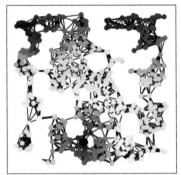

Fig. 3. Value functions on PRM graphs of 500 configurations: left) GrMPN-GN; right) GrMPN-GAT.

Analysis: Table 5 show test performance results on 2D navigation with a mobile robot. It shows that GrMPN methods not only outperform GVIN but also possess a great generalisation ability. We additionally evaluate the generalisation ability of GrMPN methods by using the trained model using Tree-like data as described in Irregular Graphs section to test on the created roadmap test set. We skip reporting on GVIN on large testing graphs (500 configuration nodes) due to its degraded performance. The trained model using Tree-like graph data could also generalise well on unseen graphs generated by PRM on different environments. In addition, they can generalise to much bigger graphs (with 500 nodes). This suggests GrMPN is able to do *zero-shot learning to plan*. The colour map in Fig. 3 also suggests that GrMPN-GAT slightly has wider value propagation, which means better generalisation for long-range planning and across task graphs.

Table 6 show the test performance results on the motion planning task with a simulated 7-DoF Baxter arm. We skip reports on GVIN due to its poor performance and scalability to learning on large graphs. The results show that GrMPN methods can do motion planning on this complex geometric domain with high accuracy and success rates.

Table 5. Motion planning problems: Test performance with varying number of configurations $\{200, 500\}$. Note that algorithms with suffix -T denote trained models using tree-like graph data.

	200 configurations			500 configurations		
	Acc	Succ	Diff	Acc	Succ	Diff
GVIN-T	62.2	58.6	0.210	–	–	–
GrMPN-GAT-T	**82.2**	**83.8**	**0.172**	61.3	72.5	**0.523**
GrMPN-GN-T	80.0	79.6	0.181	57.1	59.3	0.650
GVIN	53.9	40.6	1.326	–	–	–
GrMPN-GAT	82.2	**84.4**	**0.164**	62.3	73.1	**0.504**
GrMPN-GN	**82.7**	81.2	0.176	61.0	68.4	0.575

Table 6. Simulated 7-DoF Baxter arm motion planning: Test performance with different uses of node encoding (MLP-, GCN-, GN-) and graph-based planning (GrMPN-GAT and GrMPN-GN).

	Acc	Succ	Diff
MLP-GrMPN-GAT	**78.5**	89.1	**0.466**
GCN-GrMPN-GAT	70.4	83.4	0.875
GN-GrMPN-GAT	77.4	**90.3**	0.561
MLP-GrMPN-GN	**77.9**	89.0	**0.503**
GCN-GrMPN-GN	67.7	76.1	1.024
GN-GrMPN-GN	76.6	**89.4**	0.552

5 Conclusion

In this paper, we have proposed a general graph-based motion planning network, GrMPN. The proposed framework leverages the idea of graph neural networks to handle planning with graphs. The main idea is to integrate Graph processing modules into a differentiable planning network with the aim to capture graph isomorphism in order to achieve i) generalisation for transfer planning to graphs of arbitrary structures and ii) data-efficiency when dealing with complex graphs across task instances. Through various experiments on 2D mazes, irregular graphs and motion planning tasks, we have shown that GrMPN can improve data-efficiency significantly and improve planning task performance in comparisons to existing approaches. GrMPN outperforms baselines for regular graphs and existing approaches for irregular graphs in terms of data-efficiency and generalisation ability. For future researches, there is a promising direction in combining GrMPN with other powerful graph networks to further exploit the factored structure in planning problems, e.g. factored MDP planning or planning on high-ordered Markov models.

References

1. Atwood, J., Towsley, D.: Diffusion-convolutional neural networks. In: NIPS, pp. 1993–2001 (2016)
2. Bajpai, A.N., Garg, S., et al.: Transfer of deep reactive policies for MDP planning. In: NIPS, pp. 10965–10975 (2018)
3. Battaglia, P.W., et al.: Relational inductive biases, deep learning, and graph networks. arXiv preprint arXiv:1806.01261 (2018)
4. Battaglia, P.W., Pascanu, R., Lai, M., Rezende, D.J., Kavukcuoglu, K.: Interaction networks for learning about objects, relations and physics. In: NIPS, pp. 4502–4510 (2016)
5. Bertsekas, D.P., Bertsekas, D.P., Bertsekas, D.P., Bertsekas, D.P.: Dynamic Programming and Optimal Control, vol. 1. Athena Scientific, Belmont (1995)
6. Bruna, J., Zaremba, W., Szlam, A., LeCun, Y.: Spectral networks and locally connected networks on graphs. arXiv preprint arXiv:1312.6203 (2013)
7. Duvenaud, D.K., et al.: Convolutional networks on graphs for learning molecular fingerprints. In: NIPS, pp. 2224–2232 (2015)
8. Gilmer, J., Schoenholz, S.S., Riley, P.F., Vinyals, O., Dahl, G.E.: Neural message passing for quantum chemistry. In: ICML, pp. 1263–1272 (2017)
9. Gupta, S., Davidson, J., Levine, S., Sukthankar, R., Malik, J.: Cognitive mapping and planning for visual navigation. In: CVPR, pp. 7272–7281 (2017)
10. Hagberg, A., Swart, P., S Chult, D.: Exploring network structure, dynamics, and function using network. Technical report, Los Alamos National Lab. (LANL), Los Alamos, NM, United States (2008)
11. Heess, N., Wayne, G., Silver, D., Lillicrap, T.P., Erez, T., Tassa, Y.: Learning continuous control policies by stochastic value gradients. In: NIPS, pp. 2944–2952 (2015)
12. Henaff, M., Bruna, J., LeCun, Y.: Deep convolutional networks on graph-structured data. arXiv preprint arXiv:1506.05163 (2015)
13. Karkus, P., Hsu, D., Lee, W.S.: QMDP-Net: deep learning for planning under partial observability. In: NIPS, pp. 4694–4704 (2017)
14. Kearnes, S., McCloskey, K., Berndl, M., Pande, V., Riley, P.: Molecular graph convolutions: moving beyond fingerprints. J. Comput. Aided Mol. Des. **30**(8), 595–608 (2016). https://doi.org/10.1007/s10822-016-9938-8
15. Khan, A., Zhang, C., Atanasov, N., Karydis, K., Kumar, V., Lee, D.D.: Memory augmented control networks. In: ICLR (2018)
16. Kipf, T.N., Welling, M.: Semi-supervised classification with graph convolutional networks. In: ICLR (2017)
17. Kober, J., Bagnell, J.A., Peters, J.: Reinforcement learning in robotics: a survey. Int. J. Robot. Res. **32**(11), 1238–1274 (2013)
18. Kurutach, T., Clavera, I., Duan, Y., Tamar, A., Abbeel, P.: Model-ensemble trust-region policy optimization. In: ICLR (2018)
19. LaValle, S.M.: Planning Algorithms. Cambridge University Press, Cambridge (2006)
20. Lee, G., Hou, B., Mandalika, A., Lee, J., Srinivasa, S.S.: Bayesian policy optimization for model uncertainty. arXiv preprint arXiv:1810.01014 (2018)
21. Lee, L., Parisotto, E., Chaplot, D.S., Xing, E., Salakhutdinov, R.: Gated path planning networks. arXiv preprint arXiv:1806.06408 (2018)
22. Levine, S., Finn, C., Darrell, T., Abbeel, P.: End-to-end training of deep visuomotor policies. J. Mach. Learn. Res. **17**, 39:1–39:40 (2016)

23. Li, Y., Tarlow, D., Brockschmidt, M., Zemel, R.S.: Gated graph sequence neural networks. In: ICLR (2016)
24. Ma, T., Ferber, P., Huo, S., Chen, J., Katz, M.: Adaptive planner scheduling with graph neural networks. arXiv preprint arXiv:1811.00210 (2018)
25. Mnih, V., et al.: Human-level control through deep reinforcement learning. Nature **518**(7540), 529 (2015)
26. Niepert, M., Ahmed, M., Kutzkov, K.: Learning convolutional neural networks for graphs. In: ICML, pp. 2014–2023 (2016)
27. Niu, S., Chen, S., Guo, H., Targonski, C., Smith, M.C., Kovacevic, J.: Generalized value iteration networks: Life beyond lattices. In: AAAI, pp. 6246–6253. AAAI Press (2018)
28. Sanchez-Gonzalez, A., et al.: Graph networks as learnable physics engines for inference and control. arXiv preprint arXiv:1806.01242 (2018)
29. Scarselli, F., Gori, M., Tsoi, A.C., Hagenbuchner, M., Monfardini, G.: The graph neural network model. IEEE Trans. Neural Networks **20**(1), 61–80 (2008)
30. Schütt, K.T., Arbabzadah, F., Chmiela, S., Müller, K.R., Tkatchenko, A.: Quantum-chemical insights from deep tensor neural networks. Nat. Commun. **8**, 13890 (2017)
31. Segler, M., Preuß, M., Waller, M.P.: Towards "AlphaChem": chemical synthesis planning with tree search and deep neural network policies. arXiv preprint arXiv:1702.00020 (2017)
32. Silver, D., et al.: The predictron: End-to-end learning and planning. In: ICML, pp. 3191–3199 (2017)
33. Silver, D., et al.: Mastering the game of go with deep neural networks and tree search. Nature **529**(7587), 484–489 (2016)
34. Srinivas, A., Jabri, A., Abbeel, P., Levine, S., Finn, C.: Universal planning networks: learning generalizable representations for visuomotor control. In: ICML, pp. 4739–4748 (2018)
35. Sutton, R.S., Barto, A.G., et al.: Introduction to Reinforcement Learning, vol. 2. MIT Press, Cambridge (1998)
36. Tamar, A., Levine, S., Abbeel, P., Wu, Y., Thomas, G.: Value iteration networks. In: NIPS, pp. 2146–2154 (2016)
37. Toyer, S., Trevizan, F., Thiébaux, S., Xie, L.: Action schema networks: generalised policies with deep learning. In: AAAI (2018)
38. Velickovic, P., Cucurull, G., Casanova, A., Romero, A., Liò, P., Bengio, Y.: Graph attention networks. In: ICLR (2018)
39. Weisfeiler, B., Lehman, A.A.: A reduction of a graph to a canonical form and an algebra arising during this reduction. Nauchno-Technicheskaya Informatsia **2**(9), 12–16 (1968)
40. Zhang, S., Yao, L., Sun, A., Tay, Y.: Deep learning based recommender system: a survey and new perspectives. ACM Comput. Surv. (CSUR) **52**(1), 5 (2019)

Transfer and Multi-task Learning

Graph Diffusion Wasserstein Distances

Amélie Barbe[1,2,3(✉)], Marc Sebban[3], Paulo Gonçalves[1], Pierre Borgnat[2], and Rémi Gribonval[1]

[1] Univ Lyon, Inria, CNRS, ENS de Lyon, UCB Lyon 1, LIP UMR 5668, 69342 Lyon, France
amelie.barbe@ens-lyon.fr
[2] Univ Lyon, ENS de Lyon, UCB Lyon 1, CNRS, Laboratoire de Physique, 69342 Lyon, France
[3] Univ Lyon, UJM-Saint-Etienne, CNRS, Institut d'Optique Graduate School, Laboratoire Hubert Curien UMR 5516, 42023 Saint-Etienne, France

Abstract. Optimal Transport (OT) for structured data has received much attention in the machine learning community, especially for addressing graph classification or graph transfer learning tasks. In this paper, we present the Diffusion Wasserstein (DW) distance, as a generalization of the standard Wasserstein distance to undirected and connected graphs where nodes are described by feature vectors. DW is based on the Laplacian exponential kernel and benefits from the heat diffusion to catch both structural and feature information from the graphs. We further derive lower/upper bounds on DW and show that it can be directly plugged into the Fused Gromov Wasserstein (FGW) distance that has been recently proposed, leading - for free - to a DifFused Gromov Wasserstein distance (DFGW) that allows a significant performance boost when solving graph domain adaptation tasks.

Keywords: Optimal Transport · Graph Laplacian · Heat diffusion

1 Introduction

Many real-world problems in natural and social sciences take the form of structured data such as graphs which require efficient metrics for comparison. In this context, graphs kernels that take into account both structural and feature information have achieved a tremendous success during the past years to address graph classification tasks (see the most recent survey [12]), *e.g.* using Support Vector Machines.

Unlike standard graph kernel-based methods, we consider in this paper graphs as discrete distributions with the main objective of defining a distance on this space of probability measures, which is the main goal of Optimal Transport

Electronic supplementary material The online version of this chapter (https://doi.org/10.1007/978-3-030-67661-2_34) contains supplementary material, which is available to authorized users.

F. Hutter et al. (Eds.): ECML PKDD 2020, LNAI 12458, pp. 577–592, 2021.
https://doi.org/10.1007/978-3-030-67661-2_34

(OT) (see [15] for the original problem and [11] for a regularized version). OT has received much attention from the Machine Learning community from both theoretical and practical perspectives. It provides a natural geometry for probability measures and aims at moving a source distribution on the top of a target measure in an optimal way with respect to a cost matrix. If the latter is related to an actual distance on some geometric space, the solution of the OT problem defines a distance (the so-called p-Wasserstein distance W) on the corresponding space of probability measures.

Several recent works in OT have been devoted to the comparison of structured data such as undirected graphs. Following [16], the authors of [19] introduced the Gromov-Wasserstein (GW) distance allowing to compute a distance between two metric measures. GW can be used to catch and encode some structure of graphs, like the shortest path between two vertices. However, this distance is not able to jointly take into account both features (or attributes) at the node level and more global structural information. To address this issue, the Fused-Gromov Wasserstein (FGW) distance was introduced in [24] as an interpolation between GW (over the structure) and the Wasserstein distance (over the features). In order to better take into account the global structure of the graphs for graph comparison and alignment, and for the transportation of signals between them, the authors of [13,14] introduced GOT, as a new Wasserstein distance between graph signal distributions that resorts to the graph Laplacian matrices. This approach, initially constrained by the fact that both graphs have the same number of vertices [13] was recently extended to graphs of different sizes in [14]. However, the graph alignment and the proposed distance are still based only on the structure, and do not use the features.

In this paper, we address the limitations of the aforementioned graph OT-based methods by leveraging the notions of heat kernel and heat diffusion that are widely used to capture topological information in graphs [3,23] or graph signals [21]. Inspired from the Graph Diffusion Distance (GDD) introduced in [9], and rooted in the Graph Signal Processing (GSP) approaches [17] of graphs with features (or "graph signals" in GSP), better known under the name of *attributed graphs* in machine learning, we present the Diffusion Wasserstein (DW) distance, as a generalization of the standard Wasserstein distance to attributed graphs. While GDD is limited to graphs of the same size, does not take into account features, and would not be directly usable in the OT setting, we leverage its definition to capture in an OT problem the graph structure combined with the smoothing of features along this structure. Leveraging the properties of the heat diffusion, we establish the asymptotic behavior of our new distance. We also provide a sufficient condition for the expected value of DW to be upper bounded by the Wasserstein distance. We further show that computing DW boils down to reweighting the original features by taking into account the heat diffusion in the graphs. For this reason, DW can be plugged into FGW in place of the Wasserstein distance to get for free a family of so-called DifFused Gromov Wasserstein distances (DFGW). We will show in the experiments that DFGW significantly outperforms FGW when addressing Domain Adaptation tasks with OT (see the seminal work of [5]) whose goal here is to transfer knowledge from a source graph to a different but related target graph. Interestingly, DW alone is shown to be very competitive while benefiting from a gain in computation time.

The rest of this paper is organized as follows. Section 2 is dedicated to the main background knowledge in Optimal Transport necessary for the rest of the paper; Sect. 3 is devoted to the presentation of our new heat diffusion-based distances and the derivation of properties giving some insight into their asymptotic behavior. We perform a large spectrum of experiments in Sect. 4 that give evidence of the efficiency of our distances to address domain adaptation tasks between attributed graphs. We conclude in Sect. 5.

2 Preliminary Knowledge

We present in this section the main background knowledge in Optimal Transport as well as some definitions that will be necessary throughout this paper.

2.1 Optimal Transport

Let us consider two empirical probability measures μ and ν, called *source* and *target* distributions, and supported on two sample sets $X = \{x_i\}_{i=1}^m$ and $Y = \{y_j\}_{j=1}^n$, respectively, lying in some feature space $\mathcal{X}$ and with weights $a = (a_i)_{i=1}^m$, $b = (b_j)_{j=1}^n$ such that $\mu = \sum_{i=1}^m a_i \delta_{x_i}$ and $\nu = \sum_{j=1}^n b_j \delta_{y_j}$, where δ is the Dirac measure. If $\mathcal{X} = \mathbb{R}^r$ for some integer $r \geq 1$, a matrix representation of X (resp. of Y) is the matrix $\mathbf{X} \in \mathbb{R}^{m \times r}$ (resp. $\mathbf{Y} \in \mathbb{R}^{n \times r}$) which rows are $x_i^\top, 1 \leq i \leq m$ (resp. $y_j^\top, 1 \leq j \leq n$). Let $M = M(X,Y) \in \mathbb{R}_+^{m \times n}$ be a cost matrix, where $M_{ij} \overset{def}{=} [d(x_i, y_j)]_{ij}$ is the cost (w.r.t. to some distance function d) of moving x_i on top of y_j. Let $\Pi(a,b)$ be a transportation polytope defined as the set of admissible coupling matrices γ:

$$\Pi(a,b) = \{\gamma \in \mathbb{R}_+^{m \times n} \text{ s.t. } \gamma 1_n = a, \gamma^\top 1_m = b\},$$

where γ_{ij} is the mass transported from x_i to y_j and 1_k is the vector of dimension k with all entries equal to one. The p-Wasserstein distance $W_p^p(\mu, \nu)$ between the source and target distributions is defined as follows:

$$W_p^p(\mu, \nu) = \min_{\gamma \in \Pi(a,b)} \langle \gamma, M^p(X,Y) \rangle_F, \tag{1}$$

where $\langle ., . \rangle_F$ is the Frobenius inner product and $M^p(X,Y) := (M_{ij}^p)_{ij}$ is the entrywise p-th power of $M(X,Y)$ with an exponent $p > 0$. With $p = 2$ and d the Euclidean distance, if μ and ν are uniform ($a_i = 1/m$, $b_j = 1/n$), then the barycentric projection $\hat{\mathbf{X}}$ of $\mathbf{X}$ can be defined in closed-form [5] as follows: $\hat{\mathbf{X}} = m\gamma^\star \mathbf{Y}$, where $\gamma^\star$ is the optimal coupling of Problem (1).

2.2 Optimal Transport on Graphs

In order to be able to apply the OT setting on structured data, we need now to formally define the notion of probability measure on graphs and adapt the previous notations. Following [24], let us consider undirected and connected attributed graphs as tuples of the form $\mathcal{G} = (\mathcal{V}, \mathcal{E}, \mathcal{F}, \mathcal{S})$, where $\mathcal{V}$ and $\mathcal{E}$ are the classic sets of vertices (also called nodes) and edges of the graph, respectively. $\mathcal{F} : \mathcal{V} \to \mathcal{X}$ is a function which assigns a feature vector $x_i \in \mathcal{X}$ (also called a graph signal in [17]) to each vertex v_i of the graph (given an arbitrary ordering of the vertices). $\mathcal{S} : \mathcal{V} \to \mathcal{Z}$ is a function which associates each vertex v_i with some structural representation $z_i \in \mathcal{Z}$, e.g. a local description of the graph, the vertex and a list of its neighbors, etc. We can further define a cost function $C : \mathcal{Z} \times \mathcal{Z} \to \mathbb{R}_+$ which measures the dissimilarity $C(z, z')$ between two structural representations z, z'. Typically, $C(z, z')$ can capture the length of the shortest path between two nodes. Additionally, if the graph $\mathcal{G}$ is *labeled*, each vertex v_i is also assigned a label from some label space $\mathcal{L}$.

When each vertex of the graph is weighted according to its relative importance, the source and target graphs can be seen as probability distributions,

$$\mu = \sum_{i=1}^{m} a_i \delta_{(x_i, z_i)}, \quad \nu = \sum_{j=1}^{n} b_j \delta_{(y_j, z'_j)} \tag{2}$$

where x_i, z_i are the features/structural representations associated to the vertices of the source graph while y_j, z'_j are those associated to the target one. Equipped with these notations, we can now present the Fused Gromov-Wasserstein (FGW) distance introduced in [24] as the first attempt to define a distance that takes into account both structural and feature information in a OT problem.

Let $\mathcal{G}^s$ (resp. $\mathcal{G}^t$) be a source (resp. target) graph described by its discrete probability measure μ (resp. ν). Let $C^s \in \mathbb{R}^{m \times m}$ and $C^t \in \mathbb{R}^{n \times n}$ be the structure matrices associated with the source and target graphs respectively. The FGW distance is defined via the minimization of a convex combination between (i) the Wasserstein cost matrix which considers the features x_i, y_j associated with the nodes and (ii) the Gromov-Wasserstein cost matrix [19] which takes into account the structure of both graphs. More formally, for each $\alpha \in [0, 1]$, one can define

$$\mathrm{FGW}_p^p(\mu, \nu) = \min_{\gamma \in \Pi(a,b)} \left\{ \sum_{i,j,k,l} \left((1 - \alpha) M_{ij}^p + \alpha |C_{ik}^s - C_{jl}^t|^p \right) \gamma_{ij} \gamma_{kl} \right\} \tag{3}$$

where the summation indices are $1 \le i, k \le m$ and $1 \le j, l \le n$, and the dependency on α is omitted from the notation $\mathrm{FGW}_p(\mu, \nu)$ for the sake of concision. Note that α can be seen as a hyper-parameter which will allow FGW, given the underlying task and data, to find a good compromise between the features and the structures of the graphs. In the special case $\alpha = 0$, we recover the Wasserstein distance $\mathrm{FGW}_p(\mu, \nu \mid \alpha = 0) = W_p(\mu, \nu)$. By abuse of notation we denote $W_p(\mu, \nu) = W_p(\mu_{\mathcal{X}}, \nu_{\mathcal{X}})$, for μ, ν as in (2) and $\mu_{\mathcal{X}} := \sum_{i=1} a_i \delta_{x_i}$, $\nu_{\mathcal{X}} := \sum_j b_j \delta_{y_j}$

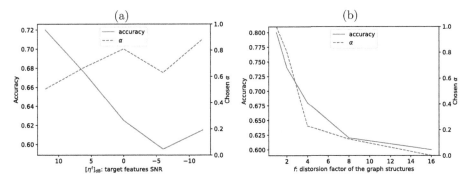

Fig. 1. DA by minimization of FGW (3). Plots display the evolutions of the classification accuracy of the target vertices (purple curves, left axis) along with the corresponding selected hyper-parameter α (dashed green curves, right axis) against two limiting regimes. (a) Features reliability. Source and target graphs structures are i.i.d. ($p_{11} = 0.02$, $p_{22} = 0.03$, $p_{12} = p_{21} = 0.01$) and $\sigma^s = 1$ (i.e. ($x_i = \pm 1 + \mathcal{N}(0, 1))_{i=1,\dots,m}$). The signal to noise ratio $[\eta^t]_{\mathrm{dB}} = 20\log_{10}(\eta^t)$ of the target features varies along the X-axis. (b) Structure reliability. Source and target features are i.i.d. ($[\eta^s]_{\mathrm{dB}} = [\eta^t]_{\mathrm{dB}} = -6\,\mathrm{dB}$). The target graph structure progressively deviates from that of $\mathcal{G}^s$ by increasing the probability of inter-class connectivity $p_{12}^t = f \cdot p_{12}^s$, with $f \geq 1$ and (p_{uv}^s) as in (a). The structure distorsion factor f evolves on the X-axis. All plots display median values estimated over 50 i.i.d realisations. The graphs' size is $n = m = 100$. (Color figure online)

their marginals on the feature space $\mathcal{X}$. The case $\alpha = 1$ corresponds to the definition of the Gromov-Wasserstein distance $\mathtt{GW}_p(\mu, \nu) := \mathtt{GW}_p(\mu_{\mathcal{Z}}, \nu_{\mathcal{Z}}) = \mathtt{FGW}_p(\mu, \nu \mid \alpha = 1)$ with $\mu_{\mathcal{Z}}, \nu_{\mathcal{Z}}$ the marginals on the structure space. Roughly speaking, the optimal coupling matrix $\gamma^\star$ will tend to associate two source and target nodes if both their feature and structural representations are similar.

Despite the fact that FGW has been shown to be efficient to address graph classification and clustering tasks [24], we claim that it might face two limitations: (i) given an underlying task, in the presence of noisy features, the best value of the hyper-parameter α in Problem (3) might be close to 1, thus focusing mainly on the structures and "forgetting" the feature information; (ii) on the other hand, if $\mathcal{G}^s$ and $\mathcal{G}^t$ are structurally very different, the optimal coupling will be likely associated with a tuned parameter α close to 0 and thus skipping the structural information by only leveraging the features.

Figure 1 illustrates these two limitations in the context of a clustering task, which aims at classifying the nodes of a target graph $\mathcal{G}^t$ by minimizing its FGW distance to a fully labelled source graph $\mathcal{G}^s$. In these experiments, α in (3) is tuned using a circular validation procedure derived from [2]. Note that we use the same following model to generate $\mathcal{G}^s$ and $\mathcal{G}^t$. The graph structures are drawn from a two-class contextual stochastic block model [10] with symmetric connectivity matrix $(p_{uv})_{(u,v)\in\{1,2\}^2}$. The vertices' features are scalar random variables $X_i \sim l(i) + \sigma\mathcal{N}(0, 1)$, with mean $l(i) = \pm 1$, depending on the class of the vertex i, and standard deviation σ. We define the signal to noise ratio $\eta = 2/\sigma$.

In Fig. 1(a), as η^t decreases, the distribution of the target features smooths out and becomes gradually less representative of the clusters. Accordingly, the Wasserstein distance between the source and the target features' distributions increases in expression (3), thus penalising the classification accuracy of FGW. To cope with this loss of feature's reliability, the tuned value of the hyper-parameter α converges towards 1, yielding a FGW based clustering that mostly relies on the graphs' structures and that necessarily undergoes a drop in performance.

In Fig. 1(b), the features distributions are now similar ($\eta^s = \eta^t$) but the structure of $\mathcal{G}^t$ differs from that of $\mathcal{G}^s$ by densifying the inter-cluster connectivity: $p_{12}^t = f \cdot p_{12}^s$, $f \geq 1$. As the distorsion factor f increases, the clusters in $\mathcal{G}^t$ tend to blur out. Accordingly, the distance C_{jl}^t between any two vertices y_j and y_l in different classes, statistically reduces in comparison with the same distance C_{ik}^s in $\mathcal{G}^s$. To lessen the penalty effect stemming from this structure mismatch, the hyper-parameter α in the distance (3) rapidly falls towards 0. Since now, the resulting FGW based clustering solely exploits the features reliability, it naturally undergoes a performance drop, as it was the case in Fig. 1-(a).

3   Diffusion and DifFused-Gromov Wasserstein Distances

To overcome the aforementioned limitations, we suggest in the following to exploit the heat diffusion in graphs and the information provided by the Laplacian matrices to design a new family of Wasserstein distances more robust to local changes in the feature and structure representations.

3.1   From Heat Diffusion to the Diffusion Wasserstein (DW) Distance

Let us consider an undirected graph $\mathcal{G}$ with m vertices and its (combinatorial) Laplacian operator $L \in \mathbb{R}^{m \times m}$. We consider a (real or complex valued) feature vector $x \in \mathbb{R}^m$ or $\mathbb{C}^m$ which contains a value for each node. A dynamical process analogous to a diffusion process is obtained by considering the heat equation on the graph: $dw(\tau)/d\tau = -Lw(\tau)$, with $\tau \in \mathbb{R}^+$ and $w(0) = x$. It admits a closed-form solution $w(\tau) = \exp(-\tau L)x$, which is the application of the heat kernel $\exp(-\tau L)$ on the initial feature vector x. Using the GSP interpretation [17], and the functional calculus on the Laplacian operator, the effect of this heat diffusion process is to smooth the features, and the larger τ, the smoother the result. The limit when $\tau \to +\infty$ is even that the solution is constant (on each connected component). It means that τ is a parameter both controlling the smoothing of features and defining a scale of analysis of the structure of the graph.

The properties of this process were used in [9] to introduce a distance between graphs (see also [23]). The original idea was to diffuse Dirac features (1 on a given node, 0, elsewhere), as illustrated in Fig. 2(a), and stack all the diffusing patterns so that $\exp(-\tau L)$ is a matrix characterizing the graph at some scale τ. Then, to compare two graphs of the same size (m nodes), given their Laplacian L_1 and L_2, the authors of [9] propose to consider $\|\exp(-\tau L_1) - \exp(-\tau L_2)\|_F$ and keep the minimum value of this quantity over all the possible τ's. While they

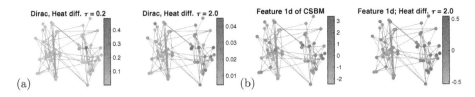

Fig. 2. Illustration of the heat diffusion of features on a graph drawn from a SBM with two blocks [10]. (a) Diffusion from a single vertex: one sees that the diffusion is first spread over the block of the vertex. (b) Diffusion of random features (drawn according to a Contextual SBM [7]): we see that the diffusion is also smoothing the features over each group, helping for their identification.

show that it is a distance, and that it captures well structural (dis)similarities between graphs, its shortcoming is that (i) it can only be used with graphs of the same size, (ii) it forgets about existing features on these graphs and (iii) it cannot be directly used in an OT setting.

To introduce our proposed Diffusion Wasserstein distance, we leverage the closed-form solution of the heat equation applied now to r features $\mathbf{X} \in \mathbb{R}^{m \times r}$ on the graph: $\exp(-\tau L)\mathbf{X}$. Each such term describes now the smoothing of all the features on the graph structure, at a specific characteristic scale τ, as seen in Fig. 2(b). Because it combines features and structure, this solution will be central in the following definition of our new distance between graphs with features.

Definition 1. *Consider a source graph $\mathcal{G}^s$, a target graph $\mathcal{G}^t$ represented through two discrete probability measures μ and ν (cf (2)) with weights vectors $a \in \mathbb{R}^m$, $b \in \mathbb{R}^n$ and Laplacian matrices $L^s \in \mathbb{R}^{m \times m}$ and $L^t \in \mathbb{R}^{n \times n}$. Let $\mathbf{X} \in \mathbb{R}^{m \times r}$, $\mathbf{Y} \in \mathbb{R}^{n \times r}$ represent the sample sets associated to the features on their vertices.*

Given parameters $0 \leq \tau^s, \tau^t < \infty$, consider the diffused sample sets $\tilde{X}, \tilde{Y}$ represented by the matrices $\tilde{\mathbf{X}} = \exp(-\tau^s L^s)\mathbf{X} \in \mathbb{R}^{m \times r}$, $\tilde{\mathbf{Y}} = \exp(-\tau^t L^t)\mathbf{Y} \in \mathbb{R}^{n \times r}$ and define $\tilde{M}(\tau^s, \tau^t) := M(\tilde{X}, \tilde{Y}) \in \mathbb{R}^{m \times n}$, a cost matrix between features that takes into account the structure of the graphs through diffusion operators. We define the Diffusion Wasserstein distance (DW) between μ and ν as:

$$\mathrm{DW}_p^p(\mu, \nu \mid \tau^s, \tau^t) = \min_{\gamma \in \Pi(a,b)} \langle \gamma, \tilde{M}^p \rangle. \tag{4}$$

Here again $\tilde{M}^p$ is the entrywise p-th power of $\tilde{M}$. The underlying distance is implicit in $M(\cdot, \cdot)$. For the sake of concision, the dependency on τ^s and τ^t will be omitted from the notation $\mathrm{DW}_p^p(\mu, \nu)$ if not specifically required.

3.2 Role of the Diffusion Parameters on DW

Denote $D^s = \exp(-\tau^s L^s) \in \mathbb{R}^{m \times m}$, $D^t = \exp(-\tau^t L^t) \in \mathbb{R}^{n \times n}$ the diffusion matrices, which depend on the (symmetric) Laplacians $L^s \in \mathbb{R}^{m \times m}$, $L^t \in \mathbb{R}^{n \times n}$ and the diffusion parameters $0 \leq \tau^s, \tau^t < \infty$. Given $1 \leq i \leq m, 1 \leq j \leq n$ let $x_i, y_j \in \mathbb{R}^r$ be the features on nodes i on $\mathcal{G}^s$ and j on $\mathcal{G}^t$, i.e. respectively the i-th row of $\mathbf{X} \in \mathbb{R}^{m \times r}$ and the j-th row of $\mathbf{Y} \in \mathbb{R}^{n \times r}$, and similarly for $\tilde{x}_i, \tilde{y}_j \in \mathbb{R}^r$ built from $\tilde{\mathbf{X}} = D^s \mathbf{X}$ and $\tilde{\mathbf{Y}} = D^t \mathbf{Y}$. Observe that $\tilde{M}(\tau^s, \tau^t)$ and $\mathrm{DW}_p^p(\mu, \nu \mid \tau^s, \tau^t)$ depend on the diffusion parameters τ^s, τ^t. When $\tau^s = \tau^t = 0$, since $D^s = I_m$ and $D^t = I_n$ we have $\tilde{M}(0,0) = M$ hence

$$\mathrm{DW}_p^p(\mu, \nu \mid 0,0) = \mathrm{W}_p^p(\mu, \nu), \tag{5}$$

i.e., DW generalizes the Wasserstein distance W.

From now on we focus on DW defined using a cost matrix $\tilde{M}$ based on the Euclidean distance and $p = 2$. Denote

$$\begin{cases} M_{ij}^2 = \|x_i - y_j\|_2^2 \\ \tilde{M}_{ij}^2 = \|\tilde{x}_i - \tilde{y}_j\|_2^2 \end{cases}$$

the squared entries of the cost matrices associated to the Wasserstein (W_2) and Diffusion Wasserstein (DW_2) distances. The next proposition establishes the asymptotic behavior of $\mathrm{DW}_2^2(\mu, \nu)$ with respect to τ^s and τ^t as well as an upper bound expressed in terms of a uniform coupling matrix. Denote $\bar{\gamma} \in \Pi(a, b) \subset \mathbb{R}_+^{m \times n}$ this (uniform) transport plan such that $\bar{\gamma}_{i,j} = 1/nm, \forall i, j$.

Proposition 1. *Consider Laplacians $L^s \in \mathbb{R}^{m \times m}$, $L^t \in \mathbb{R}^{n \times n}$ associated to two undirected connected graphs ($\mathcal{G}^s$ and $\mathcal{G}^t$) and two matrices $\mathbf{X} \in \mathbb{R}^{m \times r}, \mathbf{Y} \in \mathbb{R}^{n \times r}$ representing the sample sets $x_i \in \mathbb{R}^r, 1 \le i \le m$ and $y_j \in \mathbb{R}^r, 1 \le j \le n$ (associated to their vertices). Consider the associated measures μ, ν with flat weight vectors $a = 1_m/m$, $b = 1_n/n$. We have*

$$\lim_{\tau^s, \tau^t \to \infty} \mathrm{DW}_2^2(\mu, \nu \mid \tau^s, \tau^t) = \|\frac{1}{m} \sum_i x_i - \frac{1}{n} \sum_j y_j\|_2^2. \tag{6}$$

Moreover, the function $(\tau^s, \tau^t) \mapsto \langle \bar{\gamma}, \tilde{M}^2(\tau^s, \tau^t) \rangle$ is non-increasing with respect to τ^s and with respect to τ^t and also satisfies for each $0 \le \tau^s, \tau^t < \infty$

$$\|\frac{1}{m} \sum_i x_i - \frac{1}{n} \sum_j y_j\|_2^2 \le \mathrm{DW}_2^2(\mu, \nu \mid \tau^s, \tau^t) \le \langle \bar{\gamma}, \tilde{M}^2(\tau^s, \tau^t) \rangle \le \langle \bar{\gamma}, M^2 \rangle \tag{7}$$

$$\lim_{\tau^s, \tau^t \to \infty} \langle \bar{\gamma}, \tilde{M}^2(\tau^s, \tau^t) \rangle = \|\frac{1}{m} \sum_i x_i - \frac{1}{n} \sum_j y_j\|_2^2. \tag{8}$$

The proof is in the supplementary material.

Remark 1. The reader can check that in the proof we also establish that

$$\langle \bar{\gamma}, \tilde{M}^2(\tau^s, \tau^t) \rangle = \langle \bar{\gamma}, M^2 \rangle + \left[\sum_{i=2}^m \left(e^{-2\tau^s \lambda_i^s} - 1 \right) \|\hat{x}_i\|_2^2 + \sum_{j=2}^n \left(e^{-2\tau^t \lambda_j^t} - 1 \right) \|\hat{y}_j\|_2^2 \right].$$

Contrary to its non-increasing upper bound $\langle \bar{\gamma}, \tilde{M}^2(\tau^s, \tau^t) \rangle$, the squared Diffusion Wasserstein distance $\mathrm{DW}_2^2(\mu, \nu \mid \tau^s, \tau^t)$ may not behave monotonically with τ^s, τ^t. Even though $\mathrm{DW}_2^2(\mu, \nu \mid 0,0) = \mathrm{W}_2^2(\mu, \nu)$ we may thus have $\mathrm{DW}_2^2(\mu, \nu \mid \tau^s, \tau^t) > \mathrm{W}_2^2(\mu, \nu)$ for some values of τ^s, τ^t. The following gives a sufficient condition to ensure that (in expectation) $\mathrm{DW}_2^2(\mu, \nu \mid \tau^s, \tau^t)$ does not exceed $\mathrm{W}_2^2(\mu, \nu)$.

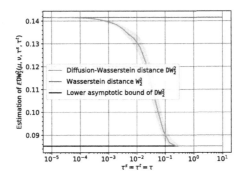

Fig. 3. Numerical illustration of Proposition 2, with distance $\mathrm{DW}_2^2(\mu, \nu \mid \tau^s, \tau^t)$ defined in Eq. (4). $\mathbb{E}\,\mathrm{DW}_2^2(\mu, \nu \mid \tau^s, \tau^t)$ is empirically estimated from 2500 independent realisations of source and target graphs drawn from the same stochastic block model, with $p_{11} = 0.32$, $p_{22} = 0.32$, $p_{12} = p_{21} = 0.02$ and $n = m = 100$. The feature vectors $\mathbf{X} \in \mathbb{R}^m$ and $\mathbf{Y} \in \mathbb{R}^n$ are arbitrarily chosen and remain fixed across all realisations, to restrict randomness only to the structures. Empirical median (solid line) and quartiles 1 and 3 (strip) of $\mathrm{DW}_2^2(\mu, \nu \mid \tau^s = \tau, \tau^t = \tau)$ are plotted against τ and compared to the Wasserstein distance $\mathrm{W}_2^2(\mu, \nu) = \mathrm{DW}_2^2(\mu, \nu \mid 0, 0)$ (upper bound) and to the asymptotic regime given in Eq. (6), when $\tau \to +\infty$ (lower plateau).

Proposition 2. *Consider integers $m, n, r \geq 1$, $a \in \mathbb{R}_+^m$, $b \in \mathbb{R}_+^n$ such that $\sum_i a_i = 1 = \sum_j b_j$, two random Laplacians $L^s \in \mathbb{R}^{m \times m}$, $L^t \in \mathbb{R}^{n \times n}$ drawn independently according to possibly distinct probability distributions, two random feature matrices $\mathbf{X} \in \mathbb{R}^{m \times r}$, $\mathbf{Y} \in \mathbb{R}^{n \times r}$, and $0 \leq \tau^s, \tau^t < \infty$. If $\mathbb{E}\,\tilde{M}_{ij}^2(\tau^s, \tau^t) \leq M_{ij}^2\ \forall (i, j)$, then $\mathbb{E}\,\mathrm{DW}_2^2(\mu, \nu \mid \tau^s, \tau^t) \leq \mathrm{W}_2^2(\mu, \nu)$.*

Remark 2. The case where the Laplacians and/or the features are deterministic is covered by considering probability distributions that are Diracs.

Proof. For brevity we omit the dependency on μ, ν.

$$\mathbb{E}\mathrm{DW}_2^2 = \mathbb{E}\inf_{\gamma \in \Pi(a,b)} \langle \tilde{M}^2, \gamma \rangle \leq \inf_\gamma \mathbb{E}\langle \tilde{M}^2, \gamma \rangle = \inf_\gamma \langle \mathbb{E}\tilde{M}^2, \gamma \rangle \leq \inf_\gamma \langle M^2, \gamma \rangle = \mathrm{W}_2^2. \qquad \Box$$

Moreover, by [18, Remark 2.19] we have $\mathrm{W}_2^2(\mu, \nu) \geq \|\frac{1}{m}\sum_{i=1}^m x_i - \frac{1}{n}\sum_{j=1}^n y_j\|_2^2$. If $\mathbf{X}$ and $\mathbf{Y}$ are such that in fact $\mathrm{W}_2^2(\mu, \nu) > \|\frac{1}{m}\sum_{i=1}^m x_i - \frac{1}{n}\sum_{j=1}^n y_j\|_2^2$ then for sufficiently large τ^s, τ^t we must have $\mathrm{DW}_2^2(\mu, \nu \mid \tau^s, \tau^t) < \mathrm{W}_2^2(\mu, \nu)$.

However we can find examples such that $\mathrm{DW}_2^2(\mu, \nu) > \mathrm{W}_2^2(\mu, \nu)$ and $\mathbb{E}\mathrm{DW}_2^2(\mu, \nu) > \mathrm{W}_2^2(\mu, \nu)$ for all $0 < \tau^s, \tau^t < \infty$. For this, it is sufficient to choose $\mathbf{X} = \mathbf{Y}$, so that $\mathrm{W}_2^2(\mu, \nu) = 0$, and deterministic or random graphs and parameters τ^s, τ^t such that $\exp(-\tau^s L^s)\mathbf{X}$ is not equal (even up to permutation) to $\exp(-\tau^t L^t)\mathbf{Y}$, so that (almost surely) $\mathrm{DW}_2^2(\mu, \nu \mid \tau^s, \tau_t) > 0$.

Figure 3 illustrates the results of Propositions 1 and 2, where we empirically estimated $\mathbb{E}\,\mathrm{DW}_2^2(\mu, \nu \mid \tau^s, \tau^t)$, and plotted its evolution against $\tau = \tau^s = \tau^t$ (experimental conditions are detailed in the legend of Fig. 3). Trivially, we verify that $\mathrm{DW}_2^2(\mu, \nu \mid 0, 0) = \mathrm{W}_2^2(\mu, \nu)$. But, more importantly, we observe that $\mathbb{E}\,\mathrm{DW}_2^2$ systematically stands below W_2^2, confirming thus the prediction of Proposition 2, and converges towards the theoretical bound given in Eq. (6) of Proposition 1, when $\tau \to \infty$. Interestingly also, although we know from the counter-example $\mathbf{X} = \mathbf{Y}$ above, that it is not true in general, the trend of $\mathbb{E}\,\mathrm{DW}_2^2$ in Fig. 3 seems to validate the conjecture whereby it is often a non-increasing function of the diffusion scale τ. However, we still lack the theoretical conditions that warrant the result of Prop. 1 on $(\tau^s, \tau^t) \mapsto \langle \bar{\gamma}, \tilde{M}^2(\tau^s, \tau^t) \rangle$ to extend to $\min_{\gamma \in \Pi(a,b)} \langle \gamma, \tilde{M}^2(\tau^s, \tau^t) \rangle$.

From an algorithmic complexity perspective, notice that compared to FGW, our new distance DW allows us to get free from the costly term in $\mathcal{O}(m^2 n^2)$ corresponding to the Gromov part of FGW (even though when $p = 2$ one can compute this term more efficiently in $\mathcal{O}(m^2 n + n^2 m)$ [19]), while still accounting for both the structure and the features of the graphs. Our study on the computational time of the state of the art methods in Sect. 4 will give evidence that DW is the cheapest way to compute a distance encompassing both sources of information.

3.3   DifFused Gromov Wasserstein (DFGW)

It is worth noticing that the heat diffusion operator, as defined in Sect. 3.1, can be seen as a reweighting scheme applied over the node features leading to the new cost matrix $\tilde{M}(\tau^s, \tau^t)$. Notice also that the latter can be precomputed during a preprocess. Therefore, by plugging $\tilde{M}(\tau^s, \tau^t)$ in place of M in FGW, we get for free a family (parameterized by $\alpha \in [0, 1]$) of so-called DifFused Gromov Wasserstein (DFGW) distances defined as follows:

Definition 2.

$$\mathrm{DFGW}_p^p(\mu, \nu) = \min_{\gamma \in \Pi(a,b)} \left\{ \sum_{i,j,k,l} \left((1-\alpha)\tilde{M}_{ij}^p + \alpha |C_{ik}^s - C_{jl}^t|^p \right) \gamma_{ij} \gamma_{kl} \right\}, \quad (9)$$

where the summation indices are $1 \le i, k \le m$ and $1 \le j, l \le n$ for a source graph $\mathcal{G}^s$ (resp. a target graph $\mathcal{G}^t$) of size m (resp. n), and the dependency on τ^s, τ^t and the considered distance d is implicit in $\tilde{M}$.

A simple lower bound on DFGW holds with arguments similar to those of [24] leading to a lower bound on FGW in terms of W and GW.

Lemma 1. *Following [24], $\forall p$, $\mathrm{DFGW}_p^p(\mu, \nu)$ is lower-bounded by the straightforward interpolation between $\mathrm{DW}_p^p(\mu, \nu)$ and $\mathrm{GW}_p^p(\mu, \nu)$:*

$$\mathrm{DFGW}_p^p(\mu, \nu) \ge (1-\alpha)\mathrm{DW}_p^p(\mu, \nu) + \alpha \mathrm{GW}_p^p(\mu, \nu)$$

Proof. By definition, for $\gamma \in \Pi(a,b)$ we have $\mathrm{GW}_p^p(\mu, \nu) \le \sum_{i,j,k,l} |C_{ik}^s - C_{jl}^t|^p \gamma_{ij} \gamma_{kl}$. Similarly, since $\sum_{k,l} \gamma_{k,l} = 1$, we get $\mathrm{DW}_p^p(\mu, \nu) \le \sum_{i,j} \tilde{M}_{ij}^p \gamma_{ij} = \sum_{i,j,k,l} \tilde{M}_{ij}^p \gamma_{ij} \gamma_{k,l}$. As a result,

$$(1-\alpha)\mathrm{DW}_p^p(\mu,\nu) + \alpha\mathrm{GW}_p^p(\mu,\nu) \leq \sum_{i,j,k,l}\left((1-\alpha)\tilde{M}_{ij}^p + \alpha|C_{ik}^s - C_{jl}^t|^p\right)\gamma_{ij}\gamma_{kl}.$$

As this holds for every $\gamma \in \Pi(a,b)$ and $\mathrm{DFGW}_p^p(\mu,\nu)$ is the infimum of the right hand side, this establishes the result. □

Just as the Diffusion Wasserstein distance, DFGW depends on the diffusion parameters τ^s, τ^t, and we have

$$\mathrm{DFGW}_p^p(\mu,\nu \mid 0,0) = \mathrm{FGW}_p^p(\mu,\nu). \tag{10}$$

Therefore, DFGW generalizes the Fused Gromov Wasserstein distance. In the same spirit as Proposition 1, the next proposition establishes the asymptotic behavior of $\mathrm{DFGW}_2^2(\mu,\nu \mid \tau^s, \tau^t)$ with respect to τ^s and τ^t.

Proposition 3. *With the notations and assumptions of Proposition 1 we have*

$$\mathrm{DFGW}_2^2(\mu,\nu \mid \tau^s, \tau^t) \geq (1-\alpha)\|\frac{1}{m}\sum_{i=1}^m x_i - \frac{1}{n}\sum_{j=1}^n y_j\|_2^2 + \alpha\mathrm{GW}_2^2(\mu,\nu), \; \forall \tau^s, \tau^t$$

$$\lim_{\tau^s,\tau^t \to \infty} \mathrm{DFGW}_2^2(\mu,\nu \mid \tau^s, \tau^t) = (1-\alpha)\|\frac{1}{m}\sum_{i=1}^m x_i - \frac{1}{n}\sum_{j=1}^n y_j\|_2^2 + \alpha\mathrm{GW}_2^2(\mu,\nu).$$

The proof is in the supplementary material.

3.4   Metric Properties of DW and DFGW

Recall that DW can be seen as a generalization of the Wasserstein distance W which leverages the diffusion operator over the features. Moreover, it is known that when the cost matrix $M_{ij} \overset{def}{=} [d(x_i, y_j)]_{ij}$ is associated to a distance d, W defines a metric. The next proposition shows that the diffusion does not change this metric property up to a natural condition.

Proposition 4. *For $p \in [1,\infty)$ and $0 \leq \tau^s, \tau^t < \infty$, the Diffusion Wasserstein $\mathrm{DW}_p(\cdot,\cdot \mid \tau^s, \tau^t)$ defines a pseudo-metric: it satisfies all the axioms of a metric, except that $\mathrm{DW}_p(\mu,\nu) = 0$ if, and only if, $\mathcal{T}(\mu) = \mathcal{T}(\nu)$. Here, $\mathcal{T}$ is the function which maps $\mu = \sum_{i=1}^m a_i\delta_{x_i,z_i}$ into $\tilde{\mu} = \mathcal{T}(\mu) = \sum_{i=1}^m a_i\delta_{\tilde{x}_i}$ where $\tilde{x}_i \in \mathbb{R}^k$ is built in a deterministic manner from the diffusion matrix D^s (which is itself a function of μ through the z_i's) and corresponds to the i-th row of $\tilde{X} = D^s X$.*

Proof. According to Def. 1, DW is defined between two probability measures $\mu = \sum_{i=1}^m a_i\delta_{(x_i,z_i)}$ and $\nu = \sum_{j=1}^n b_j\delta_{(y_j,z_j')}$ with (x_i, z_i) and (y_j, z_j') lying in some joint space $\mathcal{X} \times \mathcal{Z}$ encoding both the feature and the structure information of two source and target vertices, respectively. Since $\mathrm{DW}_p(\mu,\nu) = \mathrm{W}_p(\tilde{\mu},\tilde{\nu}) = \mathrm{W}_p(\mathcal{T}(\mu),\mathcal{T}(\nu))$, the proposition follows from the metric property of $\mathrm{W}_p(\cdot,\cdot)$. □

On the other hand, it has been shown in [24] that when C^s and C^t are distance matrices the Fused Gromov Wasserstein FGW_1 defines a metric and that FGW_p^p defines a semimetric (i.e., a relaxed version of the triangle inequality holds) when $p > 1$. Since DW is used in our DifFused Gromov Wasserstein distance in place of the Wasserstein counterpart in FGW, the same metric properties hold for DFGW.

4   Numerical Experiments

We evaluate here our diffusion distances on domain adaptation (DA) tasks between attributed graphs. We address the most complicated scenario where source and target domains are considered and labels are only available in the former. Data from the two domains are supposedly drawn from different but related distributions and the goal is to reduce this distribution discrepancy while benefiting from the supervised information from the source [20]. Note that when dealing with a DA task between attributed graphs, the divergence can come from three situations: (i) a shift in the feature representation of the source/target nodes; (ii) a difference in the graph structures; (iii) both of them. In this section, we study these three settings. Under different experimental conditions, we compare in these DA tasks the relevance of our diffusion distances DW (4) and DFGW (9), to state-of-the-art OT-based distances: W: Wasserstein (1); GW: Gromov-Wasserstein [19]; FGW: Fused Gromov-Wasserstein (3); LPL1-MM: that corresponds to the Wasserstein problem associated with a label regularization [4]; OT-LAPLACE: the same as the latter with a Laplacian regularization [5]. In the following experiments, we use the same Contextual SBM [10] and the same Gaussian mixture model as the ones described in Fig. 1, to generate the graph structures and the nodes' features, respectively. Although both source and target graphs are labelled, for OT methods implying hyper-parameters, we tune them with the circular validation procedure derived in [2] that only uses the ground truth labels on the vertices of $\mathcal{G}^s$. As for the ground truth on the vertices of $\mathcal{G}^t$, they only serve to evaluate the classification performance (accuracy) of the methods. The procedure is that each target node inherits the label of the class from which it received most of its mass by the transport plan that is solution of the optimization problem. The tuning of the hyper-parameters ($\alpha \in [0, 1]$, $\tau \in [10^{-3}, 10^{-0.5}]$ and the regularization parameters of LPL1-MM and OT-LAPLACE) and the performance evaluation are performed on two different sets of 50 i.i.d. realizations. Unless specified, we display the empirical mean values.

All codes[1] are written in Python. Graphs are generated using the Pygsp [6] library; optimal transport methods use the implementation in POT [8].

Resilience to Features' Uncertainty. We start illustrating the effect of the heat diffusion when the target features are weakly representative of the underlying clusters, leading to a divergence between the source and the target domains. This is the case in Fig. 4(a), where, as the target signal to noise ratio η^t decays, it smears out the modes of the features' distribution and makes the Wasserstein distance inefficient at discerning them. As a result, all transport methods relying uniquely on information from the features behave poorly and fail, in the limit of $\eta^t \to 0$, at inferring from $\mathcal{G}^s$, the two classes in $\mathcal{G}^t$. On the opposite, hybrid methods that also exploit similarity of the graphs' structure, naturally show better performance. Incidentally, we verified that the puzzling weak performance of Gromov-Wasserstein do not negate its capacity at clustering $\mathcal{G}^t$ correctly, but stems from the global labelling mismatch incurred by the symmetry

---

[1] https://gitlab.aliens-lyon.fr/dbarbe/graph_diffusion_wasserstein.

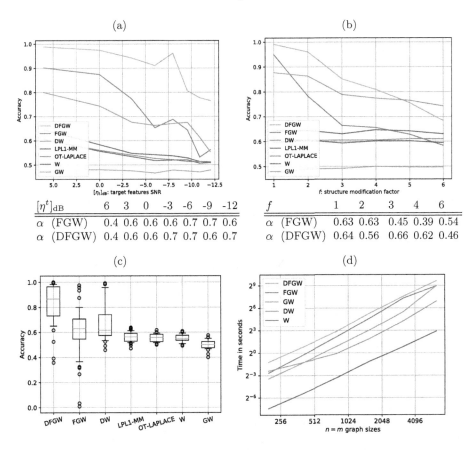

$[\eta^t]_{dB}$	6	3	0	-3	-6	-9	-12
α (FGW)	0.4	0.6	0.6	0.6	0.7	0.7	0.6
α (DFGW)	0.4	0.6	0.6	0.7	0.7	0.6	0.7

f	1	2	3	4	6
α (FGW)	0.63	0.63	0.45	0.39	0.54
α (DFGW)	0.64	0.56	0.66	0.62	0.46

Fig. 4. Comparison of OT methods in a domain adaptation task between graphs. We consider attributed graphs whose structures follow a contextual stochastic block model and attributes a mixture Gaussian model. Y-axes of plots (a)–(b) represent the classification accuracies. Hyper-parameters and mean performance are determined from two distinct sets of 50 i.i.d. realisations each. (a) Structures of $\mathcal{G}^s$ and $\mathcal{G}^t$ are identical ($p_{11}=p_{22}=0.4$, $p_{12}=p_{21}=0.05$, $n=m=250$). SNR of the source features is fixed ($[\eta^s]_{dB}=20\log_{10}(\eta^s)=6\,dB$) and σ^t of features $\mathbf{Y}_j \sim l(j)+\sigma^t\mathcal{N}(0,1)$ varies according to $[\eta^t]_{dB}$ along the X-axis. (b) Features SNR $[\eta^s]_{dB}=[\eta^t]_{dB}=6\,dB$. The target graph follows a SBM with symmetric connectivity matrix $p^t_{12}=p^s_{12}=0.05$, $p^t_{11}=p^s_{11}=0.4$ and $p^t_{22}=p^s_{22}/f$ with $p^s_{22}=0.4$ and f variable on the X-axis. Tables beneath the plots give the tuned hyper-parameters values for each case. (c) Performance when uncertainty bears on the features and on the structures simultaneously ($[\eta^t]_{dB}=0\,dB$, $f=3$). (d) Computing times wrt the size of the graphs $n=m$ ($[\eta^s]_{dB}=[\eta^t]_{dB}=0\,dB$, $f=1$).

of the SBM connectivity matrices. Now, concentrating on our difFused Gromov Wasserstein distance, it systematically and significantly[2] outperforms FGW, whatever the value of η^t. As the diffusion operator D^t estimates the conditional

---

[2] A paired Student's t-test rejects the null hypothesis with p-value equal to 2.10^{-11}.

mean value ($l(j) = \pm 1$) of each vertex j by locally averaging the features of its neighbouring nodes in the graph, DFGW turns out to be less sensitive to the noise amplitude σ_t. Notice though, that as $\tau^t \to \infty$, $D^t\mathbf{Y}$ converges to the global mean of $\mathbf{Y}$, we need to limit τ^t to the range $[10^{-3}, 10^{-0.5}]$, that is to say, before DW reaches the lower plateau in Fig. 3. As confirmed by the similar evolution of the optimized α for FGW and DFGW, the use of the Diffusion Wasserstein distance to define DFGW is responsible for the accuracy gain compared to FGW.

Interestingly, at low signal-to-noise ratio, the mere DW is able to compensate for the lack of an explicit metric between the graphs' structures, by retrieving it from the action of the diffusion operators D^s and D^t on the features.

Resilience to Structures' Uncertainty. Figure 4(b) illustrates the robustness of the diffusion distances with respect to inconsistencies in the structures of the source and target graphs. The plots display the DA performance achieved by the different OT methods, when source and target features follow the same distribution, but the graphs become less and less alike and hence, the structure information becomes less reliable in the context of an OT task.

As expected, the methods relying solely on the Wasserstein distance between the features perform constantly, with an accuracy level that is comparable to that of Fig. 4(a) at high SNRs. GW continues to get poor performance for large values of the distorsion factor f, because now, it really is unable to infer the clusters from the graphs' structures. More remarkably, the performance of FGW rapidly degrades once source and target graphs start to slightly differ. This is confirmed by the corresponding trend of α, which overall decays with f, decreasing the contribution of the GW distance in (3). In comparison, although the accuracy obtained with DFGW suffers from the growing inconsistency between graph structures too, it always remains above the curve for FGW. These results clearly demonstrate that DW bears a structure information that the solution of Eq. (9) is able to leverage and to combine with the Gromov-Wasserstein cost matrix. But the most striking result of Fig. 4(b) comes from the performance of the transport plan resulting from minimizing DW alone. Although its accuracy never outperforms that of DFGW, it not only surpasses FGW once $f > 1$, but it also degrades at a slower rate than the two competing methods. One possible explanation, is that the circular procedure used to determine α, has its own limits when $\mathcal{G}^s$ and $\mathcal{G}^t$ are drastically different, and it certainly does not yield the best compromise possible between the features and the structure modalities. DW does not entail to tune this trade-off, as it naturally embeds both modalities in its definition (4).

Resilience to Domain Divergence. The box plots of Fig. 4(c) compare the methods in a more realistic (and tricky) task of domain adaptation, when both the features and the structures in source and target spaces, differ. This scenario is a combination of the experimental settings of Fig. 4 (a) and (b). Once again, it highlights the advantage of diffusion based distances at optimizing a transport plan between the labels' nodes of attributed graphs, when these latter share some common ground but remain highly variable with regard to each other.

A situation that is most likely to occur in real world applications. Note that we also run these experiments with the convex group-lasso regularizer [5] preserving the distributions of labels. It did not change the order depicted in Fig. 4(c).

Computing Time. One major drawback of FGW is the complexity in $\mathcal{O}(m^2 n^2)$ (going down to $\mathcal{O}(m^2 n + n^2 m)$ for $p = 2$ [19]) due to the quadratic problem optimization with respect to $\gamma_{ij}\gamma_{kl}$ in its definition (3). Naturally, the same pitfall holds for DFGW. It gets even worse since in addition, it encompasses the calculation of the diffusion operator $D^s\mathbf{X}$. On the other hand, we know from [1], that the calculation cost of $D^s\mathbf{X}$, hence of the DW distance, can be done in $\mathcal{O}(m^2 r)$ for $\mathbf{X} \in \mathbb{R}^{m \times r}$. Indeed, in the experiments reported in Fig. 4(d), the computing time of the Diffusion Wasserstein distance remains one order of magnitude below that of DFGW and FGW. Then, in DFGW, the cost for computing GW always prevails over that of DW, inducing in (9), a limited overhead cost as compared to that of FGW. In this last series of experiment, our goal was not to address a domain adaptation task anymore, that is why we deemed irrelevant to compare with the computing times of LPL1-MM and OT-LAPLACE methods.

5 Conclusion

We exploit in this paper the heat diffusion in attributed graphs to design a family of Wasserstein-like distances that take into account both the feature and the structural information of the graphs. We study their asymptotic behavior and prove that they satisfy metric properties. This paper opens the door to several future lines of work. A promising solution to deal with the scalability of DFGW to large graphs would consist in relaxing the quartic constraint of the Gromov counterpart by resorting to random filtering on graphs [22] allowing for an approximation of the cost matrix. In such a case, the conservation of the metric and asymptotic properties will have to be studied. One could also use Chebyshev polynomials to approximate and speed up the heat kernel computing. Finally, other applications in graph classification or in clustering exploiting Diffusion Wasserstein barycenters can be envisioned.

Acknowledgements. Work supported by the ACADEMICS grant of the IDEX-LYON, project of the Université de Lyon, PIA operated by ANR-16-IDEX-0005.

References

1. Al-Mohy, A., Higham, N.: Computing the action of the matrix exponential, with an application to exponential integrators. SIAM J. Sci. Comp. **33**(2), 488–511 (2011)
2. Bruzzone, L., Marconcini, M.: Domain adaptation problems: a DASVM classification technique and a circular validation strategy. IEEE Trans. Pattern Anal. Mach. Intell. **32**(5), 770–787 (2010)
3. Chung, F.: The heat kernel as the pagerank of a graph. Proc. Nat. Acad. Sci. **104**(50), 19735–19740 (2007)

4. Courty, N., Flamary, R., Tuia, D.: Domain adaptation with regularized optimal transport. In: Calders, T., Esposito, F., Hüllermeier, E., Meo, R. (eds.) ECML PKDD 2014. LNCS (LNAI), vol. 8724, pp. 274–289. Springer, Heidelberg (2014). https://doi.org/10.1007/978-3-662-44848-9_18

5. Courty, N., Flamary, R., Tuia, D., Rakotomamonjy, A.: Optimal transport for domain adaptation. Trans. Pattern Anal. Mach. Intell. **39**(9), 1853–1865 (2017)

6. Defferrard, M., Martin, L., Pena, R., Perraudin, N.: Pygsp: graph signal processing in python, October 2017. https://doi.org/10.5281/zenodo.1003158

7. Deshpande, Y., Sen, S., Montanari, A., Mossel, E.: Contextual stochastic block models. In: NeurIPS 2018, Montréal, Canada, pp. 8590–8602 (2018)

8. Flamary, R., Courty, N.: Pot python optimal transport library (2017), https://github.com/rflamary/POT

9. Hammond, D.K., Gur, Y., Johnson, C.R.: Graph diffusion distance: A difference measure for weighted graphs based on the graph Laplacian exponential kernel. In: GlobalSIP, pp. 419–422. IEEE (2013)

10. Holland, P.W., Laskey, K.B., Leinhardt, S.: Stochastic blockmodels: first steps. Soc. Networks **5**(2), 109–137 (1983)

11. Kantorovich, L.: On the translocation of masses. Doklady Acad. Sci. USSR **37**, 199–201 (1942)

12. Kriege, N.M., Johansson, F.D., Morris, C.: A survey on graph kernels. Appl. Network Sci. **5**(1), 1–42 (2019). https://doi.org/10.1007/s41109-019-0195-3

13. Maretic, H.P., Gheche, M.E., Chierchia, G., Frossard, P.: GOT: an optimal transport framework for graph comparison. CoRR abs/1906.02085 (2019)

14. Maretic, H.P., Gheche, M.E., Minder, M., Chierchia, G., Frossard, P.: Wasserstein-based graph alignment. cs.LG abs/2003.06048 (2020)

15. Monge, G.: Mémoire sur la théorie des déblais et des remblais. Histoire de l'Académie royale des sciences de Paris (1781)

16. Mémoli, F.: Gromov-Wasserstein distances and the metric approach to object matching. Found. Comput. Math. **11**(4), 417–487 (2011)

17. Ortega, A., Frossard, P., Kovačević, J., Moura, J., Vandergheynst, P.: Graph signal processing: overview, challenges, and applications. Proc. IEEE **106**(5), 808–828 (2018)

18. Peyré, G., Cuturi, M.: Computational Optimal Transport. arXiv, March 2018

19. Peyré, G., Cuturi, M., Solomon, J.: Gromov-Wasserstein averaging of kernel and distance matrices. Int. Conf. on Mach. Learn. **48**, 2664–2672 (2016)

20. Redko, I., Morvant, E., Habrard, A., Sebban, M., Bennani, Y.: Advances in Domain Adaptation Theory, p. 187, Elsevier, Amsterdam, August 2019. ISBN 978-1-78-548236-6

21. Thanou, D., Dong, X., Kressner, D., Frossard, P.: Learning heat diffusion graphs. IEEE Trans. Sig. Info. Proc. Over Networks **3**(3), 484–499 (2017)

22. Tremblay, N., Puy, G., Gribonval, R., Vandergheynst, P.: Compressive spectral clustering. In: 33rd International Conference on Machine Learning, New York, USA, June 2016

23. Tsitsulin, A., Mottin, D., Karras, P., Bronstein, A., Müller, E.: Netlsd: Hearing the shape of a graph. In: ACM International Conference on Knowledge Discovery and Data Mining, London (UK), pp. 2347–2356, New York, USA (2018)

24. Vayer, T., Courty, N., Tavenard, R., Chapel, L., Flamary, R.: Optimal transport for structured data with application on graphs. Int. Conf. on Mach. Learn. **97**, 6275–6284 (2019)

Towards Interpretable Multi-task Learning Using Bilevel Programming

Francesco Alesiani$^{(\boxtimes)}$, Shujian Yu, Ammar Shaker, and Wenzhe Yin

NEC Laboratories Europe, 69115 Heidelberg, Germany
{Francesco.Alesiani,Shujian.Yu,Ammar.Shaker}@neclab.eu,
Wenzhe.Yin@stud.uni-heidelberg.de
http://www.neclab.eu

Abstract. Interpretable Multi-Task Learning can be expressed as learning a sparse graph of the task relationship based on the prediction performance of the learned models. Since many natural phenomenon exhibit sparse structures, enforcing sparsity on learned models reveals the underlying task relationship. Moreover, different sparsification degrees from a fully connected graph uncover various types of structures, like cliques, trees, lines, clusters or fully disconnected graphs. In this paper, we propose a bilevel formulation of multi-task learning that induces sparse graphs, thus, revealing the underlying task relationships, and an efficient method for its computation. We show empirically how the induced sparse graph improves the interpretability of the learned models and their relationship on synthetic and real data, without sacrificing generalization performance. Code at https://bit.ly/GraphGuidedMTL.

Keywords: Interpretable machine learning · Multi-task learning · Structure learning · Sparse graph · Transfer learning

1 Introduction

Multi-task learning (MTL) is an area of machine learning that aims at exploiting relationships among tasks to improve the collective generalization performance of all tasks. In MTL, learning of different tasks is performed jointly, thus, it transfers knowledge from information-rich tasks via task relationship [43] so that the overall generalization error can be reduced. MTL has been successfully applied in various domains ranging from transportation [9] to biomedicine [30]. The improvement with respect to learning each task independently is significant when each task has only a limited amount of training data [2].

W. Yin—Work done while at the NEC Laboratories Europe.

Electronic supplementary material The online version of this chapter (https://doi.org/10.1007/978-3-030-67661-2_35) contains supplementary material, which is available to authorized users.

© Springer Nature Switzerland AG 2021
F. Hutter et al. (Eds.): ECML PKDD 2020, LNAI 12458, pp. 593–608, 2021.
https://doi.org/10.1007/978-3-030-67661-2_35

Various multi-task learning algorithms have been proposed in the literature (see Zhang and Yang [41] for a comprehensive survey on state-of-the-art methods). Feature learning approaches [1] and low-rank approaches [7,26] assume all the tasks are related, which may not be true in real-world applications. Task clustering approaches [4,24,28] can deal with the situation where different tasks form clusters. He et al. [19] propose a MTL method that is both accurate and efficient; thus applicable in presence of large number of tasks, as in the retail sector. However, despite being accurate and scalable, these methods lack interpretability, when it comes to task relationship.

Trustworthy Artificial Intelligence, is an EU initiative to capture the main requirements of ethical AI. Transparency and Human oversight are among the seven key requirements developed by the AI Expert Group [23]. Even if MTL improves performance w.r.t to individual models, predictions made by black-box methods can not be used as basis for decisions, unless justified by interpretable models [17,31].

Interpretability can be defined locally. LIME [36] and its generalizations [33] are local method that extract features for each test sample that most contribute to the prediction and aim at finding a sparse model that describes the decision boundary. These methods are applied downstream of an independent black-box machine learning method that produces the prediction. Inpretability can also achieved globaly. For example linear regression, logistic regression and decision tree [17,31] are considered interpretable[1], since their parameters are directly interpretable as weights on the input features. Global interpretability, on the other hand, could reduce accuracy. In general, MTL methods [32,37] are not directly interpretable, unless single tasks are learned as linear models which are considered interpretable due to their simplicity [15,34,38]. This property, however, is no longer guaranteed when tasks and their relations are learned simultaneously, mainly because the relative importance of task relationship is not revealed. Since natural phenomena are often characterized by sparse structures, we explore the interpretability resulting from imposing the relationship among tasks to be sparse.

To fill the gap of interpretabilty in MTL, this paper introduces a novel algorithm, named **G**raph **G**uided **M**ulti-**T**ask regression **L**earning (GGMTL). It integrates the objective of joint interpretable (i.e. sparse) structure learning with the multi-task model learning. GGMTL enjoys a closed-form hyper-gradient computation on the edge cost; it also provides a way to learn the graph's structure by exploiting the linear nature of the regression tasks, without excessively scarifying the accuracy of the learned models. The detailed contribution of this paper is multi-fold:

Bilevel MTL Model: A new model for the joint learning of sparse graph structures and multi-task regression that employs graph smoothing on the prediction models (Sect. 3.1);

---

[1] Inteterpretability depends also on the application, where for example it may be associated with weights being integer, and can be defined thus differently.

Closed-form hyper-gradient: Presents a closed-form solution for the hyper-gradient of the graph smoothing multi-task problem (Sect. 3.4)

Interpretable Graph: The learning of interpretable graph structures;

Accurate Prediction: Accurate predictions on both synthetic and real-world datasets despite the improved interpretability (Sect. 4);

Efficient computation: of the hyper-gradient of the proposed bilevel problem (Sect. 3.6);

Efficient method: that solves the proposed bilevel problem (Sects. 3.2, 3.3);

Veracity measures: to evaluate the fidelity of learned MTL graph structure (Sect. 4.1).

2 Related Work

2.1 Multi-task Structure Learning

Substantial efforts have been made on estimating model parameters of each task and the mutual relationship (or dependency) between tasks. Usually, such relationship is characterized by a dense task covariance matrix or a task precision matrix (a.k.a., the inverse of covariance matrix). Early methods (e.g., [35]) assume that all tasks are related to each other. However, this assumption is over-optimistic and may be inappropriate for certain applications, where different tasks may exhibit different degrees of relatedness. To tackle this problem, more elaborated approaches, such as clustering of tasks (e.g., [24]) or hierarchical structured tasks (e.g., [18]) have been proposed in recent years.

The joint convex learning of multiple tasks and a task covariance matrix was initialized in Multi-Task Relationship Learning (MTRL) [42]. Later, the Bayesian Multi-task with Structure Learning (BMSL) [14] improves MTRL by introducing sparsity constraints on the inverse of task covariance matrix under a Bayesian optimization framework. On the other hand, the recently proposed multi-task sparse structure learning (MSSL) [16] directly optimizes the precision matrix using a regularized Gaussian graphical model. One should note that, although the learned matrix carries partial dependency between pairwise tasks, there is no guarantee that the learned task covariance or prediction matrix can be transformed into a valid graph Laplacian [10]. From this perspective, the learned task structures from these works suffer from poor interpretability.

2.2 Bilevel Optimization in Machine Learning

Bilevel problems [8] raise when a problem (outer problem) contains another optimization problem (inner problem) as constraint. Intuitively, the outer problem (master) defines its solution by predicting the behaviour of the inner problem (follower). In machine learning, hyper-parameter optimization tries to find the predictive model's parameters w, with respect to the hyper-parameters vector λ

that minimizes the validation error. This can be mathematically formulated as the bilevel problem

$$\min_{\lambda} F(\lambda) = E_{s \sim D^{\text{val}}} \{ f(w_{\lambda}, \lambda, s) \} \tag{1a}$$

$$\text{s.t. } w_{\lambda} = \arg\min_{w} E_{s' \sim D^{\text{tr}}} \{ g(w, \lambda, s') \}, \tag{1b}$$

The outer objective is the minimization of the generalization error $E_{s \sim D^{\text{val}}} \{ f(w_{\lambda}, \lambda, s) \}$ on the hyper-parameters and validation data D^{val}, whereas $E_{s \sim D^{\text{tr}}} \{ g(w, \lambda, s) \}$ is the regularized empirical error on the training data D^{tr}, see [12], where $D^{\text{val}} \bigcup D^{\text{tr}} = D$. The bilevel optimization formulation has the advantage of allowing to optimize two different cost functions (in the inner and outer problems) on different data (training/validation), thus, alleviating the problem of over-fitting and implementing an implicit cross validation procedure.

In the context of machine learning, bilevel optimization has been adopted mainly as a surrogate to the time-consuming cross-validation which always requires grid search in high-dimensional space. For example, [25] formulates cross-validation as a bilevel optimization problem to train deep neural networks for improved generalization capability and reduced test errors. [13] follows the same idea and applies bilevel optimization to group Lasso [40] in order to determine the optimal group partition among a huge number of options.

Given the flexibility of bilevel optimization, it becomes a natural idea to cast multi-task learning into this framework. Indeed, [29, Chapter 5] first presents such a formulation by making each of the individual hyperplanes (of each task) less susceptible to variations within their respective training sets. However, no solid examples or discussions are provided further. This initial idea was significantly improved in [11], in which the outer problem optimizes a proxy of the generalization error over all tasks with respect to a task similarity matrix and the inner problem estimates the parameters of each task assuming the task similarity matrix is known.

3 Graph Guided MTL

3.1 Bilevel Multi-tasking Linear Regression with Graph Smoothing

We consider the problem of finding regression models $\{w_i\}$ for n tasks, with input/output data $\{(X_i, y_i)\}_{i=1}^n$, where $X_i \in R^{N_i \times d}$, $y_i \in R^{N_i \times 1}$ and d is the feature size, while N_i is the number of samples for the ith task[2]. We split the data into validation $\{(X_i^{\text{val}}, y_i^{\text{val}})\}_{i=1}^n$ and training $\{(X_i^{\text{tr}}, y_i^{\text{tr}})\}_{i=1}^n$ sets and formulate the problem as a bilevel program:

$$\min_{e} \sum_{i \in [n]} ||X_i^{\text{val}} w_{e,i} - y_i^{\text{val}}||^2 + \xi ||e||_2^2 + \eta ||e||_1 + \gamma H(e) \tag{2a}$$

---

[2] In the following we assume for simplicity $N_i = N$ for all tasks, but results extend straightforward.

$$V_e = \arg\min_{V} \sum_{i \in [n]} ||X_i^{\text{tr}} w_i - y_i^{\text{tr}}||^2 + \frac{1}{2} \lambda \operatorname{tr}(V^T L_e V), \tag{2b}$$

where $V = [w_1^T, \ldots, w_n^T]^T$ is the models' vectors, $L_e = \sum_{ij \in G} e_{ij}(d_i - d_j)(d_i - d_j)^T = E \operatorname{diag}(e) E^T$ is the Laplacian matrix defined using the incident matrix E, $e = \operatorname{vec}([e_{ij}])$ is the edge weight vector with $[e_{ij}]$ being the adjacent matrix, and d_i is the discrete indicator vector which is zero everywhere except at the i-th entry. We use $[n]$ for the set $\{1, \ldots, n\}$. The regularization term in the inner problem is the *Dirichlet energy* [6]

$$\operatorname{tr}(V^T L_e V) = \sum_{ij \in G} e_{ij} ||w_i - w_j||_2^2, \tag{3}$$

where G is the graph whose Laplacian matrix is L_e. $H(e) = -\sum_{ij \in G}(|e_{ij}| \ln |e_{ij}| - |e_{ij}|)$ is the un-normalized entropy of the edge values.

The inner problem (model learning) aims at finding the optimal model for a given structure (i.e. graph), while the outer problem (structure learning) aims at minimizing a cost function that includes two terms: (1) the learned model's accuracy on the validation data, and (2) the sparseness of the graph. We capture the sparseness of the graph with three terms: (a) the ℓ_2^2 norm of the edge values, measuring the energy of the graph, (b) the ℓ_1 norm measuring the sparseness of the edges, and (c) $H(e)$ measuring the entropy of the edges. In the experiment, we limit the edges to have values in the interval $[0, 1]$, which can be interpreted as a relaxation of the *mixed integer non-linear programming* problem when $e_{ij} \in \{0, 1\}$ as defined in Eq. 2. The advantage of formulating the MTL learning as a bilevel program (Eq. 2) is the ability to derive a closed-form solution for the hyper-gradient (see Theorem 1). Moreover, for a proper choice of the regularization parameter (ℓ_1), all edge weights have a closed-form solution (see Theorem 2). For the general case, we propose a gradient descent algorithm (Algorithm 1). Entropy regularization term has superior sparsification performance to the ℓ_1 norm regularization [21], thus, the latter can be ignored during hyper-parameter search to reduce the search space at the expense of a improved flexibility.

For simplicity, we define the functions:

$$f(V_e, e) = \sum_{i \in [n]} ||X_i^{\text{val}} w_{e,i} - y_i^{\text{val}}||^2 + \xi ||e||_2^2 + \eta ||e||_1 + \gamma H(e) \tag{4a}$$

$$g(V, e) = \sum_{i \in [n]} ||X_i^{\text{tr}} w_i - y_i^{\text{tr}}||^2 + \frac{1}{2} \lambda \operatorname{tr}(V^T L_e V), \tag{4b}$$

which allow us to write the bilevel problem in the compact form:

$$\min_{e} f(V_e, e) \text{ s.t. } V_e = \arg\min_{V} g(V, e). \tag{5}$$

The proposed formulation optimally selects the sparser graph among tasks that provides the best generalization performance on the validation dataset.

Algorithm 1: GGMTL: ℓ_2-GGMTL, ℓ_2^2-GGMTL

Input   : $\{X_t, y_t\}$ for $t = \{1, 2, ..., n\}$, ξ, η, λ, ν
Output   : $V = [w_1^T, ..., w_n^T]^T, L_e$
for $i \leftarrow 1$ **to** n **do**
  $\llcorner$ Solve w_i by Linear Regression on $\{X_i, y_i\}$
Construct k-nearest neighbor graph G on V;
Construct E the incident matrix of G;
// validation-training split
$\{X_t^{\text{tr}}, y_t^{\text{tr}}\}, \{X_t^{\text{val}}, y_t^{\text{val}}\} \leftarrow$ split$(\{X_t, y_t\})$;
while *not converge* **do**
  | // compute hyper-gradient
  | compute $d_e f(V_{e^{(t)}}, e^{(t)})$ using Eq. (10), (where, with ℓ_2 norm, Eq. (10)
  | is computed using $e = e \circ l(V)$ of Eq.8 and alternating with solution
  | of Eq.9 (given by Eq.?? or in Thrm.1).;
  | // edges' values update
  | Update e: $e^{(t+1)} = [e^{(t)} + \nu d_e f(V_{e^{(t)}}, e^{(t)})]_+$;
// Train on the full datasets with alternate optimization
Solve Eq.7 on $\{X_t, y_t\}$;
return V, L_e;

3.2   The ℓ_2 Norm-Square Regularization ℓ_2^2-GGMTL Algorithm

We propose an iterative approach that computes the hyper-gradient of $f(V_e, e)$ (Eq. 4a) with respect to the graph edges (the hyper-parameters); this hyper-gradient is then used for updating the hyper-parameters based on the gradient descend method, i.e.,

$$e^{(t+1)} = e^{(t)} + \nu d_e f(V_{e^{(t)}}, e^{(t)}), \qquad (6)$$

where d_e is the hyper-gradient and ν is the learning rate. Algorithm 1 depicts the structure of the GGMTL learning method, where $[x]_+ = \max(0, x)$. The stopping criterion is evaluated on the convergence of the validation and training errors. As a final step, the tasks's models are re-learned on all training and validation data based on the last discovered edge values.

3.3   The ℓ_2 Norm Regularization ℓ_2-GGMTL Algorithm

The energy smoothing term Eq. 3 in the inner problem of Eq. 2 is a quadratic term. However, if two models are unrelated, but connected by an erroneous edge, this term grows quadratically dominating the loss. To reduce this undesirable effect, a term proportional to the distance can be achieved using not-squared ℓ_2 norm. Therefore, we extend the inner problem of Eq. (2) of the previous model to become:

$$\arg\min_{V} g(V, e) = \arg\min_{V} \sum_{i \in [n]} ||X_i^{\text{tr}} w_i - y_i^{\text{tr}}||^2 + \frac{1}{2}\lambda \sum_{ij \in G} e_{ij} ||w_i - w_j||_2, \quad (7)$$

where the regularization term in the inner problem is the non-squared ℓ_2. This can be efficiently solved using alternating optimization [19], by defining the vector of edges' multiplicative weights $l = \text{vec}([l_{ij}]) = l(V)$ such that:

$$l_{ij} = 0.5/||w_i - w_j||_2. \quad (8)$$

We can now formulate a new optimization problem equivalent to Eq. 7

$$V_e(l) = \arg\min_{V} g(V|e, l) \quad (9a)$$

$$= \arg\min_{V} \sum_{i \in [n]} ||X_i^{\text{tr}} w_i - y_i^{\text{tr}}||^2 + \frac{1}{2}\lambda \, \text{tr}(V^T L_{eol} V) + \frac{1}{4} l^{-\circ} \quad (9a)$$

where $\circ$ is the element-wise product, L_{eol} is the Laplacian matrix whose edge values are the element-wise product of e and l $(e \circ l)$, while $l^{-\circ}$ is a short notation for the element-wise inverse of l. Having fixed l, the last term of Eq. 9 can be ignored, while optimizing the inner problem w.r.t. V. The modified Algorithm 1 (ℓ_2-GGMTL), which can also be found in the supplementary material (Alg. ??), uses alternate optimization between the closed-form solution in Eq. 8 and the solution of Eq. 9 over V.

3.4   Hyper-gradient

The proposed method (Algorithm 1) is based on the computation of the hyper-gradient of Eq. 2. This hyper-gradient has a closed-form as defined by Theorem 1 and can be computed efficiently.

Theorem 1. *The hyper-gradient of problem of Eq. 2a is*

$$d_e f(V_e, e) = \xi e + \eta \, \text{sign}(e) - \gamma \, \text{sign}(e) \circ \ln e$$
$$- \lambda (V^T \otimes I_m)(B^T \otimes I_d) A^{-T} X^{val,T} (X^{val} V - Y^{val}) \quad (10)$$

where $B = [b_{11} b_{11}^T, \ldots, b_{nn} b_{nn}^T] \in R^{n \times nm}$ and B is build with only the m non-zero edges (i.e. $|\{ij|ij \in G\}| = m$).. The other variables are $b_{ij} = (d_i - d_j) \in R^{n \times 1}$, $A = \lambda L_e \otimes I_d + X^T X \in R^{dn \times dn}$ and $V = A^{-1} X^T Y$, $L_e = \sum_{ij} e_{ij} b_{ij} b_{ij}^T$, $V = [w_1^T, \ldots, w_n^T]^T \in R^{dn \times 1}$, $X = \text{diag}(X_1, \ldots, X_n) \in R^{Nn \times dn}$, $Y = [y_1, \ldots, y_n] \in R^{Nn \times 1}$. $\ln e$ is the element wise logarithm of the vector e and $\circ$ is the Hadamard product. $\text{sign}(x)$ is the element-wise sign function of x.

We notice that B and A in Theorem 1 are sparse matrices. This leads to efficient computation of the hyper-gradient, as shown in Theorem 3. All proofs are reported in the Supplementary Material (Sec. ??).

3.5   Closed-Form Hyper-edges

Alternative to applying gradient descent methods using the hyper-gradient updates, the optimal edges' values can also be directly computed. We compute e (the edge vector) as the solution of $d_e f(V_e, e) = 0$, since the optimal solution has zero hyper-gradient. In the case when ℓ_1 is the only term that has a non-zero weight in Eq. 2a , the edge vector e has a closed-form solution as proven in Theorem 2.

Theorem 2. *Let suppose $\xi = 0, \gamma = 0, \eta \neq 0$, then the hyper-edges of problem of Eq. 2 is the solution of*

$$Ue = v \tag{11}$$

where $U = (z^T \otimes (E \otimes I_d))K$, $K = [\text{vec}(\text{diag}(d_0) \otimes I_d), \ldots, \text{vec}(\text{diag}(d_{m-1}) \otimes I_d)] \in R^{m^2 d^2 \times m}$, where $d_i \in R^{m \times 1}$ is the indicator vector and $u = M^{-1} 1_m$, $v = 1/\eta C - 1/\lambda X^T X u$, $z = (E^T \otimes I_d)u$, $M = (V^T \otimes I_m)(B^T \otimes I_d) \in R^{m \times dn}$, $C = X^{val,T}(X^{val}V - Y^{val}) \in R^{nd \times 1}$ and E, V, B as in Theorem 1.

3.6   Complexity Analysis

GGMTL algorithm computes the tasks' models kNN graph, whose computational complexity can be reduced from $O(n^2)$ to $O(nd \ln n)$ [3][3], while one iteration of GGMTL algorithm computes the hyper-gradient. A naive implementation of this step requires inverting a system of dimension $nd \times nd$, whose complexity is $O((dn)^3)$. It would thus come to surprise that the actual computational complexity of the GGMTL method is $O(nd \ln n + (nd)^{1.31} + dn^2)$, where the second two terms follow from Theorem 3, while $O(dn^2)$ is the matrix-vector product which can be performed in parallel.

Theorem 3. *The computational complexity of solving hyper-gradient of Theorem. 1 is $O((nd)^{1.31} + dn^2)$ (or $O((nd) \ln^c(nd) + dn^2)$, with c constant).*

4   Experimental Results

We evaluate the performance of GGMTL against four state-of-the-art multi-task learning methodologies (namely MTRL [42], MSSL [16], BSML [14], and CCMTL [19]) on both synthetic data and real-world applications. Among the four competitors, MTRL learns a graph covariance matrix, MSSL and BSML directly learn a graph precision matrix which can be interpreted as a graph Laplacian. By contrast, CCMTL does not learn task relationship, but uses a fixed k-NN graph before learning model parameters[4].

---

[3] or, using Approximate Nearest Neighbour (ANN) methods, to $O(nd)$ [22].
[4] We performed grid-search hyper-parameter search for all methods.

4.1 Measures

Synthetic Dataset Measure for Veracity. To evaluate the performance of the proposed method on the synthetic dataset, we propose a reformulation of the measures: accuracy, recall and precision by applying the Łukasiewicz fuzzy T-norm $\top(a,b) = \max(a+b-1,0)$ and T-conorm $\bot(a,b) = \min(a,b)$ [27], where a, b represent truth values from the interval $[0, 1]$. Given the ground truth graph G_1 and the predicted graph G_2 (on n tasks) with proper adjacency matrices A_1 and A_2 (i.e., $a_{i,j}^{(1)}, a_{i,j}^{(2)} \in [0,1]$ for all $0 \leq i,j \leq n$), we define:

$$\text{recall} = \frac{\sum_{0 \leq i,j \leq n} \top(a_{i,j}^{(1)}, a_{i,j}^{(2)})}{\sum_{0 \leq i,j \leq n} a_{i,j}^{(1)}},$$

$$\text{precision} = \frac{\sum_{0 \leq i,j \leq n} \top(a_{i,j}^{(1)}, a_{i,j}^{(2)})}{\sum_{0 \leq i,j \leq n} a_{i,j}^{(2)}},$$

$$\text{accuracy} = 1 - \frac{\sum_{0 \leq i,j \leq n} \oplus(a_{i,j}^{(1)}, a_{i,j}^{(2)})}{n^2}$$

s.t $\oplus(a, b) = \top(\bot(a, b), 1 - \top(a, b))$ is the fuzzy XOR, see [5]. The definition of the F_1 score remains unchanged as the harmonic mean of precision and recall. These measures inform about the overlap between a predicted (weighted) graph and a ground truth sparse structure, in a similar way to imbalanced classification. An alternative and less informative approach would be to compute Hamming distance between the two adjacency matrices (ground truth and induced graph), provided they are both binary.

Regression Performance. The generalization performance is measured in terms of the Root Mean Square Error (RMSE) averaged over tasks.

4.2 Synthetic Data

In order to evaluate the veracity of the proposed method, we generate three synthetic datasets where the underlying structure of the relationship among tasks is known. Each task t in these datasets is a linear regression task whose output is controlled by the weight vector $\mathbf{w}_t$. Each input variable $\mathbf{x}$, for task t, is generated *i.i.d.* from an isotropic multivariate Gaussian distribution, and the output is taken by $y = \mathbf{w}_t^T \mathbf{x} + \epsilon$, where $\epsilon \sim \mathcal{N}(0, 1)$.

The first dataset `Line` mimics the structure of a line, where each task is generated with an overlap to its predecessor task. This dataset contains 20 tasks of 30 input dimensions. The coefficient vector for tasks t is $\mathbf{w}_t = \mathbf{w}_{t-1} + 0.1\mathbf{u}_{30} \odot \mathbf{b}_{30}$, where $\odot$ denotes the pointwise product, $\mathbf{u}_{30}$ is a 30-dimensional random vector with each element uniformly distributed between $[0, 1]$, $\mathbf{b}_{30}$ is also a 30-dimensional binay vector whose elements are Bernoulli distributed with $p = 0.7$, and $\mathbf{w}_0 \sim \mathcal{N}(\mathbf{1}, \mathbf{I}_{30})$.

The tasks of the second dataset `Tree` are created in a hierarchical manner simulating a tree structure such that $\mathbf{w}_t = \mathbf{w}_{t'} + 0.1\mathbf{u}_{30} \odot \mathbf{b}_{30}$, where $\mathbf{w}_{t'}$ is coefficient vector of the parent task $(t' = \lfloor (t-1)/2 \rfloor)$, and $\mathbf{w}_0 \sim \mathcal{N}(\mathbf{1}, \mathbf{I}_{30})$ is

Table 1. Results on the synthetic data of each of GGMTL, BMSL, MMSL and CCMTL (KNN) in terms of the measures described in Subsect. 4.1.

		BMSL	MSSL	CCMTL	GGMTL
Accuracy	Line	$0.384 \pm 4.1e^{-2}$	$0.180 \pm 1.1e^{-2}$	$\underline{0.631} \pm 2.1e^{-2}$	$\mathbf{0.648} \pm 3.3e^{-2}$
	Tree	$0.388 \pm 5.2e^{-2}$	$0.141 \pm 1.1e^{-2}$	$\underline{0.722} \pm 1.6e^{-2}$	$\mathbf{0.770} \pm 1.2e^{-2}$
	Star	$\underline{0.581} \pm 8.4e^{-2}$	$\mathbf{0.726} \pm 2.9e^{-2}$	$0.405 \pm 3.1e^{-2}$	$0.460 \pm 7.8e^{-2}$
Recall	Line	$0.688 \pm 5.6e^{-2}$	$\mathbf{0.958} \pm 1.5e^{-2}$	$0.288 \pm 4.4e^{-2}$	$\underline{0.726} \pm 1.0e^{-1}$
	Tree	$0.653 \pm 6.3e^{-2}$	$\mathbf{0.946} \pm 1.3e^{-2}$	$0.390 \pm 2.5e^{-2}$	$\underline{0.790} \pm 7.8e^{-2}$
	Star	$0.318 \pm 1.3e^{-1}$	$0.311 \pm 1.1e^{-1}$	$\mathbf{0.706} \pm 7.2e^{-2}$	$\underline{0.664} \pm 1.6e^{-1}$
Precision	Line	$0.100 \pm 8.1e^{-4}$	$\underline{0.100} \pm 1.5e^{-4}$	$0.083 \pm 7.1e^{-3}$	$\mathbf{0.175} \pm 2.7e^{-2}$
	Tree	$0.065 \pm 8.4e^{-4}$	$0.065 \pm 9.9e^{-5}$	$\underline{0.092} \pm 2.6e^{-3}$	$\mathbf{0.185} \pm 1.6e^{-2}$
	Star	$0.149 \pm 4.7e^{-2}$	$\mathbf{0.238} \pm 5.4e^{-2}$	$0.176 \pm 1.2e^{-2}$	$\underline{0.185} \pm 3.1e^{-2}$
F1	Line	$0.175 \pm 2.6e^{-3}$	$\underline{0.182} \pm 3.4e^{-4}$	$0.129 \pm 1.3e^{-2}$	$\mathbf{0.282} \pm 4.3e^{-2}$
	Tree	$0.117 \pm 1.9e^{-3}$	$0.121 \pm 1.8e^{-4}$	$\underline{0.149} \pm 4.0e^{-3}$	$\mathbf{0.300} \pm 2.6e^{-2}$
	Star	$0.199 \pm 6.7e^{-2}$	$0.266 \pm 7.3e^{-2}$	$\underline{0.281} \pm 2.1e^{-2}$	$\mathbf{0.288} \pm 5.e^{-2}$
RMSE	Line	6.968 ± 1.5	$4.838 \pm 7.6e^{-1}$	$\underline{4.342} \pm 7.21e^{-1}$	$\underline{4.342} \pm 7.21e^{-1}$
	Tree	7.444 ± 1.4	4.879 ± 1.3	$\underline{4.207} \pm 1.141$	$\underline{4.207} \pm 1.141$
	Star	$4.784 \pm 2.4e^{-1}$	$1.616 \pm 2.7e^{-1}$	$\underline{0.507} \pm 2.25e^{-1}$	$\mathbf{0.300} \pm 2.11e^{-1}$

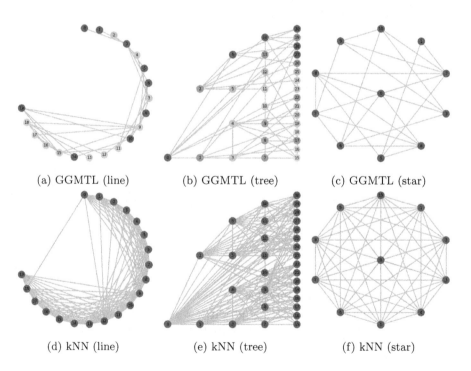

(a) GGMTL (line)   (b) GGMTL (tree)   (c) GGMTL (star)

(d) kNN (line)   (e) kNN (tree)   (f) kNN (star)

Fig. 1. The discovered graphs by GGMTL and k-NN on the synthetic datasets: Line, Tree and Star.

for the root task. In order to create a proper binary tree, we generate 31 tasks (30-dimensional) which creates a tree of five levels.

The distribution of the third dataset's tasks takes a star-shaped structure, hence called Star. The Coefficient vector of each task t is randomly created $(\mathbf{w}_t \sim \mathcal{N}(\mathbf{1}, \mathbf{I}_{20})$ for $t \in \{1, \dots, 10\})$, and the center one is a mixture of them $\mathbf{w}_0 = \sum_{t \in \{1,\dots,10\}} \mathbf{w}_t \cdot (e_{2t-1} + e_{2t})$, where $e_i \in \mathbb{R}^T$ is an indicator vector with the ith element set to 1 and the others to 0. We evaluate the performance of our method in comparison to the other methods on two aspects: (i) the ability to learn graphs that recover hidden sparse structures, and (ii) the generalization performance. For generalization error we use Root Mean Square Error (RMSE).

Table 1 depicts the results of comparing the graphs learned by GGMTL with those of CCMTL, and the covariance matrices of MSSL and BMSL when considered as adjacency matrices, after few adjustments[5]. It is apparent that GGMTL always achieves the best accuracy except on the Star dataset when MSSL performs best in terms of accuracy; this occurs only because MSSL predicts an extremely sparse matrix leading to poor recall, precision and F_1 score. Moreover, GGMTL has always the best F_1 score achieved by correctly predicting the right balance between edges (with 2nd best recall) and sparseness (always best precsion), thus, leading to correctly interpreting and revealing the latent structure of task relations. Besides the quantitative measures, interpretability is also confirmed qualitatively in Fig. 1 where the discovered edges reveal to a large extent the ground truth structures. The figure also plots the k-NN graph next to that of GGMTL, this shows how graphs of GGMTL pose a refinement of those of k-NN by removing misplaced edges while still maintaining the relevant ones among tasks. Finally, Table 1 also shows that GGMTL commits the smallest generalization error in terms of RMSE with a large margin to BMSL and MSSL.

Table 2. RMSE (mean $\pm$ std) on Parkinson's disease data set over 10 independent runs on various train/test ratios r. Best two performances underlined.

Split	MTRL	MSSL	BMSL	CCMTL	**GGMTL**
$r = 0.3$	4.147 ± 3.038	1.144 ± 0.007	1.221 ± 0.11	$\underline{1.037} \pm 0.013$	$\underline{\mathbf{1.037}} \pm 0.012$
$r = 0.4$	3.202 ± 2.587	1.129 ± 0.011	1.150 ± 0.1	$\underline{\mathbf{1.017}} \pm 0.008$	$\underline{1.019} \pm 0.01$
$r = 0.5$	1.761 ± 0.85	1.130 ± 0.009	1.110 ± 0.085	$\underline{1.010} \pm 0.008$	$\underline{1.010} \pm 0.008$
$r = 0.6$	1.045 ± 0.05	1.123 ± 0.013	1.068 ± 0.036	$\underline{\mathbf{0.998}} \pm 0.017$	$\underline{1.000} \pm 0.017$

4.3 Real-World Applications

Parkinson's Disease Assessment. Parkinson is a benchmark multi-task regression dataset[6], comprising a range of biomedical voice measurements taken

---

[5] A negative correlation is considered as a missing edge between tasks, hence, negative entries are set to zero. Besides, we normalize each matrix by division over the largest entry after setting the diagonal to zero.

[6] https://archive.ics.uci.edu/ml/datasets/parkinsons+telemonitoring.

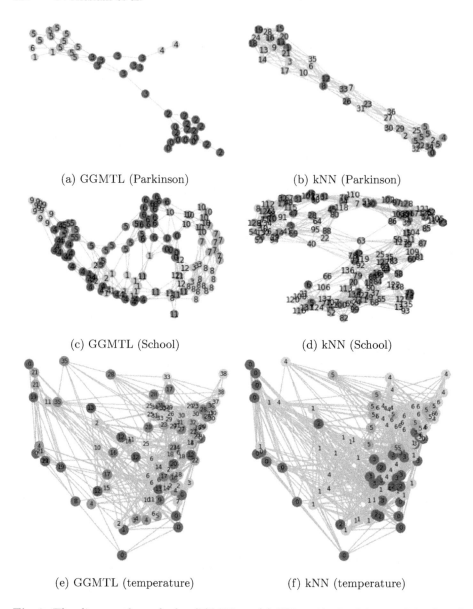

(a) GGMTL (Parkinson)　　　　　(b) kNN (Parkinson)

(c) GGMTL (School)　　　　　(d) kNN (School)

(e) GGMTL (temperature)　　　　　(f) kNN (temperature)

Fig. 2. The discovered graphs by GGMTL and k-NN on the **Parkinson**, **School** and **South** datasets.

from 42 patients with early-stage Parkinson's disease. For each patient, the goal is to predict the motor Unified Parkinson's Disease Rating Scale (UPDRS) score based 18-dimensional record: age, gender, and 16 jitter and shimmer voice measurements. We treat UPDRS prediction for each patient as a task, resulting in 42 tasks and 5,875 observations in total. We compare the generalization performance of GGMTL with that of the other baselines when different ratios of the data is used for training, ratio $r \in \{0.3, 0.4, 0.5, 0.6\}$. The results depicted in Table 2 show that GGMTL performance is close to that of CCMTL, outperforming MSSL and BMSL. However, when plotting the learned graphs, see Fig. 2(a) and Fig. 2(b), GGMTL clearly manage to separate patients into a few distinct groups unlike the heavily connected k-NN graph used by CCMTL. Interestingly, these groups are easily distinguished by Markov Clustering [39] when applied on the learned graph; this very same procedure fails to distinguish reasonable clusters when applied on the K-NN graph (35 clusters were discovered with only one task in each, and one cluster with five tasks).

Table 3. RMSE (mean ± std) on School data set over 10 independent runs on various train/test ratios r. Best two performances underlined.

	MTRL	MSSL	BMSL	CCMTL	**GGMTL**
$r = 0.2$	11.276 ± 0.103	11.727 ± 0.137	10.430 ± 0.056	$\underline{\mathbf{10.129}} \pm 0.038$	$\underline{10.137} \pm 0.034$
$r = 0.3$	10.761 ± 0.045	11.060 ± 0.045	10.223 ± 0.044	$\underline{10.078} \pm 0.038$	$\underline{\mathbf{10.072}} \pm 0.037$
$r = 0.4$	10.440 ± 0.069	10.632 ± 0.076	10.084 ± 0.067	$\underline{10.059} \pm 0.067$	$\underline{\mathbf{10.056}} \pm 0.068$
$r = 0.5$	10.267 ± 0.065	10.437 ± 0.040	10.048 ± 0.043	$\underline{\mathbf{9.994}} \pm 0.052$	$\underline{9.996} \pm 0.051$

Exam Score Prediction. School is a classical benchmark dataset in Multi-task regression [1, 28, 43]; it consists of examination scores of $15,362$ students from 139 schools in London. Each school is considered as a task and the aim is to predict the exam scores for all the students. The school dataset is available in the Malsar package [44].

Table 3 reports the RMSE on the School dataset. It is noticeable that both GGMTL and CCMTL have similar but dominating performance over the other methods. As with the Parkinson data, Fig. 2(c) and Fig. 2(d) compare the graphs induced by GGMTL and CCMTL (k-NN), and the results of applying Markov clustering on their nodes. These figures show again that graphs induced by GGMTL are easier to interpret and lead to well separated clusters with only few intercluster edges.

Table 4. RMSE (mean ± std) on US America Temperature data. The best two performances are underlined.

Hours	MTRL	MSSL	BMSL	CCMTL	**GGMTL**
8	NA	1.794	0.489	$\underline{0.101}$	$\underline{0.101}$
16	NA	1.643	0.606	$\underline{0.0975}$	$\underline{\mathbf{0.0973}}$

Temperature Forecasting in U.S. The `Temperature` dataset[7] contains hourly temperature data for major cities in the United States, collected from $n = 109$ stations for 8759 h in 2010. Data is cleaned and manipulated as described in [20]. The temperature forecasting with a horizon of 8 or 16 h in advance at each station is model as a task. We select the first 20% observations (roughly 5 weeks) to train and left the remaining 80% observations for test. We use previous 30 h of temperature as input to the model. Table 4 reports the RMSE of the methods. Learning the graph structure using GGMTL does not impact performance in term of regression error. Figure 2[e–f] shows node clustering on the graph learned on `Temerature` dataset, where the number of edges is reduced by 60% using GGMTL.

5   Conclusions

In this work, we present a novel formulation of joint multi-task and graph structure learning as a bilevel problem, and propose an efficient method for solving it based on a closed-form of hyper-gradient. We also show the interpretability property of the proposed method on synthetic and real world datasets. We additionally analyze the computational complexity of the proposed method.

References

1. Argyriou, A., Evgeniou, T., Pontil, M.: Multi-task feature learning. In: NIPS, pp. 41–48 (2007)
2. Argyriou, A., Evgeniou, T., Pontil, M.: Convex multi-task feature learning. Mach. Learn. **73**(3), 243–272 (2008)
3. Arya, S., Mount, D.M., Netanyahu, N.S., Silverman, R., Wu, A.Y.: An optimal algorithm for approximate nearest neighbor searching fixed dimensions. J. ACM (JACM) **45**(6), 891–923 (1998)
4. Bakker, B., Heskes, T.: Task clustering and gating for Bayesian multitask learning. J. Mach. Learn. Res. **4**(May), 83–99 (2003)
5. Bedregal, B.C., Reiser, R.H., Dimuro, G.P.: Xor-implications and E-implications: classes of fuzzy implications based on fuzzy Xor. Electron. Notes Theor. Comput. Sci. **247**, 5–18 (2009)
6. Belkin, M., Niyogi, P.: Laplacian eigenmaps and spectral techniques for embedding and clustering. In: Advances in Neural Information Processing Systems, vol. 14, pp. 585–591. MIT Press (2002)
7. Chen, J., Zhou, J., Ye, J.: Integrating low-rank and group-sparse structures for robust multi-task learning. In: KDD, pp. 42–50. ACM (2011)
8. Colson, B., Marcotte, P., Savard, G.: An overview of bilevel optimization. Ann. Oper. Res. **153**(1), 235–256 (2007)
9. Deng, D., Shahabi, C., Demiryurek, U., Zhu, L.: Situation aware multi-task learning for traffic prediction. In: ICDM, pp. 81–90. IEEE (2017)

---

[7] https://www.ncdc.noaa.gov/data-access/land-based-station-data/land-based-datasets/climate-normals/1981-2010-normals-data.

10. Dong, X., Thanou, D., Frossard, P., Vandergheynst, P.: Learning Laplacian matrix in smooth graph signal representations. IEEE Trans. Signal Process. **64**(23), 6160–6173 (2016)
11. Flamary, R., Rakotomamonjy, A., Gasso, G.: Learning constrained task similarities in graph regularized multi-task learning. In: Regularization, Optimization, Kernels, and Support Vector Machines, pp. 103 (2014)
12. Franceschi, L., Frasconi, P., Salzo, S., Grazzi, R., Pontil, M.: Bilevel Programming for Hyperparameter Optimization and Meta-Learning. arXiv:1806.04910 [cs, stat], June 2018
13. Frecon, J., Salzo, S., Pontil, M.: Bilevel learning of the group lasso structure. In: Advances in Neural Information Processing Systems, pp. 8301–8311 (2018)
14. Goncalves, A., et al.: Bayesian multitask learning regression for heterogeneous patient cohorts. J. Biomed. Inform. X **4**, 100059 (2019)
15. Goncalves, A.R., Das, P., Chatterjee, S., Sivakumar, V., Von Zuben, F.J., Banerjee, A.: Multi-task sparse structure learning. In: Proceedings of the 23rd ACM CIKM 2014 (2014)
16. Gonçalves, A.R., Von Zuben, F.J., Banerjee, A.: Multi-task sparse structure learning with gaussian copula models. J. Mach. Learn. Res. **17**(1), 1205–1234 (2016)
17. Guidotti, R., Monreale, A., Ruggieri, S., Turini, F., Giannotti, F., Pedreschi, D.: A survey of methods for explaining black box models. ACM Comput. Surv. **51**(5), 1–42 (2018)
18. Han, L., Zhang, Y.: Learning multi-level task groups in multi-task learning. In: AAAI, vol. 15, pp. 2638–2644 (2015)
19. He, X., Alesiani, F., Shaker, A.: Efficient and scalable multi-task regression on massive number of tasks. In: The Thirty-Third AAAI Conference on Artificial Intelligence (AAAI-19) (2019)
20. Hua, F., Nassif, R., Richard, C., Wang, H., Sayed, A.H.: Online distributed learning over graphs with multitask graph-filter models. IEEE Trans. Signal Inf. Process. Netw. **6**, 63–77 (2020)
21. Huang, S., Tran, T.D.: Sparse signal recovery via generalized entropy functions minimization. IEEE Trans. Signal Process. **67**(5), 1322–1337 (2019)
22. Hyvönen, V., et al.: Fast k-nn search. arXiv:1509.06957, 2015
23. High-Level Expert Group on Artificial Intelligence. Policy and investment recommendations for trustworthy AI. June 2019. Publisher: European Commission Type: Article; Article/Report
24. Jacob, L., Bach, F., Vert, J.P.: Clustered multi-task learning: a convex formulation. In: NIPS, pp. 745–752 (2009)
25. Jenni, S., Favaro, P.: Deep bilevel learning. In: Proceedings of the European Conference on Computer Vision (ECCV), pp. 618–633 (2018)
26. Ji, S., Ye, J.: An accelerated gradient method for trace norm minimization. In: ICML, pp. 457–464. ACM (2009)
27. Klement, E.P., Mesiar, R., Pap, E.: Triangular Norms. Kluwer Academic Publishers, Dordrecht, The Netherlands (2000)
28. Kumar, A., Daume III, H.: Learning task grouping and overlap in multi-task learning. In: ICML (2012)
29. Kunapuli G.: A bilevel optimization approach to machine learning. Ph.D. thesis, Ph.D. thesis, Rensselaer Polytechnic Institute (2008)
30. Li, L., He, X., Borgwardt, K.: Multi-target drug repositioning by bipartite block-wise sparse multi-task learning. BMC Syst. Biol. **12**, 85–97 (2018)
31. Lipton, Z.C.: The mythos of model interpretability. Commun. ACM **61**(10), 36–43 (2018)

32. Liu, P., Fu, J., Dong, Y., Qiu, X., Cheung, J.C.K.: Multi-task Learning over Graph Structures. arXiv:1811.10211 [cs], November 2018
33. Lundberg, S., Lee, S.I.: A Unified Approach to Interpreting Model Predictions. arXiv:1705.07874 [cs, stat], November 2017. arXiv: 1705.07874
34. Murugesan, K., Liu, H., Carbonell, J.G.,Yang, Y.: Adaptive Smoothed Online Multi-Task Learning. pp. 11 (2016)
35. Obozinski, G., Taskar, B., Jordan, M.I.: Joint covariate selection and joint subspace selection for multiple classification problems. Stat. Comput. **20**(2), 231–252 (2010)
36. Ribeiro, M.T., Singh, S., Guestrin, C.: "Why Should I Trust You?": Explaining the Predictions of Any Classifier. arXiv:1602.04938 [cs, stat], August 2016
37. Ruder, S.: An overview of multi-task learning in deep neural networks. arXiv preprint arXiv:1706.05098, 2017
38. Saha, A., Rai, P., Daume Iii, H., Venkatasubramanian, S.: Online Learning of Multiple Tasks and Their Relationships, pp. 9 (2011)
39. Van Dongen, S.M.: Graph clustering by flow simulation. Ph.D. thesis (2000)
40. Yuan, M., Lin, Y.: Model selection and estimation in regression with grouped variables. J. Roy. Stat. Soc.: Series B (Stat. Methodol.) **68**(1), 49–67 (2006)
41. Zhang, Y., Yang, Q.: A survey on multi-task learning. arXiv preprint arXiv:1707.08114v2, 2017
42. Zhang, Y., Yeung, D.Y.: A convex formulation for learning task relationships in multi-task learning. In: Proceedings of the Twenty-Sixth Conference on UAI, pp. 733–742 (2010)
43. Zhang, Y., Yeung, D.Y.: A regularization approach to learning task relationships in multitask learning. ACM Trans. TKDD **8**(3), 1–31 (2014)
44. Zhou, J., Chen, J., Ye, J.: Malsar: multi-task learning via structural regularization. Arizona State University, vol. 21 (2011)

Deep Learning, Grammar Transfer, and Transportation Theory

Kaixuan Zhang[1]([⊠]), Qinglong Wang[2], and C. Lee Giles[1]

[1] Information Sciences and Technology, Pennsylvania State University,
University Park, PA, USA
{kuz22,clg20}@psu.edu

[2] Alibaba Group, Building A2, Lane 55 Chuan He Road Zhangjiang,
Pudong New District, Shanghai, China
xifu.wql@alibaba-inc.com

Abstract. Despite its widespread adoption and success, deep learning-based artificial intelligence is limited in providing an understandable decision-making process of what it does. This makes the "intelligence" part questionable since we expect real artificial intelligence to not only complete a given task but also perform in a way that is understandable. One way to approach this is to build a connection between artificial intelligence and human intelligence. Here, we use grammar transfer to demonstrate a paradigm that connects these two types of intelligence. Specifically, we define the action of transferring the knowledge learned by a recurrent neural network from one regular grammar to another grammar as grammar transfer. We are motivated by the theory that there is a natural correspondence between second-order recurrent neural networks and deterministic finite automata, which are uniquely associated with regular grammars. To study the process of grammar transfer, we propose a category based framework we denote as grammar transfer learning. Under this framework, we introduce three isomorphic categories and define ideal transfers by using transportation theory in operations research. By regarding the optimal transfer plan as a sensible operation from a human perspective, we then use it as a reference for examining whether a learning model behaves intelligently when performing the transfer task. Experiments under our framework demonstrate that this learning model can learn a grammar intelligently in general, but fails to follow the optimal way of learning.

Keywords: Deep learning · Grammar transfer · Optimal transport · Category theory

1 Introduction

It is well known that deep neural architectures provide good representations for most data by using multiple processing layers that offer multiple levels of

© Springer Nature Switzerland AG 2021
F. Hutter et al. (Eds.): ECML PKDD 2020, LNAI 12458, pp. 609–623, 2021.
https://doi.org/10.1007/978-3-030-67661-2_36

abstraction. These models have dramatically improved the state-of-the-art performance in many applications such as computer vision, natural language processing, indoor localization, etc. [23] It is believed that deep learning (DL) based artificial intelligence (AI) can improve performance for many problems in several other domains [37].

However, DL has raised concerns about its innate intelligence. For example, Gary Marcus [28] doubts that DL models can learn causality, since they are often built to capture the concurrence of two events from a stateless statistical perspective. Furthermore, to capture statistics DL usually requires a very large amount of data for training. This aspect of DL brings questions to their innate functionality when the application scenario changes, especially when the captured data changes over time or the amount of data is limited. Maybe because of this, recent research has focused more on model performance and less on model intelligence. This type of "positivistic puritanism" has been observed in other fields such as physics [48] and should raise similar issues for researchers in DL.

Some ask if machines can even ever have intelligence [40]. While machines readily outperform humans in memorization and calculation, it is not evident these advantages lead to real intelligence but instead offer some narrow idiot-savant skills [29]. Indeed, the lack of reasoning in DL models has already been shown in tasks that involve abstraction, causal reasoning, and generalizing, e.g.., self-driving cars, mathematical induction, and healthcare analytics [1,6,45]. Also, many find it hard to believe that real intelligence can be generated from improved hardware and computational power. Instead of evaluating the intelligence of DL models, another approach [34,46] is to bring some self-consciousness to models by inserting in them distilled knowledge.

Our approach, instead of investigating the intelligence of DL models from a performance-oriented perspective, is to focus on its learning process. Specifically, we embed this approach as a testing method and demonstrate preliminary results obtained by applying our method to the grammar transfer problem. Our test method can be summarized as:

1. **Analogy.** Establish an analogy between machine intelligence and a reference of intelligence, e.g., human intelligence, by applying category theory [4] for building a reference. Specifically, for our built reference we provide two levels of correspondence – object correspondence and morphism correspondence – between the typical training operations in DL methods and an optimal operation obtained by using transportation theory in operations research [44]. Details are discussed in Sect. 3.
2. **Learning Process.** Investigate the learning process of a model by using the setting of grammar transfer. This is because grammar transfer facilitates the building of object correspondence as previously mentioned, and provides straightforward metrics for examining the morphism correspondence.

3. **Theoretical Basis.** Understand the critical limitation of existing methods[1] for explaining DL models sense there is not an establish theoretical connection between learned knowledge and human knowledge. In our setting, this problem is mitigated since a regular grammar (or deterministic finite automata (DFA)) has a natural correspondence with second-order recurrent neural networks [35]. Moreover, the learning cost can be measured quantitatively and compared with the transportation cost [44], which we assume to reflect human intelligence.

In summary, we propose a novel framework to investigate the mechanism of grammar transfer learning, with the intent of helping understand the foundation and intelligence of deep learning models. Please note that it is the intent of this work to open a discussion on these issues.

2 Preliminary

This section introduces mathematical tools and definitions that are used throughout the paper.

2.1 Mathematical Preliminaries

Shapley Homology. Shapley homology provides a quantitative metric for the influence of individual data samples on the topological features of a set of samples [51]. Essentially, it transforms a metric space into a discrete measure space by applying the techniques in algebraic topology and combinatorics. More specifically, the Shapley homology first constructs a simplicial complex by choosing a specific resolution, then decomposes the Betti number of the complex into the Shapley values of individual samples. Finally, the Shapley values of all samples are normalized by the L_1 norm to obtain a probability distribution. Here we applied Shapley Homology as a functor between two proposed categories. For a more detailed description as to how it works, please refer to [51].

Optimal Transport. Transportion theory studies optimal transportation and allocation of resources, and was first proposed and formalized by Gaspard Monge in 1781 [30]. The main goal is to perform transportation with the minimum cost. More formally, let X and Y be two metric spaces and $c : X \times Y \to [0, \infty]$ be a measurable function. Given probability measures μ on X and ν on Y, Monge's formulation of the transportation problem is to find a transportation map $T : X \to Y$ that realizes the infimum [43]:

$$\inf \left\{ \int_X c(x, T(x)) \mathrm{d}\mu(x) \mid T_*(\mu) = \nu \right\}, \tag{1}$$

---

[1] These methods propose to either decompose the output decision of a model into its input, which can have a linear or nonlinear form at the feature or instance level [5,22, 26], or employ interpretable models to approximate the black box DL models [13,36].

where $T_*(\mu)$ denotes the push forward of μ by T. Note that Monge's formulation can be ill-posed. As such, Kantorovich introduced a new formulation that uses a probability measure γ on $X \times Y$ that attains the infimum [43]:

$$\inf \left\{ \int_{X \times Y} c(x, y) \mathrm{d}\gamma(x, y) \mid \gamma \in \Gamma(\mu, \nu) \right\}, \tag{2}$$

where $\Gamma(\mu, \nu)$ denotes the collection of all probability measures on $X \times Y$ with marginals μ on X and ν on Y. Furthermore, the p-Wasserstein metric is defined by setting the cost function as $c(x, y) = |x - y|^p$:

$$W_p(\mu, \nu) = \left(\inf_{\gamma \in \Gamma(\mu, \nu)} \int_{X \times Y} |x - y|^p \, \mathrm{d}\gamma(x, y) \right)^{1/p}. \tag{3}$$

A more detailed introduction of this subject can be found here [44].

Category Theory. Category theory is a general theory of mathematical structures. More specifically, it studies the universal components of a family of structures of a given kind, and how structures of different kinds are interrelated in an abstract and conceptual framework [27]. There was early work on tying together category theory and neural networks [19]. Here we introduce some basic concepts that will be used to build up the framework.

Definition 1 (Category). *A category $\mathbf{C}$ consists of:*

1. *A class $\mathrm{ob}(\mathbf{C})$, whose elements are called objects;*
2. *A set $\hom_{\mathbf{C}}(a, b)$ for every pair of objects a and b, whose elements are called morphisms;*
3. *For every triad of objects a, b, and c, a function, named as composition, $\hom_{\mathbf{C}}(a, b) \times \hom_{\mathbf{C}}(b, c) \to \hom_{\mathbf{C}}(a, c)$ have its value at (f, g) denoted by $g \circ f$.*

They are required to satisfy the following conditions:

1. *(Associative) If $f : a \to b$, $g : b \to c$, and $h : c \to d$, then $h \circ (g \circ f) = (h \circ g) \circ f$;*
2. *(Identity) For every object a, a distinguished element $1_a \in \hom_{\mathbf{C}}(a, a)$, called identity on a, such that for every morphism $f : a \to b$ we have $1_b \circ f = f = f \circ 1_a$.*

We then have the following definition of a functor:

Definition 2 (Functor). *Let C and D be categories. A functor F from $\mathbf{C}$ to $\mathbf{D}$, denoted as $F : \mathbf{C} \to \mathbf{D}$, consists of:*

1. *for each object a in $\mathbf{C}$, an object $F(a)$ in $\mathbf{D}$;*
2. *for each morphism $f : a \to b$ in C, a morphism $F(f) : F(a) \to F(b)$.*

that satisfies

1. *for every object a in $\mathbf{C}$, $F(1_a) = 1_{F(a)}$;*
2. *for all morphisms $f : a \to b$ and $g : b \to c$, $F(g \circ f) = F(g) \circ F(f)$.*

Finally we say that two categories $\mathbf{C}$ and $\mathbf{D}$ are isomorphic if there exist functors $F : \mathbf{C} \to \mathbf{D}$ and $G : \mathbf{D} \to \mathbf{C}$ that are mutually inverse to each other.

2.2 Regular Grammar

A regular grammar generates a regular language – a set of strings of symbols from an alphabet, and is uniquely associated with a deterministic finite automaton (DFA) with a minimal number of states that recognizes that grammar. A DFA M can be described by a five-tuple $\{\Sigma, Q, \delta, q_0, F\}$. Σ is the input alphabet (a finite, non-empty set of symbols), and Q is a finite, non-empty set of states. $q_0 \in S$ represents the initial state, while $F \in S$ represents the set of final states. δ denotes a set of deterministic production rules. For a more detailed description of regular language and finite state machines, please refer to Hopcroft et al. [20].

2.3 Deep Learning Background

Second-Order Recurrent Neural Network. A simple second-order or tensor RNN (2-RNN) [15] has a recurrent layer updated by a weighed product of input and hidden neurons. More formally, the transition is defined by following equation:

$$h_i^t = \phi(\sum_{j,k} W_{kij} h_j^{t-1} x_k^t), \; i,j = 1, \cdots, N_h, \; k = 1, \cdots, N_x, \tag{4}$$

where t denotes the tth discrete time slot, h is the hidden units, x is the input, $W \in \mathbb{R}^{N_h \times N_h \times N_x}$ is the second-order weight, and ϕ is the activation function.

The quadratic form of the hidden activation of a 2-RNN enables a direct mapping between 2-RNN and a DFA [32], since W represents the state transition diagrams of a state process – {input, state} $\Rightarrow$ {next state}. Recent work [35] also shows the equivalence between a linear 2-RNN and weighted automata.

Please note that there are other product like neural networks [38,50]. We use this model because of its ease of representation for DFAs. Specifically, 2-RNN and DFA are expressively equivalent for representing functions defined over sequences of discrete symbols, which is not the case for other RNNs.

Transfer Learning. Transfer learning investigates the transfer of knowledge in a learning model, which has been trained with data collected from a source domain to a target domain that is different but related to the source domain. This technique can mitigate problems caused by having insufficient training data [41]. In a more formal way [33], given a source domain $\mathcal{D}_S$ and a learning task $\mathcal{T}_S$, a target domain $\mathcal{D}_T$ and a learning task $\mathcal{T}_T$, transfer learning aims to improve the learning of the target predictive function $f_T(\cdot)$ in $\mathcal{D}_T$ using the knowledge in $\mathcal{D}_S$ and $\mathcal{T}_S$.

3 Framework

3.1 Grammar Transfer Learning

Different from transfer learning, the main goal here is to examine whether the transferring process of a learning model is aligned with an optimal transportation plan. Specifically, in grammar transfer, we focus on the process of transferring the knowledge learned by 2-RNN between different regular grammars. We use strings accepted by different DFA as the data sets. These choices of data and model are based on the theoretical equivalence between a 2-RNN and DFA. In this way, we are able to build a numerical connection between the metric used in transportation theory and the metric used in grammar transfer learning. In order to adapt Eq. (3) for grammar transportation, we define a Wasserstein edit distance as follows:

$$W_e(\mu, \nu) = \inf_{\gamma \in \Gamma(\mu, \nu)} \int_{X \times Y} d_e(x, y) \mathrm{d}\gamma(x, y), \tag{5}$$

where $d_e(x, y)$ denotes the edit distance that measures the minimum number of operations needed to convert a string x into another string y [10].

3.2 Category

It is difficult to keep track of the grammar transfer process by only focusing on the models. Fortunately, category theory provides an alternative perspective. More precisely, we define three isomorphic categories to provide different representations of a regular grammar: a regular grammar category, a probability distribution category, and a learning model category.

We first start with the regular grammar category, denoted as **RG**. Here we limit our discussion to a finite number of samples since this is more realistic in practice. Suppose we have N samples. Then we can take a grammar as a definitive separation of positive and negative samples. Let ob(**RG**) denote the collection of all possible separations and hom(**RG**) contain all abstract transitions from one separation to another.

Recall that Shapley homology will assign a unique measure to a given grammar with a fixed resolution. Then we can define the probability distribution category as **PD**, of which the objects are the probability distributions associated with objects in ob(**RG**). Similarly, the morphisms contain all the transitions from one distribution to another.

Assume that a 2-RNN is trained with fixed hyper-parameters and random seeds, and the training process is stopped immediately after this 2-RNN successfully learns the underlying grammar (all strings are classified correctly). Then this 2-RNN is uniquely determined by the grammar. Hence, we denote the learning model category as ob(**LM**), which contains 2-RNNs associated with different grammars, and denote the grammar transfer learning as hom(**LM**). Based on our definitions, the cardinality of ob(**RG**), ob(**PD**), and ob(**LM**) are equal to each other. However, the morphism for each category fails to possess a well-defined correspondence. This will be addressed in the later part of this section.

3.3 Framework

Here, we introduce some necessary constraints on the morphisms to formally describe the optimal plan in our context of grammar transfer. Note that intro-

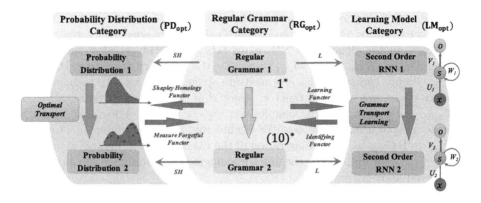

Fig. 1. Illustration of defined categories and functors.

ducing these constraints will invalidate the categories described above, since the composition in each category makes no longer makes sense. However, this will not affect our discussion. As such, we keep using the concept in category theory and the corresponding terminology in the following.

We will first construct the category $\mathbf{RG_{opt}}$ by filtering the morphisms in $\mathbf{RG}$ via optimal constraints. As such, the objects in the new category remains the same. Suppose that the N samples are enumerated as $\{x_1, x_2, \cdots, x_N\}$, and a grammar g separates the samples into two disjoint classes denoted by X_p^g and X_n^g. Given two grammars g_1 and g_2, the morphisms can be seen as a sequence of actions taken to change $X_p^{g_1}$ to $X_p^{g_2}$. Here an action is to simply flip the label of a sample. Then the optimal constraints imposed on the morphisms only allow ones with the minimal number of actions. We refer to this number as the manipulation distance and denote it as $M_e(g_1, g_2)$. It is easy to check that $M_e(g_1, g_2)$ is equal to the cardinality of symmetric difference $X_p^{g_1} \triangle X_p^2$. Note that the morphisms from g_1 to g_2 are not unique since one can change the order of actions. However, we can identify them as the same since we are only concerned with the manipulation distance instead of the entire transition process.

For the probability distribution category, we can construct $\mathbf{PD_{opt}}$ by allowing morphisms that only use the Wasserstein edit distance and identifying different transportation plans as the same one, given that one probability distribution can be transferred to another in an optimal way. As for the learning model category $\mathbf{LM_{opt}}$, we assume that M_1 and M_2 represent grammars g_1 and g_2, respectively. Then the morphisms in $\mathbf{LM_{opt}}$ contain instances of grammar transfer learning with the minimal number of training epochs, which is referred to as the learning distance and denoted as $L_e(M_1, M_2)$. Similarly, different morphisms from M_1 to M_2 with the same learning distance are considered as the same. Note that the optimal categories as defined above are not real categories, because the optimal approach from a to c is not the composition of the optimal approach from a to b and b to c in general.

It is easy to notice that functors between these categories naturally hold since the categories are isomorphic to each other. More specifically, the Shapley homology functor, denoted as SH, maps $\mathbf{RG_{opt}}$ to $\mathbf{PD_{opt}}$. Also, the learning functor L maps $\mathbf{RG_{opt}}$ to $\mathbf{LM_{opt}}$. Inversely, the transformation from $\mathbf{PD_{opt}}$ to $\mathbf{RG_{opt}}$ simply changes a discrete measure space to the set. As such, we refer to it as a measure forgetful functor, denoted as MF. Then given an object in $\mathbf{LM_{opt}}$, one can present all samples $\{x_1, x_2, \cdots, x_N\}$ to a learning model to provide a separation of the set. We denote this map identifying functor from $\mathbf{LM_{opt}}$ to $\mathbf{RG_{opt}}$ as I. An illustration of the categories and functors is provided in Fig. 1. In addition, we use the following diagram to summarize the morphisms introduced above:

$$
\begin{array}{ccccc}
\hom(\mathbf{PD_{opt}}) & \underset{SH}{\overset{MF}{\rightleftarrows}} & \hom(\mathbf{RG_{opt}}) & \underset{I}{\overset{L}{\rightleftarrows}} & \hom(\mathbf{LM_{opt}}) \\
\downarrow{\scriptstyle W_e} & & \downarrow{\scriptstyle M_e} & & \downarrow{\scriptstyle L_e} \\
\mathbb{R} & \underset{t}{\longleftarrow} & \mathbb{R} & \underset{t'}{\longleftarrow} & \mathbb{R}
\end{array}
$$

Given a morphism $f : g_1 \rightarrow g_2$ in $\hom(\mathbf{RG_{opt}})$, M_e is defined as $M_e(f) := M_e(g_1, g_2)$. Similar notations can be defined for W_e and L_e. Specifically, M_e can be obtained by calculating the cardinality of the symmetric difference of two sets. As for W_e, we can obtain its value using linear programming since we are working with discrete measures [44]. Note that W_e degenerates to M_e if one applies the uniform distribution and a constant cost. Also, L_e can be obtained empirically by conducting experiments for grammar transfer learning.

In this way, the problem of examining whether grammar transfer learning is performed in an optimal way is essentially equivalent to checking the property of function $t_1 = t'$ and $t_2 = t \circ t'$. Namely, the non-decreasingness of the functions implies the optimality in grammar transfer learning. Furthermore, we require that the domain of the functions to be the learning distance, and their range to be the manipulation distance and the Wasserstein edit distance. These two metrics provide theoretical optimal distances from two different perspectives: set theory and measure space. In summary, we provide brief procedures for our proposed framework as follows:

1. Generate strings for different regular grammar to construct $\mathbf{RG_{opt}}$ category;
2. Apply Shapley Homology functor to construct $\mathbf{PD_{opt}}$ category;
3. Train deep learning models to construct $\mathbf{LM_{opt}}$ category;
4. Obtain the distances in different categories ($\mathbf{PD_{opt}}$, $\mathbf{RG_{opt}}$, $\mathbf{LM_{opt}}$) by applying optimal transport, symmetric difference and grammar transfer learning, respectively;
5. Compare these values to see their correlation.

Next, we provide experiments to demonstrate the working mechanism of our introduced framework for examining the optimality of grammar transfer learning.

Table 1. Languages of the Tomita grammars.

G	Description
1	1^*
2	$(10)^*$
3	An odd number of consecutive 1s are always followed by an even number of consecutive 0s
4	Any string not containing "000" as a substring
5	Even number of 0s and even number of 1s [16]
6	The difference between the number of 0s and the number of 1s is a multiple of 3
7	$0^*1^*0^*1^*$

4 Experiments

We now introduce our experiment setup and provide the results of investigating the optimality of grammar transfer with discussions. Since this is the first work on studying machine intelligence based on category theory and grammar transfer, our evaluation focuses on examining the effectiveness of our proposed approach.

4.1 Experiment Setup

We use a set of seven Tomita grammars [42] to generate the string sets for evaluating transfer performance. The Tomita grammars have been widely adopted in prior work on grammatical inference and rule extraction [15,21,49]. Despite being relatively simple, these grammars cover regular languages that have a wide range of complexities [47]. Specifically, these grammars all have the alphabet of $\Sigma = \{0,1\}$, and generate infinite languages over $\{0,1\}^*$. For each grammar, we refer to its accepted strings as positive samples and its rejected strings as negative samples. In Table 1, we provide a description of the positive samples accepted by all Tomita grammars.

We generated the sets of strings by following a prior study [49]. Specifically, for the training sets, we uniformly sampled strings of various lengths $L \in \{0, 1, \ldots, 18\}$ for all seven grammars. For strings from each length, the ratio between the accepted and rejected strings was control to be 1:1 where possible, and the number of sampled strings was up to 1,000. We also constructed for each grammar a validation set consisting of up to 500 uniformly sampled strings. Each string has its lengths $L \in \{1, 4, 7, \ldots, 25\}$. On all grammars, we trained a 2-RNN with the same size of hidden layer ($N_h = 32$) and initialized with the same random seed. We monitored the training epochs needed for a model to reach 100% accuracy on the train sets and 99.9+% accuracy on a validation set.

With the generated string sets, we then computed the edit distance between string samples for each grammar and applied Shapely homology to transform the metric space into the measure space. Here we used 0-th Betti number and a resolution of 1 for decomposing the topological features of the space of sample strings. Since it is computationally intractable to apply the Shapley homology framework to a large-scale data set[2], we compute the spectrum of K subcomplexes instead of all subcomplexes as an approximation, where K is a predefined number (we set $K = 50$ in this work). More specifically, when computing the Shapley value of a specific string, we only randomly selected K subcomplexes that contains this string.

4.2   Experiment Results

Here we focus on comparing the Wasserstein edit distance and learning distance, while omitting the the the manipulation distance[3]. This is because as we mentioned earlier in Sect. 3.2, the manipulation distance is a special case of the defined Wasserstein edit distances if we neglect the measure information.

In Fig. 2, we demonstrate with heatmaps the approximated Wasserstein edit distances and the learning distances between all pairs of grammars. The latter is represented by the training epochs needed for transferring a RNN from a grammar to another. Specifically, Fig. 2a shows the ideal cases since the approximated Wasserstein distances represent the optimal transportation costs between two grammars. Clearly, it is much more difficult for performing the transportation between grammar 1 or 2 and any other grammars, since the accepted strings for these two grammars are significantly less than other grammars. Also, we can observe that grammar 5 and 6 are relatively more distinct than other grammars, since the costs needed for transporting other grammars to these two (or vice versa) are generally larger. The transportation costs shown in Figure 2a are also consistent with a prior study [47], which shows that the Tomita grammars can be categorized into different classes according to their grammar complexity. In particular, grammar 1 and 2 belong to the same class with the lowest complexity, while grammar 5 and 6 belong to another class with highest complexity.

The empirical results obtained from transferring a RNN from one grammar to another, as shown in Fig. 2b, however, only partially align with the results shown in Fig. 2a. We can first observe that the difficulty levels of being learned for grammar 3, 4, 5, and 6 are similar to the results shown in Fig. 2a. This indicates that a standard gradient descent based learning algorithm can indeed endow a RNN with a certain level of intelligence that is aligned with ideal intelligence. However, it is also clear that the symmetry in Fig. 2a is lost in Fig. 2b. This is easy to explain since the trained models may only be one of many equivalent suboptimal solutions. As such, for grammar a and grammar b, the source $RNN_{a \to b}$ (initially trained on a) and the target $RNN_{b \to a}$ (finally trained on a) do not necessarily be the same. Moreover, we can observe that grammar 1 can be easily learned as the target grammar, while grammar 2 is much harder to learn. Note that the 1,100 epochs we show in the three entries in Fig. 2b represent that the transfer learning has failed in these cases. This 1,100 epoch value is only taken for

---

[2] The algorithm introduced by [51] has exponential time complexity.
[3] We compute the manipulation distance and provide the the the results in Fig. 2c.

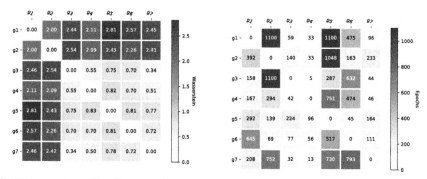

(a) Wasserstein edit distance between grammars.

(b) Learning distance between grammars.

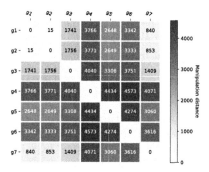

(c) Manipulation distance between grammars.

Fig. 2. Heatmaps for the grammar transfer learning performance and the estimated grammar transport learning. In both heatmaps, each entry represents the Wasserstein edit distance/learning distance/manipulation distance from the grammar indicated by its row index to the grammar indicated by its column index.

better scaling the color map used in the heatmap. The reason for the difficulty of learning grammar 2 is that its accepted strings are much too few, hence the training loss will not be significantly affected even when all the accepted strings are misclassified. This effect is also implied by noticing that it is easier to transfer a RNN initially trained on grammar 2 to other grammars. As a result, we can use the inconsistency between Fig. 2a and Fig. 2b to highlight that a standard learning algorithm cannot bring real intelligence to a model in a comprehensive way without using external guidance and reference.

5   Related Work

Measurement of Artificial Intelligence. Suggestions on how to improve AI have been sparse, especially on the measurement of artificial intelligence. This work [11] discussed the possibility of devising an intelligence quotient for machines and provided several conjectures. Recent work [7] introduced a new perspective on defining and evaluating intelligence, claiming that a measure of intelligence must account for priors, experience, and generalization difficulty, and therefore all intelligence is relative to the particular scope of application.

Category Theory in Machine Learning. Category theory has been seldom used in machine learning. Nevertheless, this does not detract the utility. It should be noted researchers in the early 2000s realized that category theory could benefit in the understanding of a neural network by applying it to its knowledge and cognition aspects [17,18]. Recent attempts to connect machine learning with category theory [24] show that a linear simplification of a recursive neural tensor network can be mapped directly onto a categorical approach. Also, a structural perspective on back-propagation [12] was provided in the language of category theory. At the same time, category theory has been widely adopted in the cognitive science [2,31]. This is because analogy, which naturally fits in category theory, is claimed to be at the core of human cognition. Motivated by this, our goal is to build a category based framework to provide a new view on analogy-related artificial intelligence.

Optimal Transport and Transfer Learning. Optimal transport theory was introduced in DL by the Wasserstein generative adversarial network (WGAN) [3]. It has also been used to improve transfer learning and domain adaptation [8,14,25,39]. In general, these studies primarily target feed-forward neural networks built for computer vision and natural language processing tasks and append the original loss function with a transport loss [25] or learn a transportation plan that matches the probability distributions of the source and target domains [9]. Different from these studies, we align the transport loss (measured by the Wasserstein edit distance) with the effort taken to transfer a learning model between different regular grammars.

6   Conclusion

We propose a category theory based framework to investigate the mechanism of grammar transfer learning. We posit that grammar transfer learning can help understand how knowledge in a deep learner can be effectively used and transferred to other machine learning systems. Furthermore, using specific examples of regular grammars can help in understanding the foundations of transfer learning and reveal actual knowledge that is learned in machine and deep learning systems. More specifically, we have constructed three isomorphic categories. Two of which represent optimal transfer approaches from a human perspective. By examining whether there are any correlations between these optimal approaches and grammar transfer learning, one can get a better understanding of the degree of intelligence demonstrated by a learning model trained with standard deep learning algorithms. Also, we demonstrated grammar transfer learning on a set of Tomita grammars to illustrate a working mechanism of our proposed framework. Future work could focus on extending this work to more sophisticated problem domains and data sets.

References

1. Abbe, E., Sandon, C.: Provable limitations of deep learning. arXiv preprint arXiv:1812.06369 (2018)
2. Arjonilla, F.J., Ogata, T.: General problem solving with category theory. arXiv preprint arXiv:1709.04825 (2017)
3. Arjovsky, M., Chintala, S., Bottou, L.: Wasserstein generative adversarial networks. In International Conference on Machine Learning, pp. 214–223 (2017)
4. Awodey, S.: Category Theory. Oxford University Press, Oxford (2010)
5. Bach, S., Binder, A., Montavon, G., Klauschen, F., Müller, K.-R., Samek, W.: On pixel-wise explanations for non-linear classifier decisions by layer-wise relevance propagation. PloS One **10**(7), e0130140 (2015)
6. Chen, D., et al.: Deep learning and alternative learning strategies for retrospective real-world clinical data. NPJ Digit. Med. **2**(1), 1–5 (2019)
7. Chollet, F.: The measure of intelligence. arXiv preprint arXiv:1911.01547 (2019)
8. Courty, N., Flamary, R., Habrard, A., Rakotomamonjy, A.: Joint distribution optimal transportation for domain adaptation. In: Advances in Neural Information Processing Systems, pp. 3730–3739 (2017)
9. Courty, N., Flamary, R., Tuia, D., Rakotomamonjy, A.: Optimal transport for domain adaptation. IEEE Trans. Pattern Anal. Mach. Intell. **39**(9), 1853–1865 (2016)
10. De la Higuera, C.: Grammatical Inference: Learning Automata and Grammars. Cambridge University Press, Cambridge (2010)
11. Falqueto, J., Lima, W.C., Borges, P.S.S., Barreto, J.M.: The measurement of artificial intelligence: an IQ for machines. In: Proceedings of The International Conference on Modeling, Identification and Control, Insbruck, Austria. Citeseer (2001)
12. Fong, B., Spivak, D., Tuyéras, R.: Backprop as functor: a compositional perspective on supervised learning. In: 2019 34th Annual ACM/IEEE Symposium on Logic in Computer Science (LICS), pp. 1–13. IEEE (2019)
13. Frosst, N., Hinton, G.E.: Distilling a neural network into a soft decision tree. In: CEx@AI*IA, Volume 2071 of CEUR Workshop Proceedings. CEUR-WS.org (2017)
14. Gayraud, N.T.H., Rakotomamonjy, A., Clerc, M.: Optimal transport applied to transfer learning for P300 detection (2017)
15. Giles, C.L., Miller, C.B., Chen, D., Chen, H.-H., Sun, G.-Z., Lee, Y.-C.: Learning and extracting finite state automata with second-order recurrent neural networks. Neural Comput. **4**(3), 393–405 (1992)
16. Giles, C.L., Sun, G.-Z., Chen, H.-H., Lee, Y.-C., Chen, D.: Higher order recurrent networks and grammatical inference. In: Advances in Neural Information Processing Systems, pp. 380–387 (1990)
17. Healy, M.J.: Category theory applied to neural modeling and graphical representations. In: Proceedings of the IEEE-INNS-ENNS International Joint Conference on Neural Networks. IJCNN 2000. Neural Computing: New Challenges and Perspectives for the New Millennium, vol. 3, pp. 35–40. IEEE (2000)
18. Healy, M.J., Caudell, T.P.: Neural networks, knowledge and cognition: a mathematical semantic model based upon category theory (2004)

19. Healy, M.J., Caudell, T.P.: Ontologies and worlds in category theory: implications for neural systems. Axiomathes **16**(1–2), 165–214 (2006)
20. Hopcroft, J.E.: Introduction to Automata Theory, Languages, and Computation. Pearson Education India (2008)
21. Hou, B.-J., Zhou, Z.-H.: Learning with interpretable structure from RNN. CoRR, abs/1810.10708 (2018)
22. Koh, P.W., Liang, P.: Understanding black-box predictions via influence functions. In: ICML, Volume 70 of Proceedings of Machine Learning Research, pp. 1885–1894. PMLR (2017)
23. LeCun, Y., Bengio, Y., Hinton, G.: Deep learning. Nature **521**(7553), 436–444 (2015)
24. Lewis, M.: Compositionality for recursive neural networks. arXiv preprint arXiv:1901.10723 (2019)
25. Lu, Y., Chen, L., Saidi, A.: Optimal transport for deep joint transfer learning. arXiv preprint arXiv:1709.02995 (2017)
26. Lundberg, S.M., Lee, S.-I.: A unified approach to interpreting model predictions. In: NIPS, pages 4765–4774 (2017)
27. Lane, S.M.: Categories for the Working Mathematician, vol. 5. Springer, Heidelberg (2013)
28. Marcus, G.: Deep learning: a critical appraisal. arXiv preprint arXiv:1801.00631 (2018)
29. Mitchell, M.: Artificial Intelligence: A Guide for Thinking Humans. Penguin UK (2019)
30. Monge, G.: Mémoire sur la théorie des déblais et des remblais. Histoire de l'Académie Royale des Sciences de Paris (1781)
31. Navarrete, J.A., Dartnell, P.: Towards a category theory approach to analogy: analyzing re-representation and acquisition of numerical knowledge. PLoS Comput. Biol. **13**(8), e1005683 (2017)
32. Omlin, C.W., Giles, C.L.: Constructing deterministic finite-state automata in recurrent neural networks. Neural Comput. **8**(4), 675–696 (1996)
33. Pan, S.J., Yang, Q.: A survey on transfer learning. IEEE Trans. Knowl. Data Eng. **22**(10), 1345–1359 (2009)
34. Papamakarios, G.: Distilling model knowledge. arXiv preprint arXiv:1510.02437 (2015)
35. Rabusseau, G., Li, T., Precup, D.: Connecting weighted automata and recurrent neural networks through spectral learning. In: AISTATS, volume 89 of Proceedings of Machine Learning Research, pp. 1630–1639. PMLR (2019)
36. Ribeiro, M.T., Singh, S., Guestrin, C.: "Why should I trust you?": explaining the predictions of any classifier. In: KDD, pp. 1135–1144. ACM (2016)
37. Russell, S.J., Norvig, P.: Artificial Intelligence: A Modern Approach. Pearson Education Limited, Malaysia (2016)
38. Schlag, I., Schmidhuber, J.: Learning to reason with third order tensor products. In: Advances in Neural Information Processing Systems, pp. 9981–9993 (2018)
39. Shen, J., Qu, Y., Zhang, W., Yu, Y.: Wasserstein distance guided representation learning for domain adaptation. arXiv preprint arXiv:1707.01217 (2017)
40. Smith, G.: The AI Delusion. Oxford University Press, Oxford (2018)
41. Tan, C., Sun, F., Kong, T., Zhang, W., Yang, C., Liu, C.: A survey on deep transfer learning. In: Kůrková, V., Manolopoulos, Y., Hammer, B., Iliadis, L., Maglogiannis, I. (eds.) ICANN 2018. LNCS, vol. 11141, pp. 270–279. Springer, Cham (2018). https://doi.org/10.1007/978-3-030-01424-7_27

42. Tomita, M.: Learning of construction of finite automata from examples using hill-climbing. RR: Regular set recognizer. Technical report, Carnegie-Mellon Univ Pittsburgh PA Dept of Computer Science (1982)
43. Villani, C.: Topics in Optimal Transportation. Number 58. American Mathematical Society (2003)
44. Villani, C.: Optimal Transport: Old and New, vol. 338. Springer, Heidelberg (2008). https://doi.org/10.1007/978-3-540-71050-9
45. Waldrop, M.M.: News feature: what are the limits of deep learning? Proc. Natl. Acad. Sci. **116**(4), 1074–1077 (2019)
46. Wang, J., Bao, W., Sun, L., Zhu, X., Cao, B., Philip, S.Y.: Private model compression via knowledge distillation. In: Proceedings of the AAAI Conference on Artificial Intelligence, vol. 33, pp. 1190–1197 (2019)
47. Wang, Q., Zhang, K., Liu, X., Giles, C.L.: Connecting first and second order recurrent networks with deterministic finite automata. arXiv preprint arXiv:1911.04644 (2019)
48. Weinberg, S.: What is quantum field theory, and what did we think it is? arXiv preprint hep-th/9702027 (1997)
49. Weiss, G., Goldberg, Y., Yahav, E.: Extracting automata from recurrent neural networks using queries and counterexamples. arXiv preprint arXiv:1711.09576 (2017)
50. Wu, Y., Zhang, S., Zhang, Y., Bengio, Y., Salakhutdinov, R.R.: On multiplicative integration with recurrent neural networks. In: Advances in Neural Information Processing Systems, pp. 2856–2864 (2016)
51. Zhang, K., Wang, Q., Liu, X., Giles, C.L.: Shapley homology: topological analysis of sample influence for neural networks. Neural Comput. **32**(7), 1355–1378 (2020)

Inductive Unsupervised Domain Adaptation for Few-Shot Classification via Clustering

Xin Cong[1,2], Bowen Yu[1,2], Tingwen Liu[1,2(✉)], Shiyao Cui[1,2], Hengzhu Tang[1,2], and Bin Wang[3]

[1] Institute of Information Engineering, Chinese Academy of Sciences, Beijing, China
{congxin,yubowen,liutingwen,cuishiyao,tanghengzhu}@iie.ac.cn
[2] School of Cyber Security, University of Chinese Academy of Sciences, Beijing, China
[3] Xiaomi AI Lab, Xiaomi Inc., Beijing, China
wangbin11@xiaomi.com

Abstract. Few-shot classification tends to struggle when it needs to adapt to diverse domains. Due to the non-overlapping label space between domains, the performance of conventional domain adaptation is limited. Previous work tackles the problem in a transductive manner, by assuming access to the full set of test data, which is too restrictive for many real-world applications. In this paper, we set out to tackle this issue by introducing a inductive framework, DaFeC, to improve **D**omain adaptation performance for **Fe**w-shot classification via **C**lustering. We first build a representation extractor to derive features for unlabeled data from the target domain (no test data is necessary) and then group them with a cluster miner. The generated pseudo-labeled data and the labeled source-domain data are used as supervision to update the parameters of the few-shot classifier. In order to derive high-quality pseudo labels, we propose a Clustering Promotion Mechanism, to learn better features for the target domain via Similarity Entropy Minimization and Adversarial Distribution Alignment, which are combined with a Cosine Annealing Strategy. Experiments are performed on the FewRel 2.0 dataset. Our approach outperforms previous work with absolute gains (in classification accuracy) of 4.95%, 9.55%, 3.99% and 11.62%, respectively, under four few-shot settings.

Keywords: Few-shot classification · Domain adaptation · Clustering

1 Introduction

Few-shot classification aims to learn a classifier to recognize unseen classes with few labeled examples. While significant progress has been made [5,13,16,17,20], most previous works are under the assumption that the samples of unseen classes should be drawn from the same domain as the training data that was used to

© Springer Nature Switzerland AG 2021
F. Hutter et al. (Eds.): ECML PKDD 2020, LNAI 12458, pp. 624–639, 2021.
https://doi.org/10.1007/978-3-030-67661-2_37

train the model. However, in the real world, the application can be used in unusual environments and novel datasets, which means that these samples are likely from different domains. Even a slight departure from a model's training domain can cause it to make spurious predictions and significantly hurt its performance. Table 1 illustrates a typical example of domain shift in the relation classification task which aims to classify the semantic relation between entities in a sentence. The training data is collected from Wikipedia, but the actual data encountered at test time comes from PubMed, a biomedical literature corpus. The new relations in the target domain such as *may_treat* are different from those in the source domain. Due to distinct domain characteristics like morphology and syntax, the performance of existing few-shot models drops drastically in such a situation.

Table 1. An example comes from FewRel 2.0, a few-shot dataset for relation classification with domain adaptation. Different colors indicate different entities, red for head entity, and blue for tail entity. Relation classification aims to determine the relation between two given entities based on their context.

	Source domain: Wikipedia		Target domain: PubMed	
Support	*member_of*	*Newton* served as the president of *the Royal Society*	*may_treat*	*Ribavirin* remains essential to *Chronic Hepatitis C* treatment
	instance_of	The *Romanian Social Party* is a left *political party* in Romania	*classified_as*	All references about a *viral infection* called *ebola haemorrhagic fever* were reviewed
Query	*Which?*	*Euler* was a member of *the Royal Swedish Academy of Sciences*	*Which?*	The *dental cysts*, especially *radicular cysts*, are compared

Unsupervised domain adaptation algorithms (UDA) aims at addressing the domain shift problem between a labeled source dataset and an unlabeled target dataset [7,18,22]. Conventional UDA methods typically assume that the target domain shares the same label space with the source domain so that the knowledge can be transferred from across domains via these same labels. However, in the few-shot settings, the source domain and target domain do not have any overlap in categories. This unique setting renders most existing UDA methods inapplicable. Previous work [4] solves this problem by making use of test data (or query set in the few shot scenario) from the target domain in the transductive manner. However, in some real-world scenarios, it is completely unrealistic to forecast test data in advance. In this paper, we work on a more realistic setting: inductive unsupervised domain adaptation for few-shot classification. It is obvious that although we do not know the ground truth classes of the target domain, some of the unlabeled target-domain data may belong to the same classes. According to the cluster hypothesis [1], the features of unlabeled data with the same latent label may cluster together in the representation space. Mining these latent cluster structures can provide auxiliary information about the target domain, which could be beneficial to improve the adaption ability of few-shot models. Based on such motivation, we design a novel framework named **DaFeC** (Unsupervised **D**omain **a**daption for **Fe**w-shot classification via **C**lustering), which effectively train the few-shot classifier with clustering-generated pseudo labels. The first step of DaFeC is the training of a representation extractor. Based on the features of unlabeled target-domain data derived from the extractor, a cluster miner is applied to group these unlabeled instances and the subsequent cluster assignments are deemed as pseudo labels. Finally, a few-shot classifier is trained based on both target-domain data with pseudo labels and source-domain training data to enable the classifier to adapt to the target domain.

Intuitively, the quality of pseudo labels significantly influences the performance of the few-shot classifier. Theoretically, if input features are well discriminative, the cluster miner can group the instances easily and assign them pseudo labels with high-confidence. Therefore, to generate high-confidence pseudo labels, we further propose a Clustering Promotion Mechanism (**CPM**) to assist in training the representation extractor to produce cluster-distributed features for unlabeled target-domain data. CPM contains three modules: First, to encourage features with the same latent class to get closer, we design a Similarity Entropy Minimization (CPM-S) objective. It calculates the Euclidean distance between each instance and others in the target domain and then minimizes the entropy of instance-wise distance vector to drive similar instances closer. Second, in our preliminary study, we observe that although the source and target domain have different label space, they may still share some similar but not identical labels. For instance (Table 1), the class *classified_as* from the target domain is semantically similar to *instance_of* from the source domain with slightly difference. Inspired by this phenomenon, we design an Adversarial Distribution Alignment method (CPM-A). It introduces a domain discriminator to play an adversarial minimax game with the representation extractor to align the distribution of similar classes cross domains. Third, we propose a Cosine Annealing Strategy (CPM-C) to support learning with CPM-S and CPM-A for achieving the optimal domain adaptation performance.

To summarize, our contributions are as following:

- For the first time, we present an inductive unsupervised domain adaptation framework, DaFeC, for few-shot classification. To the best of our knowledge, there is no similar work in few-shot classification.
- We propose a Clustering Promotion Mechanism to help the representation extractor produce cluster-distributed features for the generation of high-confidence pseudo labels.
- Our presented DaFeC is model-agnostic, which means that it can be incorporated into other models.
- Our approach achieves new state-of-the-art performance on FewRel 2.0, the currently largest unsupervised domain adaptation dataset for few-shot classification, delivering 3.99-11.62% absolute gains over previous work[1].

---

[1] The source code and data of this paper are available now and they can be obtained from https://github.com/congxin95/DaFeC.

2   Related Work

Few-shot classification aims to develop models and algorithms which are able to recognize novel classes based on few labeled instances. Recently, meta-learning has been shown to be highly effective in few-shot learning, which can be generally classified into three categories: (1) Model-based methods [12,15] design a special module such as memory to exploit meta information to make models generalize to new tasks rapidly with only a few instances. (2) Optimization-based methods [5,13] aim at learning a good initialized parameters which can achieve good performance through a few update steps. (3) Metric-based methods [16,17,20] attempt to learn a good metric function which embeds data with the same classes into adjacent distance space. Although many existing few-shot methods have achieved promising results, the performance of these methods is significantly degraded when the test data are drawn from different domains from training data, which is a quite common case in the real world.

Domain adaptation methods aim at exploiting labeled data in the source domain to perform a prediction task in the target domain. Because annotating sufficient labeled data is time-consuming and labor-intensive, unsupervised domain adaptation (no need for labeled data of the target domain), has been extensively studied recently [7,19,22]. However, all of these methods assume that the categories of the target domain are shared with the source domain, which is too restrictive to generalize to the novel classes in the few-shot classification scenario. To address this issue, [4] leverages reinforcement learning to select source data similar to the target test data to train few-shot classifiers. Nevertheless, this method works in the transductive manner, while in some real-world applications, we cannot know the test data when training. By contrast, our approach works in a more realistic inductive fashion that models cannot get information about test instances in the training phase. The only thing we can use is the unlabeled target-domain data, which can be different from the test data.

3   Task Formulation

In the few-shot classification, formally, we have two datasets: $\mathcal{D}_{meta-train}$ and $\mathcal{D}_{meta-test}$. These datasets contains a set of instances (x, y) but $\mathcal{D}_{meta-train}$ and $\mathcal{D}_{meta-test}$ have their own label space that are disjoint with each other. In the few-shot settings, $\mathcal{D}_{meta-test}$ is split into two parts: $\mathcal{D}_{support}$ and $\mathcal{D}_{query}$. If the support set contains K instances for each of N classes, this few-shot problem is called N-way-K-shot. Usually, K is really small, resulting in the poor performance when predicting $\mathcal{D}_{query}$. Therefore, models should use $\mathcal{D}_{support}$ to predict $\mathcal{D}_{query}$ labels utilizing $\mathcal{D}_{meta-train}$.

For unsupervised domain adaptation in few-shot classification, $\mathcal{D}_{meta-train}$ and $\mathcal{D}_{meta-test}$ are sampled from different domains. $\mathcal{D}_{meta-train}$ from the source domain and $\mathcal{D}_{meta-test}$ from the target domain. We rename $\mathcal{D}_{meta-train}$ as $\mathcal{D}_S$. To overcome domain discrepancy, an unlabeled target-domain dataset $\mathcal{D}_{UT} = \{x_1, x_2, \ldots, x_{UT}\}$ is provided. Our goal is to develop a model that acquires knowledge from the $\mathcal{D}_S$ and $\mathcal{D}_{UT}$, so that we can make predictions over $\mathcal{D}_{meta-test}$.

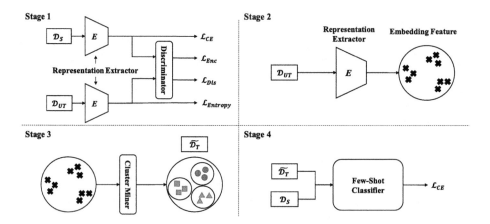

Fig. 1. The overview of our DaFeC framework. In the first stage, we train a representation extractor based on our clustering promotion mechanism and then use it to extract features for unlabeled target-domain data. Next, all unlabeled target-domain instances are grouped with a cluster miner to generate pseudo-labels. In the last stage, the few-shot classifier is trained jointly with the target-domain pseudo-labeled data and the source-domain training data.

4   Methodology

Figure 1 gives an over illustration of our framework, which operates in four stages as follows:

- **Stage 1.** Training the representation extractor with clustering promotion mechanism.
- **Stage 2.** Extracting the features of unlabeled target-domain data.
- **Stage 3.** Using the cluster miner to produce pseudo-labels for unlabeled target-domain data.
- **Stage 4.** Training the few-shot classifier based on source-domain data and target-domain data with pseudo labels.

4.1   DaFeC Framework

Representation Extractor. The Representation Extractor E is used to extract features $\mathbf{x}$ for each input instance x. Such operations is denoted as $\mathbf{x} = E(x)$. For the subsequent clustering, we use prototypical networks training method [16] and our proposed CPM (Details would be presented in Sect. 4.2) to train our representation extractor. Following [16], a Prototypical Vector representation for each class in $\mathcal{D}_S$ is generated, by averaging all the examples' representations of that label:

$$\mathbf{c}_i = \frac{1}{K} \sum_{j=1}^{K} \mathbf{x}_i^j, \quad i = 1, 2, \ldots, N, \tag{1}$$

where $\mathbf{c}_i$ refers to the prototype for class y_i and $\mathbf{x}_i^j$ represents the embedding feature of the j-th instance of class y_i. Then the probability of each class for the query instance x can be computed as:

$$P(y = y_i | x) = \frac{\exp(-d(\boldsymbol{E}(x), \mathbf{c}_i))}{\sum_{j=1}^{N} \exp(-d(\boldsymbol{E}(x), \mathbf{c}_j))}, \tag{2}$$

where $d(\cdot, \cdot)$ is the Euclidean distance. In the training phase, we expect to minimize the following objective function:

$$\min_{\theta} \mathcal{L}_{CE} = -\mathbb{E}_{x \in \mathcal{D}_S}[\log P(y|x)], \tag{3}$$

When updating the representation extractor with Eq. 3, the Euclidean distance between each instance and the prototypical vector of its class could be reduced. As a result, instances with the same class get closer to their class centroid and away from other classes. After training the representation extractor, we use it to embed all unlabeled target-domain data into embedding features for the next clustering stage.

Cluster Miner. Given the encoded features of all the target-domain instances in $\mathcal{D}_{UT}$ produced by the trained representation extractor, a cluster miner is deployed to group them into pre-defined $\tilde{N}$ distinct clusters.

Clustering has been widely studied and many approaches have been developed for a variety of circumstances. In our work, we focus on a standard clustering algorithm, k-means [11]. Same as the representation extractor, k-means get the centroid of each cluster by averaging all instances of that and use Euclidean distance to calculate the distance of every instance to their cluster centroid. Therefore, the cluster results grouped by k-Means could reveal the cluster structure generated by the representation extractor better. The subsequent cluster assignments are used as pseudo labels to guide the transformation of unlabeled $\mathcal{D}_{UT}$ to a pseudo-labeled dataset $\tilde{\mathcal{D}}_T$, which is then merged with the source-domain training set $\mathcal{D}_S$ into a new training set $\tilde{\mathcal{D}}_{meta-train} = \{\mathcal{D}_S, \tilde{\mathcal{D}}_T\}$.

Few-Shot Classifier. The few-shot classifier is trained on $\tilde{\mathcal{D}}_{meta-train}$. Because $\tilde{\mathcal{D}}_{meta-train}$ contains pseudo-labeled target-domain data, the performance of the classifier on the target domain could be improved. Our proposed DaFeC is a generally applicable and model-agnostic framework, which means it is compatible with any existing few-shot classifier. Following previous work [8], we use Proto-CNN, Proto-BERT, BERT-PAIR for the classifier backbone to demonstrate the model-agnostic property of our framework. All settings of these models are the same as the original paper.

4.2   Clustering Promotion Mechanism

Generally, the few-shot classifier learns the information of the target domain by optimizing with pseudo labels created by the cluster miner. While this seems reasonable, the inevitable label noise caused by the clustering procedure is ignored. Such noisy pseudo labels substantially hinder the model's capability to further improve the classification performance on the target domain. It is generally known that, as a typical machine learning algorithm, clustering depends heavily on the input representations, thus learning discriminative representations is fundamental to the high-confidence pseudo label generation. In order to generate features with more discriminativeness, our framework further incorporates a novel **C**lustering **P**romotion **M**echanism (CPM) into the training process. CPM is built on three components: Similarity Entropy Minimization, Adversarial Distribution Alignment, and Cosine Annealing Strategy. We describe the details of all components below.

Similarity Entropy Minimization. Obviously, only if the features of similar instances are close together, the cluster miner can assign them the same pseudo label. In order to promote this similarity without supervision, we introduce the Similarity Entropy Minimization (CPM-S) method.

We first compute the instance-wise distance vector $\mathbf{v}(\mathbf{x})$ for each target instance of $\mathcal{D}_{UT}$ as follows:

$$[\mathbf{v}(\mathbf{x}_i)]_j = \|\mathbf{x}_i - \mathbf{x}_j\|_2^2, \quad \mathbf{x}_i, \mathbf{x}_j \in \mathcal{D}_{UT}, \quad i \neq j, \tag{4}$$

where $[\cdot]_j$ means the j-th element of a vector, $\|\cdot\|_2$ means the l_2 norm and $\mathbf{x}_i, \mathbf{x}_j$ are both from $\mathcal{D}_{UT}$. To mine the latent cluster structure, we minimize the entropy of the normalized instance-wise distance vector $\mathbf{v}(\mathbf{x})$ for each target instance

$$\min_\theta \mathcal{L}_{Entropy} = \mathbb{E}_{x \sim \mathcal{D}_{UT}}[H(\text{softmax}(\mathbf{v}(\mathbf{x})/\tau))], \tag{5}$$

where $H(\cdot)$ refers to the Shannon entropy over the softmax distribution, $\tau \in \mathbb{R}^+$ is a temperature scaling parameter of the softmax distribution to control the percentage of instances we expect the target data to be similar to. Too small τ sharpens the distribution as one-hot, resulting in several pair-wise clusters while setting too large τ can smooth the distribution to be uniform, making instances get close to dissimilar ones.

Different from the conventional class-level entropy minimization [10] which calculate the entropy over the output logits of the classifier, our similarity entropy minimization over the instance-wise distance vector. Through entropy minimization, the distribution of the instance-wise distance vector will be pushed away from the uniform distribution, which means that each instance is pushed to approach its similar ones and move away from other dissimilar samples. As a result, instances are encouraged to cluster together.

Adversarial Distribution Alignment. Naturally, training the representation extractor with Eq. 3, the features of source-domain instances have been properly distributed into several distinct clusters. Although in our few-shot scenario, the target domain does not share the same label space with the source domain, we still observe that they may have some similar classes that can be leveraged to promote target-domain instances to cluster together. Recent efforts [7,19] have shown that adversarial training can align distributions of two domains, especially per-class distribution alignment. Inspired by this, we introduce the Adversarial Distribution Alignment (CPM-A) method to promote the clustering of target-domain instances by mining similar classes across domains.

First, a domain discriminator D is built to accept features encoded by the representation extractor E and classify whether a data point is drawn from the source or the target domain. Thus, D is optimized according to a standard supervised loss where the labels indicate the origin domain, defined below:

$$\min_{\phi} \mathcal{L}_{Dis} = -\mathbb{E}_{x \sim \mathcal{D}_S}[\log D_{\phi}(E(x))] - \mathbb{E}_{x \sim \mathcal{D}_{UT}}[\log(1 - D_{\phi}(E(x)))] \quad (6)$$

Second, the representation extractor E playing as the generator is optimized to maximize $\mathcal{L}_{Dis}$, which updates E to generate features to confuse discriminator. This process can be reformulated as follow:

$$\min_{\theta} \mathcal{L}_{Enc} = -\mathbb{E}_{x \sim \mathcal{D}_S}[\log(1 - D(E_{\theta}(x)))] - \mathbb{E}_{x \sim \mathcal{D}_{UT}}[\log D(E_{\theta}(x))] \quad (7)$$

Theoretically, by iterative optimization of Eq. 6 and Eq. 7, the domain discriminator D and the representation extractor E are alternated to reach the global optimality that D cannot distinguish between the features of source-domain and target-domain examples produced by E. Based on Eq. 3, the representation extractor is amended to encode instances of the source domain into cluster-distributed features. After the adversarial training phrase, the target-domain instances with similar ones in the source domain can be aligned with them as clusters.

Cosine Annealing Strategy. Unfortunately, CPM-S and CPM-A cannot work well by a simple multi-task learning strategy. In the early training phase, the representation extractor has not learned well, so the produced features are crude and inaccurate. In this time, using CPM-S to promote clustering may make instances get close to dissimilar ones, resulting in undesirable clustering results. With the training procedure going on, we can gradually increase the training weight of CPM-S to mine the latent cluster structure. CPM-A helps to align distributions of source and target domain, but over-alignment could have a detrimental effect on target-domain class separation. Since there still exist some target-domain classes completely different from classes in the source domain, the excessive alignment would mislead these classes to have inappropriate distributions for clustering, which hurts the quality of pseudo labels. Previous work [9] indicates that deep models would memorize easy instances first, and gradually adapt to hard instances as training epochs become large. Thus in the training process, our representation extractor would first align distributions of similar classes between the two domains and then those of dissimilar classes. Therefore, we can decrease the training weight of CPM-A gradually to allow model focus on the alignment of similar cross-domain classes and avoid unwanted over-alignment.

These observations motivate us to develop a Cosine Annealing Strategy (CPM-C) to adjust CPM-S and CPM-A weights in the training process. Specifically, the overall loss function $\mathcal{L}$ is designed as the combination of Eq. 3, Eq. 7, and Eq. 4 as follows:

$$\min_{\theta} \mathcal{L} = \mathcal{L}_{CE} + (1 - \lambda)\mathcal{L}_{Enc} + \lambda\mathcal{L}_{Entropy}, \tag{8}$$

where λ is the weighting parameter of $\mathcal{L}_{Entropy}$, which is designed to increase with the training epoch in the form of,

$$\lambda = \begin{cases} -\dfrac{\cos(\pi t/T) + 1}{2}, & t \leq T \\ 1, & \text{otherwise} \end{cases}, \tag{9}$$

where t is the current training epoch and T denotes a pre-defined epoch annealing hyperparameter. In the early stage of the training procedure, $\mathcal{L}_{Enc}$ has a larger weight than $\mathcal{L}_{Entropy}$, which makes the representation extractor tend to learn transferable knowledge between domains to improve the ability of encoding target-domain instances. With the training procedure going on, the weight of $\mathcal{L}_{Entropy}$ will increase continually, so CPM-S can be encouraged to promote clustering for the target domain since the representation extractor has learned enough knowledge. Compared to linear annealing, cosine function can pay more attention to CPM-A in the beginning and increase the weight of CPM-S more quickly. We experimentally found that cosine annealing outperforms linear annealing with absolute gains (in classification accuracy) of 2%–3% under different settings.

4.3   Overall Workflow

In this section, we introduce the overall working procedure of our framework DaFeC. Algorithm 1 gives the scratch.

Due to the size imbalance between $\mathcal{D}_S$ and $\mathcal{D}_{UT}$, we use the episodic paradigm proposed by [20] to train the representation extractor. In each iteration, N classes are sampled from $\mathcal{D}_S$ randomly and each class will also randomly select K instances as support instances. In this way, we can obtain the temporary support set S. And we choose other M instances from the same N classes to construct the temporary query set Q. Then the parameters of the representation extractor are optimized with Eq. 8. We use S and Q to calculate $\mathcal{L}_{CE}$. For the unlabeled target dataset $\mathcal{D}_{UT}$, we randomly sample $N \times K$ instances to construct the temporary unlabeled set U. S and U are encoded by the representation extractor into low-dimensional features. Then the discriminator takes these features as input and compute $\mathcal{L}_{Dis}$ and updates its parameters with

---

Algorithm 1. DaFeC

Input: Labeled Source-domain Dataset $\mathcal{D}_S$ and Unlabeled Target-domain Dataset $\mathcal{D}_{UT}$

Output: Few-shot Classifier $\mathbf{C}$

 1: **while** not convergence **do**
 2:    Sample S and Q from $\mathcal{D}_S$
 3:    Sample U from $\mathcal{D}_{UT}$
 4:    Update representation extractor $\boldsymbol{E}$ with Eq. 3
 5:    Update discriminator $\boldsymbol{D}$ with Eq. 6
 6:    Update representation extractor $\boldsymbol{E}$ with Eq. 7 and Eq. 5
 7: **end while**
 8: Encode $\mathcal{D}_{UT}$ into feature representations $\{\mathbf{x}_{UT}\}$ using $\boldsymbol{E}$.
 9: Run k-means on $\{\mathbf{x}_{UT}\}$ to generate clusters $\mathcal{C}_{UT}$
10: Assign each cluster in $\mathcal{C}_{UT}$ a pseudo label to construct pseudo-labeled dataset $\tilde{\mathcal{D}}_{\mathcal{T}}$
11: Merge $\tilde{\mathcal{D}}_{\mathcal{T}}$ and $\mathcal{D}_S$ into $\tilde{\mathcal{D}}_{meta-train}$
12: Train few-shot classifier $\mathbf{C}$ based on $\tilde{\mathcal{D}}_{meta-train}$
13: **return C**

Eq. 6. After that, we update the model weights of representation extractor following Eq. 8. Once the representation extractor converges, we use it to encode all the instances in $\mathcal{D}_{UT}$ into embedding features $\{\mathbf{x}_{UT}\}$. Next, we apply the k-means algorithm to mine latent cluster structure and assign them pseudo labels to construct pseudo-labeled target-domain dataset $\tilde{\mathcal{D}}_{\mathcal{T}}$, which is merged with the source-domain training set $\mathcal{D}_S$ into a new dataset $\tilde{\mathcal{D}}_{meta-train}$. Finally, the few-shot classifier is trained based on $\tilde{\mathcal{D}}_{meta-train}$.

5   Experiments

5.1   Dataset and Metric

We conduct experiments on the recently widely used benchmark FewRel 2.0 dataset introduced in [8], which is the currently largest unsupervised domain adaptation dataset for few-shot classification. It consists of 44,800 labeled instances (64 classes and 700 instances per class) from Wikipedia (source domain) as the training set and 2500 labeled instances (25 classes and 100 instances per class) from Pubmed (target domain) as the test set. It also provides SemEval-2010 task 8 as the validation set (17 classes and 8,851 instances) and unlabeled PubMed data (2500 instances) for unsupervised domain adaptation. This dataset focus on the relation classification task. Each labeled example is a single sentence, annotated with a head entity, a tail entity, and their relation. The goal is to predict the correct relation between the head and tail.

We investigate our experiments in four few-shot scenarios: 5-way-1-shot, 5-way-5-shot, 10-way-1-shot, 10-way-5-shot and report the mean and standard deviation of test accuracy according to the official evaluation scripts[2].

5.2 Implementation Details

Following [21], we implement the representation extractor E based on a convolutional neural network to encode sentences for relation classification. The window size of CNN is set to 3, and the number of filters is 230. The discriminator D is implemented as a two-layer feed forward neural network. We use the 50 dimension Glove embeddings [14] to initialize word embeddings. Following [21], we also concatenated the input word embeddings with 5-dimensional position embeddings. The model is trained using stochastic gradient descent with the learning rate of 0.1. $\tilde{N}$ (the number of clusters), τ (the temperature scaling parameter) and T (the epoch annealing parameter) are set as 10, 2 and 6000, respectively. All the hyper-parameters are tuned on the validation set. We run all experiments using PyTorch 1.1.0 on the Nvidia Tesla V100 GPU.

5.3 Baselines

We compare our model against the following 5 models proposed in [8]:

- **Proto-CNN** is a prototypical network using CNN [21] as the encoder.
- **Proto-BERT** is also a prototypical network but it uses BERT [3] as its encoder.
- **Proto-CNN-ADV** is straightforward to combine traditional domain adaptation technique, adversarial training, with few-shot model, Proto-CNN.
- **Proto-BERT-ADV** like Proto-CNN-ADV, simply utilizes adversarial training technique to augment Proto-BERT.
- **BERT-PAIR** pairs each query instance with all the supporting instances, and send the paired sequence to the BERT sequence classification model, which is the state-of-the-art on the FewRel 2.0 dataset.

5.4 Results

Table 2 reports the results of our methods (DaFeC+Proto-CNN, DaFeC+Proto-BERT and DaFeC+BERT-PAIR) against other baseline methods. From the results, we can observe that: (1) Over the previous state-of-the-art method BERT-PAIR, DaFeC+BERT-PAIR achieves substantial improvements of 4.95%, 9.55%, 3.99% and 11.62% on four few-shot settings respectively, which confirms the effectiveness and rationality of our proposed training framework. (2) Besides DaFeC+BERT-PAIR, both DaFeC+Proto-CNN and DaFeC+Proto-BERT also exceed Proto-CNN and Proto-BERT significantly. The accuracy

---

[2] https://thunlp.github.io/2/fewrel2_da.html.

Table 2. Accuracies (%) of different models on the FewRel 2.0 test set with domain adaptation. Bold marks the highest number among all models. All the results of baseline models are quoted directly from [8]. "DaFeC+" denotes our proposed method.

Model	5-Way-1-Shot	5-Way-5-Shot	10-Way-1-Shot	10-Way-5-Shot
Proto-CNN	35.09 ± 0.10	49.37 ± 0.10	22.98 ± 0.05	35.22 ± 0.06
Proto-BERT	40.12 ± 0.19	51.50 ± 0.29	26.45 ± 0.10	36.93 ± 0.01
BERT-PAIR	56.25 ± 0.40	67.44 ± 0.54	43.64 ± 0.46	53.17 ± 0.09
Proto-CNN-ADV	42.21 ± 0.09	58.71 ± 0.06	28.91 ± 0.10	44.35 ± 0.09
Proto-BERT-ADV	41.90 ± 0.44	54.74 ± 0.22	27.36 ± 0.50	37.40 ± 0.36
DaFeC+Proto-CNN	48.58 ± 0.65	65.80 ± 0.44	35.53 ± 0.67	52.71 ± 0.54
DaFeC+Proto-BERT	46.39 ± 0.68	56.32 ± 0.84	32.09 ± 0.98	40.53 ± 0.75
DaFeC+BERT-PAIR	**61.20 ± 0.91**	**76.99 ± 0.82**	**47.63 ± 1.01**	**64.79 ± 0.77**

of DaFeC+Proto-CNN and DaFeC+Proto-BERT increase 7.11% and 3.48% on average compared to Proto-CNN and Proto-BERT. This demonstrates the model-agnostic property of our framework. (3) Our models with DaFeC clearly perform better than the Proto-CNN-ADV and Proto-BERT-ADV, showing that naive applying UDA to few-shot classification is not as effective as our specifically designed framework. (4) The standard deviations of DaFeC+Proto-CNN, DaFeC+Proto-BERT and DaFeC+BERT-PAIR are slightly larger than the original models because of the pseudo-labeled noise. However, these models still outperform the original ones even considering the worst performance.

6   Analyses

6.1   Ablation Study

To study the contribution of each component in our framework, we run an ablation study (see also Table 3). From these ablations, we find that: (1) Removing the pseudo-labeled target-domain data hurts the result by 4.73%, 8.36%, 4.80% and 9.97% in four scenarios, respectively, which indicates that training the network with clustering-generated pseudo labels is vital for domain adaptation. (2) By introducing the similarity entropy minimization method (CPM-S), we can mine the latent cluster structure of unlabeled target-domain instances, which is beneficial to high-quality pseudo label generation. (3) When we remove CPM-A, the score drops by 3.81% on average, which demonstrates the effectiveness of adversarial distribution alignment over similar classes across domains. (4) When we fix the coefficients in Eq. 8 rather than use the cosine annealing strategy to adjust them, the performance declines extremely and is not on par even with only using CPM-S and CPM-A, which powerfully proves that CPM-S and CPM-A cannot work properly without CPM-C.

6.2   Effectiveness of CPM

The effectiveness of the learned discriminative feature representations through CPM can be investigated quantitatively and qualitatively.

Table 3. An ablation study of our proposed framework on the FewRel 2.0 dataset.

Model	5-Way-1-Shot	5-Way-5-Shot	10-Way-1-Shot	10-Way-5-Shot
DaFeC+BERT-PAIR	61.00	76.83	46.00	65.27
- Pseudo-labeled Data	56.27	68.47	41.20	55.30
- CPM-S	59.07	72.03	43.13	59.43
- CPM-A	58.13	71.43	43.10	60.56
- CPM-C	57.50	69.40	42.63	57.43

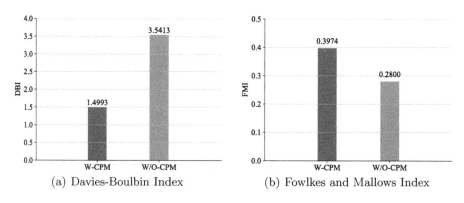

(a) Davies-Boulbin Index   (b) Fowlkes and Mallows Index

Fig. 2. Validation analysis of CPM. The Davies-Boulbin Index (DBI) is used to represent the tightness in a cluster (lower is better) and the Fowlkes and Mallows Index (FMI) is employed to indicate the clustering accuracy (higher is better).

In order to measure the cohesiveness of intra-clusters and the separation of inter-clusters, we calculate Davies-Boulbin Index (DBI) [2] on the unlabeled target-domain data. We first train two representation extractor: (1) one trained with CPM, named *W-CPM*, (2) the other trained without CPM, named *W/O-CPM*, and use them to encode the unlabeled target-domain instances from the training set respectively. Then the DBI value is calculated based on the cluster results of these features, and the lower, the better. From Fig. 2(a), we observe that W-CPM yields a considerably lower DBI value (1.4993) compared with *W/O-CPM* (3.5413).

To examine the accuracy of clustering, we first generate pseudo labels for unlabeled target-domain training data using *W-CPM* and *W/O-CPM*, respectively. Then the Fowlkes and Mallows Index (FMI) [6] score can be obtained by comparing the pseudo labels with the ground truth labels[3], and the higher, the better. The results shown in Fig. 2(b) suggest that CPM has indeed improved the clustering effect by increasing the FMI score from 0.28 to 0.3974.

In addition, we visualize the features with t-SNE projected onto 2D embedding space. Specifically, we sample 20 instances for each of 10 classes from the

---

[3] Note that FewRel 2.0 provides ground-truth labels for partial target-domain training data, but researchers are forbidden to use these labels in the training process for unsupervised domain adaption. And we only use them for experimental analysis.

target-domain training data with ground truth labels and extract their features using two representation extractors, *W-CPM* and *W/O-CPM*. Figure 3 provides the visualization of the t-SNE-transformed feature representations. We can observe that for the model without trained with CPM, the features actually are mixed and the points with the same classes are distributed in different places. Thus, the pseudo labels generated by the cluster miner may have mush noise. While for the model trained with CPM, the representation exhibits discernible clustering in the projected 2D space. Therefore, the cluster results of the cluster miner could have higher quality.

We draw the conclusion that CPM can enhance the discriminativeness of target-domain feature representations and make those instances distribute as clusters. Therefore, when the cluster miner generates pseudo labels for unlabeled target-domain data according to the features encoded by the model trained with CPM, the pseudo labels could have higher confidence, which may provide more useful target-domain information to improve the domain adaption ability of the few-shot classifier.

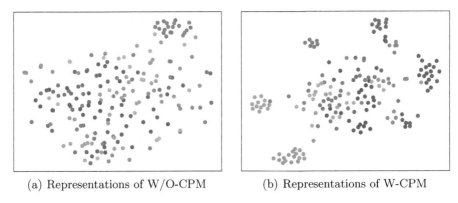

(a) Representations of W/O-CPM   (b) Representations of W-CPM

Fig. 3. A t-SNE plot of the computed feature representations of target-domain instances in the FewRel 2.0 training set. Node colors denote classes. The difference between (b) and (a) is whether the representation extractor is trained with CPM or not. (Color figure online)

6.3   Error Analysis

Although our method achieves state-of-the-art results, we still observe some phenomena which could cause failures. From Fig. 3(b), we find that five kinds of colored instances, red (*ingredient_of*), grey (*causative_agent_of*), yellow (*classified_as*), cyan (*gene_plays_role_in_process*) and blue (*biological_process-_involves_gene_product*), fail to show the cluster structure. It reveals that even trained with CPM, the representation extractor still cannot produce discriminative features for all the target-domain data. In other words, some instances belonging to different classes may have similar representations. As a result, the pseudo labels produced by the cluster miner inevitably contain noise that some instances from different classes are assigned the same pseudo label, while other instances from the same class could have distinct labels, which limits the few-shot classifier as its performance on the target domain is determined by the quality of the pseudo labels.

7   Conclusion

In this paper, we study the problem of inductive unsupervised domain adaption in the few-shot classification. We first train a representation extractor with the Clustering Promotion Mechanism. It uses Similarity Entropy Minimization to promote clustering and Adversarial Distribution Alignment to align similar class distribution across domains. Two methods are combined by the proposed Cosine Annealing Strategy. The representation extractor is used to encode unlabeled target-domain data into features, which are passed to a k-means cluster miner to generate pseudo labels. Finally, we utilize pseudo-labeled target-domain data and labeled source-domain data to train the few-shot classifier. Experimental results demonstrate that our approach achieves new state-of-the-art on FewRel 2.0 dataset. In the future, we will work on how to reduce the noise of pseudo labels to improve the domain adaption performance.

Acknowledgements. We would like to thank reviewers for their insightful comments. This work is supported in part by the National Key Research and Development Program of China (grant No. 2016YFB0801003), the Strategic Priority Research Program of Chinese Academy of Sciences (grant No. XDC02040400) and grant No. BMKY2019B04-1.

References

1. Chapelle, O., Scholkopf, B., Zien, A.: Semi-supervised learning (Chapelle, O. et al. eds.; 2006) [book reviews]. IEEE Trans. Neural Netw. **20**(3), 542 (2009)
2. Davies, D.L., Bouldin, D.W.: A cluster separation measure. IEEE Trans. Pattern Anal. Mach. Intell. **2**, 224–227 (1979)
3. Devlin, J., Chang, M., Lee, K., Toutanova, K.: BERT: pre-training of deep bidirectional transformers for language understanding. In: NAACL-HLT (2019)
4. Dong, N., Xing, E.P.: Domain adaption in one-shot learning. In: Berlingerio, M., Bonchi, F., Gärtner, T., Hurley, N., Ifrim, G. (eds.) ECML PKDD 2018. LNCS (LNAI), vol. 11051, pp. 573–588. Springer, Cham (2019). https://doi.org/10.1007/978-3-030-10925-7_35
5. Finn, C., Abbeel, P., Levine, S.: Model-agnostic meta-learning for fast adaptation of deep networks. In: Proceedings of the 34th International Conference on Machine Learning, ICML 2017 (2017)
6. Fowlkes, E.B., Mallows, C.L.: A method for comparing two hierarchical clusterings. J. Am. Stat. Assoc. **78**(383), 553–569 (1983)
7. Ganin, Y., et al.: Domain-adversarial training of neural networks. J. Mach. Learn. Res. **17**, 59:1–59:35 (2015)

8. Gao, T., et al.: FewRel 2.0: towards more challenging few-shot relation classification. In: EMNLP (2019)
9. Han, B., et al.: Co-teaching: robust training of deep neural networks with extremely noisy labels. In: Advances in Neural Information Processing Systems, pp. 8527–8537 (2018)
10. Long, M., Zhu, H., Wang, J., Jordan, M.I.: Unsupervised domain adaptation with residual transfer networks. In: Lee, D.D., Sugiyama, M., von Luxburg, U., Guyon, I., Garnett, R. (eds.) NeurIPS (2016)
11. MacQueen, J., et al.: Some methods for classification and analysis of multivariate observations. In: Proceedings of the fifth Berkeley symposium on mathematical statistics and probability, Oakland, CA, USA, vol. 1, pp. 281–297 (1967)
12. Munkhdalai, T., Yu, H.: Meta networks. In: Proceedings of the 34th International Conference on Machine Learning, ICML 2017 (2017)
13. Nichol, A., Achiam, J., Schulman, J.: On first-order meta-learning algorithms. CoRR abs/1803.02999 (2018). http://arxiv.org/abs/1803.02999
14. Pennington, J., Socher, R., Manning, C.: GloVe: global vectors for word representation. In: Proceedings of the 2014 Conference on Empirical Methods in Natural Language Processing (EMNLP), pp. 1532–1543 (2014)
15. Santoro, A., Bartunov, S., Botvinick, M., Wierstra, D., Lillicrap, T.: Meta-learning with memory-augmented neural networks. In: International Conference on Machine Learning, pp. 1842–1850 (2016)
16. Snell, J., Swersky, K., Zemel, R.: Prototypical networks for few-shot learning. In: Advances in Neural Information Processing Systems, pp. 4077–4087 (2017)
17. Sung, F., Yang, Y., Zhang, L., Xiang, T., Torr, P.H.S., Hospedales, T.M.: Learning to compare: relation network for few-shot learning. In: 2018 IEEE Conference on Computer Vision and Pattern Recognition, CVPR 2018 (2018)
18. Tzeng, E., Hoffman, J., Saenko, K., Darrell, T.: Adversarial discriminative domain adaptation. In: 2017 IEEE Conference on Computer Vision and Pattern Recognition, CVPR 2017 (2017)
19. Tzeng, E., Hoffman, J., Saenko, K., Darrell, T.: Adversarial discriminative domain adaptation. In: 2017 IEEE Conference on Computer Vision and Pattern Recognition (CVPR), pp. 2962–2971 (2017)
20. Vinyals, O., Blundell, C., Lillicrap, T., Wierstra, D., et al.: Matching networks for one shot learning. In: Advances in Neural Information Processing Systems, pp. 3630–3638 (2016)
21. Zeng, D., Liu, K., Lai, S., Zhou, G., Zhao, J.: Relation classification via convolutional deep neural network. In: COLING (2014)
22. Zou, H., Zhou, Y., Yang, J., Liu, H., Das, H.P., Spanos, C.J.: Consensus adversarial domain adaptation. In: The Thirty-Third AAAI Conference on Artificial Intelligence, AAAI 2019 (2019)

Unsupervised Domain Adaptation with Joint Domain-Adversarial Reconstruction Networks

Qian Chen, Yuntao Du, Zhiwen Tan, Yi Zhang, and Chongjun Wang[✉]

State Key Laboratory for Novel Software Technology, Nanjing University,
Nanjing, China
{chenqian,duyuntao,njuzhangyi}@smail.nju.edu.cn,
yaoyueduzhen@outlook.com, chjwang@nju.edu.cn

Abstract. Unsupervised Domain Adaptation (UDA) attempts to transfer knowledge from a labeled source domain to an unlabeled target domain. Recently, domain-adversarial learning has become an increasingly popular method to tackle this task, which bridges source domain and target domain by adversarially learning domain-invariant representations that cannot be discriminated by a domain discriminator. In spite of the great success achieved by domain-adversarial learning, most of existing methods still suffer two major limitations: (1) due to focusing only on learning domain-invariant representations, they ignore the individual characteristics of each domain and fail to extract domain-specific information that is beneficial for final classification; (2) by focusing only on performing domain-level distribution alignment to learn domain–invariant representations, they fail to achieve the invariance of representations at a class level, which may lead to incorrect distribution alignment. To address the above issues, we propose in this paper a novel model called *Joint Domain-Adversarial Reconstruction Network* (JDARN), which integrates domain-adversarial learning with data reconstruction to learn both domain–invariant and domain-specific representations. Meanwhile, we propose to employ two novel discriminators called *joint domain-class discriminators* to achieve the joint alignment and adopt a novel joint adversarial loss to train them. With both domain and class information of two domains, the two discriminators can be used to promote domain-invariant representation learning towards the class level, not only the domain level. Extensive experimental results reveal that the proposed JDARN exceeds the state-of-the-art performance on two standard UDA datasets.

Keywords: Unsupervised domain adaptation · Domain-adversarial learning · Data reconstruction · Distribution alignment

1 Introduction

Deep learning methods have achieved great success in many fields, such as computer vision and nature language process. The access to massive amounts of

F. Hutter et al. (Eds.): ECML PKDD 2020, LNAI 12458, pp. 640–656, 2021.
https://doi.org/10.1007/978-3-030-67661-2_38

labeled training data is one of the reasons for such success. Generally, deep learning methods usually train deep neural networks with a large scale labeled training dataset, and then test on a testing dataset, which has similar distribution as the training one. Nevertheless, training datasets are usually either difficult to collect, or prohibitive to annotate. Meanwhile, due to dataset bias [35] or domain shift [29], traditional deep models do not generalize well on new datasets or tasks.

Unsupervised Domain Adaptation tackles the aforementioned problems by transferring knowledge from a label-rich source domain to a label-scarce target domain whose distribution is different from the source one [25,26]. The deep UDA methods [6,19,21,32,37] have achieved remarkable performance, which leverage deep networks to learn transferable representations by embedding adaptation modules in deep architectures.

Recently, adversarial learning has been successfully embedded into deep networks to learn domain-invariant representations to reduce distribution discrepancy between source domain and target domain. Inspired by generative adversarial networks (GAN) [9], domain–adversarial learning [6] pits two networks against each other—feature extractor and domain discriminator. It plays a minimax game to learn the domain discriminator that aims to distinguish feature representations of source samples from those of target samples, and the feature extractor that aims to learn domain-invariant representations to fool the domain discriminator. Theoretically, domain alignment is achieved when the minimax optimization reach an equilibrium.

In spite of the great success achieved by domain-adversarial learning, most prior efforts still suffer two major limitations. Firstly, most of them only concentrate on learning domain–invariant representations, but ignore the individual characteristics of each domain, and are limited in extracting domain-specific information. Since domain-specific representations not only preserve the discriminability, but also contains more meaningful information of each domain, they are beneficial for the final classification. Secondly, most of them only focus on aligning the marginal distributions of two domains, which is referred as domain-level distribution alignment, but ignore the alignment of class conditional distributions across domains, which is referred as class-level distribution alignment. A perfect domain-level distribution alignment does not imply a fine-grained class-to-class overlap. With an intuitive example shown in Fig. 1, with domain-level alignment only, the learned domain–invariant representations may not only bring source domain and target domain closer, but also mix samples with different class labels together, which may cause false classification. The lack of class-level distribution alignment is a major cause of performance reduction [24]. Thus, it is necessary to pursue the class-level and domain-level distribution alignments simultaneously under the absence of target true labels.

For the first issue, we note that some UDA methods based on encoder–decoder reconstruction [3,8] use data reconstruction of source or target domain samples as an auxiliary task, which typically learn domain-invariant representations by a shared encoder and maintain domain-specific representations

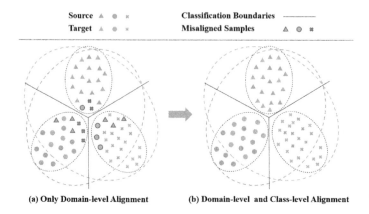

Source ▲ ● ×   Classification Boundaries ――――
Target ▲ ● ×   Misaligned Samples ▲ ● ×

(a) Only Domain-level Alignment   (b) Domain-level and Class-level Alignment

Fig. 1. Left: Only with domain-level alignment, there may exists some misaligned samples in the target domain, which will cause false classification. **Right:** With both domain-level and class-level alignment, the classification performance will be better.

by a reconstruction loss [38]. Inspired by these methods, we integrate the domain-adversarial learning with the data reconstruction to learn domain–invariant and domain-specific representations simultaneously. Besides, most non-reconstruction methods map target domain samples to source domain in the deep feature space, or map source and target domain samples to a common deep feature space, which can not promise that the structure of feature space is not distorted after mapped. On the contrary, reconstruction loss enforces the data decoded from latent representations as close to original data as possible, which is helpful for maintaining the structure of feature space and making it undistorted. In this paper, we choose variational autoencoder (VAE) [14] as the basic encoder–decoder reconstruction model. VAE is a directed graphical model with certain types of latent variables, such as Gaussian latent variables. To some degree, using VAE as the basic reconstruction model is also helpful for alleviating the distribution discrepancy between domains, since the distributions of latent representations form two domains are forced to be close to a same certain distribution by VAE.

For the second issue, to promote domain-invariant representations learning towards the class level, we propose a novel domain-adversarial learning paradigm in this paper. In previous adversarial methods, the output of the domain discriminator is a layer with 2 nodes that can only indicate the domain label of one feature representation. Different from them, in our method, the output of the discriminator is a layer with $K + 1$ (K is the number of classes) nodes, where the first K nodes represent the class of one feature representation and the last node indicates the domain of the representation. We call this new discriminator a *joint domain-class discriminator*, since it learns a joint distribution over both domain and class variables. In our model, we employ two joint domain-class discriminators in our model, and adopt a novel joint adversarial loss to train them, instead of a simple binary adversarial loss. Class labels are needed when calculating the loss, but they are not available in the target domain. Some existing methods [20, 28, 39] use pseudo-labels to make up for the lack of target domain class labels. Following this idea, we first train a source classifier using labeled source data to initialize a target classifer, and then update the target classifer by a semi-supervised learning (SSL) style [23] to provide target domain pseudo-labels.

Briefly, we summarize our contributions below.

- We propose in this paper a novel model termed Joint Domain-Adversarial Reconstruction Network for unsupervised domain adaptation. Our proposed JDARN integrates domain-adversarial learning with data reconstruction to learn both domain-invariant and domain-specific representations.
- In order to achieve the invariance of representations at the domain level and class level simultaneously, we introduce two novel discriminators called joint domain-class discriminators to our proposed model and adopt a joint adversarial loss to train them.
- We conduct careful experiments to investigate the efficacy of our proposed model. Based on commonly used basic networks (ResNet-50 [11]), our proposed model outperforms the state-of-the-art methods on two benchmark domain adaptation datasets.

2 Related Works

In this section, we briefly review recent domain adaptation methods, in particular domain-adversarial learning methods and those based on data reconstruction.

Inspired by GANs, adversarial learning has been widely adopted in domain adaptation [6, 7, 13, 18, 36, 37]. The domain-adversarial neural network (DANN) [6] uses adversarial training to find domain–invariant representations. DANN minimizes the domain confusion loss for all samples and label prediction loss only for source samples while maximizing domain confusion loss via a gradient reversal layer (GRL). The adversarial discriminative domain adaptation (ADDA) [37] summarizes that each adversarial method makes three design choices: whether to use a generative or discriminative base model, whether to tie or untie the weights, and which adversarial learning objective to use. The ADDA uses a discriminative base model without the generator and unshared weights that can better model the difference in low level features than shared ones. Besides, it use a standard GAN loss function, which has the same fixed-point properties as the minimax loss but provides stronger gradients to the target mapping.

Some recent adversarial UDA methods [20, 28, 31, 33, 39, 40] have paid attention to pursuing the class-level alignment. The moving semantic transfer network (MSTN) [39] learns semantic representations for unlabeled target samples by aligning labeled source centroid and pseudo-labeled target centroid. The multi-adversarial domain adaptation (MADA) [28] and the conditional domain adversarial network (CDAN) [20] exploit multiplicative interactions between feature representations and category predictions as highorder features to help adversarial training.

In another line of UDA methods, data reconstruction is used as an auxiliary task that simultaneously learns shared representations between domains and keeps the individual characteristics of each domain [3,8,12]. The deep reconstruction classification network (DRCN) [8] learns a shared encoding representation that provides useful information for cross-domain object recognition. DRCN minimizes the classification loss with labeled source data and the unsupervised reconstruction loss with unlabeled target data. The domain separation network (DSN) [3] explicitly and jointly models both private and shared components of the domain representations to reconstruct the images from both domains.

3   Method

3.1   Problem Setting

For clear description, we give some definitions. In unsupervised domain adaptation, we have access to a source domain $\mathcal{D}_s = \{(x_s^{(i)}, y_s^{(i)})\}_{i=1}^{N_s}$ of N_s labeled samples from $\mathcal{X}_s \times \mathcal{Y}_s$ and a target domain $\mathcal{D}_t = \{(x_t^{(i)})\}_{i=1}^{N_t}$ of N_t unlabeled samples from $\mathcal{X}_t$, where $\mathcal{X}$ represents the feature space and $\mathcal{Y}$ represents the label space. The source and target domain samples are sampled from joint distributions $P_s(x_s, y_s)$ and $P_t(x_t, y_t)$ respectively, and the i.i.d. assumption is violated as $P_s \neq P_t$. The goal of UDA is to train a deep network $f : x \to y$ which formally reduces the shifts in the data distributions across domains, such that the target error $\epsilon_t(f) = \mathbb{E}_{(x_t, y_t) \sim P_t}[f(x_t) \neq y_t]$ can be bounded by the sum of the source error $\epsilon_s(f) = \mathbb{E}_{(x_s, y_s) \sim P_s}[f(x_s) \neq y_s]$ and the distribution discrepancy $d(P_s, P_t)$.

3.2   Network Structure

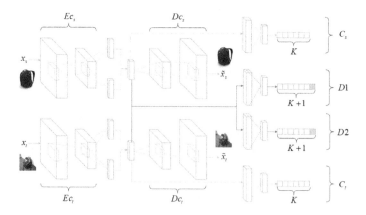

Fig. 2. Overall structure of our proposed JDARN

Our proposed JDARN is equipped with two VAEs, two joint domain-class discriminators and two task-specific classifiers. An overview of the proposed model is shown in Fig. 2. The ADDA [37] suggests that allowing independent source and target mappings is a more flexible learning paradigm as it allows more domain specific feature extraction to be learned. Therefore, following this idea, we employ two separate VAEs, one for source domain, and another for target domain. The encoders of VAEs, denoted as Ec_s for source and Ec_t for target, encode source domain or target domain samples x into latent representations $z = Ec(x)$, which aims to learn domain-invariant representations to fool the two discriminators. The decoders of VAEs, denoted as Dc_s for source and Dc_t for target, decode the latent representations z back to a reconstructed version, which aims to extract more domain-specific information and make the structure of source and target domain feature space undistorted. The two joint domain-class discriminators, denoted as $D1$ and $D2$, are not only trained to distinguish whether latent representations are from source domain or target domain, but also trained to discriminate the class of the latent representations. The two task-specific classifiers, denoted as C_s for source and C_t for target, are trained to classify the samples from two domains.

3.3   Loss Functions

In this paper, our final goal is to learn a target encoder Ec_t, and a classifier C_t that can classify target domain smaples into one of K categories correctly when testing. Due to lack of target domain class labels, it is impossible to directly learn a target encoder and a target classifier. Thereby, we instead train a source encoder, Ec_s, along with a source classifier, C_s, and then adapt that model for use in the target domain.

Classification Loss. First of all, Ec_s and C_s are pre-trained using a standard supervised loss as follows:

$$\mathcal{L}_{cls} = \mathbb{E}_{(x_s, y_s) \sim \mathcal{D}_s} \ell_{ce}(C_s(Ec_s(x_s)), y_s), \tag{1}$$

where $\ell_{ce}(f(x), y) := -\langle y, \log f(x) \rangle$ is the cross-entropy loss calculated for one-hot, ground-truth label $y \in \{0, 1\}^K$ and label prediction $f(x) \in \mathbb{R}^K$ (the output of a deep neural network with input x).

Reconstruction Loss. In this paper, data reconstruction of source or target samples is used as an auxiliary task to improve our model's ability to extract the domain-specific representations. Specifically, we use VAE as the basic reconstruction model, which first encodes a sample x to a latent representation z, and then decodes the latent representation back to data space $\tilde{x}$:

$$z \sim q(z|x) = Ec(x), \quad \tilde{x} \sim p(x|z) = Dc(z). \tag{2}$$

The VAE imposes a prior over the latent distribution $p(z)$ to regularize the encoder. Typically $z \sim N(0, I)$ is chosen. A variational lower bound of the log-likelihood is used as a surrogate objective function:

$$\log p(\boldsymbol{x}) \geq -D_{KL}[q(\boldsymbol{z}|\boldsymbol{x})||p(\boldsymbol{z})] + \mathbb{E}_{q(\boldsymbol{z}|\boldsymbol{x})}[\log p(\boldsymbol{z}|\boldsymbol{x})], \tag{3}$$

where D_{KL} is the Kullback-Leibler divergence. Thus, the VAE loss is minus the sum of the expected log likelihood (the reconstruction error) and a prior regularization term:

$$\mathcal{L}_{vae} = \mathcal{L}_{rec} + \mathcal{L}_{prior} = -\mathbb{E}_{q(\boldsymbol{z}|\boldsymbol{x})}[\log p(\boldsymbol{z}|\boldsymbol{x})] + D_{KL}[q(\boldsymbol{z}|\boldsymbol{x})||p(\boldsymbol{z})]. \tag{4}$$

With the VAE loss, we can not only extract more domain-specific information of each domain, but also allow the structure of source and target domain feature space to be undistorted. Moreover, as the distributions of latent representations form two domains are all forced to be close to a same certain distribution by VAE, the distribution discrepancy between domains might be alleviated to some degree.

Joint Adversarial Loss. As we know, in GAN, the generator G and the discriminator D are trained alternately. Similarly, domain-adversarial learning methods optimize the domain discriminator D with the feature extractor F fixed, and then optimize F with D fixed. Specifically, D and F are optimized alternately according to an adversarial objective, which is shown as follows:

$$\min_{D} \mathcal{L}_{adv-D}, \min_{F} \mathcal{L}_{adv-F}. \tag{5}$$

There are various different adversarial loss functions to choose, each of which has their own unique use cases. All adversarial losses train the D using a standard supervised loss:

$$\mathcal{L}_{adv-D} = \mathbb{E}_{\boldsymbol{x}_s \sim \mathcal{X}_s} \ell_{ce}(D(F(\boldsymbol{x}_s)), [1, 0]) + \mathbb{E}_{\boldsymbol{x}_t \sim \mathcal{X}_t} \ell_{ce}(D(F(\boldsymbol{x}_t)), [0, 1]), \tag{6}$$

where $[1, 0]$ and $[0, 1]$ are the domain labels. However, the loss used to train the F is different. Here, we only take ADDA as an example. It trains the F according to a standard loss function called GAN loss function with inverted labels, which is shown below:

$$\mathcal{L}_{adv-F} = \mathbb{E}_{\boldsymbol{x}_t \sim \mathcal{X}_t} \ell_{ce}(D(F(\boldsymbol{x}_t)), [1, 0]). \tag{7}$$

These prior efforts only align the feature distributions across domains at the domain level, but ignore the class-level alignment. To tackle the issue, we introduce a joint domain-class discriminator whose output is a layer with $K + 1$ nodes and employ a joint adversarial loss to train the target model, instead of a simple binary adversarial loss above.

To be specific, we first train a joint domain-class discriminator $D1$ according to a standard supervised loss:

$$\mathcal{L}_{adv-D1} = \mathbb{E}_{(\boldsymbol{x}_s, y_s) \sim \mathcal{D}_s} \ell_{ce}(D1(Ec_s(\boldsymbol{x}_s)), d_s^{(1)}) + \mathbb{E}_{\boldsymbol{x}_t \sim \mathcal{X}_t} \ell_{ce}(D1(Ec_t(\boldsymbol{x}_t)), d_t^{(1)}). \tag{8}$$

$d_s^{(1)}$ and $d_t^{(1)}$ are defined as follows:

$$d_s^{(1)} = \overbrace{[0,...,0,1,0,...,0}^{K},0], d_t^{(1)} = \overbrace{[0,...,0,0,0,...,0}^{K},1], \qquad (9)$$

where y_s is the one-hot, ground-truth source class label. Obviously, the discriminator $D1$ trained by minimizing Eq. 8 cannot distinguish the class of target domain samples, because the first K dimensions of $d_t^{(1)}$ are set to zero. Thereby, we introduce another joint domain-class discriminator $D2$, which is also trained according to a standard supervised loss:

$$\mathcal{L}_{adv-D2} = \mathbb{E}_{(\boldsymbol{x}_s,y_s)\sim\mathcal{D}_s}\ell_{ce}(D2(Ec_s(\boldsymbol{x}_s)),d_s^{(2)}) + \mathbb{E}_{\boldsymbol{x}_t\sim\mathcal{X}_t}\ell_{ce}(D2(Ec_t(\boldsymbol{x}_t)),d_t^{(2)}). \qquad (10)$$

Inversely, $d_s^{(2)}$ and $d_s^{(2)}$ are defined as follows:

$$d_s^{(2)} = \overbrace{[0,...,0,0,0,...,0}^{K},1], d_t^{(2)} = \overbrace{[0,...,0,1,0,...,0}^{K},0]. \qquad (11)$$

where $\hat{y}_t$ is the pseudo-label predicted by the target classifier. In this way, the two joint domain-class discriminators complement each other to align the marginal and class conditional feature distributions of two domains at the same time.

Similar to ADDA above, the target encoder Ec_t is adversarially optimized according to the loss $\mathcal{L}_{adv-Ec}$ with inverted labels to learn domain-invariant representation to fool the joint domain-class discriminators:

$$\mathcal{L}_{adv-Ec} = \mathbb{E}_{\boldsymbol{x}_t\sim\mathcal{X}_t}\ell_{ce}(D1(Ec_t(\boldsymbol{x}_t)),[\hat{y}_t,0]) + \mathbb{E}_{\boldsymbol{x}_t\sim\mathcal{X}_t}\ell_{ce}(D2(Ec_t(\boldsymbol{x}_t)),[\mathbf{0},1]), \qquad (12)$$

where $\mathbf{0}$ is a zero vector of size K. The joint adversarial loss, $\mathcal{L}_{adv-D1}$, $\mathcal{L}_{adv-D2}$ and $\mathcal{L}_{adv-Ec}$, contains both domain and class information, so the marginal and class conditional feature distributions of two domains can be aligned.

Entropy Loss. After aligning the feature distributions of two domains, UDA can be regarded as a semi-supervised learning(SSL) problem. The large amount of unlabeled target domain samples can be used to bias the decision boundaries to be in the low-density regions [16]. Applying entropy minimization for classifier training can push the decision boundaries away from the high density regions, a desired property under the cluster assumption [4]. Therefore, our method trains the target classifier C_t according to the following entropy loss:

$$\mathcal{L}_{etp} = \mathbb{E}_{\boldsymbol{x}_t\sim\mathcal{X}_t}\ell_e(C_t(Ec_t(\boldsymbol{x}_t))), \qquad (13)$$

where $\ell_e(f(\boldsymbol{x})) := -\langle f(\boldsymbol{x}), \log f(\boldsymbol{x})\rangle$. However, entropy minimization satisfies the cluster assumption only for Lipschitz classifiers [10]. Fortunately, the Lipschitz condition can be realized by virtual adversarial training(VAT) as suggested by [23]. Therefore, as with most previous methods, we apply entropy mimimization in conjunction with VAT on the target domain samples. The VAT loss is given as follows:

$$\mathcal{L}_{vat} = \mathbb{E}_{\boldsymbol{x}_t \sim \mathcal{X}_t} [\max_{\|r\| \leq \epsilon} D_{KL}(C_t(Ec_t(\boldsymbol{x}_t)) \| C_t(Ec_t(\boldsymbol{x}_t + r)))]. \tag{14}$$

The above SSL regularizations can be applied after aligning the feature distributions of two domains [34]. But, we find that applying SSL regularizations at the beginning of the training also works well.

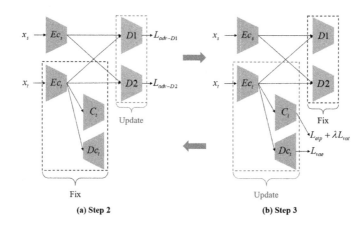

(a) Step 2   **(b) Step 3**

Fig. 3. Adversarial training steps of JDARN

3.4   Training Steps

In this section, we combine the loss functions mentioned above to train all the networks including two VAEs, two joint domain-class discriminators and two classifiers, and summarize the whole training step. Specifically, the training of JDARN is divided into three steps:

Step 1. First of all, the source VAE and source classifier are trained using all the labeled source domain samples according to the classification loss and the VAE loss together. The objective is written as follows:

$$\min_{Ec_s, Dc_s, C_s} \mathcal{L}_{cls} + \lambda_{vae}\mathcal{L}_{vae}, \tag{15}$$

where λ_{vae} is a trade-off parameter. Due to lack of target domain annotations, we use the pre-trained source model as an initialization for the target model, and then fix the source model at the later steps.

Step 2. In this step, the two joint domain-class discriminators are trained according to the joint adversarial loss to discriminate the class and domain of one latent representation, with the target VAE and classifier fixed. The objective is given as follows:

$$\min_{D1} \mathcal{L}_{adv-D1}, \min_{D2} \mathcal{L}_{adv-D2}. \tag{16}$$

Step 3. In this step, the target VAE is trained according to the joint adversarial loss to learn the domain–invariant representations to fool the discriminators, with the two discriminators fixed. In addition, VAE loss is added to the objective to help the target VAE extract more domain-specific information. The target classifier is also optimized in this step according to entropy loss and VAT loss, which can push the decision boundaries away from the high density regions. The final objective is given as follows:

$$\min_{Ec_t, Dc_t, C_t} \mathcal{L}_{adv-Ec} + \lambda_{vae}\mathcal{L}_{vae} + \lambda_t(\mathcal{L}_{etp} + \lambda_{vat}\mathcal{L}_{vat}), \tag{17}$$

where λ_{vae}, λ_t, and λ_{vat} are trade-off parameters.

Step 2 and **Step 3** are conducted in an alternating fashion, as shown in Fig. 3.

3.5   Connection to Domain Adaptation Theory

Most adversarial domain adaptation methods are based on the theoretical analysis proposed in [1,2], which states that the target error ϵ_t is bounded by four terms:

Theorem 1. *Let $\mathcal{H}$ be a hypothesis space of VC dimension d. If $\mathcal{U}_s$, $\mathcal{U}_t$ are unlabeled samples of size m each, drawn from P_s and P_t respectively, then for any $\delta \in (0,1)$, with probability at least $1 - \delta$, for every $h \in \mathcal{H}$,*

$$\epsilon_t(h) \leq \epsilon_s(h) + \frac{1}{2}d_{\mathcal{H}\Delta\mathcal{H}}(P_s, P_t) + \lambda$$
$$\leq \epsilon_s(h) + \frac{1}{2}\hat{d}_{\mathcal{H}\Delta\mathcal{H}}(\mathcal{U}_s, \mathcal{U}_t) + 4\sqrt{\frac{2d\log(2m) + \log(\frac{2}{\delta})}{m}} + \lambda, \tag{18}$$

where $\epsilon_s(h)$ is the expected error on the source domain samples that can be minimized easily with source class label information and $\lambda = \min_h[\epsilon_s(h) + \epsilon_t(h)]$ is combined error of ideal joint hypothesis. $d_{\mathcal{H}\Delta\mathcal{H}}(P_s, P_t)$ denotes the $\mathcal{H}\Delta\mathcal{H}$-distance between two domains, which is written as follows:

$$d_{\mathcal{H}\Delta\mathcal{H}}(P_s, P_t) = 2 \sup_{h,h'\in\mathcal{H}} |Pr_{x\sim P_s}[h(x) \neq h'(x)] - Pr_{x\sim P_t}[h(x) \neq h'(x)]|. \tag{19}$$

$\hat{d}_{\mathcal{H}\Delta\mathcal{H}}(\mathcal{U}_s, \mathcal{U}_t)$ is the empirical $\mathcal{H}\Delta\mathcal{H}$-distance.

Let $F_s\#P_s$ and $F_t\#P_t$ be the marginal push-forward distributions of source and target domain distribution P_s and P_t, where F_s and F_t are the feature extractors (encoders in our model). If $F_s\#P_s$ and $F_t\#P_t$ are aligned well, the $d_{\mathcal{H}\Delta\mathcal{H}}$ will vanish for any space $\mathcal{H}$ of sufficiently smooth hypotheses. However, if the corresponding class conditional distributions $F_s\#P_s(\cdot|y)$ and $F_t\#P_t(\cdot|y)$ are aligned incorrectly, λ will be large because there may not be a $h \in \mathcal{H}$ with low error in both domains. Since our proposed model aligns the marginal and class conditional distributions by the two joint domain-class discriminators and the joint adversarial loss, the above problem can be tackled well. Furthermore, as m increases, the third term will decrease. The m can increase with data augmentation. Therefore, using VAT that has the same effect of augmenting data can make the third term decrease.

4 Experiments

In this section, we evaluate our proposed model with some state-of-the-art deep UDA methods and demonstrate promising results on UDA benchmark tasks.

4.1 Experimental Setup

Office-31 [30] is a standard benchmark dataset for domain adaptation, which consists of 4,110 images spread across 31 classes in 3 domains: Amazon (**A**), Webcam (**W**), and DSLR (**D**). We focus our evaluation on six transfer tasks: **A → D, A → W, D → A, D → W, W → A** and **W → D**.

ImageCLEF-DA[1] is a benchmark dataset for ImageCLEF 2014 domain adaptation challenge, which contains three domains: Caltech-256 (**C**), ImageNet ILSVRC 2012 (**I**), and Pascal VOC 2012 (**P**). For each domain, there are 12 classes and 50 images in each class. We evaluate all methods on six transfer tasks: **C → I, C → P, I → C, I → P, P → C** and **P → I**.

We compare our proposed model with the following state-of-the-art UDA methods: Deep Adaptation Network (**DAN**) [19], Domain Adversarial Neural Network (**DANN**) [6], Adversarial Discriminative Domain Adaptation (**ADDA**) [37], Virtual Adversarial Domain Adaptation (**VADA**) [34], Maximum Classifier Discrepancy (**MCD**) [32], Conditional Domain Adversarial Network (**CDAN**) [20], and Transferable Adversarial Training (**TAT**) [17]. Besides, as a basic model, ResNet-50 [11] is also used for comparison.

We implement our model in PyTorch [27][2]. Following the commonly used experimental protocol for UDA [6,20], we use all labeled source samples and all unlabeled target samples. We repeat each domain adaptation task three times, and then compare the mean classification accuracy. ResNet-50 [11] is adopted as the based model with parameters fine-tuned from the model pre-trained on ImageNet [15].

For our proposed model, the dimension of the latent representation is set to 256; the classifier is a one-layer fully connected network (256 → K), where K is the number of classes; the discriminator consists of three fully connected layers with ReLU (256 → 3072 → 2048 → 1024 → $K + 1$). In all the tasks, the

---

[1] https://www.imageclef.org/2014/adaptation.
[2] Codes are available at https://github.com/NaivePawn/JDARN.

mini-batch stochastic gradient descent (SGD) with momentum of 0.9 is adopted. During pre-training (**Step 1**), batchsize is set to 32; learning rate is set to 10^{-4} for based ResNet-50, and 10^{-3} for the other new layers of source VAE and source classifier that are trained from scratch. During adaptation (**Step 2&3**), batchsize is set to 16; learning rate is set to 10^{-3} for two joint domain-class discriminators, and 10^{-5} for target VAE and classifier.

For hyperparameters λ_{vae} (in Eq. 15 and Eq. 17) and λ_t (in Eq. 17), we fix their value to 0.1. For hyperparameter λ_{vat}(in Eq. 17), we do search over the grid $\{1.0, 10.0\}$, which is suggested by [5]. We also search for the upper bound of the adversarial perturbation in VAT, $\epsilon \in \{0.1, 0.5, 1.0, 2.0, 4.0, 8.0\}$ (in Eq. 14), as in [5]. We train the models within 200 epochs for pre-training and 2000 epochs for adaptation.

Table 1. Classification accuracies (%) on Office-31 for unsupervised domain adaptation with ResNet-50

Method	A→D	A→W	D→A	D→W	W→A	W→D	Avg
RESNET-50 [11]	68.9 ± 0.2	68.4 ± 0.2	62.5 ± 0.3	96.7 ± 0.1	60.7 ± 0.3	99.3 ± 0.1	76.1
DAN [19]	78.6 ± 0.2	80.5 ± 0.4	63.6 ± 0.3	97.1 ± 0.2	62.8 ± 0.2	99.6 ± 0.1	80.4
DANN [6]	81.5 ± 0.4	82.6 ± 0.4	68.4 ± 0.5	96.9 ± 0.2	67.5 ± 0.5	99.3 ± 0.2	82.7
ADDA [37]	77.8 ± 0.3	86.2 ± 0.5	69.5 ± 0.4	96.2 ± 0.3	68.9 ± 0.5	98.4 ± 0.3	82.9
VADA [34]	86.7 ± 0.4	86.5 ± 0.5	70.1 ± 0.4	98.2 ± 0.4	70.5 ± 0.4	99.7 ± 0.2	85.4
MCD [32]	92.2 ± 0.2	88.6 ± 0.2	69.5 ± 0.1	98.5 ± 0.1	69.7 ± 0.3	**100.0 ± .0**	86.5
CDAN [20]	92.9 ± 0.2	94.1 ± 0.1	71.0 ± 0.3	98.6 ± 0.1	69.3 ± 0.3	**100.0 ± .0**	87.7
TAT [17]	93.2 ± 0.2	92.5 ± 0.3	73.1 ± 0.3	**99.3 ± 0.1**	72.1 ± 0.3	**100.0 ± .0**	88.4
JDARN	**93.5 ± 0.2**	**94.5 ± 0.2**	**74.2 ± 0.1**	98.9 ± 0.1	**72.9 ± 0.1**	**100.0 ± .0**	**89.0**

4.2 Results

We now discuss the experiment results. The results of baselines are directly reported from the original papers if protocol is the same.

Table 1 shows the results on office-31 dataset based on ResNet-50. The proposed model significantly outperforms all comparison methods on most transfer tasks. It is desirable that our model improves the accuracies on four hard transfer task: **A → D, A → W, D → A** and **W → A**, where the source and target domains are substantially different. Meanwhile, our model produces comparable classification accuracies on easy transfer tasks: **D → W** and **W → D**, where the source and target domain are similar.

Table 2 shows the results on ImageCLEF-DA dataset based on ResNet-50. Different from Office-31 where different domains are of different sizes, the three domains in ImageCLEF-DA are balanced, which makes it a good complement to Office-31 for more controlled experiments and avoids the class imbalance problems. Our model outperforms all comparison methods on most transfer tasks, especially on tasks **P → C** and **P → I**.

Table 2. Classification accuracies (%) on Image-CLEF for unsupervised domain adaptation with ResNet-50

Method	C→I	C→P	I→C	I→P	P→C	P→I	Avg
RESNET-50 [11]	78.0±0.2	65.5±0.3	91.5±0.3	74.8±0.3	91.2±0.3	83.9±0.1	80.7
DAN [19]	86.3±0.4	69.2±0.4	92.8±0.2	74.5±0.4	89.8±0.4	82.2±0.2	82.5
DANN [6]	87.0±0.5	74.3±0.5	96.2±0.4	75.0±0.3	91.5±0.6	86.0±0.3	85.0
CDAN [20]	91.3±0.3	74.2±0.2	**97.7±0.3**	77.7±0.3	94.3±0.3	90.7±0.2	87.7
TAT [17]	92.0±0.4	**78.2±0.4**	97.5±0.3	78.8±0.2	94.7±0.4	92.0±0.2	88.9
JDARN	**92.1±0.1**	77.5±0.1	97.0±0.2	**79.0±0.2**	**97.0±0.2**	**94.0±0.2**	**89.4**

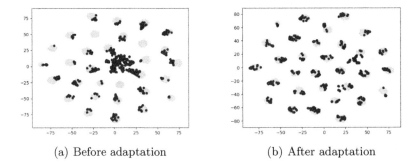

(a) Before adaptation   (b) After adaptation

Fig. 4. Features visualization. (a) and (b) show the feature representations of two domains before and after the domain adaption on the task $\mathbf{A} \rightarrow \mathbf{D}$, respectively. Yellow and purple points indicate the source and target domain samples, respectively. After adaptation, target clusters are well-aligned with source clusters. (Color figure online)

4.3   Analysis

Features Visualization. We visualize the feature representations to investigate the effectiveness of domain adaptation on the task $\mathbf{A} \rightarrow \mathbf{D}$ using t-distributed stochastic neighbor embedding (t-SNE) [22]. As shown in Fig. 4(a) and 4(b), the discrepancy between $\mathbf{A}$ and $\mathbf{D}$ is significantly reduced in the latent representation space after adaptation. Moreover, the feature representations with the same class label but from different domains are much more closer after adaptation, which demonstrates the effectiveness of the class-level distribution alignment.

Cross-Domain $\mathcal{A}$-Distance. The theory of domain adaptation [1] suggests $\mathcal{A}$-distance as a measure of distribution discrepancy. The proxy $\mathcal{A}$-distance is defined as $d_{\mathcal{A}} = 2(1 - 2\epsilon)$, where ϵ is the test error of a classifier (e.g. kernel SVM) trained to discriminate the source domain from the target domain. Table 3 shows $d_{\mathcal{A}}$ on tasks $\mathbf{A} \rightarrow \mathbf{W}$, $\mathbf{W} \rightarrow \mathbf{D}$ with features of ResNet-50, DANN, CDAN and JDARN. We observe that $d_{\mathcal{A}}$ on JDARN features is smaller than $d_{\mathcal{A}}$ on both ResNet-50, DANN and CDAN features, which implys that our model can effectively reduce the domain distribution discrepancy. As domains $\mathbf{W}$ and $\mathbf{D}$ are similar, $d_{\mathcal{A}}$ on task $\mathbf{W} \rightarrow \mathbf{D}$ is smaller than that on $\mathbf{A} \rightarrow \mathbf{W}$, implying higher accuracies.

Table 3. Cross-domain $\mathcal{A}$-distance

Method	A → W	W → D
RESNET-50 [11]	1.82	1.31
DANN [6]	1.51	1.18
CDAN [20]	1.20	**1.10**
JDARN	**1.12**	**1.10**

Table 4. Ablation experiments

Method	W → A	P → C
JDARN (w/o vae)	65.4	95.6
JDARN (w binary)	67.2	96.3
JDARN (w/o etp& vat)	71.0	96.7
JDARN	**72.9**	**97.0**

Ablation Study. To investigate the effects of each component in our model, we also try ablation study on tasks **W → A** and **P → C** with different components ablation: (i) removing the decoders and only training the encoders (w/o vae), (ii) training one domain discriminator with a binary adversarial loss (w binary), and (iii) training the target VAE without entropy loss and VAT loss (w/o etp&vat). As seen in Table 4, once one part is removed, the accuracy degrades. The results show that all the losses are designed reasonably and they all promote the classification accuracy for domain adaptation.

5   Conclusion

In this paper, we propose a novel model called Joint Domain-Adversarial Reconstruction Network for unsupervised domain adaptation. Our model mainly tackle two problems of existing domain-adversarial learning methods: ignoring the individual characteristics of each domain and ignoring the class-level alignment. JDARN uses data reconstruction as an auxiliary task to help domain-adversarial learning to learn domain-specific representations. Meanwhile, JDARN uses two joint domain-class discriminators to distinguish both class and domain of the learned representations and trains them according to a joint adversarial loss, so that the distributions of two domains can be aligned at the domain level and the class level simultaneously. Comprehensive experiments on two standard UDA datasets and tasks verify the effectiveness of our proposed model.

Acknowledgements. This paper is supported by the National Key Research and Development Program of China (Grant No. 2018YFB1403400), the National Natural Science Foundation of China (Grant No. 61876080), the Key Research and Development Program of Jiangsu (Grant No. BE2019105), the Collaborative Innovation Center of Novel Software Technology and Industrialization at Nanjing University.

References

1. Ben-David, S., Blitzer, J., Crammer, K., Kulesza, A., Pereira, F., Vaughan, J.W.: A theory of learning from different domains. Mach. Learn. 151–175 (2009). https://doi.org/10.1007/s10994-009-5152-4

2. Ben-David, S., Blitzer, J., Crammer, K., Pereira, F.: Analysis of representations for domain adaptation. In: Advances in Neural Information Processing Systems, pp. 137–144 (2007)
3. Bousmalis, K., Trigeorgis, G., Silberman, N., Krishnan, D., Erhan, D.: Domain separation networks. In: Advances in Neural Information Processing Systems, pp. 343–351 (2016)
4. Chapelle, O., Zien, A.: Semi-supervised classification by low density separation. In: AISTATS, vol. 2005, pp. 57–64. Citeseer (2005)
5. Cicek, S., Soatto, S.: Unsupervised domain adaptation via regularized conditional alignment. arXiv preprint arXiv:1905.10885 (2019)
6. Ganin, Y., Lempitsky, V.: Unsupervised domain adaptation by backpropagation (2014)
7. Ganin, Y., et al.: Domain-adversarial training of neural networks. J. Mach. Learn. Res. **17**(1), 2096–2130 (2016)
8. Ghifary, M., Kleijn, W.B., Zhang, M., Balduzzi, D., Li, W.: Deep reconstruction-classification networks for unsupervised domain adaptation. In: Leibe, B., Matas, J., Sebe, N., Welling, M. (eds.) ECCV 2016. LNCS, vol. 9908, pp. 597–613. Springer, Cham (2016). https://doi.org/10.1007/978-3-319-46493-0_36
9. Goodfellow, I., et al.: Generative adversarial nets. In: Advances in Neural Information Processing Systems, pp. 2672–2680 (2014)
10. Grandvalet, Y., Bengio, Y.: Semi-supervised learning by entropy minimization. In: Advances in Neural Information Processing Systems, pp. 529–536 (2005)
11. He, K., Zhang, X., Ren, S., Sun, J.: Deep residual learning for image recognition. In: Proceedings of the IEEE Conference on Computer Vision and Pattern Recognition, pp. 770–778 (2016)
12. Hu, L., Kan, M., Shan, S., Chen, X.: Duplex generative adversarial network for unsupervised domain adaptation. In: Proceedings of the IEEE Conference on Computer Vision and Pattern Recognition, pp. 1498–1507 (2018)
13. Kang, G., Zheng, L., Yan, Y., Yang, Y.: Deep adversarial attention alignment for unsupervised domain adaptation: the benefit of target expectation maximization. In: Ferrari, V., Hebert, M., Sminchisescu, C., Weiss, Y. (eds.) ECCV 2018. LNCS, vol. 11215, pp. 420–436. Springer, Cham (2018). https://doi.org/10.1007/978-3-030-01252-6_25
14. Kingma, D.P., Welling, M.: Auto-encoding variational bayes (2013)
15. Krizhevsky, A., Sutskever, I., Hinton, G.E.: Imagenet classification with deep convolutional neural networks. In: Advances in Neural Information Processing Systems, pp. 1097–1105 (2012)
16. Kumar, A., et al.: Co-regularized alignment for unsupervised domain adaptation. In: Advances in Neural Information Processing Systems, pp. 9345–9356 (2018)
17. Liu, H., Long, M., Wang, J., Jordan, M.: Transferable adversarial training: a general approach to adapting deep classifiers. In: International Conference on Machine Learning, pp. 4013–4022 (2019)
18. Liu, M.Y., Tuzel, O.: Coupled generative adversarial networks. In: Advances in Neural Information Processing Systems, pp. 469–477 (2016)
19. Long, M., Cao, Y., Wang, J., Jordan, M.I.: Learning transferable features with deep adaptation networks (2015)
20. Long, M., Cao, Z., Wang, J., Jordan, M.I.: Conditional adversarial domain adaptation. In: Advances in Neural Information Processing Systems, pp. 1640–1650 (2018)

21. Long, M., Zhu, H., Wang, J., Jordan, M.I.: Deep transfer learning with joint adaptation networks. In: Proceedings of the 34th International Conference on Machine Learning, vol. 70, pp. 2208–2217. JMLR. org (2017)
22. Maaten, L.V.D., Hinton, G.: Visualizing data using t-SNE. J. Mach. Learn. Res. 9(Nov), 2579–2605 (2008)
23. Miyato, T., Maeda, S.I., Koyama, M., Ishii, S.: Virtual adversarial training: a regularization method for supervised and semi-supervised learning. IEEE Trans. Pattern Anal. Mach. Intell. 41(8), 1979–1993 (2018)
24. Motiian, S., Piccirilli, M., Adjeroh, D.A., Doretto, G.: Unified deep supervised domain adaptation and generalization. In: Proceedings of the IEEE International Conference on Computer Vision, pp. 5715–5725 (2017)
25. Pan, S.J., Tsang, I.W., Kwok, J.T., Yang, Q.: Domain adaptation via transfer component analysis. IEEE Trans. Neural Netw. 22(2), 199–210 (2010)
26. Pan, S.J., Yang, Q.: A survey on transfer learning. IEEE Trans. Knowl. Data Eng. 22(10), 1345–1359 (2009)
27. Paszke, A., et al.: PyTorch: an imperative style, high-performance deep learning library. In: Advances in Neural Information Processing Systems, pp. 8024–8035 (2019)
28. Pei, Z., Cao, Z., Long, M., Wang, J.: Multi-adversarial domain adaptation. In: Thirty-Second AAAI Conference on Artificial Intelligence (2018)
29. Quiñonero-Candela, J., Sugiyama, M., Schwaighofer, A., Lawrence, N.: Covariate shift and local learning by distribution matching (2008)
30. Saenko, K., Kulis, B., Fritz, M., Darrell, T.: Adapting visual category models to new domains. In: Daniilidis, K., Maragos, P., Paragios, N. (eds.) ECCV 2010. LNCS, vol. 6314, pp. 213–226. Springer, Heidelberg (2010). https://doi.org/10.1007/978-3-642-15561-1_16
31. Saito, K., Ushiku, Y., Harada, T.: Asymmetric tri-training for unsupervised domain adaptation. In: Proceedings of the 34th International Conference on Machine Learning, vol. 70, pp. 2988–2997. JMLR. org (2017)
32. Saito, K., Watanabe, K., Ushiku, Y., Harada, T.: Maximum classifier discrepancy for unsupervised domain adaptation. In: Proceedings of the IEEE Conference on Computer Vision and Pattern Recognition, pp. 3723–3732 (2018)
33. Sener, O., Song, H.O., Saxena, A., Savarese, S.: Learning transferrable representations for unsupervised domain adaptation. In: Advances in Neural Information Processing Systems, pp. 2110–2118 (2016)
34. Shu, R., Bui, H.H., Narui, H., Ermon, S.: A DIRT-T approach to unsupervised domain adaptation (2018)
35. Torralba, A., Efros, A.A., et al.: Unbiased look at dataset bias. In: CVPR, vol. 1, p. 7. Citeseer (2011)
36. Tzeng, E., Hoffman, J., Darrell, T., Saenko, K.: Simultaneous deep transfer across domains and tasks. In: Proceedings of the IEEE International Conference on Computer Vision, pp. 4068–4076 (2015)
37. Tzeng, E., Hoffman, J., Saenko, K., Darrell, T.: Adversarial discriminative domain adaptation. In: Proceedings of the IEEE Conference on Computer Vision and Pattern Recognition, pp. 7167–7176 (2017)
38. Wang, M., Deng, W.: Deep visual domain adaptation: a survey. Neurocomputing 312, 135–153 (2018)

39. Xie, S., Zheng, Z., Chen, L., Chen, C.: Learning semantic representations for unsupervised domain adaptation. In: International Conference on Machine Learning, pp. 5419–5428 (2018)
40. Zhang, W., Ouyang, W., Li, W., Xu, D.: Collaborative and adversarial network for unsupervised domain adaptation. In: Proceedings of the IEEE Conference on Computer Vision and Pattern Recognition, pp. 3801–3809 (2018)

Diversity-Based Generalization for Unsupervised Text Classification Under Domain Shift

Jitin Krishnan[⊠], Hemant Purohit, and Huzefa Rangwala

George Mason University, Fairfax, VA, USA
{jkrishn2,hpurohit,rangwala}@gmu.edu

Abstract. Domain adaptation approaches seek to learn from a source domain and generalize it to an unseen target domain. At present, the state-of-the-art unsupervised domain adaptation approaches for subjective text classification problems leverage unlabeled target data along with labeled source data. In this paper, we propose a novel method for domain adaptation of *single-task* text classification problems based on a simple but effective idea of diversity-based generalization that does not require unlabeled target data but still matches the state-of-the-art in performance. Diversity plays the role of promoting the model to better generalize and be indiscriminate towards domain shift by forcing the model not to rely on same features for prediction. We apply this concept on the most explainable component of neural networks, the attention layer. To generate sufficient diversity, we create a multi-head attention model and infuse a diversity constraint between the attention heads such that each head will learn differently. We further expand upon our model by tri-training and designing a procedure with an additional diversity constraint between the attention heads of the tri-trained classifiers. Extensive evaluation using the standard benchmark dataset of Amazon reviews and a newly constructed dataset of Crisis events shows that our fully unsupervised method matches with the competing baselines that uses unlabeled target data. Our results demonstrate that machine learning architectures that ensure sufficient diversity can generalize better; encouraging future research to design ubiquitously usable learning models without using unlabeled target data.

Keywords: Text classification · Unsupervised domain adaptation · Natural language processing · Neural networks

1   Introduction

In natural language processing, domain adaptation of sequence classification problems has several applications ranging from sentiment analysis [4] to classifying social media posts during crisis events [1]. Knowledge learned from one domain, book reviews for instance, can be adapted to predict examples from a different domain such as reviews of electronics. Similarly, information about

© Springer Nature Switzerland AG 2021
F. Hutter et al. (Eds.): ECML PKDD 2020, LNAI 12458, pp. 657–672, 2021.
https://doi.org/10.1007/978-3-030-67661-2_39

resource-need events learned from one natural disaster can be adapted to predict events from an ongoing crisis [20]. With the publication of Amazon reviews dataset [4] consisting of around 25 different domains, cross-domain sentiment analysis became a common way to evaluate machine learning models for domain adaptation in text.

The top performing models in this line of research largely remain dependent on unlabeled target data. Although unlabeled data from the target domain tends to help, it is imperative to realize the extent of performance gain that source data alone can bring. We consider that the ideal criterion for no supervision in domain adaptation is having zero knowledge about the target domain beforehand; even if it is unlabeled. Our work can be viewed either as a strong baseline for future unsupervised cross-domain research that utilizes unlabeled target data or as a new direction in fully unsupervised domain adaptation without using any target data at all; which is necessary for tasks such as relevancy prediction for actionable information filtering in domains such as natural disasters that require timely and efficient methods. We scope our work to the following setting: **a)** *single-task transfer*, **b)** *single source and target*, and **c)** *without labeled and unlabeled target data available during training*. We compare and contrast our unsupervised methods to the existing counterparts. We do not consider any supervised or minimally supervised approaches in this work.

Contributions: a) We present a novel diversity-based generalization method using a multi-head attention model for domain adaption in unsupervised text classification tasks. **b)** To further improve the generalizability of our model and utilize additionally available unlabeled source data, we design a tri-training procedure with an additional diversity constraint between the attention heads of the tri-trained classifiers. **c)** Addressing the existing evaluation gap in component-level performance analysis, we show a systematic and incremental creation of our models by creating strong unsupervised baselines and improving upon existing work.

2   Related Work

Early works on domain adaptation such as Structural Correspondence Learning [5] make use of unlabeled target data to find a joint representation by automatically inducing correspondences among features from different domains. The importance of a good feature representation was later formally analyzed with a generalization bound by Ben-David et al. [3]. These studies realized the importance of finding commonality in features or pivots and minimizing the difference between the domains. Pan et al. [19] proposed a spectral feature alignment method to align domain-specific and domain-independent words into unified clusters via simultaneous co-clustering in a common latent space. Later, introduction of deep learning and neural networks helped remedy the problems of manual pivot selection and discrete feature representations. In order to learn better higher level representations, Stacked Denoising Autoencoders (SDA) [25] were introduced. Along with SDA, a more efficient version called marginalized

SDA [6] with low computational cost and scalability has been utilized successfully in cross-domain tasks [10,11]. Domain-Adversarial training of Neural Networks (DANN) [9] was proposed to effectively utilize unlabeled target data to create a classifier that is indiscriminate toward different domains. In their work a negative gradient (gradient reversal) from a domain classifier branch is backpropagated to promote the features at the lower layers of the network incapable of discriminating domains. DANN became an essential component in many works that followed. Recent works such as Adversarial Memory Network (AMN) [15] bring interpretability by using attention to capture the pivots. Along with attention, they effectively use gradient reversal to learn domain indiscriminate features. Hierarchical Attention Network (HATN) [14] expands upon AMN by first extracting pivots and then jointly training pivot and non-pivot networks. Interactive Attention Transfer Network (IATN) [26], another closely related work to AMN and HATN, showed the importance of attending 'aspect' information. Another line of research, that approached domain adaption through innovation in training procedure, is tri-training [22,28]. Tri-training utilizes three independently trained classifiers; of which one is trained only on unlabeled target data, pseudo-labeled by the other two. The final prediction is done by majority voting. Multi-task tri-training (MT-Tri) [21], on the other hand, introduced an orthogonality constraint between the two classifiers such that it can be trained jointly, reducing the compute time. This constraint is one of the inspirations for our work. Although diversity could be achieved through other means, we also focus on 'orthogonality'. All of these recent works used unlabeled target data for training classifiers. Our goal is to show that similar performance is achievable without using any target data at all.

Based on how the dataset is used, approaches to domain adaptation can vary from minimally-supervised to unsupervised. Minimally-supervised approaches such as Aligned Recurrent Transfer [7] utilize some *labeled* data from the target domain, while unsupervised approaches such as DANN, AMN, HATN, or IATN utilize only *unlabeled* target data making it a more realistic scenario in terms of usability where collecting labeled target data is expensive. However, many state-of-the-art unsupervised domain adaptation methods, strikingly, never compare with strong fully-unsupervised baselines where no target data is used. Newer methods have started using word vectors [18] for their input word representations. However, the baselines they compare with, utilize large 5000-dimension feature vector of the most frequent unigrams and bigrams as the input representation. In addition, many recent works present a complex system without conducting a component-wise analysis which makes it unclear as to how much each component (word vectors, gradient reversal, or attention) contributed to the performance boost as compared to a simple DANN architecture. To address these evaluation gaps, we perform a systematic and incremental construction of architectures such that individual performance gain is realized.

Advantages and Practical Utility: a) Our methods do not require any target data for training; making it out-of-the-box adaptable to any domain. **b)** We provide a method to utilize additionally available unlabeled source data. **c)** Our models are computationally cheaper (training converges quickly) when compared to the existing state-of-the-art models. **d)** Diversified attention can provide better quality of attended words which can be used for various downstream tasks such as knowledge graph construction.

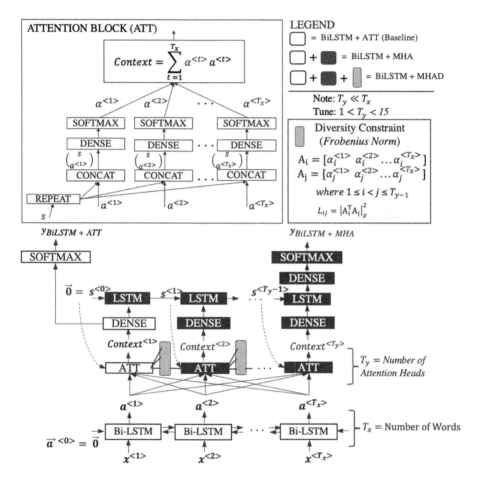

Fig. 1. Complete architecture of the multi-head attention model with diversity.

3   Methodology

3.1   Problem Definition and Notations

Given a source (D_s) and a target (D_t) domain, the goal is to train a classifier using data **only** from D_s and predict examples from the completely unseen D_t. X_s and X_t represent the set of labeled data from source and target domains respectively with their corresponding ground truth labels y_s and y_t. X_t and y_t are used for testing purposes only. X_s^u and X_t^u represent unlabeled data available from the source and target domain respectively. X_t^u (used in all of our competing models either for adversarial training or tri-training) is **never** used in our models. Finally, $[.]^{pl}$ represents data that is pseudo-labeled by the classifier. To summarize:

Input: X_s, y_s (and X_s^u for tri-training)
Output: $y_t^{pred} \leftarrow predict(X_t)$

3.2 Diversity-Based Models

We introduce 4 models with one integral concept: *diversity*. Figure 1 provides an overview of the first two models and Fig. 2 provides an overview of the last two. First is a multi-head attention baseline created to understand the naturally occurring diversity when multiple attention heads are connected. The second model enforces this diversity as a constraint such that all heads learn different features. The third model puts together three diversity-based classifiers and tri-trains them. Tri-training procedure in itself consists of an additional diversity constraint which forces two of the classifiers to learn differently. This is a one-step tri-training procedure intended for scenarios where no unlabeled source data is additionally available. When it is available, a full tri-training can be done until convergence, which is the fourth model.

Multi-head Attention for Sequence Classification (BiLSTM+MHA): BiLSTM+ATT is a standard baseline attention architecture constructed using BiLSTM [12,23] and attention mechanism [2,16]. Bidirectional Long Short-Term Memory (BiLSTM) units have been successfully used in sequence modeling tasks because of their effectiveness in representing forward and backward dependencies in a sequence. For example, meanings of words like 'good' and 'bad' can be changed when they are prefixed with 'not' or suffixed with 'but'. Attention, on the other hand, provides task-specific benefits by attending the most relevant words such as 'excellent' or 'poor' in sentiment analysis. Attention and BiLSTM have been successfully combined previously for tasks such as relation extraction [27] to capture important semantic information in a sentence.

BiLSTM+MHA is an extension of the BiLSTM+ATT baseline by adding multiple attention heads as shown in Fig. 1. This is similar to machine-translation-like architecture [16] where each attention head leads to an LSTM cell with memory carried from previous cells to predict the next word. To customize it to classification purpose, we simply use the output from the final LSTM cell. Setting the classification task this way gives more leniency for the model to learn, remember, and generalize. Multiple attention heads can learn differently and what is learned from the previous heads is transferred to the next. However, this does not guarantee diversity as we do not know if the attention heads will in fact learn differently. In order to enforce diversity, we introduce the following models.

Multi-head Attention with Diversity (BiLSTM+MHAD): In order to guarantee that these attention heads learn differently and forcing the model not to rely on the same features, we create a *diversity constraint*, an additional loss term shown below.

$$L_d = \frac{1}{k} \sum_{i=1}^{T_y-2} \sum_{j=i+1}^{T_y-1} \|A_i^T A_j\|_F^2 \; ; \text{ where } i \neq j \tag{1}$$

where $k = \frac{(T_y-2)(T_y-1)}{2}$, the total number of combinations. T_y is the total number of attention heads. A_i and A_j are i^{th} and j^{th} attention heads and $\|.\|_F^2$ is the squared Frobenius norm, similar to the orthogonality constraint used in [21]. We leave the last attention head from this loss term so that we have one layer that learns freely without any constraints. The complete architecture of this diversity-based model is shown in Fig. 1. Resulting overall loss function, consisting of a binary cross entropy loss term and the diversity loss term, for N training examples is shown below.

$$L(\theta) = -\frac{1}{N} \sum_{i=1}^{N} [y_i \log \hat{y}_i + (1 - y_i) \log(1 - \hat{y}_i)] + \gamma L_d \tag{2}$$

where γ is the hyperparameter to control how much diversity to be enforced within the model.

One-Step Diversity Tri-training (BiLSTM+MHAD-Tri-I): To further expand the concept of diversity, we tri-train the BiLSTM+MHAD models by adapting the multi-task tri-training procedure by [21]. In addition to applying the diversity constraint within each classifiers, an additional orthogonality loss is enforced between first two models m_1 and m_2. The third model m_3 is left out from the joint training. The loss term is shown below.

$$L_o = \frac{1}{k} \sum_{i=1}^{T_y} \sum_{j=1}^{T_y} \|A(m_1)_i^T A(m_2)_j\|_F^2 \tag{3}$$

where $k = \frac{(T_y-1)(T_y)}{2}$. $A(m_1)$ and $A(m_2)$ are the attention heads for models m_1 and m_2 respectively. T_y is the total number of attention heads of each model. The total tri-training diversity loss is given below.

$$L_{dtri} = \alpha L_o + \beta L_d \tag{4}$$

where α and β are the hyperparameters to control how much diversity to be enforced within and between the models.

For one-step diversity tri-training shown in Algorithm 1, we jointly train m_1 and m_2 with tri-training diversity loss L_{dtri}. m_3 is separately trained as a BiLSTM+MHAD model. For predictions, a majority voting rule is applied over the three classifiers. The overall loss function for N training examples is given below.

$$L(\theta) = -\frac{1}{N} \sum_{i=1}^{N} [y_i \log \hat{y}_i + (1 - y_i) \log(1 - \hat{y}_i)] + L_{dtri} \tag{5}$$

Algorithm 1: One-Step Diversity Tri-training

Input: X_s

Output: m_1, m_2, m_3

$m_1, m_2 \leftarrow joint\_diversity\_train\_models(X_s)$

$m_3 \leftarrow diversity\_train\_model(X_s)$

apply majority vote over m_i

Tri-training Until Convergence (BiLSTM+MHAD-Tri-II): Full tri-training, shown in Algorithm 2 and Fig. 2, utilizes additionally available unlabeled source data. While the first two classifiers m_1 and m_2 are jointly trained on labeled source data, the third classifier m_3 is solely dedicated to the training over unlabeled data that is pseudo-labeled by m_1 and m_2. Similar to [21], we define a threshold value τ such that at least one out of the two models should predict with probability greater than τ to be considered successfully pseudo-labeled. We set τ to be 0.7. Starting with second iteration, m_1 is trained jointly with m_2 using a combination of labeled source data and unlabeled source data pseudo-labeled by m_2 and m_3. During joint-training, we give priority to the primary model by setting the loss weights accordingly. For example, while joint-training m_1 with m_2, losses for the models can be minimized in a 2:1 ratio, giving priority to m_1. We continue this process until a convergence condition is met: $m_1 \approx m_2 \approx m_3$.

Algorithm 2: Tri-training [21] - Modified

Input: X_s, X_s^u

Output: m_1, m_2, m_3

while *convergence condition is not met* **do**

    for $i \in 1..3$ **do**

        $X^{pl} \leftarrow \emptyset$

        for $x \in X_s^u$ **do**

            if $p_i(x) = p_k(x)(j, k \neq i)$ **then**

                $X^{pl} \leftarrow X^{pl} \cup \{(x, p_j(x))\}$

            end

        end

        if $i = 3$ **then**

            $m_3 \leftarrow diversity\_train(X^{pl})$;   // Eq. 2

        else if $i = 1$ **then**

            $m_1 \leftarrow joint\_diversity\_train(X_s \cup X^{pl}, m_2)$;   // Eq. 5

        else

            $m_2 \leftarrow joint\_diversity\_train(X_s \cup X^{pl}, m_1)$;   // Eq. 5

        end

    end

end

apply majority vote over m_i

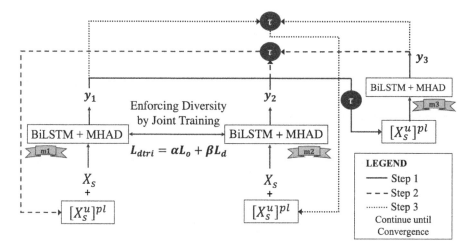

Fig. 2. Tri-training BiLSTM+MHAD models

4 Experimental Evaluation

4.1 Benchmark Dataset: Amazon Reviews

We use the standard benchmark Amazon reviews dataset[1] [4] which is widely used for cross-domain sentiment analysis. We consider four domains: Books (B), Kitchen (K), DVD (D), and Electronics (E). For a fair evaluation of the architectures, we use the exact same raw dataset[2] used by our top competitor model HATN [14], which is a part of Blitzer's original raw dataset. We also use the same 300-dimensional word vectors[3] [18]. Table 1 summarizes this dataset.

4.2 Crisis Dataset (Tweets)

Additionally, we construct a new dataset consisting of Twitter posts (tweets) collected during three hurricane crises by *CitizenHelper* [13] system: *Harvey* and *Irma* in 2017, and *Florence* in 2018. Similar to sentiment classification, our goal here is to classify whether a tweet text indicates an event or not. Using the crowd-sourcing platform Figure-Eight[4], three workers at minimum were assigned to give a binary label to each tweet. We define events to be actions that involve at least one noun/entity. Events could be past, present, or future actions. It could also be questions, news, or instructions about actions. Some examples are: '*A rescues B*', '*A is sending food to B*', '*A will move to location B*', and so on. Table 1 summarizes this dataset. Unfortunately, the labeled dataset for Florence

---

[1] http://www.cs.jhu.edu/~mdredze/datasets/sentiment/.

[2] https://github.com/hsqmlzno1/HATN/tree/master/raw_data.

[3] https://code.google.com/archive/p/word2vec/.

[4] https://www.figure-eight.com now https://appen.com/.

and Irma consists of very low number of positive events. Consequently, we set up the experiments such that we train only on *Harvey* and test on *Florence* and *Irma*.

4.3   Experimental Setup

We follow the traditional cross domain sentiment classification set up where each experiment consists of a source domain (S) and a target domain (T). A model will be trained on source data and tested on target data, represented as $S \rightarrow T$. We use all available labeled target data for testing. Crisis dataset is balanced before training and testing.

Table 1. Dataset statistics

	Positive	Negative	Unlabeled	Average number of tokens	Vocabulary
Books	3000	3000	9750	182.0	105920
DVD	3000	3000	11843	197.5	117619
Kitchen	3000	3000	13856	102.0	52972
Elec.	3000	3000	17009	119.3	72458
Harvey	1122	960	10001	17.2	23562
Florence	201	1475	10001	17.1	26380
Irma	313	596	10001	15.3	20764

4.4   Implementation Details

We use Keras deep learning library with Adam optimizer ($lr = 0.005$, $beta_1 = 0.9$, $beta_2 = 0.999$, $decay = 0.01$) for our implementations. Maximum epoch is 40 with an early stopping patience of 3. Batch size is 32 and validation split is 0.15.

We set the number of attention heads, $T_y = 5$ and number of words from each review, $T_x = 200$. To keep the model simple, we do not change this further. Dropouts are kept at 0.4. τ is kept at 0.7 and tri-training is stopped at 85% agreement. We set $\gamma = 0.01$, $\alpha = 0.05$ and $\beta = 0.01$. These values are obtained by performing a basic hyperparameter tuning using grid search.

4.5   Baselines and Modifications

Adversarial Learning Based Methods: DANN [9] introduced adversarial training by making use of unlabeled target domain data. Earlier layers of the deep neural network architecture are made domain invariant through back-propagating a negative gradient using a jointly trained domain classifier. It uses a 5000-dimensional feature vector of the most frequent unigrams and bigrams as the input representation. DAmSDA [10], on the other hand, uses mSDA [6] representation instead. We report the scores for DAmSDA and DANN from HATN [14]. For DANN, additionally, we create a customized implementation (**DANN⁺**) using BiLSTM and word vectors. This modified architecture simply consists of a shared BiLSTM layer followed by a dense layer for sentiment classification and the same BiLSTM layer followed by a gradient reversal layer and a dense layer for domain classification. Note that the accuracy for **DANN⁺** (our improved DANN) is **+3.2%** higher than what is reported in HATN.

Tri-training Based Methods: Multi-task tri-training (MT-Tri) [21] conducts tri-training on a multilayer perceptron model with an orthogonality loss between the final layers to enforce diversity between the jointly trained models. Unlabeled target data pseudo-labeled by the first two classifiers are fed to the third classifier. Three classifiers are optimized until none of the models' predictions change. We improve upon this model (**MT-Tri⁺**) by using word vectors and BiLSTM.

Attention Based Methods: Recent works such as AMN [15], HATN [14], and IATN [26] use attention to identify sentiment pivots. Utilizing unlabeled target data, gradient reversal is an essential component in their models for domain classification. AMN expands DANN to an attention-based model. HATN improves AMN further by building pivot and non-pivot networks. The pivot network (P-Net) performs the same task as AMN by extracting pivots. The non-pivot network (NP-Net) takes a transformed input that hides previously extracted pivots, which is then jointly trained with P-Net. IATN incorporates 'aspect' information in addition to the sentence attentions. At the time of writing of this paper, the open source code[5] for IATN is still being prepared by its authors. We include the reported scores for reference purpose. IATN reports a 0.8% increase in performance as compared to HATN (85.9% versus 85.1%). IATN uses the same input settings and the dataset as HATN with one difference: 200-dimensional word vectors instead of 300. Meanwhile, we use the exact same dataset and GoogleNews word vectors used by HATN for all our experiments for both reproducibility as well as blind comparison.

Strong Unsupervised Baselines: To study component-wise performance, we construct two strong unsupervised baselines from standard neural network architectures: BiLSTM and BiLSTM+ATT. BiLSTM consists of traditional BiLSTM units with the final unit making the prediction. BiLSTM+ATT, as shown in Fig. 1, adds a single attention layer on top of BiLSTM and the prediction is based on the output from the attention layer. Note that these two baselines still produce strong results and provide a reference for how much improvement following models make.

5   Results and Discussion

Tables 2, 3, and 4 show the competitive nature of our fully unsupervised methods when compared with the existing unsupervised counterparts that use unlabeled

---

[5] https://github.com/1146976048qq/IATN.

Table 2. Classification accuracy scores showing that unlabeled target data is not necessary to achieve strong performance. +: improved implementations, ◇: reproduced implementations, ♠: strong unsupervised baselines constructed from standard neural network architectures, *: reported scores from [14,26] (see description). Our scores are averaged over 5 independent runs. **Note that only the models in the bottom table do not use any unlabeled target data.**

S → T	DANN*	DAmSDA*	IATN*	DANN+	MT-Tri+	AMN◇ /P-Net	HATN◇
B → D	83.42	86.12	86.80	82.85	84.67	87.07	87.70
B → E	76.27	79.02	86.50	81.03	84.62	82.98	86.20
B → K	77.90	81.05	85.90	82.01	84.78	84.85	87.08
K → B	74.17	80.55	84.70	79.38	80.98	83.50	84.83
K → D	75.32	82.18	84.40	79.04	78.89	82.83	84.73
K → E	85.53	88.00	87.60	86.00	85.87	86.72	89.08
E → B	73.53	79.92	81.80	78.92	80.64	83.28	83.62
E → K	84.53	85.80	88.70	86.43	89.62	89.80	90.12
E → D	76.27	82.63	84.10	77.83	79.97	83.37	83.87
D → B	80.77	85.17	87.00	84.32	85.67	87.85	88.02
D → E	76.35	76.17	86.90	81.74	84.48	84.65	86.78
D → K	78.15	82.60	85.80	83.29	85.05	84.28	87.00
AVG	78.52	82.43	85.90	81.78	83.77	85.10	**86.59**

S → T	BiLSTM♠	BiLSTM +ATT♠	BiLSTM +MHA	BiLSTM +MHAD	BiLSTM +MHAD-Tri-I	BiLSTM +MHAD-Tri-II
B → D	84.19	87.44	87.29	87.54	87.76	87.46
B → E	83.61	83.90	85.36	85.63	85.75	86.08
B → K	83.87	85.21	86.04	87.06	87.34	87.68
K → B	80.52	82.15	83.11	83.70	84.19	84.23
K → D	78.28	80.17	81.50	82.27	82.11	83.34
K → E	86.33	87.30	88.60	88.81	88.98	89.22
E → B	80.58	82.10	83.55	83.67	83.96	84.33
E → K	88.07	88.19	89.61	89.96	90.07	91.05
E → D	78.08	81.93	82.77	82.93	82.87	82.81
D → B	83.93	87.72	87.77	88.22	88.51	88.74
D → E	82.98	84.57	84.75	85.93	85.79	86.21
D → K	84.38	85.45	86.50	86.73	86.74	87.37
AVG	82.90	84.68	85.57	85.98	86.17	**86.54**

Table 3. Classification accuracy scores for crisis dataset.

S → T	HATN	BiLSTM +ATT	BiLSTM +MHA	BiLSTM +MHAD	BiLSTM +MHAD-Tri-I	BiLSTM +MHAD-Tri-II
H → F	80.01	74.88	74.32	75.69	76.00	78.11
H → I	58.53	63.84	64.32	65.10	65.02	64.38

Review Text: Vornado Vortex Heat. I just had to post after reading the reviews. I have had a Vornado Vortex Heater since 1994, I paid twice what this
one costs and it has been excellent. $y_{true} = +ve$

BiLSTM+ATT: $y_{pred} = -ve$

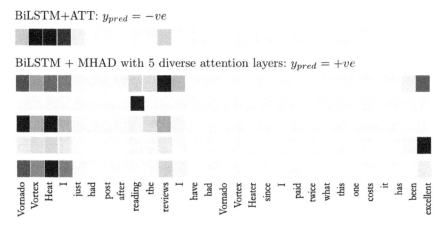

BiLSTM + MHAD with 5 diverse attention layers: $y_{pred} = +ve$

Review Text: Butter dish not tall enough. It looks nice but the lid touches the
top of the butter and sticks to it. $y_{true} = -ve$

BiLSTM+ATT: $y_{pred} = +ve$

BiLSTM + MHAD with 5 diverse attention layers: $y_{pred} = -ve$

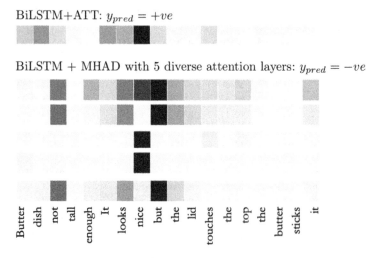

Fig. 3. Two examples of *kitchen* review predictions by BiLSTM+ATT and BiLSTM +
MHAD models trained on *book* reviews. When a single attention head fails to attend
key words like 'excellent' or 'but', at least one of the diverse heads tends to make up
for it.

target data. Our experiments showed incrementally improving results when each
component is added to the baselines. Attention with diversity improved the
single-attention baseline and tri-training with diversity improved it even further.
Using additionally available unlabeled source data proved to be fruitful for most

Table 4. Training time in d-hh:mm:ss for H→F on a Dual Intel(R) Xeon(R) Gold 5120 CPU@2.2GHz with 28 cores and 1.5TB RAM.

HATN	BiLSTM +MHA	BiLSTM +MHAD	BiLSTM +MHAD-Tri-I	BiLSTM +MHAD-Tri-II
1–08:31:09	00:50:31	01:29:28	2:07:23	6:21:18

of the domains. Note that for crisis dataset we only use *Harvey* for training because labeled data for *Florence* and *Irma* was just too low.

An implication of the diversity-based attention heads is shown in Fig. 3. Diversity pushes the model not to rely on the same features. First example shows misclassification by a single attention model that attends incorrect sentiment words like 'Vortex' and 'Heat'. However, with diversity, the model is lenient and look for alternate features. At least one of the T_y diverse heads tends to find important words like 'excellent'. These examples also show that placing diversity on attention layers, rather than on any other hidden layers, provides an explainable understanding of which words the model deems to be important and can be used for subsequent pivot extraction like in AMN or HATN.

Computational Performance: To show that our work is practically useful for all communities alike, experiments are run on a CPU. A sample training time comparison is shown in Table 4. HATN needs gradient reversal to utilize unlabeled target data for the domain classifier branch and pivot extraction for joint training; subsequently making it much slower.

Gradient Reversal: To study the impact of gradient reversal procedure with BiLSTM, we conducted experiments with unlabeled target data. The performance of BiLSTM versus DANN+ (improved DANN) models in Table 2 showed that, with a good dropout value for the BiLSTM units, gradient reversal did not help much. On a similar note, domain adversarial loss was found not to be helpful in tri-training experiments [21]. In our context, we speculate that this might be because the dropout in the BiLSTM layer drops individual words that can lead to a better generalization which is essentially the purpose of gradient reversal. This will be studied in our future work.

Table 5. Classification accuracy scores for three distinct combinations.

S → T	P-Net	BiLSTM +MHAD-Tri-II
Electronics → Yelp	88.45	89.15
Kitchen → IMDb	76.38	78.33
Yelp → IMDb	78.75	77.28

Additional Analysis: Generalizability of our models is further tested with three randomly selected experiments using very divergent domains such as Yelp[6] restaurant reviews and IMDb [17] movie reviews in addition to Amazon reviews; *Electronics (Amazon) → Yelp, Kitchen (Amazon) → IMDb*, and *Yelp → IMDb*. We randomly selected 2000 positive and negative reviews from Yelp and IMDb. Their accuracy scores on our final model when compared to PNet[7] of HATN is shown in Table 5. Once again, this shows that unlabeled target data is not always necessary; thus providing us with a fully unsupervised and computationally efficient alternative for domain adaptation in text classification tasks.

6   Future Work

Experiments shown in the additional analysis section can be expanded to a lot more datasets and divergent domains; particularly applying to domains where unlabeled target data is not readily available such as during an onset of a natural or man-made disaster event. A careful empirical study of several existing complex architectures that employ adversarial training using the gradient reversal strategy is another direction. It is crucial to understand how much of the gain claimed by the adversarial training approach can actually be brought through by good generalization of the model without using any unlabeled data from the target domain. Recent progress in deep learning and natural language processing has seen impressive performance gain across various domains by using the transformer [8,24] based models. Leveraging such models and studying the effect of encouraging diversity among the architectures appear to be a promising future direction.

7   Conclusion

Our study shows that machine learning architectures designed for achieving sufficient diversity in learning can generalize better for domain adaptation. Further, unlabeled target data, used often by state-of-the-art models, is not always necessary to produce strong performance for domain adaptation in subjective text classification problems. We introduced a novel diversity-based generalization approach for the domain shift problem using a multi-head attention model where attention heads are constrained to learn differently such that the classifier can leverage on alternative features. Experiments on the standard benchmark dataset of Amazon reviews and a newly constructed dataset of Crisis events showed that our fully unsupervised methods that completely avoid target data can indeed match the competing unsupervised baselines.

Reproducibility: Code, datasets, and documentation are available at - https://github.com/jitinkrishnan/Diversity-Based-Generalization.

---

[6] https://www.yelp.com/dataset/challenge.

[7] PNet is the first component of HATN which is computationally faster and within ∼1.5% accuracy of HATN.

Acknowledgements. Authors would like to thank U.S. National Science Foundation grant IIS-1815459 for partially supporting this research.

References

1. Alam, F., Joty, S., Imran, M.: Domain adaptation with adversarial training and graph embeddings. In: Proceedings of the 56th Annual Meeting of the Association for Computational Linguistics (Volume 1: Long Papers) (2018)
2. Bahdanau, D., Cho, K., Bengio, Y.: Neural machine translation by jointly learning to align and translate. arXiv preprint arXiv:1409.0473 (2014)
3. Ben-David, S., Blitzer, J., Crammer, K., Pereira, F.: Analysis of representations for domain adaptation. In: Advances in Neural Information Processing Systems, pp. 137–144 (2007)
4. Blitzer, J., Dredze, M., Pereira, F.: Biographies, bollywood, boom-boxes and blenders: domain adaptation for sentiment classification. In: Proceedings of the 45th Annual Meeting of the Association of Computational Linguistics, pp. 440–447 (2007)
5. Blitzer, J., McDonald, R., Pereira, F.: Domain adaptation with structural correspondence learning. In: Proceedings of the 2006 Conference on Empirical Methods in Natural Language Processing, pp. 120–128 (2006)
6. Chen, M., Xu, Z., Weinberger, K., Sha, F.: Marginalized denoising autoencoders for domain adaptation. arXiv preprint arXiv:1206.4683 (2012)
7. Cui, W., Zheng, G., Shen, Z., Jiang, S., Wang, W.: Transfer learning for sequences via learning to collocate. arXiv preprint arXiv:1902.09092 (2019)
8. Devlin, J., et al.: BERT: pre-training of deep bidirectional transformers for language understanding. In: Proceedings of the 2019 Conference of the North American Chapter of the Association for Computational Linguistics: Human Language Technologies, Vol. 1 (Long and Short Papers) (2019)
9. Ganin, Y., Lempitsky, V.: Unsupervised domain adaptation by backpropagation. In: International Conference on Machine Learning, PMLR (2015)
10. Ganin, Y., et al.: Domain-adversarial training of neural networks. J. Mach. Learn. Res. **17**(1), 2030–2096 (2016)
11. Glorot, X., Bordes, A., Bengio, Y.: Domain adaptation for large-scale sentiment classification: a deep learning approach. In: Proceedings of the 28th International Conference on Machine Learning (ICML-11), pp. 513–520 (2011)
12. Hochreiter, S., Schmidhuber, J.: Long short-term memory. Neural Compu. **9**(8), 1735–1780 (1997)
13. Karuna, P., Rana, M., Purohit, H.: Citizenhelper: a streaming analytics system to mine citizen and web data for humanitarian organizations. In: Eleventh International AAAI Conference on Web and Social Media, pp. 729–730 (2017)
14. Li, Z., Wei, Y., Zhang, Y., Yang, Q.: Hierarchical attention transfer network for cross-domain sentiment classification. In: Thirty-Second AAAI Conference on Artificial Intelligence (2018)
15. Li, Z., Zhang, Y., Wei, Y., Wu, Y., Yang, Q.: End-to-end adversarial memory network for cross-domain sentiment classification. In: IJCAI, pp. 2237–2243 (2017)
16. Luong, M.-T., Pham, H., Manning, C.D.: Effective approaches to attention-based neural machine translation. In: Proceedings of the 2015 Conference on Empirical Methods in Natural Language Processing (2015)

17. Maas, A.L., Daly, R.E., Pham, P.T., Huang, D., Ng, A.Y., Potts, C.: Learning word vectors for sentiment analysis. In: The 49th Annual Meeting of the Association for Computational Linguistics (2011)
18. Mikolov, T., Sutskever, I., Chen, K., Corrado, G.S., Dean, J.: Distributed representations of words and phrases and their compositionality. In: Advances in Neural Information Processing Systems, pp. 3111–3119 (2013)
19. Pan, S.J., Ni, X., Sun, J.T., Yang, Q., Chen, Z.: Cross-domain sentiment classification via spectral feature alignment. In: Proceedings of the 19th International Conference on World Wide Web, pp. 751–760. ACM (2010)
20. Purohit, H., Castillo, C., Imran, M., Pandey, R.: Social-EOC: serviceability model to rank social media requests for emergency operation centers. In: 2018 IEEE/ACM International Conference on Advances in Social Networks Analysis and Mining (ASONAM), pp. 119–126. IEEE (2018)
21. Ruder, S., Plank, B.: Strong baselines for neural semi-supervised learning under domain shift. In: Proceedings of the 56th Annual Meeting of the Association for Computational Linguistics (Volume 1: Long Papers) (2018)
22. Saito, K., Ushiku, Y., Harada, T.: Asymmetric tri-training for unsupervised domain adaptation. In: Proceedings of the 34th International Conference on Machine Learning, vol. 70, pp. 2988–2997. JMLR. org (2017)
23. Schuster, M., Paliwal, K.K.: Bidirectional recurrent neural networks. IEEE Trans. Signal Process. **45**(11), 2673–2681 (1997)
24. Vaswani, A., et al.: Attention is all you need. In: Advances in Neural Information Processing Systems, pp. 5998–6008 (2017)
25. Vincent, P., Larochelle, H., Lajoie, I., Bengio, Y., Manzagol, P.A.: Stacked denoising autoencoders: learning useful representations in a deep network with a local denoising criterion. J. Mach. Learn. Res. **11**, 3371–3408 (2010)
26. Zhang, K., Zhang, H., Liu, Q., Zhao, H., Zhu, H., Chen, E.: Interactive attention transfer network for cross-domain sentiment classification (2019)
27. Zhou, P., Shi, W., Tian, J., Qi, Z., Li, B., Hao, H., Xu, B.: Attention-based bidirectional long short-term memory networks for relation classification. In: Proceedings of the 54th Annual Meeting of the Association for Computational Linguistics (Volume 2: Short Papers), pp. 207–212 (2016)
28. Zhou, Z.H., Li, M.: Tri-training: exploiting unlabeled data using three classifiers. IEEE Trans. Knowl. Data Eng. **11**, 1529–1541 (2005)

Bayesian Optimization and Few-Shot Learning

On Local Optimizers of Acquisition Functions in Bayesian Optimization

Jungtaek Kim[1]($\boxtimes$) and Seungjin Choi[2]

[1] Pohang University of Science and Technology, Pohang, Republic of Korea
jtkim@postech.ac.kr
[2] Inference Lab, BARO AI, Seoul, Republic of Korea
seungjin@baroai.com

Abstract. Bayesian optimization is a sample-efficient method for finding a global optimum of an expensive-to-evaluate black-box function. A global solution is found by accumulating a pair of query point and its function value, repeating these two procedures: (i) modeling a surrogate function; (ii) maximizing an acquisition function to determine where next to query. Convergence guarantees are only valid when the global optimizer of the acquisition function is found at each round and selected as the next query point. In practice, however, local optimizers of an acquisition function are also used, since searching for the global optimizer is often a non-trivial or time-consuming task. In this paper we consider three popular acquisition functions, PI, EI, and GP-UCB induced by Gaussian process regression. Then we present a performance analysis on the behavior of local optimizers of those acquisition functions, in terms of *instantaneous regrets* over global optimizers. We also introduce an analysis, allowing a local optimization method to start from multiple different initial conditions. Numerical experiments confirm the validity of our theoretical analysis.

Keywords: Global optimization · Bayesian optimization · Acquisition function optimization · Instantaneous regret analysis

1 Introduction

Bayesian optimization provides an efficient method for finding a global optimum of an objective function $f(\mathbf{x}) : \mathcal{X} \to \mathbb{R}$, defined over a compact set $\mathcal{X} \subset \mathbb{R}^d$:

$$\mathbf{x}^\dagger = \arg\min_{\mathbf{x} \in \mathcal{X}} f(\mathbf{x}), \tag{1}$$

where, in general, $f(\mathbf{x})$ is a black-box function, i.e., its closed-form expression is not available and its gradient is not available either. The value of the function can be computed at a query point $\mathbf{x}$ but the evaluation requires a high cost. In this paper we assume that the objective function $f(\mathbf{x})$ of interest is Lipschitz-continuous. Bayesian optimization searches for a minimum of $f(\mathbf{x})$ to solve the

© Springer Nature Switzerland AG 2021
F. Hutter et al. (Eds.): ECML PKDD 2020, LNAI 12458, pp. 675–690, 2021.
https://doi.org/10.1007/978-3-030-67661-2_40

problem (1), gradually accumulating $(\mathbf{x}_t, f(\mathbf{x}_t))$ where input points $\mathbf{x}_t$ are carefully chosen and corresponding function values $f(\mathbf{x}_t)$ are calculated at $\mathbf{x}_t$. It provides an efficient approach in terms of the number of function evaluations required.

In Bayesian optimization, a global solution to the problem (1) is determined by repeating the following two procedures. At each round, we first train a probabilistic model[1] using the data observed so far to construct a surrogate function for $f(\mathbf{x})$. Then we define an acquisition function [10,12,18] over the domain $\mathcal{X}$, which accounts for the utility provided by possible outcomes drawn from the distribution determined by the surrogate model. The maximization of an acquisition function, referred to as an *inner optimization,* yields the selection of the next query point at which to evaluate the objective function. Convergence guarantees are only valid when the global optimizer of the acquisition function is found and selected as the next query point. In practice, however, local optimizers of acquisition functions are also used, since searching for the exact optimizer of the acquisition function is often a non-trivial or time-consuming task.

A recent work [22] has addressed the acquisition function optimization, elucidating gradient-based optimization of Monte Carlo estimates of acquisition functions, as well as on sub-modularity for a family of maximal myopic acquisition functions. However, so far, there is no study on what the performance loss is when a local optimizer of an acquisition function is selected as the next query point. In this paper we attempt to provide an answer to this question on the performance loss brought by local optimizers of acquisition functions over global optimizers, in terms of instantaneous regrets. To this end, we consider three different solutions to the maximization of an acquisition function: (i) a global optimizer; (ii) a local optimizer; (iii) a multi-started local optimizer. For performance analysis of local optimizers, with respect to the global optimizer, we define an *instantaneous regret difference* for a local optimizer as well as for a multi-started local optimizer and present its bound for each case. As expected, the multi-started local optimizer yields a tighter bound on the instantaneous regret difference, compared to the one for a local optimizer.

In this paper we consider three popular acquisition functions, probability of improvement (PI) [10], expected improvement (EI) [12], and Gaussian process upper confidence bound (GP-UCB) [18], each of which is calculated by posterior mean and variance determined by Gaussian process regression. The main contribution of this paper is summarized as:

- We provide an upper bound on the instantaneous regret difference between global and local optimizers, which is given in Theorem 1;
- We provide an upper bound on the instantaneous regret difference when a multi-started local optimization method is employed to search for a local maximum of the acquisition function, which is given in Theorem 2;
- Numerical experiments are provided to justify our theoretical analyses.

---

[1] Gaussian process regression is used in this paper.

2   Background

In this section, we briefly review Bayesian optimization, the detailed overview of which is referred to [2,5,17], and define instantaneous regret difference that is used as a performance measure for local optimizers of acquisition functions. We also explain global and local optimization methods that are popularly used to search for maxima of acquisition functions.

2.1   Bayesian Optimization

The Bayesian optimization strategy solves the problem (1), by gradually selecting queries $\mathbf{x}_1, \ldots, \mathbf{x}_T$ and their corresponding noisy evaluations $y_1, \ldots, y_T$ where $y_t = f(\mathbf{x}_t) + \epsilon_t$ with $\epsilon_t \sim \mathcal{N}(0, \sigma_n^2)$, such that a minimizer of $f(\mathbf{x})$ is determined from $\{\mathbf{x}_1, \ldots, \mathbf{x}_T\}$. Given the data $\mathcal{D}_{t-1} = \{(\mathbf{x}_1, y_1), \ldots, (\mathbf{x}_{t-1}, y_{t-1})\}$ observed up to round $t - 1$, the next point $\mathbf{x}_t$ is chosen as a maximizer of acquisition function $a(\mathbf{x}|\mathcal{D}_{t-1})$, i.e.,

$$\mathbf{x}_t = \arg\max \, a(\mathbf{x}|\mathcal{D}_{t-1}). \tag{2}$$

The acquisition function is the expected utility u of a query $\mathbf{x}$:

$$a(\mathbf{x}|\mathcal{D}_{t-1}) = \int u(\mathbf{x}, y) \, p(y|\mathbf{x}, \mathcal{D}_{t-1}) \, dy, \tag{3}$$

where the posterior distribution $p(y|\mathbf{x}, \mathcal{D}_{t-1})$ is calculated by Gaussian process regression using $\mathcal{D}_{t-1}$ here.

Solving (2) is another optimization problem appearing in the Bayesian optimization task given in (1). We consider three different solutions to the maximization of an acquisition function, defined in detail below.

Definition 1 (Global optimizer).   *We denote by* $\mathbf{x}_{t,g}$ *the optimizer of the acquisition function* $a(\mathbf{x}|\mathcal{D}_{t-1})$ *at round t, determined by a global optimization method, given a time budget* τ:

$$\mathbf{x}_{t,g} = \overset{global}{\arg\max}_{\mathbf{x} \in \mathcal{X}} \, a(\mathbf{x}|\mathcal{D}_{t-1}). \tag{4}$$

$\mathbf{x}_{t,g}$ *is referred to as a global optimizer.*

Definition 2 (Local optimizer).   *We denote by* $\mathbf{x}_{t,l}$ *the optimizer of the acquisition function* $a(\mathbf{x}|\mathcal{D}_{t-1})$ *at round t, determined by an iterative (local) optimization method where the convergence meets* $\|\mathbf{x}_{t,l}^{(\tau)} - \mathbf{x}_{t,l}^{(\tau-1)}\|_2 \leq \epsilon_{opt}$ *for iteration* τ:

$$\mathbf{x}_{t,l} = \overset{local}{\arg\max}_{\mathbf{x} \in \mathcal{X}} \, a(\mathbf{x}|\mathcal{D}_{t-1}). \tag{5}$$

$\mathbf{x}_{t,l}$ *is referred to as a local optimizer.*

Definition 3 (Multi-started local optimizer). *Suppose that* $\{\mathbf{x}_{t,l_1}, \ldots,$ $\mathbf{x}_{t,l_N}\}$ *is a set of N local optimizers, each of which is determined by a local optimization method (5), starting from a different initial condition. The multi-started local optimizer, denoted by $\mathbf{x}_{t,m}$, is the one at which $a(\mathbf{x}|\mathcal{D}_{t-1})$ achieves the maximum:*

$$\mathbf{x}_{t,m} = \arg\max_{\mathbf{x}\in\mathcal{X}}^{m\text{-}local} a(\mathbf{x}|\mathcal{D}_{t-1}). \tag{6}$$

With solutions to (2), defined in (4), (5), and (6), we define *instantaneous regret* for each of these solutions and *instantaneous regret difference* for each of local solutions below.

Definition 4 (Instantaneous regret). *Suppose that $\mathbf{x}^\dagger$ is the true global minimum of the objective function in (1). Denote by $\mathbf{x}_t$ a maximum of acquisition function $a(\mathbf{x}|\mathcal{D}_{t-1})$ at round t, determined by either a global or local optimization method. The instantaneous regret r_t at round t is defined as*

$$r_t = f(\mathbf{x}_t) - f(\mathbf{x}^\dagger). \tag{7}$$

Depending on an optimization method (i.e., one of global, local, and multi-started local optimization methods) used to search for a maximum of the acquisition function, we define the following instantaneous regrets: $r_{t,g} = f(\mathbf{x}_{t,g}) - f(\mathbf{x}^\dagger)$, $r_{t,l} = f(\mathbf{x}_{t,l}) - f(\mathbf{x}^\dagger)$, and $r_{t,m} = f(\mathbf{x}_{t,m}) - f(\mathbf{x}^\dagger)$.

Definition 5 (Instantaneous regret difference). *With Definition 4, we define instantaneous regret differences for an local optimizer $\mathbf{x}_{t,l}$ and for a multi-started local optimizer $\mathbf{x}_{t,m}$:*

$$|r_{t,g} - r_{t,l}| = |f(\mathbf{x}_{t,g}) - f(\mathbf{x}_{t,l})|, \tag{8}$$

$$|r_{t,g} - r_{t,m}| = |f(\mathbf{x}_{t,g}) - f(\mathbf{x}_{t,m})|, \tag{9}$$

which measures a performance gap with respect to the one induced by $\mathbf{x}_{t,g}$, at round t.

Henceforth, instantaneous regret and instantaneous regret difference are simply called to *regret* and *regret difference*, respectively.[2]

2.2   Maximization of Acquisition Functions

As described earlier, we may consider either global or local solutions to (2). Famous global optimization methods include DIRECT [8] and CMA-ES [6]. DIRECT is a deterministic Lipschitzian-based derivative-free partitioning

---

[2] In Bayesian optimization, a cumulative regret is usually used to analyze the performance of convergence quality. By Lemma 5.4 of [18] and Theorem 3 of [3], our analysis on instantaneous regret differences can be expanded into the analysis on cumulative regrets. However, to concentrate the scope of this work on the behavior of local optimizers, these analyses are not included in this paper.

method where it observes function values at the centers of rectangles and divides the rectangles without the Lipschitz constant iteratively. CMA-ES is a stochastic derivative-free method based on evolutionary computing. In this paper we use DIRECT to determine $\mathbf{x}_{t,g}$ in (4).

A local optimization method we used to determine $\mathbf{x}_{t,l}$ or $\mathbf{x}_{t,m}$, given in (5) or (6) is the Broyden-Fletc.her-Goldfarb-Shanno (BFGS) algorithm, which is a quasi-Newton optimization technique. A limited memory version, referred to as L-BFGS [11] and a constrained version known as L-BFGS-B are widely used in the Bayesian optimization literature [5,13,20]. Multi-started optimization methods are also widely used [2,7], where a local optimization method starts from N distinct initializations and such N local solutions started from N distinct initializations are combined to determine the best local solution.

Compared to our work, [22] introduces a reparameterization form to allow differentiability of Monte Carlo acquisition functions to integrate them and query in parallel, which is not related to the topics covered in this paper.

3   Performance Analysis

In this section we present our main contribution on the performance analysis for the local optimizer and the multi-started local optimizer, given in (5) and (6).

3.1   Main Theorems

Before introducing the lemmas used to prove the main theorems, we explain the main theorems and their intuition first. Our theorems are described as follows.

Theorem 1. *Given $\delta_l \in [0,1)$ and $\epsilon_l, \epsilon_1, \epsilon_2 > 0$, the regret difference for a local optimizer $\mathbf{x}_{t,l}$ at round t, $|r_{t,g} - r_{t,l}|$ is less than ϵ_l with a probability at least $1 - \delta_l$:*

$$\mathbb{P}\big(\,|r_{t,g} - r_{t,l}| < \epsilon_l\big) \geq 1 - \delta_l, \tag{10}$$

where $\delta_l = \frac{\gamma}{\epsilon_1}(1-\beta_g) + \frac{M}{\epsilon_2}$, $\epsilon_l = \epsilon_1\epsilon_2$, $\gamma = \max_{\mathbf{x}_i,\mathbf{x}_j \in \mathcal{X}} \|\mathbf{x}_i - \mathbf{x}_j\|_2$ is the size of $\mathcal{X}$, β_g is the probability that a local optimizer of the acquisition function collapses with its global optimizer, and M is the Lipschitz constant explained in Lemma 8.

Theorem 1 is extended for a multi-started local optimizer.

Theorem 2. *Given $\delta_m \in [0,1)$ and $\epsilon_m, \epsilon_2, \epsilon_3 > 0$, a regret difference for a multi-started local optimizer $\mathbf{x}_{t,m}$, determined by starting from N initial points at round t, is less than ϵ_m with a probability at least $1 - \delta_m$:*

$$\mathbb{P}\big(\,|r_{t,g} - r_{t,m}| < \epsilon_m\big) \geq 1 - \delta_m, \tag{11}$$

where $\delta_m = \frac{\gamma}{\epsilon_3}\left(1 - \beta_g\right)^N + \frac{M}{\epsilon_2}$, $\epsilon_m = \epsilon_2\epsilon_3$, $\gamma = \max_{\mathbf{x}_i,\mathbf{x}_j \in \mathcal{X}} \|\mathbf{x}_i - \mathbf{x}_j\|_2$ is the size of $\mathcal{X}$, β_g is the probability that a local optimizer of the acquisition function collapses with its global optimizer, and M is the Lipschitz constant explained in Lemma 8.

As shown in Theorem 1, $|r_{t,g} - r_{t,l}|$ is smaller than ϵ_l with a probability $1 - \delta_l$. It implies the probability $1 - \delta_l$ is controlled by three statements related to γ, β_g, and M: the probability is decreased (i) as γ is increased, (ii) as β_g is decreased, and (iii) as M is increased. If $\mathcal{X}$ is a relatively small space, γ is naturally small. Moreover, β_g is close to one if converging to global optimum by Definition 3 is relatively easy for some reasons: (i) a small number of local optima exist, or (ii) a global optimum is easily reachable.

Theorem 2 suggests the implications that are similar with Theorem 1 in terms of the control factors of $1 - \delta_m$. The main difference of two theorems is that δ_m is related to the number of initial points in Definition 3, N. Because $0 \leq 1 - \beta_g < 1$ is given, N can control the bound of (11). Additionally, by this difference, we theoretically reveal how many runs for a multi-started local optimizer are needed to obtain the sufficiently small regret difference over a global optimizer.

3.2   Lemmas

Next, we prove two statements (i) how different the global and local optimizers are (see Lemma 1 to Lemma 7), and (ii) how steep the slope between the global and local optimizers is (see Lemma 8). First, the Lipschitz continuities of acquisition functions are proved in the subsequent lemmas.

Lemma 1 (Lipschitz continuity of PI). *The PI criterion $a(\mathbf{x}|\mathcal{D}_{t-1})$, formed by the posterior distribution calculated by Gaussian process regression on $\mathcal{D}_{t-1}$ is Lipschitz-continuous.*

Proof. $\mathcal{X}$ is a compact subset of d-dimensional space $\mathbb{R}^d$. In this paper, we analyze our theorem with Gaussian process regression as a surrogate function. If we are given $t - 1$ covariates $\mathbf{X} = [\mathbf{x}_1 \cdots \mathbf{x}_{t-1}]^\top$ obtained from $\mathcal{X}$ and their corresponding responses $\mathbf{y} = [y_1 \cdots y_{t-1}] \in \mathbb{R}^{t-1}$, posterior mean and variance functions, $\mu(\mathbf{x})$ and $\sigma^2(\mathbf{x})$ over $\mathbf{x} \in \mathcal{X}$ can be computed, using Gaussian process regression [14]:

$$\mu(\mathbf{x}) = \mathbf{k}(\mathbf{x}, \mathbf{X})\tilde{\mathbf{K}}^{-1}\mathbf{y}, \tag{12}$$

$$\sigma^2(\mathbf{x}) = k(\mathbf{x}, \mathbf{x}) - \mathbf{k}(\mathbf{x}, \mathbf{X})\tilde{\mathbf{K}}^{-1}\mathbf{k}(\mathbf{X}, \mathbf{x}), \tag{13}$$

where $k(\cdot, \cdot)$ is a covariance function, $\tilde{\mathbf{K}} = \mathbf{K}(\mathbf{X}, \mathbf{X}) + \sigma_n^2 \mathbf{I}$, and σ_n is an observation noise. $\mathbf{k}(\cdot, \cdot)$ accepts a vector and a matrix as two arguments (e.g., $\mathbf{k}(\mathbf{x}, \mathbf{X}) = [k(\mathbf{x}, \mathbf{x}_1) \cdots k(\mathbf{x}, \mathbf{x}_{t-1})]$). Similarly, $\mathbf{K}(\cdot, \cdot)$ can take two matrices (e.g., $\mathbf{K}(\mathbf{X}, \mathbf{X}) = [\mathbf{k}(\mathbf{X}, \mathbf{x}_1) \cdots \mathbf{k}(\mathbf{X}, \mathbf{x}_{t-1})]$). Before showing the Lipschitz continuity of the acquisition function, we first show the derivatives of (12) and (13). It depends on the differentiability of covariance functions, but the famous covariance functions, which are used in Bayesian optimization are usually at least once differentiable (e.g., squared exponential kernel[3] and Matérn kernel[4]). Thus, the derivatives of (12) and (13) are

---

[3] Squared exponential kernel is infinite times differentiable.
[4] Matérn kernel is $\lceil \nu \rceil - 1$ times differentiable.

$$\frac{\partial \mu(\mathbf{x})}{\partial \mathbf{x}} = \frac{\partial \boldsymbol{k}(\mathbf{x}, \mathbf{X})}{\partial \mathbf{x}} \tilde{\boldsymbol{K}}^{-1} \mathbf{y}, \tag{14}$$

$$\frac{\partial \sigma^2(\mathbf{x})}{\partial \mathbf{x}} = -2 \frac{\partial \boldsymbol{k}(\mathbf{x}, \mathbf{X})}{\partial \mathbf{x}} \tilde{\boldsymbol{K}}^{-1} \boldsymbol{k}(\mathbf{X}, \mathbf{x}), \tag{15}$$

using vector calculus identities. To show (14) and (15) are bounded, each term in both equations should be bounded. For all $i \in \{1, \ldots, t-1\}$, $\mathbf{y}$ and $\boldsymbol{k}(\mathbf{X}, \mathbf{x})$ are obviously bounded:

$$|y_i| < \infty \quad \text{and} \quad |k(\mathbf{x}_i, \mathbf{x})| < \infty, \tag{16}$$

but $\partial \boldsymbol{k}(\mathbf{x}, \mathbf{X})/\partial \mathbf{x}$ and $\tilde{\boldsymbol{K}}^{-1}$ should be revealed.

First of all, the bound of $\partial \boldsymbol{k}(\mathbf{x}, \mathbf{X})/\partial \mathbf{x}$ would be proved in Lemma 4. For the latter one, all entries of $\tilde{\boldsymbol{K}}^{-1}$ are bounded by the Kantorovich and Wielandt inequalities [15]. As a result, by (16), Lemma 4, and the Kantorovich and Wielandt inequalities, the following inequalities are satisfied:

$$\left| \frac{\partial \mu(\mathbf{x})}{\partial x_i} \right| < \infty \quad \text{and} \quad \left| \frac{\partial \sigma^2(\mathbf{x})}{\partial x_i} \right| < \infty, \tag{17}$$

for all $i \in \{1, \ldots, d\}$. Thus, we can say

$$\left| \frac{\partial \mu(\mathbf{x})}{\partial x_i} \right| < M_\mu \quad \text{and} \quad \left| \frac{\partial \sigma^2(\mathbf{x})}{\partial x_i} \right| < M_{\sigma^2}, \tag{18}$$

for some $M_\mu, M_{\sigma^2} < \infty$. It implies that $\mu(\mathbf{x})$ and $\sigma^2(\mathbf{x})$ are Lipschitz-continuous with the Lipschitz constants M_μ and M_{σ^2} to each axis direction, respectively. Thus, (12) and (13) are Lipschitz-continuous with the Lipschitz constants dM_μ and dM_{σ^2}, where d is a dimensionality of $\mathbf{x}$, by the triangle inequality.

PI is written with $z(\mathbf{x}) = \left(f(\mathbf{x}^\ddagger) - \mu(\mathbf{x})\right)/\sigma(\mathbf{x})$ if $\sigma(\mathbf{x}) > \sigma_n$, and 0 otherwise, where $\mathbf{x}^\ddagger$ is the current best observation which has a minimum of $\mathbf{y}$. Given the PI criterion $a_{\mathrm{PI}}(\mathbf{x}) = \Phi(z(\mathbf{x}))$, where $\Phi(\cdot)$ is a cumulative distribution function of standard normal distribution, the derivative of PI criterion is

$$\frac{\partial a_{\mathrm{PI}}(\mathbf{x})}{\partial \mathbf{x}} = \phi(z(\mathbf{x})) \frac{\partial z(\mathbf{x})}{\partial \mathbf{x}} = \phi(z(\mathbf{x})) \left(\frac{\mu(\mathbf{x}) - f(\mathbf{x}^\ddagger)}{\sigma^2(\mathbf{x})} \frac{\partial \sigma(\mathbf{x})}{\partial \mathbf{x}} - \frac{1}{\sigma(\mathbf{x})} \frac{\partial \mu(\mathbf{x})}{\partial \mathbf{x}} \right), \tag{19}$$

where $\phi(\cdot)$ is a probability density function of standard normal distribution. By (12), (13), and (18), we can show (19) is bounded, $\|\partial a_{\mathrm{PI}}(\mathbf{x})/\partial \mathbf{x}\|_2 < \infty$. $\qquad \square$

Lemma 2 (Lipschitz continuity of EI). *The EI criterion $a(\mathbf{x}|\mathcal{D}_{t-1})$, formed by the posterior distribution calculated by Gaussian process regression on $\mathcal{D}_{t-1}$ is Lipschitz-continuous.*

Proof. EI expresses with $z(\mathbf{x}) = \left(f(\mathbf{x}^\ddagger) - \mu(\mathbf{x})\right)/\sigma(\mathbf{x})$ if $\sigma(\mathbf{x}) > \sigma_n$, and 0 otherwise, where $\mathbf{x}^\ddagger$ is the current best observation which has a minimum of $\mathbf{y}$. For the EI criterion:

$$a_{\mathrm{EI}}(\mathbf{x}) = \left(f(\mathbf{x}^\ddagger) - \mu(\mathbf{x})\right) \Phi(z(\mathbf{x})) + \sigma(\mathbf{x})\phi(z(\mathbf{x})), \tag{20}$$

the derivative of (20) is

$$
\frac{\partial a_{\mathrm{EI}}(\mathbf{x})}{\partial \mathbf{x}} = \left(f(\mathbf{x}^{\ddagger}) - \mu(\mathbf{x})\right)\phi(z(\mathbf{x}))\frac{\partial z(\mathbf{x})}{\partial \mathbf{x}} - \frac{\partial \mu(\mathbf{x})}{\partial \mathbf{x}}\Phi(z(\mathbf{x}))
$$
$$
+ \sigma(\mathbf{x})\phi'(z(\mathbf{x}))\frac{\partial z(\mathbf{x})}{\partial \mathbf{x}} + \frac{\partial \sigma(\mathbf{x})}{\partial \mathbf{x}}\phi(z(\mathbf{x})). \tag{21}
$$

Similar to (19), the following inequality,

$$
\left\|\frac{\partial a_{\mathrm{EI}}(\mathbf{x})}{\partial \mathbf{x}}\right\|_2 < \infty, \tag{22}
$$

is satisfied.   □

Lemma 3 (Lipschitz continuity of GP-UCB). *GP-UCB* $a(\mathbf{x}|\mathcal{D}_{t-1})$, *formed by the posterior distribution calculated by Gaussian process regression on* $\mathcal{D}_{t-1}$ *is Lipschitz-continuous.*

Proof. GP-UCB [18] and its derivative are

$$
a_{\mathrm{UCB}}(\mathbf{x}) = -\mu(\mathbf{x}) + \alpha\sigma(\mathbf{x}), \tag{23}
$$

$$
\frac{\partial a_{\mathrm{UCB}}(\mathbf{x})}{\partial \mathbf{x}} = -\frac{\partial \mu(\mathbf{x})}{\partial \mathbf{x}} + \alpha\frac{\partial \sigma(\mathbf{x})}{\partial \mathbf{x}}, \tag{24}
$$

where α is a coefficient for balancing exploration and exploitation. By (18), the following inequality,

$$
\left|\frac{\partial a_{\mathrm{UCB}}(\mathbf{x})}{\partial x_i}\right| = \left|-\frac{\partial \mu(\mathbf{x})}{\partial x_i} + \alpha\frac{\partial \sigma(\mathbf{x})}{\partial x_i}\right| \leq \left|-\frac{\partial \mu(\mathbf{x})}{\partial x_i}\right| + \alpha\left|\frac{\partial \sigma(\mathbf{x})}{\partial x_i}\right|
$$
$$
= \left|\frac{\partial \mu(\mathbf{x})}{\partial x_i}\right| + \alpha\left|\frac{\partial \sigma(\mathbf{x})}{\partial x_i}\right| \leq M_\mu + \alpha\sqrt{M_{\sigma^2}}, \tag{25}
$$

is bounded for $i \in \{1, \ldots, d\}$. Therefore, (24) is bounded.   □

Lemma 4. *Given a stationary covariance function* $k(\cdot, \cdot)$ *that is widely used in Gaussian process regression [4, 21],* $\partial \mathbf{k}(\mathbf{x}, \mathbf{X})/\partial \mathbf{x}$ *is bounded where* $\mathbf{X} \in \mathbb{R}^{n\times d}$ *and* $\mathbf{k}(\mathbf{x}, \mathbf{X}) = [k(\mathbf{x}, \mathbf{x}_1) \cdots k(\mathbf{x}, \mathbf{x}_n)]$.

Proof. The well-known stationary covariance functions such as squared exponential (SE) and Matèrn kernels are utilized in Gaussian process regression [4, 21]. Since such kernels are additive or multiplicative [4], this lemma can be generalized to most of kernels applied in Gaussian process regression. In this paper, we analyze the cases of SE and Matèrn 5/2 kernels. Because the cases of Matèrn 3/2 and periodic kernels can be simply extended from the cases analyzed, it is omitted. The SE and Matèrn 5/2 kernels are at least one time differentiable, thus $\partial \mathbf{k}(\mathbf{x}, \mathbf{X})/\partial \mathbf{x}$ can be computed. Before explaining in detail, $\partial \mathbf{k}(\mathbf{x}, \mathbf{X})/\partial \mathbf{x}$ is written as

$$
\frac{\partial \mathbf{k}(\mathbf{x}, \mathbf{X})}{\partial \mathbf{x}} = \left[\frac{\partial k(\mathbf{x}, \mathbf{x}_1)}{\partial \mathbf{x}} \cdots \frac{\partial k(\mathbf{x}, \mathbf{x}_n)}{\partial \mathbf{x}}\right], \tag{26}
$$

for $\mathbf{X} = [\mathbf{x}_1 \cdots \mathbf{x}_n]$. Furthermore, we can define

$$d(\mathbf{x}_1, \mathbf{x}_2) = \sqrt{(\mathbf{x}_1 - \mathbf{x}_2)^\top \boldsymbol{L}^{-1}(\mathbf{x}_1 - \mathbf{x}_2)}, \tag{27}$$

where $\boldsymbol{L}$ is a diagonal matrix of which entries are lengthscales for each each dimension. For simplicity, $\mathbf{x}_1 - \mathbf{x}_2$ is denoted as $\mathbf{s}_{12}$. The derivative of (27) is

$$\frac{\partial d(\mathbf{x}_1, \mathbf{x}_2)}{\partial \mathbf{x}_1} = \left(\boldsymbol{L}^{-1}\mathbf{s}_{12}\right)\left(\mathbf{s}_{12}^\top \boldsymbol{L}^{-1}\mathbf{s}_{12}\right)^{-\frac{1}{2}}. \tag{28}$$

The derivative of $d^2(\mathbf{x}_1, \mathbf{x}_2)$ is $\frac{\partial d^2(\mathbf{x}_1, \mathbf{x}_2)}{\partial \mathbf{x}_1} = 2\boldsymbol{L}^{-1}\mathbf{s}_{12}$.

First, the SE kernel is $k(\mathbf{x}_1, \mathbf{x}_2) = \sigma_s^2 \exp(-\frac{1}{2}d^2(\mathbf{x}_1, \mathbf{x}_2))$ where σ_s is a signal scale. The derivative of each $\partial k(\mathbf{x}, \mathbf{x}_i)/\partial \mathbf{x}$ is

$$\begin{aligned}
\frac{\partial k(\mathbf{x}, \mathbf{x}_i)}{\partial \mathbf{x}} &= \frac{\partial}{\partial \mathbf{x}}\left(\sigma_s^2 \exp\left(-\frac{1}{2}d^2(\mathbf{x}, \mathbf{x}_i)\right)\right) = k(\mathbf{x}, \mathbf{x}_i)\frac{\partial}{\partial \mathbf{x}}\left(-\frac{1}{2}d^2(\mathbf{x}_1, \mathbf{x}_2)\right) \\
&= -\frac{k(\mathbf{x}, \mathbf{x}_i)}{2}\left(2\boldsymbol{L}^{-1}(\mathbf{x} - \mathbf{x}_i)\right) = -k(\mathbf{x}, \mathbf{x}_i)\left(\boldsymbol{L}^{-1}(\mathbf{x} - \mathbf{x}_i)\right). \tag{29}
\end{aligned}$$

Because all the terms of (29) are bounded in $\mathcal{X}$, $\|\partial k(\mathbf{x}, \mathbf{x}_i)/\partial \mathbf{x}\|_2 < \infty$ is satisfied for all $i = \{1, \ldots, n\}$. Note that (27) and (28) are bounded, because $\mathbf{x}_1, \mathbf{x}_2 \in \mathcal{X}$.

The Matérn 5/2 kernel is

$$k(\mathbf{x}_1, \mathbf{x}_2) = \sigma_s^2 \left(1 + \sqrt{5}d(\mathbf{x}_1, \mathbf{x}_2) + \frac{5}{3}d^2(\mathbf{x}_1, \mathbf{x}_2))(-\sqrt{5}d(\mathbf{x}_1, \mathbf{x}_2)\right), \tag{30}$$

where σ_s is a signal scale. Its derivative is

$$\frac{\partial k(\mathbf{x}, \mathbf{x}_i)}{\partial \mathbf{x}} = -\frac{5\sigma_s^2}{3}\left(1 + \sqrt{5}d(\mathbf{x}, \mathbf{x}_i)\right)\exp(-\sqrt{5}d(\mathbf{x}, \mathbf{x}_i))\boldsymbol{L}^{-1}(\mathbf{x} - \mathbf{x}_i). \tag{31}$$

Since the derivative is bounded, $\|\partial k(\mathbf{x}, \mathbf{X})/\partial \mathbf{x}\|_2 < \infty$ is satisfied. Similarly, other kernels can be straightforwardly proved. Thus, this lemma is concluded. $\square$

From now, we show the number of local optima is upper-bounded, using the condition involved in a frequency domain.

Lemma 5. *Let $\mathcal{X} \subset \mathbb{R}^d$ be a compact set. Given some sufficiently large $|\hat{\boldsymbol{\xi}}| > 0$, a spectral density of stationary covariance function for Gaussian process regression is zero for all $|\boldsymbol{\xi}| > |\hat{\boldsymbol{\xi}}|$ with very high probability. Then, the number of local maxima at iteration t, ρ_t is upper-bounded.*

Proof. By Sard's theorem [16] for a Lipschitz-continuous function [1], critical points (i.e., the points whose gradients are zero) do not exist almost everywhere. Since the number of local maxima ρ_t is upper-bounded by the number of critical points, it can be a starting point to bound ρ_t. Especially, by Lemma 1, Lemma 2, and Lemma 3, the number of local maxima ρ_t can be restrained in the compact set $\mathcal{X}$. Since it cannot express the upper-bound of ρ_t, we transform a stationary covariance function using a Fourier transform and obtain the spectral density of each covariance function [14, Chapter 4]. Because a spectral density of stationary covariance function is naturally a light-tail function by Bochner's theorem [19], a spectral density of covariance function for Gaussian process regression is zero for all $|\boldsymbol{\xi}| > |\hat{\boldsymbol{\xi}}|$ with very high probability (i.e., exponentially saturated probability over $|\boldsymbol{\xi}|$ due to the form of stationary kernels [14, Chapter 4]), given some sufficiently large $|\hat{\boldsymbol{\xi}}|$. Then, a function has finite local maxima, which implies that the number of local maxima at iteration t is upper-bounded. □

Based on Lemma 5, we can prove the ergodicity of local maxima that are able to be discovered by the local optimization method and coincided with the local optimizers started from different initial points.

Lemma 6. *Let the number of local maxima of acquisition function at iteration t be ρ_t. Since local optimizers which are started from some initial conditions $\in \mathcal{X}$ are ergodic to all the local maxima, the probability of reaching each solution is $\beta_1, \ldots, \beta_{\rho_t} > 0$ such that $\Sigma_{i=1}^{\rho_t} \beta_i = 1$.*

Proof. If we start from some different initial conditions $\in \mathcal{X}$, it is obvious that all the local solutions are reachable. Therefore, all the local optimizers are ergodic, and the probability of reaching each solution is larger than zero and they sum to one: $\sum_{i=1}^{\rho_t} \beta_i = 1$, where $\beta_1, \ldots, \beta_{\rho_t} > 0$. □

We now prove the distance between two points acquired by Definition 1 and Definition 2 is bounded with a probability.

Lemma 7. *Let $\mathcal{X} \subset \mathbb{R}^d$ be a compact space where $\gamma = \max_{\mathbf{x}_1, \mathbf{x}_2 \in \mathcal{X}} \|\mathbf{x}_1 - \mathbf{x}_2\|_2$. Then, for $\gamma > \epsilon_1 > 0$, we have*

$$\mathbb{P}\big(\|\mathbf{x}_{t,g} - \mathbf{x}_{t,l}\|_2 \geq \epsilon_1\big) \leq \frac{\gamma}{\epsilon_1}(1 - \beta_g), \tag{32}$$

where β_g is the probability that some local optimizer is coincided with the global optimizer of the acquisition function.

Proof. By Markov inequality for $\epsilon_1 > 0$, we have

$$\mathbb{P}\big(\|\mathbf{x}_{t,g} - \mathbf{x}_{t,l}\|_2 \geq \epsilon_1\big) \leq \frac{1}{\epsilon_1}\mathbb{E}\big[\|\mathbf{x}_{t,g} - \mathbf{x}_{t,l}\|_2\big]. \tag{33}$$

Following from Lemma 6 and (33), the expectation in the right-hand side of (33) is calculated as

$$\frac{1}{\epsilon_1}\mathbb{E}\big[\|\mathbf{x}_{t,g} - \mathbf{x}_{t,l}\|_2\big]$$

$$= \frac{1}{\epsilon_1}\beta_g\|\mathbf{x}_{t,g} - \mathbf{x}_{t,l}\|_2\Big|_{\mathbf{x}_{t,g}=\mathbf{x}_{t,l}} + \frac{1}{\epsilon_1}(1 - \beta_g)\|\mathbf{x}_{t,g} - \mathbf{x}_{t,l}\|_2\Big|_{\mathbf{x}_{t,g}\neq\mathbf{x}_{t,l}}$$

$$\leq \frac{1}{\epsilon_1}(1 - \beta_g)\gamma, \tag{34}$$

which completes the proof. □

The lower-bound of $\frac{|f(\mathbf{x}_{t,1}) - f(\mathbf{x}_{t,2})|}{\|\mathbf{x}_{t,1} - \mathbf{x}_{t,2}\|_2}$ can be expressed with a probability as follows.

Lemma 8. *Given any $\epsilon_2 > 0$, the probability that $\frac{|f(\mathbf{x}_{t,g}) - f(\mathbf{x}_{t,l})|}{\|\mathbf{x}_{t,g} - \mathbf{x}_{t,l}\|_2} \geq \epsilon_2$ is less than $\frac{M}{\epsilon_2}$:*

$$\mathbb{P}\left(\frac{|f(\mathbf{x}_{t,g}) - f(\mathbf{x}_{t,l})|}{\|\mathbf{x}_{t,g} - \mathbf{x}_{t,l}\|_2} \geq \epsilon_2\right) \leq \frac{M}{\epsilon_2}. \tag{35}$$

Proof. By Markov's inequality, we can express

$$\mathbb{P}\left(\frac{|f(\mathbf{x}_{t,g}) - f(\mathbf{x}_{t,l})|}{\|\mathbf{x}_{t,g} - \mathbf{x}_{t,l}\|_2} \geq \epsilon_2\right) \leq \frac{1}{\epsilon_2}\mathbb{E}\left[\frac{|f(\mathbf{x}_{t,g}) - f(\mathbf{x}_{t,l})|}{\|\mathbf{x}_{t,g} - \mathbf{x}_{t,l}\|_2}\right] \leq \frac{M}{\epsilon_2}, \tag{36}$$

where M is the Lipschitz constant of function f, because f is M-Lipschitz continuous and $\mathbf{x}_{t,g}, \mathbf{x}_{t,l} \in \mathcal{X}$. $\qquad\square$

3.3 Proof of Theorem 1

Now we present the proof of Theorem 1 here.

Proof. The probability of $|r_{t,g} - r_{t,l}| < \epsilon_l$ can be written as

$$\mathbb{P}\big(|r_{t,g} - r_{t,l}| < \epsilon_l\big) = \mathbb{P}\big(|(f(\mathbf{x}_{t,g}) - f(\mathbf{x}^\dagger)) - (f(\mathbf{x}_{t,l}) - f(\mathbf{x}^\dagger))| < \epsilon_l\big)$$

$$= \mathbb{P}\left(\|\mathbf{x}_{t,g} - \mathbf{x}_{t,l}\|_2 \cdot \frac{|f(\mathbf{x}_{t,g}) - f(\mathbf{x}_{t,l})|}{\|\mathbf{x}_{t,g} - \mathbf{x}_{t,l}\|_2} < \epsilon_l\right). \tag{37}$$

We define two events:

$$E_1 = \big(\|\mathbf{x}_{t,g} - \mathbf{x}_{t,l}\|_2 < \epsilon_1\big) \quad \text{and} \quad E_2 = \left(\frac{|f(\mathbf{x}_{t,g}) - f(\mathbf{x}_{t,l})|}{\|\mathbf{x}_{t,g} - \mathbf{x}_{t,l}\|_2} < \epsilon_2\right). \tag{38}$$

Then, (37) can be expressed as

$$\mathbb{P}\big(|r_{t,g} - r_{t,l}| < \epsilon_l\big) = \mathbb{P}\big(E_1 \cap E_2\big), \tag{39}$$

where $\epsilon_l = \epsilon_1\epsilon_2$. Thus, (37) can be written as

$$\mathbb{P}\big(E_1 \cap E_2\big) = 1 - \mathbb{P}\big(E_1^c \cup E_2^c\big) \geq 1 - \mathbb{P}\big(E_1^c\big) - \mathbb{P}\big(E_2^c\big), \tag{40}$$

since $\mathbb{P}\big(E_1^c \cup E_2^c\big) \leq \mathbb{P}\big(E_1^c\big) + \mathbb{P}\big(E_2^c\big)$ by Boole's inequality. Then, we have

$$\mathbb{P}\big(E_1 \cap E_2\big) \geq 1 - \mathbb{P}\big(\|\mathbf{x}_{t,g} - \mathbf{x}_{t,l}\|_2 \geq \epsilon_1\big) - \mathbb{P}\left(\frac{|f(\mathbf{x}_{t,g}) - f(\mathbf{x}_{t,l})|}{\|\mathbf{x}_{t,g} - \mathbf{x}_{t,l}\|_2} \geq \epsilon_2\right)$$

$$\geq 1 - \frac{\gamma}{\epsilon_1}(1 - \beta_g) - \frac{M}{\epsilon_2}, \tag{41}$$

where Lemma 7 and Lemma 8 are used to arrive at the last inequality. Therefore, the proof is completed:

$$\mathbb{P}\big(|r_{t,g} - r_{t,l}| < \epsilon_l\big) \geq 1 - \delta_l, \tag{42}$$

where $\delta_l = \frac{\gamma}{\epsilon_1}(1 - \beta_g) + \frac{M}{\epsilon_2}$. $\qquad\square$

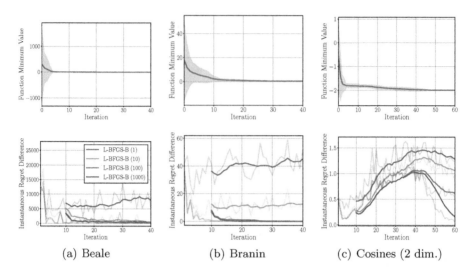

Fig. 1. Empirical results on Theorem 1 and Theorem 2. The caption of each figure indicates a target function. The upper panel of each figure is an optimization result for a global optimizer, and the lower panel is the regret differences between a global optimizer and four types of local optimizer (i.e., multi-started local optimizers found by starting from {1, 10, 100, 1000} initial points). Some legends of the lower panels are missed not to interfere with the graphs, but all the legends are same. For the lower panels, transparent lines are observed instantaneous regret differences and solid lines are moving average (10 steps) of the transparent lines. All experiments are repeated 50 times.

As described above, Theorem 1 implies that the regret difference is basically controlled by γ, β_g, and M. For example, if ρ_t is close to one, the regret difference is tight with high probability. On the other hand, if ρ_t goes to infinity, the difference is tight with low probability.

3.4   On Theorem 2

We extend Theorem 1 into the version for a multi-started local optimizer defined in Definition 3. To prove the next theorem, we need to prove Lemma 9.

Lemma 9. *Let the number of initial points for a multi-started local optimizer be N. A global optimizer and a multi-started local optimizer are different with a probability:*

$$\mathbb{P}(\mathbf{x}_{t,g} \neq \mathbf{x}_{t,m}) = (1 - \beta_g)^N, \tag{43}$$

where $\mathbf{x}_{t,m}$ is determined by (6).

Proof. Since each initial condition of local optimizer is independently sampled, N local optimization methods started from different initial points are independently run. Therefore, the proof is obvious. □

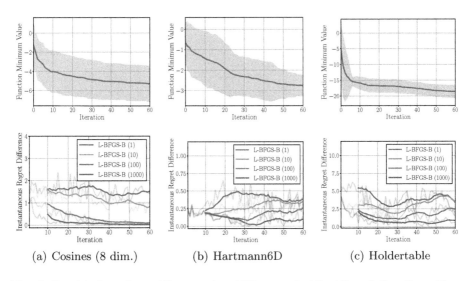

(a) Cosines (8 dim.)   (b) Hartmann6D   (c) Holdertable

Fig. 2. Empirical results on Theorem 1 and Theorem 2. All settings follow the settings described in Fig. 1.

Because $N \geq 1$ and (43) is less than one, (43) is decreased as N is increased. For instance, these are satisfied:

$$(1 - \beta_g)^N \leq 1 - \beta_g \quad \text{and} \quad \lim_{N \to \infty} (1 - \beta_g)^N = 0. \tag{44}$$

By Lemma 9, we can prove the theorem for the local optimization method started from multiple initial points. Before introducing Theorem 2, we simply derive Corollary 1.

Corollary 1. l_2 *distance between the acquired points* $\mathbf{x}_{t,g}$ *and* $\mathbf{x}_{t,m}$ *from (4) and (6) at iteration* t *is larger than any* $\gamma > \epsilon_3 > 0$ *with a probability:*

$$\mathbb{P}\big(\|\mathbf{x}_{t,g} - \mathbf{x}_{t,m}\|_2 \geq \epsilon_3\big) \leq \frac{\gamma}{\epsilon_3}(1 - \beta_g)^N. \tag{45}$$

Proof. Because it can be proved in the same manner of Lemma 7, it is trivial. □

We provide the proof of Theorem 2 with the above lemmas.

Proof. It is an extension of Theorem 1. By Lemma 8 and Corollary 1, it is proved in the same way. □

As we mentioned before, because the equations in (44) are satisfied, we can emphasize a lower-bound on the probability of the case using a multi-started local optimizer is tighter than the case using a local optimizer. It implies an appropriate multi-started local optimizer can produce a similar convergence quality with the global optimizer without expensive computational complexity.

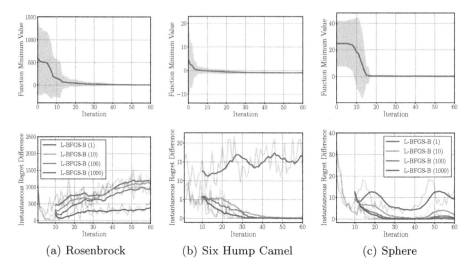

(a) Rosenbrock   (b) Six Hump Camel   (c) Sphere

Fig. 3. Empirical results on Theorem 1 and Theorem 2. All settings follow the settings described in Fig. 1.

4   Empirical Analysis

We present empirical analyses for Theorem 1 and Theorem 2, demonstrating the acquisition function optimization with global, local, and multi-started local optimizers on various examples: Beale, Branin, Cosines (2 dim. and 8 dim.), Hartmann6D, Holdertable, Rosenbrock, Six Hump Camel, and Sphere functions, which are widely used as benchmark functions in the Bayesian optimization literature. We use Gaussian process regression with Mátern 5/2 kernel as a surrogate function and EI as an acquisition function. In addition, the hyperparameters (e.g., signal scale and lengthscales) of Gaussian process regression are found by maximizing the marginal likelihood. All the experiments are implemented with bayeso [9].

To show regret differences, we need to sync up the historical observations for all the methods. The Bayesian optimization results via DIRECT are compared to the results via local optimization methods, considering each of the results by DIRECT as a true global solution for the acquisition function given. At each iteration, four L-BFGS-B algorithms started from $\{1, 10, 100, 1000\}$ different initial points find the next point to measure a regret difference. The points found by L-BFGS-B are only used to compute the regret differences. The transparent lines described in the lower panels of Fig. 1 and Fig. 2 are the observed regret differences, and the solid lines are the moving averages of the transparent lines, each of which is computed as the unweighted mean of the previous 10 steps.

As shown in Fig. 1 and Fig. 2, the regret difference at each iteration is decreased as N is increased, which supports the main theorems. For some cases, the regret differences are slightly increased as the optimization step is repeated.

Table 1. Time (sec.) consumed in optimizing acquisition functions.

	1(a)	1(b)	1(c)	2(a)	2(b)	2(c)	3(a)	3(b)	3(c)
DIRECT	3.434	2.987	2.306	2.508	0.728	2.935	13.928	4.639	10.707
L-BFGS-B (1)	0.010	0.004	0.052	0.023	0.026	0.017	0.005	0.010	0.030
L-BFGS-B (10)	0.096	0.036	0.515	0.224	0.253	0.177	0.050	0.100	0.311
L-BFGS-B (100)	0.977	0.363	5.173	2.224	2.533	1.760	0.504	0.969	3.048
L-BFGS-B (1000)	9.720	3.633	51.818	22.306	25.305	17.629	5.049	9.682	30.764

It means that the acquisition function at the latter iteration has relatively many local optima, which is usually observed in the Bayesian optimization procedures. Furthermore, Table 1 shows Bayesian optimization with a multi-started optimizer is a fair and efficient choice for most of the cases. However, the cases using 1000-started optimizer tends to be slower than the ones using DIRECT, which implies choosing the adequate number of initial conditions for a multi-started local optimizer is significant and it should be carefully selected.

5   Conclusion

In this paper, we theoretically and empirically analyze the upper-bound of instantaneous regret difference between two regrets occurred by global and local optimizers for an acquisition function. The probability on this bound becomes tighter, using a multi-started local optimizer instead of the local optimizer. Our experiments show our theoretical analyses can be supported.

References

1. Barbet, L., Dambrine, M., Daniilidis, A., Rifford, L.: Sard theorems for Lipschitz functions and applications in optimization. Israel J. Math. **212**(2), 757–790 (2016). https://doi.org/10.1007/s11856-016-1308-7
2. Brochu, E., Cora, V.M., de Freitas, N.: A tutorial on Bayesian optimization of expensive cost functions, with application to active user modeling and hierarchical reinforcement learning. arXiv preprint arXiv:1012.2599 (2010)
3. Chowdhury, S.R., Gopalan, A.: On kernelized multi-armed bandits. In: Proceedings of the International Conference on Machine Learning (ICML), pp. 844–853, Sydney, Australia (2017)
4. Duvenaud, D.: Automatic model construction with Gaussian processes. Ph.D. thesis, University of Cambridge (2014)
5. Frazier, P.I.: A tutorial on Bayesian optimization. arXiv preprint arXiv:1807.02811 (2018)
6. Hansen, N.: The CMA evolution strategy: A tutorial. arXiv preprint arXiv:1604.00772 (2016)
7. Hutter, F., Hoos, H.H., Leyton-Brown, K.: Sequential model-based optimization for general algorithm configuration. In: Coello, C.A.C. (ed.) LION 2011. LNCS, vol. 6683, pp. 507–523. Springer, Heidelberg (2011). https://doi.org/10.1007/978-3-642-25566-3_40

8. Jones, D.R., Perttunen, C.D., Stuckman, B.E.: Lipschitzian optimization without the Lipschitz constant. J. Optim. Theory Appl. **79**(1), 157–181 (1993). https://doi.org/10.1007/BF00941892

9. Kim, J., Choi, S.: Bayeso: a Bayesian optimization framework in Python (2017). http://bayeso.org

10. Kushner, H.J.: A new method of locating the maximum point of an arbitrary multipeak curve in the presence of noise. J. Basic Eng. **86**(1), 97–106 (1964)

11. Liu, D.C., Nocedal, J.: On the limited memory BFGS method for large scale optimization. Math. Program. **45**(3), 503–528 (1989). https://doi.org/10.1007/BF01589116

12. Moćkus, J., Tiesis, V., Žilinskas, A.: The application of Bayesian methods for seeking the extremum. Towards Glob. Optim. **2**, 117–129 (1978)

13. Picheny, V., Gramacy, R.B., Wild, S., Le Digabel, S.: Bayesian optimization under mixed constraints with a slack-variable augmented Lagrangian. In: Advances in Neural Information Processing Systems (NeurIPS), vol. 29, pp. 1435–1443, Barcelona, Spain (2016)

14. Rasmussen, C.E., Williams, C.K.I.: Gaussian Processes for Machine Learning. MIT Press, Cambridge (2006)

15. Robinson, P.D., Wathen, A.J.: Variational bounds on the entries of the inverse of a matrix. IMA J. Numer. Anal. **12**(4), 463–486 (1992)

16. Sard, A.: The measure of the critical values of differentiable maps. Bull. Am. Math. Soc. **48**(12), 883–890 (1942)

17. Shahriari, B., Swersky, K., Wang, Z., Adams, R.P., de Freitas, N.: Taking the human out of the loop: a review of Bayesian optimization. Proc. IEEE **104**(1), 148–175 (2016)

18. Srinivas, N., Krause, A., Kakade, S., Seeger, M.: Gaussian process optimization in the bandit setting: no regret and experimental design. In: Proceedings of the International Conference on Machine Learning (ICML), pp. 1015–1022, Haifa, Israel (2010)

19. Stein, M.L.: Interpolation of Spatial Data: Some Theory for Kriging. Springer, New York (1999). https://doi.org/10.1007/978-1-4612-1494-6

20. Wang, Z., Gehring, C., Kohli, P., Jegelka, S.: Batched large-scale Bayesian optimization in high-dimensional spaces. In: Proceedings of the International Conference on Artificial Intelligence and Statistics (AISTATS), pp. 745–754, Lanzarote, Spain (2018)

21. Wilson, A.G., Adams, R.P.: Gaussian process kernels for pattern discovery and extrapolation. In: Proceedings of the International Conference on Machine Learning (ICML), pp. 1067–1075, Atlanta, Georgia, USA (2013)

22. Wilson, J.T., Hutter, F., Deisenroth, M.P.: Maximizing acquisition functions for Bayesian optimization. In: Advances in Neural Information Processing Systems (NeurIPS), vol. 31, pp. 9906–9917, Montreal, Quebec, Canada (2018)

Bayesian Optimization with Missing Inputs

Phuc Luong[(⊠)], Dang Nguyen, Sunil Gupta, Santu Rana,
and Svetha Venkatesh

Applied Artificial Intelligence Institute (A2I2), Deakin University, Waurn Ponds,
Geelong, VIC 3216, Australia
{pluong,d.nguyen,sunil.gupta,santu.rana,svetha.venkatesh}@deakin.edu.au

Abstract. Bayesian optimization (BO) is an efficient method for optimizing expensive black-box functions. In real-world applications, BO often faces a major problem of missing values in inputs. The missing inputs can happen in two cases. First, the historical data for training BO often contain missing values. Second, when performing the function evaluation (e.g., computing alloy strength in a heat treatment process), errors may occur (e.g., a thermostat stops working) leading to an erroneous situation where the function is computed at a random unknown value instead of the suggested value. To deal with this problem, a common approach just simply skips data points where missing values happen. Clearly, this naive method cannot utilize data efficiently and often leads to poor performance. In this paper, we propose a novel BO method to handle missing inputs. We first find a probability distribution of each missing value so that we can impute the missing value by drawing a sample from its distribution. We then develop a new acquisition function based on the well-known Upper Confidence Bound (UCB) acquisition function, which considers the uncertainty of imputed values when suggesting the next point for function evaluation. We conduct comprehensive experiments on both synthetic and real-world applications to show the usefulness of our method.

Keywords: Bayesian optimization · Missing data · Matrix factorization · Gaussian process

1 Introduction

Bayesian optimization (BO) [20] is a powerful tool to optimize expensive black-box functions. Typically, at each iteration BO first models the black-box function via a statistical model, e.g., a Gaussian process (GP) based on historical data (*observed data*) and then seeks out the next point (*suggestion*) for function evaluation by maximizing an *acquisition function*. BO has been successfully applied to a wide range of practical applications such as hyper-parameter tuning, automated machine learning, material design, and robot exploration [6,15,16,22].

© Springer Nature Switzerland AG 2021
F. Hutter et al. (Eds.): ECML PKDD 2020, LNAI 12458, pp. 691–706, 2021.
https://doi.org/10.1007/978-3-030-67661-2_41

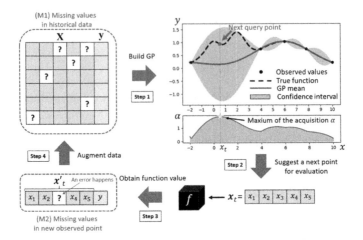

Fig. 1. Four main steps in BO: (1) build a GP from the historical data $[\mathbf{X}, \mathbf{y}]$, (2) maximize an acquisition function α to get a suggested point $\boldsymbol{x}_t$, (3) evaluate the suggested point $\boldsymbol{x}_t$ with the true black-box function and obtain a function value y, and (4) augment the historical data with the new observed point. With the presence of missing values in (**M1**) *historical data* and (**M2**) *new observed point*, BO faces two significant problems: it cannot build the GP at Step-1 and it cannot use the new observed point at Step-4.

In real-world applications, BO often faces a significant problem that is *missing values in inputs*. As shown in Fig. 1, missing values in input can happen in two cases. First, similar to other machine learning models, the historical data for training BO may contain missing values (*missing values in historical data*). Without imputing these missing values, we cannot model the black-box function using GP. Second, when performing the function evaluation at the suggested point, if an error happens (e.g., failure of devices), we obtain the function value at an unknown random point (*missing values in new observed point*). Missing values in input can lead to many crucial failures in BO optimization such as erroneous calculation, and difficulties in interpretation and representation of information [21].

To address the missing input problem in BO, one approach is to apply imputation methods, e.g., mean/mode imputation and k-nearest neighbors (KNN) [1,3] to fill missing values first and then apply a traditional BO method. Although these imputation methods can predict missing values, their performance is non-optimal since mean/mode methods do not consider the correlation between missing values and non-missing values while KNN strongly depends on the current available data and distance metric [5]. Recently, more complex imputation methods have been introduced, including using random forest [24] and deep neural network [26]; however, these methods require extensive training data, which is unrealistic in the BO context where the historical data is quite limited. Another approach is to simply apply BO to non-missing data where points with missing values are removed [11]. As mentioned before, this approach does not use the data efficiently, leading to poor performance in optimization. Oliveira et al. proposed a BO method for uncertain inputs [16], where they observe the difference between the actual input value and the one recommended by BO, and they estimate the variance needed to build the probability distribution of input values. However, in the case of missing values, this variance is unknown (i.e., the noise level added to the actual input value is unknown), their method cannot approximate missing values well. To the best of our knowledge, there is no BO method that can directly handle missing values in input.

Our Method. To overcome the disadvantages of existing methods, we propose a novel method (named **BOMI**) for optimizing black-box functions with missing values in input. In particular, we first adapt the idea of Bayesian probabilistic matrix factorization (BPMF) [19] to find the distribution of each missing value for imputation. Note that none of the imputation methods discussed above use the distributions of missing values for imputation even though these distributions are essential since they represent a certain level of noise in the actual values. By adapting the idea of BPMF, these distributions are built using one of the collaborative filtering technique so that the correlation between values in the data is taken into account. We then propose a new acquisition function, based on the widely used UCB acquisition function [23], to achieve greater confidence in modeling the black-box function. Our new acquisition function differs from a traditional acquisition function in a sense that it does not use one single GP built from imputed data but leverages multiple GPs to take into account the uncertainty of predicted values. By doing this, our method achieves an agreement on the imputed values that results in a higher confidence in the posterior predictive distribution. As a result, it improves the optimization performance when the black-box functions involve missing inputs.

To summarize, we make the following contributions.

- Develop _Bayesian Optimization with Missing Inputs_ (**BOMI**) to optimize black-box functions with missing values in input.
- Propose a new acquisition function that takes into account the distributions of missing values when suggesting the next point for function evaluation.
- Demonstrate the usefulness of **BOMI** in both synthetic and real-world applications, and show that it outperforms well-known state-of-the-art baselines.

2   Background

2.1   Bayesian Optimization

Bayesian optimization (BO) is an efficient method for automatically finding the optimum of an expensive black-box function within a small number of function evaluations [4,13]. Given an unknown function $f : \mathcal{X} \to \mathbb{R}$, our goal is to find the optimal input $x^* = \arg\max_{x \in \mathcal{X}} f(x)$, where $\mathcal{X}$ is a bounded domain in $\mathbb{R}^d$. Since the objective function f is expensive to evaluate, BO attempts to model f via a surrogate model, e.g., Gaussian process (GP) [17]. The function f is assumed to be drawn from the GP, i.e., $f(x) \sim \mathcal{GP}(\mu(x), k(x, x'))$, where $\mu : \mathcal{X} \to \mathbb{R}$ and $k : \mathcal{X} \times \mathcal{X} \to \mathbb{R}$ are mean and covariance functions. Normally, $\mu(x)$ is assumed to be zero and k is the *squared exponential* kernel (Eq. (1)):

$$k(x, x') = \sigma^2 exp(-\frac{1}{2l^2}\|x - x'\|^2) \tag{1}$$

where σ^2 is a parameter dictating the uncertainty in f, and l is a length scale parameter which controls how quickly f can change.

Given the historical data up to iteration t, $\mathcal{D}_t = \{(x_i, y_i)\}_{i=1}^t$ that contains inputs x_i and their evaluations $y_i = f(x_i) + \epsilon_i$ for $i = 1, 2, \ldots, t$ where $\epsilon_i \sim \mathcal{N}(0, \sigma_\epsilon^2)$, we obtain the predictive distribution $f(x) \mid \mathcal{D}_t \sim \mathcal{N}(\mu_t(x), \sigma_t^2(x))$ with $\mu_t(x)$ and $\sigma_t^2(x)$ as:

$$\mu_t(x) = \mathbf{k}^T(\mathbf{K} + \sigma_\epsilon^2\mathbf{I})^{-1}\mathbf{y} \tag{2}$$

$$\sigma_t^2(x) = k(x, x) - \mathbf{k}^T(\mathbf{K} + \sigma_\epsilon^2\mathbf{I})^{-1}\mathbf{k} \tag{3}$$

where $\mathbf{y} = (y_1, ..., y_t)$ is a vector of function evaluations, $\mathbf{k} = [k(x_i, x)]_{\forall x_i \in \mathcal{D}_t}$ is the covariance between a new input x and all observed inputs x_i, $\mathbf{K} = [k(x_i, x_j)_{\forall x_i, x_j \in \mathcal{D}_t}]$ is the covariance matrix between all inputs, $\mathbf{I}$ is an identity matrix with the same dimension as $\mathbf{K}$, and σ_ϵ^2 is a measurement noise.

BO uses the predictive mean and standard deviation in Eqs. (2) and (3) in an *acquisition function* $\alpha(x)$ to find the next point to evaluate. The acquisition function uses the predictive distribution to balance two contrasting goals: sampling where the function is expected to have a high value vs. sampling where the uncertainty about the function value is high. Some well-known acquisition functions are *Probability of Improvement* (PI) [12], *Expected Improvement* (EI) [10], *Upper Confidence Bound* (UCB) [23], and *Predictive Entropy Search* (PES) [9].

Since we use UCB as a base to develop a new acquisition function (see Sect. 3.3) for the optimization problem with missing inputs, we describe it in detail in the next section.

2.2   Upper Confidence Bound Acquisition Function

The UCB acquisition function is a weighted sum of predictive mean and variance from Eqs. (2) and (3), computed as:

$$\alpha_t^{UCB}(x) = \mu_t(x) + \sqrt{\beta_t}\sigma_t(x) \tag{4}$$

where β_t is the exploitation-exploration trade-off factor. Following [23], β_t is calculated as $\beta_t = 2\log\left(t^2 2\pi^2/3\delta\right) + 2d\log\left(t^2 dbr\sqrt{\log(4da/\delta)}\right)$ to guarantee an upper bound on the cumulative regret with probability greater than $1 - \delta$ in the search space $\mathcal{X} \subseteq [0, r]^d$, where $r > 0$ and $a, b > 0$ are constants.

To suggest a next point for the black-box objective function evaluation, we maximize the UCB acquisition function in Eq. (4) as follows:

$$x_{t+1} = \arg\max_{x \in \mathcal{X}} \alpha_t^{UCB}(x) \tag{5}$$

3 Framework

3.1 Problem Definition

Before formally defining the problem of Bayesian optimization (BO) with missing inputs, we provide two cases when missing values occur in inputs.

Case 1. (**Missing values in historical data**) Given a point $\boldsymbol{x} = \{x_1, \ldots, x_d\}$ in historical data, $\boldsymbol{x}$ contains missing values if $\exists x_i \in \boldsymbol{x}$ ($i \in \{1, ..., d\}$ and d is the input dimension) such that x_i is unobserved (i.e., missing), and we denote x_i by a question mark '?'.

Case 2. (**Missing values in the next suggested point**) At iteration t, when we intend to evaluate the black-box function f at a suggested point $\boldsymbol{x}_t = \{x_1, \ldots, x_d\}_t$ to obtain the function value y_t, two scenarios may arise. (1) Due to an error in the evaluation, the function may actually be evaluated at $\boldsymbol{x}'_t$ instead of intended point $\boldsymbol{x}_t$. In general, we denote an element x'_i of $\boldsymbol{x}'_t$ using the corresponding element of $\boldsymbol{x}_t$ as $x'_i = x_i \pm \eta$ where η is an unknown noise amount. (2) In case of <u>no error</u>, $\boldsymbol{x}'_t$ is same as $\boldsymbol{x}_t$.

We present the problem of BO with missing inputs. Given a historical data $[\mathbf{X}, \mathbf{y}]$ and a *black-box* function $f : \mathcal{X} \rightarrow \mathbb{R}$ ($\mathcal{X}$ is the input domain), $\mathbf{X}$ may contain missing values as mentioned in Case 1, and if we query a point $\boldsymbol{x}_t \in \mathcal{X}$ to compute the function value $y_t = f(\boldsymbol{x}_t)$, then we may obtain $y_t = f(\boldsymbol{x}'_t)$ as mentioned in Case 2. Our goal is to find the optimal point $\boldsymbol{x}^*$ that maximizes the black-box function f, as follows:

$$\boldsymbol{x}^* = \arg\max_{x \in \mathcal{X}} f(\boldsymbol{x}) \tag{6}$$

3.2 Building a Probability Distribution for Each Missing Value

Let $\boldsymbol{x}_o$ and $\boldsymbol{x}_m$ be non-missing and missing values. An observation is denoted as $\boldsymbol{x} = \{\boldsymbol{x}_o, \boldsymbol{x}_m\}$. To solve the optimization problem in Eq. (6), one simple approach is to omit observations having missing values $\boldsymbol{x}_m$ and then apply a standard BO to observations containing only non-missing values. As discussed in Sect. 1, this method may perform poorly since it has too few data points to build a good model. To overcome this, we propose to use the distribution of a missing value so that we can both impute it as well as utilize the uncertainty in its prediction. Therefore, instead of directly substituting $x = c$ ($x \in \boldsymbol{x}_m$ and c is a single constant value), we assume that $x \sim p(x)$, where $p(x)$ is an unknown probability distribution of x, and our goal is to find $p(x)$ for each $x \in \boldsymbol{x}_m$.

We represent the observed data $[\mathbf{X}, \mathbf{y}]$ as a matrix $R = [\mathbf{X}, \mathbf{y}] \in \mathbb{R}^{N \times M+1}$, where N is the number of rows (data points) and M is the number of columns (features). Let x_{ij} be a missing value at row i and column j, and x_{ij} is assumed to be sampled from a normal distribution $p(x_{ij}) = \mathcal{N}\left(\mu_{x_{ij}}, \sigma^2_{x_{ij}}\right)$. To find the distribution $p(x_{ij})$, we adapt the idea of Bayesian probabilistic matrix factorization (BPMF) [19].

Our goal is to decompose the partially-observed matrix $R \in \mathbb{R}^{N \times M+1}$ into a product of two smaller matrices $U \in \mathbb{R}^{N \times K}$ and $V \in \mathbb{R}^{K \times M+1}$ such that $R \approx UV$, i.e., we find two matrices U and V whose product is as close as possible to the original matrix R.

We first construct the prior distributions on U and V as follows:

$$p\left(U \mid \mu_U, \Lambda_U\right) = \prod_{i=1}^{N} \mathcal{N}\left(U_i \mid \mu_U, \Lambda_U^{-1}\right)$$
$$p\left(V \mid \mu_V, \Lambda_V\right) = \prod_{j=1}^{M+1} \mathcal{N}\left(V_i \mid \mu_V, \Lambda_V^{-1}\right) \tag{7}$$

where $\Theta_U = \{\mu_U, \Lambda_U\}$ and $\Theta_V = \{\mu_V, \Lambda_V\}$ are hyper-parameters of the priors. We learn them using Gibbs sampling [14].

Next, we sample U and V from their distributions, as in Eq. (8):

$$U_i^{l+1} \sim p\left(U_i \mid R, V^l, \Theta_U^l\right) \text{ for } i = 1, \dots, N \text{ rows}$$
$$V_j^{l+1} \sim p\left(V_i \mid R, U^{l+1}, \Theta_V^l\right) \text{ for } j = 1, \dots, (M+1) \text{ columns} \tag{8}$$

where l is the number of iterations used in Gibbs sampling.

Finally, we reconstruct $R \approx UV$ and the missing value $x_{ij} = R_{ij}$ is filled by a linear combination of matrix product, i.e., $x_{ij} = U_{i,:}.V_{:,j}$, where $U_{i,:}$ is the row i of U and $V_{:,j}$ is the column j of V.

Although we can impute a missing value using $x_{ij} = U_{i,:}.V_{:,j}$, using a single predicted value is not effective. Thus, we go a step further to obtain the distribution $p\left(x_{ij}\right)$ of x_{ij}. In particular, following [19] we use the Monte Carlo approximation [14] to approximate $p\left(x_{ij}\right)$ as:

$$p\left(x_{ij}\right) \approx p\left(x_{ij} \mid U_{i,:}.V_{:,j}, \xi\right),$$

where $\xi = \sigma_{R_{ij}}^2$ (called *precision factor*) is the "width" of distribution covering the actual value of x_{ij}. To fill/predict a missing value x_{ij}, we simply draw a sample from its distribution $\tilde{x}_{ij} \sim p\left(x_{ij}\right)$ and set $x_{ij} = \tilde{x}_{ij}$.

3.3   Bayesian Optimization with Missing Inputs (BOMI)

In Sect. 3.2, we find a distribution $p\left(x_{ij}\right)$ for each missing value x_{ij}. To optimize the black-box function $f(\boldsymbol{x})$, we can simply draw a sample $\tilde{x}_{ij} \sim p\left(x_{ij}\right)$ to fill the missing value x_{ij}, and then apply a standard BO to the new non-missing data. We call this method *Imputation-BPMF*. However, the performance of this approach heavily depends on the quality of $\tilde{x}_{ij} \sim p\left(x_{ij}\right)$. In other words, it does not consider the uncertainty of $\tilde{x}_{ij}$.

We propose a novel BO method to optimize black-box functions with missing inputs, called *Bayesian Optimization with Missing Inputs* (**BOMI**). Our method has three main steps, which illustrated in Fig. 2. **Step 1:** from the observed data with missing values, **BOMI** learns a distribution for each missing value (see Sect. 3.2), then uses these distributions to impute and generate Q new non-missing data. **Step 2:** for each new non-missing data, **BOMI** builds a GP and computes the acquisition function UCB (see Eq. (4)). **Step 3: BOMI**

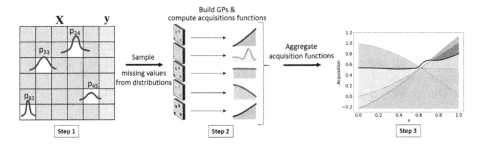

Fig. 2. Three main steps in our method **BOMI**: (1) sample missing values from their distributions, (2) build GPs and compute UCB acquisition functions (Eq. (4)), and (3) develop a new acquisition function based on aggregated information.

aggregates the information from Q acquisition functions to come up with a new acquisition function that takes into account the uncertainty of imputed values. The new acquisition function (called **UCB-MI**) is described next.

Upper Confidence Bound acquisition function for Missing Inputs (UCB-MI). Our new acquisition function UCB-MI aggregates the information from Q standard UCB acquisition functions computed at Step-2, as follows:

$$\alpha^{UCB-MI}(x) = \mu_\alpha\left(\boldsymbol{\alpha}^{UCB}(x)\right) + \beta_\alpha \sigma_\alpha\left(\boldsymbol{\alpha}^{UCB}(x)\right) \tag{9}$$

$$= \frac{1}{Q}\sum_{q=1}^{Q}\left(\alpha_q^{UCB}(x)\right) + \beta_\alpha\sqrt{\frac{\sum_{q=1}^{Q}\left(\alpha_q^{UCB}(x) - \frac{1}{Q}\sum_{q=1}^{Q}\alpha_q^{UCB}(x)\right)^2}{Q-1}}$$

where α^{UCB} is the UCB acquisition function (see Eq. (4)).

Our acquisition function α^{UCB-MI} is based on the commonly used UCB acquisition function, but it incorporates the posterior predictive information from different GPs. It is described as a summation of the mean of Q acquisition values $\mu_\alpha(\alpha)$ and their standard deviation $\sigma_\alpha(\alpha)$ multiplied by a trade-off factor β_α. This acquisition function quantifies the level of agreement between Q individual acquisition functions to determine the confidence in predicting the outcome of an input. As a result, we have more information about the variance of one point $\boldsymbol{x}$ and more certainty about its outcome. To suggest a next point for evaluation, we maximize the acquisition function α^{UCB-MI}:

$$x_{t+1} = \underset{x\in\mathcal{X}}{\operatorname{argmax}} \, \alpha^{UCB-MI}(x) \tag{10}$$

Discussion. We can see that when Q is set to a small value, our acquisition function α^{UCB-MI} is close to the standard acquisition function UCB α^{UCB}. For example, with $Q = 1$, the standard deviation $\sigma_\alpha\left(\boldsymbol{\alpha}^{UCB}\right) = 0$ and $\alpha^{UCB-MI}(x) = \alpha_1^{UCB}(x) + \beta_\alpha 0 = \alpha^{UCB}(x)$.

When $Q > 1$, the first term $\frac{1}{Q}\sum_{q=1}^{Q}\left(\alpha_q^{UCB}(x)\right)$ in Eq. (9) represents the average among different acquisition functions, which can be considered as an agreement on different acquisition functions. In contrast, the second term represents the disagreement on acquisition values since it is the standard deviation measuring how much acquisition functions differ from their mean (agreement). The trade-off factor β_α is used to control the balance between agreement and disagreement.

Our proposed method **BOMI** is summarized in Algorithm 1.

Algorithm 1: The proposed **BOMI** algorithm.

Input: Observed data D_0, # iterations T, # new non-missing data Q

1 **begin**
2   **for** $t = 0, ..., T$ **do**
3   **for** $q = 1, ..., Q$ **do**
4   Sample $U_{(q)} \sim p\left(U \mid R, V, \Theta_U\right)$ and $V_{(q)} \sim p\left(V \mid R, U, \Theta_V\right)$
5   Generate new non-missing data $R_{(q)} = U_{(q)} V_{(q)}$
6   Build GP $GP_{(q)} \leftarrow R_{(q)}$
7   Compute acquisition function $\alpha_q^{UCB} \leftarrow \textsc{Acquisition}(GP_{(q)})$
8   **end**
9   Compute α^{UCB-MI} using Equation (9)
10   Suggest a next point $x_{t+1} = \arg\max_{x \in \mathcal{X}} \alpha^{UCB-MI}(x)$
11   Evaluate the objective function $y_{t+1} = f(x_{t+1})$
12   **if** <u>missing event</u> **then**
13             $x_{t+1} \rightarrow x_{t+1}'$ (see Case 2)
14             $y_{t+1} = f(x_{t+1}')$
15   **end**
16   Augment $D_{t+1} = \{D_t, (x_{t+1}, y_{t+1})\}$
17   **end**
18 **end**

---

4   Experimental Results

We evaluate our proposed method **BOMI** in both synthetic and real-world applications. For synthetic experiments, we test our method with four benchmark synthetic functions to show its optimization performance and stability. For real-world experiments, we test the performance of our method in two real-world applications, namely, a robot exploration simulation and a heat treatment process. In these two applications, missing inputs often occur since the failures of robots and thermostat are unmanageable.

Baselines. We compare **BOMI** with six state-of-the-art baselines that use different ways to deal with missing values. They are categorized into two groups:

- **Imputation-based methods:** These methods first use imputation methods to predict missing values and then simply apply a standard BO method to optimize the black-box functions. Here, we use three well-known imputation methods in machine learning, namely, mean, mode, and KNN [3]. The mean method (called *Imputation-Mean*) replaces a missing value by the mean of its feature column. The mode method (called *Imputation-Mode*) replaces a missing value by the mode of its feature column. The KNN method (called *Imputation-KNN*) replaces a missing value by the mean value of its k nearest points. We also compare with *Imputation-BPMF*, where missing values are imputed using the BPMF method (see Sect. 3.3).
- **BO-based methods:** Since standard BO methods cannot directly deal with missing inputs, we consider two variants of BO. *DropBO* – whenever a data point containing missing values occurs in historical data or new observed point, this method simply skips that data point and applies a standard BO method to non-missing data [11]. *SuggestBO* – similar to DropBO this method removes data points containing missing values in historical data; however when a new observed point contains missing values, instead of skipping this new observation this method still uses it but substitutes missing values by the values suggested by the acquisition function. We also compare with *BO-uGP* [16] – a recent BO method proposed for optimizing black-box functions with uncertain inputs. This method assumes that there is no missing values but all of them are noisy. It first maps all points into distributions and then builds a surrogate model over the distributions of points.

Implementation Details. We implement our method **BOMI** and all baselines using GPyTorch [7] to accelerate matrix multiplication operations in GP inference. For a fair comparison, in our experiments we use the same kernel (*squared exponential* kernel) and identical initial points for all methods. For Imputation-KNN, we use the number of neighbors $k = 5$ and the Euclidean distance, following [2]. For BO-uGP, we use the same hyper-parameter setting, as mentioned in the paper. For our method **BOMI**, we set the dimension K of matrices U and V to 15, the precision $\xi = 0.01$, the number of new non-missing data $Q = 5$, and the number of iterations in Gibbs sampling $l = 40$. We repeat each method 10 times and report the average result along with the standard error.

4.1 Synthetic Experiments

We test our method and baselines with four benchmark synthetic functions where their characteristics are summarized in Table 1.

Performance Comparison. The first experiment illustrates how our method **BOMI** outperforms other methods in terms of optimization result.

Experiment Settings. We initialize 30 data points (historical data) for each function and keep them the same for all methods. To see the effect of missing values, we allow 80% of historical data to have missing values. When evaluating

Table 1. Characteristics of synthetic functions.

Function	Dimension	Range
Eggholder	2	$x_1, x_2 \in [-512, 512]$
Schubert	4	$x_i \in [-10, 10]$ for $i = 1, \ldots, 4$
Alpine	5	$x_i \in [-10, 10]$ for $i = 1, \ldots, 5$
Schwefel	5	$x_i \in [-500, 500]$ for $i = 1, \ldots, 5$

a suggested point, there is a probability ρ (called *missing rate*) that the new observed point has missing values (i.e., an error occurs, see Case 2). With this probability ρ, an amount of noise η (called *missing noise*) is added to the suggested value, which is calculated as $x_i' = x_i \pm \eta r_i$, where x_i is the actual value suggested by the acquisition function, r_i is the value range of x_i, and x_i' is a random unknown value. In our experiments, we set $\rho = 0.25$ and $\eta = 0.05$ for all functions.

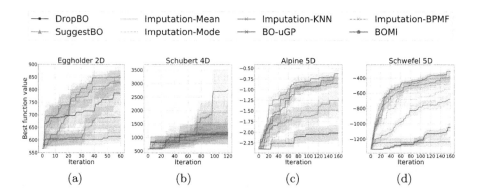

Fig. 3. Optimization results for four synthetic functions in Table 1.

Results and Discussion. Figure 3 shows the optimization results for four synthetic functions in Table 1. We can see that our method **BOMI** generally outperforms other methods. On the 2d-function *Eggholder* (Fig. 3(a)), **BOMI** and Imputation-BPMF are the two best methods, where they are slightly better than SuggestBO. When the dimension is increased up to 4 and 5 (Fig. 3(b-d)), **BOMI** is always the best method, and especially it significantly outperforms other baselines on the 4d-function *Schubert* (Fig. 3(b)).

Imputation-based methods (mean, mode, KNN, and BPMF) work fairly well; however, their performance is not very consistent. Mean and mode imputations often fall behind KNN since they suffer from biases. Imputation-BPMF is often better or comparable with other imputation methods, which verifies our intuition about the importance of distributions of missing values, as mentioned in Sect. 3.2.

BOMI is often better than Imputation-BPMF. This clearly proves that our proposal of using probability distributions to impute missing values along with the new acquisition function is more effective, as discussed in Sect. 3.3.

DropBO underperforms on most functions since it throws away many observations when they contain missing values, which leads to too few data to train a good GP model. In contrast, SuggestBO is always better than DropBO since it has more observations by replacing the missing value with the value of suggested point. On the *Eggholder* function (Fig. 3(a)), SuggestBO achieves a very good performance, where it is the second-best method. However, on other functions SuggestBO only achieves fair results since these functions vary very quickly even with a small change in the input values. As expected, BO-uGP unsuccessfully optimizes most functions due to its lack of the ability to handle missing values.

Stability Comparison. The second experiment illustrates how different values of three factors *missing rate ρ, missing noise η,* and *maximum number of missing values v* affect to our method and other baselines. Note that ρ and η were defined in the first experiment setting, while v indicates how many dimensions in a data point contain missing values.

Experiment Settings. We show the optimization result on the 5d-function *Schwefel* as a function of one chosen factor while the others are fixed to their default values. We sequentially set up three separate settings as follows:

1. **Missing Rate.** We fix $\eta = 0.05$ and $v = 1$, then let $\rho \in [0.25, 0.65]$ with a step of 0.1.
2. **Missing Noise.** We fix $\rho = 0.25$ and $v = 1$, then let $\eta \in [0.1, 0.9]$ with a step of 0.1.
3. **Maximum number of missing values.** We fix $\rho = 0.5$ and $\eta = 0.05$, then allow v in a range of $[1, d - 1]$, where $d = 5$ is the dimension of function.

Results and Discussion. From Fig. 4(a), we can see that when ρ increases, the performance of DropBO drastically declines. This can be explained by the fact that the number of observations in DropBO is inversely proportional to ρ, which leads to too few data to train a good optimization model. Similarly, BO-uGP also faces the same problem as DropBO since it has no mechanism to handle missing values. Meanwhile, SuggestBO seems to be unstable, where its performance drops at $\rho = 0.35$ and 0.45 but increases at $\rho = 0.55$ before going down again at $\rho = 0.65$. In contrast, imputation-based methods and our method **BOMI** are stable and robust to the missing rate, where the performance is just slightly changed with different values for ρ.

From Fig. 4(b), when η increases SuggestBO heavily drops since it imputation error increases in proportional to η. Interestingly, the performance of DropBO does not change since the noise is only applied to observations with missing values and DropBO does not consider these observations. BO-uGP is the worst method in this experiment since it computes wrong probability distributions of very noisy values. Imputation-based methods except KNN wiggles a lot, indicating that

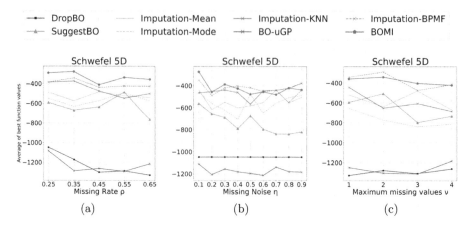

Fig. 4. Optimization results of our method **BOMI** and other methods on the 5d-fucntion *Schwefel* with different values for (a) missing rate ρ, (b) missing noise η, and (c) maximum number of missing values v.

they suffer from an over-fitting. In contrast, our method **BOMI** can maintain a good performance even with very high values for missing noise (e.g., $\eta = 0.8, 0.9$).

Finally, when the number of missing values increases (Fig. 4(c)), all methods trend to decrease, as expected. When more values are missing, the correlation vanishes that, in turn, reduces the optimization performance. Our method is still the best method, where it significantly outperforms other methods. Similar to Figs. 4(a-b), DropBO and BO-uGP perform poorly in this experiment, where they are the two worst methods.

4.2   Real-World Experiments

We also demonstrate the benefits of our method in two real-world applications, namely, robot exploration simulation [18] and heat treatment process [8].

Robot Exploration Simulation. We use the simulation software named CoppeliaSim[1] v.4 to simulate an environment for a robot to explore and measure the concentration of copper in the soil [18]. The environment is created by using the dataset Brenda Mines[2], which includes a textured terrain, trees, and bumps. *Our goal is to find the best configuration for the robot to obtain the highest percentage of copper.*

Figure 5(a) visualizes the copper percentage in the dataset Brenda Mines, where the highest copper percentage is 1.024. This map is matched with the area shown in Fig. 5(b), where the robot needs to explore. Figure 5(b) shows the starting location of the robot, its target location (i.e., its *next location*), and also

---

[1] https://www.coppeliarobotics.com/.
[2] http://www.kriging.com/datasets/.

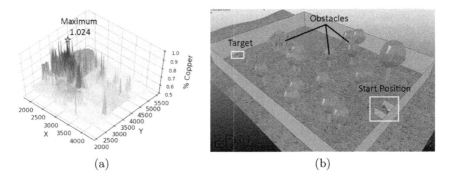

Fig. 5. (a) A visualization of copper density in the dataset Brenda Mines – the darker blue indicates a location with more copper, and the highest copper percentage is 1.024. (b) A screenshot of the robot simulation using CoppeliaSim v.4 software [18] – the figure shows the starting location of the robot, its target location (i.e., *the next location it needs to move to*), and obstacles in the environment, e.g., trees and bumps.

obstacles (e.g., trees and bumps). A 2-wheel robot is allowed 10 s to move along a pre-calculated path to a specific next location. If the robot is unable to reach the target location within 10 s, it takes the measurement at the current location before the simulation stops. On the way to the target location, many errors such as overturn or being stuck can happen and prevent the robot from reaching the target location. Whenever these errors occur, a certain noise is added to the current location of the robot by the simulation software.

In this experiment, we tune four parameters X, Y, Z, and *velocity* of the robot. $X \in [1899.94, 4301.51]$ and $Y \in [2177.37, 5400.19]$ are the coordinates of the next location where the robot needs to move to. $Z \in [4330.96, 5467.46]$ is the depth underground that the robot needs to drill to measure the copper percentage at the location (X, Y). *velocity* $\in [200, 700]$ is the speed of robot moving; the value range of velocity is chosen according to the simulation and path finding algorithm. We use the missing rate $\rho = 0.5$ and the number of missing variables $v = 1$ (i.e., one of two coordinators of the robot can be missing). We do not set value for the missing noise η since the noise is automatically added by the simulation when errors happen.

From the result in Fig. 6(a), we can see our method **BOMI** performs the best, where it significantly outperforms other methods after 70 iterations. Two imputation-based methods Imputation-KNN and Imputation-Mode perform well in this experiment, where Imputation-KNN is the second-best method. Interesting, BO-uGP shows a good performance in this application, where it is better than SuggestBO and two other imputation-based methods. Again, the performance of DropBO is very poor.

Heat Treatment Process. This is a process of heating an alloy to achieve a desired strength. In particular, an Al-Sc alloy is posed to the heat in four stages, where each stage has a different temperature and a different time duration. *Our*

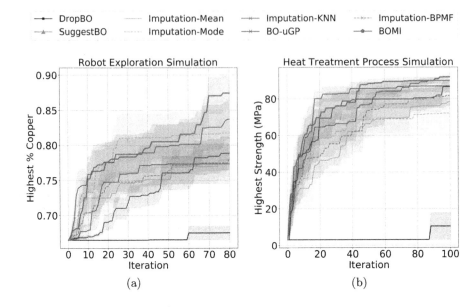

Fig. 6. Optimization results on two real-world applications: (a) robot exploration simulation and (b) heat treatment process.

goal is to choose which temperature and how long to heat the alloy at each stage to maximize its strength.

To simulate the heat treatment process for Al-Sc alloy, we use the Kampmann-Wagner model [25], same as in [8]. At each stage, there are two values to set, temperature te (in $°C$) and time ti (in second). In total, we tune eight parameters, including $te_1 \in [1, 100]$, $te_2 \in [1, 1000]$, $te_3 \in [1, 1000]$, $te_4 \in [1, 1000]$, and the heating times $ti_k \in [1, 21600]$ for $k \in \{1, 2, 3, 4\}$. Since both temperature and heating time can be missing, we set the missing rate and missing noise $\rho_1 = 0.35$, $\eta_1 = 0.8$ for temperature and $\rho_2 = 0.25$, $\eta_2 \in [0.7, 0.9]$ for heating time.

Figure 6(b) shows the optimization result for the heat treatment process. We can see our method **BOMI** is the best method, where it slightly outperforms the second-best method Imputation-KNN. Two BO-based methods SuggestBO and BO-uGP perform well and become the third-best method. It can be seen that BO methods are generally better than imputation-based methods in this experiment. We also see DropBO performs very poorly; it can be concluded that this method is not favorable in practice.

The results in Fig. 6 again confirm the real benefits of our method not only in synthetic applications but also in real-world applications when optimizing black-box functions with missing inputs.

5   Conclusion

We have presented a novel BO method **BOMI** to optimize expensive black-box functions with missing values in inputs. Our method computes the distributions of missing values for imputation and develops a new acquisition function that takes into account the uncertainty of imputed values to suggest the next point with more confidence. We demonstrate the efficiency of **BOMI** with several benchmark synthetic functions and two real-world applications in robot exploration simulation and heat treatment process. The empirical results show that **BOMI** has a better and more stable performance compared to state-of-the-art baselines, especially in experiments with high missing rates. Our future work will focus on improving the prediction of missing values, which can help to improve the performance of our method.

Acknowledgements. This research was partially funded by the Australian Government through the Australian Research Council (ARC). Prof Venkatesh is the recipient of an ARC Australian Laureate Fellowship (FL170100006).

References

1. Ambler, G., Omar, R.Z., Royston, P.: A comparison of imputation techniques for handling missing predictor values in a risk model with a binary outcome. Stat. Methods Med. Res. **16**(3), 277–298 (2007)
2. Beretta, L., Santaniello, A.: Nearest neighbor imputation algorithms: a critical evaluation. BMC Med. Inform. Decis. Making **16**(3), 74 (2016). https://doi.org/10.1186/s12911-016-0318-z
3. Bertsimas, D., Pawlowski, C., Zhuo, Y.D.: From predictive methods to missing data imputation: an optimization approach. J. Mach. Learn. Res. **18**(1), 7133–7171 (2017)
4. Brochu, E., Cora, V.M., De Freitas, N.: A tutorial on Bayesian optimization of expensive cost functions, with application to active user modeling and hierarchical reinforcement learning. arXiv preprint arXiv:1012.2599 (2010)
5. Donders, A.R.T., Van Der Heijden, G.J., Stijnen, T., Moons, K.G.: A gentle introduction to imputation of missing values. Clin. Epidemiol. **59**(10), 1087–1091 (2006)
6. Frazier, P.I., Wang, J.: Bayesian optimization for materials design. In: Lookman, T., Alexander, F.J., Rajan, K. (eds.) Information Science for Materials Discovery and Design. SSMS, vol. 225, pp. 45–75. Springer, Cham (2016). https://doi.org/10.1007/978-3-319-23871-5_3
7. Gardner, J., Pleiss, G., Weinberger, K.Q., Bindel, D., Wilson, A.G.: GPytorch: Blackbox matrix-matrix Gaussian process inference with GPU acceleration. In: NIPS. pp. 7576–7586 (2018)
8. Gupta, S., Shilton, A., Rana, S., Venkatesh, S.: Exploiting strategy-space diversity for batch Bayesian optimization. In: Artificial Intelligence and Statistics (AISTATS), pp. 538–547 (2018)
9. Hernández-Lobato, J.M., Hoffman, M.W., Ghahramani, Z.: Predictive entropy search for efficient global optimization of black-box functions. In: Advances in Neural Information Processing Systems, pp. 918–926 (2014)

10. Jones, D.R., Schonlau, M., Welch, W.J.: Efficient global optimization of expensive black-box functions. J. Global Optim. **13**(4), 455–492 (1998). https://doi.org/10.1023/A:1008306431147

11. Kang, H.: The prevention and handling of the missing data. Korean J. Anesthesiol **64**(5), 402–406 (2013)

12. Kushner, H.J.: A new method of locating the maximum point of an arbitrary multipeak curve in the presence of noise. Basic Eng. J. **86**(1), 97–106 (1964)

13. Mockus, J., Tiesis, V., Zilinskas, A.: The application of Bayesian methods for seeking the extremum. Towards Glob. Optim. **2**(117–129), 2 (1978)

14. Neal, R.M.: Probabilistic inference using Markov chain Monte Carlo methods. Department of Computer Science, University of Toronto Toronto, ON, Canada (1993)

15. Nguyen, D., Gupta, S., Rana, S., Shilton, A., Venkatesh, S.: Bayesian optimization for categorical and category-specific continuous inputs. In: AAAI (2020)

16. Oliveira, R., Ott, L., Ramos, F.: Bayesian optimisation under uncertain inputs. arXiv preprint arXiv:1902.07908 (2019)

17. Rasmussen, C.E.: Gaussian processes in machine learning. In: Bousquet, O., von Luxburg, U., Rätsch, G. (eds.) ML -2003. LNCS (LNAI), vol. 3176, pp. 63–71. Springer, Heidelberg (2004). https://doi.org/10.1007/978-3-540-28650-9_4

18. Rohmer, E., Singh, S.P., Freese, M.: V-rep: A versatile and scalable robot simulation framework. In: Intelligent Robots and Systems (IROS), pp. 1321–1326. IEEE (2013)

19. Salakhutdinov, R., Mnih, A.: Bayesian probabilistic matrix factorization using Markov chain Monte Carlo. In: ICML, pp. 880–887 (2008)

20. Shahriari, B., Swersky, K., Wang, Z., Adams, R.P., De Freitas, N.: Taking the human out of the loop: a review of Bayesian optimization. Proc. IEEE **104**(1), 148–175 (2016)

21. Śmieja, M., Struski, Ł., Tabor, J., Marzec, M.: Generalized RBF Kernel for incomplete data. Knowl.-Based Syst. **173**, 150–162 (2019)

22. Snoek, J., Larochelle, H., Adams, R.P.: Practical Bayesian optimization of machine learning algorithms. In: NIPS, pp. 2951–2959 (2012)

23. Srinivas, N., Krause, A., Kakade, S.M., Seeger, M.W.: Information-theoretic regret bounds for Gaussian process optimization in the bandit setting. IEEE Trans. Inf. Theory **58**(5), 3250–3265 (2012)

24. Tang, F., Ishwaran, H.: Random forest missing data algorithms. Stat. Anal. Data Min. ASA Data Sci. J. **10**(6), 363–377 (2017)

25. Wagner, R., Kampmann, R., Voorhees, P.W.: Homogeneous second phase precipitation. Phase Transform. Mater. **5**, 213–303 (1991)

26. Yoon, J., Jordon, J., Van Der Schaar, M.: Gain: Missing data imputation using generative adversarial nets. arXiv preprint arXiv:1806.02920 (2018)

Confusable Learning for Large-Class Few-Shot Classification

Bingcong Li[1,2(✉)], Bo Han[2], Zhuowei Wang[3], Jing Jiang[3], and Guodong Long[3]

[1] School of Automation, Guangdong University of Technology, Guangzhou, China
bingcongli@qq.com
[2] Department of Computer Science, Hong Kong Baptist University,
Kowloon Tong, Hong Kong SAR, China
[3] Australian Artificial Intelligence Institute, University of Technology Sydney,
Ultimo, Australia

Abstract. Few-shot image classification is challenging due to the lack of ample samples in each class. Such a challenge becomes even tougher when the number of classes is very large, i.e., the large-class few-shot scenario. In this novel scenario, existing approaches do not perform well because they ignore confusable classes, namely similar classes that are difficult to distinguish from each other. These classes carry more information. In this paper, we propose a biased learning paradigm called *Confusable Learning*, which focuses more on confusable classes. Our method can be applied to mainstream meta-learning algorithms. Specifically, our method maintains a dynamically updating confusion matrix, which analyzes confusable classes in the dataset. Such a confusion matrix helps meta learners to emphasize on confusable classes. Comprehensive experiments on *Omniglot*, *Fungi*, and *ImageNet* demonstrate the efficacy of our method over state-of-the-art baselines.

Keywords: Large-class few-shot classification · Meta-learning · Confusion matrix

1 Introduction

Deep Learning has made significant progress in many areas recently, but it relies on numerous labeled instances. Without enough labeled instances, deep models usually suffer from severe over-fitting, while a human can easily learn patterns from a few instances. By incorporating this ability, meta-learning based few-shot learning has become a hot topic [1,2]. Mainstream meta-learning methods obtain meta-knowledge from a base dataset containing a large number of labeled instances, and employ meta-knowledge to classify an meta-testing dataset.

Preliminary work was done during an internship at Hong Kong Baptist University (HKBU). This research was partially funded by the Australian Government through the Australian Research Council (ARC) under grants LP180100654, HKBU Tier-1 Start-up Grant and HKBU CSD Start-up Grant.

© Springer Nature Switzerland AG 2021
F. Hutter et al. (Eds.): ECML PKDD 2020, LNAI 12458, pp. 707–723, 2021.
https://doi.org/10.1007/978-3-030-67661-2_42

Recent progress in few-shot learning focuses on small-class few-shot scenario [3]. These methods consist of three directions: model initialization based methods, metric learning methods and data augmentation. In particular, model initialization based methods include learning a good model initialization [1]. Metric learning methods assume that there exists an embedding for any given dataset, where the representation of instances drawn from the same class is close to each other [2,4]. The predictions of these methods are conditioned on distance or metric to few labeled instances during meta-training. Data augmentation generates new instances based on existing "seed" instances [5].

Table 1. Difficulties of different learning scenarios.

	Few-shot	Many-shot
Small-class	Hard [1]	Easy [24]
Large-class	**Hardest** [11]	Medium [12]

Most existing methods are evaluated on tasks with less than 50 classes [6–9]. However, in practice, we are often asked to classify thousands of classes, which naturally brings large-class few-shot scenario [10,11]. As shown in Table 1, this scenario is challenging to conquer. In this kind of scenario, some classes are more difficult for the model to classify. Performance of the model on these classes will suffer from a relatively low accuracy. Meanwhile, experiments in large-class many-shot scenario also provides the same conclusion [12].

To tackle the large-class few-shot scenario, we propose a biased learning paradigm called *Confusable Learning*. Our key idea is to focus on confusable classes in meta-training dataset, which can improve model robustness in meta-testing dataset. In each iteration, we uniformly sample a few classes and denote them as *target classes*. For each target class, our paradigm selects several similar classes, which the model has difficulty in distinguishing from their target class. We call these classes *distractors*[1]. The model is then trained by a meta-learning algorithm to recognize instances of target class from those of distractors. Note that distractors are dynamically changing: when the model fits the distractors in each iteration, they become less confusable; while other classes become relatively more confusable and have higher chance to be selected as distractors. In this way, the model goes through every class in meta-training dataset dynamically. We briefly show how *Confusable Learning* works in Fig. 1, where *Confusable Learning* is presented as a framework agnostic to different meta-learners. In the experiment, we build our method on the top of several state-of-the-art meta-learning methods, including Prototypical Network, Matching Network, Prototypical Matching Network and Ridge Regression Differentiable Discriminator [2,4,13,14]. *Confusable Learning* is a training framework applied only in meta-training stage. In meta-test stage, these models are evaluated in

---

[1] Distractors defined in our paper are different from those defined by Ren [17].

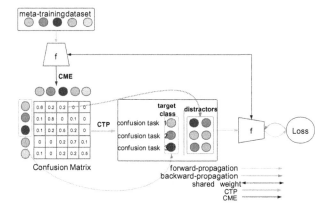

Fig. 1. Demonstration of one episode of *Confusable Learning* as specified in Algorithm 1. In this example, there are five classes in total in the meta-training dataset, which are represented by five circles with different colors. In the Confusion Matrix, the entry $\mathbf{C}_{i,j}$ (i $\neq$ j) is the average probability for instances of class i to be misclassified as class j. f denotes the meta-learning algorithm, where *Confusable Learning* is applied. Specifically, we use *Confusion Task Processor* in Algorithm 2 to sample three classes (red box) as our *target classes* and then locate two *distractors* for each target class (grey box). For example, for the orange class, we choose the purple and blue classes in the grey box as its *distractors*, because they have a higher probability in the Confusion Matrix. Second, *Confusion Matrix Estimation (CME)* in Algorithm 3, represented by the blue arrow, is applied to update the Confusion Matrix using probabilities calculated by the meta-learner. Confusable classes change as the model updates. (Color figure online)

the same way their authors originally did [2,4,13,14]. We evaluate our method on datasets that have more than a thousand classes, including *Omniglot* [15], *Fungi*[2], and *ImageNet* [16]. The empirical results show that the models with *Confusable Learning* has better generalization in unseen meta-testing datasets than the models without *Confusable Learning* on large-class few-shot tasks.

2   Related Literature

In the following, we discuss representative few-shot learning algorithms organized into three main categories: model initialization and learnable optimizer based methods, metric learning methods, and data augmentation. The first category is learning an optimizer or a specified initialization. For example, Wichrowska *et al.* [18] used an RNN network to replace the gradient descent optimizer. Finn *et al.* [1] proposed a gradient-based method to backward-propagate through the learning process of a task, which finds a good initialization for the meta-learner. This method takes a very illuminating perspective on learning which is very

---

[2] https://github.com/visipedia/fgvcx_fungi_comp.

effective. However, as reported by Chen *et al.* [5], this method can be easily hindered by large shifts between training and testing domain.

Metric learning methods assume that there exists an embedding for any given dataset, where the representation of instances drawn from the same class is close to each other. For example, Koch *et al.* [19] addressed one-shot problem by comparing distances from embedding of query instances to a single support instance. Vinyals *et al.* [4] proposed to focus attention on features that help solve a particular task. Instead of comparing distances to the embedding of labeled instances, Snell *et al.* [2] compared the distance to the prototype of each class, which is computed by averaging the embedding of support instances belonging to that class. In contrast with previous methods using handcraft metrics like Euclidean distance and cosine similarity, Garcia and Bruna [20] used a Graph Neural Network to learn the metric specific to a given dataset. Liu *et al.* [21,22] proposed a novel graph structure to tackle the similar problem.

Data augmentation methods address few-shot problem by generating new instances based on existing "seed" instances. Chen *et al.* [5] showed that applying traditional data augmentation methods like crop, flip, and color jitter in traditional many-shot scenario can produce a competitive result. To capture more realistic variation for generating new instances, Hariharan and Girshick [10] learned to transfer the variation of the meta-training dataset to the meta-testing dataset. Instead of concerning the authenticity of those generated instances, Wang *et al.* [13] proposed to use a hallucinator to produce additional training instances. Data augmentation methods are orthogonal with other few-shot methods and can be considered as the pre-processing procedure for other few-shot methods. Thus, we do not consider these methods in this work.

On the other hand, researchers have studied large-class problem under many-shot and few-shot setting. Deng *et al.* [12] studied the ability of deep models to classify ImageNet subsets with more than 10000 classes of images. They found that only a small portions of classes are truly confusable and suggested to focus on these confusable classes to improve performance. Gupta *et al.* [23] treated confusable classes as the noise of lower priority and prevented the model from updating classes that are consistently misclassified. However, the assumption of ignoring confusable classes will cause performance degradation. Li *et al.* [11] proposed a novel large-class few-shot learning model by learning transferable visual features with the class hierarchy, which encodes the semantic relations between source and target classes. These works admit the difficulty brought by large-class setting. Deng *et al.* [12] motivates us to focus on, instead of ignoring, confusable classes in large-class setting to improve classification performance. Meanwhile, it is necessary to point out that Deng *et al.* [12] did not propose a method to deal with confusable classes, neither did they study how the learning of confusable classes of meta-training dataset influences the model performance on meta-testing dataset.

3   Algorithm

Here we clarify the notations that will be used later. Given a meta-training dataset with K classes called $\mathbb{D} = \{(\mathbf{x}^1, y^1), ..., (\mathbf{x}^N, y^N)\}$, we denote $\mathbb{D}_k$ as a set containing all instances within $\mathbb{D}$ that belong to the kth class. For conventional meta-learning methods, a few classes, represented by their indices in our work, are drawn from K classes in each episode. A support set and a held-out query set are sampled from instances of these classes in $\mathbb{D}$.

Algorithm 1: Confusable Learning

Input: Training set $\mathbb{D} = \{(\mathbf{x}^1, y^1), ..., (\mathbf{x}^N, y^N)\}$ (denote $\mathbb{D}_k$ as a set containing all instances of $\mathbb{D}$ that belong to kth class); Learnable weights Θ; Number of times performing CME M.

Parameter: ρ

1   Initialize confusion matrix $\mathbf{E}$ as a matrix of shape (K, K), with each entry initialized as $1/K$;

2   **while** *Model does not converge* **do**

3   $\quad$ $\mathbf{C} \leftarrow \mathbf{E}$;
   $\quad$ // Algorithm 2

4   $\quad$ $\Theta \leftarrow$ ConfusionTaskprocessor($\mathbb{D}, \Theta, \mathbf{C}$);

5   $\quad$ **foreach** i *in* $\{1, ..., M\}$ **do**
   $\quad\quad$ // Algorithm 3

6   $\quad\quad$ $\mathbf{E} \leftarrow$ ConfusionMatrixEstimation($\mathbb{D}, \Theta, \mathbf{E}$);

7   $\quad$ **end**

8   **end**

To focus on confusable classes, we propose our method called *Confusable Learning*. Specifically, *Confusable Learning* utilizes the confusion matrix $\mathbf{C}$, which is designed as a square matrix of K rows and K columns for a meta-training dataset with K classes. This matrix is calculated by training a model and then setting the entry $\mathbf{C}_{i,j}$ as the count of the test instances of class i that are misclassified as class j [25], as formally shown below:

$$\mathbf{C}_{i,j} = \sum_{(\mathbf{x}, y) \in \mathbb{Q}^i} \mathbf{1}_{j = \arg\max_k P(\hat{y} = k | \mathbf{x})}, \tag{1}$$

in which $\mathbb{Q}$ denotes a query set, and $P(\hat{y} = k | \mathbf{x})$ denotes the prediction given by a meta-learning model.

The confusion matrix is often used to describe the performance of a classification model on a test dataset with true labels. We make use of the confusion matrix to find out the confusable classes and then learn from these confusable classes to update parameters in the model. We need to calculate this matrix in each episode because when parameters in the model change, the corresponding confusion matrix changes as well.

In the case where classes are imbalanced, it is useful to normalize the confusion matrix by dividing each entry of the confusion matrix by the sum of the entries of the corresponding row [27]. The normalized confusion matrix $\mathbf{C}^n$ is formally given by:

$$\mathbf{C}^n_{i,j} = \frac{\mathbf{C}_{i,j}}{\sum_{k=1}^{K} \mathbf{C}_{i,k}}. \tag{2}$$

As a result, the sum of entries in each row of $\mathbf{C}^n$ is equal to 1.

To learn more about the confusable classes, we propose the definition of soft confusion matrix, which is slightly different from the traditional one. Instead of counting the times of classifying class i as class j, we utilize the probability of classifying instances of class i as class j:

$$\mathbf{C}^p_{i,j} = \frac{1}{|\mathbb{Q}^i|} \sum_{(\mathbf{x},y)\in\mathbb{Q}^i} P(\hat{y} = j|\mathbf{x}). \tag{3}$$

It is easy to observe that $\sum_{j=1}^{K} \mathbf{C}^p_{i,j}$ is always equal to 1. Thus, the soft confusion matrix defined by Eq. (3) is normalized itself. We will show in the experiment section that by using such a definition *Confusable Learning* exploit more detailed information about the error the model made for each instance.

We show the framework of our method in Algorithm 1. Algorithm 1 shows that *Confusable Learning* consists of 2 steps: Algorithm 2 and Algorithm 3, which will be introduced in detail in Sect. 3.1 and Sect. 3.2 respectively.

3.1   Confusion Task Processor

Before delving into the calculation of confusion matrix, we assume that we have a confusion matrix at the start of each episode. Next, *Confusion Learning* constructs several tasks called confusion tasks according to the confusion matrix. Then the meta-learning method is trained on these tasks.

As shown in Algorithm 2, to construct the confusion tasks, we first uniformly sample N_T^c target classes $\bar{v}^c = \{\bar{v}_1^c, ..., \bar{v}_{N_T^c}^c\}$. For target class $k \in \bar{v}^c$, we use a multinomial distribution to sample N_D classes as distractors $\bar{w}^k = \{\bar{w}_1^k, ..., \bar{w}_{N_D}^k\}$. That is:

$$\bar{w}^k \sim \text{Multinomial}(N_D, \mathbf{p}_1^k, ..., \mathbf{p}_{k-1}^k, \mathbf{p}_{k+1}^k, ..., \mathbf{p}_K^k), \tag{4}$$

in which:

$$\mathbf{p}_j^k = \frac{\mathbf{C}_{k,j}}{\sum_{h=\{1,2,...,k-1,k+1,...,K\}} \mathbf{C}_{k,h}}. \tag{5}$$

$\mathbf{C}$ is the confusion matrix given by either Eq. (2) or Eq. (3). We can see here that the classes with higher confusion are more likely to be sampled. By exposing the target class and the coupling distractors to the model, *Confusion Learning* enables the model to learn better.

Combining all target classes and their corresponding distractors, we have N_T^c pairs of target classes and distractors $\{(\bar{v}_1^c, \bar{w}^1), ..., (\bar{v}_{N_T^c}^c, \bar{w}^{N_T^c})\}$. For the k-th pair $(\bar{v}_k^c, \bar{w}^k)$, our learning paradigm samples N_S instances for each distractor

---

Algorithm 2: ConfusionTaskProcessor($\mathbb{D}$, Θ,**C**)

Input: Confusion Matrix **C**; Training set $\mathbb{D}$; Learnable weights Θ.

Parameter: Number of support instances for each class N_S; Number of query instances for each class N_Q; Number of distractors for each target class N_D; Number of confusable tasks N_T^c.

1   $\bar{v}^c \leftarrow$ RandomSample($\{1,...,K\}, N_T^c$) ;   // Sample target classes

2   **foreach** k *in* $\{\bar{v}_1^c,...,\bar{v}_{N_T^c}^c\}$ **do**

3   Sample distractors $\bar{w}^k$ with Eq. (4);

4   **foreach** l *in* $\{\bar{w}_1^k,...,\bar{w}_{N_D}^k\}$ **do**

         // Sample instances for distractors

5          $\mathbb{S}_{k,l}^{\text{distractor}} \leftarrow$ RandomSample($\mathbb{D}_l, N_S$);

6   **end**

7      $\mathbb{S}_k^{\text{distractor}} \leftarrow \bigcup_{l=\{\bar{w}_1^k,...,\bar{w}_{N_D}^k\}} \mathbb{S}_{k,l}^{\text{distractor}}$;

8      $\mathbb{S}_k^{\text{target}} \leftarrow$ RandomSample($\mathbb{D}_k, N_S$); // Sample instances for target class

9      $\mathbb{Q}_k \leftarrow$ RandomSample($\mathbb{D}_k, N_Q$) ;   // Sample instances for query set

10 **end**

11 $J \leftarrow 0$;

12 **foreach** k *in* $\{\bar{v}_1^c,...,\bar{v}_{N_T^c}^c\}$ **do**

13   **foreach** (x,y) *in* $\mathbb{Q}_k$ **do**

14   Calulate loss for any base meta-learning algorithm using the support set given by $\mathbb{S}_k^{\text{distractor}} \cup \mathbb{S}_k^{\text{target}}$;

15          $J \leftarrow J + \frac{1}{N_T^c N_Q} \log \mathrm{P}(\hat{y} = k|\mathbf{x})$;

16   **end**

17 **end**

18 Update Θ by back-propagating on J;

---

and target class, generating a support set $\mathbb{S}$ with $N_S \times (N_D + 1)$ instances, and samples N_Q instances of the target class $\bar{v}_k^c$, generating a query set $\mathbb{Q}$ with N_Q instances. Combining $\mathbb{S}$ and $\mathbb{Q}$, we have a confusion task. Predictions $\mathrm{P}(\hat{y} = k|\mathbf{x} \in \mathbb{Q})$ are computed by the meta-learning algorithm using N_T^c confusion tasks. By maximizing $\mathrm{P}(\hat{y} = k|\mathbf{x})$ and applying stochastic gradient descent, we can optimize parameters in the model to distinguish target classes better in the next episode until they become less confusable, after which the model switches its attention to relatively more confusable classes as the confusion matrix updates dynamically. The pseudo-code is demonstrated in Algorithm 2.

3.2 Confusion Matrix Estimation

The traditional way of the confusion matrix calculation requires inferences of instances drawn from K classes, which can be extremely slow and require high

RAM usage when K is large. To bypass this hinder, we propose a novel iterative way called Confusion Matrix Estimation (CME) to estimate the confusion matrix, the difficulty of which is much smaller than the traditional one.

Given a trained model, we regard the confusion matrix $\mathbf{C}$ calculated by evaluating the model on K classes using the traditional method as the ideal confusion matrix. The purpose of CME is to estimate $\mathbf{C}$ in an incremental yet less computation-consuming way. CME initializes a matrix $\mathbf{E}$ of size $K \times K$, which is the same as the ideal confusion matrix. Each entry of $\mathbf{E}$ is initialized with a positive constant. Since the summary of each row of $\mathbf{C}$ is equal to 1, $1/K$ is usually good for the initialization of $\mathbf{E}$. The purpose of CME is to update $\mathbf{E}$ in multiple steps to make it closer to the ideal confusion matrix $\mathbf{C}$. In each step, N_T^e classes are uniformly sampled from meta-training dataset. Let us mark their indices among the K meta-training classes as $\bar{v}^e = \{\bar{v}_1^e, ..., \bar{v}_{N_T^e}^e\}$. By performing inference on $\bar{v}^e$, we are able to obtain a smaller confusion matrix named $\mathbf{E}'$ with the size of $N_T^e \times N_T^e$. Clearly, $\mathbf{E}'$ contains the information on how a class is confused with other classes among $\bar{v}^e$. $\mathbf{E}'$ is somehow like an observation of $\mathbf{C}$ through a small "window". To incorporate the observation into $\mathbf{E}$, in each step, CME updates $\mathbf{E}$ by:

$$\mathbf{E}_{\bar{v}_i^e, \bar{v}_j^e} = \rho \mathbf{E}_{\bar{v}_i^e, \bar{v}_j^e} + (1 - \rho)\mathbf{E}'_{i,j} Z, \tag{6}$$

in which ρ is a hyperparameter between 0 and 1, and Z is used to scale the observation to make sure the summary of the current observation is constant with the result of previous observations:

$$Z = \sum_{k=1,...,N^e} \mathbf{E}_{\bar{v}_i^e, \bar{v}_k^e}. \tag{7}$$

To combine CME with previously introduced *Confusable Learning* framework, we can initialize $\mathbf{E}$ and perform multi-steps CME at each *Confusable Learning* episode to get a reliable estimation of the confusion matrix. However, it is not necessary to make a fresh start in each episode. Denoting the ideal confusion matrix of the model at episode i as $\mathbf{C}_i$, since the ability of the model will not change a lot between successive episodes, $\mathbf{C}_{i+1}$ should be close to $\mathbf{C}_i$. Intuitively, we do not need to initialize a new $\mathbf{E}$ in each episode. Instead, to calculate the estimation $\mathbf{E}_{i+1}$ of episode $i + 1$, we can perform CME update based on the estimation $\mathbf{E}_i$ of episode i. As such, $\mathbf{E}_0$ is initialized only once when the meta-learning model is initialized. Then at each episode, CME updates for M steps. In our experience, by setting M to 1, the performance of CME is good enough. Algorithm 3 shows how the confusion matrix is updated in an episode.

To demonstrate the training dynamic of our method using Algorithm 1, Fig. 2 shows that *Confusable Learning* first spreads its attention to many classes, and then turns to the classes that are more difficult to distinguish. In the early stage (top 100 rows) of meta-training, attention is dynamically spread to many classes. Later, most of these classes are well fitted by the model and get less attention. Meanwhile, other classes get more attention because they are inherently intractable to distinguish. We found that these intractable classes are visually similar (Fig. 2(c)).

Algorithm 3: ConfusionMatrixEstimation($\mathbb{D}$, Θ, $\mathbf{E}$)

Input: Training set $\mathbb{D}$; Learnable weights Θ; Estimation of confusion matrix $\mathbf{E}$.

Parameter: Number of support instances for each class N_S; number of query instances for each class N_Q; number of classes to use in each CME step N_T^e.

1 Initialize $\mathbf{E}'$ as a zero matrix of shape (N_T^e, N_T^e), with each entry initialized as 0;

2 $\bar{v}^e \leftarrow$ RandomSample($\{1, ..., K\}, N_T^e$);

3 **foreach** k *in* $\{\bar{v}_1^e, ..., \bar{v}_{N_T^e}^e\}$ **do**

4     $\mathbb{S}_k \leftarrow$ RandomSample($\mathbb{D}_k, N_S$);

5     $\mathbb{Q}_k \leftarrow$ RandomSample($\mathbb{D}_k, N_Q$);

6 **end**

7 **foreach** m *in* $\{1, ..., N_T^e\}$ **do**

8   **foreach** (x, y) *in* $\mathbb{Q}_{\bar{v}_m^e}$ **do**

9   **foreach** n *in* $\{1, ..., N_T^e\}$ **do**

10   Calculate $\mathrm{P}(\hat{y} = \bar{v}_n^e | \mathbf{x})$ with any base meta-learning algorithm using the support set given by $\bigcup_{k=\{\bar{v}_1^e, ..., \bar{v}_{N_T^e}^e\}} \mathbb{S}_k$;

11             $\mathbf{E}'_{m,n} \leftarrow \mathbf{E}'_{m,n} + \mathrm{P}(\hat{y} = \bar{v}_n^e | \mathbf{x})$;

12   **end**

13   **end**

14 **end**

15 $\mathbf{E}' \leftarrow \frac{\mathbf{E}'}{N_Q}$;

16 Update $\mathbf{E}$ using Eq. (6);

17 **return** $\mathbf{E}$;

---

4 Experiment

4.1 Experiment Setup

To empirically prove the efficacy of our method, we conduct experiments on three real-world datasets: *Omniglot*, *Fungi*, and *ImageNet*. We choose these datasets because they contain more than 1000 classes, unlike prior works on meta-learning which experiment with smaller images and fewer classes.

Confusable Learning can be easily applied to various mainstream meta-learning algorithms. To demonstrate the efficacy of our method, we choose four state-of-the-art meta-learning algorithms as our base meta-learning methods in our experiments: Prototypical Network (PN) [2], Matching Network (MN) [4], Prototype Matching Network (PMN) [13] and Ridge Regression Differentiable Discriminator (R2D2) [14]. We will show that by applying *Confusable Learning* to these methods, we can easily improve their performance in large-class few-shot learning setting. For notation, we denote our method by attaching a

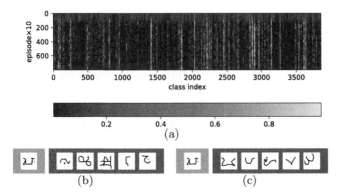

Fig. 2. (a) A Prototype Network is trained with *Confusable Learning* on *Omniglot*. Here, each pixel of the image denotes the frequency for a class to be focused, i.e., to be selected as a distractor in 10 episodes. For better visual effect, dilation has been applied to the image. (b) Target class (green) and distractors that get no attention (red) in the last 100 episodes. (c) Target class and distractors that get the most attention in the last 100 episodes. (Color figure online)

"w/CL" behind each of them. For example, PN w/CL means prototype networks with *Confusable Learning*.

In few-shot learning, N-shot K-way classification tasks consist of N labeled instances for each K classes. As stated in PN [2], it can be extremely beneficial to train meta-learning models with a higher way than that will be used in meta-testing dataset. Particularly, for PN, to train a model for 5/20 ways tasks, the author used training tasks with 60 classes [2]. The same setting is also mentioned in other mainstream few-shot methods [14]. However, for the large-class few-shot problem which has a large number of classes, it becomes impractical to build even larger support sets due to the limitation of memory and the exponentially growing load of computation. Therefore, in our experiment, the models are all trained in meta-training tasks with fewer ways than the meta-testing task. We evaluate the models using query sets and support sets constructed in the same way as their authors originally did [2,4,13,14].

Omniglot. *Omniglot* contains 1623 handwritten characters. As what Vinyals [4] has done, we resize the image to 28×28 and augment the dataset by rotating each image by 90, 180, and 270 degrees. For a more challenging meta-testing environment, we employ the split introduced by Lake [15], constructing our meta-training dataset with 3856 classes and meta-testing dataset with 2636 classes. In meta-testing stage, we use all 2636 classes in every single meta-testing task.

For PN, MN and PMN, learning rate is set to $1e-4$. For R2D2, learning rate is set to $5e-5$. For our method, We set N_D to 40, N_T^e to 500. In our experiments of all datasets, N_T^c is always set to $(N_T^e \times 2)/(N_D+2)$, which is 23 here. ρ is set to 0.9. M is set to 1.

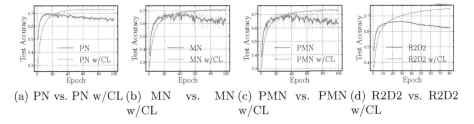

(a) PN vs. PN w/CL (b) MN   vs.   MN (c) PMN   vs.   PMN (d) R2D2 vs. R2D2
                                w/CL   w/CL   w/CL

Fig. 3. Test accuracy vs. number of epoch of 4 kinds of meta-learning method without vs. with *Confusable Learning* in *Omniglot*.

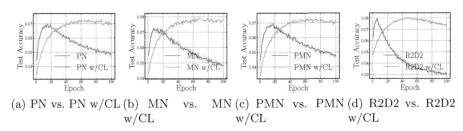

(a) PN vs. PN w/CL (b) MN   vs.   MN (c) PMN   vs.   PMN (d) R2D2 vs. R2D2
                                w/CL   w/CL   w/CL

Fig. 4. Test accuracy vs. number of epoch of 4 kinds of meta-learning method without vs. with *Confusable Learning* in *fungi*.

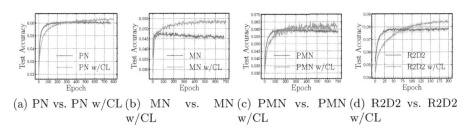

(a) PN vs. PN w/CL (b) MN   vs.   MN (c) PMN   vs.   PMN (d) R2D2 vs. R2D2
                                w/CL   w/CL   w/CL

Fig. 5. Test accuracy vs. number of epoch of 4 kinds of meta-learning methods without vs. with *Confusable Learning* in *Imagenet*.

Fungi. *Fungi* is originally introduced by the 2018 FGVCx Fungi Classification Challenge. We randomly sample 632 classes to construct meta-training dataset, 674 classes to construct meta-testing dataset and 88 classes to construct validation dataset. All 674 meta-testing classes are used in every meta-testing task.

Learning rate is set to $5e - 4$ for all methods. For our method, We set N_D to 5. N_T^e is set to 70, and thus N_T^c is set to 20. ρ is set to 0.9. M is set to 1. For R2D2 w/CL, the temperature of the softmax is set to 0.1 in CME.

ImageNet64x64. We also conduct experiments on *ImageNet64x64* dataset [26], which is a downsampled version of the original *ImageNet* used in ILSVRC with images resized to 64×64 pixels. The reason why we do not perform experiments on regular few-shot datasets like *mini*ImageNet and *tiered*ImageNet is that neither of them holds enough number of classes for our large-class setting. Although *ImageNet64x64* is not regularly used to evaluate few-shot learning methods, it contains 1000 classes in total. Here we merge its original training dataset and original testing dataset together and then split it by class again into a new meta-training dataset, a validation dataset, and a new meta-testing dataset according to the category of the classes. All classes that belong to the category **Living Thing** are used for meta-training stage and all classes that belong to the category **Artifact** are used for meta-testing stage. This gives us a meta-training dataset with 522 classes and a meta-testing dataset with 410 classes. The rest 68 classes belong to the validation dataset.

For MN and MN w/CL, we set the learning rate to $1e - 3$. For the other settings, the learning rate is set to $1e - 4$. In our experiment, We set N_D to 5. N_T^e is set to 90, and thus N_T^c is set to 25. ρ is set to 0.9. M is set to 1.

4.2   Result

We perform five repeated experiments with different random seeds. In Table 2, the averaged accuracy for each method is shown. It can be seen all meta learners coupled with our method outperform original ones in all three datasets.

To delve into the training dynamic, Figs. 3, 4 and 5 show how testing accuracy changes with respect to the number of epochs, with the standard deviation shown with the shaded area. The meta-learning algorithm with *Confusable Learning* has a better generation than those without it. As shown in Fig. 4, it is interesting to find out that *Confusable Learning* shows a great ability to resist over-fitting than corresponding original methods in *Fungi*, which contains very similar mushroom classes and thus leads to over-fitting easily.

Table 2. 5-shot classification test accuracies of PN, PN w/CL, MN, MNw/CL, PMN, PMNw/CL and R2D2, R2D2 w/CL in *Omniglot* (2636-way), *Fungi* (674-way) and *ImageNet64x64* (410-way).

Algorithm	PN	PN w/CL	MN	MN w/CL	PMN	PMN w/CL	R2D2	R2D2 w/CL
Omniglot	69.90%	**73.21%**	68.04%	**71.07%**	69.73%	**73.68%**	66.12%	**74.72%**
Fungi	6.96%	**7.29%**	7.26%	**7.92%**	6.96%	**7.28%**	7.92%	**8.04%**
ImageNet	6.02%	**6.27%**	4.79%	**5.41%**	5.90%	**6.41%**	7.78%	**8.46%**

To further demonstrate the ability of *Confusable Learning* to resist over-fitting, we visualize the co-relationship between the training loss and the testing loss of PN and PN w/CL. Note that we estimate the training loss of the model using query sets and support sets constructed in the same way as Snell [2], but including all classes in the meta-training dataset in each task. To be specific, we estimate training losses in *Omniglot* (3856-way 5-shot), *Fungi* (632-way 5-shot), and *ImageNet64x64* (522-way 5-shot). Figure 6 shows that under the same training loss, *Confusable Learning* has smaller testing loss. We believe that PN over-fits the regular patterns and ignores confusable patterns, while *Confusable Learning* focuses on these confusable patterns.

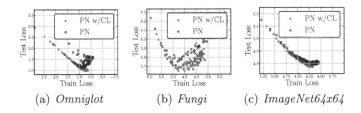

(a) *Omniglot*   (b) *Fungi*   (c) *ImageNet64x64*

Fig. 6. (a) Training loss vs. testing loss in *Omniglot* dataset. The triangles or circles on the right have smaller training loss, and thus usually represent the performance of the meta-learning algorithm in the late stage of training (b) Training loss vs. testing loss in *Fungi* dataset. (b) Training loss vs. testing loss in *ImageNet64x64* dataset.

Table 3. 5-shot classification test accuracies on *Omniglot* for *Confusable Learning* using the traditional confusion matrix and the proposed soft confusion matrix.

Algorithm	PN	MN	PMN	R2D2
Baseline	69.90 ± 0.16%	68.04 ± 0.23%	69.73 ± 0.21%	66.12 ± 0.16%
w/CLN	**73.23 ± 0.10%**	70.63 ± 0.09%	73.24 ± 0.09%	74.65 ± 0.14%
w/CL	73.21 ± 0.04%	**71.07 ± 0.19%**	**73.68 ± 0.21%**	**74.72 ± 0.52%**

4.3 Ablation Study

In this section, we will discuss the influence of proposed soft confusion matrix and Confusion Matrix Estimation in *Confusable Learning*.

Influence of Calculating Confusion Matrix with Probability. *Confusable Learning* adopts a novel definition of confusion matrix called soft confusion matrix as shown in Eq. (3). To demonstrate its benefit, we implement *Confusable Learning* using the traditional definition of confusion matrix given in Eq. (2). To implement such a setting, we simply replace $P(\hat{y} = \bar{v}_n^e|\mathbf{x})$ in line 11 of Algorithm 3 with $\mathbf{1}_{\bar{v}_n^e=\arg\max_k P(\hat{y}=k|(\mathbf{x},y))}$. We denote this setting by attaching a "w/CLN" behind the name of each meta-learning algorithm. Results of *Omniglot* and *Fungi* are shown in Table 3 and Table 4 respectively. It can be seen that the accuracy of the model using soft confusion matrix, marked with "w/CL", is higher than the model marked as "w/CLN" on *Fungi*. However, on *Omniglot* they are almost equal. This is because the model is more confident about its prediction on *Omniglot*, making $P(\hat{y}|(\mathbf{x},y))$ close to either 0 or 1. In this case, the proposed soft confusion matrix is equivalent to the traditional one.

Influence of Confusion Matrix Estimation (CME). To demonstrate the performance and the efficiency of CME, we implement *Confusable Learning* with the traditional way of confusion matrix calculation, which is performing the meta-learning algorithm on a K-way task and calculating confusion matrix with Eq. (3). In fact, the traditional confusion matrix calculation can be seen as a

Table 4. 5-shot classification test accuracies on *Fungi* for *Confusable Learning* using the traditional confusion matrix and the proposed soft confusion matrix.

Algorithm	PN	MN	PMN	R2D2
Baseline	6.96 ± 0.07%	7.26 ± 0.09%	6.96 ± 0.12%	7.92 ± 0.13%
w/CLN	7.09 ± 0.06%	7.89 ± 0.02%	7.14 ± 0.05%	8.04 ± 0.14%
w/CL	**7.29 ± 0.09%**	**7.92 ± 0.04%**	**7.28 ± 0.11%**	8.04 ± 0.12%

special case of CME, in which ρ is set to 0 and N_T^e is set to K. We denote this setting by attaching a "w/CLT" behind the name of meta-learning algorithm.

Table 5. 5-shot classification test accuracies, averaged elapsed time in each iteration and memory in need for confusion matrix calculation in *Omniglot* (2636-way).

	PN w/CL-1	PN w/CL-2	PN w/CL-4	PN w/CL-8	PN w/CLT
Accuracy	73.21%	73.58%	73.46%	73.59%	73.60%
Elapsed time	0.073 s	0.146 s	0.289 s	0.570 s	1.432 s
GPU memory	2463 MiB	2479 MiB	2511 MiB	2567 MiB	12457 MiB

In Table 5, we compare the results of *Confusable Learning* with the traditional confusion matrix calculation and the result of *Confusable Learning* with our proposed CME. Here, we denote *Confusable Learning* with M steps of CME by attaching a "w/CL-M". The experiment is conducted on a machine with a Tesla P100 GPU. It can be concluded that the CME settings achieve almost the same accuracy with the setting using the traditional confusion matrix calculation but largely decrease the elapsed time and memory requirement. It is noteworthy that *Omniglot* is a small dataset, so we are able to implement *Confusable Learning* with the traditional confusion matrix calculation easily. When training with a larger dataset like *fungi* and *Imagenet64x64*, traditional confusion matrix calculation will require much longer time and larger memory.

4.4   Parameter Sensitivity Analysis

Confusable Learning contains 5 parameters: N_D, N_T^e, N_T^c, M and ρ. Table 5 already shows that larger M yields a better result. In this section, we discuss how sensitive the other 4 parameters are.

Based on the parameters we use in *Omniglot* PN w/CL experiment in Sect. 4.1, we adjust each of the 4 parameters at a time. The results are shown in Fig. 7. Performance of the model is very stable with ρ changing from 0 to 0.9. It is not surprising that the accuracy drops below 70% when ρ is set to 1. In such a case, the confusion matrix will not be updated and thus, *Confusable Learning* can not obtain any useful information from the confusion matrix. Increasing N_D

and N_T^e always help improve accuracy but requires longer elapsed time and more memory. When increasing N_T^c, as shown in 7(d), accuracy firstly increases and then decreases. The optimal N_T^c is about 10. Regardless of this observation, the accuracy is over 73% within a large range of N_T^c, significantly higher than the accuracy of the PN model without $ConfusableLearning$ reported in Table 2. It can be concluded that $Confusable\ Learning$ can yield a great performance without any elaborate tuning.

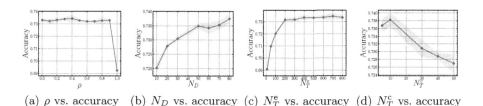

(a) ρ vs. accuracy   (b) N_D vs. accuracy   (c) N_T^e vs. accuracy   (d) N_T^c vs. accuracy

Fig. 7. Accuracy of PN w/CL vs. parameters in *Omniglot*.

5   Conclusion

We have presented an approach to locate and learn from confusable classes in large-class few-shot classification problem. We show significant gains on top of multiple meta-learning methods, achieving state-of-the-art performance on three challenging datasets. Future work will involve constructing better confusion tasks to learn confusable classes better.

References

1. Finn, C., Abbeel, P., Levine, S.: Model-agnostic meta-learning for fast adaptation of deep networks. In: International Conference on Machine Learning (ICML), pp. 1126–1135 (2017)
2. Snell, J., Swersky, K., Zemel, R.: Prototypical networks for few-shot learning. In: Advances in Neural Information Processing Systems (NeurIPS), pp. 4077–4087 (2017)
3. Wang, Y., Yao, Q.: Few-shot learning: A survey. arXiv preprint arXiv:1904.05046 (2019)
4. Vinyals, O., Blundell, C., Lillicrap, T., Wierstra, D.: Matching networks for one shot learning. In: Advances in Neural Information Processing Systems (NeurIPS), pp. 3630–3638 (2016)
5. Chen, W., Liu, Y., Zsolt, K., Wang, Y., Huang, J.: A closer look at few-shot classification. In: The International Conference on Learning Representations (ICLR), (2019)
6. Zhang, R., Che, T., Ghahramani, Z., Bengio, Y., Song, Y.: MetaGAN: an adversarial approach to few-shot learning. In: Advances in Neural Information Processing Systems (NeurIPS), pp. 2365–2374 (2018)

7. Oreshkin, B., Rodríguez López, P., Lacoste, A.: TADAM: task dependent adaptive metric for improved few-shot learning. In: Advances in Neural Information Processing Systems (NeurIPS), pp. 721–731 (2018)

8. Liu, Y., Lee, J., Park, M., Kim, S., Yang, E., Hwang, S., Yang, Y.: Learning to propagate labels: transductive propagation network for few-shot learning. In: Conference on Computer Vision and Pattern Recognition (CVPR) (2018)

9. Franceschi, L., Frasconi, P., Salzo, S., Grazzi, R., Pontil, M.: Bilevel programming for hyperparameter optimization and meta-learning. In: International Conference on Machine Learning (ICML) (2018)

10. Hariharan, B., Girshick, R.: Low-shot visual recognition by shrinking and hallucinating features. In: International Conference on Computer Vision (ICCV), pp. 3018–3027 (2017)

11. Li, A., Luo, T., Lu, Z., Xiang, T., Wang, L.: Large-scale few-shot learning: knowledge transfer with class hierarchy. In: Conference on Computer Vision and Pattern Recognition (CVPR), pp. 7212–7220 (2019)

12. Deng, J., Berg, A.C., Li, K., Fei-Fei, L.: What does classifying more than 10,000 image categories tell us? In: Daniilidis, K., Maragos, P., Paragios, N. (eds.) ECCV 2010, Part V. LNCS, vol. 6315, pp. 71–84. Springer, Heidelberg (2010). https://doi.org/10.1007/978-3-642-15555-0_6

13. Wang, Y., Girshick, R., Hebert, M., Hariharan, B.: Low-shot learning from imaginary data. In: Conference on Computer Vision and Pattern Recognition (CVPR), pp. 7278–7286 (2018)

14. Bertinetto, L., Henriques, J., Torr, P., Vedaldi, A.: Meta-learning with differentiable closed-form solvers. In: The International Conference on Learning Representations (ICLR) (2019)

15. Lake, B., Salakhutdinov, R., Tenenbaum, J.: Human-level concept learning through probabilistic program induction. Science **350**(6266), 1332–1338 (2015)

16. Deng, J., Dong, W., Socher, R., Li, L., Li, Kai., Li, F.: ImageNet: a large-scale hierarchical image database. In: Conference on Computer Vision and Pattern Recognition (CVPR), pp. 248–255 (2009)

17. Ren, M., et al.: Meta-learning for semi-supervised few-shot classification. In: The International Conference on Learning Representations (ICLR) (2018)

18. Wichrowska, O., et al.: Learned optimizers that scale and generalize. In: International Conference on Machine Learning (ICML), pp. 3751–3760 (2017)

19. Koch, G., Zemel, R., Salakhutdinov, R.: Siamese neural networks for one-shot image recognition. In: International Conference on Machine Learning Deep Learning Workshop (ICML) (2015)

20. Garcia, V., Bruna, E.: Few-shot learning with graph neural networks. In: The International Conference on Learning Representations (ICLR) (2018)

21. Liu, L., Zhou, T., Long, G., Jiang. J., Yao, L., Zhang, C.: Prototype propagation networks (PPN) for weakly-supervised few-shot learning on category graph. In International Joint Conferences on Artificial Intelligence (IJCAI) (2019)

22. Liu, L., Zhou, T., Long, G., Jiang. J., Zhang, C.: Learning to propagate for graph meta-learning. In Advances in Neural Information Processing Systems (NeurIPS), pp. 1037–1048 (2019)

23. Gupta, R., Bengio, S., Weston, J.: Training highly multiclass classifiers. J. Mach. Learn. Res. (JMLR) **15**(1), 1461–1492 (2014)

24. He, K., Zhang, X., Ren, S., Sun, J.: Deep residual learning for image recognition. In: Conference on Computer Vision and Pattern Recognition (CVPR), pp. 770–778 (2016)

25. Griffin, G., Perona, P.: Learning and using taxonomies for fast visual categorization. In: Conference on Computer Vision and Pattern Recognition (CVPR), pp. 1–8 (2008)
26. Chrabaszcz, P., Loshchilov, I., Hutter, F.: A Downsampled Variant of ImageNet as an Alternative to the CIFAR datasets. arXiv preprint arXiv:1707.08819 (2017)
27. Giannakopoulos, T., Pikrakis, A.: Introduction to Audio Analysis: a MATLAB® Approach. Academic Press, Orlando (2014)

Inductive Generalized Zero-Shot Learning with Adversarial Relation Network

Guanyu Yang[1], Kaizhu Huang[1(✉)], Rui Zhang[1], John Y. Goulermas[2], and Amir Hussain[3]

[1] Xi'an Jiaotong-Liverpool University, SIP, Suzhou 215123, China
{Guanyu.Yang,Kaizhu.Huang,Rui.Zhang02}@xjtlu.edu.cn
[2] Department of Computer Science, University of Liverpool, Liverpool L69 3BX, UK
J.Y.Goulermas@liv.ac.uk
[3] School of Computing, Edinburgh Napier University, Edinburgh EH11 4BN, UK
A.Hussain@napier.ac.uk

Abstract. We consider the inductive Generalized Zero Shot Learning (GZSL) problem where test information is assumed unavailable during training. In lack of training samples and attributes for unseen classes, most existing GZSL methods tend to classify target samples as seen classes. To alleviate such problem, we design an adversarial Relation Network that favors target samples towards unseen classes while enjoying robust recognition for seen classes. Specifically, through the adversarial framework, we can attain a robust recognizer where a small gradient adjustment to the instance will not affect too much the classification of seen classes but substantially increase the classification accuracy on unseen classes. We conduct a series of experiments extensively on four benchmarks i.e., AwA1, AwA2, aPY, and CUB. Experimental results show that our proposed method can attain encouraging performance, which is higher than the best of state-of-the-art models by 10.8%, 14.0%, 6.9%, and 1.9% on the four benchmark datasets, respectively in the inductive GZSL scenario. (The code is available on https://github.com/ygyvsys/AdvRN-with-SR)

Keywords: Zero-shot learning · Adversarial examples · Gradient penalty

1 Introduction

Deep learning models have made significant achievements in image classification under the support of a large number of labeled training samples [9,19,25]. However, data collection and labeling can become time-consuming and difficult, especially when the data are from different modalities or domains. Moreover, conventional classification models are inflexible in that they can only be applied to the categories existing during training. To overcome such difficulties, zero-shot learning (ZSL), i.e. performing classification without supporting samples in some categories, is inspired by human's cognition and knowledge transferability.

© Springer Nature Switzerland AG 2021
F. Hutter et al. (Eds.): ECML PKDD 2020, LNAI 12458, pp. 724–739, 2021.
https://doi.org/10.1007/978-3-030-67661-2_43

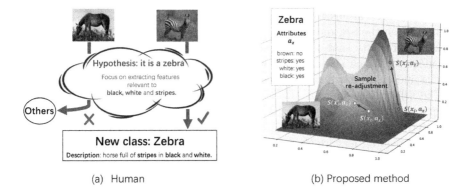

(a) Human   (b) Proposed method

Fig. 1. Processes of the new class recognition: (a) for human, (b) for the proposed method. Human make hypothesis based on previous knowledge and then make judgement. Similarly, the proposed method re-adjusts the instances based on descriptive attributes before recognition.

In a similar way that human can construct the concepts of categories through language descriptions without any visual information, most of the ZSL try to recognize the unknown categories (unseen classes) by transferring the learned knowledge of source categories (seen classes) through the side information. Currently, zero-shot image classification is the most common ZSL task where semantic attributes and word vectors are widely used as the side information [6,13,28,29]. Moreover, the most stringent and practical ZSL task is defined as inductive generalized zero-shot learning (inductive GZSL) where all information about unseen classes is unavailable during training, and the generalized task requires that the targets contain both seen and unseen categories [3]. Researchers have proposed various zero-shot methods, most of which can be summarized as generative methods [20,21,30] and embedding methods [12,15,32,33]. However, both types of these methods have drawbacks in handling the inductive GZSL task. The generative methods need to utilize the attributes of the target classes during training, which actually breaks the principle of 'unknown'; on the other hand, the embedding methods are commonly not discriminative enough and tend to classify target samples as seen classes.

In this paper, we propose an adversarial framework called adversarial Relation Network (advRN) based on an embedding method for the inductive GZSL scenario. Inspired by the human recognition process [4] revealing that brain will make hypothesis about incoming data based on previous knowledge, we design a sample re-adjustment (SR) process during test to enhance the awareness of the unseen classes attributes. In this process, the adversarial noises [8] are reversely exploited to support the hypothesis that samples are from unseen classes. Namely, a gradient based worst-case perturbation is subtracted from the visual features in order to encourage high response of the unseen classes. On the other hand, starting from the adversarial training method, we further show that a robust classifier can be built with a gradient penalty (GP) regularization that proves insensitive to small perturbations on samples of seen classes. Overall, when the SR and GP are both applied, the resulting advRN enjoys the appealing property that a small gradient adjustment on the target samples will not affect too much the classification of seen classes but substantially increase the classification accuracy on unseen classes.

The schematic diagram of the recognition process is illustrated in Fig. 1. Given an instance x, the proposed method will adjust x to better fit the unseen class attributes. Applying the strategies of both SR and GP, the proposed model has the appealing feature that small sample adjustment imposed by our model can lead to high response to unseen classes while not affect the recognition of seen classes due to the robust adversarial training. Namely, compared with the adjusted instance from seen classes, such as horse, only the true zebra instance x'_j will gain after adjustment a high confidence $S(x'_j, a_z)$ due to the robust nature of the recognizer (incurred by the gradient penalty). Detailed justification can be seen in Sect. 3.

In summary, the proposed GP and SR actually form an adversarial framework which makes the model sensitive enough to the unseen classes due to the true unseen hypotheses, while being simultaneously sufficiently robust to avoid misleading by the false unseen hypotheses. The main contributions of this paper can be listed in the following:

1. A gradient penalty, derived from the adversarial training and constraining the derivatives of the relation scores w.r.t input samples, is constructed as a regularization term during training to strengthen the robustness of recognizer for seen classes.
2. A sample re-adjustment is designed based on the gradient during test to make the recognizer more inclined to the unseen classes.
3. To the best of our knowledge, this is the first work using the framework of adversarial examples to obtain significant improvement on inductive GZSL. The proposed adversarial relation network achieves the state-of-the-art performance and is competitive compared to those methods taking advantages of unseen class attributes during training.

2   Related Work

Among the current ZSL ideas, the generative methods can usually achieve better results [20,21,30] in which the task is divided into two steps. First, a generator is trained to generate pseudo samples for unseen classes based on semantic attributes. Then a simple classifier is trained via the combination of exact seen samples and pseudo unseen samples. VAE [11] or GAN [7] based structures with several restrictions on distribution make these models perform more generalized to both seen and unseen classes. However, in practice, the attributes of unseen classes are usually not available till test, which is exactly the setting as defined in the inductive GZSL. Without these attributes, the pseudo samples for training the classifier cannot be generated, thereby the generative methods cannot be implemented in the inductive GZSL.

In the inductive scenario, the most common idea is to map the semantic attributes and the extracted image features into a same target space, thereby achieving the classification via comparing distribution relationships in this space. In some early studies, the semantic attribute space was selected as the target space [18]. However, this kind of methods were shown aggravating the hubness phenomenon which harms the performance [22]. Based on this idea, the opposite projection direction from the semantic space to the visual space was proposed that can alleviate hubness [33]. In fact, both of the project directions could be used to form a reconstruction process for learning a better embedding function [12,32]. To train a more generalized embedding function, models rectifying prototypes of unseen classes [15], learning shared representation via VAE [24] and maintaining the similarities in the hidden space [1] were also designed. Moreover, to make the model more discriminative, a latent guided attention module was designed to deal with semantic ambiguous objects [14], an orthogonal mapping was exploited to push the inter-class distances [32], and a similarity metric was further learned in a meta learning process [23]. Although these existing methods have concentrated on how to adapt the limited information, the class-level over-fitting problem still makes the model lack of awareness for some unseen classes. There were also some embedding methods which trained more discriminative recognizers with the assistance of unseen attributes [10,14]. Similar to the the generative methods, they become ineffective under the inductive setting.

In recent years, adversarial examples [26] defined as the worst imperceptible perturbation on input that could however deceive the learning models, have become an active research topic. Recent studies indicate that most deep neural networks (DNNs) are vulnerable to these small perturbations in various tasks [2, 5,31,34,36]. To promote the robustness of DNNs, researches have developed a series defense methods. Among these approaches, adversarial training (AT) [8,17, 27] is a typical defense strategy that proves effective in improving the recognition robustness. In the training process of AT, different types of adversarial examples are generated and utilized for optimization to reduce the impact of perturbations on samples. On the other hand, a family of gradient regularization techniques were also proposed, which could also enhance the robustness of learning models. Though adversarial examples are widely adopted in many fields, it is rarely applied in the ZSL. It is noted that adversarial training was applied to the ZSL model in [35]; however, this work focused on evaluating the robustness of the model under several adversarial attacks. Differently, in our work, we propose an adversarial framework focusing on improving the performance for ZSL task instead of surviving under the attack.

3   Methodology

In this section, we will first define the ZSL task formally and then detail the proposed adversarial relation network.

Table 1. Schematic diagrams of data splits for several types of ZSL

3.1   Problem Definition

A brief overview of different ZSL classification tasks is shown in Table 1 where the colors blue and green are used to distinguish between seen and unseen classes. In this paper, we focus on the most challenged scenario, i.e. inductive GZSL, where the specific information of the targets is always limited till test and the models are required not only to recognize the unseen classes, but also to be discriminative enough between the seen and unseen classes.

Following the idea of Relation Network (RN) [23], the classification is performed by comparing the relation scores of the sample and the semantic attributes for all the classes. Denote $X = \{x_1, ..., x_N\}$, $Y = \{y_1, ..., y_N\}$, $A = \{a_1, ..., a_K\}$ as visual features, class labels, and semantic attributes sets respectively, with the total sample number N and class number K. We define $N = N_s + N_u$ and $K = K_s + K_u$ to divide the whole dataset into a source set (seen set) $X_s \times Y_s \times A_s = \{(x_i, y_i, a_{y_i}), i = 1, 2, ..., N_s\}$ and a target set (unseen set) $X_u \times Y_u \times A_u = \{(x_m, y_m, a_{y_m}), m = 1, 2, ..., N_u\}$ constructed in terms of 3-tuple. Here $X_s \cap X_u = \emptyset$ and $\forall y_i \in Y_s : y_i \leq K_s$ are required in the inductive scenario. With a trained relation score function $S(x, a)$ based on the source set, the ZSL target is to achieve the classification by satisfying $S(x_m, a_{y_m}) > S(x_m, a_k)$ for all $k \neq y_m$ where $0 < k \leq K$ for GZSL and $K_s < k \leq K$ for conventional zero-shot learning (CZSL).

3.2   Adversarial Relation Network

Relation Network [23]. In RN, the visual features X are extracted as training samples by a pre-trained DNN. A learned embedding function $f_\theta(\cdot)$ will project the semantic attribute vectors into the embedded space (visual feature space) as

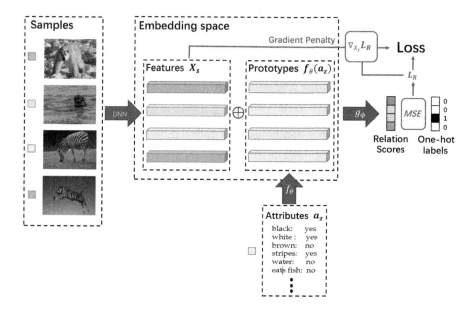

Fig. 2. Flowchart of the training process in advRN. As an example, the semantic attribute of Zebra is embedded into the visual feature space and copied several times used to calculate the relation scores. With these relation scores the loss function for RN and the further defined gradient penalty can be obtained. The DNN denotes the pre-trained feature extractor, $\oplus$ represents the concatenating operations, f_θ and g_ϕ are the embedded and metric function to be learned, respectively.

the prototypes of the corresponding classes. After a feature and a prototype are concatenated, their relation score will be calculated via a learned metric function $g_\phi(\cdot)$. By calculating the mean squared error (MSE) between the one-hot labels and the relation scores, a loss function for optimizing parameters θ and ϕ is designed as follows:

$$L_R(\boldsymbol{X}_s, \boldsymbol{Y}_s, \boldsymbol{A}_s; \theta, \phi) = \frac{1}{N_s} \frac{1}{K_s} \sum_{i=1}^{N_s} \sum_{k=1}^{K_s} [g_\phi(x_i \oplus f_\theta(a_k)) - r(x_i, a_k)]^2, \quad (1)$$

$$r(x_i, a_k) = \begin{cases} 0 & k \neq y_i, \\ 1 & k = y_i, \end{cases} \quad (2)$$

where $\oplus$ represents the concatenation operation and $r(x_i, a_k)$ indicates whether x_i and a_k represent the same class. The $g_\phi(x_i \oplus f_\theta(a_k))$ can be regarded as the required relation score function $S(x_i, a_k)$.

AdvRN with Gradient Penalty. In order to achieve a robust perception of seen classes, we design a robust training process encouraging that any small perturbation on instances of seen classes would not affect the network's output:

$$\min_{\theta,\phi} \max_{\epsilon:\|\epsilon\|_p \le \sigma} L_R(\boldsymbol{X}_s + \epsilon, \boldsymbol{Y}_s, \boldsymbol{A}_s; \theta, \phi), \tag{3}$$

where the robustness against any small perturbation defined as ϵ is required in the inner problem.

Since the minmax problem is usually difficult to solve, the first order Taylor expansion at feature point x can be used to approximately solve this non-convex problem. The inner problem can then be relaxed as follows:

$$\max_{\epsilon} L_R(\boldsymbol{X}_s, \boldsymbol{Y}_s, \boldsymbol{A}_s; \theta, \phi) + \nabla_{\boldsymbol{X}_s} L_R^{\mathsf{T}} \epsilon \quad \text{s.t.} \quad \|\epsilon\|_p \le \sigma. \tag{4}$$

To solve the optimization problem, we first introduce Lemma 1 from [16].

Lemma 1. *The optimization problem* $\max_{\epsilon} \nabla_{\boldsymbol{X}_s} L_R^{\mathsf{T}} \epsilon$ *s.t.* $\|\epsilon\|_p \le \sigma$ *has a closed form solution* $\epsilon = \sigma \, \text{sign}(\nabla L_R)(\frac{|\nabla L_R|}{\|\nabla L_R\|_{p^*}})^{\frac{1}{p-1}}$ *where p^* is the dual of p, i.e.,* $\frac{1}{p^*} + \frac{1}{p} = 1$.

Proof. Since L_R is independent of ϵ, the optimal ϵ should have a norm of σ. Then the maximum problem can be solved by Lagrangian multiplier method with defined $f(\epsilon) \equiv \nabla_{\boldsymbol{X}_s} L_R^{\mathsf{T}} \epsilon$ and $g(\epsilon) \equiv \|\epsilon\|_p = \sigma$. Set $\nabla f(\epsilon) = \lambda \nabla g(\epsilon)$, we have

$$\nabla f(\epsilon) = \lambda \nabla g(\epsilon) \tag{5}$$

$$\nabla_{\boldsymbol{X}_s} L_R = \lambda \frac{\epsilon^{p-1}}{p(\sum_i \epsilon_i^p)^{1-\frac{1}{p}}} \tag{6}$$

$$\nabla_{\boldsymbol{X}_s} L_R = \frac{\lambda}{p}(\frac{\epsilon}{\sigma})^{p-1} \tag{7}$$

$$\nabla_{\boldsymbol{X}_s} L_R^{\frac{p}{p-1}} = (\frac{\lambda}{p})^{\frac{p}{p-1}}(\frac{\epsilon}{\sigma})^p \tag{8}$$

By summing over two sides, we get

$$\sum \nabla_{\boldsymbol{X}_s} L_R^{\frac{p}{p-1}} = \sum (\frac{\lambda}{p})^{\frac{p}{p-1}}(\frac{\epsilon}{\sigma})^p \tag{9}$$

$$\|\nabla L_R\|_{p^*}^{p^*} = (\frac{\lambda}{p})^{p^*} * 1 \tag{10}$$

$$(\frac{\lambda}{p}) = \|\nabla L_R\|_{p^*} \tag{11}$$

By combining Eq. (7) and (11), we can easily obtain

$$\epsilon = \sigma \, \text{sign}(\nabla L_R)(\frac{|\nabla L_R|}{\|\nabla L_R\|_{p^*}})^{\frac{1}{p-1}} \tag{12}$$

This completes the proof. $\qquad\qquad\square$

Using Lemma 1, we can obtain a closed form solution of problem (4). Then, with this solution and a settled $p = 2$, the approximation of the minmax problem becomes

$$\min_{\theta,\phi} L_R(\boldsymbol{X}_s, \boldsymbol{Y}_s, \boldsymbol{A}_s; \theta, \phi) + \sigma \|\nabla L_R\|_2. \tag{13}$$

Here, instead of only minimizing the loss function (1), we additionally minimize a gradient penalty $\sigma \|\nabla L_R\|_2$. Adding this gradient penalty into the loss as a regularization term is approximately equivalent to injecting adversarial noises to the samples during training as shown in the above. As a result, the trained model will be less sensitive to any small changes in samples for seen classes.

In practice, to make the model more resistant to over-fitting, the regularization term $\|\theta\|^2$ can be usually applied. Consequently, the final optimization loss function for training the advRN can be summarized as below:

$$L_{total} = L_R + \sigma \|\nabla L_R\|_2 + \alpha \|\theta\|^2, \tag{14}$$

where σ and α are the trade-off parameters.

For illustration, we also plot Fig. 2 which briefly shows the proposed training process.

3.3   Sample Re-adjustment with Gradient Guidance

As an imitation of the hypothesis testing cognition, we design our test process in the following. The test process consists of three parts: gradient calculation with the unseen classes hypothesis, sample re-adjustment, and classification as shown in Fig. 3.

Similar to hypothesis testing cognition, we first make an unseen-against-seen hypothesis that each test instance x_m belongs to unseen categories. Then we can construct an objective function L_{Gm} for this hypothesis as follows:

$$L_{Gm} = \frac{1}{K} \sum_{k=1}^{K} (g_\phi(x_m \oplus f_\theta(a_k)) - l(a_k))^2, \tag{15}$$

$$l(a_k) = \begin{cases} 0 & 0 < k \leqslant K_s, \\ 1 & K_s < k \leqslant K, \end{cases} \tag{16}$$

where $l(a_k)$ indicates whether a_k is an unseen class attribute. This objective function actually measures the total cost with the hypothesis that each test sample was from unseen categories.

Inspired by the adversarial idea as discussed in Sect. 3.2 that gradient based perturbations on samples tend to deceive the classification, we introduce a similar gradient, but adjust in an opposite direction the instances. Namely, the gradient of the loss function w.r.t. the input instance $\nabla_{x_m} L_{Gm}$ presents the worst-case perturbation which would change the output of classifier. On the contrary, adjusting the sample in a reversed gradient direction would then lead to more confidence on the original unseen-against-seen hypothesis.

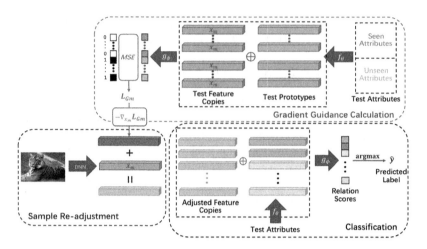

Fig. 3. Flowchart of the proposed test process. A gradient is calculated which makes the model tend to recognize the feature as an unseen class; then the instance feature is re-adjusted with the gradient. The DNN denotes the pre-trained feature extractor, $\oplus$ represents the concatenating operations, f_θ and g_ϕ are the need to be learned embedded and metric function, respectively.

As a result, the adjusted features become:

$$x'_m = x_m - \lambda \nabla_{x_m} L_{Gm}, \tag{17}$$

where λ denotes the adjustment rate. After the feature instance is adjusted, we can directly predict its label as the category with the maximum relation score between the adjusted feature and the semantic attributes:

$$\hat{y}_m = \arg\max_{k \in K} g_\phi(x'_m \oplus f_\theta(a_k)) \tag{18}$$

In a short summary, by applying the SR, the model is adapted so that high response can be obtained for small adversarial perturbations to unseen classes, consequently enhancing the awareness of the unseen classes. On the other hand, robust training with GP has the property that any small perturbations on the samples of seen classes would not change the network's output (as discussed in the previous subsection). This would greatly alleviate the limitation of previous inductive GSZL methods, i.e., they are commonly not discriminative enough and tend to classify target samples as seen classes.

Remarks on Entire Adversarial Framework. The training and test processes form an entire adversarial framework for the inductive GZSL. Adding the gradient penalty during training can be deemed as an adversarial defense for the seen classes, making the classifier sufficiently robust to small sample perturbations; on the other hand, small sample re-adjustment in the test phase can be regarded as an adversarial attack that tends to mislead the classifier to choose the unseen classes. In summary, applying the strategies of both SR and GP, the proposed model has the appealing feature that small sample re-adjustment can lead to high response to unseen classes while not affect the recognition of seen classes due to the robust adversarial training.

4 Experiment

4.1 Setup

Dataset. To evaluate the performance of the proposed method, we select four commonly used ZSL benchmark datasets: AwA1 [13], AwA2 [29], aPY [6] and CUB [28]. Detailed statistic information for each benchmark is shown in Table 2. AwA1 is a medium-scale coarse-grained dataset containing 50 classes of animals with 85 attributes. AwA2 keeps the same setting as AwA1 but contains different samples. aPY is a small-scale coarse-grained dataset with 32 object classes and 64 attributes, and CUB is a medium-scale fine-grained dataset with 200 bird classes described in 312 attributes.

Table 2. Statistic information for ZSL benchmark in terms of scale, granularity, attribute dimension, class size and sample size.

Benchmark	Scale	Granularity	Attribute dimension	Class size		Sample size		
				Train (seen)	Unseen	Train	Test_{seen}	Test_{unseen}
AwA1	Medium	Coarse	85	27 + 13	10	19832	4958	5685
AwA2	Medium	Coarse	85	27 + 13	10	23527	5882	7913
aPY	Small	Coarse	64	15 + 5	12	5932	1483	7924
CUB	Medium	Fine	312	100 + 50	50	7057	1764	2967

Table 3. Specific experiment settings for ZSL benchmarks in terms of layer structures, learning rate, batch size, regularization weight, and adjustment rate.

Benchmark	Layer structures		Learning rate	Batch size	Regularization weight		Adjustment rate λ
	f_θ	g_ϕ			σ	α	
AwA1	$85 \times 1600 \times 2048$	$4096 \times 400 \times 1$	1×10^{-5}	32	1.5	1×10^{-5}	10000/32
AwA2	$85 \times 1600 \times 2048$	$4096 \times 400 \times 1$	1×10^{-5}	32	1.5	1×10^{-5}	10000/32
aPY	$65 \times 1200 \times 2048$	$4096 \times 400 \times 1$	1×10^{-4}	32	1.5	1×10^{-5}	5000/32
CUB	$312 \times 1200 \times 2048$	$4096 \times 1200 \times 1$	1×10^{-5}	32	1.5	1×10^{-5}	2500/32

Setting. All quantitative evaluations were conducted in the inductive GZSL scenario. We follow the attributes, features and train/test splits proposed in [29] (GBU setting), to avoid overlapping of categories between the test set and the set for training feature extractor. The average per-class top-1 accuracy is selected to measure the performance. Specifically, for GZSL, criteria ACC_S and ACC_U denote accuracies for the seen and unseen categories, respectively, and H is defined as the harmonic mean of ACC_S and ACC_U to evaluate the overall performance as follows:

$$H = \frac{2 \times ACC_S \times ACC_U}{ACC_S + ACC_U}. \tag{19}$$

The learned embedding function f_θ is designed as a two-layer neural network with the activation function *relu*. The learned metric function g_ϕ engages a similar setting to that of f_θ, but the activation function of the last layer is replaced by *sigmoid*. The specific values of hyperparameters for each benchmark are shown in Table 3 where the adjustment rate λ is roughly chosen from $\{2500/32, 5000/32, 10000/32\}^1$ by the validation set.

Table 4. Comparisons between the proposed **advRN** and the other state-of-the-art methods. Values are accuracies in %.

Method	AwA1			AwA2			aPY			CUB		
	ACC_U	ACC_S	H	ACC_U	ACC_S	H	ACC_U	ACC_S	H	ACC_U	ACC_S	H
SAE [12]	1.8	77.1	3.5	1.1	82.2	2.2	0.4	80.9	0.9	7.8	54.0	13.6
PSR-ZSL [1]	–	–	–	20.7	73.8	32.3	13.5	51.4	21.4	24.6	54.3	33.9
DEM [33]	32.8	84.7	47.3	30.5	86.4	45.1	11.1	75.1	19.4	19.6	57.9	29.2
Triple Verification Net [32]	27.0	67.9	38.6	–	–	–	16.1	66.9	25.9	26.5	62.3	37.2
LFGAA+Hybrid [14]	–	–	–	27.0	**93.4**	41.9	–	–	–	36.2	**80.9**	50.0
MIVAE [24]	35.9	87.8	51.0	32.9	90.4	48.2	15.3	**87.4**	26.0	31.5	60.4	41.4
RN [23]	31.4	**91.3**	46.7	30.0	**93.4**	45.3	21.2	68.8	32.4	38.1	61.1	47.0
advRN(ours)	**50.6**	79.5	**61.8**	**49.3**	84.0	**62.2**	**28.0**	66.0	**39.3**	**44.3**	62.6	**51.9**

4.2   Results

Since we are focus on inductive GZSL tasks, here we only compared the proposed methods with the models designed under inductive scenario. It is observed that the proposed model achieves the most significant performance for unseen accuracy and harmonic mean compared to the state-of-the-art methods as shown in Table 4. As the aim of our method is to learn a more generalized recognition model, it is reasonable that the accuracies for seen classes are not the highest. Compared to the baseline RN, while ACC_S was decreased by 11.8%, 9.4% and 2.8%, ACC_U has been significantly promoted by 19.2%, 19.3% and 6.8% in AwA1, AwA2 and aPY, respectively. More importantly, overall, the proposed

---

^{1} This division by 32 is due to the effect of batch size during the derivation in test process.

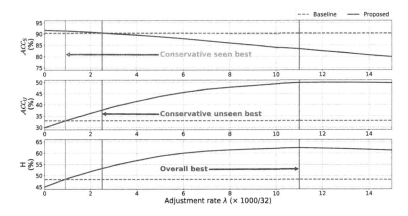

Fig. 4. Accuracies ($\mathcal{ACC}_S$, $\mathcal{ACC}_U$ and H) vs. adjustment ratio λ for the proposed method on AwA2. RN without sample re-adjustment is also provide as the baseline.

model attains the best harmonic mean in all the four datasets. Taking a closer examination on CUB, we observe an increase in both $\mathcal{ACC}_S$ and $\mathcal{ACC}_U$ over RN, showing that the proposed advRN model obtains a more generalized recognizer. It is noted that, even without using the test information during training, our model leads to comparable performance with most of the semantic transductive GZSL methods.

4.3 Further Analysis

In order to further demonstrate the effectiveness, we take AwA2 as one illustrative example to have a closer examination. We first plot in Fig. 4 the performance of advRN as the adjustment rate λ increases. As observed, when sample re-adjustment is not applied (i.e., $\lambda = 0$), the gradient penalty improves the accuracy of known or seen classes, but it actually leads to an over-fitting on the seen classes and consequently causes the decline in $\mathcal{ACC}_U$. However, after combining the sample re-adjustment, the proposed advRN method improves the recognition of unseen classes significantly. A small adjustment (with a small λ) leads to a sharp increase in $\mathcal{ACC}_U$ and a slight decrease in $\mathcal{ACC}_S$; this alleviates the aforementioned over-fitting problem. As we observed in Fig. 4, for the proposed method, the conservative seen best node corresponds to the performance level with $\mathcal{ACC}_U$ the same as that of the baseline, the conservative unseen best node corresponds to the performance level with $\mathcal{ACC}_S$ the same as that of the baseline, and the overall best node denotes the performance with the highest H. When λ is in the region between the two conservative nodes, both $\mathcal{ACC}_S$ and $\mathcal{ACC}_U$ are beyond the baseline, indicating that a more discriminative recognizer can be achieved.

For a more intuitive view, classification accuracies for three confusing classes in AwA2 are demonstrated in Fig. 5 where bobcat is one of the unseen classes,

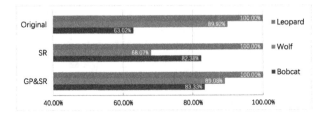

Fig. 5. Accuracy histogram for bobcat, wolf and leopard in AwA2.

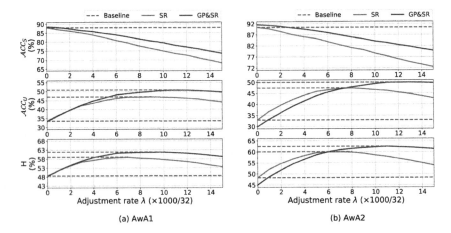

Fig. 6. Accuracies ($\mathcal{ACC}_S$, $\mathcal{ACC}_U$ and H) vs. adjustment ratio λ on (a) AwA1 and (b) AwA2. GP largely increases the robustness of the recognizer especially on seen classes. SR means that only sample re-adjustment is applied, GP&SR denotes both gradient penalty and sample re-adjustment are applied and the dashed line represents the baseline performance as reference.

leopard and wolf are two seen classes. applying SR, 20.31% more bobcat samples can be correctly recognized than without applying SR. Simultaneously, the adversarial training prevents a 21.01% accuracy drop caused by the misleading of sample re-adjustment for wolf.

To illustrate the importance of the gradient penalty, we also compare the performances of the proposed advRN with and without gradient penalty. As seen in Fig. 6, we can inspect an obvious difference between the models. For the model trained without the gradient penalty, the learned recognizer appears much sensitive to perturbations especially for seen classes. As a result, $\mathcal{ACC}_S$ falls sharply when λ increases and the overall performance is mediocre. In comparison, with the gradient penalty, the learned recognizer shows high robustness to seen classes. Moreover, while promoting the recognition for the seen classes, this gradient penalty can also increase the model's generalization to unseen classes, as the upper bound of the unseen accuracy is also slightly increased in Fig. 6.

5   Conclusion

In this work, we have designed a sample re-adjustment process for GZSL to increase the model's awareness of unseen class. However, this adversarial perturbation based adjustment may mislead the correct judgment about the seen class. Therefore, with the idea of adversarial examples, we have also designed a gradient penalty during training which makes the model less sensitive to the input fluctuations for those samples of seen classes. Applying both the sample re-adjustment and gradient penalty, we attain a more generalized model that has achieved superior performance compared to the state-of-the-art methods. Extensive quantitative experimental results have validated the effectiveness of our work on four benchmark datasets.

Acknowledgements. The work was partially supported by the following: National Natural Science Foundation of China under no.61876155; Natural Science Foundation of Jiangsu Province BK20181189; Key Program Special Fund in XJTLU under no. KSF-A-01, KSF-T-06, KSF-E-26, KSF-P-02 and KSF-A-10.

References

1. Annadani, Y., Biswas, S.: Preserving semantic relations for zero-shot learning. In: Proceedings of the IEEE Conference on Computer Vision and Pattern Recognition, pp. 7603–7612 (2018)
2. Carlini, N., et al.: Hidden voice commands. In: 25th {USENIX} Security Symposium ({USENIX} Security 16). pp. 513–530 (2016)
3. Chao, W.-L., Changpinyo, S., Gong, B., Sha, F.: An empirical study and analysis of generalized zero-shot learning for object recognition in the wild. In: Leibe, B., Matas, J., Sebe, N., Welling, M. (eds.) ECCV 2016, Part II. LNCS, vol. 9906, pp. 52–68. Springer, Cham (2016). https://doi.org/10.1007/978-3-319-46475-6_4
4. Eagleman, D.: Incognito: The Secret Lives of the Brain. Vintage Books, New York (2013)
5. Ebrahimi, J., Rao, A., Lowd, D., Dou, D.: Hotflip: White-box adversarial examples for text classification. arXiv preprint arXiv:1712.06751 (2017)
6. Farhadi, A., Endres, I., Hoiem, D., Forsyth, D.: Describing objects by their attributes. In: IEEE Conference on Computer Vision and Pattern Recognition, pp. 1778–1785 (2009)
7. Goodfellow, I., et al.: Generative adversarial nets. In: Advances in Neural Information Processing Systems, pp. 2672–2680 (2014)
8. Goodfellow, I.J., Shlens, J., Szegedy, C.: Explaining and harnessing adversarial examples. arXiv preprint arXiv:1412.6572 (2014)
9. Huang, K., Hussain, A., Wang, Q.F., Zhang, R.: Deep Learning: Fundamentals, Theory and Applications. Springer, Cham (2019). ISBN 978-3-030-06072-5
10. Jiang, H., Wang, R., Shan, S., Chen, X.: Transferable contrastive network for generalized zero-shot learning. In: Proceedings of the IEEE International Conference on Computer Vision, pp. 9765–9774 (2019)
11. Kingma, D.P., Welling, M.: Auto-encoding variational bayes. arXiv preprint arXiv:1312.6114 (2013)

12. Kodirov, E., Xiang, T., Gong, S.: Semantic autoencoder for zero-shot learning. In: IEEE Conference on Computer Vision and Pattern Recognition, pp. 3174–3183 (2017)
13. Lampert, C.H., Nickisch, H., Harmeling, S.: Attribute-based classification for zero-shot visual object categorization. IEEE Trans. Pattern Anal. Mach. Intell. **36**(3), 453–465 (2014)
14. Liu, Y., Guo, J., Cai, D., He, X.: Attribute attention for semantic disambiguation in zero-shot learning. In: Proceedings of the IEEE International Conference on Computer Vision, pp. 6698–6707 (2019)
15. Luo, C., Li, Z., Huang, K., Feng, J., Wang, M.: Zero-shot learning via attribute regression and class prototype rectification. IEEE Trans. Image Process. **27**(2), 637–648 (2018)
16. Lyu, C., Huang, K., Liang, H.N.: A unified gradient regularization family for adversarial examples. In: 2015 IEEE International Conference on Data Mining, pp. 301–309. IEEE (2015)
17. Madry, A., Makelov, A., Schmidt, L., Tsipras, D., Vladu, A.: Towards deep learning models resistant to adversarial attacks. arXiv preprint arXiv:1706.06083 (2017)
18. Mikolov, T., Le, Q.V., Sutskever, I.: Exploiting similarities among languages for machine translation. arXiv preprint arXiv:1309.4168 (2013)
19. Russakovsky, O., et al.: ImageNet large scale visual recognition challenge. Int. J. Comput. Vis. **115**(3), 211–252 (2015)
20. Sariyildiz, M.B., Cinbis, R.G.: Gradient matching generative networks for zero-shot learning. In: Proceedings of the IEEE Conference on Computer Vision and Pattern Recognition, pp. 2168–2178 (2019)
21. Schonfeld, E., Ebrahimi, S., Sinha, S., Darrell, T., Akata, Z.: Generalized zero-and few-shot learning via aligned variational autoencoders. In: Proceedings of the IEEE Conference on Computer Vision and Pattern Recognition, pp. 8247–8255 (2019)
22. Shigeto, Y., Suzuki, I., Hara, K., Shimbo, M., Matsumoto, Y.: Ridge regression, hubness, and zero-shot learning. In: Appice, A., Rodrigues, P.P., Santos Costa, V., Soares, C., Gama, J., Jorge, A. (eds.) ECML PKDD 2015, Part I. LNCS (LNAI), vol. 9284, pp. 135–151. Springer, Cham (2015). https://doi.org/10.1007/978-3-319-23528-8_9
23. Sung, F., Yang, Y., Zhang, L., Xiang, T., Torr, P.H., Hospedales, T.M.: Learning to compare: relation network for few-shot learning. In: IEEE Conference on Computer Vision and Pattern Recognition, pp. 1199–1208 (2018)
24. Suzuki, M., Iwasawa, Y., Matsuo, Y.: Learning shared manifold representation of images and attributes for generalized zero-shot learning (2018). https://openreview.net/pdf?id=Hkesr205t7
25. Szegedy, C., et al.: Going deeper with convolutions. In: Proceedings of the IEEE Conference on Computer Vision and Pattern Recognition, pp. 1–9 (2015)
26. Szegedy, C., Zaremba, W., Sutskever, I., Bruna, J., Erhan, D., Goodfellow, I., Fergus, R.: Intriguing properties of neural networks. arXiv preprint arXiv:1312.6199 (2013)
27. Tramèr, F., Kurakin, A., Papernot, N., Goodfellow, I., Boneh, D., McDaniel, P.: Ensemble adversarial training: Attacks and defenses. arXiv preprint arXiv:1705.07204 (2017)
28. Wah, C., Branson, S., Welinder, P., Perona, P., Belongie, S.: The caltech-ucsd birds-200-2011 dataset (2011)
29. Xian, Y., Lampert, C.H., Schiele, B., Akata, Z.: Zero-shot learning-a comprehensive evaluation of the good, the bad and the ugly. IEEE Transactions on Pattern Analysis and Machine Intelligence (2018)

30. Xian, Y., Lorenz, T., Schiele, B., Akata, Z.: Feature generating networks for zero-shot learning. In: IEEE Conference on Computer Vision and Pattern Recognition, pp. 5542–5551 (2018)
31. Xie, C., Wang, J., Zhang, Z., Zhou, Y., Xie, L., Yuille, A.: Adversarial examples for semantic segmentation and object detection. In: Proceedings of the IEEE International Conference on Computer Vision, pp. 1369–1378 (2017)
32. Zhang, H., Long, Y., Guan, Y., Shao, L.: Triple verification network for generalized zero-shot learning. IEEE Trans. Image Process. **28**(1), 506–517 (2019)
33. Zhang, L., Xiang, T., Gong, S.: Learning a deep embedding model for zero-shot learning. In: IEEE Conference on Computer Vision and Pattern Recognition, pp. 2021–2030 (2017)
34. Zhang, S., Huang, K., Zhang, R., Hussain, A.: Generalized adversarial training in riemannian space. In: 2019 IEEE International Conference on Data Mining (ICDM), pp. 826–835 (2019)
35. Zhang, X., Gui, S., Zhu, Z., Zhao, Y., Liu, J.: Atzsl: Defensive zero-shot recognition in the presence of adversaries. arXiv preprint arXiv:1910.10994 (2019)
36. Zügner, D., Akbarnejad, A., Günnemann, S.: Adversarial attacks on neural networks for graph data. In: Proceedings of the 24th ACM SIGKDD International Conference on Knowledge Discovery and Data Mining, pp. 2847–2856 (2018)

Author Index

Printed in the United States
By Bookmasters